C0-AZG-712

The Materials Science of Microelectronics

The Materials Science of Microelectronics

Klaus J. Bachmann

Klaus J. Bachmann
Department of Materials Science and Engineering
and
Department of Chemical Engineering
North Carolina State University
Raleigh, North Carolina

This book is printed on acid-free paper. ∞

Library of Congress Cataloging-in-Publication Data

Bachmann, Klaus J.
The materials science of microelectronics / Klaus J. Bachmann.
p. cm.
Includes bibliographical references and indexes.
ISBN 0-89573-280-7
1. Microelectronics--Materials. 2. Semiconductors. 3. Materials--Electric properties. I. Title.
TK7874.B32 1994
621.381--
93-7878
CIP

Printed in the United States of America

ISBN 0-89573-280-7 VCH Publishers, Inc.

Printing History:
10 9 8 7 6 5 4 3 2

Published jointly by

VCH Publishers, Inc.
220 East 23rd Street
New York, New York 10010

VCH Verlagsgesellschaft mbH
P.O. Box 10 11 61
69451 Weinheim, Germany

VCH Publishers (UK) Ltd.
8 Wellington Court
Cambridge CB1 1HZ
United Kingdom

Preface

Microelectronics is a highly competitive industrial enterprise that is backed by a multidisciplinary research effort, running at a high pace. The purpose of this book is to provide a concise overview of the materials science and engineering that supports this industrial development. It is based on courses on materials chemistry and physics taught by the author during the past decade at North Carolina State University and is laid out for teaching a multidisciplinary audience.

Chapters 1–4 of this book provide relevant background information from chemistry, physics, electrical and computer engineering, and materials science and engineering. Depending on the background of the reader, they may be either skipped or supplemented by further reading. A selection of books and original literature is listed at the end of each chapter to facilitate such supplementary studies. Chapters 5–8 cover special topics of the materials science of microelectronics, that is, the purification of raw materials, the preparation of bulk crystals for the fabrication of substrates, chemical vapor deposition and epitaxy, pattern definition and doping, and oxidation and metallization. Brief tutorial sections on selected methods of characterization are mixed into this text to add to the review of properties and methods of characterization given in Chapter 3. It is hoped that the selected formal background information, discussions of specialized topics, and tutorials on methods of characterization will convey the essential linkages of fundamental ideas and experimental techniques of different disciplines in the development of microelectronics, which make the participation in research in this field such challenging and rewarding task. The book closes with a look at the materials science of optical electronics that is the topic of another book in preparation.

I would like to thank the management and staff of the North Carolina State University Library service for providing me with an office and valuable assistance in creating this book.

I am indebted to Dr. Edmund Immergut for his steadfast support to Ms. Camille Pecoul, Director of Production, and to the staff of VCH for their help in producing this book. Also, thanks to Professor Sigurd Wagner of Princeton University and Professor Kenneth A. Jackson of the University of Arizona for helpful comments and suggestions. Furthermore, I would like to thank all my colleagues who have contributed original figures from their publications, and their publishers who permitted the reproduction of these figures.

Klaus J. Bachmann
Raleigh, North Carolina
June 1994

Contents

CHAPTER

1

Introduction

Microelectronics is a special field of engineering that has evolved over the past 35 years. It encompasses the design, fabrication, and testing of electronic circuits based, at present, on semiconductor structures of submicrometer feature size. Through its entire history, the development of microelectronics was driven by a rapid succession of inventions that permitted revolutionary improvements in the performance and reliability of electronic circuits accompanied by a drastic reduction of their cost. The progress in this field is intimately tied to multidisciplinary research, with equally important inputs from chemistry, physics, applied mathematics, and electrical engineering. Chemistry provides for novel and/or improved materials and processing methods. Physics establishes the principles of device operation and provides for the methods of characterization that are needed for process diagnostics and the analysis of reliability problems. Electrical engineering, in conjunction with inputs from applied mathematics, provides for circuit design and for the testing of circuits as well as for their assembly into systems. Depending on their end use, this may involve further inputs from other engineering disciplines or medicine. Therefore, the success of research and development in microelectronics is invariably tied to the interactions of large teams spanning a wide range of expertise.

The origin of microelectronics may be traced back to the inventions of the point contact transistor by John Bardeen and Walter Brattain in 1947 [1] and of the junction transistor by William Shockley in 1950 [2] at Bell Telephone Laboratories. These devices were originally made from germanium, which has

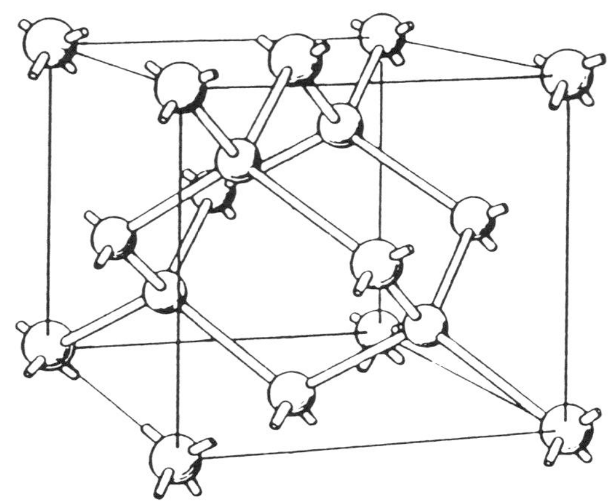

Figure 1.1 Schematic representation of the position and bonding of atoms in the diamond structure.

been almost totally replaced by silicon. Both silicon and germanium crystallize in the diamond structure shown in Figure 1.1.

The three-dimensional periodicity of the atom positions in crystal lattices is important for the understanding of the high mobility of electrical carriers in semiconductors, such as silicon and germanium. Upon ordering in a crystal lattice, the discrete energy levels that characterize the electronic properties of separated atoms or molecules broaden into bands of closely spaced states of allowed energy that are separated by gaps of forbidden energy. This is illustrated in Figure 1.2, which shows schematically the spreading of the energy eigenvalues of discrete atoms with decreasing interatomic distance for a linear periodic array of atoms. The tightly bound core states of these atoms are

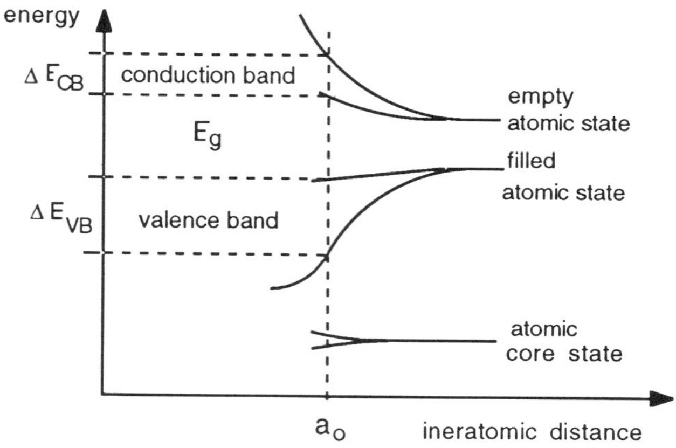

Figure 1.2 Schematic representation of the broadening of the discrete energy levels of atoms upon forming an equally spaced array. a_0 = lattice parameter.

affected very little by the proximity of other atoms and consequently do not broaden very much in energy. Therefore, the core states may be utilized for element identification in analytical evaluations of crystalline solids. As in the ground state of molecules, the electrons of the constituent atoms fill the molecular orbitals up to a specific energy, the core states and bands of a crystalline solid are filled by electrons up to a specific energy, and the states above this energy are empty. Since there are many allowed states within each band, the situation may arise that the highest occupied band is only partially filled. In this case, the electrons in this band can rise to higher energy states, that is, can gain kinetic energy upon acceleration in an external field, thus contributing to the flow of an electrical current. The band that provides for the eletronic conduction is referred to as the *conduction band*. In the case of metals there are many filled and unfilled states in the conduction band, resulting in their low specific resistivity ρ. For example, in the case of copper the density of mobile electrons is $6.3 \times 10^{22}\,\mathrm{cm}^{-3}$ and $\rho(\mathrm{Cu}) = 1.67 \times 10^{-6}\,\Omega\,\mathrm{cm}$ at room temperature. On the other side of the conductivity scale are the insulators. They have a completely occupied uppermost filled band (valence band) separated by a bandgap of several eV from the next higher unoccupied band. Since the bandgap is large as compared to the thermal energy, $k_B T = 26\,\mathrm{meV}$ at room temperature, the thermal excitation of electrons across the gap is negligible, and the material is thus essentially devoid of mobile carriers.

In between these extremes there exists the class of semiconducting materials. They usually have small concentrations of electron in the conduction band and holes (unfilled states) in the valence band, respectively. With the exception of low-gap materials, they exhibit thus in ultrapure form high specific resistivities, such as $\rho(\mathrm{Si}) = 1.1 \times 10^{5}\,\Omega\,\mathrm{cm}$ at room temperature. However, by the addition of impurities that introduce states with energy close to the band edges, the resistivity of semiconductors may be lowered and controlled within narrow margins. Shallow impurity states in semiconductors are thermally ionized at room temperature and, depending on their nature, either donate electrons into the conduction band or accept electrons from the valence band. If both donor and acceptor impurities are present, their effects on the conductivity compensate. Usually a high compensation level is undesirable, and extremely pure semiconductors are needed for device fabrication. An excess of either a donor or an acceptor impurity is added to these ultrapure starting materials so that the conduction is carried primarily by electrons in the conduction band (*n*-type conduction) or by holes in the valence band (*p*-type conduction). Typical background impurity levels in high-quality semiconductors are in the parts per trillion to parts per billion range. The doping levels in semiconductor devices are typically 10^{14} to $10^{18}\,\mathrm{cm}^{-3}$, corresponding to resistivities in the 100 to $0.01\,\Omega\,\mathrm{cm}$ range, depending on the carrier mobility. The ability to control the doping of semiconductors closely with high spatial resolution provides for the fabrication of a variety of semiconductor devices, such as diodes and transistors, which are discussed in more detail in Chapter 4.

Figure 1.3 shows a schematic representation of a bipolar transistor that

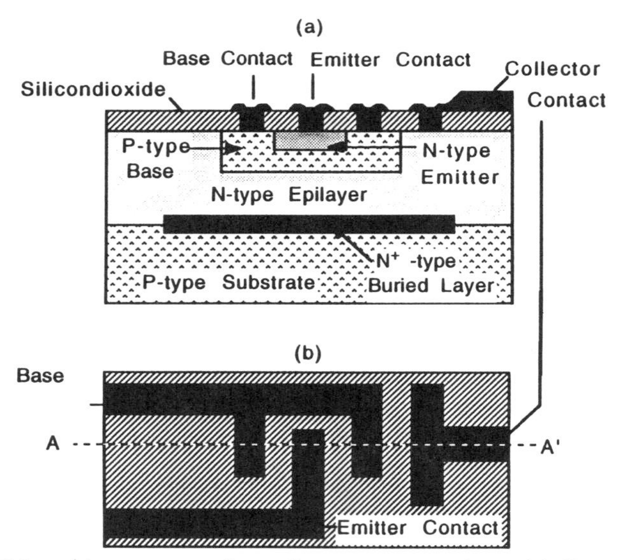

Figure 1.3 Schematic representations of an n–p–n transistor. (a) Cross section A–A′; (b) top view of the contact pattern.

utilizes the electrical manipulation of the barriers to current flow in either n–p–n or p–n–p structures for the switching and amplification of electrical currents. For example, in the case of an n–p–n transistor the current flow from an n-type emitter to an n-type collector is controlled by the current into the p-type base. Upon a positive bias of the base with respect to the emitter, electrons are injected from the emitter into the base region. Also, a small hole current flows from the base into the emitter. The electrons entering the base are accelerated toward the collector, which is held at a yet more positive potential than the base. Small changes in the base current cause large changes in the collector current, which is the basis for the use of transistors for amplification and switching operations.

In 1958 Jack H. Kilby of Texas Instruments added to the invention of the transistor the concept of the integrated circuit (IC) [3], setting the stage for modern microelectronics [4]. An IC consists of internally interconnected transistors and passive elements, such as resistors and capacitors, and is fabricated on a single semiconductor chip, keeping the length of input and output lines to a minimum. This provides for high reliability and the economic use of materials, minimum resistive power loss, and minimum delay by the minimization of the RC constants associated with the device elements and connecting metal lines. There are two classes of ICs, linear and digital. Linear ICs relate a continuous range of input signals to a continuous range of output signals. Figure 1.4 shows as an example of a small-scale linear IC the circuit diagram of an operational amplifier, employing five bipolar transistors. The operational amplifier was an important early product of solid-state electronics because its

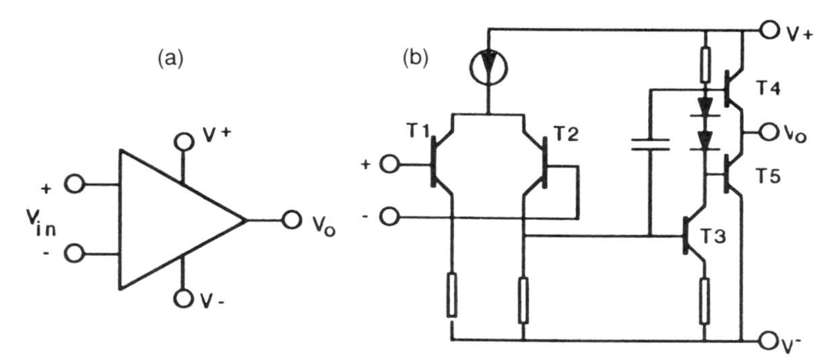

Figure 1.4 Symbolic representation (a) and circuit diagram (b) of a simple operational amplifier. After Wojslaw [5].

high input resistance and low output resistance made it a useful building block in a variety of applications.

Digital ICs perform arithmetic and logic functions, representing the zeros and ones of binary coded information or the true and false values of logic decisions by the high (logic 1) and low (logic 0) states of the input and output terminals. For conventional bipolar transistor technology, they are 0 and +3.3 V for 0 and 1, respectively. The relation between the input variables and output functions of these variables is represented by truth tables and the corresponding expressions of Boolean algebra. For example, the AND function

$$X = A \cdot B \cdot C \cdot \ldots$$

sets the output X to a high state if the inputs A, B, $C, \ldots$ are all high. Otherwise X is set to a low state. Figures 1.5(a) and (b) show the truth table and a circuit diagram for the NAND (NOT AND) operation, where the output X is set to a logic low state if all inputs are high and to a logic high state otherwise.

Several small-scale logic circuits may be combined to larger IC blocks for other tasks. For example, two cross-connected two-input NAND gates make an RS flip-flop, as illustrated in Figure 1.5(b), which shows the symbolic representation of such a circuit. Figures 1.5(c) and (d) show a circuit diagram and schematic view of a cross section of a bipolar inverter circuit, using integrated injection logic (I^2L). The fan out into two collector contacts is provided as part of the output transistor design.

The RS flip-flop has two inputs, set (S) and reset (R), and two outputs labeled $\bar{X}_1$ and $\bar{X}_2$, respectively. When the R input is high and the S input is low, the inverted output $\bar{X}_1$ is high, and the $\bar{X}_2$ output is low. This is so because adding another low input on the NAND gate with the low S input forces a high output, and the cross connection of this output to the other NAND gate establishes there two high inputs in accordance with its low output. Conversely, a high input at S and a low input at R flips the output at $\bar{X}_1$ and $\bar{X}_2$. However, if the inputs at R and S are the same, a conflict may arise. Therefore, more complex circuits are used that avoid this problem.

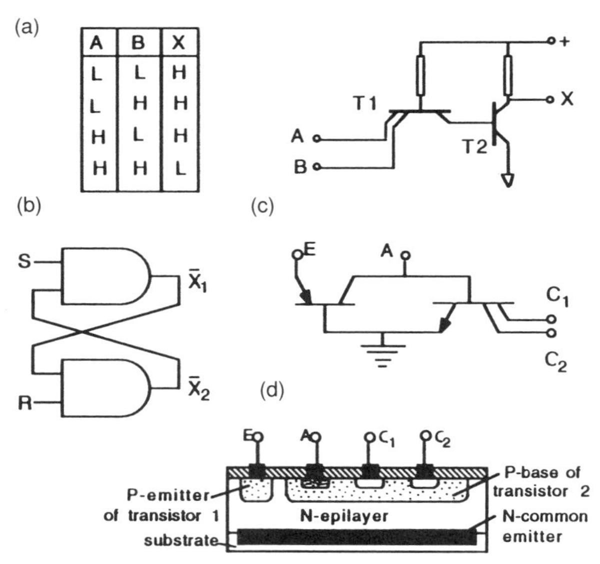

Figure 1.5 (a) Truth tables, logic symbol, and bipolar circuit for the NAND operation. b) Logic diagram for the RS flip-flop. (c) Bipolar inverter circuit. (d) Implementation of I^2L fanout into two collectors by a planar design. After Wojslaw [5].

For example, three cross-connected RS flip-flops make a D flip-flop, which may be controlled by an external clock that delivers a train of equally spaced pulses of logic one height to the clock input at the device. The D flip-flop thus can be set and reset in a timed fashion, that is, has the attributes of a memory cell. Other types of simpler memory cells are described in Chapter 4. For storing many *bi*nary digi*ts* (contracted to bits), many memory cells must be combined into registers. If the information is entered into and retrieved from such a register sequentially bit by bit, it is called a *shift register*. On the other hand, if all the flip-flops are set and read simultaneously, the register is called a *parallel register*. In modern computers, a variety of building blocks are added, for example, demultiplexer/decoder circuits for addressing and the timing of interactions between the arithmetic logic unit and input output devices. A discussion of the design and the function of these circuits is outside the scope of this book. However, an understanding of the design and function of the devices that are used in their implementation is essential for the selection of materials and processing options, which is an important point in the discussion of the materials science of microelectronics.

Figure 1.6 shows a photograph of an IC representing the state of the art of very large-scale integration (VLSI) in the early 1980s. A number of such circuits are produced on a single semiconductor wafer, which is subsequently diced into chips, each containing a complete integrated circuit. The particular circuit shown in Figure 1.6 incorporates 100,000 transistors and is based on a processing technology that achieves 1 μm resolution in the device dimensions.

Figure 1.6 The BELLMAC™-32 microprocessor photographed against the background of a postage stamp to illustrate its size. Reprinted by permission of AT&T Bell Laboratories from the brochure Bell Labs 1980; copyright © 1980, AT&T Bell Laboratories, Murray Hill, NJ.

The level of integration in VLSI technology is typically 10^4 internal interconnects per external connection so that most of the data handling is performed internally maximizing the efficiency and reliability of the circuit. Although the progress in the development of silicon microelectronics from its initial stage represented by Figure 1.4 to the VLSI circuit represented by Figure 1.6 is staggering, the push toward ultra-large-scale integration (ULSI) that is, beyond one million transistors on a chip, is at present being pursued with unabated vigor and success.

The above-discussed advance of transistor technology has been greatly facilitated by the development of the field-effect transistor (FET). The function of the FET is based on the opening and closure of a conducting channel between heavily doped regions below two contacts called the *source* and the *drain*, respectively. A gate contact in between the source and the drain controls the carrier concentration in the channel by either attracting or repelling mobile electrical charges in the low-doped surface region below the gate contact. Although this principle of operation was known before the invention of the junction transistor [6], bipolar technology dominated the initial phase of the history of microelectronics. The primary reason for this initial difficulty was the lack of an understanding of the surface chemistry of silicon, leading to problems with the reliability and yield of early FET circuits. These problems were overcome in the early 1960s by the development of the metal oxide semiconductor (MOS) FET.

Figure 1.7 shows a schematic representation of a cross section through an advanced *n*-channel MOSFET of 100 nm channel length. The gate is isolated from the semiconductor by a 4.5-nm-thick gate oxide layer that is grown by

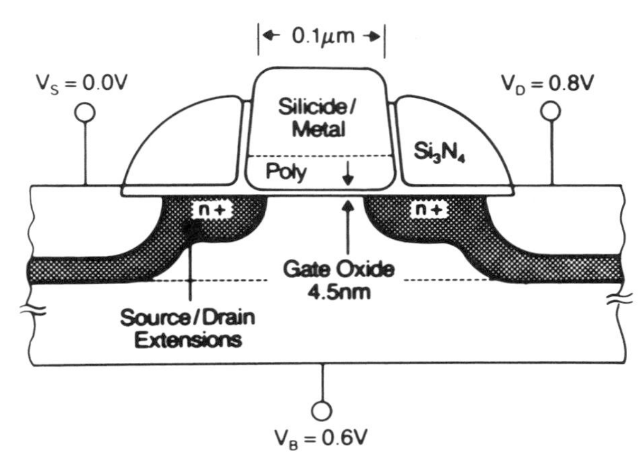

Figure 1.7 Schematic representation of a cross section through an advanced *n*-channel MOSFET device. After Kern [7]; copyright © 1989, Springer Verlag, Berlin.

thermal oxidation of the silicon surface at ~1000°C. The isolation of the metallization grid is provided by a thicker field oxide layer that is usually deposited by chemical vapor deposition methods. Since the source and drain regions must be heavily doped, the IC fabrication process still involves ion implantation and annealing steps, as in the case of bipolar technology. However, it does usually require fewer processing steps and is thus more reliable and less expensive than the bipolar process. Depending on the type of conduction in the channel, the potential at which a MOSFET conducts may be either high or low.

Figure 1.8 shows designs of (a) inverter, (b) NOR, and (c) NAND circuits using complementary MOS (CMOS) and MESFET transistor technology, respectively. The combination of *n*-channel and *p*-channel FETs in Figure 1.8(a) results in a substantial simplification, since, for a low gate bias, a high impedance of the bottom NMOS transistor and a low impedance of the top PMOS transistor is obtained, yielding a high output close to the drain voltage V_{DD}. Conversely, a high input at both gates results in high impedance for the top PMOS transistor and low impedance for the bottom NMOS transistor, yielding a low output close to ground. Because one of the two transistors is always nonconducting, the power consumption of this circuit is very small.

In the NOR circuit shown in Figure 1.8(b), two normally conducting driver transistors are combined with a normally nonconducting load transistor. The circuit has been implemented using MESFET technology, which employs GaAs as the semiconductor and uses metal gate contacts without any gate oxide layer for the switching operations [7]. Since a high input at either of the driver transistors results in a low impedance, a high output is only obtained if both inputs are low. For the NAND circuit of Figure 1.8(c), a low output is obtained only if the gates *A* and *B* on the two normally nonconducting transistors are high, that is, if both transistors conduct.

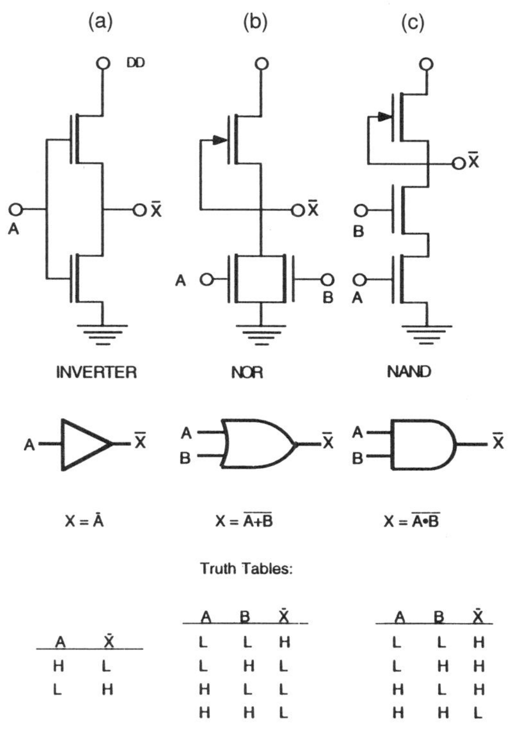

Figure 1.8 FET circuits, symbols, and truth tables for the inversion, NOR and NAND operations.

Although the current design goals of ULSI can be reached by known device concepts, significant changes in the processing technology are necessary for realizing submicrometer resolution. For example, since any thermal treatment expands the boundaries of previously established doped regions of the IC by interdiffusion, the tightening of the tolerances beyond VLSI designs requires fundamental changes in the processing conditions of epitaxy, oxidation, doping, and metallization. Low-temperature processes, utilizing plasma excitation or light to enhance the reaction rate, as well as rapid annealing techniques are favored because they limit or at least localize the cumulative effects of thermal cycles. The kinetics of these low-thermal-budget processes differs substantially from that encountered in conventional VLSI processing. Current efforts devoted to ULSI routinely produce circuits with 0.25 μm resolution. Research and development efforts aim currently at 0.1 μm resolution.

Figure 1.9 shows the progression of the number of transistors per chip over the past 2 decades, which doubled on average every 3 years. Concomitantly the computing power has drastically increased, and the cost per operation has significantly declined. As a consequence of this development, the cost of the products of the electronics industry has dramatically decreased. For example,

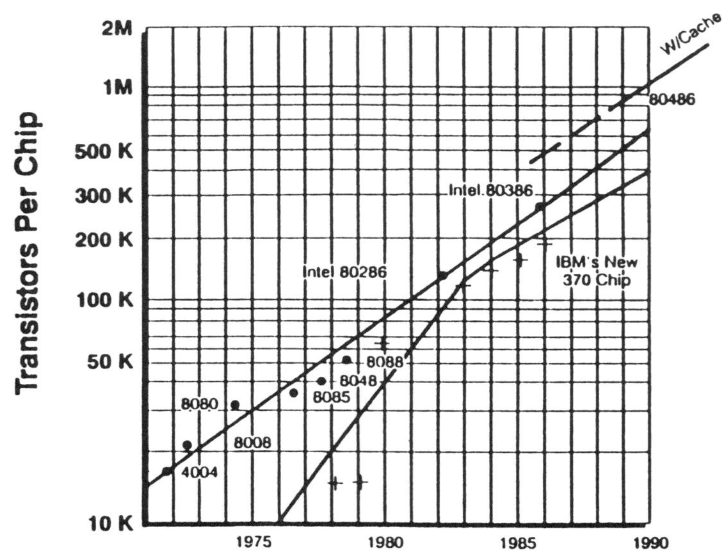

Figure 1.9 Number of transistors per chip for a variety of ICs. After Siu [8]; copyright © 1989, The Electrochemical Society, Pennington, NJ.

the cost of electronic calculators declined between 1960 and 1980 by approximately 3 orders of magnitude. In the past decade, the U.S. semiconductor industry has grown 5–6 times faster than the real GNP, and similar developments have been observed in Europe and Asia. The world market value of semiconductor ICs alone is predicted to reach $150 billion in 1996. At the beginning of the twenty-first century, the total annual value of the electronics industry is expected to reach $2 trillion and will thus become the world's largest industry. Production of 256 Mbit DRAMs and 4 Mbit SRAMs in an experimental stage of development appears to be realistic prediction for this time frame [9].

As shown in Figure 1.10, the delay achieved by conventional Si circuits tends toward saturation, but new devices and circuit designs allow for significant advances in speed if the RC time constant limitations that are associated with the electrical interconnects between circuit elements can be overcome. Past experience shows that new markets open up for Si ICs with enhanced storage/switching speed, once the technology exists. Therefore, it is likely that the progression of the complexity of silicon circuits based on MOSFET technology will continue to be supported by new applications.

However, novel processing approaches must be introduced because at feature sizes $\leq 0.1\,\mu$m, uv photolithography is no longer adequate for the patterning of the ICs because of the wavelength limit to the resolution. Also, the limitation of the switching speed and the increasing problem of power dissipation cannot be addressed with conventional approaches. In particular, new interconnect technologies and the exploration of alternative semiconduc-

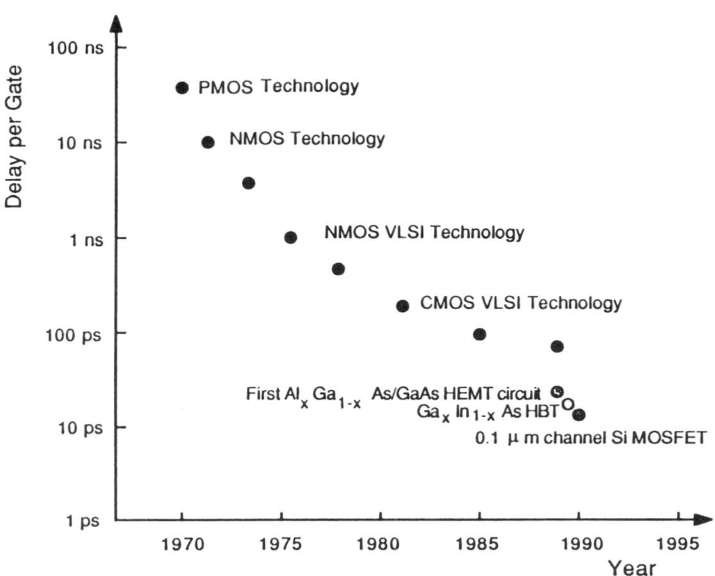

Figure 1.10 Decrease in the propagation delay of ICs in the past 25 years. ● Data of Siu [8].

tors with large carrier mobility and thermal conductivity are important. For example, optical interconnects, transferring information between circuit elements with the speed of light, are an attractive option and constitute a driving force for the utilization of compound semiconductors, which will be discussed further in Chapter 9.

Table 1.1 shows a section of the periodic table of the elements. The diamond structure elemental semiconductors silicon and germanium considered thus far belong to group IVB. As shown in Figure 1.1, each atom in the diamond structure is bonded to 4 other atoms of the same kind occupying the corners

Table 1.1 Section of the periodic table of the elements pertaining to the formation of the most important classes of compound semiconductors

IB	IIB	IIIB	IVB	VB	VIB
		13	14	15	16
		Al	Si	P	S
29	30	31	32	33	34
Cu	Zn	Ga	Ge	As	Se
47	48	49	50	51	52
Ag	Cd	In	Sn	Sb	Te

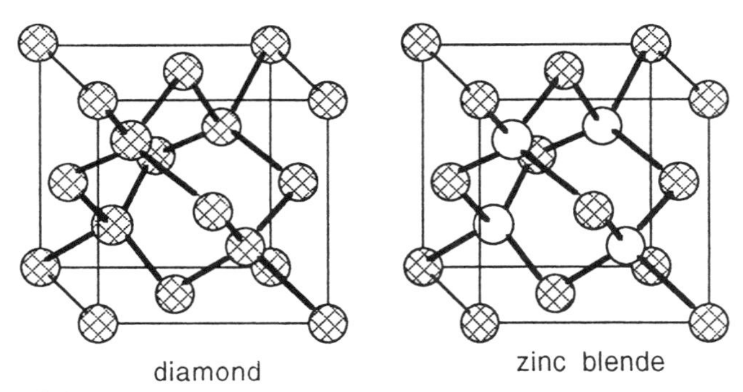

Figure 1.11 Schematic representation of the position and chemical bonding of atoms in the zincblende structure.

of a tetrahedron about the central atom. H. Welker of Siemens AG recognized in 1952 that an important class of compound semiconductors of general type *AB* is related to the diamond structure by replacing every second group IV atom by *A* and *B* atoms, respectively [10]. The compound structure thus created is shown in Figure 1.11 and is named the *zincblende* (sphalerite) structure after the mineral form of ZnS. Each *A* atom is tetrahedrally coordinated to *B* atoms and vice versa. An essential requirement for retaining this structure is that on average eight electrons must be provided to form the 4 covalent bonds between the atoms. This requirement is satisfied if the *A* atoms are chosen from the groups IIIB and IIB, respectively, and the *B* atoms are selected from the groups VB and VIB, respectively. Thus two families of compounds are defined that are the III–V compounds (e.g., GaAs) and the II–VI compounds (e.g., ZnS). Since gallium arsenide provides for a higher carrier mobility than silicon, it is currently a preferred compound semiconductor for high-speed ICs.

Unfortunately, the fabrication of GaAs circuits is complicated by the formation of native point defects on both the anion and cation sublattices of compounds of general composition *AB*. For example, there may exist vacancies on the *A* and *B* sublattices, *A* atoms on *B* sites and *B* atoms on *A* sites. The concentrations of these native point defects generally differ substantially from each other, depending on the processing conditions. As a net effect, this leads to deviations from the exactly stoichiometric composition of the compound. Native point defects often are electrically active and also affect the optical properties of the material. In addition, their interactions with impurities further complicate the electrical and optical properties. The great variety of chemical species present at the interfaces of compound semiconductors structures and the strong gradients in composition frequently result in interfacial interactions. The point defect chemistry near such interfaces then becomes rather complex, leading to variations in the yield and reliability of circuits. Although the degree to which this can be controlled for selected compounds and structures at this time is amazing, it often represents a serious limitation. For example, the adverse interfacial chemistry at the GaAs/native oxide interface as compared to the Si/SiO_2 interface prevents to date the extension of MOSFET technology to GaAs ICs. The level of integration of today's GaAs circuits is well below that of Si circuits, that is, $\sim 10^5$ transistors per chip. Nevertheless, the high speed of GaAs ICs provides an incentive for their use in advanced systems for both military and civilian applications [11].

A summary of the current state of the art and projections of the improvement in the performance expected of ULSI are presented in the form of a plot propagation delay versus power consumption per gate in Figure 1.12. The power–delay product is a figure of merit of digital ICs since both the speed of operation and the heat dissipation in high-density ICs are critical limiting aspects. VLSI circuits with 2–3 μm feature sizes achieve routinely power–delay products of 0.1–10 pJ at typically $\sim$1 ns delay for silicon circuits and 1–10 ps delay for GaAs MESFETs (see Section 8.4). In addition to the higher mobility of carriers in selected compound semiconductors as compared to silicon, further improvement of the power–delay product of ICs is attained by the use of heterostructures combining different semiconductors. An example of such a heterostructure is the combination of GaAs with $Al_xGa_{1-x}As$, that is, with a solid solution formed from GaAs and AlAs ($0 \leqslant x \leqslant 1$). It has a larger band gap than pure GaAs and matches the lattice constant of GaAs very well. Because of the small difference in the lattice constants of GaAs and AlAs, the

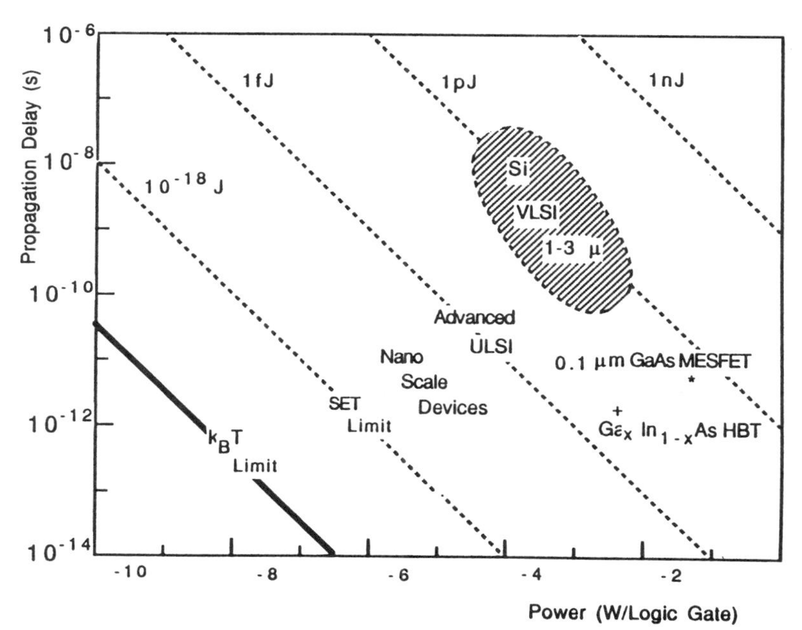

Figure 1.12 Propagation delay vs. power dissipation per gate for modern digital circuits.

growth of oriented single crystalline (epitaxial) films of $Al_xGa_{1-x}As$ on GaAs substrates proceeds with minimum strain. In heterostructure base transistors (HBTs), a narrow-gap base is combined with a wide-gap emitter, allowing a reduction of the propagation delay in HBT digital ICs at low power consumption per gate, that is, power–delay products of $\sim 10\,fJ$ (see Section 4.2). Advanced ULSI circuits are expected to reach 1 fJ and nanoscale devices may push the power–delay product beyond this limit. Initial steps towards such devices have been made in the context of confined heterostructures for which the dimensions of the potential wells, defined by modulations the conduction- and valence-band edges, become comparable to the de Broglie wavelength of the electrical carriers. For semiconductors, these parameters can be engineered within wide margins, which opens up opportunities for the development of novel device structures, such as high electron mobility transistors (HEMTs) (see Section 4.6).

For example, Figure 1.13(a) shows a high-resolution transmission electron microscopy image of a cross section through a $Al_xGa_{1-x}As/GaAs$ multiple heterostructure. The band-edge offsets in this heterostructure confine the electrons and holes inside two-dimensional quantum wells in the GaAs separated by $Al_xGa_{1-x}As$ barriers. As shown on the basis of a quantum-mechanical model in Chapter 2, the position relative to the band edges and spacings of the set of energy levels in the quantum well depend on the width and depth of the potential modulation. This is illustrated in Figure 1.13(b). By doping the $Al_xGa_{1-x}As$ layers heavily, leaving the GaAs layers undoped, mobile carriers spill from the doped barrier layers into the undoped quantum wells. Thus they are separated from the scattering dopant ions, providing for unusually high electron mobility.

The high-resolution heterostructures needed for the fabrication of HBTs and HEMTs are made by vapor deposition, which is discussed in detail in Chapter 6. In molecular beam epitaxy (MBE), ballistic beams of the precursors to epitaxial crystal growth are employed that mix at the surface of the substrate. Alternatively the surface of the substrate may be exposed to a vapor atmosphere at higher pressure, where gas-phase collisions provide for efficient exchange of energy and the precursors to epitaxial growth, formed by gas-phase reactions, diffuse across a boundary layer to the surface of the solid. Both molecular beam epitaxy and chemical vapor deposition processes currently achieve epitaxial layer growth with atomic resolution. Also, digital etching methods have been realized that achieve the monolayer by monolayer removal of atoms and permit thus the fine tuning of etching processes with atomic resolution [12].

The pattern definition processes discussed in Chapter 7 permit today in a manufacturing environment the resolution of $0.25\,\mu m$ features. However, advanced methods of patterning achieve in a research environment a resolution of $< 10\,nm$. In combination with modern techniques of vapor deposition, they provide access to the engineering of solid surfaces on a molecular scale. Simultaneously highly sensitive surface analysis and imaging methods achiev-

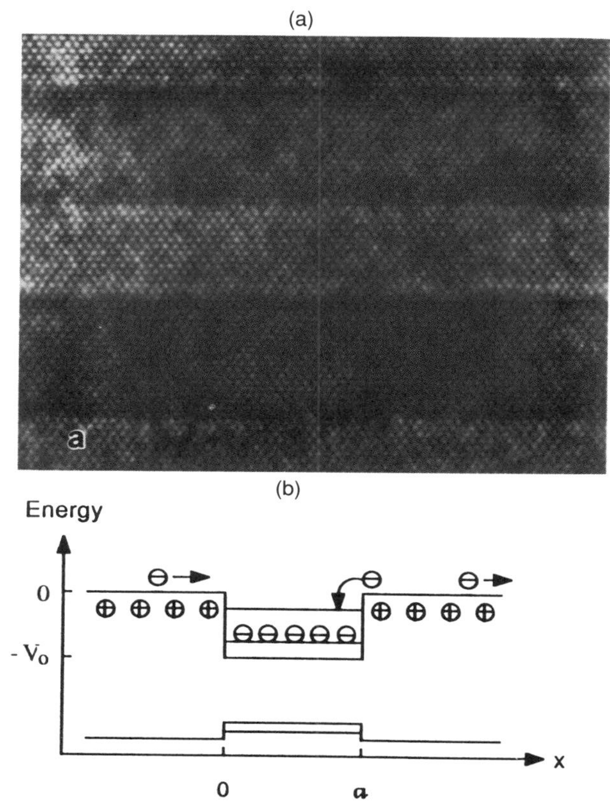

Figure 1.13 (a) High-resolution transmission electron micrograph of an AlAs/GaAs MQW heterostructure. (b) Schematic representation of the band-edge positions and carrier separation from the ionized donors in a modulation-doped single quantum well $Al_xGa_{1-x}As/GaAs/Al_xGa_{1-x}As$ heterostructure. (a) After Jeng et al. [12]; copyright © 1985, American Institute of Physics, New York.

ing atomic resolution have been developed that greatly enhance our ability to characterize nanoscale surface structures. New insights into the mechanisms of thermal oxidation of semiconductors and into the interactions of semiconductor surfaces with metal overlayers have been gained by these methods.

Further progress in the understanding of the structure–property relations of solids is essential for carrying the development of microelectronics to a scale below 100 nm. At this scale, the width of the depletion layers associated with electrical junctions becomes large compared to the device dimensions and discontinuities in the bands at heterojunctions become of increased interest in the control of device properties. Also, the electrical transport becomes largely ballistic, requiring new approaches to the modeling of device properties. Future advances with regard to the performance and reliability of nanoscale devices and circuits depend on the understanding of the local bonding environment during all processing steps, which are required to control the atomic structure at the critical interfaces. Chemistry thus plays an increasingly important role in advanced solid-state electronics research and development, which motivates the discussion of chemical bonding in Chapter 2.

References

1. J. Bardeen and W. H. Brattain, *Phys. Rev.* **74**, 230 (1948); U.S. Patent No. 2,524,035, October, 1950.
2. W. Shockley, *Proc. I.R.E.* **40**, 1365 (1952).
3. J. S. Kilby, *Miniaturized electronic circuits*, U.S. Patent No. 3,138,743, 1964.
4. Stan Augarten, State of the art, *A Photographic History of the Integrated Circuit*, Ticknor and Fields, New Haven and New York, 1983.
5. C. F. Wojslaw, *Integrated Circuits Theory and Applications*, Reston Publishing Company, Inc., Reston, VA, 1978.
6. J. E. Lilienfield, U.S. Patent No. 1,745,175, 1930.
7. D. Kern, *Sub-0.1 Micron MOSFETS, ESSDERC '89*, A. Heuberger, H. Ryssel, and P. Lange, eds., Springer Verlag, Berlin, 1989, p. 631.
8. W. M. Siu, in *ULSI Science and Technology*, C. M. Osburn and J. M. Andrews, eds., The Electrochemical Society, Pennington, NJ, 1989, p. 3.
9. I. Hayashi, *Jpn, J. Appl. Phys.* **32**, 266 (1993).
10. H. Welker, *Z. Naturforschg.* **7A**, 744 (1952).
11. W. Kellner, *GaAs Electronic Devices, ESSDERC '89*, A. Heuberger, H. Ryssel, and P. Lange, eds., Springer Verlag, Berlin, 1989.
12. Y. Horiike, T. Hashimoto, K. Asami, J. Amamoto, Y. Todokoro, H. Sakaue, S. Shingunbara, and H. Shindo, *Microelectronic Engineering* **13**, 417 (1991).
13. S. J. Jeng, C. M. Wayman, J. J. Coleman, and G. Costrini, *Materials Lett.* **3**, 89 (1985).

CHAPTER

2

The Electronic Structure of Atoms and Molecules

2.1 Early Concepts of Atomic Structure and Bonding

Since the electronic properties of solids are intimately linked to the chemical bonding and electronic properties of the constituent atoms and molecules, we review in this chapter selected topics of atomic structure, molecular symmetry, and chemical bonding. The first steps towards the present understanding of atomic structure and bonding were strongly influenced by investigations of the macroscopic symmetry of crystals, which arises from their strongly anisotropic growth in a supersaturated nutrient phase. Planes that are perpendicular to the directions of slowest growth are represented in the largest dimensions on the stationary growth form of a crystal.

Due to limitations of the growth kinetics, distortions of this ideal growth habit usually occur, which is illustrated in Figure 2.1, presenting schematically the growth of a crystal in two dimensions. Initially the crystal has a hexagonal shape. If there exist no kinetically induced differences in the addition of atoms or molecules in a supersaturated nutrient phase at the six equivalent faces labeled a–f, they advance with a uniform rate, maintaining the same shape for the full duration of growth, as shown in Figure 2.1a. However, due to the adsorption of impurities or differences in the defect density and in the mass transport to specific faces, at some time during growth, the rates at the initially equivalent faces may become different (e.g., increase at the faces A and D and decrease at the faces B and E). Thus the stationary shape of the crystal changes as demonstrated in Figure 2.1(b). As a consequence of such distortions, crystals of the same material occur in nature in a great variety of habits. This delayed

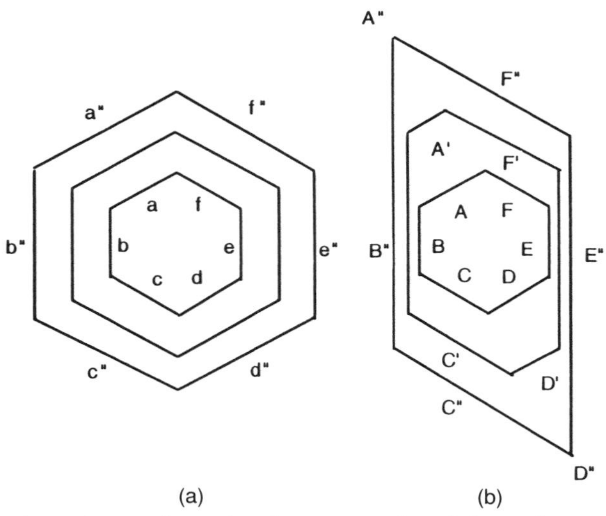

Figure 2.1 Schematic two-dimensional representation of the stationary growth form of crystals.

for many years the recognition of their fundamental symmetry properties. In 1669 the Danish crystallographer Nicolas Steno discovered that the angles between specific crystal faces were constant irrespective of their relative size on the stationary growth form. This discovery opened the door to the systematic evaluation of the macroscopic symmetry of crystals (see Chapter 3) and represented a significant first step in the development of modern solid-state sciences. It was followed within a decade by yet another pivotal discovery.

In an address to the French Royal Academy concerning the refraction of light in crystals, the Dutch physicist Christian Huygens presented in 1678 the following observations [1]:

> Rock crystal grows ordinarily in hexagonal bars and diamonds are found which occur with a square point and polished surfaces.... It seems that in general the regularity which occurs in these productions comes from the arrangement of small indivisible equal particles of which they are composed.... I will not undertake to say anything touching the way in which so many corpuscles all equal and similar are generated nor how they are set in such a beautiful order.... To develop truth so recondite there would be needed a knowledge of nature much greater than that we have. I will add only that these little spheroids could well contribute to form the spheroids of waves of light, here above supposed, these as well as those being similarly situated, and with their axes parallel.

Figure 2.2 shows a reproduction of Huygens' drawing of the arrangement of the elementary building blocks of a crystal [1]. It represents a remarkably foresighted conception that predates by more than a century Haüy's structure theory of crystals [2] and the formulation of the atomic theory of matter by John Dalton in 1803. In fact, Dalton's atomic theory, which was based on

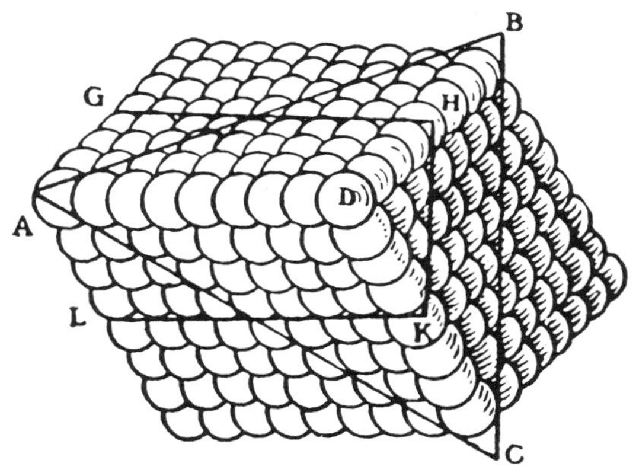

Figure 2.2 Drawing of the atomic arrangement and of the structure of the surface of a crystal. After Huygens [1].

strictly chemical arguments, remained hotly disputed well into the second half of the nineteenth century [3]. It was Mendeleev's formulation of the periodic table of the elements and his spectacular success in predicting the properties of unknown elements [4] that changed this perception. The atomic theory of matter became thus accepted and led to rapid progress in the understanding of the structure of solids.

The first theory of chemical bonding based on the electrostatic interactions between ions was formulated by Davy in 1806. Spatial models of the arrangements of atoms in molecules were developed by Laurent and Dumas in 1839, who were strongly influenced by Haüy's theory of crystal structures. In 1857 Kekulé von Stradonitz introduced the concept of covalent bonding between tetravalent carbon atoms, establishing the basis for modern organic structure theory, which was further advanced by his conception of the ring structure of benzene in 1861.

In addition to relating for the first time the physical properties of solids to their atomic structure, Huygens explained in his *traite de la lumiere* the phenomenon of birefringence. The birefringence of calcite provided the basis for the generation of polarized light, which enabled a number of important subsequent discoveries. For example, certain optically active molecules interact with polarized light, rotating the plane of polarization either to the left or to the right. In 1874 van t'Hoff and Le Bel postulated that the valence bonds of a carbon atom point to the corners of a tetrahedron and linked the optical activity of organic molecules to their symmetry (i.e., the presence of a carbon atom bonded to four different atoms or chemical groups). Although this explanation had to be modified on the basis of subsequent observations [5], it stimulated research on the symmetry of molecules well into the twentieth century. Also, the observation of the change in the optical activity upon the conversion of cane sugar into fructose and glucose enabled the first study of the rate of a chemical reaction. The enhancement of the rate of a chemical reaction by the presence of materials that are themselves not changed was investigated by Berzelius, who coined the name *catalysis* for this phenomenon in 1836. A further advance in the understanding of vapor–solid surface interactions was made in 1873 by van der Waals' investigation of the weak bonding forces between molecules that cause the nonideal behavior of gases at high density [6] and enable the adsorption of gases at the surface of solids. They also establish, in conjunction with hydrogen bonding [7], the secondary and tertiary structures of organic macromolecules that coil or fold up under the influence of weak intramolecular forces and determine thus to a large degree the physical properties of polymers.

At the turn of the century a series of discoveries and bold advances in theoretical physics set the stage for the development of modern concepts of the electronic structure of atoms and molecules. In particular, the discovery of radioactivity by Becquerel in 1896, and its subsequent investigation by Rutherford and others, led to new insights into the electronic structure of atoms. Based on his experimentation concerning the scattering of α particles by the atoms of a gold foil, Rutherford concluded that an atom consists primarily of empty space with most of its mass located in a positively charged nucleus that is surrounded by an equal electronic charge. The success of Planck's formulation [8] of the radiation law in explaining the properties of blackbody radiation, based on the assumption of quantized energy exchange between the walls of a cavity and the enclosed radiation field, established in 1900 the first quantum theory. It was followed by Bohr's interpretation of Rutherford's model of the atom on the basis of quantized angular momentum [9], which explained Ritz's principle, that is, the relation of the wave numbers of the line spectra of atoms to differences between spectroscopic terms.

Bohr's theory made a significant impact on the development of the theory of chemical bonding. In 1916 Lewis [10] and Kossel proposed the octet rule [11], explaining the formation of ionic bonds by the transfer of electrons from atoms with an excess of electrons to atoms that have a deficiency of electrons as compared to the stable 8 outer electron configuration of the noble gases. It explains qualitatively why the elements on the left and right sides of the periodic table tend to form cations and anions, respectively, and the elements located in the center of the periodic table tend to form covalent bonds by the sharing of electrons. The representation of covalent chemical bonds by straight lines between the element symbols, introduced by Cooper in 1858, was thus interpreted as the pairing of electrons. Also, the donation of an electron pair from an ion or molecule possessing lone pairs of electrons (Lewis base) to an ion or molecule having a gap in its octet shell (Lewis acid) explained the formation of complex bonds. However, although the Kossel–Lewis model provided an interpretation of the earlier chemical bonding symbolism in terms of the electronic structure of atoms and ions, numerous exceptions and modifications to the octet rule were soon required to keep up with the experimental evidence. Also, the selection rules governing the atomic and molecular spectra could not be understood on the basis of Bohr's postulate and demonstrated the need for a more sophisticated theory.

This theory was developed in the 1930s and 1940s, building upon the concepts of wave mechanics introduced by Schrödinger in 1926 [12]. The basis for this development was laid by a series of observations of particlelike behavior of waves [13]–[15]. They were summarized by de Broglie's hypothesis [16] that all waves are associated with quasiparticles of momentum $\mathbf{p}$ and all particles are associated with waves of wavelength λ so that

$$|\mathbf{p}|\lambda = m|\mathbf{v}_g|\lambda = |\mathbf{k}|\hbar\lambda = h, \tag{2.1}$$

where h is Planck's constant, m is the mass and $\mathbf{v}_g$ is the group velocity of the quasiparticle, λ is the wavelength, and $\mathbf{k} = 2\pi/\lambda$ is the wave vector of the associated wave [8]. The wave character of particles was confirmed by Davisson and Germer in 1927 by the observation of the diffraction of a beam of electrons on a nickel foil [17]. For further details of these exciting developments in the history of chemistry and physics, the reader is referred to Refs. [18] and [19], respectively.

2.2 Elements of Quantum Mechanics

According to the correspondence principle, any physical observable is represented in quantum mechanics by a linear operator acting on the wave function Ψ. In particular, the stationary electronic states of atoms, ions, molecules, and solids are the solutions of the time-independent Schrödinger equation

$$H\Psi = (T + V)\Psi = \left(\frac{\mathbf{p}^2}{2m} + V\right)\Psi = E\Psi, \tag{2.2}$$

where H is the Hamiltonian, which is composed of the kinetic energy operator T and potential energy operator V. We consider as an example the electronic states in the quantum-well heterostructure described in Figure 1.12b. They are the solutions to Equation (2.2) for a rectangular potential well of depth $-V_0$ and width a (see Fig. 1.12). Since the momentum $\mathbf{p}$ corresponds to the operator $(\hbar/i)\nabla$ [20], for the one-dimensional problem, Equation (2.2) assumes inside the well the form

$$-\frac{\hbar^2}{2m}\frac{\partial^2}{\partial x^2}\Psi - (E + V_0)\Psi = 0. \tag{2.3}$$

Substituting $V = 0$ in the ranges $x < 0$ and $x > a$, we get

$$\frac{\partial^2}{\partial x^2}\Psi - k_0^2\Psi = 0, \qquad k_0 = \frac{\sqrt{(-2mE)}}{\hbar} \tag{2.4a}$$

outside the well and

$$\frac{\partial^2}{\partial x^2}\Psi + k^2\Psi = 0, \qquad k = \frac{\sqrt{[2m(E + V_0)]}}{\hbar} \tag{2.4b}$$

inside the well. Since $-V_0 < E < 0$, $|\mathbf{k}| = k$, and $|\mathbf{k}_0| = k_0$ are real. With the boundary condition of vanishing wave function at infinite distance from the well, the solutions to Equations (2.4a) and (2.4b) are

$$\Psi = \exp(k_0 x), \qquad x \leqslant 0 \tag{2.5a}$$

$$\Psi = A\exp(-k_0 x), \qquad x \geqslant a \tag{2.5b}$$

and

$$\Psi = B\exp(ikx) + C\exp(-ikx), \qquad 0 \leqslant x \leqslant a. \tag{2.5c}$$

The continuity of both Ψ and $\partial\Psi/\partial x$ at $x = 0$ and $x = a$ results in the conditions

$$B + C = 1, \qquad x = 0, \tag{2.6a}$$

$$k_0 = ikB - ikC, \qquad x = 0, \tag{2.6b}$$

$$A\exp(-k_0 a) = B\exp(ika) + C\exp(-ika), \qquad x = a, \tag{2.6c}$$

$$-k_0 A\exp(-k_0 a) = ikB\exp(ika) - ikC\exp(-ika), \qquad x = a, \tag{2.6d}$$

which are satisfied simultaneously only for specific values of the energy

$$E_n = -V_0(1 - \alpha_n^2). \tag{2.7}$$

In Equation (2.7),

$$\alpha_n = \left(\frac{E_n + V_0}{V_0}\right)^{1/2} \tag{2.8}$$

and n is an integer labeling the individual solutions. They are determined by the compatibility condition

$$\gamma\alpha_n = (n - 1)\pi + 2\cos^{-1}\alpha_n \tag{2.9}$$

derived from Equation (2.6) [21] with

$$\gamma = \left(\frac{2mV_0 a^2}{\hbar^2}\right)^{1/2}. \tag{2.10}$$

As illustrated in Figure 2.3(a), each specific value of γ determines a set of values α_n, E_n, and the associated wave functions Ψ_n. Solving Equation (2.6) for A, B, and C results in

$$\Psi_n = \begin{cases} \dfrac{1}{\alpha_n}\sin\left[k_n\left(x - \dfrac{a}{2}\right) + \dfrac{n\pi}{2}\right], & 0 \leqslant x \leqslant a \quad (2.11a) \\ \exp(k_{0n}x), & x < 0 \quad (2.11b) \\ (-1)^{n+1}\exp[-k_{0n}(x - a)], & x > a \quad (2.11c) \end{cases}$$

Figure 2.3(b) shows Ψ_n for $n = 3$. We note that, for any γ, the lowest value of α_n (i.e., the largest negative value of E_n) corresponds to $n = 1$ and that with

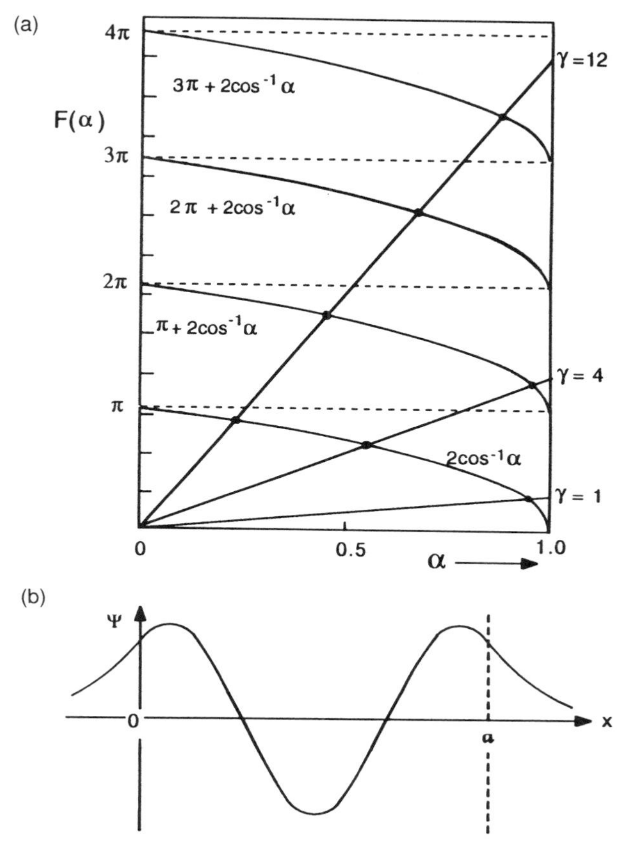

Figure 2.3 (a) Graphical solution of Equation (2.9) for $\gamma = 1$, 4, and 12. (b) Representation of Ψ_n for $n = 3$. After Powell and Craseman [21].

increasing γ the number of energy levels inside the well increases. In particular, for $V_0 \to \infty$, the line $F(\alpha) = \gamma\alpha$ approaches the ordinate of Figure 2.3, resulting in an infinite number of intersections $ka = n\pi$ and consequently in a countably infinite set of energy levels and wave functions.

Figure 2.4 shows the sets of energy levels for $\gamma = 1$, 4, and 12 using for convenience a variable scale as indicated by the well depth values

$$V_0 = \gamma^2(\hbar^2/2ma^2). \tag{2.12}$$

As will be shown in Chapter 3, in a crystalline solid the electron mass must be replaced by an effective mass. Otherwise the theory holds to a lower limit of a. A discussion of very thin quantum wells will be provided in Chapter 4.

For bound states, that is, when the classical motion is confined by the potential, the solutions $\psi_i (i = 1, 2, \ldots)$ to Equation (2.2) are always quantized; that is, the set of energy values, for which solutions to Equation (2.2) exist, is discrete. Consequently, the set of solutions defines a vector space (Hilbert

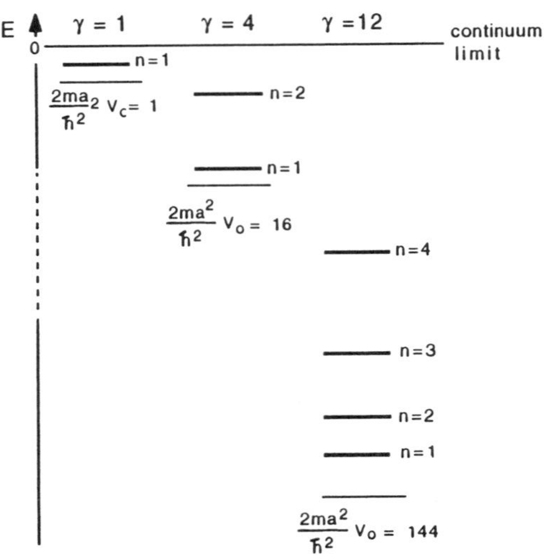

Figure 2.4 Energy eigenvalues for a square potential well of depth-V_0 and width a at selected values of γ. After Powell and Craseman [21].

space). Any arbitrary vector ϕ_l' in Hilbert space may be represented by a linear combination of the set of basis vectors; that is, for a countably infinite set

$$\phi_l' = \sum_{i=1}^{\infty} c_i \psi_i \tag{2.13}$$

As there exists an infinite number of sets of three noncoplanar vectors that define a three-dimensional space, the definition of Hilbert space by a specific set of wave vectors may be replaced by other sets of linearly independent basis vectors related to the original one by transformation matrices. A set of wave vectors ψ_i is said to be linearly independent when Equation (2.12) vanishes if and only if all $c_i = 0$. The electronic wave vectors solving Equation (2.2) are functions of spatial and spin coordinates. Scalar products between wave vectors are defined as

$$\langle \psi_i | \psi_k \rangle = \int_{\text{all space}} \psi_i^* \psi_k \, d\tau \tag{2.14}$$

where ψ_i^* stands for the complex conjugate of ψ_i and $d\tau$ is a volume element of the coordinate space in which ψ_i is defined. The length or norm of the wave vector ψ_i is represented by $\langle \psi_i | \psi_i \rangle^{1/2}$. Vectors of unit norm are said to be *normalized*. For orthogonal vectors $\langle \psi_i | \psi_k \rangle = 0$. It is always possible to construct from a complete linearly independent set ψ_i a complete set of linearly independent orthonormal vectors Δ_j so that $\langle \Delta_i | \Delta_k \rangle = \delta_{ik}$, where $\delta_{ik} = 1$ if $i = k$ and $\delta_{ik} = 0$ if $i \neq k$ (see Ref. [20] for more information).

If the action of the linear operator on a wave vector merely changes its length, the solutions are said to be *state vectors*, each belonging to a well-defined *eigenvalue* of the observable. Since H is a linear operator and E is a constant, Equation (2.2) is an example of such an eigenvalue equation. If d distinct state vectors ψ_l solve the eigenvalue equation for the same eigenvalue of the observable, this eigenvalue is said to be d-fold degenerate. The average value or expectation $\langle\psi_i|A|\psi_i\rangle$ of an observable for a system in state ψ_i is given by

$$\langle\psi_i|A|\psi_i\rangle = \int_{\text{all space}} \psi_i^* A\psi_i \, d\tau \tag{2.15a}$$

or if the observable is linked to a coupling of states ψ_i and ψ_k

$$\langle\psi_i|A|\psi_k\rangle = \int_{\text{all space}} \psi_i^* A\psi_k \, d\tau \tag{2.15b}$$

Representing the set of solutions ψ_l to an eigenvalue equation in the form of a column vector $\{\Psi\}$, the operation of the general linear operator A on Ψ may be expressed as

$$\{A\Psi\} = [A]\{\Psi\}. \tag{2.16}$$

where $[A]$ is the matrix

$$[A] = \begin{bmatrix} A_{11} & A_{12} & A_{13} & \cdots \\ A_{21} & A_{22} & A_{23} & \cdots \\ A_{31} & A_{32} & A_{33} & \cdots \\ \cdot & \cdot & \cdot & \cdots \end{bmatrix} \tag{2.17}$$

relating the components $(A\Psi)_k$ of the column vector $\{A\Psi\}$ to the components ψ_l of the column vector $\{\Psi\}$ according to

$$(A\Psi)_k = \sum_{l=1}^{\infty} A_{kl}\Psi_l \tag{2.18}$$

for a countably infinite set of wave vectors. Let $[M] = [M_{ik}]$ denote a matrix with generally complex elements M_{ik} and $[M]^+ = [M_{ki}^*]$ denote its Hermitian conjugate, that is, a matrix where all elements are replaced by their complex conjugates M_{ik}^* and the rows and columns are scrambled. If the rows and columns are mutually orthogonal so that $[M][M]^+ = [\delta_{ik}]$, the matrix is said to be unitary. In particular, for unitary matrices

$$[U][U]^+ = [U]^+[U] = [1] \tag{2.19}$$

or

$$[U]^+ = [U]^{-1}. \tag{2.20}$$

The unitary transformation

$$\begin{aligned} [U]^+([H]\Psi) = [U]^+(E\Psi) &= E[U]^+\Psi = E\Delta \\ &= [U]^+[H][U][U]^+\Psi = [H']\Delta, \end{aligned} \tag{2.21}$$

that rotates and scales the arbitrary set of wave vectors $\{\Psi\}$ into the orthonormal set $\{\Delta\}$ is thus associated with the transformation of the Hamiltonian matrix $[H]$ operating on $\{\Psi\}$ into the matrix $[H'] = [U]^{+}[H][U]$ operating on $\{\Delta\}$. Since

$$\langle \Delta_i | H' | \Delta_k \rangle = E_k \langle \Delta_i | \Delta_k \rangle = E_k \delta_{ik}, \tag{2.22}$$

in the absence of degeneracies, the matrix $[H']$ is diagonal, that is, is of the form

$$[H'] = \begin{bmatrix} H_{11} & 0 & 0 & 0 & 0 & \cdots \\ 0 & H_{22} & 0 & 0 & 0 & \cdots \\ 0 & 0 & H_{33} & 0 & 0 & \cdots \\ 0 & 0 & 0 & H_{44} & 0 & \cdots \\ \cdot & \cdot & \cdot & \cdot & \cdot & \cdots \end{bmatrix} \tag{2.23}$$

Since chemical applications of quantum mechanics frequently are too complicated to allow solving Equation (2.2) exactly, approximate solutions must be found. This is done by replacing the complete expansion of the wave vectors by a linear combination of a reduced number n of basis functions

$$\Psi_{\approx} = \sum_{i=1}^{n} c_i \phi_i. \tag{2.24}$$

The energy calculated with the approximate wave function is thus

$$E_{\approx} = \frac{\int_{\text{all space}} \Psi_{\approx}^{*} H \Psi_{\approx} \, d\tau}{\int_{\text{all space}} \Psi_{\approx}^{*} \Psi_{\approx} \, d\tau} > E \tag{2.25}$$

which is less negative than the value obtained for the exact solution. The deviation of the approximated energy from the true value is a good measure of the degree to which the approximate wave function approaches the exact solution to Equation (2.2). Based on this assessment, the variational principle is formulated that achieves the best approximation of Equation (2.13) by the minimization of $E_{\approx}$ through the variation of the c_i in Equation (2.21). Let

$$S_{kl} = \int \phi_k \phi_l \, d\tau \tag{2.26a}$$

$$H_{kl} = \int \phi_k H \phi_l \, d\tau \tag{2.26b}$$

denote the overlap integral and the Hamiltonian matrix element, respectively, between states ϕ_k and ϕ_l of the set ϕ_i that is used in the approximation Equation (2.24). Then upon substitution of Equation (2.26) into Equation (2.25)

$$E_{\approx} = \frac{\Sigma_k \Sigma_l c_k c_l H_{kl}}{\Sigma_k \Sigma_l c_k c_l S_{kl}} \tag{2.27}$$

or

$$\sum_k \sum_l c_k c_l (H_{kl} - E_{\approx} S_{kl}) = 0. \tag{2.28}$$

By differentiation of Equation (2.28) we get

$$2\sum_l c_l (H_{kl} - E_{\approx} S_{kl}) - \sum_k \sum_l c_k c_l S_{kl} \frac{\partial E_{\approx}}{\partial c_k} = 0 \tag{2.29}$$

For $E_{\approx}$ to have a minimum, the condition $\partial E_{\approx}/\partial c_k = 0$ must be satisfied, which results in a set of n secular equations

$$\sum_l c_l (H_{kl} - E_{\approx} S_{kl}) = 0, \tag{2.30}$$

one for each c_k $(k = 1, 2, \ldots, n)$. The set of nontrivial solutions to Equation (2.30) is represented by the secular determinant

$$|H_{kl} - E_{\approx} S_{kl}| = 0. \tag{2.31}$$

For example, in case of an expansion using only two basis vectors, the secular determinant is

$$\begin{vmatrix} H_{11} - E_{\approx} & H_{12} - E_{\approx} S_{12} \\ H_{21} - E_{\approx} S_{21} & H_{22} - E_{\approx} \end{vmatrix} = 0. \tag{2.32}$$

For $H_{12} = H_{21}$ and $S_{12} = S_{21}$, this determinant corresponds to the quadratic equation

$$E_{\approx}^2 - E_{\approx}(H_{11} + H_{22} - 2H_{12}S_{12})/(1 - S_{12}^2) + (H_{11}H_{22} - H_{12}^2) = 0, \tag{2.33}$$

which has two solutions, $E_{\approx 1}$ and $E_{\approx 2}$, respectively. Inserting these solutions into the secular equations results in the coefficients c_1 and c_2 that determine $\Psi_{\approx 1}$ and $\Psi_{\approx 2}$.

2.3 Atomic Wave Functions and Energy Eigenvalues

Electrons moving in the central field of the nucleus of an atom or ion are described by the N-electron Hamiltonian

$$H = \sum_{i=1}^{N} \left(-\frac{\hbar^2}{2m}\Delta_i - \frac{Zq^2}{r_i} + \sum_{j<i}^{N} \frac{q^2}{r_{ij}} + \xi(r_i)\mathbf{l}_i \cdot \mathbf{s}_i \right) + H' \tag{2.34}$$

The four terms in the large parentheses represent the kinetic energy of the electrons, the nuclear–electronic attraction, electron–electron repulsion, and spin–orbit coupling, respectively, and H' is a term accounting for a number of small corrections. For the hydrogen atom ($Z = N = 1$) and one-electron ions

($Z \neq 1, N = 1$), the Hamiltonian may be represented by the first two terms in the large parentheses of Equation (2.34), resulting in exact solutions to Equation (2.2) of the form

$$\Psi_{nlm} = R_{nl}(r)\Theta_{lm}(\theta)\Phi_{m}(\phi) \tag{2.35}$$

where n, l, and m are quantum numbers. The detailed properties of these functions are discussed in textbooks of atomic physics [19] and spectroscopy [22]. Here we review only a few selected features that are needed in the subsequent discussions of chemical bonding and the analytical evaluation of materials and processes by atomic spectroscopies.

The eigenvalues of energy associated with the wavefunctions of the hydrogen atom are quantized according to

$$E_n = Z^2 \frac{mq^4}{2\hbar^2 n^2} = Z^2 \frac{E_B}{n^2} \tag{2.36}$$

where $E_B = -13.6\,\text{eV}$ is the energy of the hydrogen atom in the ground state. An analogous relation governs the energy eigenvalues associated with shallow impurity states in semiconductors that are discussed in Chapter 3. The length of the angular momentum vector $\mathbf{l}$ and its component with regard to one distinct direction in space, which is usually chosen to be parallel to the z axis, are quantized according to the eigenvalue equations

$$|\mathbf{l}|^2\Psi = l(l+1)\hbar^2\Psi \tag{2.37}$$

$$l_z\Psi = m\hbar\Psi \tag{2.38}$$

Since for a given principal quantum number n, the azimuthal quantum number l ranges from 0 to $n - 1$, and there are $2l + 1$ values of the magnetic quantum number m for each value of l, there exist

$$\sum_{l=0}^{n-1} (2l + 1) = n^2 \tag{2.39}$$

distinct orbital wavefunctions Ψ_{nlm} for each eigenvalue E_n. Also, the electron spin is quantized according to the eigenvalue equation

$$s_z u = m_s \hbar u \tag{2.40}$$

with $m_s = \pm 1/2$. Therefore, two spin orbitals $u_{nlmm_s} = \sigma_\alpha \Psi_{nlm}$ and $u_{nlmm_s} = \sigma_\beta \Psi_{nlm}$ are formed from each orbital wavefunction Ψ_{nlm} by multiplication with spin functions σ_α for $m_s = +\frac{1}{2}$ and σ_β for $m_s = -\frac{1}{2}$. Consequently the eigenvalues of the hydrogen atom are $2n^2$-fold degenerate.

Since the function $R_{nl}(r)$ is isotropic, the angular symmetry of the wavefunctions is represented by the spherical harmonics

$$Y_l^m = \Theta(\theta)\Phi(\phi) \tag{2.41}$$

that are listed in Table 2.1 for $l \leqslant 2$. For $m \neq 0$ the wavefunctions are complex. However, since any linear combination of the set of degenerate wavefunctions

Table 2.1 Spherical harmonics for $l \leqslant 2$

$Y_0^0 = 1/\sqrt{4\pi}$	$Y_1^0 = \sqrt{3/4\pi}\cos\theta$
$Y_1^1 = \sqrt{3/8\pi}\sin\theta\exp(i\phi)$	$Y_1^{-1} = \sqrt{3/8\pi}\sin\theta\exp(-i\phi)$
$Y_2^0 = (\sqrt{5}/4\sqrt{\pi})(3\cos^2\theta - 1)$	$Y_2^1 = \sqrt{15/8\pi}\sin\theta\cos\theta\exp(i\phi)$
$Y_2^{-1} = \sqrt{15/8\pi}\sin\theta\cos\theta\exp(-i\phi)$	$Y_2^2 = (\sqrt{15}/4\sqrt{2\pi}\sin^2\theta\exp(2i\phi)$
$Y_2^{-2} = (\sqrt{15}/4\sqrt{2\pi})\sin^2\theta\exp(-2i\phi)$	

also solves the Schrödinger equation for the same degenerate eigenvalue of energy as the original set, it is possible to construct from the set of degenerate complex wavefunctions an equivalent set of real wavefunctions. In particular, it is possible to rewrite the wavefunctions as products $F(r)G(r, \theta, \phi)$ of an isotropic real function $F(r)$ and a real function $G(r, \theta, \phi)$ that is equal to products of the Cartesian coordinates x, y, z of order l. These basis functions represent the angular symmetry of the wavefunctions (see [41] and Section 2.6 for further discussions). By convention, the electronic wavefunctions are labeled by the letters *s*, *p*, *d*, *f*,... for $l = 0, 1, 2, 3, \ldots$ adding the basis functions as right-hand subscripts (e.g., p_x, p_y, p_z for $l = 1$ or d_{xy}, d_{xz}, d_{yz}, $d_{x^2-y^2}$, d_{z^2} for $l = 2$).

For multielectron atoms or ions, no analytical solutions to Equation (2.2) exist. However, accurate numerical solutions can be obtained by a method introduced by D. R. Hartree in 1927 [23]. This method treats the electrons of the atom as separate states solving Equation (2.2), one electron at a time, for the field created by the nucleus and the charge-density distribution of the other electrons. Starting with a finite set of trial functions that retain the angular symmetry properties of the hydrogen wavefunctions, but are modified in the radial part, first the spherically averaged potential U is calculated. This potential is then used to determine a new set of improved wavefunctions. The variational principle is applied in the iterative process until the energy does not change within a chosen error between cycles. The field of the average space-charge distribution given by the one-electron wavefunctions determined in this fashion must be the same as the field used in their calculation. Therefore, the method is called the *self-consistent-field* (SCF) *method*.

Table 2.2 shows the one-electron energies for the first 36 elements of the perodic table. Because of the spherical averaging of the potential, the angular symmetry of the wavefunctions is still represented by the spherical harmonics Y_l^m; that is, there are $2l + 1$ orbital wavefunctions belonging to the same eigvalue of the energy. However, the n^2-fold degeneracy of the energy eigenvalues of the hydrogen atom is no longer valid for multielectron atoms.

According to the Bohr–Stoner "Aufbauprinzip" [24], the positions of the atoms in the periodic table may be understood by placing electrons into spin orbitals in ascending order of their energies. The electron configuration of the element with nuclear charge Z is thus related to that of the element with nuclear charge $Z - 1$ by the addition of one electron. Table 2.3 shows the

Table 2.2 One-electron energies (Rydberg) for the first 36 elements (after Slater [25])

Element	1*s*	2*s*	2*p*	3*s*	3*p*	3*d*	4*s*	4*p*
H	1.00							
He	1.81							
Li	4.77	0.40						
Be	8.9	0.69						
B	14.5	1.03	0.42					
C	21.6	1.43	0.79					
N	30.0	1.88	1.03					
O	39.9	2.38	1.17					
F	51.2	2.95	1.37					
Ne	64.0	3.56	1.59					
Na	79.4	5.2	2.80	0.38				
Mg	96.5	7.0	4.1	0.56				
Al	115.3	9.0	9.0	0.83	0.44			
Si	135.9	11.5	7.8	1.10	0.57			
P	158.3	14.1	10.1	1.35	0.72			
S	182.4	17.0	12.5	1.54	0.86			
Cl	208.4	20.3	15.3	1.86	1.01			
Ar	236.2	24.2	18.5	2.15	1.16			
K	266.2	28.2	22.2	3.0	1.81		0.32	
Ca	297.9	32.8	26.1	3.7	2.4		0.45	
Sc	331.1	37.3	30.0	4.2	2.6	0.59	0.55	
Ti	366.1	42.0	34.0	4.8	2.9	0.68	0.52	
V	402.9	46.9	38.3	5.3	3.2	0.74	0.55	
Cr	441.6	51.9	43.0	6.0	3.6	0.75	0.57	
Mn	482.2	57.7	47.8	6.6	4.0	0.57	0.50	
Fe	524.3	63.0	52.8	7.3	4.4	0.64	0.50	
Co	568.3	69.0	58.2	8.0	4.9	0.66	0.53	
Ni	614.1	75.3	63.7	8.7	5.4	0.73	0.55	
Cu	662.0	81.3	69.6	9.6	6.1	0.79	0.57	
Zn	712.0	88.7	76.2	10.5	7.0	1.28	0.69	
Ga	764.0	96.4	83.0	11.8	7.9	1.6	0.93	0.44
Ge	818.2	104.6	90.5	13.5	9.4	2.4	1.15	0.55
As	874.5	113.0	98.5	15.4	10.8	3.4	1.30	0.68
Se	932.6	122.1	106.8	17.3	12.2	4.5	1.54	0.80
Br	993.0	131.7	115.6	19.9	13.8	5.6	1.80	0.93
Kr	1055.5	142.0	124.7	22.1	15.9	7.1	2.00	1.03

Bayley–Thomsen–Bohr step-pyramid representation of the periodic table of the elements, which illustrates this principle explicitly. However, caution is in order with regard to this description of the electronic structure by a particular electron configuration, because the construction of the atomic wavefunctions from one-electron wavefunctions generally requires the consideration of a number of different electron configurations.

According to the Pauli principle, the atomic wavefunction Ψ_a must be antisymmetric [26]; that is, Ψ_a must change sign upon simultaneous exchange

Table 2.3 The modified Bayley–Thomsen–Bohr step-pyramid periodic table (After Jensen [29])

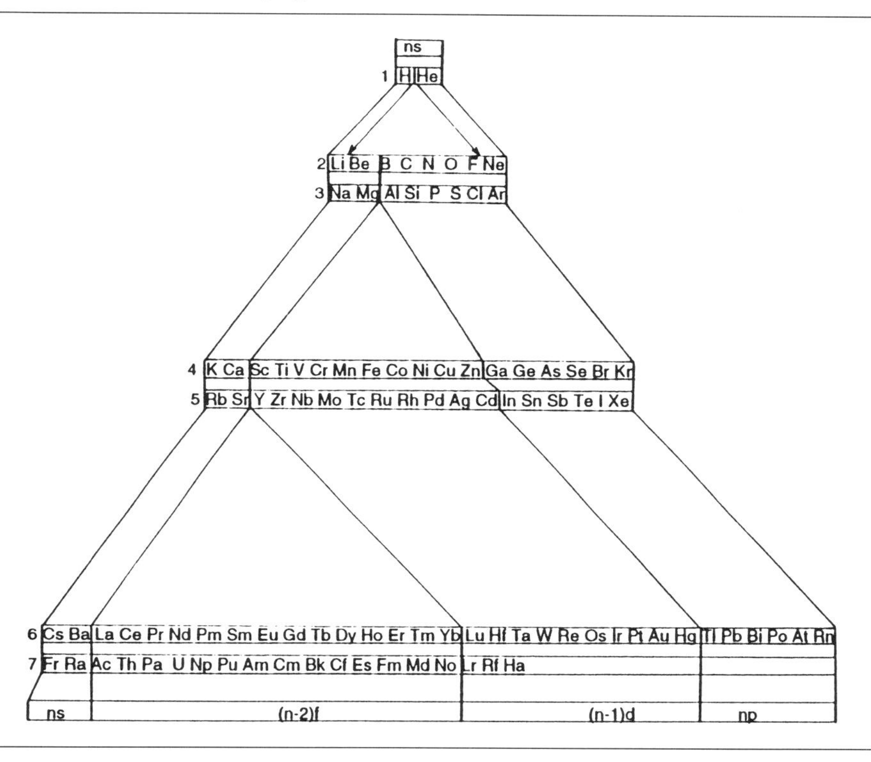

of the spatial and spin coordinates of two electrons. For closed-shell atoms, an appropriate representation of Ψ_a in terms of the one-electron spin orbitals u_i was introduced by Fock [27] and Slater [28] in the form of the determinant

$$\Psi_a = |S| = \frac{1}{\sqrt{N!}} \begin{vmatrix} u_1(1) & u_2(1) & . & u_N(1) \\ u_1(2) & u_2(2) & . & u_N(2) \\ \vdots & \vdots & . & \vdots \\ u_1(N) & u_2(N) & . & u_N(N) \end{vmatrix} \tag{2.42}$$

where the factor $1/\sqrt{N!}$ assures the normalization of Ψ_a if the u_i are normalized. The numbers in parentheses in Equation (2.42) represent the different sets of spatial and spin coordinates of the N electrons of the atom. Since they are indistinguishable, a permutation of all electrons and spin orbitals is required, which is provided in the determinantal form of Equation (2.42). In accord with the Pauli principle, the Slater determinant $|S|$ changes sign upon exchange of two rows. Also, it vanishes if the spatial and spin coordinates of two electrons are equal, that is, if two rows are identical. Therefore, two electrons with same spin cannot exist at the same location in space. Neglecting spin–orbit coupling and the term H' in Equation (2.43), the Hartree–Fock energy is given by

$$\begin{aligned}\int \Psi^* H \Psi \, d\tau = \sum_{i=1}^{N} \int u_i^*(1) \left(-\frac{\hbar^2}{2m} \Delta_i - \frac{Zq^2}{r_i} \right) u(1) \, d\tau \\ + \sum_{i>j}^{N} \iint u_i^*(1) u_j^*(2) \frac{q^2}{r_{12}} u_i(1) u_j(2) \, d\tau_1 \, d\tau_2 \\ - \sum_{i>j;\|}^{N} \iint u_i^*(1) u_j^*(2) \frac{q^2}{r_{12}} u_i(1) u_j(2) \, d\tau_1 \, d\tau_2 \end{aligned} \tag{2.43}$$

Note that Equation (2.43) contains an exchange term that sums over all electrons with parallel spins. Since this term lowers the energy, the ground state of an atom generally is represented by an electron configuration maximizing the number of parallel spins (Hund's rule).

For open-shell atoms, spin orbitals of higher energy than indicated by the electron configuration derived by the Aufbauprinzip must be included in the calculation of the energy. This was first realized by Fock [28], who introduced the use of a linear combination of Slater determinants to represent the atomic wavefunction; that is,

$$\Psi_a(1, 2, \ldots, N) = \sum_{S=1}^{\binom{N_S}{N}} c_S |S| \tag{2.44}$$

The upper limit on the sum in Equation (2.44) represents the fact that there are $N_s!/N!(N_s - N)!$ possible arrangements of N electrons into N_s available spin orbitals; that is, each Slater determinant Ψ_S represents a specific electron configuration by a combination of a specific set of spin orbitals. The Bohr–Stoner principle determines the positions of the elements in the periodic table correctly because the selected electron configurations make the largest contributions to the ground-state energies. Nevertheless they correspond to a considerable simplification. In accurate calculations of the ground-state energies of atoms by the Hartree–Fock (HF) method all the possible electron configurations for an open-shell atom are taken into account. However, since there are correlations of the electrons that go beyond the correlation due to the Pauli principle, the HF ground-state energies still deviate by a small correlation energy correction from the true values, corresponding to the term H' in Equation (2.34).

2.4 Molecular Wave Functions and Energy Eigenvalues

The calculation of the electron wave functions and energies for molecules requires solving Equation (2.2) with a Hamiltonian of the form

$$H_t = T_N + T_E + V_{NN} + V_{EN} + V_{EE} \tag{2.45}$$

where T_N and T_E, V_{NN} and V_{EE}, and V_{EN} represent, respectively, energies of the nuclei and electrons, the nuclear and electronic repulsions, and the attraction between the electrons and the nuclei of the molecule. Since the electrons move in the field of the nuclei of all atoms belonging to the molecule, shielded by the charge distribution of all electrons, the task of solving Equation (2.2) with the Hamiltonian (2.45) is complex and usually requires several approximations in managing the problem. Because the masses of the nuclei are large as compared to the electronic mass, in the Born–Oppenheimer approximation, the assumption is made that the electrons can adjust instantaneously to any change in the nuclear positions. This allows a separation of the effects of the nuclear motion. First the electronic contributions to the total energy are calculated for fixed nuclei using the simplified Hamiltonian

$$H = H_T - T_N, \tag{2.46}$$

where, for a given internuclear distance R_N, V_{NN} is a constant, and all other terms depend only on the coordinates of the electrons. Solving Equation (2.2) with the Hamiltonian given by Equation (2.46) results in the electronic molecular wavefunctions Ψ_{Ei} and the associated energy eigenvalues E_{Ei}. The calculation of E_{Ei} as a function of the internuclear distance defines the potential energy function that governs the nuclear motion in the molecular state represented by Ψ_{Ei}, e.g., the electronic ground state. Solving Equation (2.2) with the Hamiltonian $H = T_N + E_{Ei}(R_N)$ that depends only on the nuclear coordinates results the nuclear wavefunctions Ψ_{Nki} and the associated energy eigenvalues E_{Nki}. Since additive terms in the Hamiltonian correspond to

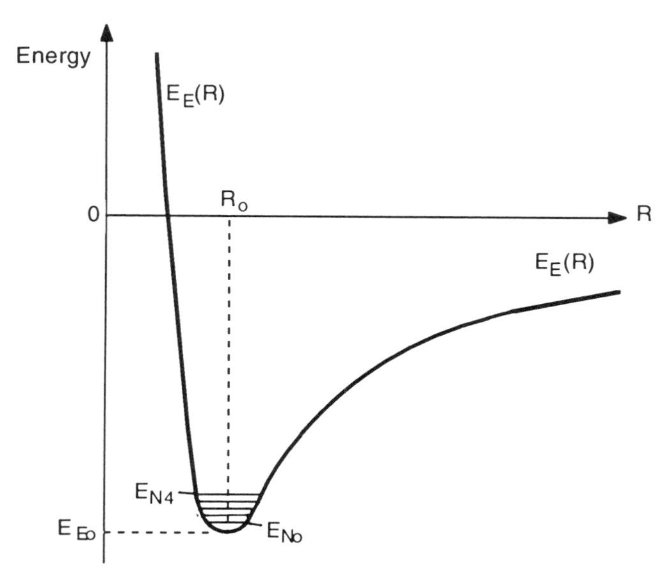

Figure 2.5 Schematic representation of the superposition of the nuclear contributions to the energy.

product wave functions [30], they combine with the electronic wave functions to the complete solutions of the Schrödinger equation

$$\Psi_t = \Psi_{Ei}\Psi_{Nki} \tag{2.47}$$

Figure 2.5 shows schematically the curve $E_{Ei}(R_n)$ and the superimposed set of eigenvalues E_{Nki}, which are quantized and represent the vibrational and rotational motions of the molecule.

Depending on the type of further simplifications made after the Born–Oppenheimer approximation, several types of calculations of molecular wavefunctions and energy eigenvalues may be distinguished. Valence-bond (VB) theory adheres to the two-center bonding concept of classical chemistry by an a priori distinction between bonding electrons and electrons that are not involved in the bonding. It evolved from a calculation of Heitler and London for the H_2 molecule, which was the first quantum chemical calculation addressing covalent bonding [30]. In the molecule, the electrons of the two hydrogen atoms interact with both nuclei and are indistinguishable. Therefore, the molecular wavefunction must be expressed as a linear combination of product functions, that is, $\Psi = c_1\Psi_{1a}\Psi_{2b} + c_2\Psi_{2a}\Psi_{1b}$, where for example, Ψ_{1b} refers to the association of electron 1 with nucleus *b* (compare Figure 2.6). In this labeling the assumption is made that if electron 1 is associated with nucleus *a*, electron 2 is associated with nucleus *b* and vice versa. Ionic configurations, where both electrons are associated with the same nucleus, are thus deliberately neglected. In more advanced VB calculations this restriction is removed. Upon determination of the constants c_1 and c_2, two solutions to Equation (2.2) are obtained:

$$\Psi_+ = \frac{\Psi_{1a}\Psi_{2b} + \Psi_{2a}\Psi_{1b}}{(2 + 2S^2)^{1/2}} \tag{2.48}$$

$$\Psi_- = \frac{\Psi_{1a}\Psi_{2b} - \Psi_{2a}\Psi_{1b}}{(2 - 2S^2)^{1/2}} \tag{2.49}$$

$S = \langle \Psi_a | \Psi_b \rangle$ represents the overlap integral for the H_2^+ molecule, for which exact solutions to Equation (2.2) exist.

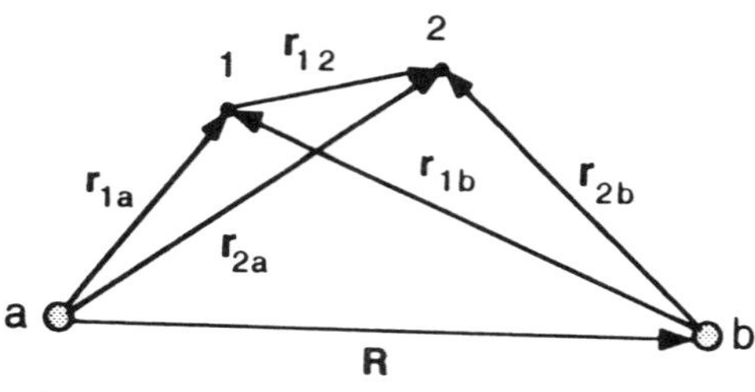

Figure 2.6 Schematic representation of the positions of the nuclei and electrons in the H_2 molecule.

In the case where both electrons have same spin ($\alpha_1\alpha_2$ or $\beta_1\beta_2$), antisymmetric spin orbitals are obtained only for multiplication with Ψ_-. If the two electrons have opposite spins, linear combinations of spin functions must be used to account for the indistinguishability of the electrons. Antisymmetric spin orbitals are obtained upon multiplication of Ψ_- by $\alpha_1\beta_2 + \alpha_2\beta_1$ and of Ψ_+ by $\alpha_1\beta_2 - \alpha_2\beta_1$. Therefore the molecular spin orbitals (MSOs) formed from Ψ_+ and Ψ_- are a singlet and a triplet, respectively. The energy eigenvalues associated with Ψ_+ and Ψ_- are

$$E_{\pm} = \frac{E_{1s} + H_1 \pm H_2}{1 \pm S^2}, \tag{2.50}$$

$$H_1 = -\int \frac{\Psi_{1a}^2}{r_{1b}}\, d\tau_1 - \int \frac{\Psi_{2b}^2}{r_{2a}}\, d\tau_2 + \frac{1}{R} + \iint \frac{\Psi_{1a}^2 \Psi_{2b}^2}{r_{12}}\, d\tau_1\, d\tau_2 \tag{2.51}$$

$$H_2 = -S \int \frac{\Psi_{1a}^* \Psi_{1b}}{r_{1b}}\, d\tau_1 - S \int \frac{\Psi_{2a}^* \Psi_{2b}}{r_{2a}}\, d\tau_2 + \frac{S^2}{R} + \iint \frac{\Psi_{1a}^* \Psi_{1b} \Psi_{2b}^* \Psi_{2a}}{r_{12}}\, d\tau_1\, d\tau_2 \tag{2.52}$$

are the direct and the exchange integrals, respectively. H_1 represents the Coulomb attractions of electron 1 to nucleus b and of electron 2 to nucleus a the nuclear repulsion and the repulsion between the two electrons, respectively. H_2 is characterized by integrands associating the electrons simultaneously with both nuclei. This is in accord with the Kossel–Lewis concept of covalent bonding as a sharing of the bonding electron pair by the two nuclei between which the bond is established. However, no classical interpretation is possible of the various terms in H_2, which plays an important role in determining the dissociation energy D.

In extensions of the Heitler–London (HL) model to other polyatomic molecules [31]–[33], valence bonds are set up between the valence electrons of the constituent atoms using HL-type wavefunctions. In the perfect pairing approximation the assumption is made that the interaction of individual valence electron pairs forming a bond is not affected by the formation of the other bonds in the molecule. If this approximation holds, the heat of formation of a compound with purely covalent bonding is simply the sum of the enthalpies associated with the various bonds minus the enthalpy of atomization of the constituent elements. Therefore, bonding enthalpies can be estimated from thermodynamic data, or conversely heats of formations of compounds can be estimated from known bonding enthalpies and enthalpies of atomization.

A selection of bond energies is listed in Table 2.4. Note that these energies are identical to the dissociation energies only for biatomic molecules. Otherwise the dissociation energies are different from the average bond energies. For example, the O–O bond energy shown in Table 2.4 is derived from the enthalpy of formation of H_2O_2, assuming that the O–H bond energies are the same as the average O–H bond energy of the water molecule. In heteronuclear molecules the polarization of the molecule due to differences in the attraction of the electronic charge by the two nuclei must be taken into account. In classical VB theory, bond polarities are evaluated on the basis of electronegativity scales. For example, Mulliken defined an electronegativity scale by averaging the ionization energy and the electron affinity. There are other definitions that are discussed in more detail in Refs. [34] and [35]. However, although these scales are useful in qualitative evaluations, they are not a reliable guide for the quantitative determination of the charge distribution in molecules.

The energy of polyatomic molecules in VB theory is minimized by the provision of maximum overlap of the wavefunctions of the individual bonding electron pairs and minimum overlap of the other wavefunctions in the region of this bond. The atomic orbitals (AOs) calculated for the spherically symmetric charge distribution of isolated atoms generally do not conform to this requirement. However, it is possible to construct by an appropriately chosen linear combination of these AOs hybridized wavefunctions of the atom in a hypothetical valence state that results in maximum overlap along the bonds. In classical VB theory, specific valence states corresponding to specific mixes of the atomic orbitals (AOs) are used and define characteristic sets of bond angles.

Table 2.4 Average energy values for bonds (kcal/mol), $1\text{eV} = 3.829 \times 10^{-20}$ cal (after Pauling [33])

Bond	Energy	Bond	Energy	Bond	Energy
H–H	104.2	C–H	98.8	C–F	105.4
C–C	83.1	Si–H	70.4	Si–F	129.3
C=C	147	N–H	93.4	N–F	64.5
C≡C	193.8	P–H	76.4	As–F	111.3
Si–Si	42.2	As–H	58.6	C–Cl	78.5
Ge–Ge	37.6	O–H	110.6	Si–Cl	85.7
N–N	38.4	S–H	81.1	Ge–Cl	97.5
N=N	111.4	Se–H	66.1	N–Cl	47.7
N≡N	226.1	H–F	134.6	P–Cl	79.0
P–P	51.3	H–Cl	103.2	F–F	36.6
As–As	32.1	C–Si	69.3	Cl–Cl	58.0
O–O	33.2	C–N	69.7	C–O	84.0
S–S	50.9	C=N	147.2	C=O	174.2
Se–Se	44.0	C≡N	213.2	Si–O	88.2

For example, the sp^3 hybrid valence-shell wavefunctions of the group IVB elements are mixtures of the type

$$\begin{aligned}
\Psi_1 &= (s + p_x + p_y + p_z)/2, \\
\Psi_2 &= (s - p_x - p_y + p_z)/2, \\
\Psi_3 &= (s + p_x - p_y - p_z)/2 \\
\Psi_4 &= (s - p_x + p_y - p_z)/2
\end{aligned} \tag{2.53}$$

For the purpose of discussion, we show in Figures 2.7(a) and (b) the geometry of the sp^3 hybrid wavefunctions in reference to a Cartesian coordinate system and a $(\bar{1}10)$ cut (see Section 3.1 for an explanation of the Miller indices) through the functions Ψ_1 and Ψ_2, respectively. The bond directions point to the corners of a tetrahedron so that the pairs of hybrid functions Ψ_1, Ψ_3 and Ψ_3, Ψ_4 lie on the $(\bar{1}10)$ and (110) planes, respectively. They are thus perpendicular to each other.

The considerable asymmetry in the hybridized wavefunctions between the

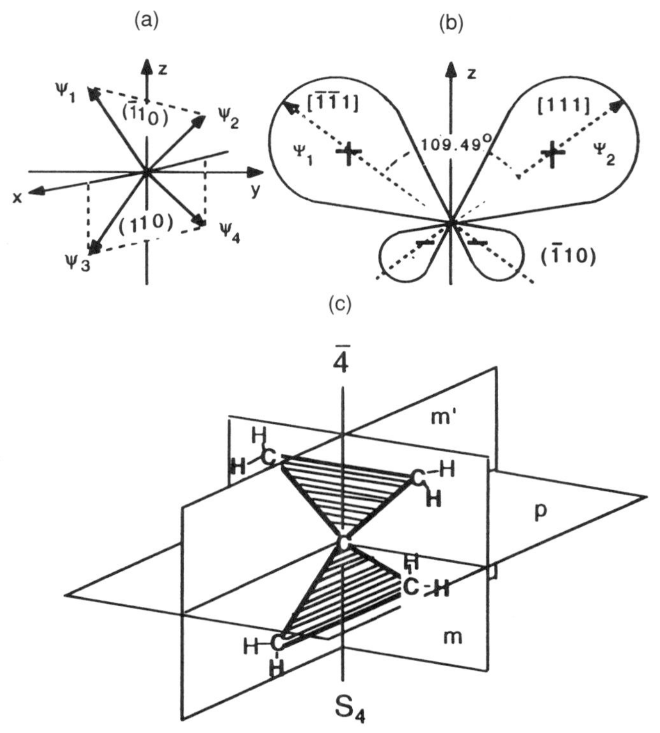

Figure 2.7 (a) Schematic illustrations of the directions of the sp^3 hybrid wavefunctions, (b) of a $(\bar{1}10)$ cut through Ψ_1 and Ψ_2, and (c) of the spiropentane molecule.

positive and negative lobes assures a substantial gain in the overlap in the formation of bonds provided there are no prohibitive structural limitations. Figure 2.7(c) shows a schematic representation of the spiropentane molecule, which illustrates the importance of structural constraints. The bond angles in each of the three-membered rings in this molecule cannot be achieved by any real combination of *s* and *p* AOs to hybrid valence-bond functions; that is, the C–C bonds in these rings are bent. In the language of VB theory, this reduces the energy per bond as compared to an unstrained C–C bond because of a loss in the achievable overlap. However, even in the absence of bond strain, the bond angles in many molecules deviate substantially from the ideal bond angles predicted on the basis of the hybridized wavefunctions of VB theory.

An attempt at reconciling these observations with VB theory is made in the valence-shell electron-pair repulsion (VSEPR) model by including into the consideration of molecular geometry the repulsions exerted by nonbonding lone pairs of electrons onto the bonding pairs of the molecule [34]. Since lone pairs take up more space than bonding electron pairs, the following rank order exists in the repulsions: lone pair–lone pair > lone pair–bonding pair > bonding pair > bonding–pair > single electron–bonding pair. Also, multiple bonds take more space than single bonds, and increasing electronegativity of the elements participating in the bonding reduces the space taken by the bonding electron pair and consequently its repulsion by other pairs. Thus the bond angles in the molecules $AsCl_3$, $AsBr_3$, and AsI_3 increase in the stated order. A systematic representation of predicted molecular shapes based on the VSEPR model is shown in Table 2.5. Although the VSEPR model and other general principles developed on the basis of classical VB theory have helped chemists to rationalize many empirically established relations and stimulated chemical research, they are modifications of a model "valence" state of the atoms that is entirely fictitious. Therefore, it is more satisfying to look for modifications of the theory that retain the two-center bond concept of chemistry, but do not depend on an a priori determined mixing scheme. Such a theory is provided by the generalized valence-bond (GVB) method [35]. In contrast to classical VB theory, the orbitals are determined self-consistently so that the molecular geometry and the details of the bonding evolve as a result of the minimization of the energy.

However, the expense in GVB computing is substantial, favoring self-consistent molecular orbital (MO) calculations that are based on expansions of the molecular wavefunction by a limited set of basis functions. This is similar to the calculations for multielectron atoms of Section 2.2, but with the complications of the Hamiltonian Equation (2.45). Since atomic orbitals must emerge upon complete dissociation of the molecule into atoms, and the inner electrons close to the individual atomic cores in the molecule resemble closely the corresponding atomic wavefunctions, it is reasonable to start out with a linear combination of atomic orbitals Ψ_i (LCAO approximation)

$$\Psi_{MO} = \sum_{i=1}^{v} c_i \Psi_i \tag{2.54}$$

Table 2.5 Structures of simple molecules predicted by the VSPER model (after Bestgen et al. [36])

Number of Pairs	Number of bond Explanation	Molecular Shape	Number of Pairs	Number of bond Explanation	Molecular Shape
2	2 AX_2 linear	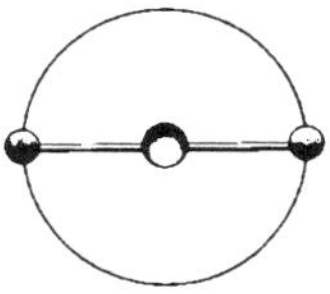	5	4 AX_4E distorted tetrahedron	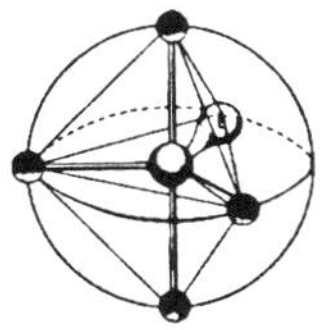
3	3 AX_3 planar triangular	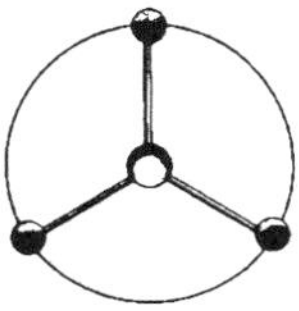	5	3 AX_3E_2 T-shaped	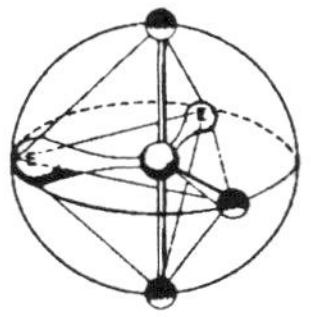
3	2 AX_2E V-shaped	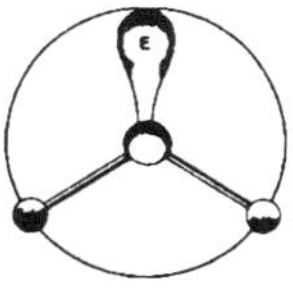	5	2 AX_2E_3 linear	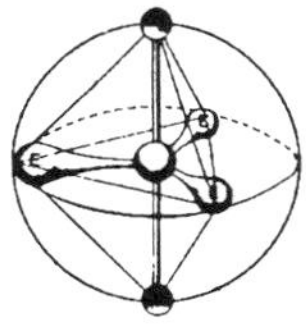
4	4 AX_4 tetrahedral	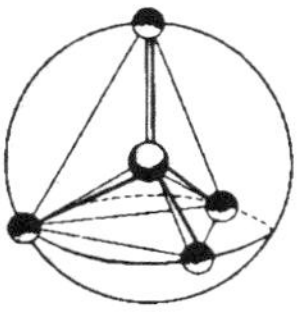	6	6 AX_6 octahedral	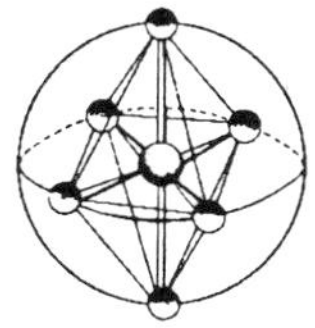
4	3 AX_3E trigonal-bipyramidal	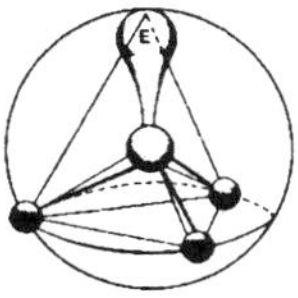	6	5 AX_5E quadratic bipyramidal	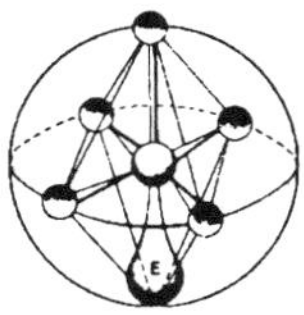
4	2 AX_2E_2 V-shaped	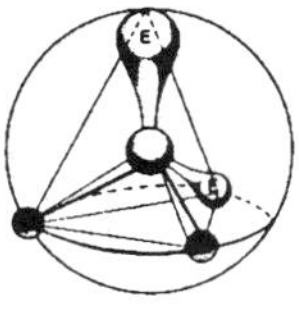	6	4 AX_4E_2 planar quadratic	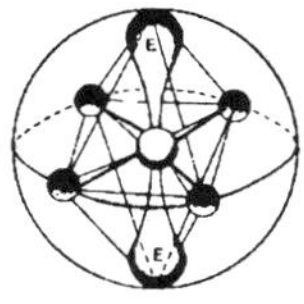
5	5 AX_5 trigonal bipyrimidal	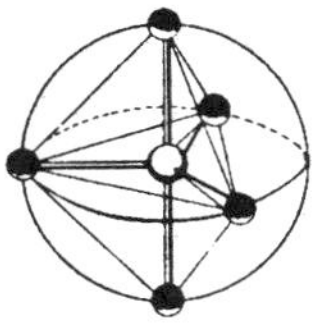			

utilizing the formalism developed for the optimization of the coefficients and the calculation of the energies that are associated with the set of approximate solutions to Equation (2.2). For the H_2 molecule, according to Equation (2.33), two MOs are formed from the two 1*s* AOs. They are associated with energies located below and above the energy of the hydrogen atom. Placing the two *s* electrons into the MO with lower energy results in a lowering of the energy for the molecule as compared to the two hydrogen atoms in separation. Conversely, placing the two electrons into the MO having higher energy results in a more negative energy for the two hydrogen atoms in separation. Therefore, these two MOs are called *bonding* and *antibonding* orbitals, respectively.

The two-basis function approximation may be extended to other homonuclear biatomic molecules, resulting in a pair of bonding and antibonding orbitals [37]–[40] for each pair of electrons of same energy as illustrated in Figure 2.8. In analogy to the Aufbauprinzip for atoms, electron configurations for molecules may be generated by filling the MOs with electrons in ascending order of their energies allowing for two electrons with antiparallel spins per MO at maximum. The labeling of the MOs in Figure 2.8 is based on their symmetry with regard to rotations about the internuclear axis, which is described by the function $\Phi'(\phi) = \exp(\pm il\phi)$, where l is an integer. Similar to the labeling of atomic orbitals according to their quantum numbers l, the one-electron MOs are labeled by the lower case Greek letters $\sigma, \pi, \delta, \ldots$ for $l = 0,\ 1,\ 2, \ldots$, respectively. The σ bonds have cylindrical symmetry about the internuclear axis, while π bonds are characterized by a nodal plane containing the internuclear axis as illustrated in Figure 2.9. The MOs for entire molecules are labeled by the corresponding capital letters, that is, Σ, Λ, and Δ for $l = 0$, 1,

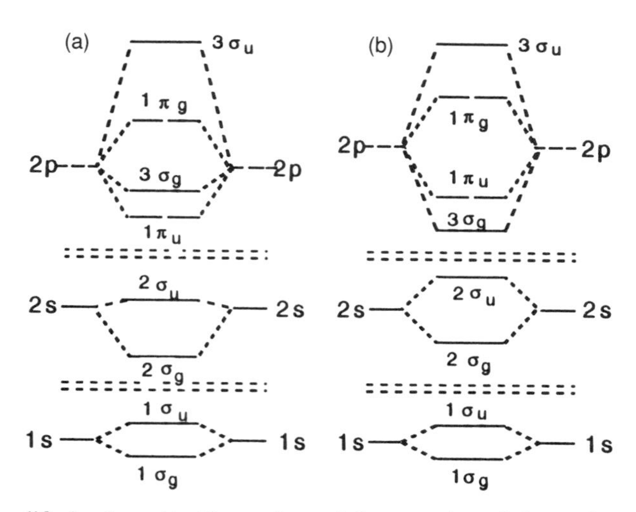

Figure 2.8 Simplified schematic illustration of the energies of the MOs of biatomic homonuclear molecules.

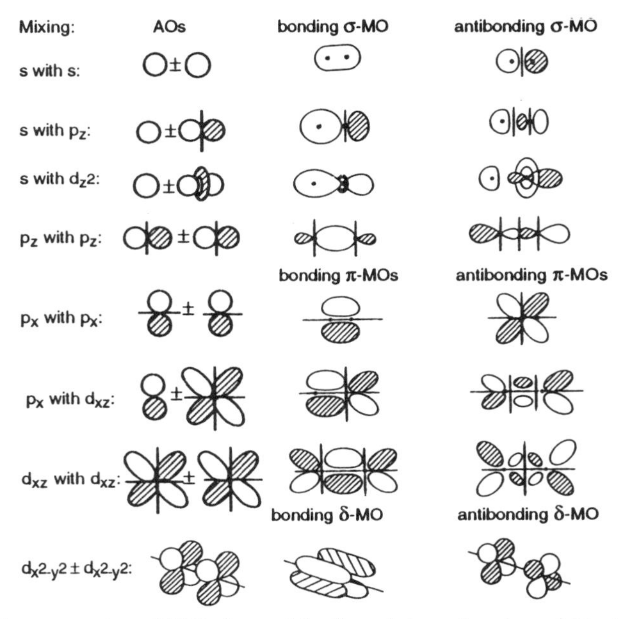

Figure 2.9 The symmetry of MOs formed by the mixing of various AOs. After Douglas and Hollingsworth [22].

and 2, respectively. Right-hand superscripts + (−) and subscripts $g(u)$ on the MO symbols refer to even (uneven) behavior of the MO with regard to reflections on a symmetry plane containing the internuclear axis and to inversion on the midpoint between the nuclei on the internuclear axis, respectively. The spin multiplicity is listed as a left-hand superscript.

Table 2.6 shows the electron configurations for the biatomic molecules and selected molecular ions of the first 10 elements of the periodic table. An excellent correlation exists between the dissociation energies and internuclear distances, respectively, on the one side and the differences between the number of electrons N_b and N_a in bonding orbitals and antibonding orbitals, respectively, on the other side. However, the experimentally observed dissociation energies D_{exp} consistently exceed the results D_{HF} of HF calculations employing a minimum basis of atomic wavefunctions. This problem in the prediction of dissociation energies is largely due to the fact that the HF calculation does not account for the interatomic correlation of the electrons that tend to separate their charge distributions. These correlations are:

1. The "in–out correlation" corresponding to the tendency of one electron to reside primarily in the outer region of the molecule if the other electron is located primarily in the interior region.

2. The "left–right correlation" corresponding to the tendency of the electrons to separate to the vicinity of different nuclei.
3. The "angular correlation" corresponding to the tendency of electrons being confined in approximately the same region to separate by the angle.

As in the case of the electronic states of atoms, corrections to the deficiency of molecular HF theory with regard to interatomic electron correlations can be made by incorporating configuration interactions into the calculations. This is the basis of multiconfiguration self-consistent-field (MCSCF) calculations. They include the optimization of the set of basis functions as part of the SCF calculation [41]. The results listed in Table 2.6 suggest that these corrections are far more important for molecules than the corrections for intra-atomic electron correlation to the HF calculations for atoms. MCSCF calculations that neglect configuration-interaction (CI) corrections for the intraatomic electron correlations beyond the correlation due to the Pauli principle, but account for the interatomic electron correlations result in optimized valence configuration (OVC) wavefunctions. An important aspect of making such a calculation as economical as possible is the selection of an appropriate set of electron configurations that contribute the most to the energy of the molecule. In addition to the configuration representing the ground state, which is determined in a zero-order approximation to the correlated wavefunctions by a HF calculation, configurations corresponding to various excited states must be considered. A discussion of the selection criteria for these excited states is outside the scope of this book, and the author refers the reader to Ref. [41] for a concise review of this topic.

Table 2.6 Electron configurations of the biatomic homonuclear molecules and of selected molecular ions of the first 10 elements of the periodic table (after Kutzelnigg [40])

Molecule	Configuration	$N_b - N_a$	R_n (Å)	D_{HF} (eV)	D_{exp} (eV)
H_2^+	$1\sigma_g$	1	1.06	–	2.7
H_2	$(1\sigma_g)^2$	2	0.74	–	4.7
He_2^+	$(1\sigma_g)^2 1\sigma_u$	1	1.08	–	2.5
He_2	$(1\sigma_g)^2(1\sigma_u)^2$	0	–	–	–
Li_2^+	$(1\sigma_g)^2(1\sigma_u)^2 2\sigma_g$	1	3.14	–	1.3
Li_2	$(1\sigma_g)^2(1\sigma_u)^2(2\sigma_g)^2$	2	2.67	0.17	1.05
Be_2	$(1\sigma_g)^2(1\sigma_u)^2(2\sigma_g)^2(2\sigma_u)^2$	0	2.45	–	0.09
B_2	$[Be_2](1\pi_u)^2$	2	1.59	0.89	3 ± 0.5
C_2	$[Be_2](1\pi_u)^4$	4	1.24	0.79	6.36
N_2^+	$[Be_2](1\pi_u)^4 3\sigma_g$	5	1.12	3.13	8.86
N_2	$[Be_2](1\pi_u)^4(3\sigma_g)^2$	6	1.10	5.18	9.90
O_2^+	$[Be_2](1\pi_u)^4(3\sigma_g)^2 1\pi_g$	5	1.12	3	6.55
O_2	$[Be_2](1\pi_u)^4(3\sigma_g)^2(1\pi_g)^2$	4	1.21	1.28	5.21
F_2	$[Be_2](1\pi_u)^4(3\sigma_g)^2(1\pi_g)^4$	2	1.44	−1.37	1.68
Ne_2	$[Be_2](1\pi_u)^4(3\sigma_g)^2(1\pi_g)^4(3\sigma_u)^2$	0	–	–	–

Careful studies of the selection of the basis functions were initiated in the mid-1960s [42], [43] using normalized complex Slater-type orbitals (STOs), that is, basis functions of the type

$$\Psi_{n'lm} = (2\zeta)^{n'+1/2}[(2n')!]^{-1/2} r^{n'-1} \exp(-r\zeta) Y^{m_l}(\theta, \phi) \tag{2.55}$$

in calculations for biatomic molecules. Note that although the form of the STOs resembles the hydrogenic wavefunctions and the angular part is indeed identical to the spherical harmonics introduced in Section 2.3, both the effective principal quantum number n' and the parameter ζ in the radial part of the STO are adjustable and are optimized as part of the SCF calculation to minimize the energy in a particular calculation.

As the accuracy of the calculations improved, extensive tabulations of data for specific simple molecules were generated. The extraction of general trends and comparisons between details in the bonding of different molecules requires considerable thought and is not as easily displayed as the results of the less sophisticated early VB calculations. A valuable representation that reveals sufficient details, but still permits a straight forward interpretation, is the plot of electron density difference maps of the molecules [39]. In such plots the electron density of the constituent atoms in the same nuclear positions as in the molecule are subtracted from the molecular electron density distribution. Figures 2.10 and 2.11 show the electron density difference maps of N_2 and LiF generated by ab initio calculations. N_2 is covalently bonded transferring extra charge into the bonding region between the nuclei. Thus an attracting force is established that binds the nuclei in the molecular complex. Note that a significant fraction of the redistributed atomic charge in the nitrogen molecule

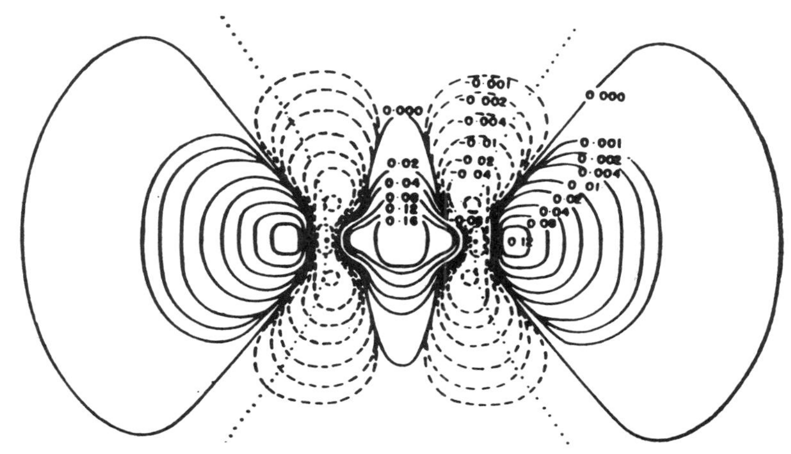

Figure 2.10 Electron-density difference map of nitrogen. After Bader, Henneker, and Cade [43]; copyright © 1967, American Institute of Physics, New York.

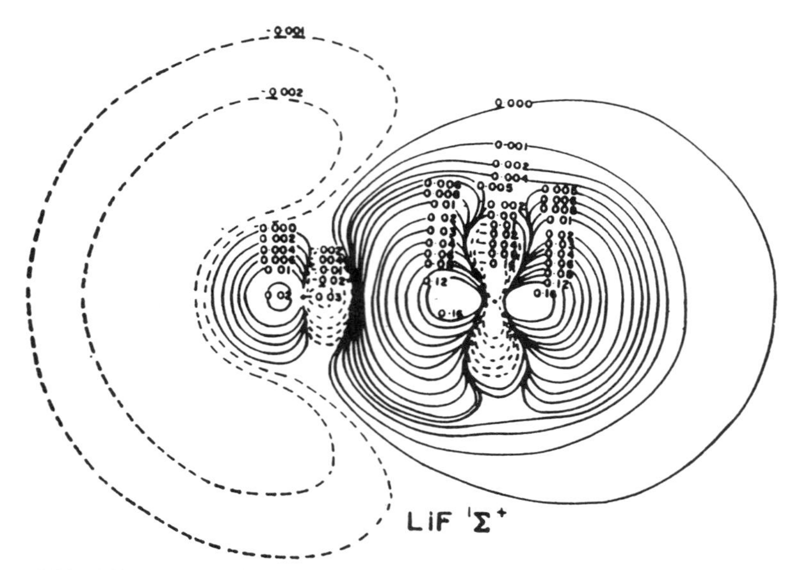

Figure 2.11 Electron-density difference map of lithium fluoride. After Bader, Henneker, and Cade [43]; copyright © 1967, *J. Chem. Phys.*, New York.

is channeled into the antibonding region of the molecule which is separated from the bonding region by the two dotted lines shown in Figure 2.10. However, this charge is smaller and far more diffuse than the charge concentrated in the bonding region near to the interatomic axis so that the N_2 molecule is stabilized by strong covalent bonding (compare Tables 2.4 and 2.6). In the case of the LiF molecule, clearly a substantial transfer of charge from the Li atom to the F atom occurs leading to an essentially ionic bond. Based on similar calculations the charge distributions have been analyzed for many molecules [44].

In recent years, ab initio quantum chemical calculations have targetted increasingly complex molecules for which all-electron HF calculations with appropriate Cl corrections are often prohibitive. Since the most important electronic properties of large molecules and solids depend primarily on the valence electrons of the atoms from which they are formed, the computations can be considerably simplified by a separation of the core-electron problem from the calculation of the interactions between the outer electrons. In order to accomplish such a separation, the condition of orthogonality between the core and valence electrons must be dropped. This can be done at the cost of adding a term, called the *pseudopotential*, to the Hamiltonian. To show this we follow the argument made by J. C. Phillips and L. Kleinman [45]: Let $\psi_1, \psi_2, \ldots, \psi_N$ denote the set of wavefunctions of the atomic core and ψ_n denote the wavefunction associated with a single valence electron, all being orthogonal with the exception of degenerate subsets. Also, let ϕ_n denote a nonorthogonal wavefunction associated with the same valence electron, which may be represented as

$$\psi_n = \phi_n - \sum_{i=1}^{N} c_i \psi_i \tag{2.56}$$

where the coefficients

$$c_i = \langle \psi_i | \phi_n \rangle \tag{2.57}$$

measure the nonorthogonality between ϕ_n and the ψ_i. Equation (2.2) leads to

$$H\psi_i = E_i \psi_i \tag{2.58}$$

and, in particular,

$$H\psi_n = E_n \psi_n \tag{2.59}$$

By replacing ψ_n by ϕ_n

$$(H + V_p)\phi_n = E_n \phi_n \tag{2.60}$$

where

$$V_p = \sum_{i=1}^{N} c_i \frac{(E_n - E_i)\psi_i}{\phi_n} \tag{2.61}$$

The pseudopotential method that allows considerably more freedom in the choice of the type of basis function than in the work of Ref. [45] originally was utilized to advantage in matching the core states and plane-wave-type extended states in the calculation of the band structure of solids (see Chapter 3). However, the method is of more general importance and is currently employed in studies that address molecules.

Faster convergence of the calculations is achieved by replacing the STOs by nodeless Gaussian-type orbitals (GTOs), that is, orbitals of the type $N_{lmn\zeta} x^l y^m z^n \exp(-\zeta r^2)$, where $N_{lmn\zeta}$ is a normalization constant. The most direct form of such a replacement is the fitting of the STOs by a linear combination of GTOs, but such a linkage of the GTOs to STOs is not essential for their use. Generally more GTOs than STOs are required to achieve the same accuracy in a particular molecular calculation, but the overall computational effort is still reduced. GTOs of the form $\exp(-\zeta r^2)$, $z\exp(-\zeta r^2)$, and $xy\exp(-\zeta r^2)$ are called *s*-, *p*-, and *d*-GTOs in reference to the analogous form of the *s*-, *p*-, and *d*-AOs.

Figure 2.12 shows the accuracy of an MCSCF calculation for the O_2 molecule using an increasing number of *s*- and *p*-like GTOs. The economy of the use of such sets of basis functions can be further improved by contracting them into specific linear combinations. More details on this topic and a list of references concerning the choice of basis functions for molecular calculations are provided in Refs. [46]–[48]. For simple molecules, calculations with a sufficiently large set of parameters are capable of achieving an accuracy

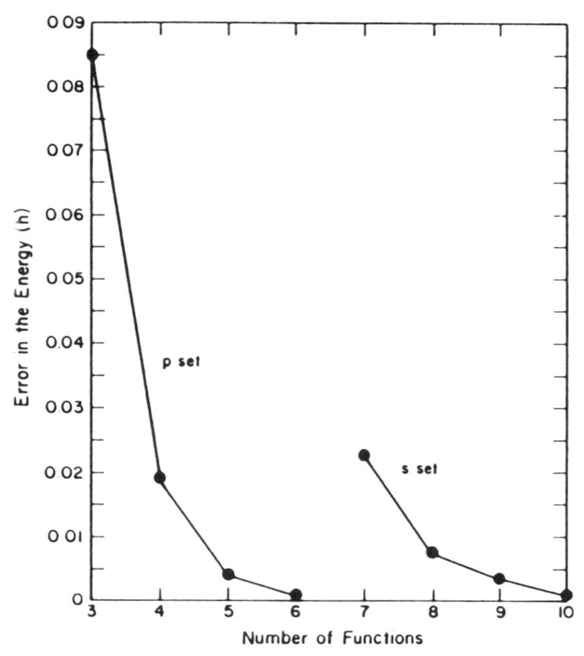

Figure 2.12 Accuracy of a pseudopotential calculation as a function of the number of *s*- and *p*-GTOs used as a basis set. After Dunning, Jr., and Hay [46]; copyright © 1977, Plenum Press, New York.

exceeding that of the best available experimental data. Therefore, the traditional supporting role of theoretical chemistry in explaining existing experimental facts is beginning to change into a leading role, providing theoretical motivation and guidance for critical experiments. An early demonstration of this point was the 100-parameter calculation of Kolos and Wolniewicz for H_2. This calculation predicted a dissociation energy $D = 36117.4\ \text{cm}^{-1}$ [49]. Since this value was $3.8\ \text{cm}^{-1}$ larger than the existing experimental value at the time of this prediction, it forced a reevaluation leading to a revision of the spectroscopic data [50].

Experimental information concerning the energy position of the MOs can be obtained rather directly from photoemission spectra. In photoemission spectroscopy, energetic photons are used to excite electrons from the filled MOs to energies exceeding the ionization limit. The energy of the free electrons is then analyzed. Knowing the energy of the incident photons, it is thus possible to allocate structure in the photoemission spectra to the energy positions of the MOs. This provides an opportunity to fix critical parameters of the calculation to match the experimental results. Depending on whether or not all or only part of the parameters of the calculation are obtained in this way, it is called an *empirical* or *semiempirical calculation*. Further details of the experimental approach and of the analysis of photoemission data are added in Chapters 7 and 8 in the context of gas molecule and metal atom interactions with solid surfaces.

In principle, the ab initio methods of quantum chemistry have the advantage that by an appropriate extension of the set of basis functions and of the number of configuration interactions incorporated into the calculations, it is always possible to improve their accuracy. However, the gains made in such calculations must be paid for in computer time. Therefore, in the predictions of molecular properties and in the analysis of molecular reaction mechanisms involving complex molecules, where the limitations in affordable computer time force restrictions in the choice of the basis function set and in the selection of the configurations, it is still advantageous to use a semiempirical approach. Considerable simplifications in MO calculations arise from symmetry restrictions of the various integrals, which are exact. Since these simplifications are also employed to advantage in the complex task of calculating the electronic properties of crystalline solids and in defining the selection rules that govern atomic and molecular spectra, we consider in Section 2.5 the symmetry of molecules in more detail. Because the description of symmetry is most concisely presented in the language of group theory, we also include in Section 2.5 a brief review of group theory and its applications to MO calculations.

2.5 Molecular Symmetry

In this section, we develop a more precise description of the symmetry of molecules, which we then apply in the discussion of atomic and molecular properties. Figure 2.13 shows as examples the symmetry of the water and the ammonia molecules. Because of the lone pairs on the oxygen and nitrogen atoms, according to the VSEPR rules, the H_2O molecule is nonlinear (bond angle H–O–H 104.45°), and the HN_3 molecule is nonplaner (bond angle H–N–H 107.3°). The atomic arrangement of the H_2O is invariant to a *rotation* by 180° about a vertical axis through the center of the oxygen atom, which rotates the hydrogen atom labeled H1 into the position of the hydrogen atom labeled H2 and vice versa. Two succesive rotations about this axis by 180° bring the atoms of the molecule into coincidence with the original positions, which constitutes the *identity* operation. Therefore, the rotational symmetry of the H_2O molecule is characterized by a twofold *axis of rotation* that is collinear to the z axis. The NH_3 molecule is invariant to rotations by 120° about a symmetry axis that is collinear to the z axis. Since the identity operation corresponds to three such rotations in succession, this axis is called a *threefold axis of rotation*.

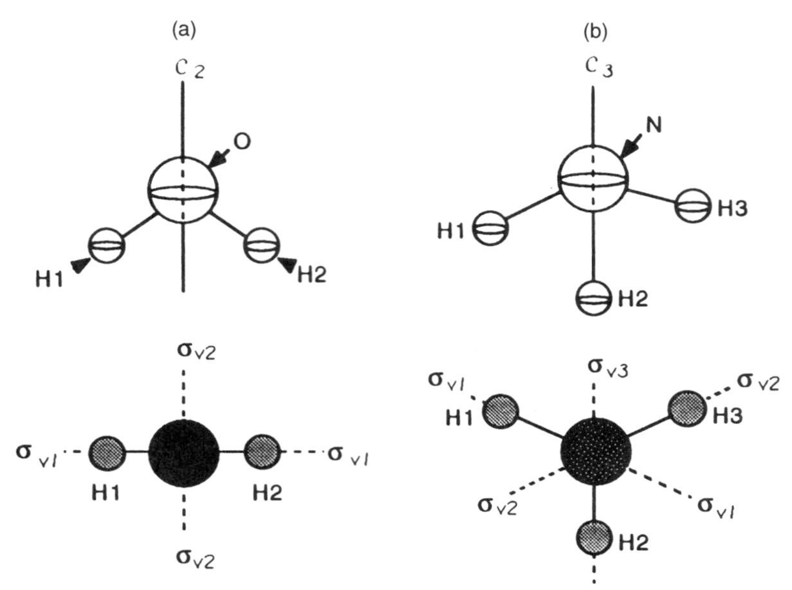

Figure 2.13 Schematic representations of the water molecule (a) and the ammonia molecule (b); top: side views, bottom: top views.

Note that there are geometrical symmetry elements (e.g., axes of rotation) and symmetry operations (e.g., rotations about symmetry axes) that re distinguished by the use of italics for the latter. In the international notation that is exclusively used in crystallography, a symmetry axis of order n (rotations by $2\pi/n$) is denoted by the number n. For example, twofold and threefold axes of rotation are represented by 2 and 3, respectively. We also introduce the equivalent specialized symbols of the Schönfließ notation, which is maintained with great tenacity in chemistry. In the Schönfließ notation, n-fold symmetry axes are denoted by C_n.

The atom positions of the H_2O molecule shown in the top of Figure 2.13(a) are also invariant to a *reflection* on the plane of the paper and to a plane that is perpendicular to the plane of the paper bisecting the oxygen atom. Such *symmetry planes* are labeled m (*mirror symmetry*) in the international notation, and are denoted by the Greek letter σ in the Schönfließ notation. The subscript v denotes that the symmetry planes contain a vertical axis of rotation. The traces of the two symmetry planes labeled σ_{v1} and σ_{v2} are indicated by dashed lines in the bottom part of Figure 2.13(a). Similarly three vertical symmetry planes σ_{v1}, σ_{v2}, and σ_{v3} are identified for the NH_3 molecule in Figure 2.13(b).

As a further example, we consider the spiropentane molecule shown in Figure 2.7. The z axis is collinear to a twofold symmetry axis, and the molecule has mirror symmetry with regard to both planes m and m'. Clearly a reflection on the plane, which is perpendicular to m and m', does not bring the molecule into an equivalent position, but rotation about the z axis by $2\pi/4$ followed by a reflection on this plane does. The composite symmetry operation of a rotation by $2\pi/n$ followed by a reflection on a plane that is perpendicular to the axis of the rotation is called a *rotatory reflection*, and the corresponding geometrical symmetry element is called a *mirror axis* and is labeled S_n in the Schönfließ notation. The z axis in Figure 2.7 is thus collinear to an S_4 mirror axis. There is no analogous symmetry element to S_n in the international notation.

Figure 2.14 shows a schematic representation of the SF_6 molecule, in which the sulfur atom is octahedrally coordinated by the fluorine atoms and of the cyclopentadienyl complexes of divalent metal ions. The metal ion, to which the

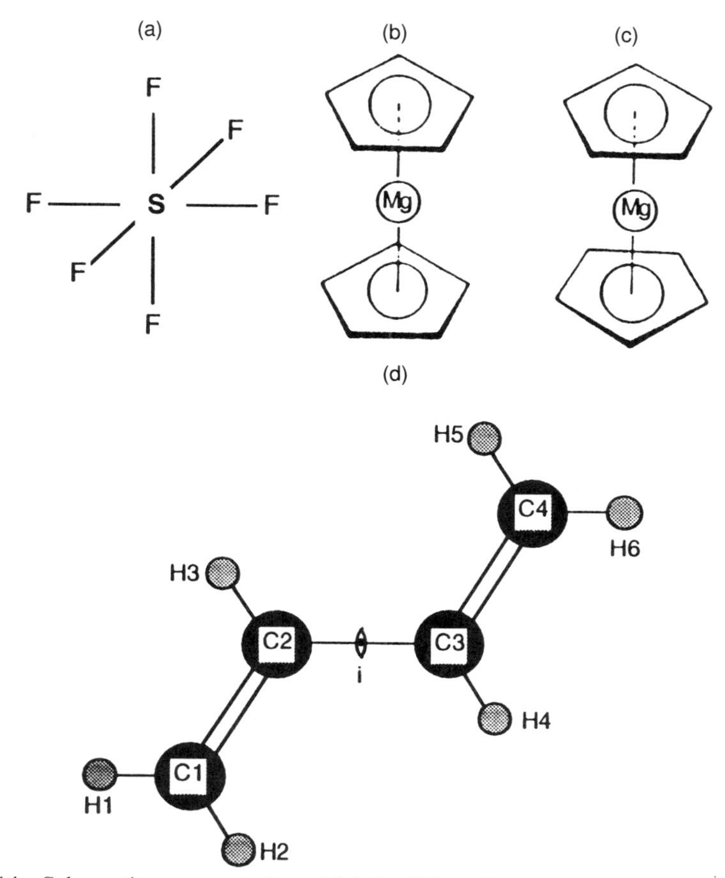

Figure 2.14 Schematic representation of (a) the SF_6 molecule and of cyclopentadienyl complexes of divalent metal ions in the eclipsed (b) and staggered (c) arrangements of the rings. (d) Schematic representation of the trans-butadiene molecule.

complex bonding is established by the two C_5H_5 rings, can be, for example, Mg^{2+} (magnesocene) or Fe^{2+} (ferrocene). Since the cyclopentadienyl complexes are volatile, they are useful dopant sources for chemical vapor deposition processes. The operation of replacing the coordinates of a point $[[x, y, z]]$ by the coordinates $[[-x, -y, -z]]$ is called an *inversion* on the origin of the coordinate system. A geometrical form that is invariant to inversion has an *inversion center*, which is labeled i in the Schönfließ convention. For the SF_6 molecule an inversion center exists at the position of the sulfur atom. Note that the cyclopentadienyl complexes have an inversion center on the metal ion in the staggered configuration (c), but not in the eclipsed position (b) of the cyclopentadienyl rings. The trans-butadiene molecule has an inversion center at the midpoint of the internuclear axis of the carbon atoms C2 and C3, which is also intersected by a twofold symmetry axis that is perpendicular to the paper plane. The molecule is planar and is thus invariant to reflections at the mirror plane σ_h coinciding with the plane of the paper.

In the international convention the inversion operation is compounded with rotations to the *rotoinversion* operation. The rotoinversion axes are labeled $\bar{n}$; that is, the inversion center is represented by $\bar{1}$. There exists for each rotatory reflection an equivalent rotoinversion operation, but their orders are not necessarily the same. For example, the cyclopentadienyl complex shown in Figure 2.14(c) has a $\bar{5}$ rotoinversion axis corresponding to an S_{10} rotatory reflection axis. Generally rotoinversion axes of odd order correspond to mirror axes of doubled order. Rotoinversion axes of even order divisible by 4 correspond to mirror axes of same order. There are no rotinversion axes of even order divisible by 2, but not by 4, since these axes correspond to axes of rotation C_n.

Let $(A, B, C, \ldots)$ denote the set of general symmetry operations of a molecule, including the identity operation E, which leaves the molecule in the original position. This set forms a group $\underset{\sim}{G}$ in the sense of mathematical group theory [51] because it has the following properties:

1. The multiplication of any two elements of the set is another element in the same set.

$$A \circ B = C, \qquad A, B, C \in \underset{\sim}{G} \tag{2.62}$$

2. There exists the identity element E, that is,

$$R \circ E = R \tag{2.63}$$

3. The associative law holds for all elements; that is,

$$A \circ (B \circ C) = (A \circ B) \circ C \tag{2.64}$$

4. For each element R there is an inverse element M so that

$$R \circ M = E \tag{2.65}$$

that is, $R = M^{-1}$.

In the case of molecular symmetry discussed here, the multiplication symbol $\circ$ stands for the successive application of the symmetry operations connected by it. To give a specific example, the rotation $C_3^{(1)}$ about a threefold symmetry axis by $2\pi/3$ in the mathematical convention (i.e., counterclockwise) and the rotation $C_3^{(2)}$ by $2 \times 2\pi/3$ about the same axis in the same direction are distinct elements of G. Since $C_3^{(2)}$ cancels the action of $C_3^{(1)}$ (i.e., $C_3^{(1)} \circ C_3^{(2)} = E$), $C_3^{(1)}$ and $C_3^{(2)}$ are inverse operations. If the order of the operations does not affect the outcome so that

$$A \circ B = B \circ A \tag{2.66}$$

the group is called an *Abelian group*. The elements $P \in \underline{G}$ and $Q \in \underline{G}$ are said to be *conjugated* if

$$P = NQN^{-1} \rightarrow Q = N^{-1}PN \tag{2.67}$$

where $N \in \underline{G}$. Note that we have dropped in Equation (2.67) the multiplication sign, which can be done if the nature of the multiplication is uniquely defined. The collection of mutually conjugate elements constitute a class. Classes are disjoint. For Abelian groups $NQN^{-1} = QNN^{-1} = Q$ so that each element is in a class by itself. Since E commutes with all elements in any group, it is always in a class by itself. The identity class is labeled E.

Symmetry groups are labeled by their highest-order axis of rotation, which by convention is taken to be vertical. Information on the orientation of symmetry planes with regard to this axis is added by a second right-hand subscript. For example, the group of symmetry operations describing the trans-butadience molecule is C_{2h}: $\{E, C_2, i, \sigma_h\}$. The heteronuclear biatomic molecules of general type AB (e.g., LiF or HCl) shown in Figure 2.15(a) have an internuclear axis of infinite order C_∞ that is contained in an infinite number of vertical symmetry planes σ_v. In conjunction with the identity element, the symmetry operations associated with this axis and the infinite set of symmetry

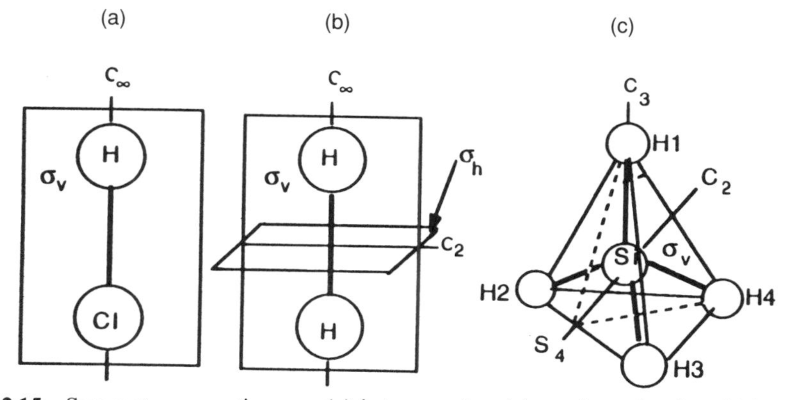

Figure 2.15 Symmetry operations on (a) heteronuclear biatomic molecules, (b) homonuclear biatomic molecules, and (c) tetrahedrally coordinated molecules.

planes constitute the group $C_{\infty v}$. The homonuclear molecules A_2 (e.g., N_2) shown in Figure 2.15(b), and the linear molecules of the type AB_2 (e.g., CO_2) have, in addition to the elements of $C_{\infty v}$, mirror operations on a horizontal symmetry plane σ_h and rotations about twofold axes C_2 contained in σ_h. Groups that have a set of twofold symmetry axes perpendicular to the axis of highest order are called *dihedral groups*. Therefore, the A_2 and the linear AB_2 molecules belong to the dihedral group $D_{\infty h}$.

The molecules of type AB_4 [e.g., the SiH_4 molecule shown in Figure 2.14(c)] have four threefold axes of rotation each associated with two distinct operations, $C_3^{(1)}$ and $C_3^{(2)}$, respectively. They form the class C_3 that thus contains 8 elements. Also, there are three twofold axes and three collinear S_4 rotatory reflection axes bisecting opposite edges of the tetrahedron. Since $S_4^{(2)} = C_2^{(1)}$ and $S_4^{(4)} = E$, two distinct symmetry elements, $S_4^{(1)}$ and $S_4^{(3)}$, are associated with each of the S_4 axes. Furthermore there are six symmetry planes, each containing one edge (e.g., H_1–H_4) and bisecting the opposite edge (e.g., H_2–H_3), as illustrated in Figure 2.15(c). Each of these planes corresponds to a single symmetry element. Therefore, the group of the tetrahedrally coordinated AB_4 compounds corresponds to the set of symmetry elements $(E, 8C_3, 6S_4, 3C_2, 6\sigma)$. It is designated by the special label T_d. Also, the symmetry group describing octahedrally coordinated structures [e.g., the SF_6 molecule shown in Figure 2.15(a) has a special label, which is O_h (see Table 2.10).

The symmetries of the ammonia molecule shown in Figure 2.13(b) and of the trichlorosilane molecule shown in Figure 2.16 are represented by the same group, which is C_{3v}: $(E, 2C_3, 3\sigma_v)$. Table 2.7 shows the results of all possible multiplications of two elements of this group. Such a table is called a *multiplication table*. Note that the $SiHCl_3$ molecule is derived from the SiH_4 molecule by substituting three hydrogen atoms by chlorine atoms. Because C_{3v} has fewer symmetry elements than the group T_d it is said to be of lower symmetry. Substitution generally reduces the symmetry of a molecule.

Since the symmetry operations discussed correspond to coordinate transformations in a three-dimensional space, each symmetry element may be represented by a 3 × 3 transformation matrix $[R]$ so that the set of coordinates in column vector form $\{x', y', z'\}$ after application of the symmetry operation

Table 2.7 Multiplication table of the elements of the group C_{3v}

C_{3v}	E	C_3^1	C_3^2	σ_v	σ_v'	σ_v''
E	E	C_3^1	C_3^2	σ_v	σ_v'	σ_v''
C_3^1	C_3^1	C_3^2	E	σ_v'	σ_v''	σ_v
C_3^2	C_3^2	E	C_3^1	σ_v''	σ_v	σ_v'
σ_v	σ_v	σ_v''	σ_v'	E	C_3^2	C_3^1
σ_v'	σ_v'	σ_v	σ_v''	C_3^1	E	C_3^2
σ_v''	σ_v''	σ_v'	σ_v	C_3^2	C_3^1	E

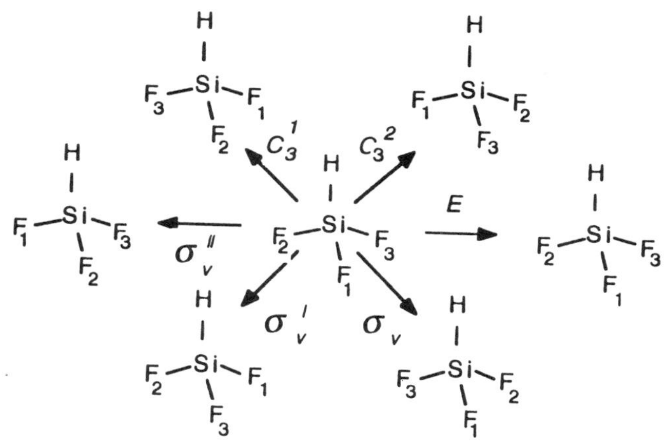

Figure 2.16 Complete set of symmetry operations on a substituted tetrahedrally coordinated (trifluorosilane) molecule of C_{3v} symmetry.

is related to the original set $\{x, y, z\}$ according to the transformation $\{x', y', z'\} = [R]\{x, y, z\}$. Figure 2.17 shows two examples with respect to a Cartesian coordinate system. Of course, other coordinate axes may be chosen that lead to different matrix representations. In fact, the same group may be represented by a set of matrices of more than three dimensions $\Gamma = ([A], [B], [C], [D], \ldots)$. This representation is said to be a *faithful representation* of the original group if for each element in the latter set there exists an equivalent element in the former set so that the multiplications between these elements reproduce the multiplication table of the original set. It is always possible to

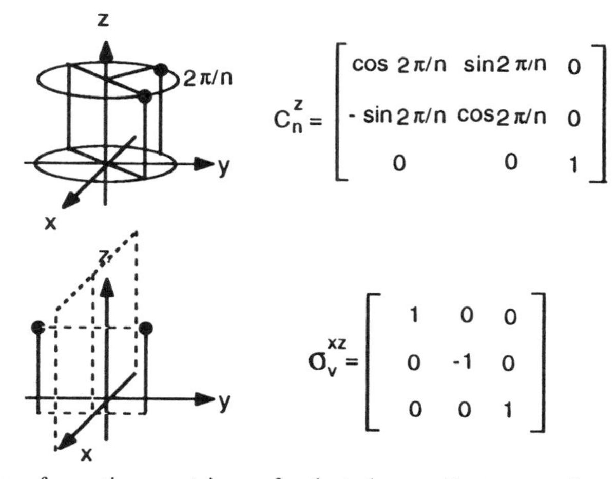

Figure 2.17 Transformation matrices of selected coordinate transformations. Top: rotation about the z axis by $2\pi/n$. Bottom: reflection on the xz plane.

generate from a faithful representation of a group other faithful representations by similarity transformations

$$[R'] = [S]^{-1}[R][S] \tag{2.68}$$

where $[R]$ represents a general element of the group of symmetry elements. In particular, block diagonalization by a unitary transformation results in a representation where each element is in block form; that is,

$$[R] = [U]^{-1}[R][U] = \begin{bmatrix} R_1 & 0 & 0 & . \\ 0 & R_2 & 0 & . \\ 0 & 0 & R_3 & . \end{bmatrix} \tag{2.69}$$

It can be shown that each of the sets of matrices $\Gamma_1 = ([A_1], [B_1], [C_1], [D_1], \ldots)$, $\Gamma_2 = ([A_2], [B_2], [C_2], [D_2], \ldots), \ldots$ formed from the blocks is a faithful representation of the group. Since these representations are of smaller dimensions than the original representation Γ, the latter is called a *reducible representation.*

In particular, the set of representations $\Gamma_i = ([A_i], [B_i], [C_i], [D_i], [E_i], \ldots, [R_i], \ldots)$, $i = 1, 2, 3, \ldots, N_i$ of smallest dimensions that can be obtained by simultaneous block diagonalization of all elements of a reducible representation is called the set of *irreducible representations* (IRs) of the group. Of course, by similarity transformations $[S]^{-1}[R_\alpha][S] = [R_\beta]$ of all elements of the IR Γ_α another IR Γ_β may be generated; that is, any IR of the group is only defined within a similarity transformation. However, since the trace χ of a matrix is invariant to a similarity transformation, the trace of each matrix belonging to Γ_α is the same as the trace of the corresponding matrix belonging to Γ_β; that is,

$$\chi([R_\alpha]) = \sum_{k=1}^{l_\alpha} R_{\alpha kk} = \chi([R_\beta]) = \sum_{k=1}^{l_\beta} R_{\beta kk} \tag{2.70}$$

where l_α, l_β, $R_{\alpha kk}$, and $R_{\beta kk}$ are the dimension and the diagonal elements of the matrix $[R_\alpha]$ of the set Γ_α and the corresponding matrix $[R_\beta]$ of the set Γ_β, respectively.

The set of numbers $\chi([E_i])$, $\chi([A_i])$, $\chi([B_i])$, $\chi([C_i]), \ldots$ is thus unique and is called the *character* of the IR. Since all elements in a given class are related by similarity transformations, to which χ is invariant, only one value needs to be listed per class to specify the character of an IR completely. The characters of all IRs of a group are listed in its character table, which consists of N_i rows labeled by the various IRs Γ_i and columns labeled by the classes of the group. Usually the class label is preceded by the number of elements n_k in this class. The number of classes of a group is labeled N_k. It can be shown that there are as many classes as IRs [52]; that is, $N_k = N_i$. Since the diagonal elements of the identity matrix are all ones, the traces listed under the identity class represent the dimensionalities l_i of the various IRs of the group. There always exists the identity representation where all symmetry elements are represented by the number 1. It appears in every character table as one of the IRs of the group, which is usually listed as the first row.

Further details of the construction of the character tables follow from Wigner's orthogonality theorem [53]

$$\sum_{\text{all } R} [R_\alpha]^*_{ik}[R_\beta]_{jl} = \frac{g}{\sqrt{(l_\alpha l_\beta)}} \delta_{\alpha\beta}\delta_{ij}\delta_{kl} \tag{2.71}$$

which provides a general relation between the matrix elements of the two IRs Γ_α and Γ_β of the same group. Let $k = i$ and $l = j$ and sum over i and j to get

$$\begin{aligned}\sum_i \sum_j \sum_R [R_\alpha]_{ii}[R_\beta]_{jj} &= \sum_R \chi([R_\alpha])\chi([R_\beta]) \\ &= \sum_i \sum_j \frac{g}{\sqrt{l_\alpha l_\beta}} \delta_{\alpha\beta}\delta_{ij}\delta_{ij}\end{aligned} \tag{2.72a}$$

For $i = j$

$$\begin{aligned}\sum_R \chi([R_\alpha])\chi([R_\beta]) &= \sum_{i=1}^{l_\alpha} \frac{g}{\sqrt{l_\alpha l_\beta}} \delta_{\alpha\beta}\delta_{ii} \\ &= g\frac{l_\alpha}{\sqrt{l_\alpha l_\beta}} \delta_{\alpha\beta}\end{aligned} \tag{2.72b}$$

The rows of the character table are thus orthogonal and normalized to g. For $\alpha = \beta$ we get from Equation (2.72)

$$\sum_R |\chi([R_\alpha])|^2 = \sum_{k=1}^{N_k} |\chi_\alpha(C_k)|^2 n_k = g \tag{2.73}$$

This reduces under the identity representation to

$$\sum_{k=1}^{N_k} n_k = g \tag{2.74}$$

Also, the columns in the same IR are orthogonal and are normalized to g/n_k; that is,

$$\sum_{\alpha=1}^{N_i} \chi_\alpha(C_k)\chi^*_\alpha(C_l) = \frac{g}{n_k} \delta_{kl} \tag{2.75}$$

which holds for all classes, including the identity class, where all the diagonal elements are equal to one so that $\chi_\alpha(E) = l_\alpha$

$$\sum_{\alpha=1}^{N_i} l_\alpha^2 = g \tag{2.76}$$

Because of Equation (2.75) all IRs of a group consisting of classes that contain just one symmetry operation each (e.g., C_{2h}) must be one dimensional.

Therefore, the three-dimensional transformation matrices representing the symmetry of the group C_{2h}

$$E: \begin{bmatrix} 1 & 0 & 0 \\ 0 & 1 & 0 \\ 0 & 0 & 1 \end{bmatrix}, \quad C_2: \begin{bmatrix} -1 & 0 & 0 \\ 0 & -1 & 0 \\ 0 & 0 & 1 \end{bmatrix}$$

$$i: \begin{bmatrix} -1 & 0 & 0 \\ 0 & -1 & 0 \\ 0 & 0 & -1 \end{bmatrix}, \quad \sigma_h \begin{bmatrix} 1 & 0 & 0 \\ 0 & 1 & 0 \\ 0 & 0 & -1 \end{bmatrix}$$

are reducible representations. Since they are in block-diagonal form, some of the IRs of this group can be determined by inspection. The diagonal 1×1 blocks of the first, second, and third rows of these reducible representation correspond to the IRs $\Gamma_\alpha = \{[1], [-1], [-1], [1]\}$, $\Gamma_\alpha = \{[1], [-1], [-1], [1]\}$, and $\Gamma_\beta = \{[1], [1], [-1], [-1], [-1]\}$, respectively. Since the IRs are representations of the group, they must obey the multiplication rules; that is,

$$\begin{aligned}
\Gamma_\alpha \times \Gamma_\beta &= \{[1] \times [1], [-1] \times [1], [-1] \times [-1], [1] \times [-1]\} \\
&= \{[1], [-1], [1], [-1]\} = \Gamma_\gamma \\
\Gamma_\alpha \times \Gamma_\alpha &= \{[1] \times [1], [-1] \times [-1], [-1] \times [-1], [1] \times [1]\} \\
&= \Gamma_\delta = \Gamma_\beta \times \Gamma_\beta = \Gamma_\gamma \times \Gamma_\gamma \\
\Gamma_\alpha \times \Gamma_\gamma &= \{[1] \times [1], [-1] \times [-1], [-1] \times [1], [1] \times [-1]\} \\
&= \{[1], [1], [-1], [-1]\} = \Gamma_\beta \\
\Gamma_\beta \times \Gamma_\gamma &= \{[1] \times [1], [1] \times [-1], [-1] \times [1], [-1] \times [-1]\} \\
&= \{[1], [-1], [-1], [1]\} = \Gamma_\alpha
\end{aligned} \tag{2.77}$$

Note that the multiplications among the four IRs, thus determined, result in all the other IRs of the same group; that is, there are four classes and four IRs. Since all matrices are one dimensional so that the traces equal the matrix elements, this permits the construction of the character table of C_{2h}, which is shown in Table 2.8. Note that the basis functions x and y transform as Γ_α, z transforms as Γ_β, xz and yz transform as Γ_γ, and xy, x^2, y^2, and z^2 transform as Γ_δ. Also, the rows and columns are orthogonal and normalized to g and g/n_k, respectively.

There exist two frequently used conventions in the labeling of IRs, the Wigner notation, where all IRs are labeled Γ, and the Mulliken notation, which uses different capital letters to convey information concerning the dimensionalities and nature of the IRs. Subscripts and superscripts are added in both notations to convey additional information. In the Mulliken notation, the IRs

Table 2.8 Character table of the group C_{2h}

Class: IR	E	C_2	i	$\sigma_h(xy)$	Basis functions
A_g	1	1	1	1	x^2, y^2, z^2, xy
A_u	1	1	-1	-1	z
B_g	1	-1	1	-1	xz, yz
B_u	1	-1	-1	1	x, y

of finite groups are labeled A or B for $l_i = 1$. Bases belonging to A and B representations, respectively, do not and do change sign under the highest-order rotational operation. The subscripts 1 or 2 and single or double prime superscripts, respectively, indicate that the bases do not or do change sign under a C_2 operation in dihedral groups and under σ_v if there is no C_2 axis perpendicular to the symmetry axis of highest order. Multidimensional IRs are labeled E, T, G, H for $l_i = 2, 3, 4, 5$, respectively. Subscripts g or u indicate that the bases do not or do change sign under the inversion operation. In the character tables of the infinite rotational groups $C_{\infty v}$ and $D_{\infty h}$ the Mulliken labels are often replaced by the Greek letters that specify the angular symmetry about the internuclear axis, that is, Σ for cylindrical symmetry, and Π, Δ, and Φ, respectively, for one, two, and three symmetry planes containing the internuclear axis.

In order to apply group theory to problems of quantum chemistry, it is necessary to relate the results of symmetry operations on the spatial coordinates $[[x, y, z]]$ to the action of linear operators on functions of the variables x, y, z. The desired relation is established by introducing a group of operators $(A, B, C, \ldots)$ with the property that each of these operators acting on functions of the spatial coordinates x, y, z exactly compensates for the effects of an associated coordinate transformation matrix from the group $([A], [B], [C], [D], \ldots)$ on the variables x, y, z, that is,

$$AF([A]\{xyz\}) = F(x, y, z) = F([A]^{-1}[A]\{xyz\}) \tag{2.78}$$

where $\{xyz\}$ represents a column vector. Since there is for each product $A \circ B$ of operators an associated product $[A]^{-1} \circ [B]^{-1} = [A \circ B]^{-1}$ of elements in the group of coordinate transformation matrices, the two groups are isomorphic (i.e., have the same IRs).

Because of the spherical symmetry of the atomic potential, the symmetry of the Hamiltonian is described by the spherical group K_h; that is, H is invariant to all operations of K_h. The group of symmetry operators to which the Hamiltonian is invariant is called the group of the Schrödinger equation. Let R denote a general operator in this group. Since H is invariant to R, the two operators commute so that

$$RH\Psi_i = HR\Psi_i = H\Psi_i' = RE_i\Psi_i = E_iR\Psi_i = E_i\Psi_i' \tag{2.79}$$

The state vectors Ψ_i and Ψ_i' generated in this manner for a specific eigenvalue

E_i are either identical or not identical. In the former case E_i is nondegenerate while in the latter it is degenerate. If all eigenvectors solving Equation (2.2) for the same energy eigenvalue are generated by operations of the elements of the group of the Schrödinger equation on a particular solution Ψ_i, the resulting degeneracy is called *normal*. All other degeneracies are called *accidental*.

In the absence of accidental degeneracies, the action of the operator on the complete degenerate set of state vectors $\{\Psi_i\}$ then simply creates another state vector belonging to the same eigenvalue of energy as already discussed in Section 2.2. In matrix notation,

$$\{\Psi_i'\} = [R]\{\Psi_i\} \tag{2.80}$$

so that

$$\Psi_k = \sum_{j=1}^{l_i} R_{kj}\Psi_j \tag{2.81}$$

Since the complete degenerate set of state vectors is used in the expansion Equation (2.81), the dimension of the matrix $[R]$ must be equal to the degeneracy d_i of this set so that $l_i = d_i$. In particular, for nondegenerate eigenvalues $l_i = 1$. Also, each of the state vectors belonging to a complete set associated with a degenerate eigenvalue of energy is linearly independent of the other members of the same set. Thus there is no smaller set, and the matrices $[R]$ representing the symmetry operations, to which H is invariant, form for each eigenvalue E_i an irreducible representation of the group of the Schrödinger equation. This can be generalized: Each eigenvalue of any linear operator defining an eigenvalue equation is associated with a unique irreducible representation of the associated group of symmetry operations, to which this operator is invariant. The solutions associated with each eigenvalue of the operator are basis functions to the corresponding IR (i.e., transform accordingly).

For example, the IRs of the rotational group $R_{\text{h}}(3)$, which is identical to K_h, are labeled by the symbol $D_{\text{g}}^{(l)}$ or $D_{\text{u}}^{(l)}$ depending on the sign of the character under the inversion class (see Table 2.9). Since the radial parts of the atomic wavefunctions are isotropic, they have the spherical harmonics $Y_l^m(\theta, \phi)$ as basis functions so that their dimension is $2l + 1$. Since there is no upper limit to l, $R_{\text{h}}(3)$ is an infinite group. Application of all operations of $R_{\text{h}}(3)$ leave the s electron totally unchanged; that is, the s orbitals transform as the completely symmetric representation $D_{\text{g}}^{(0)}$. The p-electron orbitals are threefold degenerate and change sign under the inversion operation. Therefore, they belong to the irreducible representation $D_{\text{u}}^{(1)}$. The d-electron orbitals that are fivefold degenerate do not change sign upon inversion; that is, they transform as $D_{\text{g}}^{(2)}$. See Table 2.9.

An important consequence of the relation between the degeneracies of the energy eigenvalues and the IRs of the appropriate group of the Schrödinger equation is that a reduction in the symmetry of a molecule may force a splitting of certain energy levels. For example, if a transition metal ion becomes incorporated into a complex of O_{h} symmetry (e.g., the Fe^{3+} ion in the complex

Table 2.9 Character table of the group $R_h(3)$

Class: IR	E	$C(\phi, x, y, z)$	i	$S(-\phi, x, y, z)$	σ
$D_g^{(0)}$	1	1	1	1	1
$D_g^{(1)}$	3	$1 + 2\cos\phi$	3	$1 - 2\cos\phi$	-1
$D_g^{(2)}$	5	$1 + 2\cos\phi + 2\cos 2\phi$	5	$1 - 2\cos\phi + 2\cos 2\phi$	1
...	...	...	...	...	...
$D_g^{(k)}$	$2k + 1$	$1 + \sum_{l=1}^{k} 2\cos l\phi$	$2k + 1$	$1 + \sum_{l=1}^{k} 2\cos l\phi$	$(-1)^k$
$D_u^{(0)}$	1	1	-1	-1	-1
$D_u^{(1)}$	3	$1 + 2\cos\phi$	-3	$-1 + 2\cos\phi$	1
$D_u^{(2)}$	5	$1 + 2\cos\phi + 2\cos 2\phi$	-5	$-1 + 2\cos\phi - 2\cos 2\varphi$	-1
...	...	...	...	...	...
$D_u^{(k)}$	$2k + 1$	$1 + \sum_{l=1}^{k} 2\cos l\phi$	$-(2k + 1)$	$-1 - \sum_{l=1}^{k} (-1)^k 2\cos l\phi$	$-(-1)^k$

$[Fe(CN)_6]^{3-}$), the potential is no longer spherical since the 6 ligands interact with the central ion. Due to the O_h symmetry of the field created by the presence of the ligands for octahedrally coordinated complexes, the group of the Schrödinger equation is O_h. There are at most three-dimensional IRs (see Table 2.10) and consequently threefold degenerate eigenvalues permitted for a molecule of O_h symmetry. Therefore, the fivefold orbital degeneracy permitted

Table 2.10 Character table of the group O_h^a

Class: IR	E	$3C_4^2$	$6C_4$	$6C_2$	$8C_3$	i	$3\sigma_h$	$6S_4$	$6\sigma_d$	$8S_6$
A_{1g}	1	1	1	1	1	1	1	1	1	1
A_{1u}	1	1	1	1	1	-1	-1	-1	-1	-1
A_{2g}	1	1	-1	-1	1	1	1	-1	-1	1
A_{2u}	1	1	-1	-1	1	-1	-1	1	1	-1
E_g	2	2	0	0	-1	2	2	0	0	-1
E_u	2	2	0	0	-1	-2	-2	0	0	1
T_{1g}	3	-1	1	-1	0	3	-1	1	-1	0
T_{1u}	3	-1	1	-1	0	-3	1	-1	1	0
T_{2g}	3	-1	-1	1	0	3	-1	-1	1	0
T_{2u}	3	-1	-1	1	0	-3	1	1	-1	0
Γ_r	5	1	-1	1	-1	5	1	-1	1	-1

[a]Groups of high symmetry may be formed by the multiplication of smaller subgroups, in particular, products containing the subgroups $C_i = \{E, i\}$ and $C_s = \{E, \sigma\}$ as factors. They are either direct products $G = G_1 \times G_2$ or semidirect products $G = G_1 \wedge G_2$. Direct products are formed if all elements of the subgroups G_1 and G_2 commute. In this case

$$G = G_1 \times G_2 = G_2 \times G_1 = (A_1, B_1, C_1, D_1, \ldots, E) \times (A_2, B_2, C_2, D_2, \ldots, E)$$
$$= (A_1 \circ A_2, A_1 \circ B_2, \ldots, A_1, B_1 \circ A_2, B_1 \circ B_2, \ldots, B_1, \ldots, E)$$

results a group of order $g = g_1 g_2$.

for the d-electron wavefunctions of the free ion under $D_{\mathrm{g}}^{(2)}$ is lifted; that is, the energy level of the d functions of the free ion must split upon the formation of the complex. The magnitude of this ligand-field splitting cannot be specified by group theory, but the specific nature of the IRs under O_h into which the five-dimensional IR under $R_{\mathrm{h}}(3)$ decomposes is easily established on the basis of the already-introduced group-theoretical principles.

Let $\chi(R)$ denote the trace of the reducible matrix representation Γ_{r} of the symmetry element R, and f_α specify how often the matrix $[R_\alpha]$ belonging to the IR Γ_α appears in the block-diagonalized matrix $[R]$. Then

$$\chi([R]) = \sum_{\alpha} f_\alpha \chi([R_\alpha]) \tag{2.82}$$

so that

$$\sum_{R} \chi([R])\chi([R_\beta]) = \sum_{\alpha} f_\alpha \sum_{R} \chi([R_\alpha])\chi([R_\beta]) = \sum_{\alpha} f_\alpha g \delta_{\alpha\beta} \tag{2.83}$$

that is, for $\alpha = \beta$

$$\sum_{R} \chi(R)\chi([R_\alpha]) = f_\alpha g \tag{2.84}$$

which gives

$$f_\alpha = \frac{1}{g} \sum_{R} \chi([R])\chi([R_\alpha]) \tag{2.85}$$

A look at the character tables of $R_{\mathrm{h}}(3)$ and O_{h} allows us to work out the traces of the reducible representation into which $D_{\mathrm{g}}^{(2)}$ converts under O_{h}. In accord with the fivefold dimension the trace under E is 5. With $\phi = 2\pi/3$, the trace under C_3 is $1 + 2\cos(2\pi/3) + 2\cos(4\pi/3) = 1 - 1 - 1 = -1$. Similarly the traces under C_2 and C_4 are 1 and -1, respectively. The trace under i remains 5. The trace under S_4 is $1 - 2\cos(\pi/2) + 2\cos\pi = 1 - 0 - 2 = -1$. Similarly we get for S_6 the trace -1. The traces under σ_{h} and σ_{d} are 1. These traces are listed as the row labeled Γ_{r} below Table 2.10. Based on Equation (2.85) we get: $f_{A_{1g}} = f_{A_{2g}} = f_{T_{1g}} = f_{A_{2u}} = f_{A_{2u}} = f_{E_u} = f_{T_{1u}} = f_{T_{2u}} = 0$. For example,

$$\begin{aligned} f_{A_{1g}} &= \{1/48\}\{1(5)[1] + 3(1)[1] + 6(-1)[1] + 6(1)[1] + 8(-1)[1] \\ &\quad + 1(5)[1] + 3(1)[1] + 6(-1)[1] + 6(1)[1] + 8(-1)[1]\} \\ &= \{1/48\}\{28 - 28\} = 0, \end{aligned}$$

and $f_{E_g} = f_{T_{2g}} = 1$

$$\begin{aligned} f_{E_g} &= \{1/48\}\{1(5)[2] + 3(1)[2] + 6(-1)[0] + 6(1)[0] + 8(-1)[-1] \\ &\quad + 1(5)[2] + 3(1)[2] + 6(-1)[0] + 6(1)[0] + 8(-1)[-1]\} \\ &= \{1/48\}\{48\} = 1 \end{aligned}$$

and

$$f_{T_{2g}} = \{1/48\}\{1(5)[3] + 3(1)[-1] + 6(-1)[-1] + 6(1)[1] + 8(-1)[0]$$

$$+ 1(5)[3] + 3(1)[-1] + 6(-1)[-1] + 6(1)[1] + 8(-1)[0]\}$$

$$= \{1/48\}\{54 - 6\} = 1.$$

Therefore, E_g and T_{2g} appear in the block-diagonalized form of Γ_r just once so that $D_g^{(2)}$ in $R_h(3)$ correlates with $E_g + T_{2g}$ in O_h. As a consequence of this correlation, the fivefold orbital degeneracy of the energy eigenvalue associated with the d-electron wavefunctions permitted under $R_h(3)$ is lifted, and the energy eigenvalue must split under O_h into one twofold- and one threefold-degenerate energy level. The wavefunctions $d_{x^2-y^2}$ and d_{z^2} have e_g symmetry, while the d_{xy}, d_{xz}, and d_{yz} wavefunctions have t_{2g} symmetry. Note that by convention the orbitals are labeled by lower-case plain letters, reserving the Mulliken symbols using capital letters for the IRs. The correlation between the IRs of the various groups and their subgroups has been worked out [52] and is shown for a number of atomic orbitals in Table 2.11. Of course, the knowledge of the Cartesian basis functions that describe the angular symmetry of the atomic wavefunctions permits one to determine this information directly from the character tables [e.g. the p-electron wavefunctions that belong in $R_h(3)$ to $D_u^{(1)}$ belong in $D_{\infty h}$ to IRs of Σ_u and Π_u symmetry (see Table 2.12) so that in the formation of the homonuclear biatomic molecules from the atoms the threefold-degenerate atomic energy levels split into a singly degenerate σ MO and a doubly degenerate π MO as shown in Figure 2.8.

From the invariance of a linear operator A to all symmetry operations of the associated group (e.g., the invariance of the Hamiltonian operator to the operations of the group of the Schrödinger equation), follows that the operator belongs to the identity representation Γ_1. Therefore, the action of this operator onto a wavefunction Ψ_i being a basis function to the IR Γ_i leaves its character unchanged. This may be expressed by the direct product $\Gamma_1 \times \Gamma_i = \Gamma_i$. Furthermore, from the orthogonality of the basis functions follows that the matrix elements $\langle \Psi_k | A | \Psi_i \rangle$ vanishes if Ψ_k and Ψ_i belong to different IRs. In more general terms, for matrix elements to assume nonzero values, the

Table 2.11 Correlation of the irreducible representations of atomic wavefunctions in $R_h(3)$, O_h, and T_d

Orbitals	$R_h(3)$	O_h	T_d
s	$D_g^{(0)}$	A_{1g}	A_1
p	$D_u^{(1)}$	T_{1u}	T_2
d	$D_g^{(2)}$	$E_g + T_{2g}$	$E + T_2$
f	$D_u^{(3)}$	$A_{2u} + T_{1u} + T_{2u}$	$A_2 + T_1 + T_2$

Table 2.12 Character table of the group $D_{\infty h}(3)$

Class: IR	E	$2C_{\infty}(\phi)$	∞C_2	i	$2S_{\infty}(-\phi)$	$\infty\sigma_v$
Σ_g^+	1	1	1	1	1	1
Σ_g^-	1	1	-1	1	1	-1
Π_g	2	$2\cos\phi$	0	2	$-2\cos\phi$	0
Δ_g	2	$2\cos 2\phi$	0	2	$2\cos 2\phi$	0
…	…	…	…	…	…	…
Σ_u^+	1	1	1	-1	-1	-1
Σ_u^-	1	1	-1	-1	-1	1
Π_u	2	$2\cos\phi$	0	-2	$2\cos\phi$	0
Δ_u	2	$2\cos 2\phi$	0	-2	$-2\cos 2\phi$	0
…	…	…	…	…	…	…

integrand in Equation (2.15) must transform as the completely symmetric representation or, if represented by a reducible representation, must decompose into a set of representations containing the completely symmetric representation at least once. This is the basis of selection rules for the optical transitions in atoms and molecules eliminating symmetry-forbidden lines or bands in the spectra. It also provides for a considerable simplification of MO calculations by the use of symmetry-adapted linear combinations of atomic orbitals (SALCAOs).

As an example, we consider a simple Hückel calculation [54] for the trans-butadiene molecule shown in Figure 2.14(d) and the cyclobutadiene molecule. As in the case of the trans-butadiene molecule, after establishing the σ-bonded backbone of the molecule, one p_z electron remains on each of the 4

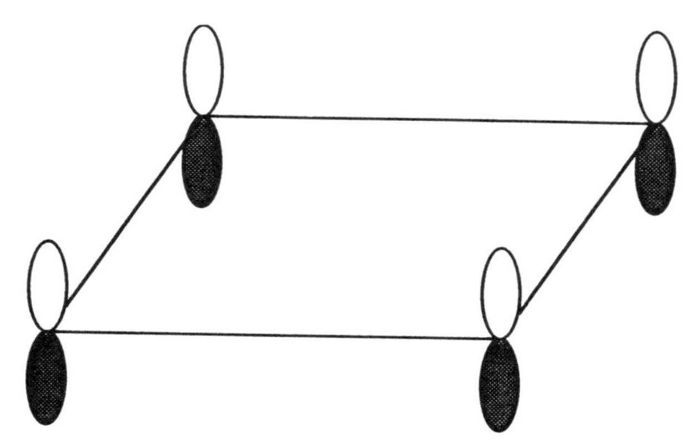

Figure 2.18 Schematic representation of the σ-bonding system and the p_z electrons forming the π-bonding system of the cyclobutadiene molecule.

carbon atoms of the cyclobutadiene molecule as shown in Fig. 2.18. Interactions of these electrons result in the formation of the π-bonding system. In the Hückel model the Hamiltonian of the σ- and π-bonding systems are separated; that is,

$$H = H_\sigma + H_\pi \tag{2.86}$$

so that

$$\Psi = \Psi_\sigma \Psi_\pi. \tag{2.87}$$

Since the p_z-electron wavefunctions on each of the carbon atoms are the same, all Coulomb integrals and exchange integrals are taken to have same values α and β, respectively. Exchange contributions arise only from atoms that are directly connected by a σ bond (i.e., $\beta = 0$ for all interactions not involving directly bonded atoms). Also, all overlap integrals between p_z-electron wave functions on different atoms are set to zero; that is, $S_{ij} = \delta_{ij}$, which is a drastic simplification. Forming from the four p_z wavefunctions $\phi_1, \phi_2, \phi_3, \phi_4$ on the carbon atoms C_1, C_2, C_3, C_4 in Figure 2.14(d) linear combinations that transform as the IRs of the group C_{2h}, we get

$$A_u: \Phi_1 = \tfrac{1}{2}(\phi_2 + \phi_3), \qquad \Phi_2 = \tfrac{1}{2}(\phi_1 + \phi_4) \tag{2.88}$$

$$B_g: \Phi_3 = \tfrac{1}{2}(\phi_2 - \phi_3), \qquad \Phi_4 = \tfrac{1}{2}(\phi_1 - \phi_4) \tag{2.89}$$

No other symmetry-adapted linear combinations exist since the atom positions 1 and 3 and 2 and 4 are not transformed into each other by any of the symmetry operations of C_{2h}.

Applying the rules of the Hückel model, we get with the symmetry-adapted orbitals Φ_1 and Φ_2

$$\begin{aligned} H_{11} &= \frac{1}{2}\int (\phi_2^* + \phi_3^*) H (\phi_2 + \phi_3)\, d\tau = \frac{1}{2}\int \phi_2^* H \phi_2\, d\tau \\ &\quad + \frac{1}{2}\int \phi_2^* H \phi_3\, d\tau + \frac{1}{2}\int \phi_3^* H \phi_2\, d\tau + \frac{1}{2}\int \phi_3^* H \phi_3\, d\tau \\ &= \frac{1}{2}(\alpha_{22} + \beta_{23} + \beta_{32} + \alpha_{33}) = \alpha + \beta \end{aligned} \tag{2.90}$$

$$H_{22} = \tfrac{1}{2}(\alpha_{11} + \beta_{14} + \beta_{41} + \alpha_{44}) = \alpha \tag{2.91}$$

$$H_{12} = H_{21} = \tfrac{1}{2}(\beta_{21} + \beta_{24} + \beta_{31} + \beta_{34}) = \beta \tag{2.92}$$

because $\beta_{14} = \beta_{41} = \beta_{24} = \beta_{31} = 0$. Similarly one obtains with the symmetry-adapted orbitals Φ_3 and Φ_4

$$H_{33} = \tfrac{1}{2}(\alpha_{22} - \beta_{23} - \beta_{32} + \alpha_{33}) = \alpha - \beta \tag{2.93}$$

$$H_{44} = \tfrac{1}{2}(\alpha_{11} - \beta_{14} - \beta_{41} + \alpha_{44}) = \alpha \tag{2.94}$$

$$H_{34} = H_{34} = \tfrac{1}{2}(\beta_{23} - \beta_{24} - \beta_{31} + \beta_{34}) = \beta \tag{2.95}$$

Since the matrix elements formed between symmetry-adapted orbitals from different IRs vanish, the secular determinant for the trans-butadiene molecule is of the form

$$\begin{vmatrix} H_{11}-E & H_{12} & 0 & 0 \\ H_{21} & H_{22}-E & 0 & 0 \\ 0 & 0 & H_{33}-E & H_{34} \\ 0 & 0 & H_{43} & H_{44}-E \end{vmatrix}$$

$$= \begin{vmatrix} \alpha+\beta-E & \beta & 0 & 0 \\ \beta & \alpha-E & 0 & 0 \\ 0 & 0 & \alpha-\beta-E & \beta \\ 0 & 0 & \beta & \alpha-E \end{vmatrix} = 0 \tag{2.96}$$

Solving this determinant block by block, results in bonding MOs at the energy eigenvalues

$$E_1(A_u) = \alpha + 1.62\beta \tag{2.97}$$

$$E_2(B_g) = \alpha + 0.62\beta \tag{2.98}$$

and antibonding MOs at the energy eigenvalues

$$E_3(A_u) = \alpha - 0.62\beta \tag{2.99}$$

$$E_4(B_g) = \alpha - 1.62\beta \tag{2.100}$$

where $\beta < 0$. In the ground state, the four electrons forming the π-bonding system of trans-butadiene are accomodated in the two bonding orbitals. The contribution of the p system to the bonding of the molecule is thus $4\alpha + 4.48\beta$. The cyclobutadiene molecule has the symmetry D_{4h} and results with the symmetry-adapted orbitals

$$A_{2u}\colon\ \Phi_1 = \tfrac{1}{2}(\phi_1 + \phi_2 + \phi_3 + \phi_4) \tag{2.101}$$

$$B_{1u}\colon\ \Phi_2 = \Phi(A_{2u}) = \tfrac{1}{2}(\phi_1 - \phi_2 + \phi_3 - \phi_4) \tag{2.102}$$

$$E_g\colon\ \Phi_3 = \Phi(E_{1g}) = \frac{1}{\sqrt{2}}(\phi_1 - \phi_3) \tag{2.103}$$

$$E_g\colon\ \Phi_4 = \Phi'(E_{1g}) = \frac{1}{\sqrt{2}}(\phi_2 - \phi_4) \tag{2.104}$$

in the secular determinant

$$\begin{vmatrix} H_{11} - E & 0 & 0 & 0 \\ 0 & H_{22} - E & 0 & 0 \\ 0 & 0 & H_{33} - E & H_{34} \\ 0 & 0 & H_{43} & H_{44} - E \end{vmatrix}$$

$$= \begin{vmatrix} \alpha + 2\beta - E & 0 & 0 & 0 \\ 0 & \alpha - 2\beta - E & 0 & 0 \\ 0 & 0 & \alpha - E & 0 \\ 0 & 0 & 0 & \alpha - E \end{vmatrix} = 0 \qquad (2.105)$$

It consists of two 1×1 blocks along the diagonal for the integrals involving the SALCAOs belonging to the one-dimensional representations A_{2u} and B_{1u} and one 2×2 block involving the two SALCAOs that are basis functions of the E_g representation. However, since $H_{34} = H_{43} = 0$, the secular determinant is diagonalized, resulting in

$$E(A_{2u}) = \alpha + 2\beta \qquad (2.106)$$

$$E(E_g) = \alpha \qquad (2.107)$$

$$E(B_{1u}) = \alpha - 2\beta \qquad (2.108)$$

Therefore, the electron distribution in the ground-state $a_{2u}^2 e_g^2$ corresponds to a contribution of the π-bonding system to the energy of $4\alpha + 4\beta$.

Of course, any distortion of the square planar molecule, leading to a molecular shape described by a group having only one-dimensional IRs, would lift the orbital degeneracy of the e_g orbital, lowering the total energy of the π-bonding system even further, as shown in Figure 2.19. This effect is called a Jahn–Teller distortion and is illustrated for the example of a distortion from

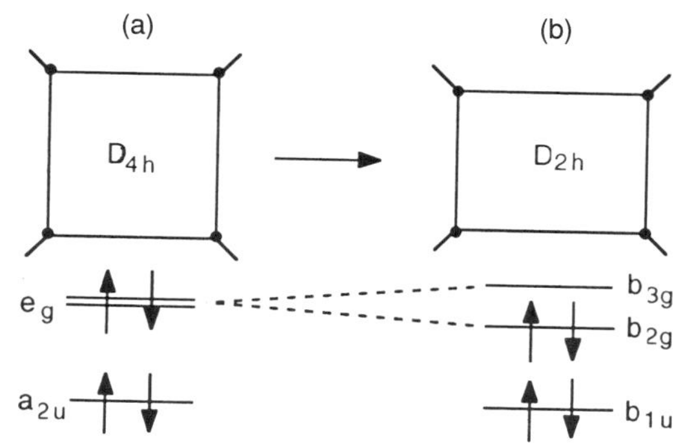

Figure 2.19 Example of a Jahn–Teller distortion of the cyclobutadiene molecule from (a) D_{4h} to (b) D_{2h} and the associated lowering of the energy of the π-bonding system.

D_{4h} to D_{2h} in Figure 2.19(b). Note that, although the cyclobutadiene molecule is susceptible to a Jahn–Teller distortion, one cannot decide whether this distortion actually occurs on the basis of group theory. In order to make this decision, a quantitative evaluation is required of not only the gain in energy resulting from the splitting of the degenerate eigenvalue of the π-bonding system, but also of the associated change in the energy of the σ-bonding system. Therefore, group theory is just a qualitative, albeit valuable, guide in the theoretical determination of the structure of molecules, providing suggestions that must be backed by calculations to decide on the energetically most favorable configuration.

References

1. C. Huygens, *Traite de la Lumière*, Leiden, 1692. English Translation by S. P. Thompson, The University of Chicago Press, Chicago, Illinois, 1945.
2. Rene Just Haüy, *Essay d'une théorie sur la structure des cristeaux, appliquée à plusieurs genres de substances crystallisées*, Paris, 1784.
3. A. W. Williamson, *J. Chem. Soc. London* **7**, 328 and 433 (1869).
4. D. I. Mendeleev, Faraday Lecture, *J. Chem. Soc.* **55**, 634 (1889).
5. A. Werner, *Ber.* **47**, 3087 (1914).
6. van der Waals, *Thesis*, Universität Leipzig, 1873.
7. G. C. Pimentel and A. L. McClellan, *The Hydrogen Bond*, W. H. Freeman, San Francisco, CA, 1959.
8. M. Planck, *Ann. Physik* **4**, 553 (1901).
9. N. Bohr, *Phil. Mag.* **26**, 1, 476 (1913).
10. G. N. Lewis, *J. Am. Chem. Soc.* **38**, 762 (1916).
11. W. Kossel, *Ann. Physik* **49**, 229 (1916).
12. E. Schrödinger, *Ann. Physik* **79**, 489 (1926); **80**, 437 (1926).
13. A. Einstein, *Ann. Physik* **17**, 132 (1905).
14. J. Franck and G. Hetz, *Verhandl. deut. phys. Ges.* **16**, 457 and 512 (1914).
15. A. H. Compton, *Phys. Rev.* **21**, 483 (1923); *ibid.* **22**, 409 (1923).
16. L. de Broglie, *These de doctorat*, Paris, Masson 1924; *J. phys. et radium* **7**, 1 and 321 (1926).
17. C. Davisson and L. H. Germer, *Nature* **119**, 558 (1927); *Phys. Rev.* **30**, 705 (1927).
18. C. Bugge, *Das Buch der Grossen Chemiker*, Verlag Chemie, Weinheim, 1984.
19. J. C. Slater, *Quantum Theory of Atomic Structure*, McGraw–Hill, New York, 1960.
20. H. S. Green, *Matrix Mechanics*, P. Noordhoff Ltd., Groningen, 1965.
21. J. L. Powell and B. Craseman, *Quantum Mechanics*, Addison–Wesley, Palo Alto, CA, 1965.
22. B. D. Douglas and C. E. Hollingsworth, *Symmetry in Bonding and Spectra*, Academic Press, Orlando, FL, 1985.
23. D. R. Hartree, *Proc. Cambridge Phil. Soc.* **24**, 89 (1927).

24. E. Stoner, *Phil. Mag.* **48**, 719 (1924).
25. J. C. Slater, *Phys. Rev.* **98**, 1039 (1955).
26. H. Pauli, *Z. Phys.* **31**, 765 (1925); W. Heisenberg, *ibid.* **39**, 499 (1926).
27. V. Fock, *Z. für Physik* **61**, 126 (1930); (b) **62**, 795 (1930).
28. J. C. Slater, *Phys. Rev.* **34**, 481 (1929).
29. W. A. Jensen, *Comp. & Maths. with Applications* **12A**, 487 (1986).
30. W. Heitler and F. London, *Z. Physik* **44**, 455 (1927).
31. J. C. Slater, *Phys. Rev.* **38**, 1109 (1931).
32. L. Pauling, *JACS* **53**, 1367 (1931).
33. L. Pauling, *The Nature of the Chemical Bond*, Cornell University Press, Ithaca, NY, 1960.
34. R. J. Gillespie, *J. Chem. Educ.* **47**, 18 (1970).
35. W. J. Hunt, P. J. Hay, and W. A. Goddard III, *J. Chem. Phys.* **57**, 738 (1972).
36. J. Bestgen, K. Holland-Moritz, D. O. Hummel, H. Meisenheimer, and G. Trafara, *Aufgaben zur Physikalischen Chemie*, Walter De Gruyter, Berlin, 1989.
37. P. E. Cade, K. Sales, and A. C. Wahl, *J. Chem. Phys.* **44**, 1973 (1966).
38. C. A. Coulson, *Valence*, The Clarendon Press, Oxford, 1954.
39. C. A. Coulson, *The Shape and Structure of Molecules*, revised by R. McWeeny, Clarendon Press, Oxford, 1982.
40. W. Kutzelnigg, Einführung in die Theoretische Chemie, Verlag Chemie, Weinheim, 1978.
41. A. C. Wahl and G. Das, in *Methods of Electronic Structure Theory*, H. F. Schaefer III, Ed., Plenum Press, New York, 1977, pp. 51–78.
42. A. C. Wahl, *J. Chem. Phys.* **41**, 2600 (1964).
43. R. F. Bader, W. H. Henneker, and P. E. Cade, *J. Chem. Phys.* **46**, 3341 (1967).
44. K. Ohno and K. Morokuma, *Quantum Chemistry Literature Data Base*, Elsevier Scientific Publ. Corp., Amsterdam, 1982.
45. J. C. Phillips and L. Kleinman, *Phys. Rev.* **116**, 287 (1959); **118**, 1153 (1960).
46. T. D. Dunning, Jr., and P. J. Hay, in *Methods of Electronic Structure Theory*, H. F. Schaefer III, Ed., Plenum Press, New York, 1977, pp. 1–27.
47. R. F. Stewart, *J. Chem. Phys.* **52**, 431 (1970).
48. J. C. Slater, *The Calculation of Molecular Orbitals*, John Wiley & Sons, New York, 1979.
49. W. Kolos and L. Wolniewicz, *J. Chem. Phys.* **49**, 404 (1968).
50. G. Herzberg, *J. Mol. Spectrosc.* **33**, 147 (1970).
51. A. F. Cotton, *Chemical Applications of Group Theory*, Wiley–Interscience, New York, 1971.
52. R. L. Flury, Jr., *Symmetry Groups*, Prentice–Hall, Englewood Cliffs, 1980.
53. E. P. Wigner, *Gruppentheorie und ihre Anwendung auf die Quantenmechanik der Atomspektren*, Friedich Vieweg und Sohn, Braunschweig, Germany, 1931; translated and revised edition, Academic Press, New York, 1959.
54. E. Hückel, *Z. Physik* **70**, 204 (1931); **72**, 310 (1931); **76**, 628 (1932).

CHAPTER

3

The Electronic Structure of Crystalline Solids

3.1 The Structure and Symmetry of Crystals

The macroscopic symmetry of crystals is defined by the symmetry operations that leave their external shape unchanged. This is illustrated in Figure 3.1a for a cubic crystal where three fourfold, four threefold, and six twofold symmetry axes plus an inversion center and sets of perpendicular symmetry planes are identified. No new symmetry elements are added in this description to those already introduced in the context of molecular symmetry. Planes in a crystal lattice are identified by their *Miller indices* (hkl), which are the reciprocals of the cut-offs of the plane on the crystal axes converted into integers by multiplication with a common factor (see Figure 3.1b). Directions are simply labeled by their smallest integer components on the crystal axis [uvw] while points are labeled by their coordinates [[mnp]], as illustrated in Figure 3.1c and d, respectively. Equivalent faces are represented by the Miller indices of any one of them enclosed in curly brackets, for example, {100} and {111} for the six cube faces and the eight octahedral faces on a cubic crystal, respectively. Planes that contain a common axis are said to form a *zone* and the axis is called a *zone axis*. If a plane (hkl) belongs to the zone [uvw] then $hu + kv + lw = 0$. In Figure 3.1, a *clinographic projection*, that is, a method that projects parallel edges as parallel lines, is utilized for the visualization of the crystal symmetry. A frequently used alternative is the *stereographic projection*. It is illustrated in Figure 3.2 using as an example the cuboctahedron, which is derived from a cube by truncation of all corners.

For the purpose of generating a sterographic projection of the faces of a crystal or of a general geometric body, it is placed inside a sphere so that its

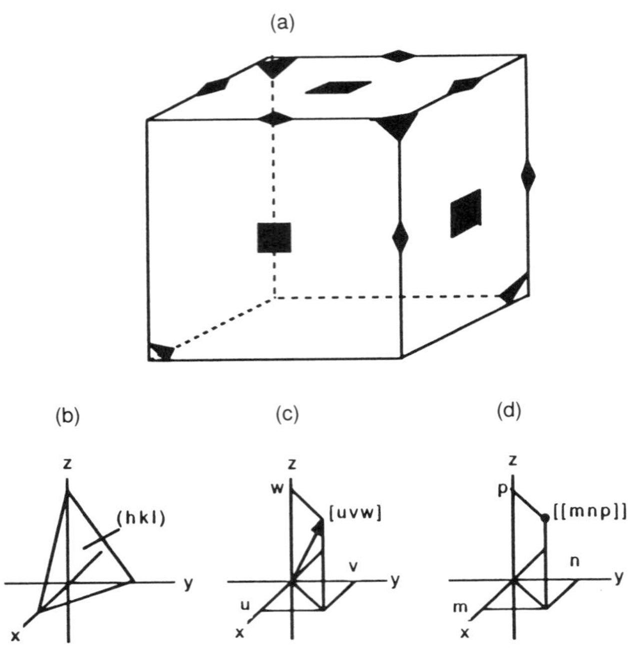

Figure 3.1 (a) Symmetry elements of a cubic crystal; (b)–(d) illustrations of the labeling of planes, directions, and coordinates in a crystal lattice.

center coincides with the center of the sphere. Orthogonal lines to the faces drawn from the center intersect the sphere in poles. To construct the stereographic projection, the poles of planes that are located on the northern hemisphere are projected from the south pole into the equatorial plane and are represented as full circles. The poles that are locate on the southern hemisphere are projected from the north pole into the equatorial plane and are represented by open circles. If both projections are superimposed on the same equatorial plane it is called a *Gadolin projection.*

Figure 3.2 shows the stereographic projection of the cuboctahedron representing the cube faces by squares and the faces perpendicular to the body diagonals of the cube as triangles with two of the cube faces parallel to the equator and two of the other four cube faces being perpendicular to the east–west axis. Since there are an infinite number of possible relative orientations of the cuboctahedron with respect to these axes, there exist an infinite number of projections of the same body. However, the angular distance between the poles is invariant to changes in the orientation. They can be read conveniently by the aide of a Wulff's net, and the stereographic projections can be rotated into standard configurations with a symmetry axis or any other outstanding symmetry element in its center, as described in more detail in textbooks of crystallography [1]. Note that directions in space (e.g., symmetry

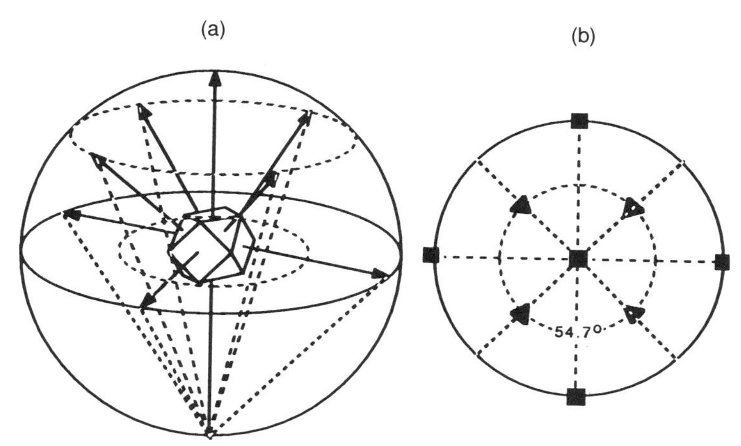

Figure 3.2 (a) Illustration of the generation of a stereographic projection of a cuboctahedron; (b) (001) stereographic projection.

axes) may be plotted in a stereographic projection in a similar fashion as the poles of crystal faces by projecting their points of intersection with the sphere from the north or south poles into the equatorial plane. Also, note that symmetry planes passing through the center of the crystal cut the surface of the sphere in large circles. Symmetry planes that contain the north and south poles are projected as straight lines.

There are altogether 32 different point groups that describe the macroscopic symmetry of crystals listed in Table 3.1 under the corresponding stereographic projections of their symmetry elements. These groups are called by crystallographers the 32 *crystal classes*, which should not be confused with the group-theoretical definition of classes of symmetry elements. They are ordered into 7 *crystal systems* (see Table 3.2). Crystals belonging to the isometric and tetragonal systems have four triad axes and a single tetrad axis, respectively. Point groups of crystals having three diad axis and no axis of higher symmetry belong to the orthorhombic system, while crystals having one triad or hexad axis belong to the hexagonal system, which may be subdivided into the trigonal and hexagonal systems, respectively (see Ref. [1] for a more complete discussion). The remaining point groups belong either to the monoclinic system if they have diad axes or to the triclinic system if they have only monad axes. The monoclinic point groups are shown twice corresponding to two different orientations of the stereographic projections of symmetry elements. The international notations and the corresponding Schönfließ symbols are listed in the lower right and left of each box, respectively.

As intuitively predicted by Huygens in 1678, the atomic or molecular motifs from which a crystal is built occupy the positions of a point lattice. At present several experimental methods have been established that resolve the atoms at the surface of a crystal lattice (see Section 3.6). The most important property

Table 3.1 Point groups obtained from symmetry elements of order 1, 2, 3, 4, and 6 (after Henry and Lonsdale [3])

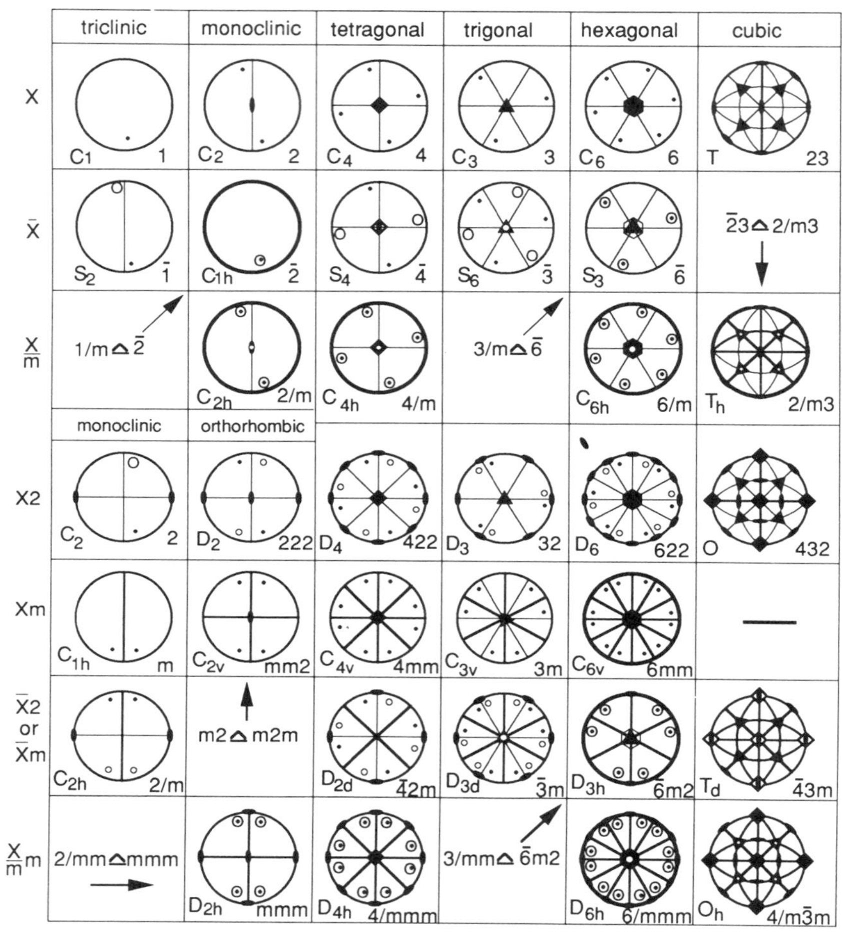

	triclinic	monoclinic	tetragonal	trigonal	hexagonal	cubic
X	C_1 1	C_2 2	C_4 4	C_3 3	C_6 6	T 23
$\bar{X}$	S_2 $\bar{1}$	C_{1h} $\bar{2}$	S_4 $\bar{4}$	S_6 $\bar{3}$	S_3 $\bar{6}$	$\bar{2}3$ △ 2/m3 ↓
$\frac{X}{m}$	1/m △ $\bar{2}$ ↗	C_{2h} 2/m	C_{4h} 4/m	3/m △ $\bar{6}$ ↗	C_{6h} 6/m	T_h 2/m3
	monoclinic	orthorhombic				
X2	C_2 2	D_2 222	D_4 422	D_3 32	D_6 622	O 432
Xm	C_{1h} m	C_{2v} mm2	C_{4v} 4mm	C_{3v} 3m	C_{6v} 6mm	—
$\bar{X}2$ or $\bar{X}m$	C_{2h} 2/m	↑ m2 △ m2m	D_{2d} $\bar{4}2m$	D_{3d} $\bar{3}m$	D_{3h} $\bar{6}m2$	T_d $\bar{4}3m$
$\frac{X}{m}m$	2/mm △ mmm →	D_{2h} mmm	D_{4h} 4/mmm	3/mm △ $\bar{6}m2$ ↗	D_{6h} 6/mmm	O_h 4/m$\bar{3}$m

of a point lattice is its *translational symmetry*, that is, any property that depends on the position in the crystal lattice is invariant to a translation by a lattice vector

$$\mathbf{T} = n\mathbf{a} + m\mathbf{b} + p\mathbf{c} \tag{3.1}$$

where n, m, p are integers and **a**, **b**, **c** are noncoplanar vectors pointing from the origin to neighboring lattice positions. The origin is usually chosen to coincide with a lattice point. The parallelepiped defined by **a**, **b**, **c** represents the *unit cell* of the lattice. The entire lattice is generated by integer translations of the unit cell on the 3 axes defined by unit vectors $\mathbf{a}/a$, $\mathbf{b}/b$, and $\mathbf{c}/c$, where

Table 3.2 Definition of the seven crystal systems in terms of the unit cell parameters

Triclinic	$a \neq b \neq c$	$\alpha \neq \beta \neq \gamma$	P
Monoclinic	$a \neq b \neq c$	$\alpha = \gamma = 90^\circ \neq \beta$	P, C
Orthorhombic	$a \neq b \neq c$	$\alpha = \beta = \gamma = 90^\circ$	P, C, I, F
Tetragonal	$a = b \neq c$	$\alpha = \beta = \gamma = 90^\circ$	P, I
Trigonal	$a = b = c$	$\alpha = \beta = \gamma = 120^\circ$	P
Hexagonal	$a = b \neq c$	$\alpha = \beta = 90^\circ,\ \gamma = 120^\circ$	P
Isometric or cubic	$a = b = c$	$\alpha = \beta = \gamma = 90^\circ$	P, I

$a = |\mathbf{a}|$, $b = |\mathbf{b}|$, $c = |\mathbf{c}|$. Denoting the angles between **b** and **c**, **a** and **c**, and **a** and **b** as α, β, and γ, respectively, the seven crystal systems may be defined in terms of the unit cell parameters. Usually the axes **a**, **b**, **c** are chosen to minimize the unit cell volume

$$V_{\mathrm{T}} = \mathbf{a} \cdot \mathbf{b} \times \mathbf{c} = abc \tag{3.2}$$

but exceptions are made in some cases where a unit cell of larger dimensions achieves higher symmetry. Figure 3.3 shows the 14 distinct translational unit cells that are either primitive (P), basis centered (C), body centered (I for German "innenzentriert"), or face centered (F). Since they were conceived in 1848 by Bravais, they are called *Bravais lattices*. Note that, for example, in the orthorhombic system there exist 3 types of faces on the unit cell. After choosing the labeling of the axes a, b, and c, respectively, the faces perpendicular to these axes may be labeled A, B, and C so that A, B, and C centering may be distinguished. However, since there exists no a priori labeling of these faces, they all constitute the same Bravais lattice type.

In crystalline materials, where the points of the Bravais lattice are occupied by atoms or ions, the crystal structure is often derived from a *close packing of spheres*. For example, many metals crystallize in close-packed form. There exist three different atom positions in close packings of spheres and consequently two fundamental repeat sequences that are illustrated in Figure 3.4. Figure 3.4(a) shows a top view of a close packing of spheres in the $\cdots ABABAB \cdots$ stacking sequence, where every two layers the atoms are located on top of each other. This stacking corresponds to a hexagonal unit cell, as illustrated in Figure 3.5(b) and is sometimes represented by the abbreviation H. The three-layer stacking $\cdots ABCABC \cdots$ shown in Figure 3.5(c) corresponds to an fcc unit cell, as illustrated in Figure 3.5(b) and is sometimes abbreviated C. By changes in the repeat pattern, a large variety of *polytypes* can be created. Since they differ only slightly in free energy, several polytypes may be present in one crystal, which poses a challenge in the growth of certain compound semiconductors of technological interest. For example, there are more than 40 known polytypes of α-SiC. Figure 3.5 shows the interface between the α-SiC 6H

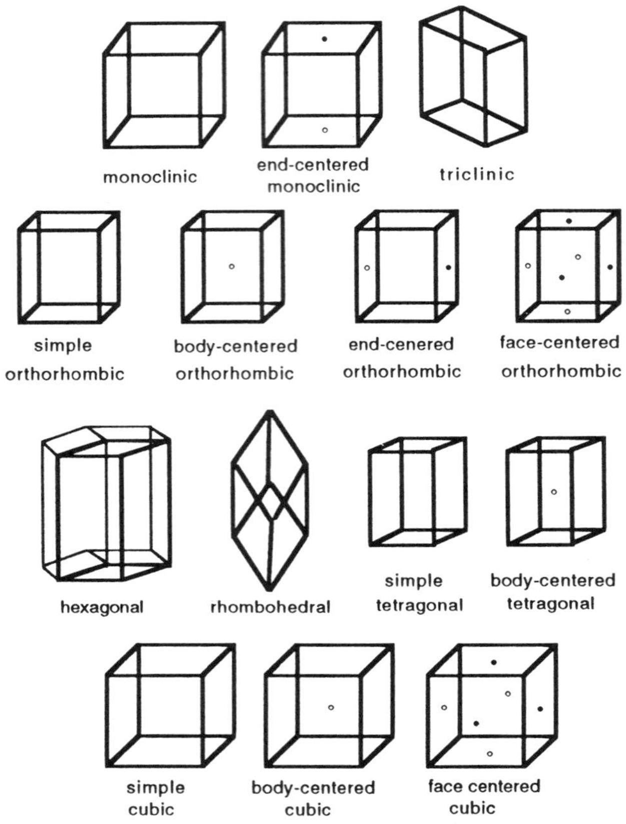

Figure 3.3 The 14 Bravais lattices.

polytype (bottom), which has a $\cdots ABABABBABABA \cdots$ stacking (i.e., a six-layer repeat distance), and cubic β-SiC (top).

For ionic crystals more open structures are favored that minimize the distance between ions carrying opposite charges, but simultaneously maximize the distance between ions that carry like charges. Many of these structures may be formally related to the hcp and ccp structures by the filling of the interstices between the close-packed spheres. There are two sorts of holes that may be filled by smaller spheres, *octahedral* and *tetrahedral holes.* Figure 3.6(a) shows the positions of these holes, which are either in between six or four close-packed spheres corresponding thus to octahedral [Figure 3.6(b)] or tetrahedral [Figure 3.6(c)] coordination. Note that there are twice as many tetrahedral as octahedral holes. The ratio of the radii of the spheres occupying the interstices to the radii of the close-packed spheres is 0.414 and 0.225 for ideal filling of the octahedral and tetrahedral holes, respectively.

Filling of all octahedral holes of the hcp structure generates the niccolite (NiAs) structure. Actually the nickel–nickel distance in niccolite is similar to

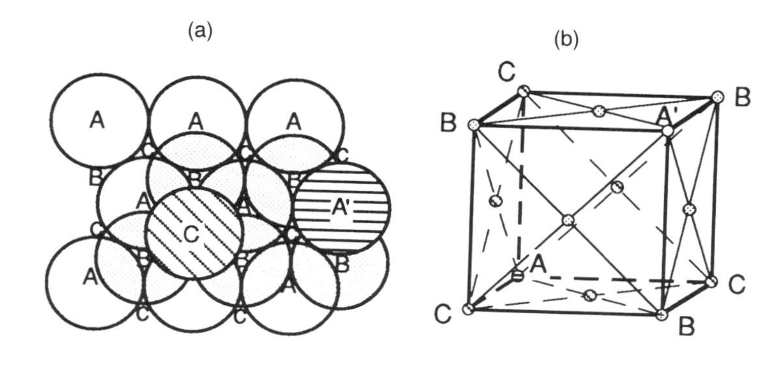

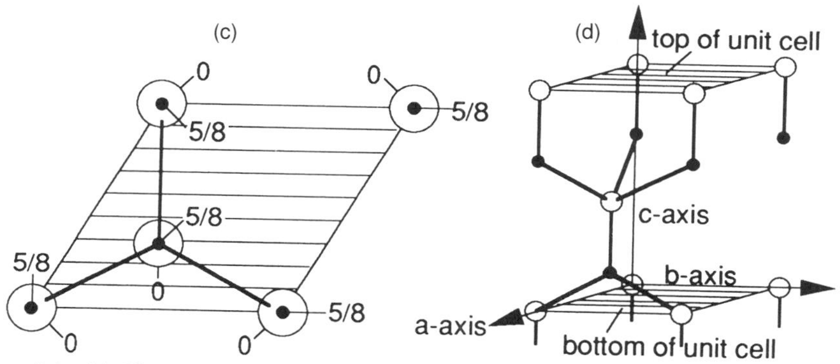

Figure 3.4 (a) Illustration of the close-packing of spheres with sides A, B and C. (b) Unit cell of the fcc lattice. (c) and (d) hexagonal close-packed (hcp) structures. After Smith [1].

the As–As distance. Therefore, the anions are pushed apart by the cations, which is desirable for minimizing their mutual repulsion. Removing every second row of cations from the niccolite structure results in the related $CaCl_2$ and rutile (TiO_2) structures, which differ by distortions of the ideal close-packed anion sublattice. Filling of all octahedral holes in the ccp structure results in the NaCl structures. The ratio of cation to anion radii is $\sim$0.55, so that in this case an open structure is also generated. Alternatively the NaCl structure thus may be considered as a *superstructure* of the primitive cubic lattice, as illustrated in Figure 3.7(b). In such a superstructure part of the symmetry-related sites of the lattice are occupied by different motifs. Figure 3.7(a) shows the unit cell of the CsCl structure lattice, which is a superstructure of the bcc lattice.

Often the motifs associated with the lattice sites are molecules, adding to the complexity of the structure. However, regardless of the complexity of the motifs, it is always possible to describe the structure by a combination of the

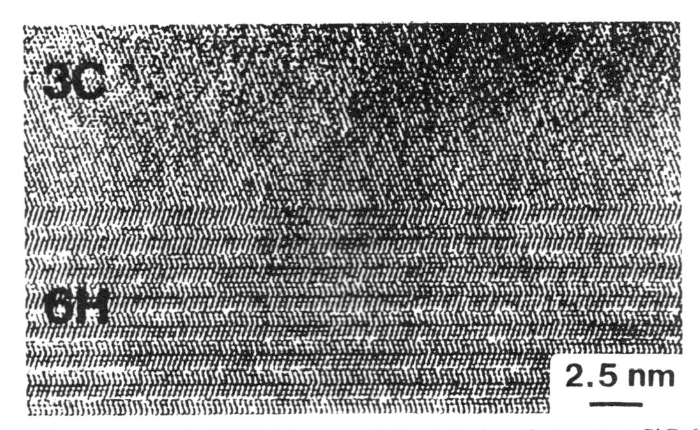

Figure 3.5 Cross sectional HREM image of the interface between α-SiC $6H$ poltype and β-SiC; courtesy of R. F. Davis, Department of Materials Science and Engineering, North Carolina State University.

translational symmetry operations of the Bravais lattices with the symmetry operations of the 32 point groups that characterize the symmetry of the finite structures that are compatible with a particular point lattice. Using a notation introduced by F. Seitz, the combination of a point group symmetry operation R with the translational operation t of the Bravais lattice is expressed as $\{R \mid t\}$; that is, operating on a general point x

$$\{R \mid t\}x = Rx + t \tag{3.3}$$

The combination of a Bravais lattice with point groups belonging to the same

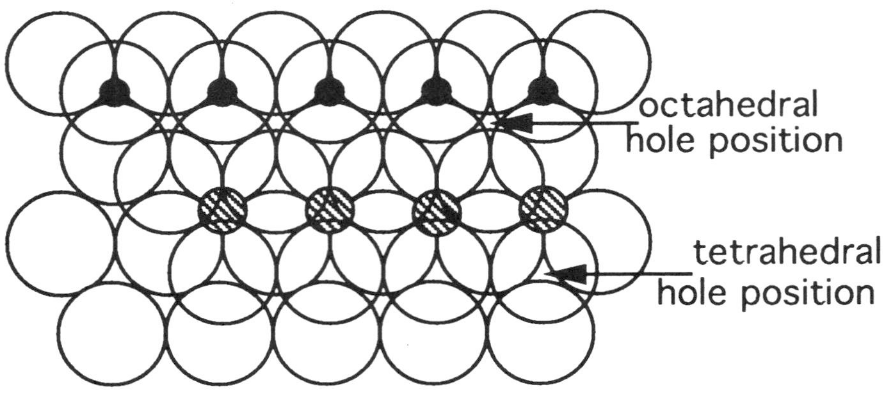

Figure 3.6 Tetrahedral and octahedral holes in close-packed structures.

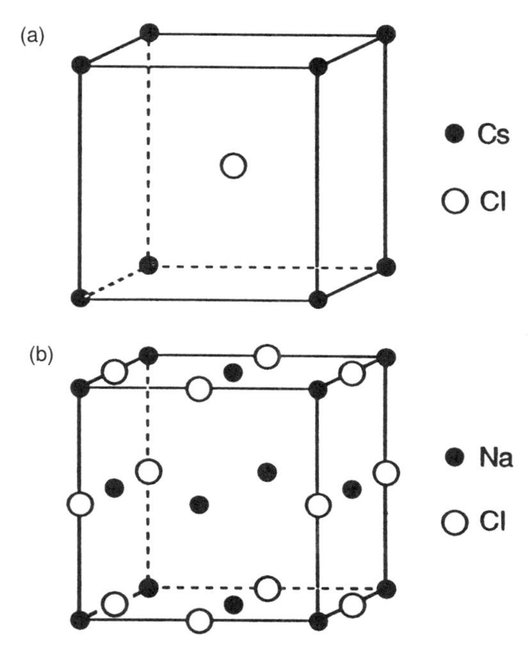

Figure 3.7 The crystal structures of (a) CsCl and (b) NaCl.

crystal system are called *symmorphic space groups.* For example, there are two tetragonal Bravais lattices and 7 tetragonal point groups resulting in 14 space groups. Sixty-six space groups can be formed in this simple manner. However, several alternatives in the positioning of specific point groups on certain Bravais lattices exist so that there are altogether 73 symmorphic space groups. In their labeling, using the system of the *International Tables of X-Ray Crystallography* [3], the Bravais lattice type is listed first, followed by the international point group symbol, for example, *P4mm* corresponding to a primitive tetragonal Bravais lattice combined with a motif of 4*mm* symmetry.

This is illustrated in Figures 3.8(a) and (b). Each motif of equivalent positions associated with the corners of the unit cell has 4*mm* symmetry, that is, a tetrad axis perpendicular to the plane of the paper plus one set of mirror planes *m* perpendicular to the directions of the *a* and *b* axes plus a second set of mirror planes *m* at 45° to the first set. The space group symmetry is shown in Figure 3.8(b) and contains, in addition, a set of *diagonal glide planes.* The reflection of the motifs on the glide planes, combined with a translation along the diagonal by $\frac{1}{2}$ of the repeat distance, brings them into equivalent positions. In this particular case, the glide planes are the result of the combination of integer lattice translations on the Bravais lattice with the point group symmetry and do not require explicit specification. Adding a body-centered position at $[[\frac{1}{2}\frac{1}{2}\frac{1}{2}]]$ associated with another set of general points (space group *I4mm*) generates another set of diagonal glide planes where the translation is of the

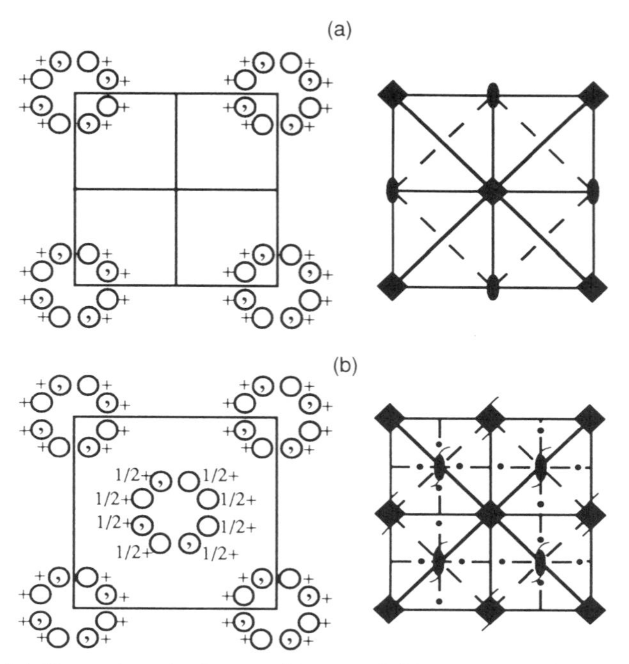

Figure 3.8 (a) General equivalent positions and (b) symmetry elements for the space groups *P4mm* and *I4mm*. After Henry and Lonsdale [3].

form $(\pm a \pm b \pm c)/2$. Diagonal glide planes are specified in the space group symbols by the general label n (e.g., $P4/n$ for a tetragonal lattice with one diagonal glide plane perpendicular to the tetrad axis). *Axial glide planes* (i.e., glide planes where the translations occur along one of the crystal axes by $a/2$, $b/2$, or $c/2$) are labele a, b, and c, respectively, for example, *Pbca* for an

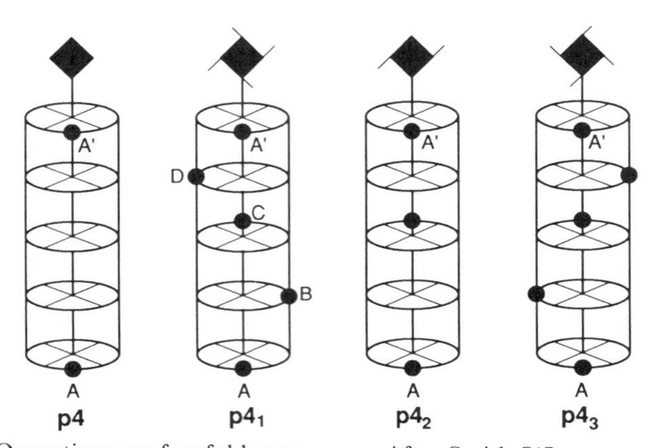

Figure 3.9 Operations on fourfold screw axes. After Smith [1].

orthorhombic lattice with 3 axial glide planes, one along each axis.

In addition to diagonal glide planes, a set of new symmetry elements is automatically created in the group *I4mm*, which are called *screw axes*. A screw axis n_i in the direction $\mathbf{c}/|\mathbf{c}|$ represents rotations by $2\pi/n$ coupled with simultaneous translations by $|\mathbf{c}|i/n$, where n and i are integers. Therefore, after n applications the position of a general point is related to its initial position by a lattice vector. For example, in the operation $4_1^{(1)}$ a general point **A** is rotated by 90° and is translated by $c/4$ along the axis into position **B** as illustrated in Figure 3.9 which also depicts the operations of p4, $p4_3$, $p4_2$ and $p\bar{4}$. The operations $4_1^{(2)}$ and $4_1^{(3)}$ repeat the rotation by 90° and translation by c/4, moving the general point into positions **C** and **D**, respectively. $4_1^{(4)}$ moves the point **A** into the equivalent lattice position **A′**.

Figure 3.10 represents views along the *c*-axes and the associated symmetry elements of (a) the diamond structure and (b) the zincblende structure. The lattice positions are labeled by their distances from the base in fractions of *c*. In the diamond structure there are screw tetrad axes that move a general point associated with the positions labeled 0 on the base of the cell through the

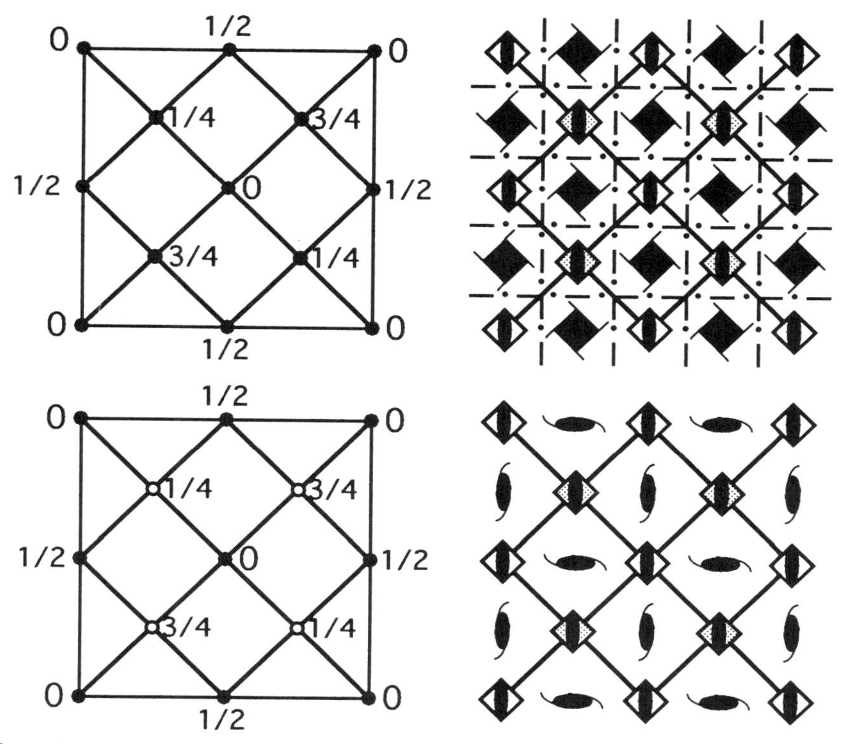

Figure 3.10 Lattice positions in the unit cells and symmetry elements of (a) the diamond and (b) the zincblende structures. After Smith [1].

positions $\frac{1}{4}$, $\frac{1}{2}$, $\frac{3}{4}$ into the equivalent positions labeled 1 on the top of the unit cell. Also, the strings of atoms parallel to the c axis are associated with fourfold rotoinversion axes. Furthermore, a new kind of glide plane exists, the traces of which are shown in Figure 3.10(a) by dash-dotted lines containing arrows. The translations on these diamond glide planes are of the type $(\pm a \pm b \pm c)/4$. The space group symbol for the diamond structure is $Fd3m$. Since the explicit specification of a symmetry element outside the point group symmetry is needed, in this case, the space group is *asymmorphic*. There are 157 asymmorphic space groups so that there exist all together 230 space groups that describe all possible crystal structures. As discussed already in the context of molecules, substitution lowers the symmetry so that the space groups of the zincblende structure are a subgroup of $Fd3m$. In the zincblende structure the order of the screw axes is reduced to 2 and the diamond glide planes are removed, resulting in the space group $F\bar{4}3m$.

Experimentally the space groups of crystals are determined from diffraction data. Such data are most conveniently interpreted in *reciprocal space*, defined by three noncoplanar vectors that are related to the lattice vectors **a**, **b**, **c** as

$$\mathbf{a}^* = \frac{\mathbf{b} \times \mathbf{c}}{abc}, \qquad \mathbf{b}^* = \frac{\mathbf{c} \times \mathbf{a}}{abc}, \qquad \mathbf{c}^* = \frac{\mathbf{a} \times \mathbf{b}}{abc} \tag{3.4}$$

having the properties of being reciprocal to the lattice vectors, so that

$$\mathbf{a}^* \cdot \mathbf{a} = \mathbf{b}^* \cdot \mathbf{b} = \mathbf{c}^* \cdot \mathbf{c} = 1 \tag{3.5}$$

and of vanishing other dot products among the reciprocal vectors and **a**, **b**, and **c** (e.g., $\mathbf{a}^* \cdot \mathbf{b} = \mathbf{a}^* \cdot \mathbf{c} = 0$). Thus $\mathbf{a}^*$ is perpendicular to **b** and **c**, $\mathbf{b}^*$ is perpendicular to **a** and **b**, and $\mathbf{c}^*$ is perpendicular to **a** and **b**. The unit cell volume V_{R} in the reciprocal space defined by $\mathbf{a}^*$, $\mathbf{b}^*$, and $\mathbf{c}^*$ is

$$V_{\mathrm{R}} = \mathbf{a}^* \cdot \mathbf{b}^* \times \mathbf{c}^* = \frac{\mathbf{a}^* \cdot \mathbf{b}^* \times (\mathbf{a} \times \mathbf{b})}{abc} = \frac{1}{abc} = \frac{1}{V_{\mathrm{T}}} \tag{3.6}$$

[The cross product of 3 vectors is $\mathbf{m} \times (\mathbf{n} \times \mathbf{p}) = (\mathbf{m} \cdot \mathbf{p})\mathbf{n} - (\mathbf{m} \cdot \mathbf{n})\mathbf{p}$.] Usually the unit cell in reciprocal space is defined by planes that bisect the reciprocal lattice vectors and that are perpendicular to $\mathbf{a}^*$, $\mathbf{b}^*$, and $\mathbf{c}^*$. The cell thus created is called the *first Brillouin zone* of the reciprocal lattice. A unit cell defined in the same way in the original lattice is called a *Wigner–Seitz cell*. Figure 3.11 shows the Brillouin zones that are associated with a selection of Bravais lattices.

Figure 3.12 illustrates the condition of constructive interference of x rays on a one-dimensional lattice. It is given by

$$\mathbf{a} \cdot (\mathbf{s} - \mathbf{s}_0) = H\lambda \tag{3.7}$$

so that for three dimensions

$$\mathbf{a} \cdot (\mathbf{s} - \mathbf{s}_0) = H\lambda, \qquad \mathbf{b} \cdot (\mathbf{s} - \mathbf{s}_0) = K\lambda, \qquad \mathbf{c} \cdot (\mathbf{s} - \mathbf{s}_0) = L\lambda \tag{3.8}$$

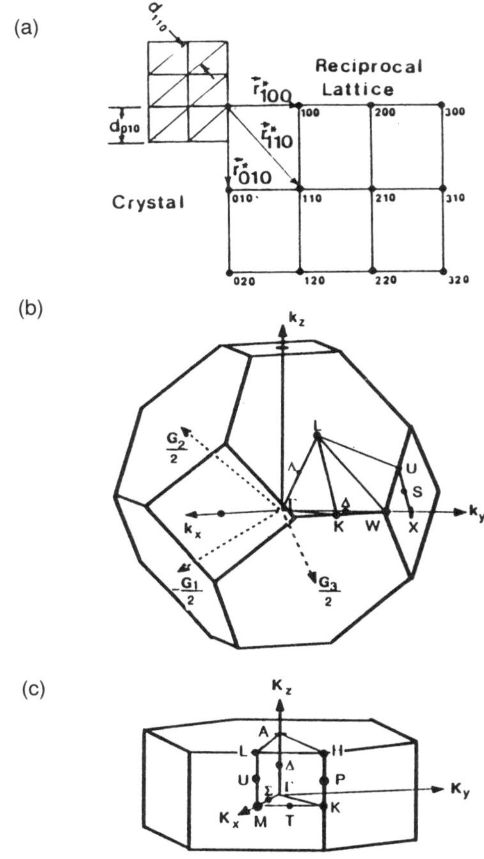

Figure 3.11 (a) The relation between lattice planes and reciprocal lattice points. After Spruiell and Clark [7]; copyright © 1980, Academic Press, Inc., New York, NY. (b) and (c), the first Brillouin zones of the fcc/diamond structure and the hcp structure, respectively.

where λ is the x-ray wavelength, the Laue indices H, K, and L are integers, and $\mathbf{s}$ and $\mathbf{s}_0$ are unit vectors in the directions of the diffracted beam and the incident beam corresponding to wave vectors

$$\mathbf{k} = \frac{2\pi}{\lambda}\mathbf{s} \qquad \text{and} \qquad \mathbf{k}_0 = \frac{2\pi}{\lambda}\mathbf{s}_0 \tag{3.9}$$

respectively. With Equation (3.9), the Laue equations (3.8) may be rewritten as

$$\mathbf{a}\cdot\Delta\mathbf{k} = 2\pi H, \qquad \mathbf{b}\cdot\Delta\mathbf{k} = 2\pi K, \qquad \mathbf{c}\cdot\Delta\mathbf{k} = 2\pi L \tag{3.10}$$

Since $|\mathbf{k}|$ and $|\mathbf{k}_0|$ have the dimension of a reciprocal length, $\Delta\mathbf{k} = \mathbf{k} - \mathbf{k}_0$ is a vector in reciprocal space. Based on a simple consideration of the quadruple

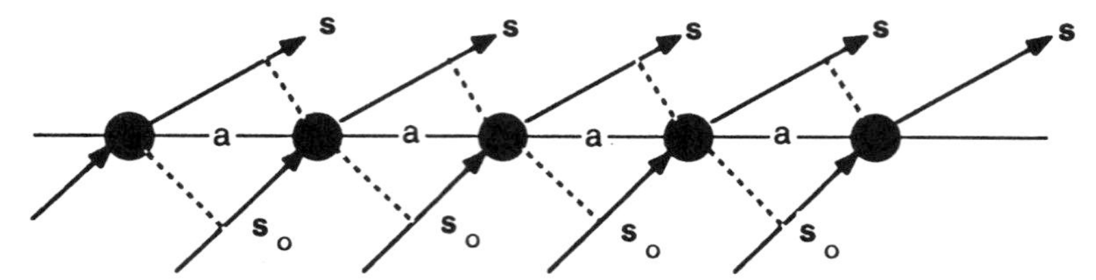

Figure 3.12 Illustration of the constructive interference for an incident wave with wave vector $(2\pi/\lambda)\mathbf{s}_c$ on a one-dimensional lattice.

vector product $\mathbf{a}^* \times \mathbf{b}^* \times \mathbf{c}^* \times \mathbf{d}^*$, it can be shown that a general vector $\mathbf{d}^*$ in reciprocal space can be expressed as

$$\mathbf{d}^* = (\mathbf{a}\cdot\mathbf{d}^*)\mathbf{a}^* + (\mathbf{b}\cdot\mathbf{d}^*)\mathbf{b}^* + (\mathbf{c}\cdot\mathbf{d}^*)\mathbf{c}^*$$

Thus the vector $\Delta\mathbf{k}$ is related to the basis vectors of the reciprocal lattice by the equation

$$\begin{aligned}\Delta\mathbf{k} &= (\Delta\mathbf{k}\cdot\mathbf{a})\mathbf{a}^* + (\Delta\mathbf{k}\cdot\mathbf{b})\mathbf{b}^* + (\Delta\mathbf{k}\cdot\mathbf{c})\mathbf{c}^* \\ &= 2\pi(H\mathbf{a}^* + K\mathbf{b}^* + L\mathbf{c}^*) = 2\pi\mathbf{G}^*_{HKL}\end{aligned} \tag{3.11}$$

In order to avoid carrying the factor 2π, it is frequently included in the definition of the reciprocal lattice vectors. In the following text this is recognized by replacing $\mathbf{G}^*_{HKL}$ by $\mathbf{G} \equiv 2\pi\mathbf{G}^*$. The difference between the wave vectors of the diffracted and incident x-ray beams is thus a vector from the common origin of the original lattice and the reciprocal lattice to the point $[[HKL]]$ in reciprocal space. This is the basis of the Ewald construction of the x-ray diffraction patterns, which is shown in Figure 3.13. Note that the lattice

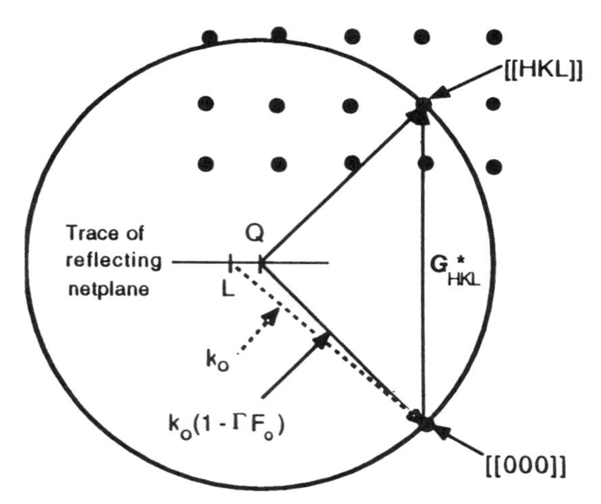

Figure 3.13 Ewald construction of the x-ray diffraction for the case of an ideal mosaic crystal.

planes (HKL) are perpendicular to $\mathbf{G}^*_{HKL}$ so that they bisect the angle between $\mathbf{k}_{HKL}$ and $\mathbf{k}_0$. Therefore, the diffracted beams may be regarded formally as reflections of the incident beam on the net planes of the crystal lattice. Also, note that the wave vectors inside the crystal must be corrected for the average refractive index. This moves the point Q off the Laue point L, as illustrated in Figure 3.13. The correction term $\delta_c = (r_e|^2\pi V)F_0$ is a very small number $(\sim 10^{-5})$ because the electron radius $r_e = 2.8 \times 10^{-13}$ cm $\ll |\mathbf{a}|$, while $\lambda \approx |\mathbf{a}|$ (see [7] for details).

In the kinematical theory of x-ray diffraction, the assumption is made that the x rays once diffracted penetrate the crystal without being diffracted back into the direction of the incident beam. This would be the case for an ideal mosaic crystal, that is, a crystal composed of small blocks that are tilted with regard to each other. Since the x rays reflected on the net planes of a particular mosaic block would be outside the glancing angle for other regions of the crystal they penetrate under ordinary absorption conditions. In this case, the intensity of the diffracted beam is

$$I_{HKL} = I_T L_{HKL} |F_{HKL}|^2 \tag{3.12}$$

where I_T and L_{HKL} are the Thomson scattering intensity for a free electron and the Laue interference function, respectively, and

$$F_{HKL} = \sum_k f_k \exp(i\mathbf{G}_{HKL} \cdot \mathbf{r}_k) \tag{3.13}$$

is the structure factor, which sums over all atoms with atomic scattering factors f_k at positions $\mathbf{r}_k$ in the unit cell. The electron density distribution in the unit cell of the crystal lattice is related to the structure factor by an inverse Fourier integral [5]. In view of the fact that the intensity peaks sharply in the vicinity of the reciprocal lattice points and decays rapidly with increasing order of the reflections, this integral can be approximated quite accurately by the Fourier sum over a sufficiently large set of experimentally determined structure factors

$$\rho(r) = \frac{1}{VF_T} \sum_H \sum_K \sum_L F_{HKL} \exp(-2\pi i \mathbf{G}^*_{HKL} \cdot \mathbf{r}) \tag{3.14}$$

Unfortunately the integrated intensity

$$R_{HKL} = \int\int I_{HKL} d^2 \, d\Omega \, dt \tag{3.15}$$

measured at distance d from the crystal integrated over the intercepted solid angle Ω during the time interval t is usually recorded for imperfect crystals approaching the kinematical behavior. Thus R_{HKL} is proportional to $|F_{HKL}|^2$. In order to relate the electron density distribution to the experimental data, it is thus necessary to replace the Fourier transformation over the position $\mathbf{r}$ in the unit cell by the Patterson function over the interatomic distance $\mathbf{r}_p$

$$P(r_p) = \int_{\text{unit cell}} \rho(r)\rho(r + r_p)\,d\tau(\mathbf{r}) = \sum_H \sum_K \sum_L F^2_{HKL} \exp(-2\pi i \mathbf{G}^*_{HKL} \cdot \mathbf{r}_p) \tag{3.16}$$

which sums over $|F_{HKL}|^2$. Because the Patterson function is the convolute of $\rho(r)$ with itself, and for $r_p = 0$ each atom in the unit cell is placed into the origin, the Patterson function represents a superposition of images that is difficult to unravel. However, with access to fast digital computers, it is at present possible to generate detailed high-resolution 3D images of the atomic positions and electron density distributions from diffraction data even for very complex macromolecules. For the simple crystal structures that are of interest in the context of solid-state electronics, accurate information on the electron

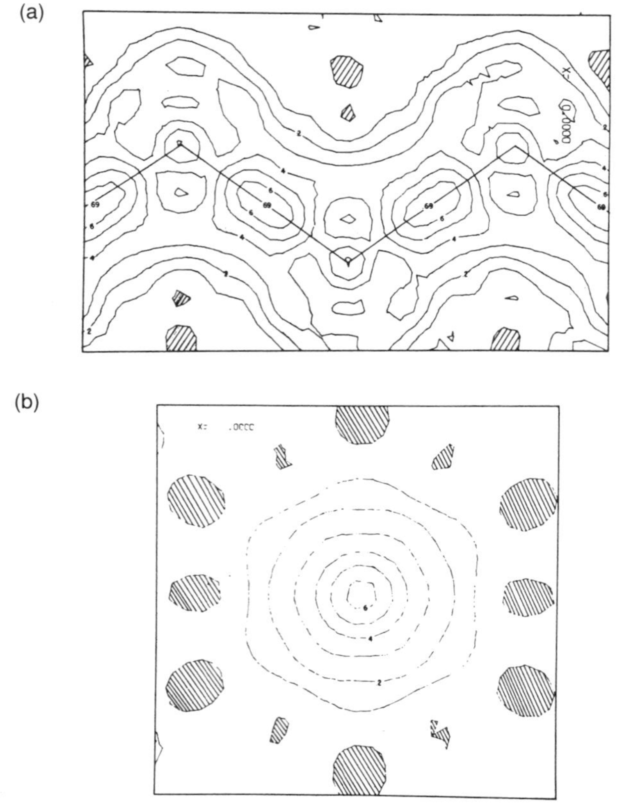

Figure 3.14 Valence charge-density contours ($0.1q\,\text{Å}^{-3}$) for crystalline silicon (a) along the Si bonds (b) perpendicular to the bonds. After Yang and Coppens [5]; copyright © 1974, Pergamon Press, Ltd, Oxford.

density distributions is obtained from x-ray data with relatively little effort. Therefore, reliable information on the bonding can be obtained from x-ray diffraction data. For example, Figure 3.14 shows the electron density difference maps along (a) the bond chain of Si and (b) perpendicular to the bond at the midpoint between the nuclei [5]. Note that with the exception of the first contour in Figure 3.14(b), the electron distribution is cyclindrically symmetric about the bond. Figure 3.14(a) shows that there is substantial excess charge channeled in between the atoms of the crystal lattice as compared to the charge distribution of noninteracting free Si atoms at the same positions corresponding to strong covalent bonding.

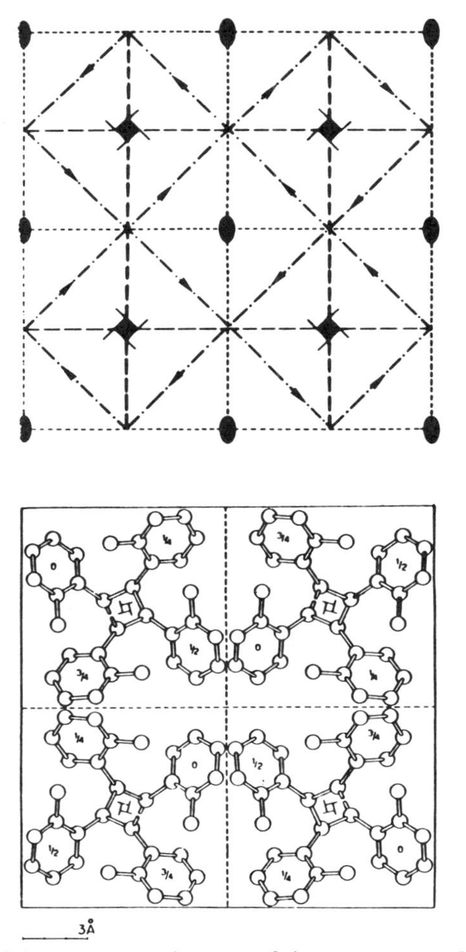

Figure 3.15 Some of the symmetry elements of the space group $I4_1cd$ and a model of the structure of poly(o-methylstyrene) projected onto (001). After Corradini and Ganis [7]; copyright © 1960, Nuovo Cimento, Rome.

In addition to ionic bonding and covalent bonding, there exist molecular crystals where the lattice is held together primarily by weak bonding forces. A wide variety of crystal structures ranging from highly symmetric structures to structures of low symmetry exist in this class of compounds. Of particular interest in the context of solid-state electronics are x-ray structure determinations on polymers that reveal details about their secondary structure, as discussed in Section 2.1 [6]. Figure 3.15 shows as an example the structural model of the carbon backbones of 4 chain units and the associated symmetry elements projected onto (001) of a crystal of poly(*o*-methylstyrene) that crystallizes in the space group $I4_1cd$. The screw tetrad axes coincide with the axes of the polymer chains, and the repeat distance on the chains defines the c-axis lattice parameter of the tetragonal unit cell.

In the case of the technologically important semiconductors, the art of crystal growth is currently in such an advanced state that large perfect single crystals are produced routinely. Therefore, it is necessary to widen the scope of our discussion of diffraction phenomena to include perfect crystals and nearly perfect crystals. They differ in their diffraction properties from the predictions of the kinematical theory in several important points. For example, the geometrical diffraction conditions are no longer set by the Ewald construction for a single excitation point Q, but are defined by the excitation of tie points on the two branches of a dispersion surface. A detailed discussion of the properties of this dispersion surface and dynamic x-ray diffraction is outside the scope of this book, and the reader is referred to the comprehensive review in Ref. [7] for further studies. However, two particular x-ray diffraction conditions, the symmetric Laue case, where the entrance and exit surfaces for the incoming and diffracted x rays differ, and the symmetric Bragg reflection case, where the entrance and exit surfaces on the crystal are identical, are briefly discussed in the following.

Considering the Laue case first, we note that, in a defect-free crystal under diffraction conditions, the x-ray beam is diffracted back and forth on the perfectly parallel net planes. The wave field in the crystal excited in this manner extends over the entire Borrmann delta of Poynting vectors between $\mathbf{s}$ and $\mathbf{s}_0$, as illustrated in Figure 3.16(a). At the exit surface this wave field breaks up into diffracted beam R and a forward-diffracted beam R_0 as illustrated in Fig. 3.16(b). They are recorded on a film, as separate lines. The absorption inside the crystal depends on the direction of the energy flow. In the directions of the incident and diffracted beams the absorption coefficient is the usual value for the ideal mosaic crystal μ_0, but for energy flow parallel to the diffracting net planes it is substantially reduced to the value μ_s, as illustrated in Figure 3.16(a). The cause for this behavior is the generation of standing wave patterns inside the crystal with planes of constant intensity parallel to the diffracting net planes. One of the standing wave patterns has nodal planes coinciding with the atom positions on the net planes, while the other standing wave pattern has nodal planes in between the net planes. Since in the former case the maximum electron density in the crystal coincides with the nodal planes, the waves of this

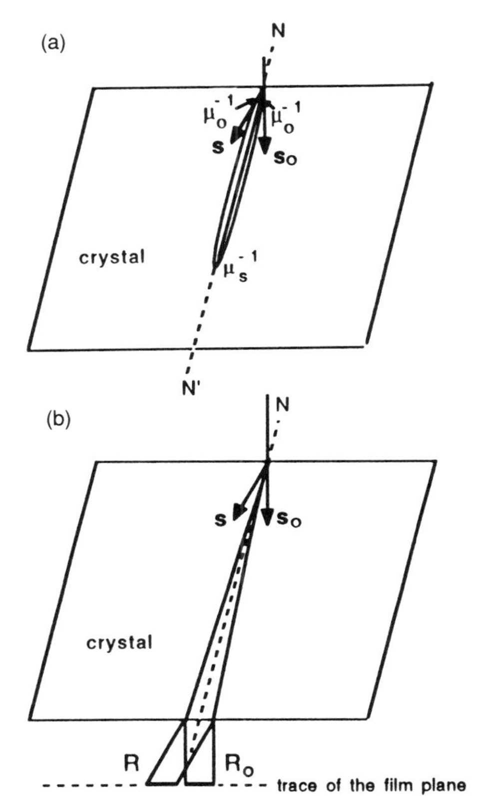

Figure 3.16 Schematic representation of anomalous transmission of x rays through a perfect crystal. (a) The variation of the absorption coefficient with the direction of energy flow; (b) breakup of the wave field that is generated inside the crystal at the exit surface. N − N′ trace of the reflecting netplane. After Borrmann, Hildebrandt, and Wagner [9].

pattern are very little attenuated, while the wave pattern corresponding to coincidence of the planes of maximum intensity with the positions of maximum electron density leads to strong absorption.

For a thick crystal (i.e., absorption coefficient thickness product $\mu t > 1$), only the lightly absorbed wave pattern survives, giving rise to the anomalous transmission of x rays through perfect crystals. This effect is called the *Borrmann effect* [8]. Since the fan of surviving x rays narrows with increasing thickness of the crystal, the Borrmann delta attains a smaller halfwidth than twice the Bragg angle. Therefore, the beams R and R_0 are separated from the primary beam and are recorded on a film as distinct lines, the separation of which depends on the entrance position of the primary beam and the distance of the film from the exit surface. If there are defects in the crystal lattice that destroy locally the perfect order of the net planes and with it the anomalously transmitted wave pattern, the intensity in their vicinity varies casting shadows of the defects. This is the basis for the imaging of crystal defects by x-ray topography.

Because of the deviation of the refractive index for x rays from 1, the diffraction peak shown in Figure 3.17(a) for the symmetric Bragg reflection case is shifted off the position determined by Bragg's law by

$$\Delta\theta = (r_e \lambda^2/\pi V) F_0/\sin 2\theta \tag{3.17}$$

The width of the diffraction curve is given by

$$\Delta\theta_{\text{FWHM}} = |\mathbf{k}|\,|\mathscr{P}|(r_e \lambda^2/\pi V)|F_{HKL}| \sec\theta \tag{3.18}$$

Therefore, $\Delta\theta_{\text{FWHM}}$ decreases with increasing energy of the x-ray radiation and

(a)

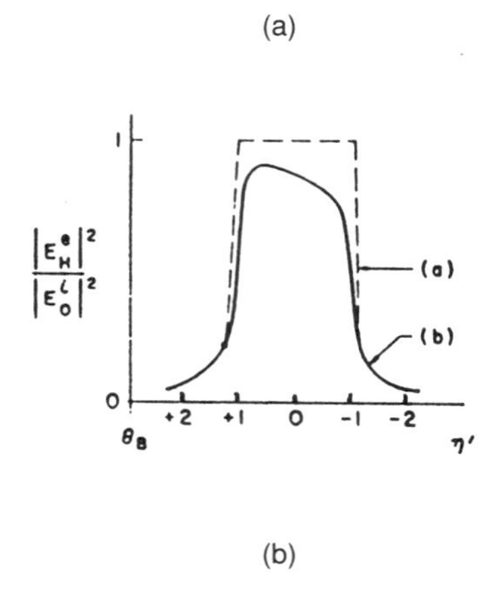

(b)

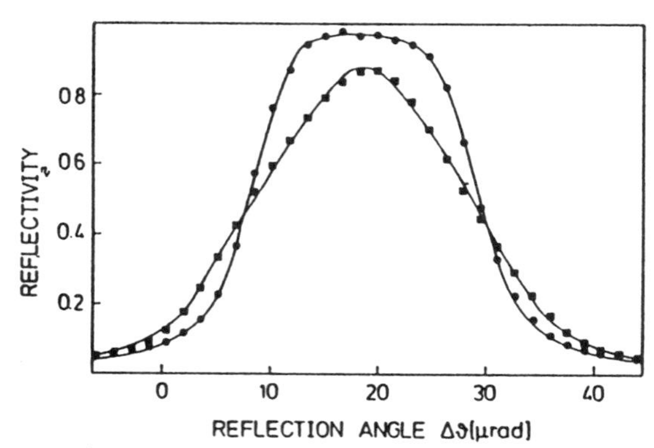

Figure 3.17 (a) Normalized intensity versus angular position for the Bragg reflection of x rays on a perfect crystal with (full line) and without (dashed line) absorption. (b) Reflectivity of the (111) Si surface for x rays measured in a triple crystal spectrometer. ● Perfect crystal, ■ reflectivity curve after sample sputtering. After (a) Batterman and Cole [7] and (b) Funke and Materlik [9]; copyrights (a): © 1964, *Reviews of Modern Physics*, New York, and (b): © 1987, North Holland Publishing Company, Amsterdam.

decreasing structure factor of the reflection and is typically of the order of seconds of an arc. An example of this behavior is given in Table 3.3, which presents the values of $\Delta\theta$ for perfect crystals of Cu at selected x-ray wavelengths and diffraction vectors $\mathbf{G}^*_{HKL}$. $\mathscr{P} = \cos 2\theta$ is the polarization factor. The penetration depth of the x rays into the crystal is proportional to the inverse of the absorption coefficient and is under kinematic diffraction conditions typically 10–100 μm. Under dynamic diffraction conditions, the effective absorption coefficient is significantly smaller because it is determined by extinction effects. It is related to the linear absorption coefficient for kinematic diffraction conditions μ according to the equation

$$\mu_s = \left(1 + \mathscr{P}\sqrt{u}\,\frac{F'_{HKL}\sin\varphi + F''_{HKL}\cos\varphi}{F''_0}\right)\frac{\mu}{\sin\theta_B} \tag{3.19}$$

where F'_{HHL} and F''_{HKL} are the real and imaginary parts of the complex structure factor for the (HKL) reflection and u is the reflectivity of the crystal. For a perfect (111) silicon surface the penetration depth is of the order of 0.1–1 μm. As the angle of incidence θ is advanced through the diffraction peak, the phase angle φ varies from π to 0, corresponding to nodes and antinodes of the standing x-ray waves at the diffracting net planes. This shift in the standing-wave pattern relative to the net planes can be utilized for a measurement of the registration of adsorbed atoms on a crystal surface relative to the underlying lattice. For crystals containing defects the width of the diffraction curve increases rapidly and can thus be utilized for characterizing nondistructively the perfection of a crystal. Since for dynamic x-ray diffraction the integrated intensity is proportional to F_{HKL} [4], while it is proportional to F^2_{HKL} for the ideal mosaic crystal, the integrated intensity also provides information concerning the crystal perfection.

Table 3.3 Calculated rocking curve halfwidth $\Delta\theta$ and integrated intensity R for perfect crystals and ideal mosaic crystals of copper (after Bachmann, Baldwin, and Young, Jr. [10])

λ (Å)	HKL	$\Delta\theta_{perf}$ (sec)	$R_{perf} \times 10^7$	$R_{imperf} \times 10^3$	R_{perf}/R_{imperf}
0.56	111	10.9	448	3315	7.3
	222	3.4	128	610	4.7
	333	1.3	42	109	2.6
	444	0.7	20	36	1.8
0.71	111	15.4	542	2930	5.4
	222	4.5	148	473	3.2
	333	1.8	48	104	2.2
	444	1.0	29	43	1.5
1.54	111	28.8	1095	8280	7.6
	222	11.0	435	435	5.5

Table 3.3 lists the halfwidth of the x-ray rocking curve of perfect crystals of copper for selected values of the x-ray wavelength and the Laue indices. Figure 3.18 shows the experimental variation of $\Delta\theta$ with increasing defect density N_d measured for a set of Cu crystals deformed to selected points on the stress–strain curve. Considerable broadening is observed with increasing deformation. Note that the x-ray rocking curve measurements cover the range of defect densities in between the low- and high-density ranges that are suitable for x-ray topography and transmission electron microscopy studies, respectively.

3.2 Energy Bands

The electronic structure of solids can be predicted by theoretical modeling, which, similar to MO calculations, relies on solving Equation (2.2) for the entire assembly of atoms in the crystal with a one-electron Hamiltonian that incorporates the effects of correlation and exchange. In view of the large number of atoms, the computational effort is substantial and requires taking full advantage of the simplifications that arise from the symmetry of the crystal lattice. In order to connect the LCAOMO calculations for small molecules to calculations of the electronic structure of solids, we consider first the energy eigenvalues of the π-bonding system of conjugated polyenes, that is, com-

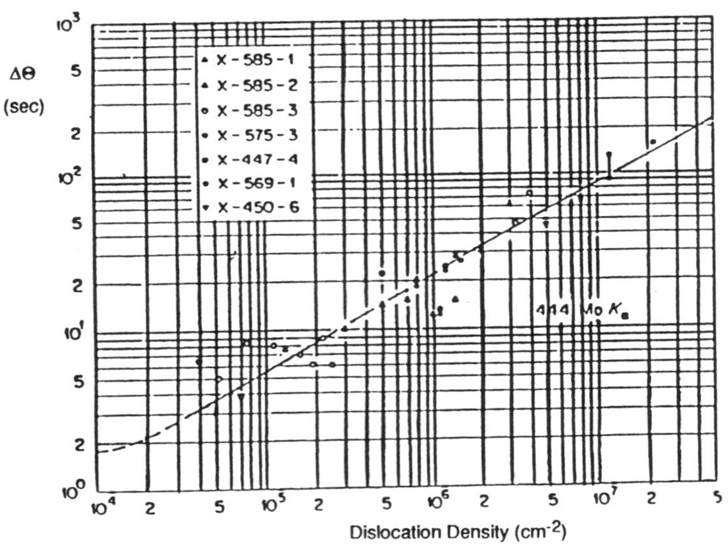

Figure 3.18 Change in the FWHM of the diffraction curve with increasing density of dislocations in a copper crystal. After Bachmann, Baldwin, and Young, Jr. [10]; copyright © 1970, The American Institute of Physics, New York.

pounds of the general formula $H[-CH = CH-]_nH$, as a function of increasing chain length parameter n. For very large n these chains approach the structure of a one-dimensional crystal. A simple Hückel calculation for the π-bonding system of conjugated polyenes results in the general expression

$$E_j = \alpha + 2\beta \cos \frac{j\pi}{n+1} \tag{3.20}$$

where j is an integer $\leqslant n$ [11], [12]. Figure 3.19 shows the energy levels and, schematically, the orbitals up to $n = 4$. Since $|\cos(j\pi/n + 1)| \leqslant 1$, these levels are restricted to fall within the limits $\alpha \pm 2\beta$, so that with increasing chain length the spacings of the energy levels become closer and closer, forming in the limit $n \to \infty$ a quasicontinuous band of allowed energy. Since there is one p_z electron per atom and the band contains one bonding and one antibonding level per p_z-electron pair, there are as many states as atoms per unit volume. Since two electrons are accomodated by each bonding orbital, the band formed from the p_z electrons should be half filled. Accordingly, polyacetylene should behave like a metal. In reality, it is a semiconductor because the π-bonding system is not evenly distributed over the chain, causing uneven spacings of pairs of atoms that affect the band structure.

Although the number of repeat units in crystals is usually very large, a conceptional problem in energy band-structure calculations is the termination of the periodic structure at its ends. A solution to this problem is the provision

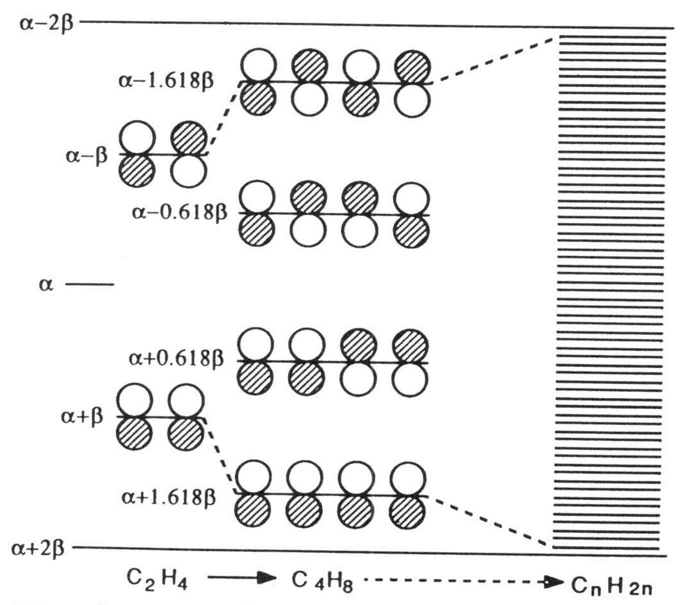

Figure 3.19 Schematic representation of the energy levels and of the associated combinations of p_z orbitals for conjugated π-bonded chains up to butadiene.

of cyclic boundary conditions assuming that the crystal is bent so that its end faces touch. For the example of the polyenes, this can be accomplished by chemical closure at their ends to form large conjugated ring systems. This is illustrated in Figure 3.20 for the ring closure of octatetraene to cyclooctatetraene under abstraction of a hydrogen atom at each of the terminal carbon atoms of the chain. We have already evaluated the energy levels of one member of this class of compounds, cyclobutadiene, in Section 2.5. An extension of these calculations to cyclic conjugated polyenes containing N carbon atoms [12] results in

$$E_j = \alpha + 2\beta \cos \frac{2\pi i}{N} = \alpha + 2\beta \cos(\mathbf{k} \cdot \mathbf{a}) \tag{3.21}$$

where $j = 0, \pm 1, \pm 2, \pm N/2$ and $\mathbf{k} = (j/N)\mathbf{a}^*$ is a reciprocal lattice vector with values in the range $-\pi/a \leq |\mathbf{k}| \leq +\pi/a$, that is, within the first Brillouin zone of the 1D crystal. Figure 3.21 shows the evolution of the $E(\mathbf{k})$ dispersion from discrete values of $E(j)$ for 5- and 15-membered conjugated rings into a quasi-continuous band at very large N. Note that in the limit $N \to \infty$ the bond strain associated with the ring closure vanishes. Therefore, the cyclic boundary condition does not limit the accuracy of band structure calculations for sufficiently large crystal dimensions.

Figure 3.22 shows the band structure of polyethylene, which is a σ-bonded system. It depends sensitively on the conformation of the molecule in the helical chain representing its minimum-energy configuration. The angle ω specifies the rotation of adjacent CH_2 groups about the C–C axis, for example, 180° in the trans configuration and 60° in the gauche configuration. Note that for any finite number of atoms a nonzero spacing of the states in a band results and that the density of states in an energy interval $[E_n, E_n + dE]$ generally varies with the value of E_n. This is shown on the right side margins in Figure 3.22(a) for the nonane molecule, which represents a small-size analog to the polyethylene chain so that the spacings between the energy levels is still fairly large. Nevertheless, the much simpler calculation for the nonane molecule already outlines the variations in the density of states distribution within the bands, which is further illustrated in Figure 3.22(b).

The consideration of the energy levels of the π-bonding electrons in the

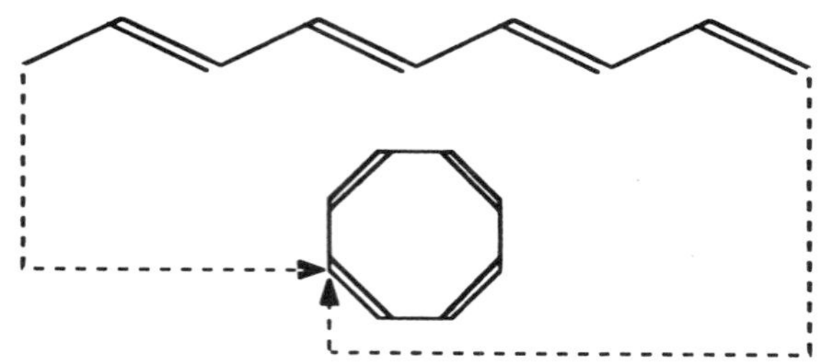

Figure 3.20 A visualization of cyclic boundary conditions for a 1D crystal: The formation of a conjugated chain ring system by the closure of a conjugated chain on its ends.

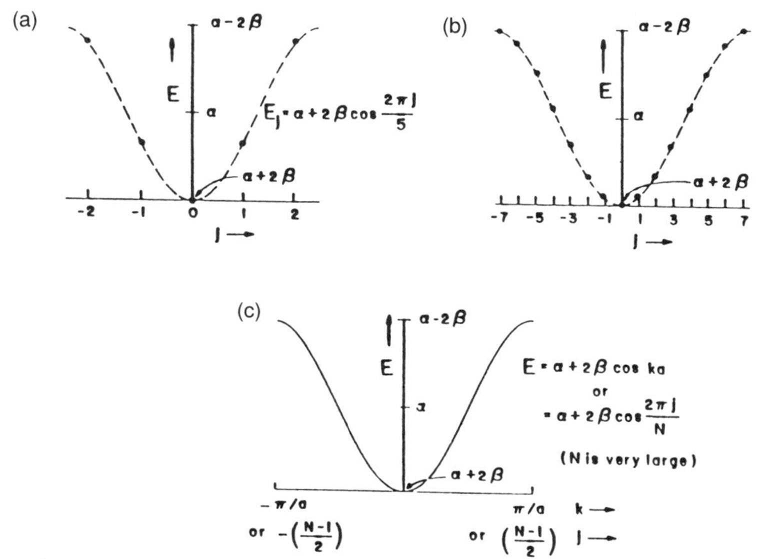

Figure 3.21 The evolution of a quasicontinuous band of energy levels from the discrete energy levels of conjugated ring systems with increasing ring size: (a) cyclopentadiene, (b) 15-annulene, (c) infinite loop. After Burdett [12]; copyright © 1984, Pergamon Press, Oxford.

Hückel approximation can be extended to band structures in three dimensions within the framework of tight-binding calculations. However, although the tight-binding model is very satisfying in maintaining a simple relation between the bonding of the atoms and the band structure, it is difficult to handle mathematically. Therefore, we next consider alternative approaches to the modeling of the band structure of 3D solids, based on perturbation theory and cellular methods, respectively.

It can be shown [13] that the translational symmetry of the crystal lattice in conjunction with cyclic boundary conditions requires the solutions of Equation (2.2) for crystals to conform to Bloch's theorem

$$\Psi_q(\mathbf{r} + \mathbf{T}) = \exp(i\mathbf{q}\cdot\mathbf{T})\Psi_q(\mathbf{r}) \tag{3.22}$$

where $\mathbf{q}$ is a vector in reciprocal space. Bloch's theorem may be rewritten in the equivalent form

$$\Psi_q(\mathbf{r} + \mathbf{T}) = \exp(i\mathbf{q}\cdot\mathbf{r})u_q(\mathbf{r}) \tag{3.23}$$

where $u_q(\mathbf{r})$ has the periodicity of the lattice; that is, $\Psi_q(\mathbf{r})$ is a plane wave traveling in the direction of $\mathbf{q}$ modulated by a lattice periodic function. With

$$\mathbf{q} = \mathbf{G}_{HKL} + \mathbf{k} \tag{3.24}$$

$$\begin{aligned}\exp(i\mathbf{q}\cdot\mathbf{T})\Psi_q(\mathbf{r}) &= \exp(i\mathbf{G}_{HKL}\cdot\mathbf{T})\exp(i\mathbf{k}\cdot\mathbf{T})\Psi_q(\mathbf{r})\\ &= \exp(i\mathbf{k}\cdot\mathbf{T})\Psi_q(\mathbf{r})\end{aligned} \tag{3.25}$$

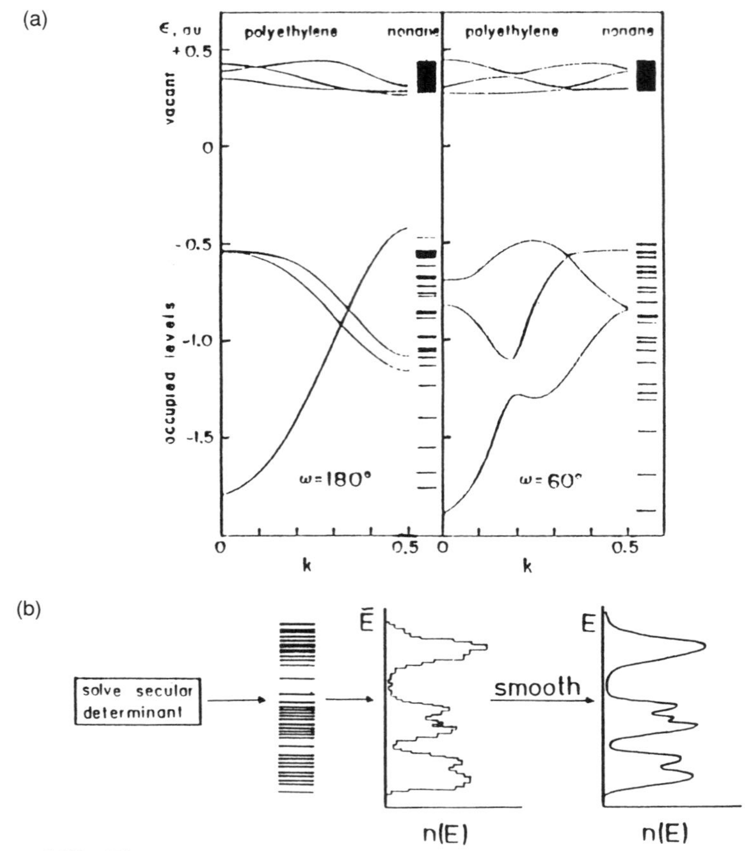

Figure 3.22 The energy band structure of polyethylene (a) and density of states versus energy (b). (a) After Morokuma [14]; (b) After Burdett [12], reprinted by permission of the publishers; copyrights (a) © 1976, Elsevier Science Publishers B.V., Amsterdam, and (b) © 1984, Pergamon Press, Oxford.

that is, $\Psi_q(\mathbf{r})$ satisfies Bloch's theorem with the reduced wave vector $\mathbf{k} = \mathbf{q} - \mathbf{G}_{HKL}$. By convention, $|\mathbf{k}|$ is made as small as possible, so that its range extends over the first Brillouin zone. The set $\mathbf{k}, \mathbf{G}_{HKL}$ suffices to label all one-electron states $\Psi_q(\mathbf{r})$ of the crystal.

In an empty lattice [i.e., a lattice minus the motifs, $V(\mathbf{r}) = 0$], $u_q(\mathbf{r}) = 1$ and Equation (2.2) reduces to

$$-\frac{\hbar^2}{2m}\Delta\Psi = E\Psi \tag{3.26}$$

so that

$$E_q = \frac{\hbar^2\mathbf{q}^2}{2m} \tag{3.27}$$

Figure 3.23 shows a plot of E_q versus **k** for a 1D lattice of period $|\mathbf{a}|$. The first Brillouin zone (BZ) of a 1D lattice extends from $-\pi/|\mathbf{a}|$ to $+\pi/|\mathbf{a}|$. The parts of the parabolic curve defined by Equation (3.27) that are outside these limits in **q**, shown as a dashed line in Figure 3.23, are folded back into this zone, as indicated by the full lines.

Adding a motif to the points of the lattice generates the periodic potential $V(\mathbf{r})$ that determines the details of the electronic structure of the crystals. If the electrons are nearly free, that is, $V(\mathbf{r})$ is small as compared to the kinetic energy part of the Hamiltonian, the wave functions Ψ_q can be calculated by first-order perturbation theory, that is, can be expanded in terms of the set of orthogonal unperturbed plane waves

$$\Psi^0_{q'} = A \exp(i\mathbf{q}' \cdot \mathbf{r}) \tag{3.28}$$

normalized to unit volume. Since $V(\mathbf{r})$ has the periodicity of the lattice, it can be expressed by a Fourier integral,

$$V(\mathbf{r}) = \frac{1}{(2\pi)^{3/2}} \int \tilde{V}(\mathbf{G}) \exp(i\mathbf{G} \cdot \mathbf{r})\, d\mathbf{G} \tag{3.29}$$

with

$$\tilde{V}(\mathbf{G}) = \frac{1}{(2\pi)^{3/2}} \int V(\mathbf{r}) \exp(-i\mathbf{G} \cdot \mathbf{r})\, d\mathbf{r} \tag{3.30}$$

The translational periodicity of $V(\mathbf{r})$

$$V(\mathbf{r}) = V(\mathbf{r} + \mathbf{T}) \tag{3.31}$$

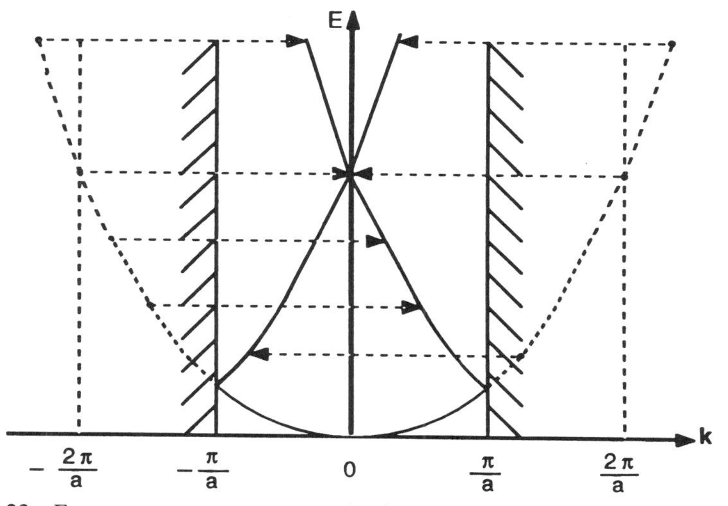

Figure 3.23 Energy versus wave vector plot for a linear lattice of period **a**.

imposes the condition $\exp(i\mathbf{G}\cdot\mathbf{r}) = \exp[i\mathbf{G}\cdot(\mathbf{r} + \mathbf{T})]$ on the Fourier components so that

$$\mathbf{G}\cdot\mathbf{T} = j2\pi \tag{3.32}$$

where j is an integer. This is only the case if $\mathbf{G} = \mathbf{G}_{HKL}$ (i.e., $\mathbf{G}$ is a vector pointing from the origin to a reciprocal lattice point). Therefore, the integral in Equation (3.29) can be replaced by a sum over all reciprocal lattice vectors. Using this Fourier series for $V(\mathbf{r})$,

$$V(\mathbf{r}) = \sum_{HKL} \tilde{V}(\mathbf{G}_{HKL}) \exp(i\mathbf{G}_{HKL}\cdot\mathbf{r}) \tag{3.33}$$

results in the matrix elements

$$\langle\mathbf{q}'|H|\mathbf{q}\rangle = \frac{\hbar^2\mathbf{q}^2}{2m}\int \exp[i(\mathbf{q} - \mathbf{q}')\cdot\mathbf{r}]\,d\mathbf{r} + \sum_{HKL} \tilde{V}(\mathbf{G}_{HKL})\int \exp[i(\mathbf{q} - \mathbf{q}' + \mathbf{G}_{HKL})\cdot\mathbf{r}]\,d\mathbf{r} \tag{3.34}$$

Since the set of basis vectors is orthogonal and normalized and nonzero Fourier coefficients exist only if $\mathbf{q} - \mathbf{q}'$ equals a reciprocal lattice vector

$$\langle\mathbf{q}'|H|\mathbf{q}\rangle = \frac{\hbar^2\mathbf{q}^2}{2m}\,\delta_{\mathbf{q}\mathbf{q}'} + \tilde{V}(\mathbf{G}_{HKL}) \tag{3.35}$$

for $\mathbf{q}' - \mathbf{q} = \mathbf{G}_{HKL}$ and $\langle\mathbf{q}'|H|\mathbf{q}\rangle = 0$, for $\mathbf{q}' - \mathbf{q} \neq \mathbf{G}_{HKL}$. The Hamiltonian thus couples only states that are directly above each other in the reduced zone representation.

Considering two such states

$$\psi_1^0 = \exp[i(\mathbf{k} + \mathbf{G}_{H_1K_1L_1})] \quad \text{and} \quad \psi_2^0 = \exp[i(\mathbf{k} + \mathbf{G}_{H_2K_2L_2})],$$

for $\mathbf{q} - \mathbf{q}' = 0$, results in

$$H_{11} = \frac{\hbar^2(\mathbf{k} + \mathbf{G}_{H_1K_1L_1})^2}{2m} + \tilde{V}(\mathbf{0}) \tag{3.36}$$

$$H_{22} = \frac{\hbar^2(\mathbf{k} + \mathbf{G}_{H_2K_2L_2})^2}{2m} + \tilde{V}(\mathbf{0}) \tag{3.37}$$

and for $\mathbf{q} \neq \mathbf{q}'$,

$$H_{12} = \tilde{V}_G = \tilde{V}(\mathbf{G}_{H_1K_1L_1} - \mathbf{G}_{H_2K_2L_2}) \tag{3.38}$$

The solutions of the secular Equation (2.33) are thus

$$E_{\pm} = \frac{H_{11} + H_{22}}{2} \pm \left(\frac{H_{11} - H_{22}}{2} + |\tilde{V}_G|^2\right)^{1/2} \tag{3.39}$$

$\tilde{V}_G$ is the Fourier coefficient for the reciprocal lattice vector $\mathbf{G} = \mathbf{G}_{H_1K_1L_1} - \mathbf{G}_{H_2K_2L_2}$. In Equations (3.36) and (3.37), the first term represents the empty lattice $E(\mathbf{k})$ relation, and the second term represents a constant shift that does not affect the evaluation of the band gaps and the effective masses. At points of high symmetry in the first BZ, where the empty lattice $E(\mathbf{k})$ relation exhibits degeneracies (i.e., $H_{11} = H_{22}$),

$$E_{\pm} = H_{11} \pm |V_G| \tag{3.40}$$

opening up bandgaps of magnitude $2|V_G|$. For $H_{11} - H_{22} \gg |V_G|$,

$$E_{+} = H_{11} + \frac{|V_G|}{H_{11} - H_{22}} |V_G| \tag{3.41}$$

$$E_{-} = H_{22} - \frac{|V_G|}{H_{11} - H_{22}} |V_G| \tag{3.42}$$

Because the factor in front of $|V_G|$ is very small and the potential energy is itself small as compared to the kinetic energy, E_+ and E_- are essentially the free-electron values. Therefore, the influence of the potential is significant only in the vicinity of the high-symmetry points, where the branches of the empty lattice $E(\mathbf{k})$ relation associated with different $\mathbf{G}_{HKL}$ meet, that is, where the wave functions coupled by the Hamiltonian have similar energies.

In Figure 3.24 the empty-lattice $E(\mathbf{k})$ relation and the band structure calculated by the nearly free-electron model are compared with a more accurate band-structure calculation for aluminum [16] based upon a method introduced by Kohn and Rostoker [15]. In this method, the wavefunction at any location $\mathbf{r}$ in the crystal is calculated by solving the integral equation

$$\Psi(\mathbf{r}) = \int_{\tau_{uc}} \mathscr{G}(\mathbf{r} - \mathbf{r}')V(\mathbf{r}')\Psi(\mathbf{r})\, d\mathbf{r}' \tag{3.43}$$

where the integral extends over the unit cell volume τ_{uc} and the Green's function

$$\mathscr{G}(\mathbf{r} - \mathbf{r}') = -\frac{1}{\tau_{uc}} \sum_{HKL} \frac{\exp[i(\mathbf{G}_{HKL} + \mathbf{k})\cdot(\mathbf{r} - \mathbf{r}')]}{(\mathbf{G}_{HKL} + \mathbf{k})^2 - \mathrm{E}} \tag{3.44}$$

extends over all vectors of the reciprocal lattice [16(b)].

In the regions of the first Brillouin zone that are far removed from the high-symmetry points, the Green's function calculation essentially reproduces the empty-lattice $E(\mathbf{k})$ relation, but in the vicinity of the high-symmetry points bandgaps open up as expected from the nearly free-electron model. Using the potential $V_{[111]} = 0.023$ Ry and $V_{[200]} = 0.043$ Ry in a nearly free-electron calculation reproduces in good approximation the bandgaps and curvature of the bands at the L and X points, respectively. Since the $E(\mathbf{k})$ curves are symmetric to the energy axis, it suffices to show only half of the BZ. The points and directions of high symmetry are labeled as in Figure 3.11(c).

Because the diamond structure group IV semiconductors are of particular interest in the context of solid-state electronics, we consider next the evolution of their band structure from the empty-lattice $E(\mathbf{k})$ plot. This is done in Figure 3.25 for cuts through the BZ along the directions Δ and Λ. The first two reciprocal lattice vectors in the Fourier expansion, $\mathbf{G}_{000}$ of length 0 and

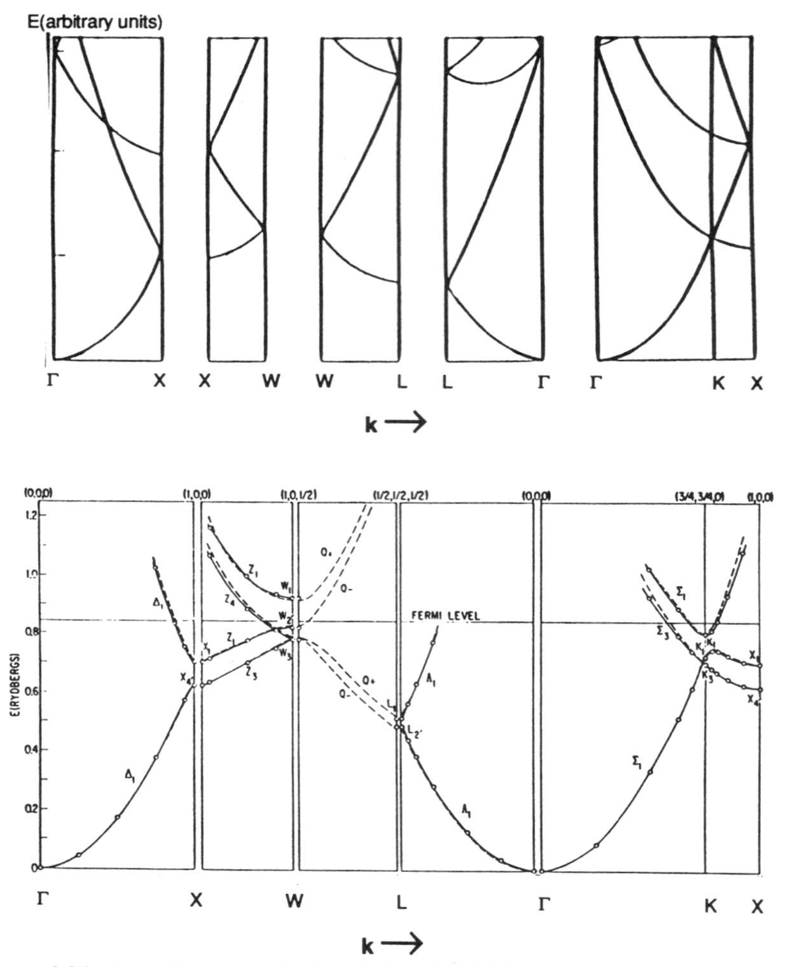

Figure 3.24 Top: Empty lattice bands for the FCC structure. Bottom: Band structure of aluminum (solid lines and circles calculated by the Greensfunction method; dashed lines calculated by the nearly-free electron method). After Segal [16(a)]; Copyright © 1961 *The American Physical Society*, New York.

$\mathbf{G}_{\pm1\pm1\pm1}$ of length $2\pi\sqrt{3}/a$, are associated with one parabolic band $E_{0\mathbf{k}}$ and 8 bands $E_{\langle 111\rangle\mathbf{k}}$. The next larger reciprocal lattice vector $\mathbf{G}_{\langle 200\rangle\mathbf{k}}$ of length $4\pi/a$ is associated with 6 bands and so on. The degeneracies of the bands are noted in Figure 3.25(a) in square brackets.

Figure 3.25(b) shows the symmetry-induced splitting into bands under the influence of the lattice symmetric potential in analogy to the crystal-field splitting of the energy levels of transition metal complexes discussed in Section 2.5. For example, at the zone center the full symmetry of the group O_h is established. The highest possible degeneracy is 3, corresponding to the irreduc-

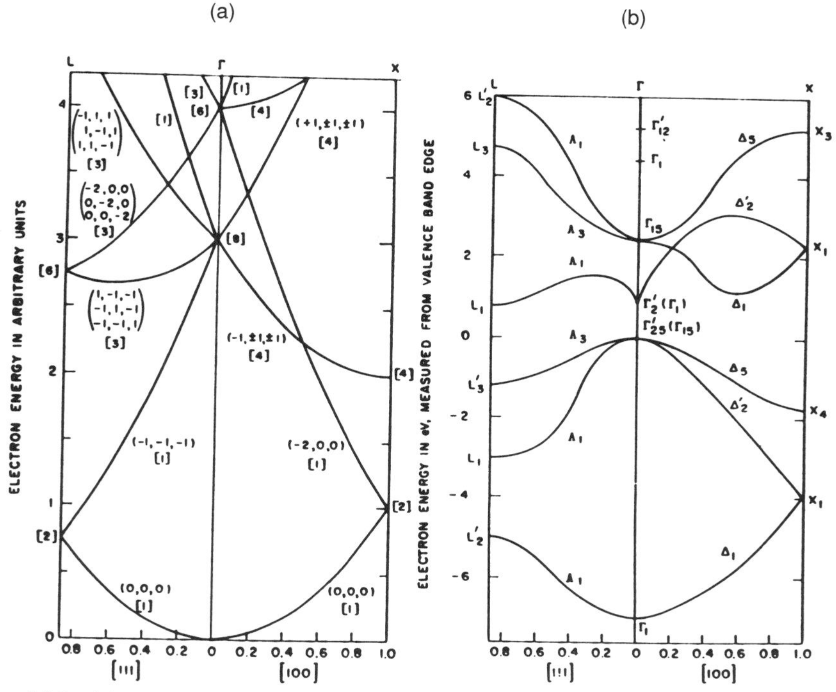

Figure 3.25 (a) Empty lattice E_{Gk} versus **k** curves for a diamond structure lattice, energy in arbitrary units, wave vector in units of $|\mathbf{k}_{100}|$. (b) Crystal-field-split bands without spin–orbit interactions, energy in eV measured from the valence-band edge, wave vector in units of $|\mathbf{k}_{100}|$. After Long [17]; copyright © 1968, Interscience Publishers, John Wiley and Sons, New York.

ible representations of T symmetry. The band of $A_{1\mathrm{g}} = \Gamma_1$ symmetry associated with $\mathbf{G}_{[000]\mathbf{k}}$ is unaffected, but the eightfold degeneracy of the bands associated with $\mathbf{G}_{[\pm1\pm1\pm1]\mathbf{k}}$ at Γ is lifted. The 8×8 reducible representation Γ_{r} of the symmetry operations of O_{h}, corresponding to the matrices operating on the set of the 8 plane-wave basis functions of the empty lattice

$$\exp[i(2\pi|a_0)(\pm x \pm y \pm z)]$$

has traces:

E	$8C_3$	$3C_2$	$6C_4$	$6C_2'$	i	$8S_6$	$3\sigma_{\mathrm{h}}$	$6S_4$	$6\sigma_{\mathrm{d}}$
8	2	0	0	0	0	0	0	0	4

Using Equation (2.85), we get the decomposition of this representation

$$\Gamma_{\mathrm{r}} = A_{1\mathrm{g}} + A_{2\mathrm{u}} + T_{1\mathrm{u}} + T_{2\mathrm{g}} \tag{3.45}$$

that is, two nondegenerate and two threefold degenerate bands. In the Wigner notation, this sum corresponds to $\Gamma_{\mathrm{r}} = \Gamma_1 + \Gamma_2' + \Gamma_{15} + \Gamma_{25}'$. The order of the splitting in energy shown in Figure 3.25(b) corresponds to the case of Ge, where the band of Γ'_{25} symmetry is lowest in energy, corresponding to the valence band, and the band of Γ'_2 symmetry contains the lowest conduction-band minimum. The triply degenerate band of Γ_{15} symmetry and the non-degenerate band of Γ_1 symmetry are higher-lying conduction bands. Note that off the Γ point, the symmetry is lowered, causing further splittings of the bands. For example, a general point in the direction Δ to the X point [see Figure 3.11(b)] has the symmetry C_{4v}. Therefore, the threefold-degenerate Γ'_{25} representation at the zone center splits for $|\mathbf{k}| \neq 0$ along the Δ direction into one nondegenerate band Δ'_2 and one twofold-degenerate band Δ_5. For the zinc-blende structure the symmetry at the zone center is T_d and the valence-band maximum is of Γ_{15} symmetry, while the lowest conduction-band minimum is of Γ_1 symmetry.

Thus far the discussion has been limited to orbital degeneracies of the bands. If the electron spin is included, the number of elements in the groups is doubled; that is, double groups are formed. Several new classes and IRs are added to the double groups. For O_h, the added IRs are in the Mulliken notation $E_{1/2g}$, $E_{5/2g}$, $G_{3/2g}$, $E_{1/2u}$, $E_{5/2u}$, and $G_{3/2u}$, corresponding in the Wigner notation to Γ_6^+, Γ_7^+, Γ_8^+, Γ_6^-, Γ_7^-, and Γ_8^-. The relations between single and double group representations are: $A_{1g} \rightarrow E_{1/2g}$, $A_{2g} \rightarrow E_{5/2g}$, $E_g \rightarrow G_{3/2g}$, $T_{1g} \rightarrow E_{1/2g} + G_{3/2g}$, $T_{2g} \rightarrow E_{5/2g} + G_{3/2g}$, $A_{1u} \rightarrow E_{1/2u}$, $A_{2u} \rightarrow E_{5/2u}$, $E_u \rightarrow G_{3/2u}$, $T_{1/2u} \rightarrow E_{1/2u} + G_{3/2u}$, $T_{2u} \rightarrow E_{5/2u} + G_{3/2u}$. Therefore, the triple orbital degeneracy of the valence bands in the diamond structure at the Γ point is lifted; that is, the T_{2g} valence band separates into a band of $G_{3/2g}$ symmetry and a band of $E_{3/2g}$ symmetry, which is lowered in energy relative to the top valence-band edge by the spin–orbit splitting energy Δ_0.

Although the nearly free-electron model describes remarkably well the properties associated with the conduction and upper valence bands of semiconductors, as will be discussed, the effective electron masses differ significantly from the free-electron mass, and the core states cannot be represented adequately by a plane-wave expansion. This is addressed by cellular models of band-structure calculations that evolved from the early recognition by Wigner and Seitz that upon the construction of planes about the atoms of fcc or bcc metals, bisecting the vectors to their nearest neighbors, almost spherical cells are obtained [18]. Approximating these Wigner–Seitz cells by spheres of an effective radius $r_{\text{effective}}$ to result in the same volume and solving the SE for one of these spheres, in combination with the translational symmetry of the crystal, provided information about its electronic properties.

The method was improved upon by the introduction of the muffin-tin potential, which is illustrated in Figure 3.26. About each nucleus, a spherical region is defined in which the potential is spherically averaged to represent the atomic core states. In the interstitial spaces outside these regions, a constant potential is chosen to represent the free-electron-like behavior of the outer electrons. The augmented plane waves (APW), obtained by joining the solutions to the SE in the two regions at their junctions, have generally

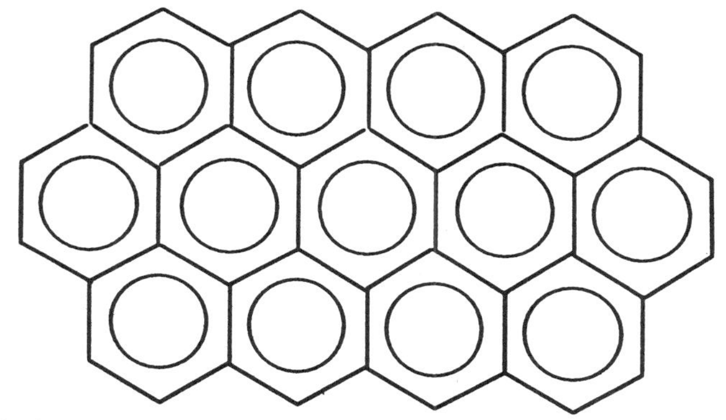

Figure 3.26 Schematic representation of the muffin-tin potential associated with a two-dimensional lattice.

different derivatives, resulting in kinks. Therefore, a linear combination of a set of augmented plane waves is needed to obtain a smooth approximation to the wavefunction of the solid and requires a considerable computational effort.

In 1940, C. Herring [19] introduced the orthogonalized plane-wave (OPW) method, which represents the core wave functions of the crystal in terms of the functions Φ_c known from calculations for the free atoms. These core functions are coupled in the form of a Bloch function summing over all atoms of the solid; that is,

$$\Phi_c(\mathbf{r}) = \sum_{(n,m,p)} \exp[i\mathbf{k}\cdot(\mathbf{r} - \mathbf{T})]\Phi(\mathbf{r} - \mathbf{T}) \tag{3.46}$$

where $\mathbf{T} = n\mathbf{a} + m\mathbf{b} + p\mathbf{c}$ is a lattice translation vector from the nucleus placed at the origin to all points of the lattice. An OPW functions is then constructed by subtracting this normalized Bloch function from the free-electron wavefunction at same $\mathbf{r}$

$$\Psi_{\text{OPW}} = \exp(i\mathbf{k}\cdot\mathbf{r}) - a_c\Phi_c(\mathbf{r}) \tag{3.47}$$

The coefficient a_c is defined as

$$a_c = \int \Phi_c^* \exp(i\mathbf{k}\cdot\mathbf{r})\, d\tau \tag{3.48}$$

so that

$$\int \Phi_c^* \Psi_{\text{OPW}}\, d\tau = \int \Phi_c^* \exp(i\mathbf{k}\cdot\mathbf{r})\, d\tau - \int \Phi_c^* \exp(i\mathbf{k}\cdot\mathbf{r})\, d\tau \int \Phi_c^* \Phi_c\, d\tau$$
$$= 0 \tag{3.49}$$

that is, the OPW and the Bloch function associated with the atomic cores are orthogonal. Usually there will be more than just one core state, so that

$$\Psi_{\mathrm{OPW}} = \exp(i\mathbf{k}\cdot\mathbf{r}) - \sum_{c} a_c \Phi_c(\mathbf{r}) \tag{3.50}$$

Substituting the OPW into the SE results

$$H \exp(i\mathbf{k}\cdot\mathbf{r}) - \sum_{c} a_c H\Phi_c(\mathbf{r}) = E \exp(i\mathbf{k}\cdot\mathbf{r}) - E \sum_{c} a_c \Phi_c(\mathbf{r}) \tag{3.51}$$

so that with $H\Phi_c(\mathbf{r}) = E_c\Phi_c(\mathbf{r})$ satisfied exactly we get

$$(H + V_{\mathrm{p}}) \exp(i\mathbf{k}\cdot\mathbf{r}) = E \exp(i\mathbf{k}\cdot\mathbf{r}) \tag{3.52}$$

with

$$V_{\mathrm{p}} \exp(i\mathbf{k}\cdot\mathbf{r}) = \sum_{c} (E - E_c) a_c \Phi_c \tag{3.53}$$

In this manner, the potential V in Equation (2.2) is replaced by an effective potential

$$V_{\mathrm{eff}} = V + V_{\mathrm{p}} \tag{3.54}$$

The addition of the pseudopotential V_p cancels the negative oscillatory part of V almost exactly, so that nearly free-electron-like solutions result [20].

A distinction is made between local pseudopotential calculations, where V_{p} is a function of position only, and nonlocal pseudopotential calculations, where V_{p} is a function of energy. Excellent agreement of local pseudopotential calculations was achieved with experimental results concerning the upper valence bands and lower conduction bands of solids [21]. However, discrepancies became evident with the results of ultraviolet and x-ray photoemission studies, which probe the lower lying valence bands and core levels. Also, electron-density distributions derived from x-ray diffraction data were in conflict with the predictions of local pseudopotential calculations [6]. Therefore, nonlocal pseudopotential calculations were introduced to improve upon the calculations and were carried out for the technologically most important semiconductors in the mid to late 1970s.

Figure 3.27(a), (b) shows the results of such calculations for Si and GaAs, respectively. Note that for GaAs the minimum separation between the top valence band and the lowest conduction band occurs at the Γ point between states of Γ_8 and Γ_6 symmetry. Interband transitions in this case thus connect states that are directly on top of each other in the BZ. Therefore, the fundamental gap of GaAs is called a *direct bandgap*. For Ge and Si, the highest valence-band maximum also occurs at the Γ point, but the lowest conduction-band minima are located at the L and close to the X points, respectively. In this case, the interband transitions across the fundamental gap are associated with a change in the momentum, that is, require the participation of a phonon for momentum conservation. Such transitions are referred to as *indirect transitions*, and the fundamental gaps of Si and Ge are called

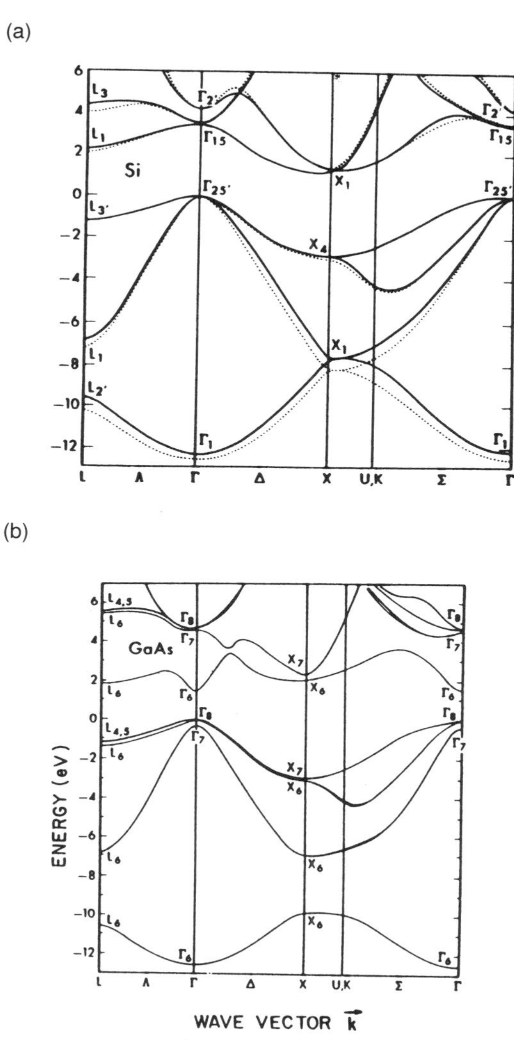

Figure 3.27 The band structures of (a) Si and (b) GaAs. After Chelikowsky and Cohen [22]; copyright © 1976, *The American Physical Society*, New York.

indirect bandgaps.

Figure 3.28 shows the calculated density of states function for the upper valence bands and the conduction band of Si. The energy distribution curve obtained by XPS measurements that probe the density of states below the upper valence band edge ($E_{VBE} = 0$) is shown in the upper part of the figure for comparison.

Figure 3.29 shows the valence electron-density difference map of Si calculated by the nonlocal pseudopotential method in comparison to the experimental result of Ref. [6]. Note that in Figure 3.29 the electron-density

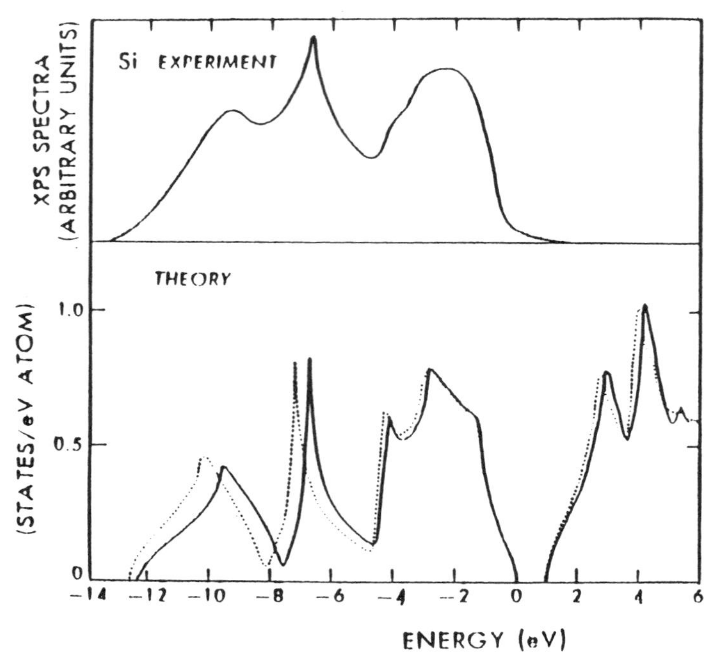

Figure 3.28 Density of states function for the upper valence and lowest conduction bands of Si: --- nonlocal pseudopotential calculation; --- local pseudopotential calculation. Top: XPS energy distribution curve. After Chelikowsky and Cohen [22]; copyright © 1976, *The American Physical Society*, New York.

distributions of the ionic cores are subtracted from the measured electron-density data to reveal the valence-electron contributions. This differs from Figure 3.14, where the electron distributions of neutral Si atoms were subtracted from the experimental electron-density map to reveal the redistribution of the charge caused by the covalent bonding of the Si atoms. The position of the lower valence bands of the group IV semiconductors and III–V compounds is in good agreement with the results of XPS measurements (see Chapter 8 for further discussion).

Although the pseudopotential method explains in a formal way why the conduction electrons of metals and semiconductors are nearly free, it does not address the physical reasons for this behavior, which is done more explicitly by scattering theories. Scattering theories construct the electronic wavefunctions at some point **r** by adding the partial waves scattered to this point by all the atoms of the solid. Thus they must focus on the question of why the scattering of the electronic wavefunctions by the atomic cores is diminished. There are two reasons for this behavior: electron correlation effects that screen the nuclear potential and the Ramsaur effect, which is a tunnelinglike behavior leaving an incident electronic wave unperturbed by the nuclear potential upon passing. This may be understood qualitatively by evaluating the simplified problem of a free electron passing over a square potential well of depth $-V_0$

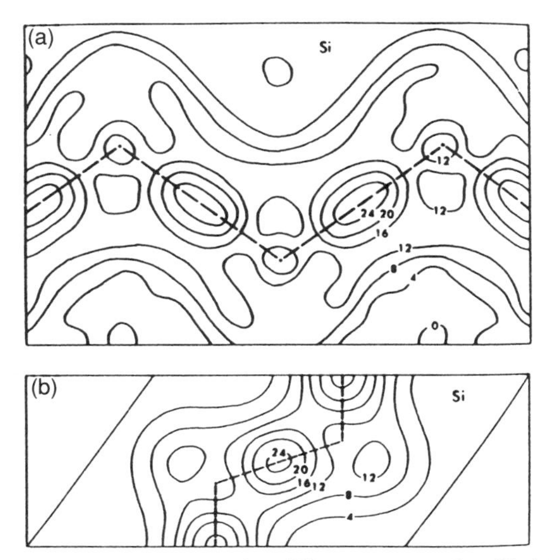

Figure 3.29 Valence electron-density difference maps for Si. (a) Experimental data of Ref. [6]; (b) nonlocal pseudopotential calculation. After Chelikovsky [22], copyright © 1976, The American Physical Society, New York, NY.

and width a. In this case, solving Equation (2.2) with the boundary conditions that the wavefunctions and their derivatives are continuous at $x = 0$ and a results in the transmission coefficient

$$T = \left(1 + \frac{V_0^2}{4V_0(V_0 + E)} \sin^2 \frac{2\pi a}{\lambda_e}\right)^{-1} \tag{3.55}$$

which takes on the value 1 for integer multiples of the electron wavelength $n\lambda_e = 2a$.

The screened potential in a solid differs significantly from the unscreened potential of the free atoms, because the mobile electrons buildup a negative charge about the positively charged atomic cores, that is, correlate their motions to minimize scattering. Since this correlation and exchange contribution of the electrons to the energy is a cooperative phenomenon, its representation by a one-electron Hamiltonian is not easily justified. However, for a fairly uniform distribution of the electron density in the solid, correlation effects may be incorporated into the effective potential by adding a term

$$V_{xc}(\mathbf{r}) = -\tfrac{1}{3}q^2[3\pi^2 n(\mathbf{r})]^{1/3} \tag{3.56}$$

that depends on the local electron density $n(\mathbf{r})$, making the contribution $E_{xc}(n[\mathbf{r}])$ to the energy. The method is called the *functional method* because the energy is a function of a function [23].

Although all-electron first-principles calculations in the local-density approximation have led to further advances of the modeling of the electronic structure of crystalline solids, they were impaired by a serious underestimation of the energy gaps. Recently dynamical screening was incorporated into first-principles calculations that eliminate this deficiency [24]. Figure 3.30 shows the results of such calculations for the band structure of diamond, demonstrating the considerable difference between conventional and time-dependent screened Hartree–Fock (TDSHF) calculations. Figure 3.31 shows the result of a TSCF calculation of the valence-band structure of Ge in comparison with experimental data. There thus exist at present first-principles methods of band-structure calculations, which provide detailed and sufficiently accurate information for the evaluation of the electronic properties pertaining to solid-state electronics applications.

For example, the velocity of the carriers is represented by the group velocity of the wave package

$$\mathbf{v}_g = \nabla_{\mathbf{k}}\omega = \frac{1}{\hbar}\nabla_{\mathbf{k}}E \tag{3.57}$$

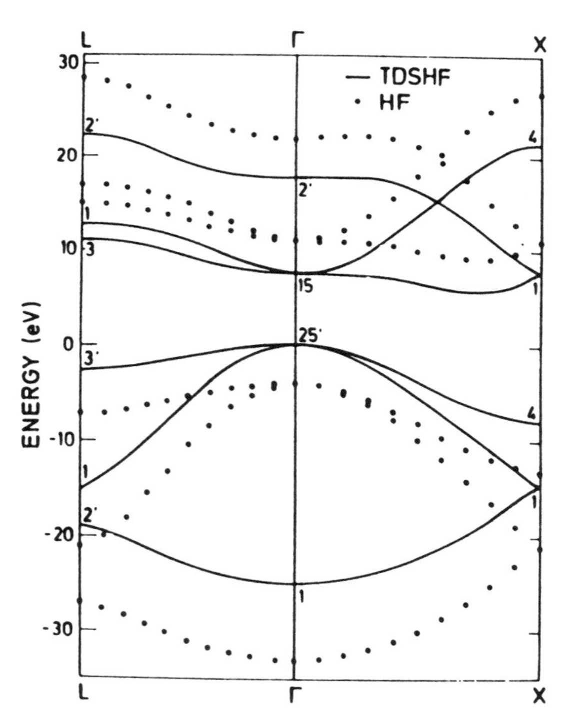

Figure 3.30 A comparison between the HF and TDSHF band structures of diamond. After Hanke, Mattausch, and Strinati [24]; copyright © 1983, Plenum Press, New York.

In an electrical field exerting a force

$$\mathbf{F} = \frac{d\mathbf{p}}{dt} = \hbar \frac{d\mathbf{k}}{dt} \tag{3.58}$$

onto the carriers their group velocity changes with time t according to

$$\frac{d\mathbf{v}_g}{dt} = \frac{1}{\hbar} \frac{\partial^2 E}{\partial \mathbf{k}\, \partial t} = \frac{1}{\hbar^2} \frac{\partial^2 E}{\partial \mathbf{k}^2} \frac{\partial \hbar \mathbf{k}}{\partial t} \tag{3.59}$$

$$= \frac{1}{\hbar^2} \begin{bmatrix} \dfrac{\partial^2 E}{\partial k_x^2} & \dfrac{\partial^2 E}{\partial k_x \partial k_y} & \dfrac{\partial^2 E}{\partial k_x \partial k_z} \\ \dfrac{\partial^2 E}{\partial k_y \partial k_x} & \dfrac{\partial^2 E}{\partial k_y^2} & \dfrac{\partial^2 E}{\partial k_y \partial k_z} \\ \dfrac{\partial^2 E}{\partial k_z \partial k_x} & \dfrac{\partial^2 E}{\partial k_z \partial k_y} & \dfrac{\partial^2 E}{\partial k_z^2} \end{bmatrix} \begin{bmatrix} \dfrac{d}{dt}(\hbar k_x) \\ \dfrac{d}{dt}(\hbar k_y) \\ \dfrac{d}{dt}(\hbar k_z) \end{bmatrix}$$

Since

$$\frac{d\mathbf{v}_g}{dt} = \frac{1}{m^*} \frac{\partial \mathbf{p}}{\partial t} \tag{3.60}$$

the effective mass of the carriers is related to the curvature of the bands by the relation

$$m^* = \frac{\hbar^2}{\nabla_{\mathbf{k}}^2 E} \tag{3.61}$$

Since the electrons fill states of lowest energy, they reside under equilibrium conditions in the lowest conduction-band minimum, while holes (i.e., unfilled electronic states) reside in the topmost valence-band maximum. The extrema in the bands usually occur at or close to high-symmetry points. Therefore, it suffices to calculate the details of the curvature of the bands in the vicinity of these points to obtain the relevant effective masses. This can be done most conveniently on the basis of perturbation theory, for example, a $\mathbf{k} \cdot \mathbf{p}$ calculation (see Ref. [17]). In the case of parabolic bands, the constant-energy surfaces in $\mathbf{k}$ space are spherical. Consequently, there exists one unique effective mass m_e^* for such semiconductors. However, for Si and Ge, the conduction-band minima near the X and L points, respectively, are associated with ellipsoidal constant-energy surfaces. Therefore, two effective masses m_l^* and m_t^* corresponding to the different curvatures of the bands along the major axes of the ellipsoids must be specified. The conductivity effective mass m_c^* is related to m_l^* and m_t^* as

$$\frac{1}{m_c^*} = \frac{1}{3}\left(\frac{1}{m_l^*} + \frac{2}{m_t^*}\right) \tag{3.62}$$

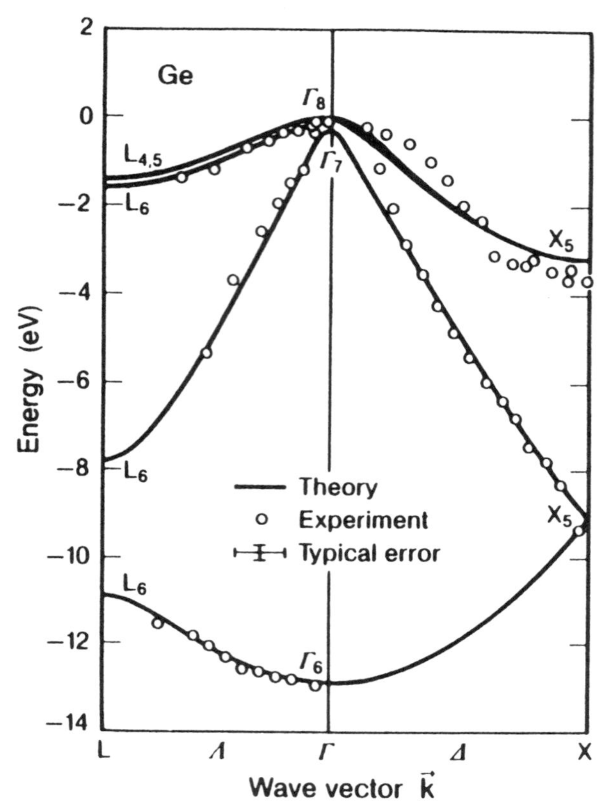

Figure 3.31 A comparison between calculated (curve) and experimental data (open circles) for the upper valence bands of Ge. After Hybertsen and Louie [25]; copyright © 1986, American Institute of Physics, New York.

The top valence bands of Si, Ge, and the III–V compounds at the Γ point are parabolic, but two effective masses m_{hh}^* and m_{lh}^* exist because of the degeneracy of the heavy- and light-hole bands at the zone center.

For selected well-behaved materials, trends in the effective masses can be evaluated on the basis of the nearly free-electron model. Consider a high-symmetry point in the first BZ [i.e., $\mathbf{k}_0 + \mathbf{G}_1 = -(\mathbf{k}_0 + \mathbf{G}_2)$]. Let $\Delta\mathbf{k} = \mathbf{k} - \mathbf{k}_0$ define a small region of $\mathbf{k}$ in the vicinity of $\mathbf{k}_0$. Then, under conditions described by the nearly free-electron model,

$$E_\pm = \frac{\hbar^2 \Delta\mathbf{k}^2}{2m} \pm |V_G| \pm \frac{\hbar^2 \Delta\mathbf{k}^2}{2m} \frac{\hbar^2 G^2}{4m|V_G|} + \frac{\hbar^2}{2m}(\mathbf{k}_0 + \mathbf{G}_1)^2 \tag{3.63}$$

Comparing Equation (3.63) with a Taylor expansion of E about $\mathbf{k}_0$

$$E = E_0 + \nabla_k E(\mathbf{k} - \mathbf{k}_0) + \tfrac{1}{2}\nabla_k^2 E(\mathbf{k} - \mathbf{k}_0)^2 + \cdots \tag{3.64}$$

results with Equation (3.61) at the band extrema where $\nabla_k E = 0$

$$\frac{m^*}{m_0} \approx \frac{4m_0|V_G|}{\hbar^2 G^2} \tag{3.65}$$

Since $E_g \sim |V_G|$, within a particular class of semiconductors, the effective masses should scale approximately with the energy gaps. This is indeed the case, for example, $m_e^*(\text{InSb}) = 0.0116$, $E_g(\text{InSb}) = 0.17\,\text{eV}$ versus $m_e^*(\text{InAs}) = 0.025$, $E_g(\text{InAs}) = 0.35\,\text{eV}$, and $m_e^*(\text{InP}) = 0.073$, $E_g(\text{InP}) = 1.29\,\text{eV}$ at room temperature. Note that while the effective masses of the III–V compounds differ substantially from the mass of a free electrons, the effective mass of the nearly free-electron calculation of Ref. [16] for aluminum $m^*/m_0 = 1.03$ is very close to the free-electron mass.

3.3 Defects in Crystalline Solids

The discussion of the band structure of crystalline solids has been based on the assumption of a perfect crystal lattice. However, defects generally exist in

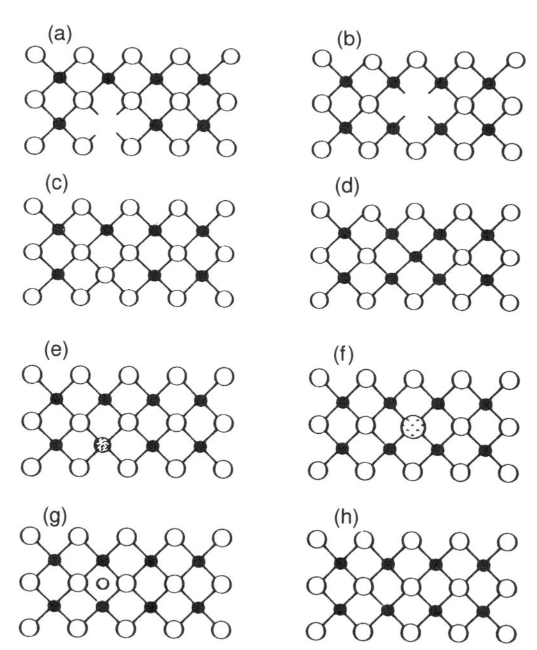

Figure 3.32 Point defects in a crystal lattice: (a) cation vacancy; (b) anion vacancy; (c) B_A antisite defect; (d) A_B antisite defect; (e) substitutional impurity on the cation sublattice; (f) substitutional impurity on the anion sublattice; (g) interstitial impurity; (h) perfect lattice.

thermodynamic equilibrium with the crystal lattice and the environment. These defects are associated with electronic states that are located in energy either within the bands or within the bandgaps. Figure 3.32(a)–(d) shows various native point defects, for example, vacancies, interstitials, and, in the case of compounds, antisite defects. The concentration of such defects existing in equilibrium with the lattice are well defined by thermodynamic principles and are functions of temperature. At low temperatures, where the kinetics of establishing equilibrium is slow, quenched-in nonequilibrium concentrations of these defects may exist. Also, nonequilibrium point defect distributions may be created by radiation damage. The latter is a side effect of many processing steps in the manufacturing of microelectronic circuits exposing the semiconductor to radiation. For example, the semiconductor surface is exposed to uv light, electrons, focused ion beams, or x rays during high-resolution lithography; to uv light, electron, and ion bombardment in plasma deposition and etching processes; and to high-energy ions in ion implantation processes. These processes will be considered in detail in subsequent chapters. Here we focus on the electronic states associated with defects that for one reason or another exist in the lattice and that affect, often critically, the electrical transport and optical properties of the solid. Depending on the distance of the energy levels associated with the defects from the band edges as compared to kT, the distinction is made between shallow defect states that are to a large fraction thermally ionized at room temperature and deep defect states that act as carrier traps. Often defects interact, forming complexes that are associated with energy levels differing from the energies of the component defects. This complicates significantly the analysis of the electrical properties of solids. Usually several traps exist simultaneously, but by an appropriate choice of the crystal growth conditions, one particular trap may be made to dominate (see Chapters 5 and 8 for further details). For example, the dominant native defect in As-rich GaAs is the EL2 trap. It is related to As_{Ga} antisite defects, albeit not in isolation, but as a complex with the As interstitial, as shown in Figure 3.33 [26].

In addition to the native point defects, impurities may be introduced either in substitutional or interstitial lattics sites, perturbing the periodicity of the potential in their vicinity, which is reflected in the calculations by a perturbation of the potential energy part of the Hamiltonian

$$V = V^0 + U \tag{3.66}$$

where V^0 denote the unperturbed potential of the perfect crystal and U is the defect potential. For weak perturbations (i.e., shallow impurities that have ground-state energies close to the band edges), effective-mass theory (EMT) represents a good approximation. In the EMT the effects of V^0 are incorporated formally into an effective kinetic energy operator T_{eff} so that the eigenvalue problem becomes

$$[T_{\text{eff}} + U(\mathbf{r})]F(\mathbf{r}) = EF(\mathbf{r}) \tag{3.67}$$

For the weakly bound electrons of a shallow donor impurity, which are

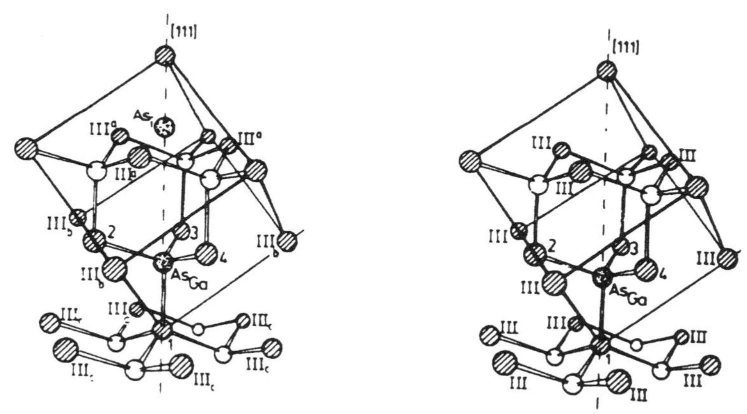

Figure 3.33 Model of the EL2 trap in As-rich GaAs as As_{Ga}–As_i pair defect (left) and of the isolated As_{Ga} defect (right). After Spaeth, Fockele, and Krambrock [26]; copyright © 1992, Elsevier Science Publishers B.V., Amsterdam.

expected to be located at a distance $\langle \mathbf{r} \rangle$ well removed from the substitutional donor ion, the impurity potential can be approximated as

$$U(r) = -\frac{q^2}{\varepsilon r} \tag{3.68}$$

where the factor $1/\varepsilon$ accounts for the shielding of the charge in the crystal medium with dielectric constant ε. By expanding the wavefunctions $\Psi(\mathbf{r})$ of the donor states in terms of the wavefunctions $\Psi_{n\mathbf{k}}(\mathbf{r})$ of the conduction band of the perfect crystal; that is,

$$\Psi(\mathbf{r}) = \sum_{\mathbf{k}} C_{\mathbf{k}} \Psi^0_{n\mathbf{k}} \tag{3.69}$$

and expanding the energy to second order in $\mathbf{k}$ about $\mathbf{k} = 0$, one can show that the function $\Psi(\mathbf{r})$ is a solution to the eigenvalue equation

$$\left(-\frac{\hbar^2}{2m^*} \Delta - \frac{q^2}{\varepsilon r} \right) \psi(\mathbf{r}) = E\Psi(\mathbf{r}) \tag{3.70}$$

which is similar to the Schrödinger equation of the hydrogen atom discussed in Section 2.3 and has solutions

$$\Psi(\mathbf{r}) = R_{nl}(\mathbf{r}) Y_l^m(\theta, \phi) \tag{3.71}$$

with energy eigenvalues

$$E_n = -\frac{m^* q^4}{2\varepsilon^2 \hbar^2 n^2} \tag{3.72}$$

measured from the conduction band edge (CBE), which coincides with the continuum limit $n \rightarrow \infty$.

The position of acceptor energy levels above the top valence band edge (VBE) can be calculated by the EMT in a similar manner, taking into account the valence-band degeneracy at Γ, which requires one to incorporate into the expansion of $\Psi(\mathbf{r})$ states from several bands. Since for $m^* \ll m_0$

$$\langle \mathbf{r} \rangle_n = n^2 \frac{m_0}{m^*} a_B \gg a_B \tag{3.73}$$

the premise on which we started this discussion of hydrogenic EMT calculations seems to be satisfied for many semiconductors, particularly for excited states where $n > 1$. Indeed excellent agreement is observed between the predictions of EMT calculations and experimentally determined energies of excited states associated with shallow donors and acceptors in a variety of semiconductors. This is illustrated for several acceptor states in Si and Ge in Figure 3.34. However, for the groundstate the model achieves only a poor match to the experimental ionization energies, which differ for different impurities, while the hydrogenic model results the same value for all impurities in a given semiconductor. There exists an inherent difficulty with the model for s-like wavefunctions that, in contrast to the excited states with $l > 0$, have nonzero amplitude at the substitutional lattice site. Corrections to the potential Equation (3.68) have been proposed to account for the screening conditions in

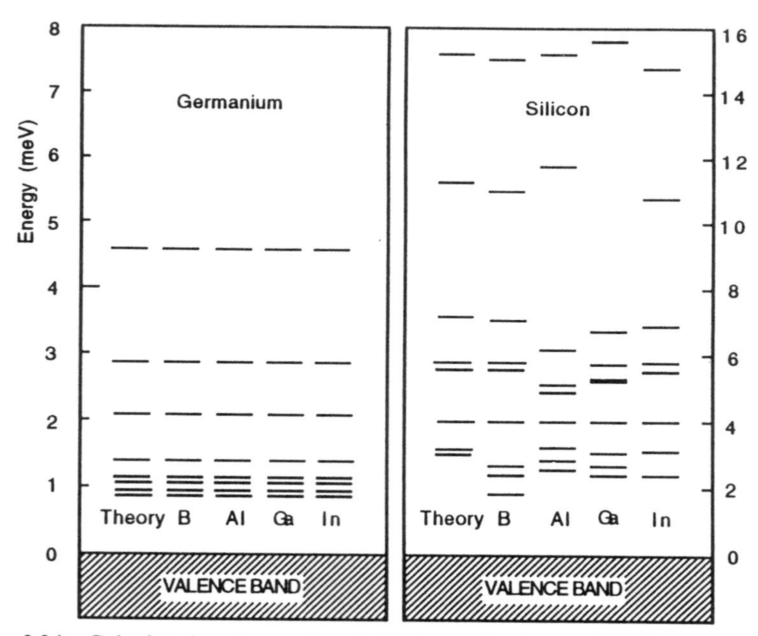

Figure 3.34 Calculated and experimentally determined energies of acceptor states in Ge and Si. After Pantelides [27].

the core region of the donor ion. A critical evaluation of such corrections is provided in Ref. [27].

The occupation of the defect states and bands in semiconductors is governed by the Fermi–Dirac distribution function

$$f(E) = \left[1 + \exp\left(\frac{E - E_F}{kT}\right)\right]^{-1} \tag{3.74}$$

which is a function ranging from 1 in the limit $E - E_F \to -\infty$ to 0 in the limit $E - E_F \to \infty$ and which has the value $\frac{1}{2}$ at the Fermi energy $E = E_F$.

Figure 3.35 shows this function in relation to the band edges for several values of E_F: (a) $E_F \approx (E_c + E_v)/2 \gg kT$, (c) $E_F = E_c - \Delta E$, and (b) $E_F = E_v + \Delta E$. In the intrinsic case (a) the electron concentration in the conduction band equals the concentration of holes in the valence band

$$n_i = \int_0^\infty N(E) f(E)\, dE = p_i \tag{3.75}$$

where

$$N(E) = \frac{1}{4\pi^3} \int_{S(E)} \frac{dS}{\hbar v_g} \tag{3.76}$$

is the density of states, $S(E)$ is a constant-energy surface in $\mathbf{k}$ space, and the energy is measured from the CBE. For a parabolic $E(\mathbf{k})$ relation

$$N(E) = \frac{(2m^*/m_0)^{3/2}}{2\pi^2\hbar^3} \sqrt{E} \tag{3.77}$$

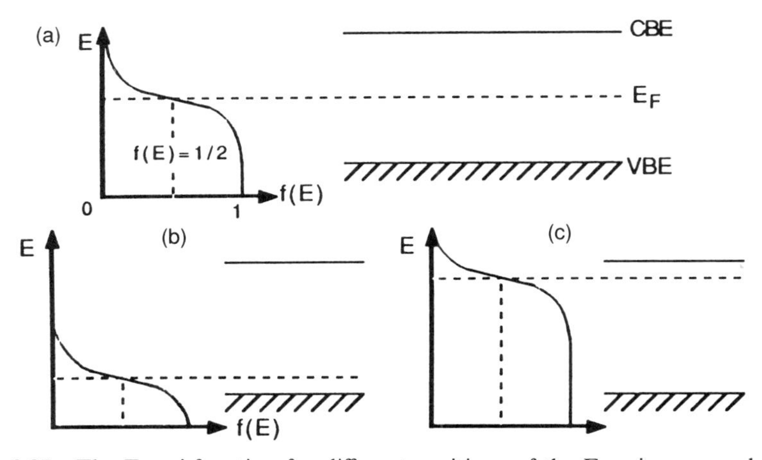

Figure 3.35 The Fermi function for different positions of the Fermi energy relative to the band edges of (a) an intrinsic semiconductor, (c) an n-type semiconductor, and (b) a p-type semiconductor.

so that

$$n_{\mathrm{i}} = 2\left(\frac{m_n^* kT}{2\pi m_0 \hbar^2}\right)^{3/2} \exp\left(\frac{E_{\mathrm{F}} - E_{\mathrm{C}}}{kT}\right) \tag{3.78}$$

$$p_{\mathrm{i}} = 2\left(\frac{m_{\mathrm{p}}^* kT}{2\pi m_0 \hbar^2}\right)^{3/2} \exp\left(-\frac{E_{\mathrm{F}} - E_{\mathrm{V}}}{kT}\right) \tag{3.79}$$

Therefore, the intrinsic carrier concentration product is

$$n_{\mathrm{i}}^2 = n_{\mathrm{i}} p_{\mathrm{i}} = 4\left(\frac{kT}{2\pi\hbar^2}\right)^3 \left(\frac{m_n^* m_p^*}{m_0^2}\right)^{3/2} \exp\left(-\frac{E_{\mathrm{g}}}{kT}\right) \tag{3.80}$$

At room temperature the intrinsic carrier concentrations for Ge, Si, and GaAs are 2.4×10^{13}, 1.6×10^{10}, and $1.1 \times 10^{7}\ \mathrm{cm}^{-3}$, respectively.

For doped semiconductors in equilibrium the product of the carrier concentrations of holes and electrons is still represented by n_{i}^2, but $n \neq p$ because of the donation of electrons from shallow donors into the conduction band and of the removal of electrons from the valence band by the ionization of shallow acceptors. The conductivity type depends on the sign of $N_A - N_D$, that is, on the difference between the concentrations of ionized acceptors N_A and donors N_D. In reference to the energy of the CBE, the minority hole concentration in n-type material is thus

$$p_n = \frac{n_{\mathrm{i}}^2}{N_D - N_A} = n_{\mathrm{i}} \exp\left(-\frac{E_{\mathrm{i}} - E_{\mathrm{F}}}{kT}\right) \tag{3.81}$$

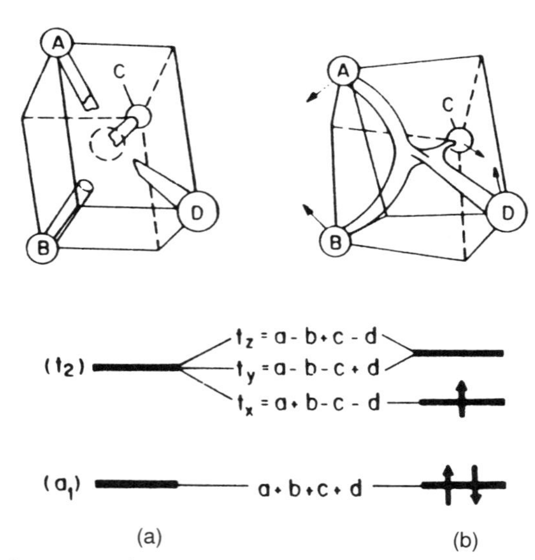

Figure 3.36 One-electron orbitals associated with the vacancy in the diamond structure lattice: (a) undistorted and (b) after Jahn–Teller lattice distortion. After Baraff, Kane, and Schluter [29]; copyright © 1980, The American Physical Society, New York.

and the minority electron concentration in p-type material is

$$n_p = \frac{n_i^2}{N_A - N_D} = n_i \exp\left(\frac{E_i - E_F}{kT}\right) \tag{3.82}$$

where

$$E_i = -\frac{E_g}{2} + kT \ln\left(\frac{m_p^*}{m_n^*}\right)^{3/4} \tag{3.83}$$

is the Fermi level of the intrinsic semiconductor. According to Eq. (3.83), the intrinsic Fermi level is close to the midpoint between the VBE and the CBE if $E_g \gg kT$.

As in the case of native defects, certain impurities represent deep traps. The calculation of the energy levels associated with deep states is far more complex than for shallow impurities, since the strong perturbation of the periodic potential by the defect makes the use of the EMT inappropriate. The first calculations of the energy eigenvalues of a deep native defect state were made on the grounds of MO theory by Coulson and Kearsley [28], who treated the vacancy in diamond as a "defect molecule" expanding the MOs of the defect in terms of the sp^3 hybridized orbitals of the four nearest-neighbor atoms surrounding the defect. The calculation includes Cl corrections and results for the undistorted vacancy shown in Figure 3.36 (a) in 4 MOs: a_1, t_x, t_y, and t_z, corresponding to a singlet and a triplet state of A_1 and T_2 symmetry, respectively.

Note that the ideally terminated dangling bond structure of Figure 3.36(a) is subject to Jahn–Teller distortion, that is, a lowering of the total energy of the defect molecule by reducing the symmetry from T_d to D_{2d}. The energy level associated with the triply degenerate t_2 orbital under T_d splits under D_{2d} into a doubly degenerate eigenvalue of E symmetry and a singly degenerate eigenvalue of B_2 symmetry (compare the correlation table for T_d in Appendix 2). Since the b_2 orbital has a lower energy than the t_2 orbital, the distortion lowers the energy for both the neutral and positively charged vacancies. This is depicted for V^+ in Figure 3.36(b). For negative charging of the vacancy the added electrons must be accomodated in the higher-lying e orbital, which diminishes the driving force of the Jahn–Teller distortion. Since lifting of the degeneracy of this orbital allows for a lowering of the total energy, the defect may undergo a Jahn–Teller distortion to C_{2v}. A look at the correlation table reveals that the e orbital splits into two b orbitals. The b orbital with the lower energy accommodates the excess electrons of the negatively charged vacancy up to the doubly charged state.

The position of the energies associated with vacancies in Si and Ge were evaluated recently by a perturbative Green's function method [29]. Figure 37 shows contour plots on (110) of the total defect potential $U(\mathbf{r})$ for the undistorted (a) and the distorted (b) vacancy in Si. Although the defect potential for the undistorted defect decays within nearest neighbours, the defect

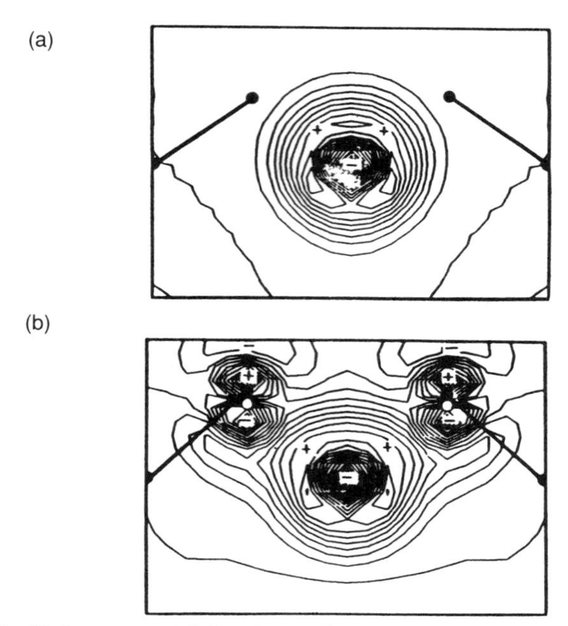

Figure 3.37 Defect potential for the undistorted (a) and distorted (b) vacancy in Si. After Baraff, Kane, and Schluter [29]; copyright © 1980, The American Institute of Physics, New York.

potential and charge distribution associated with the distorted defect spreads well beyond second-nearest neighbors. The value for the energy associated with the defect states, taking into account the readjustments of the lattice, differs from the result of a one-electron HF calculation for the undistorted defect, which does not describe the defect in its most stable configuration and is thus not suitable for direct comparison with the experimental data. Since the charge state of the defect, and consequently the lattice distortion, depends on the position of the Fermi level, the energy eigenvalues associated with the defect may differ for *n*- and *p*-type material.

Note that, as in the case of native point defects, the interpretation of the effects of impurities on the electrical and optical properties of a crystal depend on their interactions with other point defects. For example, replacing one of the corners of the vacancy defect molecule of Figure 3.37 by a donor atom (e.g., P, As, and Sb in Si) results a defect of C_{1h} symmetry [30]. In this case, all IRs are one dimensional, and this defect cannot distort any further as the result of a Jahn–Teller instability. However, other impurity–vacancy complexes may undergo Jahn–Teller distortions, so that their energy position and consequently their electrical activity become dependent of the Fermi level position. Also, interstitials may undergo Jahn–Teller distortions, as suggested by the partially filled degenerate energy eigenvalues for various interstitial positions in the diamond structure that are shown in Figure 3.38. Interstitial impurities

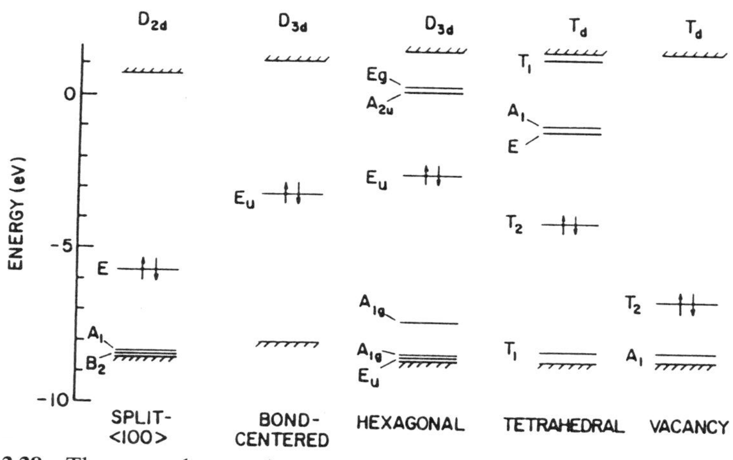

Figure 3.38 The ground-state electronic states of the neutral split-⟨100⟩, bond-centered, hexagonal, and tetrahedral interstitials in a diamond structure semiconductor. After Corbett and Bourgoin [30]; copyright © 1975, Plenum Press, New York, NY.

frequently form complexes with vacancies, as, for example, interstitial oxygen in silicon, which attains an energetically favored position by partially filling a vacancy site.

Depending on the kinetics of the capture and emission of carriers by defects, they may act as traps for either electrons or holes or, if the successive capture/emission of electrons and holes by the same defect proceeds efficiently with approximately the same rate, they may act as recombination centers for carriers (see Figure 3.39 and Section 8.2). In analogy to the kinetics for a chemical reaction, the rate is proportional to the concentrations of the participating states. For example, the rate of electron capture is proportional to the concentrations of filled states in the conduction band n and empty trap states $c_t[1 - f(E_t)]$; that is,

$$v_{nt} = v_{th}\sigma_n n c_t[1 - f(E_t)] \tag{3.84}$$

where E_t is the trap energy, v_{th} is the thermal velocity of the carriers, which is typically 10^7 cm/s, and σ_n is the electron capture cross section of the trap. The rate of the electron emission process is proportional to the concentration of filled traps

$$v_{tn} = P_{tn} c_t f(E_t) \tag{3.85}$$

where P_{tn} is the electron emission probability. In thermal equilibrium $v_{nt} = v_{tn}$, so that

$$P_{tn} = v_{th}\sigma_n n_i \exp\left(\frac{E_i - E_t}{kT}\right) \tag{3.86}$$

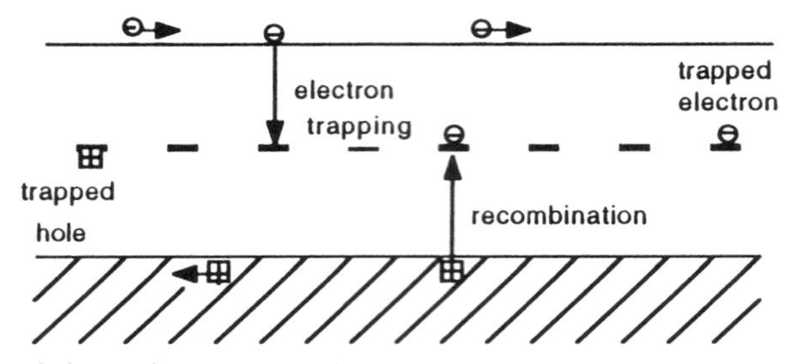

Figure 3.39 Schematic representation of the recombination at midgap states.

Similarly, the rates of hole capture and emission are

$$v_{pt} = v_{th}\sigma_p n c_t f(E_t) \tag{3.87}$$

$$v_{tp} = P_{tp} c_t [1 - f(E_t)] \tag{3.88}$$

with

$$P_{tp} = v_{th}\sigma_p n_i \exp\left(\frac{E_t - E_i}{kT}\right) \tag{3.89}$$

Under nonequilibrium conditions in steady state $v_{nt} - v_{tn} = v_{pt} - v_{tp}$. This allows the determination of

$$f(E_t) = \frac{n\sigma_n + pn_i \exp\left[\frac{E_i - E_t}{kT}\right]}{\sigma_n\left(n + n_i \exp\left[\frac{E_t - E_i}{kT}\right]\right) + \sigma_p\left(p + n_i \exp\left[\frac{E_i - E_t}{kT}\right]\right)} \tag{3.90}$$

resulting in a net recombination rate

$$U = \frac{v_{th}\sigma_n\sigma_p c_t [pn - n_i^2]}{\sigma_n\left(n + n_i \exp\left[\frac{E_t - E_i}{kT}\right]\right) + \sigma_p\left(p + n_i \exp\left[\frac{E_i - E_t}{kT}\right]\right)} \tag{3.91}$$

For devices operating close to thermal equilibrium, $np = n_i^2$, which implies that the net recombination rate is zero. In this case the recombination of carriers at traps is exactly compensated by the generation of carriers at traps in the entire volume of the semiconductor.

Extended defects may be created either as a consequence of errors during crystal growth or as a consequence of stress. Because of the high activation energy, the plastic deformation of a crystalline solid under stress does not occur in an abrupt mode where all bonds along the plane of slip are separated and then reformed simultaneously. Instead, the deformation proceeds by the

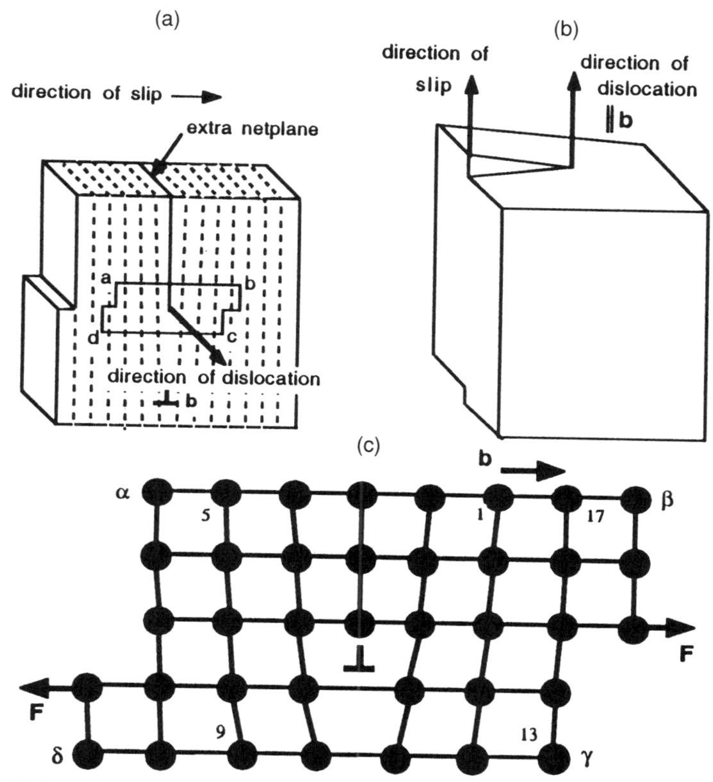

Figure 3.40 Dislocations in a crystal lattice: (a) edge dislocation, (b) screw dislocation, (c) magnified view of (a), showing atom positions and the Burgers cycle.

motion of dislocations in the crystal lattice, which is illustrated in Figure 3.40(a) and (c). Slip has occurred to the position of the dislocation line. Further motion of the dislocation under a force **F** causes the already-stretched bonds along the row of atoms to move to the left so that the slip progresses with minimum activation energy. A path in the lattice about the dislocation consists of an arbitrarily chosen number of steps, which, in this example, starts at point 1 and proceeds 4 lattice spacings to the left, downwards, to the right and upwards to that, in a perfect lattice, the points 1 and 17 coincide. In the imperfect lattice, the addition of the Burgers vector **b** is required for closure. In the examples shown in Figure 3.40(a) and (b), the Burgers vectors are perpendicular and parallel to the dislocation lines corresponding to a pure edge and a pure screw dislocation, respectively. Frequently intermediate cases are realized in which **b** is composed of edge and screw components.

Impurities that do not fit into the perfect crystal lattice tend to segregate in the dislocation core, relieving part of the strain [31]. This decoration mechanism applies to all crystal defects that are associated with long-range stress. It is utilized in the gettering of impurities from nearly perfect regions of the lattice into deliberately introduced neighboring defective regions, thus improving the purity of active device regions placed in the perfect part of the structure (see Chapter 4). On the other hand, the decoration of dislocations by impurities generally degrades the electrical properties of the defective crystal so that dislocations must be kept out of the active device region. Since the presence of impurities and the strained bonds in the vicinity of the core region affects the dissolution rate in the vicinity of the emergence points of dislocations during etching of the crystal surface, dislocation etch pits or hillocks are generated that reveal the density of dislocation intersecting the surface (see Section 7.4 and Figure 3.40 for further details). However, there are a variety of causes for changes in the surface relief upon etching that are unrelated to dislocations. Therefore, independent tests are required to confirm the validity of equating the density of pits or hillocks on the surface to the dislocation density in the crystal.

Quantitative information on the density and type (Burgers vector) of dislocations is obtained by x-ray topography and electron microscopy. The former is useful for dislocation densities $N_d \leqslant 10^4\,\mathrm{cm}^{-2}$, while the latter is suitable for $N_d \geqslant 10^6\,\mathrm{cm}^{-2}$. Figure 3.41 shows a schematic representation of the experimental arrangement for x-ray topography. A high-intensity monochromatic x-ray beam generated by a rotating anode x-ray source or synchrotron source impinges onto the crystal in the diffraction angle. The transmitted beam is delimited by a slit in a lead screen, and the crystal and the photographic plate are scanned simultaneously to generate an image of the crystal. For reasons discussed in Section 3.1, the dislocation lines are imaged, provided that the associated displacement in the net planes locally detroys the dynamical diffraction conditions. For diffraction vectors $\mathbf{g}$ of orientations associated with the condition $\mathbf{g}\cdot\mathbf{b} = 0$, no displacement results, that is, no

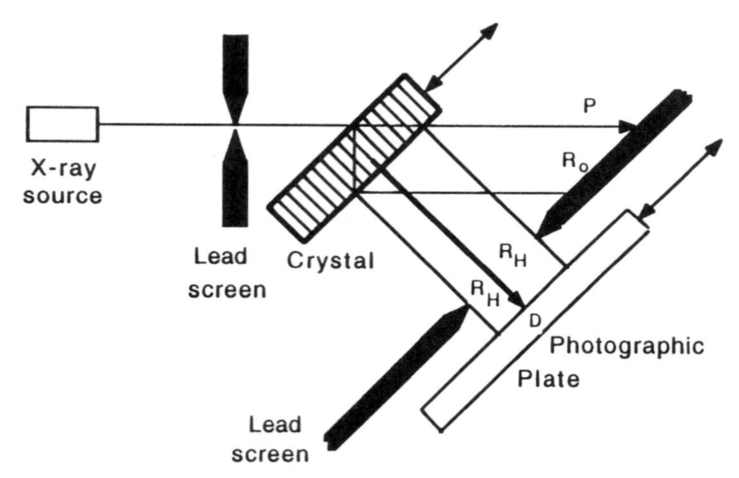

Figure 3.41 Schematic representation of the experimental arrangement for x-ray topography.

contrast is observed, which provides a means for determining the Burgers vectors of individually imaged dislocations. Since the resolution of the method is restricted by the minimum grain size of the photographic emulsion ($\sim 10^{-4}$ cm), the method is restricted to the characterization of nearly perfect crystals, (e.g., Fig. 5.36).

At high dislocation densities, transmission electron microscopy is a suitable characterization method. Figure 3.42 illustrates the conditions of image formation in a transmission electron microscope. An electron beam of typically several hundred keV energy impinges onto a thin foil of the material to be studied. According to

$$\lambda_e(\text{Å}) = \sqrt{[150/E(\text{eV})]} \tag{3.92}$$

its wavelength is $\lambda_e \leqslant 0.1$Å, thus resulting in diffraction on the atoms of the crystal. The diffracted beams and the transmitted beam are brought into focus at the back focal plane and can be recorded there as a diffraction pattern that contains information about the crystal structure and the orientation of the foil. Under bright-field imaging conditions, the diffracted beams are blocked by an aperture that is centered on the transmitted beam. A dark-field image may be produced by centering the aperture on a selected diffraction spot. Thus certain features (e.g., defects in a microstructure) contributing to this particular diffraction spot may be highlighted. As in the case of x-ray diffraction, defects are revealed by the contrast due to distortions of the lattice in their vicinity. Elastic distortions of a perfect crystal foil that bring the net planes in certain parts of the foil into strong diffraction conditions result in extinction contours. Under the conditions of either bright- or dark-field imaging by a single beam, the contrast is produced by local variations of the amplitude. By widening the

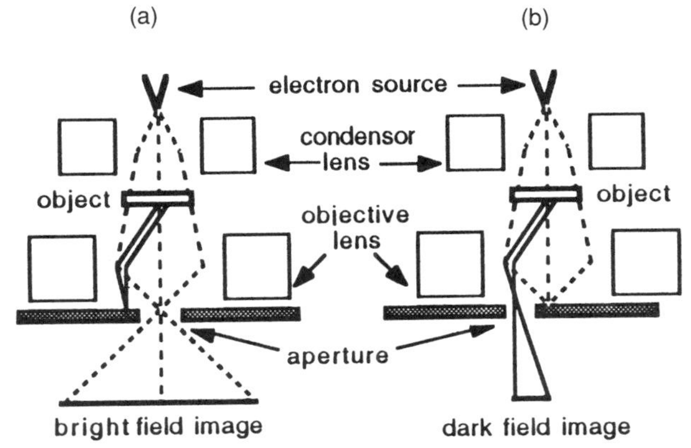

Figure 3.42 Schematic representation of the conditions of (a) bright- and (b) dark-field imaging in the transmission electron microscope. As an alternative to shifting the apperture, the dark-field image may be produced by tilting.

aperture to permit several beams to contribute to the imaging, phase contrast is produced. Since the smallest features in real space are the farthest apart in reciprocal space, including beams in the off-center part of the diffraction pattern into the image formation thus allows for higher resolution than under single beam imaging conditions. It can be shown that the resolution is given by

$$\delta_r = k_r \Sigma^{1/4} \lambda_e^{3/4} \tag{3.93}$$

where $k_r \approx 1$ and Σ is the spherical aberration coefficient that depends on the operating conditions of the microscope. Thus specialized high-voltage instruments that provide for small Σ are required for high-resolution electron microscopy (HREM). Because the contrast depends critically on a set of parameters (e.g., the sample thickness and the position of the focus with regard to the imaging plane) that is often not known with sufficient accuracy, the interpretation of HREM images, such as Figure 1.15, requires a comprehensive simulation effort to fit the experimental data to a particular model.

Figure 3.43 shows transmission electron microscopy images of (top) dislocations in a crystal of InP and (bottom) etch pits at the emergence points of dislocations on the polar faces of an InP wafer of (111) orientation (see Section 7.3 for further explanations). Dislocations are formed in crystals due to stress or the occurrence of attachment errors during growth. Stress may be generated during crystal growth either due to thermal gradients or due to lattice mismatch and/or mismatch of the thermal expansion coefficients at heterojunctions. If the stress exceeds a critical limit, dislocation halfloops form at the crystal surface that move in the stress field on the glide planes toward the interface, as illustrated in Figure 3.44a. Thus a network of dislocations is formed at the mismatched interface, relieving the stress. The critical thickness has been related to the materials properties on the basis of simplified models [34], [35]. For elastically isotropic media, neglecting nonlinear effects and assuming the formation of a network of noninteracting dislocations, Matthews et al. [34] derived the following implicit expression for the critical thickness scaled by the absolute value of the Burgers vector of the misfit dislocations

$$\frac{t_c}{|\mathbf{b}|} \ln\left(\frac{t_c}{|\mathbf{b}|} + 1\right) = \frac{aG_1}{4\pi\, \Delta a(G_1 + G_2)(1 + \nu)} \tag{3.94}$$

where G_1 and G_2 are the shear moduli of the materials joined at the interface, **b** is the Burgers vector, $\Delta a/a$ is the lattice mismatch between materials 1 and 2, and ν is the Poisson ratio. Although this equation represents an oversimplification, and more complex models have been developed for specific systems, it has been used as a general guide for estimating the critical thickness under the conditions of III–V alloy heteroepitaxy. Substantial deviations from the predicted behavior have been observed.

Dislocations interact with each other and with point defects so that their distribution and structure usually are quite complex. Also, they may dissociate into partials delimiting a stacking fault, as illustrated for a dissociated edge

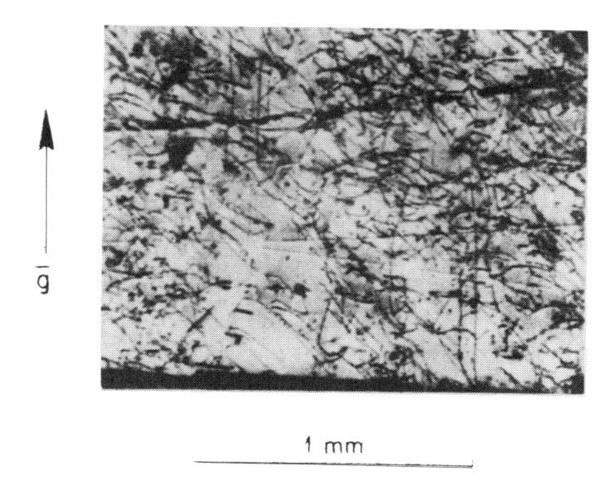

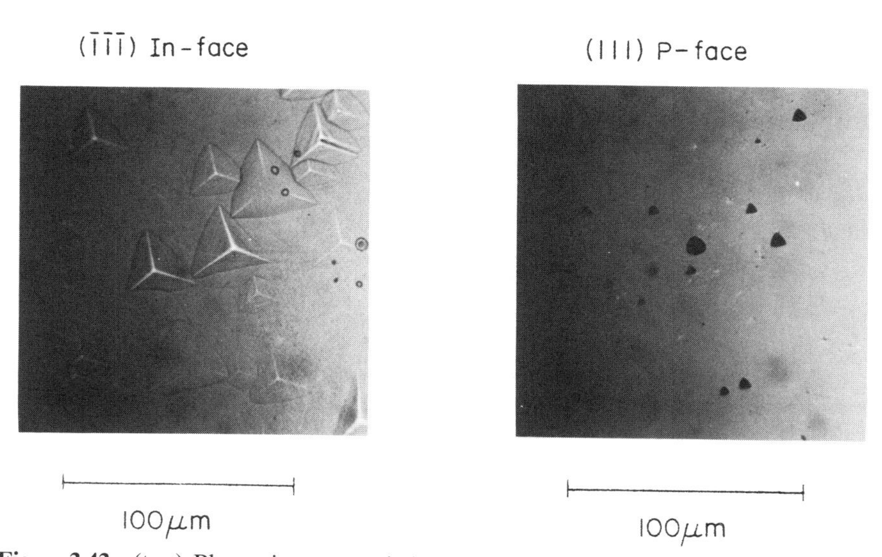

Figure 3.43 (top) Plane view transmission electron microscopy image of dislocations in a single crystal of InP. (bottom) Dislocation etch pits on the (111) In- and (111) P-faces of a single crystal of InP.

dislocation in the cubic $ABC \cdot BC$ stacking sequence in Figure 3.44(e) shows the TEM image of a stacking fault (SF). SFs are called *extrinsic* or *intrinsic* depending on whether an extra plane of atoms is inserted into or a plane of atoms is removed from the space defined by the partial dislocation loop [see Figure 3.44(c) and (d)]. Since a dislocation represents a disturbance of the three-dimensional periodicity of the lattice, it is associated with defect states that may be located inside the bandgap. The actual situation depends strongly on reconstruction, minimizing the number of broken bonds along the dislocation core.

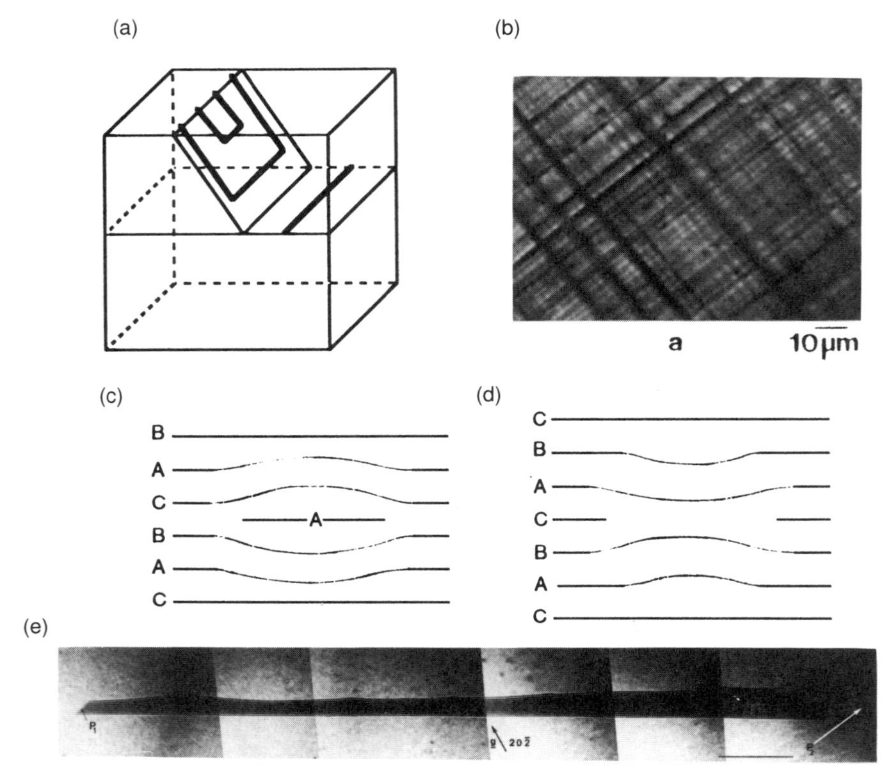

Figure 3.44 (a) Formation of misfit dislocations by nucleation of half-loops at the surface and glide to the interface. (b) Cathodoluminescence image of a misfit dislocation network at the interface of a 3500 Å thick $Ga_xIn_{1-x}As$ epilayer on a low-dislocation-density GaAs substrate wafer. (c) and (d) Schematic representations of extrinsic and intrinsic stacking faults in a crystal lattice. (e) Image of a stacking fault in InP. (b) After Fitzgerald et al. [36]; copyright © 1989, The Electrochemical Society, Pennington, NJ. (e) After Mahajan et al. [32]; copyright © 1981, American Institute of Physics, New York.

Figure 3.45(a)–(d) shows this for a 90° dislocation and a 30° dislocation in silicon. Reconstruction causes the formation of unusual (e.g., pentagonal) rings corresponding to straining of the bonds (angles and lengths). However, this is nevertheless energetically favorable as compared to the unreconstructed dislocation core structure containing dangling bonds, which introduce defect states into the bandgap. This is illustrated in Figure 3.46(b). It shows the density of states distribution that is associated with a 30° partial dislocation. While for the unrelaxed partial dislocation a high density of midgap states is generated (dashed line), no such states are present after reconstruction (dotted line) [37]. Therefore, the observed association of dislocations with deep states has been attributed, to kinks and jogs in the dislocation lines. They represent short offsets in the dislocation that connect parallel segments, which, in the case of

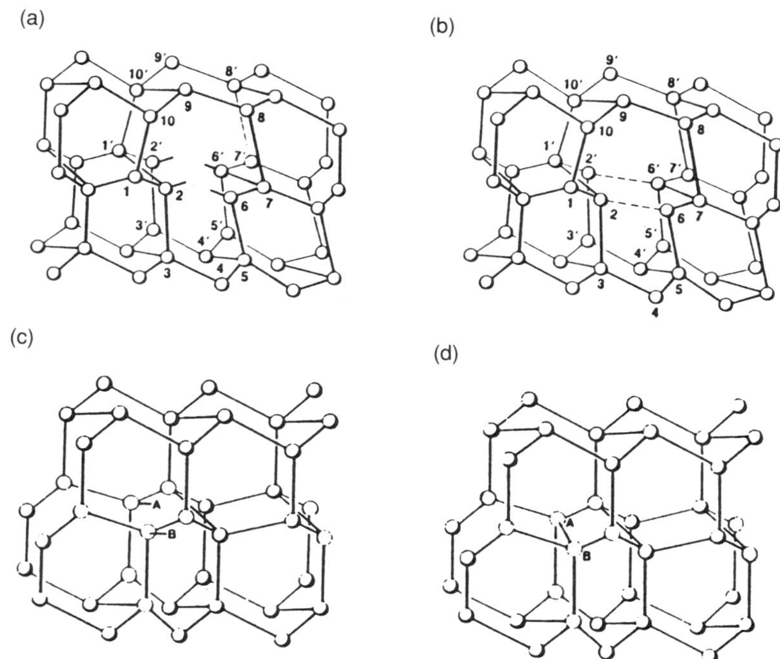

Figure 3.45 Schematic representation of unreconstructed (a), (c) and reconstructed (b), (d) 90° (top) and 30° (bottom) dislocations in silicon. After Chelicowsky [37] and Chelicowsky and Spence [38]; copyright © 1982 and 1984, American Institute of Physics, New York.

kinks, lie within one and the same glide plane, while, in the case of jogs, they lie on parallel glide planes, as illustrated in Figure 3.46(c) and (d).

Arrays of dislocations form grain boundaries. This is easily visualized for low-angle tilt and twist boundaries composed of arrays of pure edge and screw dislocations, respectively, and also holds for large-angle boundaries [39].

The energy and properties of grain boundaries vary with the angle between the adjacent grains, assuming a minimum at certain angles where a large number of coincidence sites exist. Figure 3.47(a) shows schematically the maximization of coincidence sites in a 22° tilt boundary for a simple square lattice. As discussed for the case of dislocations, generally the atoms in the boundary region relax, so that the structure of the reconstructed boundary differs from the idealized picture of Figure 3.47(a). Nevertheless, minima in the grain boundary energy are observed in the vicinity of coincidence angles determined by simple geometric relations for the ideally terminated bulk lattices [40].

Figure 3.47(b) shows etch pits at the emergence points of the dislocations that are generated in the volume swept by the grain boundary upon annealing of an InP bicrystal that is oriented along the $\langle 111 \rangle$ axis. Some of the dislocations form small angle subboundaries. Note that, in addition to dislo-

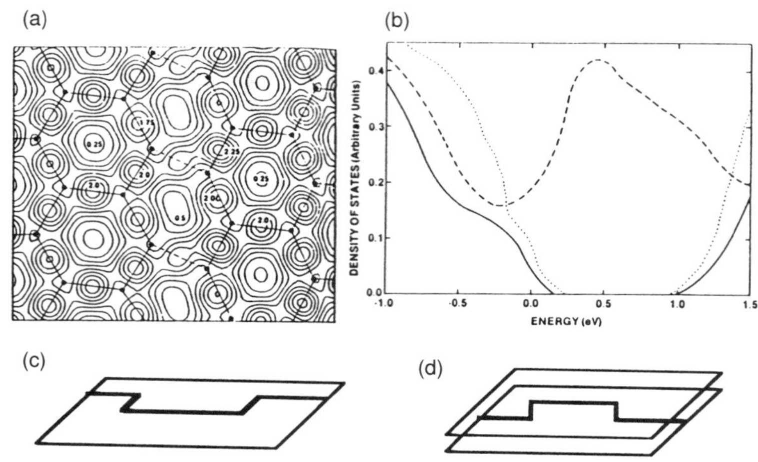

Figure 3.46 (a) Schematic representation of the charge-density distribution about a reconstructed 30° dislocation in silicon and (b) density of states for the same dislocation (dotted curve) in relation to the conduction and valence band edges (solid line). The dashed line corresponds to the unreconstructed 30° dislocation. (c) and (d) Schematic representation of kinks and jogs in a dislocation line. (a), (b) After Chelikowsky [37]; copyright © 1982, American Institute of Physics, New York.

cations, microtwins are formed under the conditions of grain growth at the growth interface, which is discussed further in Chapter 5.

3.4 Crystal Surfaces

Since many solid-state circuits are built in near-surface regions, the disruption of the three-dimensional periodicity of crystalline solids at their surfaces and its effects on the electronic properties are an important aspect of device engineering. Surfaces prepared by cutting, polishing, and etching are not atomically clean; that is, they are covered by adsorbed impurities. Frequently they are covered by a native oxide layer or other reaction products of the chemical processing. These reaction products, and changes in the point defect chemistry in the subsurface region of the semiconductor caused by surface reactions, may dominate the electronic properties (see Chapters 8 and 9 for details). Therefore, in situ cleaning steps are essential for the successful fabrication of devices and circuits.

Under ultra-high-vacuum (UHV) conditions (i.e., at $<10^{-10}$ Torr pressure) atomically clean surfaces can be made and maintained for many hours by cleaving of the crystal. The disadvantage of this method is its restriction to specific surface orientations at which cleavage occurs. These orientations depend on the crystal structure. For example, Si and Ge cleave on $\{111\}$, while zincblende structure materials cleave on $\{110\}$. A generally applicable cleaning

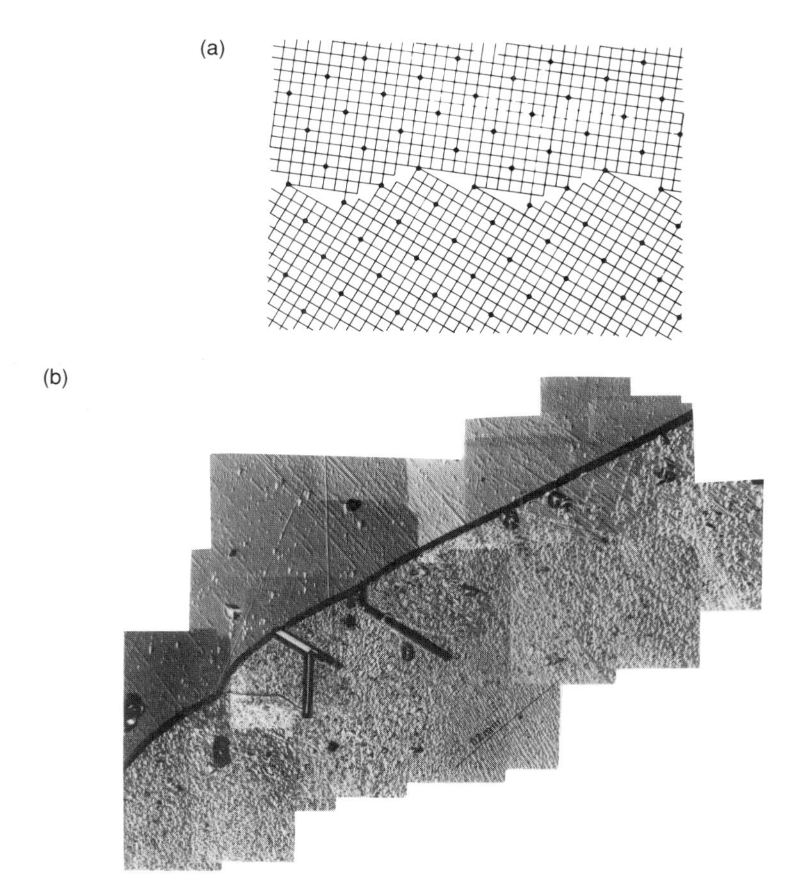

Figure 3.47 Schematic representations of (a) a 22° tilt boundary in a square lattice, and (b) etch pit pattern on (111) delineating the emergence points of dislocations and growth twins that are created upon motion of a grain boundary in an InP bicrystal of ⟨111⟩ orientation. (b) After Gleiter et al. [33b], copyright © 1980, Pergamon Press, Oxford.

process in vacuum that is not limited to a specific surface orientation is sputter cleaning: the removal by ion bombardment of the contaminants and of a number of surface layers of the semiconductor. However, the damage to the crystal lattice near the surface caused by this process requires subsequent UHV annealing. For compound semiconductors that exhibit high decomposition pressures, low temperature annealing for many hours may be necessary, requiring extremely reliable control of the residual gas atmosphere in the UHV chamber.

For Si, clean surfaces can be prepared conveniently by heating the surface in UHV to $\geqslant 1000°C$ after an oxidizing etch and subliming off the oxide film due to the volatility of silicon monoxide at this temperature. Similar

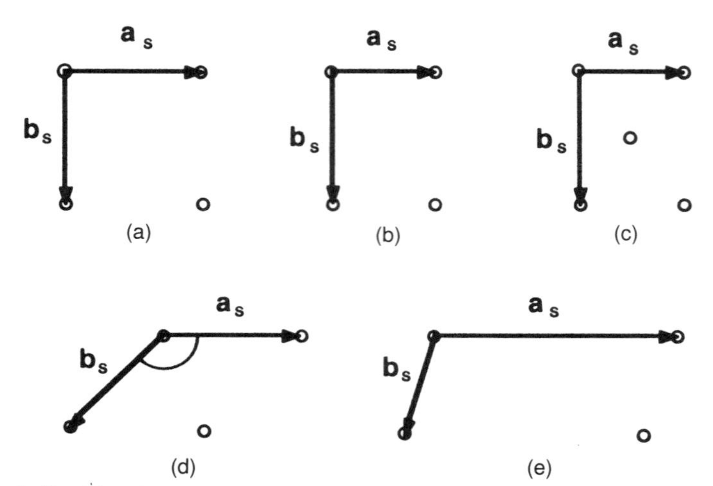

Figure 3.48 The five surface unit cells: (a) $\mathbf{a}_s = \mathbf{b}_s$, 90°; (b) $\mathbf{a}_s \neq \mathbf{b}_s$, 90°; (c) as (b), centered; (d) $\mathbf{a}_s = \mathbf{b}_s$, 120°; (e) $\mathbf{a}_s \neq \mathbf{b}_s$, $\neq$ 90°, 120°.

procedures have been worked out at lower annealing temperatures for GaAs and InP. Under the conditions of chemical vapor deposition and liquid-phase epitaxy, a combination of in situ chemical surface cleaning and sealing steps are essential for producing high-quality interfaces with well-controlled reproducible properties. These methods usually are more rapid and closer to conditions of a manufacturing process than UHV methods but require special recipes for each material, which we will discuss in the following chapters in the context of epitaxy and dielectric thin-film deposition.

Another important aspect of preparation of the crystal surface for subsequent processing is its reconstruction. A look at Figure 1.1 shows that there are two broken bonds per surface atom on the ideally terminated (001) surface and one broken bond per surface atom on the ideally terminated (111) surface of a diamond structure material. Generally the ideally terminated surface does not represent minimum free energy; that is, the surface atoms relax from their bulk lattice positions, healing the defective surface as far as possible.

The reconstruction of the lattice on the surface leads to a different periodicity of the surface mesh of atoms as compared with the bulk net planes of equal orientation. As there is a limited number of 3D space-filling parallelepipeds, there is also a limited number of surface-filling tesselations defining 5 types of surface unit cells, which are shown in Figure 3.48. The surface lattice vectors $\mathbf{a}_s$ and $\mathbf{b}_s$ thus defined are related to the bulk lattice vectors $\mathbf{a}$ and $\mathbf{b}$ by a transformation matrix $[M]$

$$\begin{Bmatrix} \mathbf{a}_s \\ \mathbf{b}_s \end{Bmatrix} = \begin{bmatrix} M_{11} & M_{12} \\ M_{21} & M_{22} \end{bmatrix} \begin{Bmatrix} \mathbf{a} \\ \mathbf{b} \end{Bmatrix} \tag{3.95}$$

The unit cell area of the surface mesh is

$$A_s = |\mathbf{a}_s \times \mathbf{b}_s| \tag{3.96}$$

In order to characterize the various possible surface structures, one may use $[M]$ or another label agreed upon by convention. A frequently used notation is $(|\mathbf{a}_s|/|\mathbf{a}| \times |\mathbf{b}_s|/|\mathbf{b}|)\alpha''$, where α'' is the angle of rotation of the surface mesh in relation to the bulk lattice plane. If one chooses the unit cell vectors of the surface mesh to be parallel to the corresponding bulk unit cell vectors, α'' may be omitted.

Figure 3.49 shows schematic representations of top views of the Si(100) (a) 1×1 and (b) 2×1 surface structures. Figure 3.50 shows a model of the Si(111)-1×1 and -2×1 surface structures. The (111)-2×1 structure corresponds to the as-cleaved Si{111} surface. It achieves a further lowering of the surface energy by buckling, as illustrated in Figure 3.50. Some of the surface atoms of the original (1×1) surface are pushed down into the second layer, and the atoms of the original second layer relax laterally. These relaxed and unrelaxed positions are shown as superimposed shaded and filled black circles, respectively. Note that the cleaved Si(111)-2×1 surface exists at room

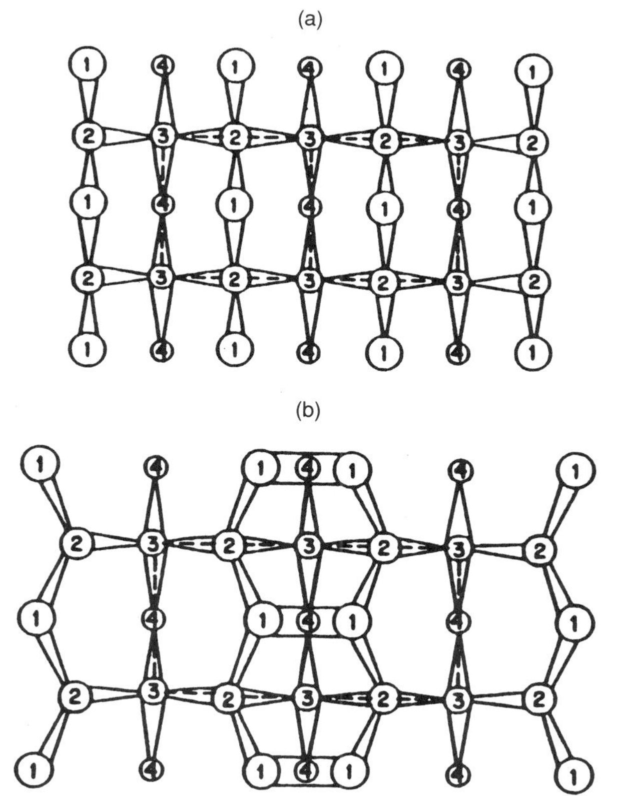

Figure 3.49 Top views of (a) the ideally terminated and (b) the dimer model of the reconstructed (001) surface of a diamond structure crystal. After Appelbaum, Baraff, and Hamann [41]; copyrights © 1976, The American Physical Society, New York.

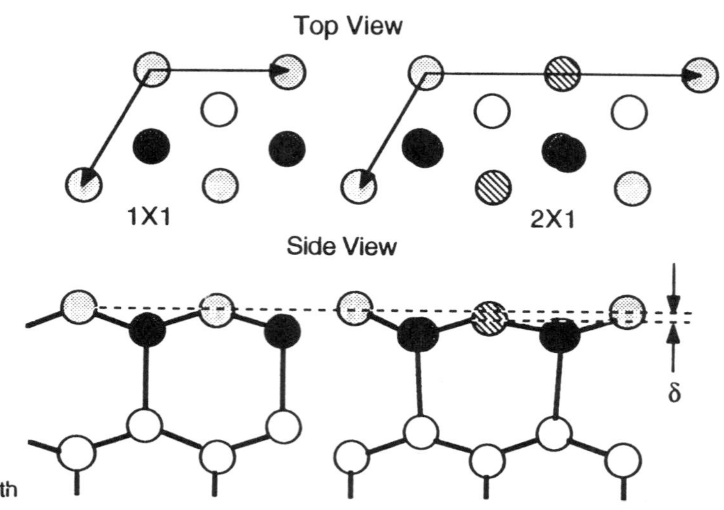

Figure 3.50 The geometric relations between surface meshes and the underlying bulk net planes for the Si{111}-(1 × 1) and -(2 × 1) surfaces. After Taloni and Haneman [42].

temperature in metastable equilibrium. It reconstructs into the stable (7 × 7) surface upon heating to a temperature above $350 \leqslant T_c \leqslant 425°C$. The reason for the range in the critical transformation temperature T_c for the Si(111)-(2 × 1) surface is that cleaving does not necessarily produce a perfectly flat (111) surface, but often generates a vicinal surface that contains a number of monatomic steps that may be evenly or unevenly spaced. T_c decreases from the high-temperature limit in proportion to $1/\tan \alpha_c$, where α_c is the average angle of misorientation between the cleavage face and (111). Also, other surface properties (e.g., the reactivity of the surface) are strongly affected by the density and distribution of steps. The stable high-temperature Si(111) surface configuration is (1 × 1) and is formed at temperatures above 840°C.

Experimental information about the geometry and real structure of solid surfaces is obtained by diffraction techniques, such as reflection high-energy electron diffraction (RHEED), low-energy electron diffraction (LEED), transmission electron diffraction, and direct imaging of the surface atoms, for example, by scanning tunneling microscopy (STM), atomic force microscopy (AFM), or field-ion microscopy (FIM). In order to discuss the geometry of the electron diffraction patterns obtained under the conditions of RHEED and LEED experiments, we consider next the modifications in the Ewald construction for the diffraction of electrons by the surface atoms. Since the scattering is elastic, the diffracted electrons have the same energy as the incident electron beam; that is, $|\mathbf{k}| = |\mathbf{k}_0|$. In analogy to the definition of the three Miller indices of the net planes in a 3D crystal in terms of their cutoffs on the axes of the crystal lattice, the orientations of the various rows of atoms in the surface of a crystal are labeled by two Miller indices derived from their cutoffs on the $\mathbf{a}_s$

and $\mathbf{b}_s$ axes. A reciprocal lattice can be defined by the vectors $\mathbf{a}_s^*$ and $\mathbf{b}_s^*$ that have the properties $\mathbf{a}_s \cdot \mathbf{a}_s^* = \mathbf{b}_s \cdot \mathbf{b}_s^* = 1$ and $\mathbf{a}_s \cdot \mathbf{b}_s^* = \mathbf{b}_s \cdot \mathbf{a}_s^* = 0$, respectively. However, since the lattice is two dimensional, the reciprocal lattice is extended to infinity in the third dimension so that to each row with indices (hk) belongs one infinitely extended reciprocal lattice line that is perpendicular to the crystal surface. The diffraction condition is

$$\mathbf{k}_{\parallel} = \mathbf{k}_{0\parallel} + \mathbf{G}_{hk} \tag{3.97}$$

where $\mathbf{k}_{\parallel}$ and $\mathbf{k}_{0\parallel}$ are the components of $\mathbf{k}$ and $\mathbf{k}_0$ parallel to the surface and

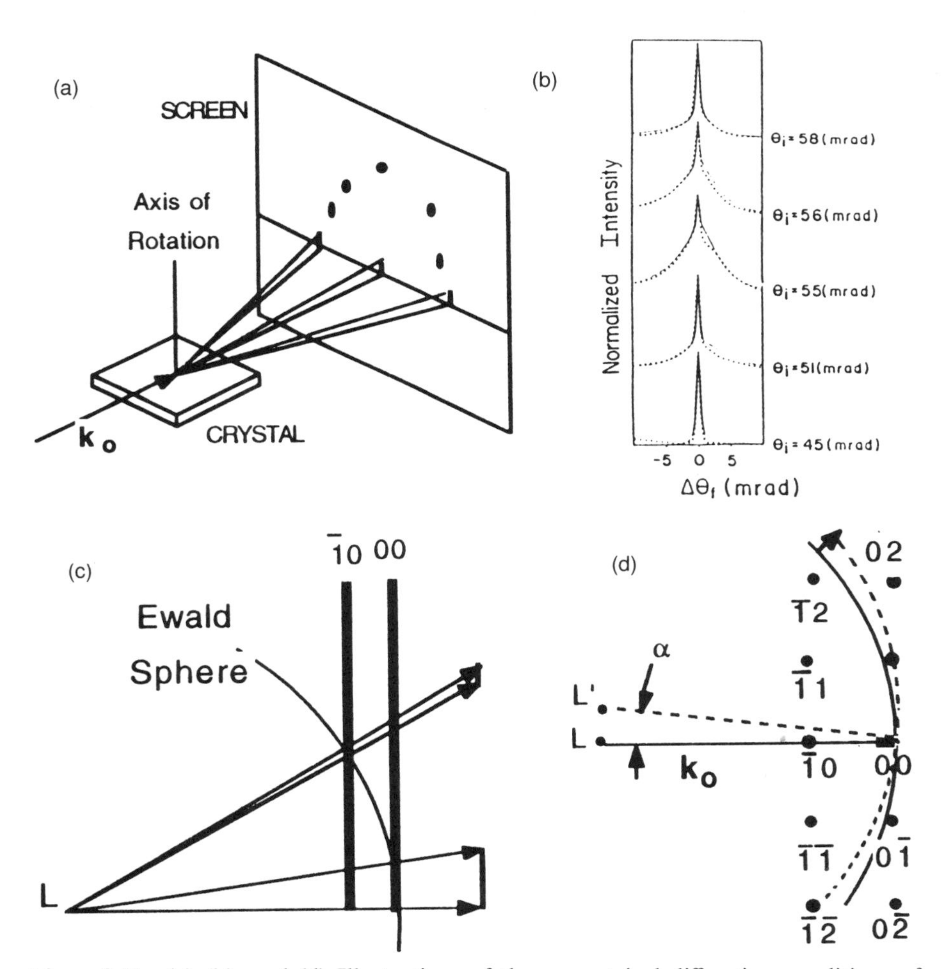

Figure 3.51 (a), (c), and (d) Illustrations of the geometrical diffraction conditions of RHEED. (b) Spot profile analysis for a GaAs surface after deposition of a small amount of GaAs at $T = 450°C$. (b) After Lent and Cohen [43]; copyright © 1986, American Institute of Physics, New York.

$\mathbf{G}_{hk}$ is a vector from the origin to the points of intersection between the reciprocal lattice line (hk) and the Ewald sphere.

Figure 3.51(a) shows schematically the experimental arrangement for RHEED. A beam of electrons, typically of 100 keV energy, impinges onto the surface of the crystal in glancing incidence. Because of the small angle the penetration depth is very small, and scattered electrons collected on a fluorescence screen emerge from only the topmost layer of the crystal. Figure 3.51(c, d) illustrate the geometric conditions that lead to the diffraction pattern. Since the de Broglie wavelength of the electrons is $\lambda \approx 0.04$ Å, the radius $|\mathbf{k}_0|$ of the Ewald sphere is large as compared to $\mathbf{a}_s^*$ and $\mathbf{b}_s^*$. The rows of the reciprocal lattice lines perpendicular to $\mathbf{k}_{0\|}$ cut the Ewald sphere on small half-circles about (00). The spot diameter is affected by the width of the diffraction curve and instrumetal broadening, which is due to the divergence of the electron beam and the energy distribution. Therefore, the reciprocal lattice lines and the Ewald sphere have a nonzero width. A flat screen or photographic film captures the diffracted beams that project the points of interjection of the Ewald sphere with the reciprocal lattice rods from the Laue point onto the film, as illustrated in Figure 3.51(a). The lower half of the Ewald sphere does not contribute to the diffraction pattern because it is shaded by the crystal. Figure 3.51(c) shows a perpendicular cut through the Ewald sphere containing the incident beam plus two of the reciprocal lattice rods ($0K$). For perfectly flat surfaces, the spots are elongated in the vertical direction, with maximum streaking for diffracted beams that are located in the equatorial plane, where the reciprocal lattice rods intersect the Ewald sphere tangentially. Azimuthal

Figure 3.52 LEED patterns of Si(100)-2 × 1 pattern for a flash-annealed surface at 60 eV energy. Courtesy of S. Habermehl and G. Lucovsky [46].

rotation of the sample about a vertical axis moves the Ewald sphere through the reciprocal lattice rods. This is illustrated in Figure 3.51(d), which shows a top view onto the crystal surface. The rotation of the Laue point by the angle α moves the rod (01) into the diffraction condition on the equator Also, rod $(\bar{1}\bar{2})$, which intersects the sphere originally above the equator, is moved simultaneously down into a tangential position to the Ewald sphere at the equator.

If the surface is not perfectly flat, the incident electron beam will penetrate through the three-dimensional hillocks on the surface, resulting in a 3D diffraction pattern removing the streaking. Quantitative information on the surface profile is obtained by an analysis of the angular intensity distribution in the diffraction spots as a function of the angle of incidence. Figure 3.51(b) shows a series of scans of the specular beam intensity for an initially step-free GaAs surface after depositing ~ 0.5 monolayers of GaAs at angles of incidence in the range $45 \leqslant \theta_i \leqslant 58$ mrad [43]. The intensity distribution consists of a central spike due to the long-range order of the surface and a diffuse contribution that is due to the step disorder. The solid line represents to a theoretical fit to the dotted experimental curves, corresponding to an average terrace length of 300 Å. It is based on the kinematical theory of electron diffraction, which is not strictly applicable since multiple scattering contributes to the line shape. However, under the conditions of incidence used in the work of Ref. [43], the deviations due to dynamical effects are small. More detailed analyses of the RHEED intensities for the example of the Si(111)-7 $\times$ 7 surface are provided in Refs. [44] and [45]. Because RHEED provides information of both the surface structure and roughness, it is a valuable tool for the characterization of crystal growth and etching processes, which will be discussed in more detail in Sections 6.4 and 7.4, respectively.

Figure 3.52 shows the LEED patterns of an Si(100)-2 $\times$ 1 surface. Under the conditions of LEED, an electron beam of 10–500 eV energy corresponding to $4 > \lambda > 0.5$ Å at nearly normal incidence is diffracted off the surface of the crystal. The experimental arrangement is shown in Figure 3.53(a). Due to their low energy the diffracted electrons emerge from the top layer and thus give information concerning the surface structure. The intensity of the diffracted beams under the conditions of LEED is governed by dynamical diffraction phenomena [47]. Figure 3.53(b) shows the I–V spectra measured for the (00) spot of the Si(111)-2 $\times$ 1 surface in the as-cleaved (top) and annealed (300°C, middle) states and of the Si(111)-7 $\times$ 7 surface (bottom). An analysis of these I–V spectra confirmed the Haneman model shown in Figure 3.50 with an experimental value for the buckling of the top layer $\delta = 0.24 \pm 0.02$ Å. Although the quantitative analysis of LEED patterns mandates extensive dynamical calculations, qualitative information on the surface structure can be obtained by LEED spot analysis based on the results of kinematic calculations, which has been applied in the investigation of crystal growth [49]. Figure 3.54 shows the spot profile for the (00) spot of the LEED pattern of a clean Si(111)-7 $\times$ 7 surface after deposition of half of a double layer of Si at

(a)

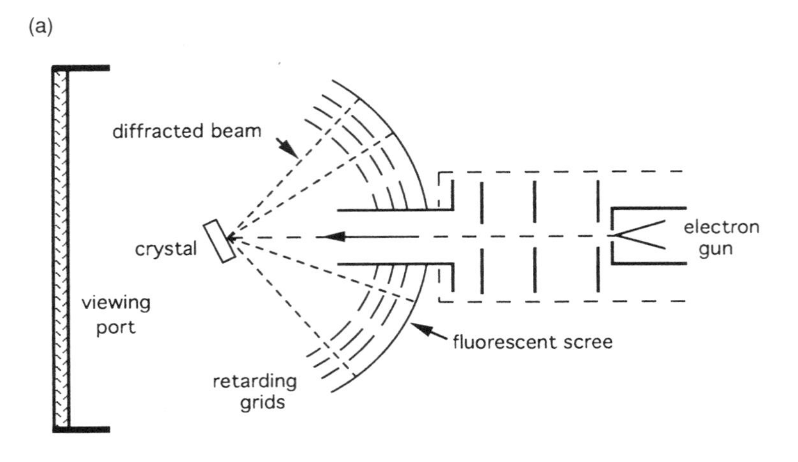

(b)

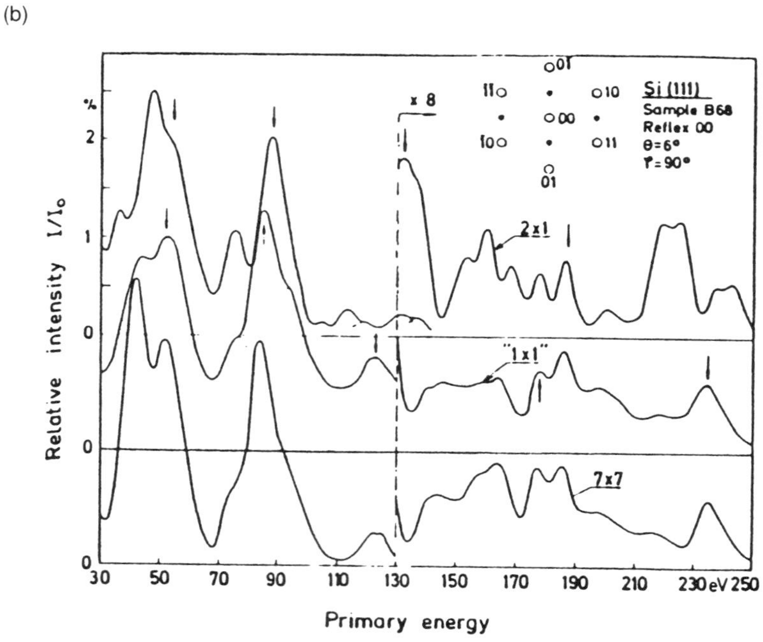

Figure 3.53 (a) Schematic representation of a LEED experiment. After Forstmann [47]. (b) Plot of relative intensity versus primary energy for the Si{111} surface. After Mönch and Auer [48]; copyright © 1978, American Institute of Physics, New York.

600°C. The analysis of the intensity distribution in the (00) spot at various coverages shows that new Si double-layer islands are formed after completion of the first double layer.

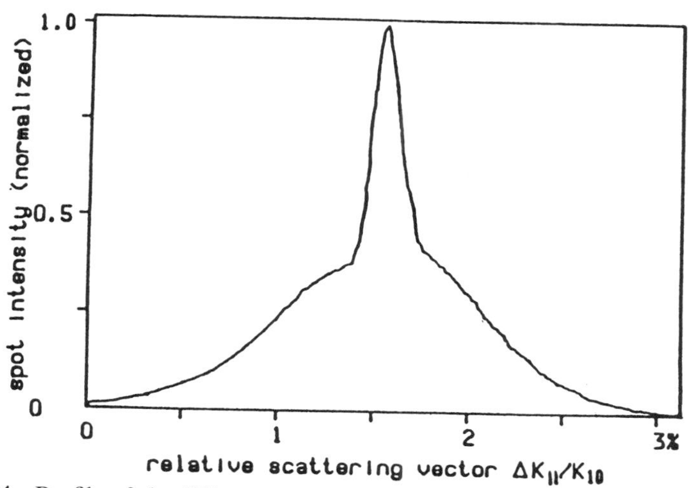

Figure 3.54 Profile of the (00) spot of the LEED pattern for a Si(111)-7 × 7 surface partially overgrown by a double layer of Si atoms at 600°C. After Henzler [49]; copyright © 1983, Elsevier Science Publishers BV, Amsterdam.

An important contribution to the understanding of the surface structure of silicon has been made by transmission electron diffraction studies. In transmission electron diffraction patterns the surface structure is reflected in superlattice spots about the fundamental diffraction spots, as shown for the Si(111)-7 × 7 surface in Figure 3.55. Based on an analysis of the intensities in the diffraction pattern, a model for the reconstructed Si(111)-7 × 7 surface has been proposed by Takayanagi et al. [50], which is widely accepted by the scientific community. Figure 3.56(b) shows a top view (top) and a side view projection on $(10\bar{1})$ (bottom) of this model. The atoms represented by open circles belong to the reconstructed surface layer that extends over 3 atomic layers and connects to atoms in bulk lattice positions represented by full circles. In the corners of the 7 × 7 unit cell, vacancies are formed. In the triangular right half of the unit cell the stacking sequence is *(bCcA)aB*, while the left half is stacked according to the sequence *(bCcA)a|C* containing a stacking fault indicated by a vertical bar. Consequently the 12 top atoms represented by large circles are grouped into two clusters of 6 adatom positions, which differ slightly in height. In the next layer down (a layer = small circles), dimers are formed all around on the outer perimeter of the unit cell.

An important recent addition to the experimental methods of surface physics is scanning tunneling microscopy (STM) [52], which is based on the tunneling of electrons between an atomically sharp protrusion at the tip of the probe needle and either filled or empty surface states. STM is supplemented by atomic force microscopy (AFM) [53], which measures the deflection of a cantilever, carrying at its end a sharp stylus that experiences a force by the surface atoms, as illustrated in Figure 3.56. While AFM is suitable for the measurement of surface reliefs of both conductors and insulators, STM relies

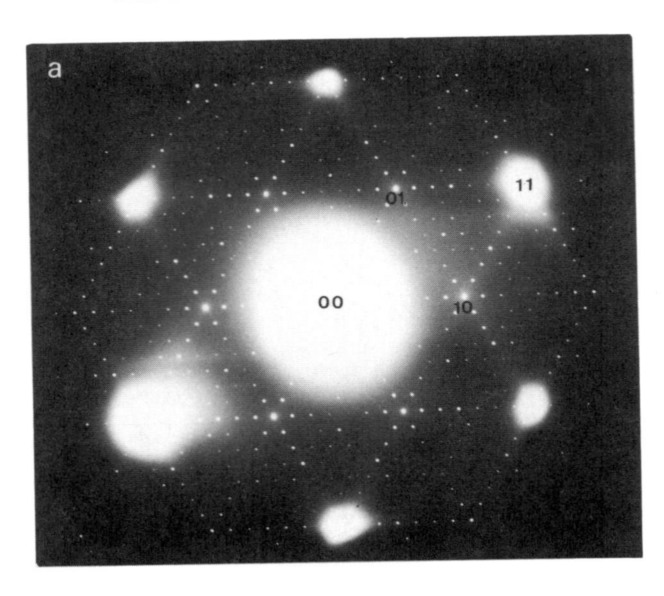

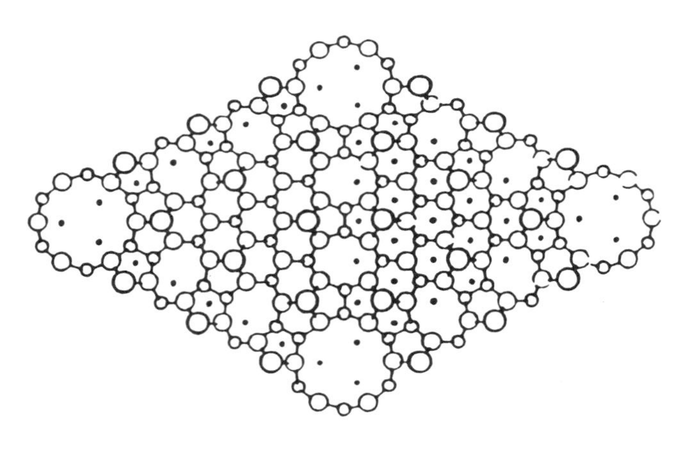

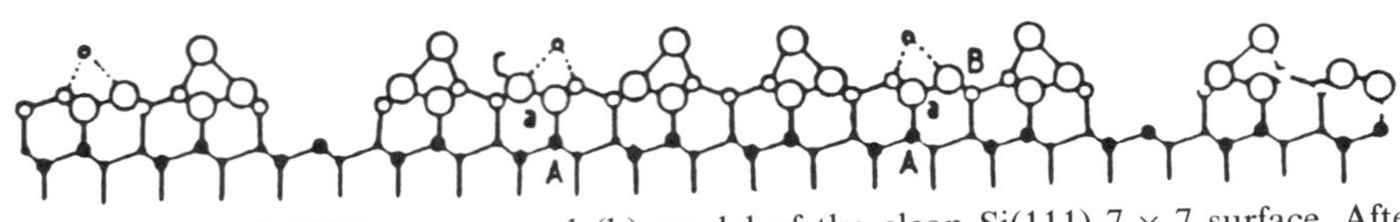

Figure 3.55 (a) TED pattern and (b) model of the clean Si(111)-7 × 7 surface. After (a) Kajiyama et al. [51] and (b) Takayanagi et al. [50]; copyrights (a) © 1989, Elsevier Science Publishers BV, Amsterdam; (b) © 1985, American Institute of Physics, New York.

on conducting samples and either a conducting cantilever stylus assembly or a sharp metallic needle that is kept by the z-drive in a position, resulting in a constant tunneling current as the sample surface is scanned by means of the

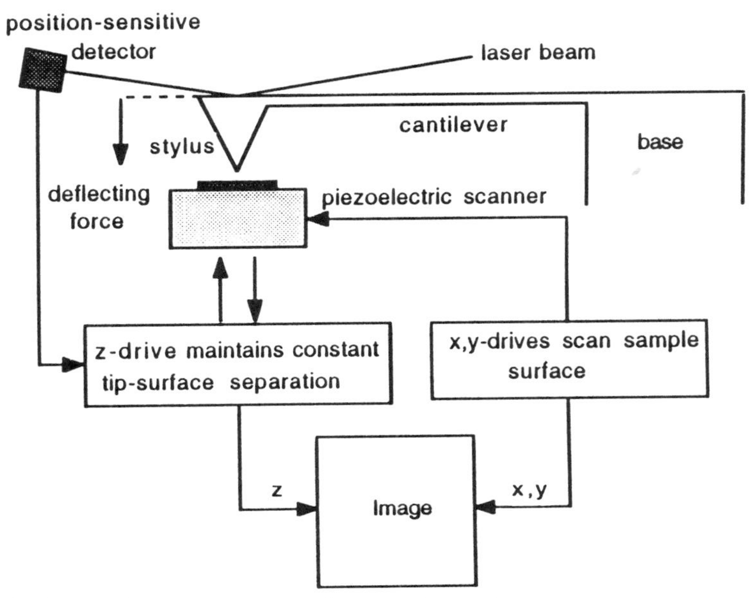

Figure 3.56 Schematic representations of atomic force microscopy.

x, y-drives. The distance of the probe needle that is scanned over the surface is only a few Å, requiring extraordinary positioning control and precautions with regard to vibration-free operation [54]. Although special thinning techniques provide for a very sharp radius of the tip surface, on an atomic scale, the tip generally terminates in a small cluster of atoms of which ideally only one is active. Multiple images may be superimposed under unfavorable conditions if the tunneling current is carried by several active protrusions on the tip or if the surface is rough, that is, contains features with a radius of curvature being comparable to or smaller than the radius of curvature of the tip of the needle. The formation of such ghost images on a rough surface is illustrated in Figure 3.57.

However, with proper precautions, both STM and AFM provide reliable information on the structure and electronic properties of clean surfaces and on their interactions with adsorbates during crystal growth and etching processes. For example, Figure 3.58 shows an STM image of the Si(111)-7 $\times$ 7 surface, which is in excellent agreement with the model of Ref. [50]. STM studies of the Si(100) surface reveal domains with the expected 2 $\times$ 1 reconstruction, coexisting with domains in which the dimer rows are buckled [i.e., a $c(4 \times 2)$ unit cell is established] [56]. The STM image of the Si(111)-7 $\times$ 7 surface shown in Fig. 3.58 also contains surface defects that appear as vacancy clusters and contaminants. However, one should keep in mind that the absence of contrast merely indicates the absence of a tunneling current, which may or may not be due to the absence of atoms under the tip. The interpretation of STM images thus requires careful modeling and verification, where possible, by independent methods.

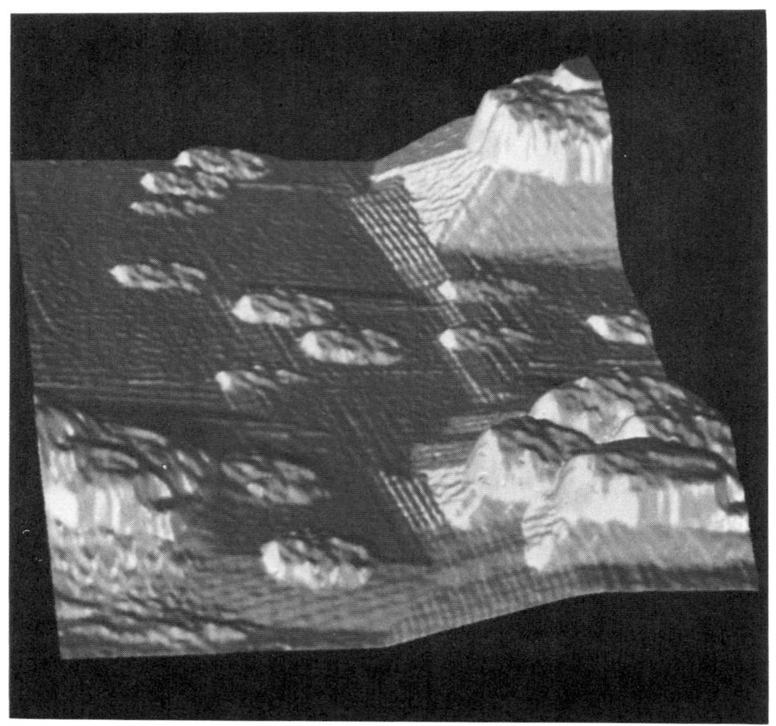

Figure 3.57 Ghost images of the probe tip in the STM image of a 200 × 228 nm^2 area of a Si(110) surface. After van Loenen et al. [55]; copyright © 1990, American Institute of Physics, New York.

Figure 3.59 shows a schematic representation of the tunneling from the probe tip into surface states and out-of-surface states into the probe tip as well as the relation between the density of surface states distribution and the steplike onsets in the current–voltage (I–V) characteristics. The specific features in the density of states distribution that contribute to the tunneling current thus depend on the bias voltage. Since the correlation of the density of states distribution to specific surface states depends on the position in the reconstructed surface, selected bonds to specific surface atoms can be imaged by choosing an appropriate bias voltage for the acquisition of the image. This is illustrated in Fig. 3.60 which shows images of the Si(111)-7 × 7 unit cell acquired at different bias voltages. In compound semiconductors, the discrimination of the imaging with regard to a specific feature in the J-V curve associated with the bonding to a particular type of surface atom in conjunction with optical excitation permits the use of the STM for elemental surface analysis with atomic resolution [57].

Also, the STM images provide for valuable information concerning the topography of the surface of semiconductors that is essential for the understanding of crystal growth processes and surface properties. Figure 3.61 shows

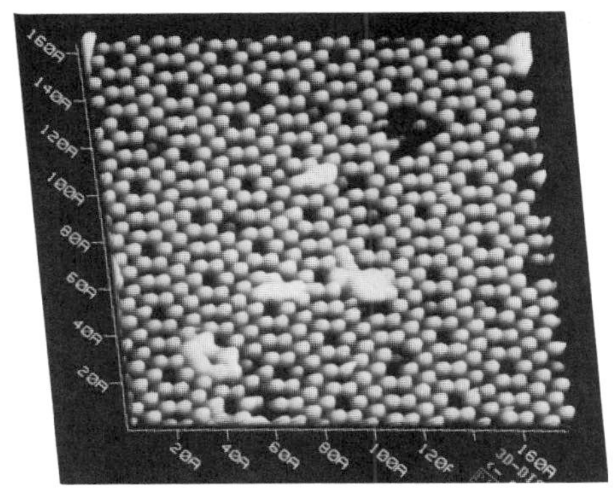

Figure 3.58 STM image of a Si(111)-7 × 7 surface. Reprinted by permission of Omicron GmbH, Taunusstein, Germany.

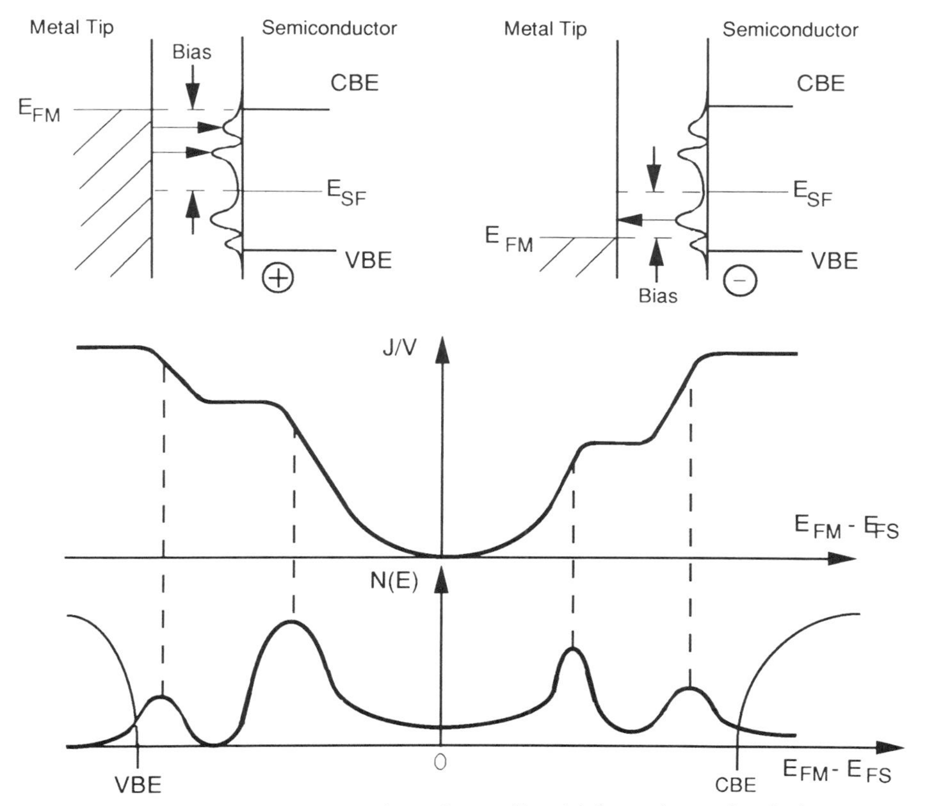

Figure 3.59 Schematic representation of tunneling (a) from the probe tip into empty surface states and (b) from filled surface states into the probe tip. (c) The relation between the onsets in the conductance and the surface states. After Tromp, Hammers and Demuth [58].

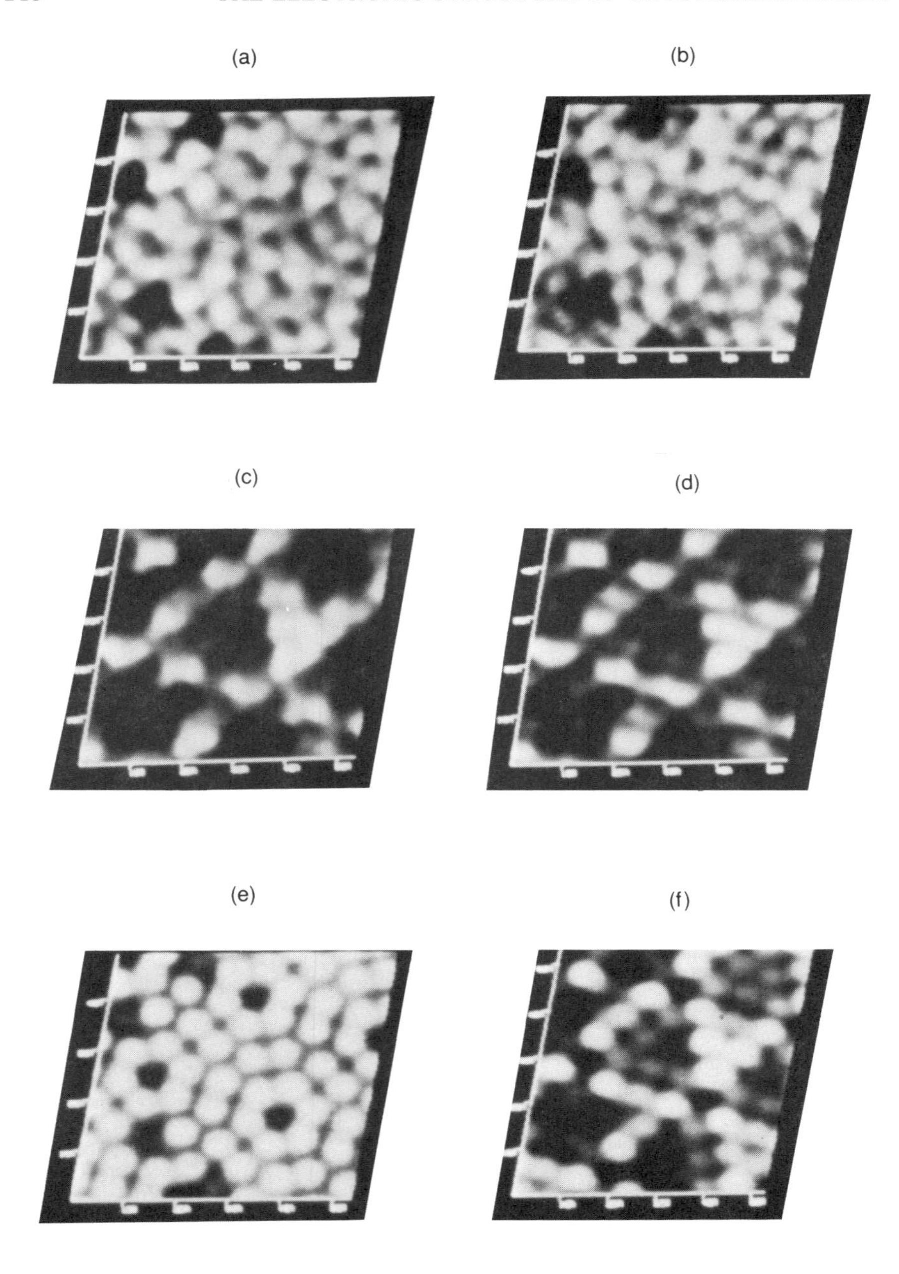

Figure 3.60 I/V spectroscopy maps of the Si (111)-7 × 7 surface. (a)–(d) Current maps at selected bias voltages: (a) +0.8 V, (b) +0.4 V, (c) −1.2 V, (d) −1.8 V. (e) and (f) surface images at +2 V forward bias and −2 V reverse bias, respectively. Reprinted by permission of Omicron Vakuumphysik GmbH, Taunusstein, Germany.

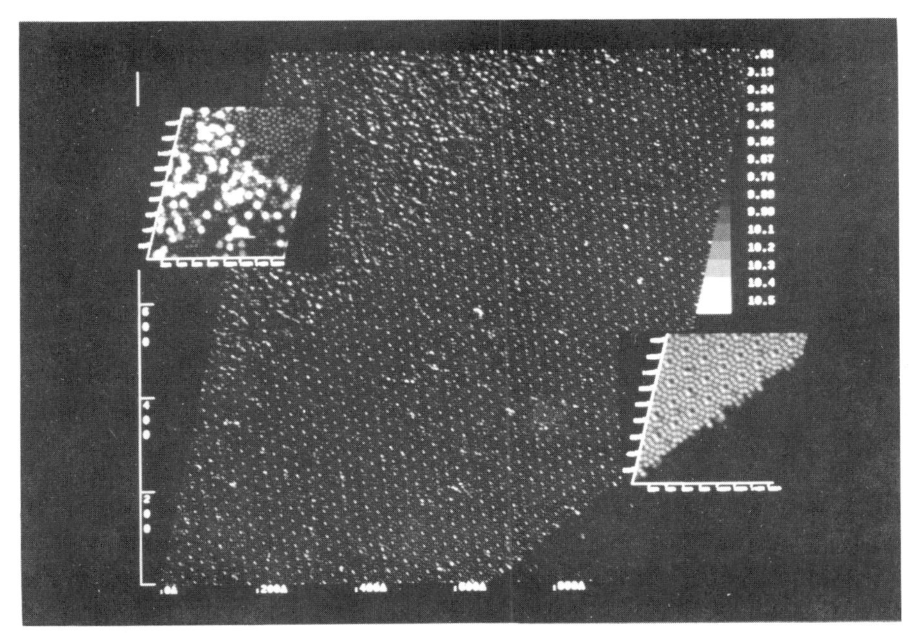

Figure 3.61 STM image of Sl (111)-7 × 7 surface containing in the lower right a mono-step and in the upper left a band of distortions. Magnified views of these features are shown as insets and reveal details of the step structure and local order in the distorted part of the surface, respectively. Reprinted by permission of Omicron Vakuumphysik GmbH, Taunusstein, Germany.

an STM image of the (111)-7 × 7 surface of a boron doped silicon crystal containing in the lower right a mono-step and in the upper left a band of distortions. A magnified image of the step is shown as an inset at the lower right edge of the figure. The step contains straight segments extending over several surface unit cells, but also kinks and atoms in positions at the step edge. A magnified image of the surface distortions is shown as an inset at the upper left edge of the figure. It reveals local order in the distorted region that differs from the 7 × 7 reconstruction of the remaining (111) surface, that is, $\sqrt{3} \times \sqrt{3}$ R90°, 2 × 2 and c(2 × 4) surface domains. STM images of GaAs surfaces have been utilized to investigate by direct inspection the step structure on UHV cleaved (1$\bar{1}$0) surfaces. Figure 3.62 shows three step features revealed by line scans of such a surface. Steps (a) are two atomic layers in height or multiples thereof. They run in the [001] direction corresponding to step facets of ($\bar{1}$10) orientation. Steps (b) running in the [112] direction are one atomic layer in height or multiples thereof and expose a (1$\bar{1}$1) ledge which is As terminated.

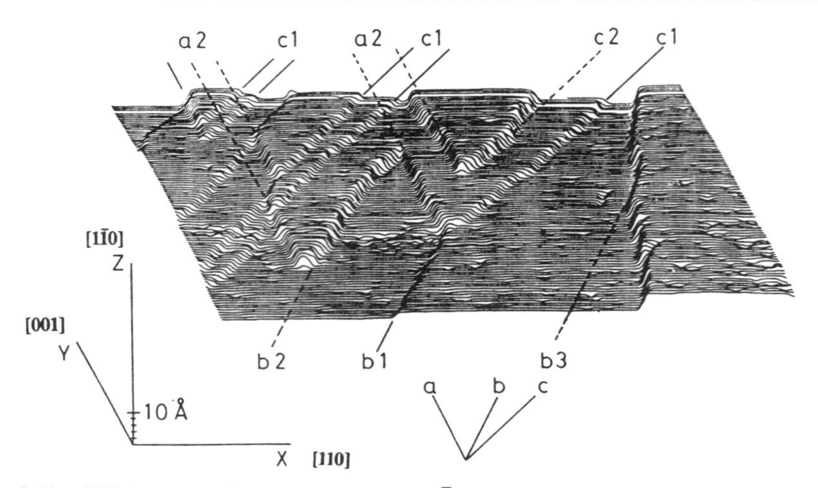

Figure 3.62 STM scans of a UHV-cleaved (1$\bar{1}$0) surface of GaAs. The size of the scanned area is 840 × 840 Å^2. The step height is listed as an index on the labeling of the different step directions by lower-case letters. After Möller et al. [59]; copyright © 1989, North Holland Publishing Company, Amsterdam.

Steps (c) run in the [114] direction and are one atomic layer in height or multiples thereof and expose a (22$\bar{1}$) facet.

For device applications generally the GaAs(001) surface is preferred, which, depending on conditions, reconstructs in the 2 × 1 or 2 × 4 structures. The mechanism of the As-dimer formation on the surface is illustrated in Figure 3.63(a), and the geometries of the (001)-2 × 1 and (001)-2 × 4 surfaces are shown in Figure 3.63(b). As in the discussion of the bulk properties, the electronic structure of the crystal surface is most conveniently represented by a plot of the energy bands in the surface Brillouin zones. Figure 3.63(d) shows the BZ of the (001) surface inscribed into the BZ of the bulk lattice. Figure 3.63(c) shows the effects of reconstruction on the (001) surface BZ. Because of the larger spacings of the 2 × 1 and 2 × 4 reconstructed surface meshes, the associated BZs are of smaller dimensions. As in the description of the band structure of the bulk crystal lattice along high-symmetry directions in the three-dimensional Brillouin zone, the labeling of the surface Brillouin zones corresponds to high-symmetry points and directions in reciprocal space.

Figure 3.63(e) shows as full lines the result of a calculation for the position of the bands associated with the surface atoms [60]. They are superimposed onto the projections of the bulk bands of GaAs into the two-dimensional Brillouin zone, which are represented as vertically dashed regions. The surface bands are either inside these bands, giving rise to resonance, or are within the gap regions. The bands labeled D_{up} and D_{down} are associated with the up and down As–As dimer atoms of Figure 3.63(a). The bands labeled B_{up} and B_{down} correspond to the back bonds of these atoms. The features related to the As–As dimers and the back bonding in Figure 3.63(e) are similar to those obtained for the fully reconstructed 2 × 4 surface both theoretically and by

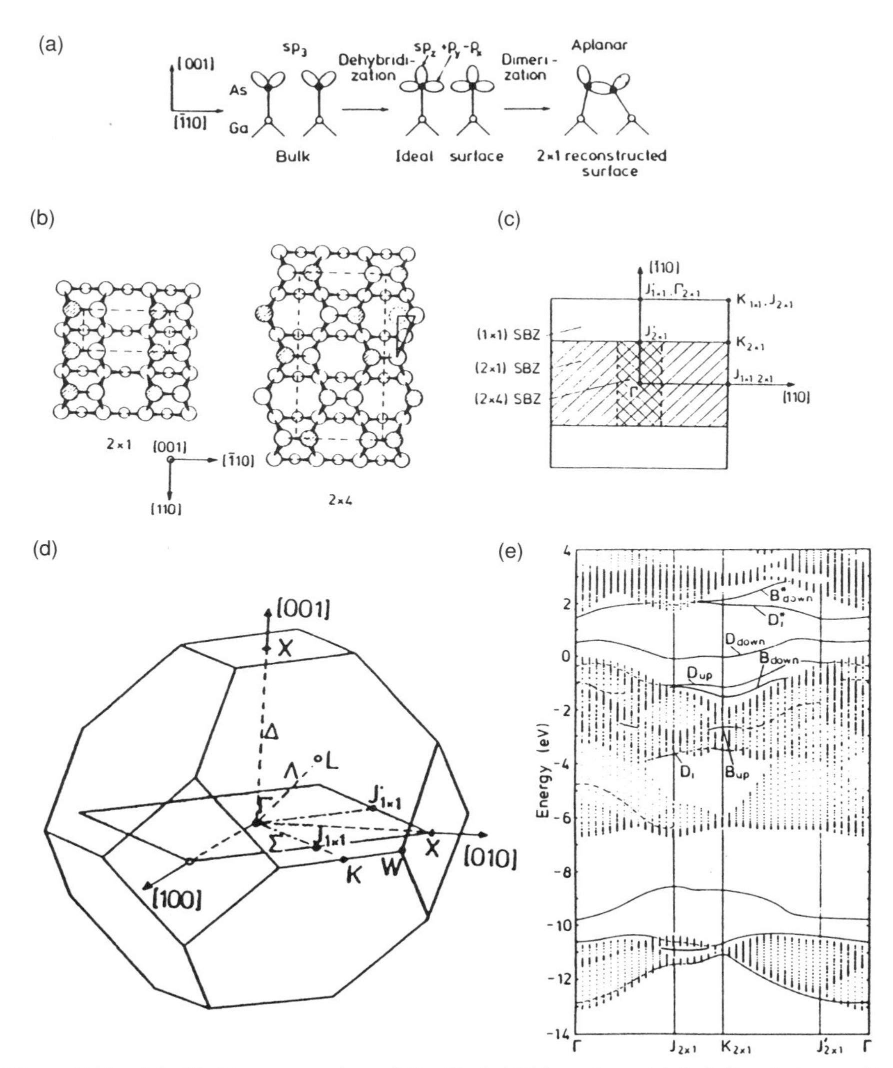

Figure 3.63 (a), (b) Reconstruction of the GaAs(001) surface. (c) Relation between the 1 × 1, 2 × 1, and 2 × 4 surface BZs. (d) 2D Brillouin zone for the (001) surface inscribed into the 3D Brillouin zone of the zincblende structure. (e) Surface band structure for the (001) GaAs-2 × 1 asymmetric dimer model. After Larsen et al. [60]; copyright © 1982, American Institute of Physics, New York.

experimental evaluations. In particular, angle-resolved photoemission studies (see Chapter 8) of the surface band structure associated with the GaAs(001)-2 × 4 surface reveal that there are no surface states located inside the gap. Also, both theoretical and experimental investigations of the (110) surfaces show that there are no intrinsic surface states in the band gaps of GaAs and InP, which are currently the most widely utilized III–V compounds. However, extrinsic surface states may be generated at the surface of these semiconductors by chemical and physical interactions. Thus the density of surface states depends critically on the processing conditions. Note that similar considerations apply also to solid–solid interfaces. For heterostructures employing materials with nearly perfect matching of the lattice parameters and chemistry, the interface state density can be very low. However, for incommensurate structures and solids that enter into chemical interactions at the interface, high densities of interface states in the energy gap may result in a concomitant degradation of the electronic properties of the interface.

An alternative method providing for atomic resolution of solid surfaces is the field ion microscope (FIM). It images the tip of a sharp needle by either noble gas ions [61] or field desorption of adsorbates or the tip material itself [62]. A schematic representation of the FIM imaging process and a schematic outline of a field ion microscope are shown in Figure 3.64. By application of a large positive potential to the needle, a high electric field is generated. In the FIM, this field causes the polarization and attraction of noble gas atoms to the surface. Upon desorption these atoms are ionized and accelerated toward a screen, where they generate an image of the surface. The resolution of this image is

$$\delta_{\text{FIM}} = \left[\delta_0^2 + \frac{4\hbar}{c_1} \left(\frac{r_t}{2c_2 mqE_{\text{bi}}} \right)^{1/2} + \left(\frac{16 r_t kT}{c_1^2 c_2 q E_{\text{bi}}} \right) \right]^{1/2} \tag{3.98}$$

where $c_1 \approx 0.6$ and $4 \leqslant c_2 \leqslant 8$ are numerical factors, r_t is the tip radius, E_{bi} is the best imaging field that is proportional to the applied voltage

$$E_{\text{bi}} = \frac{V_a}{c_2 r_t} \tag{3.99}$$

and δ_0 is the diameter of the ionization disc within which ionization occurs at a distance x_c above the tip surface. It is related to the ionization energy of the imaging gas atoms I_i, the work function of the metal Φ_{Me}, and the local field E_l as $x_c = (I_i - \Phi_{\text{Me}})/qE_l$. In the field desorption microscope (FDM), the atoms of the tip surface or adsorbed atoms or molecules detach under the influence of a field exceeding the threshold for field evaporation V_{th}. They are ionized and accelerated toward the screen, where they form an image of the tip surface. Ions of different mass, desorbed from the surface simultaneously upon application of the high-voltage pulse to $V > V_{\text{th}}$, are characterized by different times of flight from the tip to the image-intensifying microchannel plates. Since ions of the same mass-to-charge ratio arrive at the channel plates at the same time, ion-specific images of the surface of the tip can be created with atomic resolution by switching the channel plates on and off with a controlled delay. Injection of a sample of the accelerated ions into a time-of-flight mass spectrometer permits the identification of the surface chemistry [64]. The recording of field evaporations over a period of time permits the reconstruction of the subsurface structure and composition.

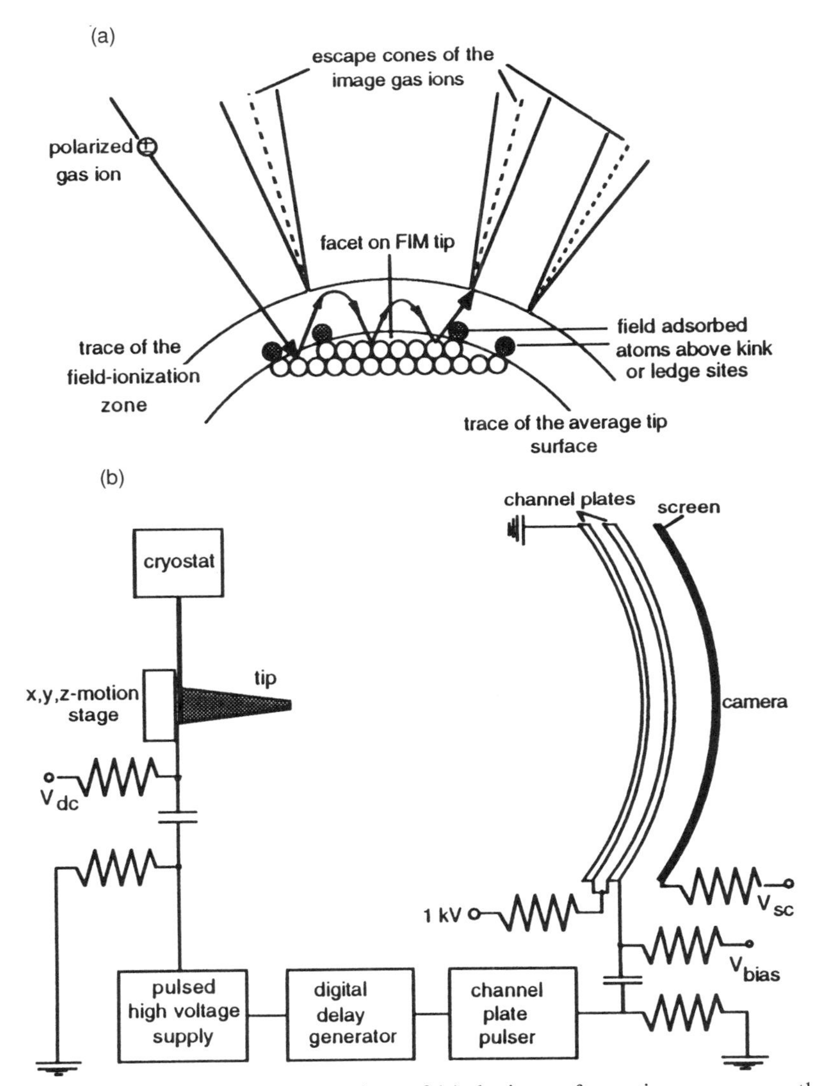

Figure 3.64 Schematic representations of (a) the image formation process on the tip of an FIM probe and (b) the most important components of a field ion microscope. After Wagner [63].

The recording of a timed sequence of FIM images without field evaporation (i.e., at $E_{bi} < E_{th}$), reveals the motion of surface atoms. For example, Figure 3.65 shows the FIM images of a (110) facet created by field evaporation on the tip of an iridium needle, revealing the relative motions of three Ir atoms sequentially forming a dimer (b) and a trimer (d) cluster [65]. The temperature dependence of the jump frequency and of the average lifetime of dimers provides information concerning the activation energy for surface diffusion and

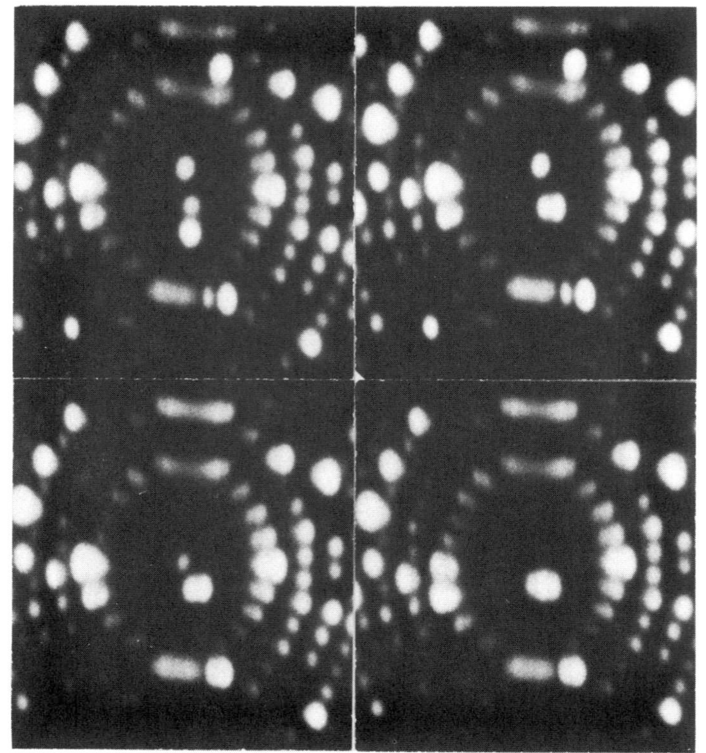

Figure 3.65 Sequence of FIM images of a (110) facet on an iridium needle revealing the relative motion and clustering of Ir atoms. After Tsong and Chen [65]; copyright © 1991, Elsevier Science Publishers BV, Amsterdam.

the energy of bonding. However, these data are not necessarily indicative of the surface properties in the absence of the field. On the other hand, the use of the FIM is particularly appropriate for the characterization of field emitter tips, which are widely employed in the construction of electron microscopes and may become useful in microwave applications, as discussed in more detail in Chapter 7.

References

1. J. V. Smith, *Geometrical and Structural Crystallography*, John Wiley & Sons, New York, 1982.

2. J. E. Spruiell and E. S. Clark in *Methods of Experimental Physics: Polymers*, R. E. Fava, ed., Vol. 16B, Academic Press, New York, NY, 1980, p. 1.

3. N. F. M. Henry and K. Lonsdale, *International Tables of X-Ray Crystallography*, Kynoch Press, Birmingham, England, 1952.

4. G. N. Ramachandran and R. Srinivasan, *Fourier Methods in Crystallography*, Wiley-Interscience, New York, 1970.

5. Y. W. Yang and P. Coppens *Solid State Commun.* **15**, 1555 (1974).

6. P. Corradini and P. Ganis, Nuovo Cimento, Suppl 15, 96 (1960).

7. B. W. Batterman and H. Cole, *Reviews of Modern Physics* **36**, 681 (1964).

8. G. Borrmann, G. Hildebrandt, and H. Wagner, *Zeitsch. f. Physik* **142**, 406 (1955).

9. P. Funke and G. Materlik, *Surface Sci.* **188**, 378 (1987).

10. K. J. Bachmann, T. O. Baldwin, and F. W. Young, *J. Appl. Phys.* **41**, 4783 (1970).

11. E. Hückel, *Z. Physik* **70**, 204 (1931); **72**, 310 (1931); **76**, 628 (1932).

12. J. K. Burdett, *Prog. Solid St. Chem.* **15**, 173 (1984).

13. F. Bloch, *Z. Physik* **52**, 555 (1928).

14. K. Morokuma, *Chem. Phys. Lett.* **6**, 186 (1970).

15. W. Kohn and N. Rokstoker, *Phys. Rev.* **94**, 1111 (1954).

16. (a) B. Segall, *Phys. Rev.* **124**, 1797 (1961); (b) F. S. Ham and B. Segall, *ibid.*, 1786.

17. D. Long, *Energy Bands in Semiconductors*, Interscience Publishers, New York, 1968.

18. E. Wigner and F. Seitz, *Phys. Rev.* **43**, 804 (1933); *ibid.* **45**, 509 (1934).

19. C. Herring, *Phys. Rev.* **57**, 1169 (1940).

20. V. Heine, in *Solid State Physics*, H. Ehrenreich, F. Seitz, and D. Turnbull, eds., Vol. 24, Academic Press, New York, 1970, p. 1.

21. M. L. Cohen and T. K. Bergstresser, *Phys. Rev.* **141**, 789 (1966).

22. J. R. Chelikowsky and M. L. Cohen, *Phys. Rev. B* **14**, 556 (1976).

23. P. Hohenberg and W. Kohn, *Phys. Rev.* **136**, B864 (1964).

24. W. Hanke, H. J. Mattausch, and G. Strinati, in *Electron Correlations in Solids, Molecules and Atoms*, J. T. Devreese and F. Brosens, eds., Plenum Press, NY 1983, p. 289.

25. M. S. Hybertsen and S. G. Louie, *Phys. Rev. B* **34**, 5390 (1986).

26. J.-M. Spaeth, M. Fockele, and K. Krambrock, in *Non-Stoichiometry in Semiconductors*, K. J. Bachmann, H.-L. Hwang, and C. Schwab, eds., Elsevier Science Publ. BV, Amsterdam, 1992, p. 193.

27. S. T. Pantelides, *Reviews of Modern Physics* **50**, 797 (1978).

28. C. A. Coulson and M. J. Kearsley, *Proc. Roy. Soc. A* **241**, 433 (1957).

29. G. A. Baraff, E. O. Kane, and M. Schlüter, *Phys. Rev. B* **21**, 5662 (1980).

30. J. W. Corbett and J. C. Bourgain, in *Point Defects in Solids*, J. H. Crawford, Jr., and L. M. Slifkin, eds., Plenum Press, New York, 1975, p. 1.

31. G. D. Watkins, *Radiation Defects and Damage in Semiconductors—1972*, J. E. Whitehouse, ed., The Institute of Physics, London, 1973; R. P. Messmer and G. D. Watkins, *Phys. Rev. B* **7**, 2568 (1973).

32. S. Mahajan, K. J. Bachmann, D. Brasen, and E. Buehler, *J. Appl. Phys.* **49**, 245 (1978).

33. H. Gleiter, S. Mahajan, and K. J. Bachmann, *Acta Metallurgica* **28**, 1603 (1980).

34. J. H. Van der Merwe, *J. Appl. Phys.* **34**, 123 (1962).

35. J. W. Matthews, A. E. Blakeslee, and S. Mader, *Thin Solid Films* **33**, 253 (1976).

36. E. A. Fitzgerald, G. P. Watson, P. D. Kirchner, R. E. Proano, G. D. Pettit, J. M. Woodall, and D. G. Ast, in *Proc. Symp. Heteroepitaxial Approaches in Semiconductors*, R. People and G. P. Schwartz, eds., The Electrochemical Society, Pennington, NJ 1988, p. 21.

37. J R. Chelikowsky, *Phys. Rev. Lett.* **49**, 1569 (1982).

38. J. R. Chelikowsky and J. C. H. Spence, *Phys. Rev. B* **30**, 694 (1984).

39. J. Heydenreich, Synthesis, *Crystal Growth and Characterization*, K. Lal, ed., North Holland Publishing Company, Amsterdam, 1982, p. 339.

40. H. Gleiter and B. Chalmers, in *Progr. in Mater. Sci.*, B. Chalmers, J. W. Christian, and T. B. Massalski, eds., Pergamon Press, New York, 1972, Vol. 16, p. 1.

41. J. A. Appelbaum, G. A. Baraff, and D. R. Hamann, *Phys. Rev. B* **14**, 588 (1976).

42. A. Taloni and D. R. Haneman, *Surface Science* **10**, 215 (1968).

43. C. S. Lent and P. I. Cohen, *Phys. Rev.* **33**, 8329 (1986).

44. S. Ino, *Japan. J. Appl. Phys.* **19**, 1277 (1980).

45. Z. C. Wu and L. J. Schowalter, *J. Vac. Sci. Technol. B* **6**, 1457 (1988).

46. S. Habermehl and G. Lucovsky, unpublished results.

47. F. Forstmann, *Festkörperprobleme* **13**, 275 (1973).

48. W. Mönch and P. P. Auer, *J. Vac. Sci. Technol.* **15**, 1230 (1978).

49. M. Henzler, *Surface Science* **132**, 82 (1983).

50. K. Takayanagi, Y. Tanishiro, M. Takahashi, and S. Takahashi, *J. Vac. Sci. Technol. A* **3**, 1502 (1985).

51. K. Kajiyama, Y. Tanishiro, and K. Takayanagi, *Surf. Saci.* **222**, 47 (1989).

52. G. Binnig and H. Rohrer, *Helv. Phys. Acta* **55** 726 (1982).

53. G. Binnig, C. F. Quate, and C. Gerber, *Phys. Rev. Lett.* **56**, 930 (1986).

54. S. Chiang, R. J. Wilson, C. Gerber, and V. M. Hallmark, *J. Vac. Sci. Technol. A* **6**, 386 (1988).

55. E. J. van Loenen, D. Dijkkamp, A. J. Hoeven, J. M. Lenssinck, and J. Dieleman, *Appl. Phys. Lett.* **56**, 1755 (1990).

56. R. J. Hammers, R. M. Tromp, and J. E. Demuth, *Phys. Rev. B* **34**, 5343 (1986).

57. L. L. Kazmerski, Paper 12aA1, ITCM-9, August 8–12, 1993, Yokohama, Japan, *Jpn. J. Appl. Phys.* **32**, 25 (1993).

58. R. M. Tromp, R. J. Hammers, and J. E. Demuth, *Science* **234**, 304 (1986).

59. R. Möller, R. Coenen, B. Koslowski, and M. Rauscher, *Surf. Sci.* **217**, 189 (1989).

60. P. K. Larsen, J. F. van der Veen, A. Mazur, J. Pollmann, J. H. Neave, and B. A. Joyce, *Phys. Rev. B* **15**, 3222 (1982); P. K. Larsen, J. H. Neave, and B. A. Joyce, *J. Phys. C* **14**, 167 (1981).

61. E. W. Müller, *Z. Physik* **131**, 136 (1951).

62. J. A. Panitz, *Progr. Surf. Sci.* **8**, 219 (1978).

63. R. Wagner, *Field Ion Microscopy*, Springer-Verlag, Berlin, 1982.

64. E. W. Müller and S. V. Krishnaswamy, *Rev. Sci. Instr.* **46**, 1053 (1978).

65. T. T. Tsong and C-L. Chen, *Surface Science* **246**, 13 (1991).

CHAPTER

4

Semiconductor Devices and Circuits

4.1 *p–n* Junctions

Building upon the concepts introduced in the preceding two chapters, in this chapter we discuss in more detail the control of carrier transport by electronic devices and the general features of an IC processing sequence. In this section, we consider the conditions of equilibrium at p–n junctions and electrical transport in semiconductors, both in the absence and presence of a built-in electrical field. In thermal equilibrium the Helmholtz free energy

$$F = U - TS \tag{4.1}$$

has a minimum so that its total differential $dF = 0$. Since F is a state function, its total differential may be separated into partial terms,

$$dF = -P\,dV - S\,dT + \sum_{i=1}^{c} \mu_i' \, dn_i + \sum_{m=1}^{f} \varepsilon_m \, dO_m \tag{4.2}$$

where T, V, and S are the temperature, molar volume, and entropy, ε_m is the specific surface free energy of the crystal face labeled m of area O_m, n_i is the number of moles in the multicomponent system of the constituent C_i labeled by the index i. Its chemical potential μ_i depends on the activity a_i of C_i according to

$$\mu_i = \mu_i^0 + RT \ln a_i = \mu_i^0 + RT \ln \gamma_i x_i, \tag{4.3}$$

where γ_i is the activity coefficient of C_i and μ_i^0 refers to the chemical potential under standard conditions, where $a_i = 1$. If charged constituents are present, μ_i is replaced by the electrochemical potential

$$\mu_i' = \mu_i + z_i F \phi_G \tag{4.4}$$

The electrochemical potential differs from the purely chemical potential by the reversible work $z_i F\phi$ for adding the charge $z_i F$ associated with one mole of C_i to the system, which is in its interior at the Galvani potential ϕ_G. Consider two phases, a and b, which have constant volume, temperature, and external shape and are in contact across an interface that allows the exchange of constituents. If $\mu_{i'}(a) > \mu_{i'}(b)$, the free energy of the system decreases upon transport of constituent C_i from phase a into phase b. Due to the concentration dependence of μ_i, in this process, $\mu_{i'}(a)$ is decreased and $\mu_{i'}(b)$ is increased. Equilibrium is reached when the electrochemical potentials in both phases are the same for all exchangeable constituents.

In the special case of thermal equilibrium between the two conducting phases a and b where electrons are the only exchangable constituent, we have thus the condition $\mu_e'(a) = \mu_e'(b)$. Let N_{na} be the number of electronic states available in the energy interval $[E_n, E_n + \Delta E]$. Then the probability of N_{n0} out of a total of N electrons occupying states in this energy interval leaving $N_{na} - N_{n0}$ states empty is

$$W_n = \binom{N_{na}}{N_{n0}} = \frac{N_{na}!}{N_{n0}!(N_{na} - N_{n0})!} \tag{4.5}$$

The addition of electrons to the system such that the number of filled states in the energy interval $[E_n, E_n + \Delta E]$ increases by unity and the number of empty states in the same interval decreases by unity results in the change

$$\begin{aligned} d \ln W_n &= \ln \frac{N_{na}!}{(N_{n0} + 1)!(N_{na} - N_{n0} - 1)!} - \ln \frac{N_{na}!}{N_{n0}!(N_{na} - N_{n0})!} \\ &= \ln(N_{na} - N_{n0}/N_{n0} + 1) \end{aligned} \tag{4.6}$$

Adding a number of electrons that increases the occupancy of states in the energy interval $[E_n, E_n + \Delta E]$ by 2 results for $N_{na} > N_{n0} \gg 1$ in the change $d \ln W_n' = \ln[(N_{na} - N_{n0})(N_{na} - N_{n0})/N_{n0} + 2] \approx \ln[(N_{na} - N_{n0})^2/N_{n0} + 1] = 2 \ln(N_{na} - N_{n0}/N_{n0} + 1)$. Therefore, a small arbitrary change dN_e in the total number of electrons in the system causing N_{n0} to change by dN_{n0} corresponds to a change in the occupation probability

$$d \ln W_n \approx dN_{n0} \ln \frac{N_{na} - N_{n0}}{N_{n0} + 1} \tag{4.7}$$

Since $N_{n0} \gg 1$, this reduces to

$$d \ln W_n \approx dN_{n0} \ln \left(\frac{N_{na}}{N_{n0}} - 1 \right) \tag{4.8}$$

Substituting this result into Equation (3.74), and using

$$W = \sum_{n=1}^{\infty} W_n \tag{4.9}$$

results in the relation

$$kT\,d\ln W = \sum_{n=1}^{\infty} (E_n - E_F)\,dN_{n0}. \tag{4.10}$$

Based on the Boltzmann relation for the entropy

$$S = k\ln W, \tag{4.11}$$

equation (4.10) is identified with the term $T\,dS$ in the total differential

$$dF = dU - T\,dS \tag{4.12}$$

so that

$$dF_T = \sum_{n=1}^{\infty} E_n\,dN_{n0} - \sum_{n=1}^{\infty} (E_n - E_F)\,dN_{n0} = E_F\,dN_e. \tag{4.13}$$

Consequently

$$E_F = \left(\frac{\partial F}{\partial N_e}\right)_{V,T,O_m} = \mu_e' \tag{4.14}$$

that is, the Fermi level is identified with the electrochemical potential of the electrons in a conductor. Therefore, in thermal equilibrium between two conducting phases, the Fermi level must be the same on both sides of the junction (see A. Vander Ziel, *Solid State Physical Electronics*, Prentice Hall, 1968).

Based on this equilibrium condition, it is possible to construct the energy-band diagrams for various semiconductor junctions at zero bias. This is illustrated in Figure 4.1 for a homojunction between *n*- and *p*-type portions of the same semiconductor. For the purpose of discussion, the assumption is made that the doping level in the *n*-type material is much larger than in the *p*-type material, which is indicated by the plus superscript. Consequently the Fermi level in the n^+-type material is much closer to the majority carrier band edge than in the *p*-type material with a Fermi level separation ΔE_F, as shown in Figure 4.1(a) for the two doping levels in separate pieces of the semiconductor. Upon establishing intimate contact between the n^+- and *p*-type portions of the junction, initially a steep gradient in the concentrations of the mobile carriers exists at the interface. Consequently a diffusion current results in electrons from the n^+ side into the *p*-type part and in holes from the *p* side into the n^+ part of the semiconductor. These diffusing carriers recombine with the respective majority carriers at the other side of the junction. This lowers the Fermi level of the n^+ materials and raises the Fermi level of the *p*-type material until at equilibrium the Fermi level is constant across the junction.

As a consequence of the recombination of the mobile carriers in the vicinity

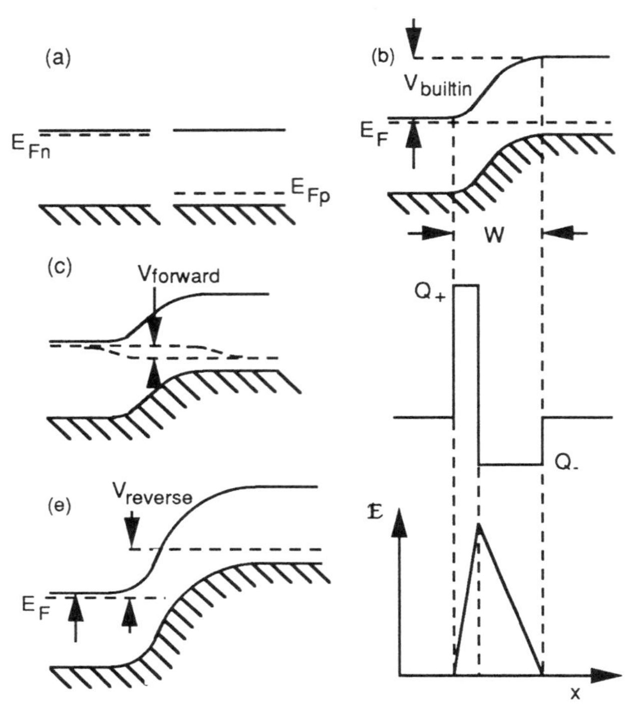

Figure 4.1 (a) n^+- and p-type portions of a semiconductor in separation; (b) n^+p homojunction; (c) homojunction under forward bias; (d) charge distribution; (e) homojunction under reverse bias; (f) electrical field distribution.

of the junction, a zone is established that is denuded of mobile carriers. It is called the *depletion* or *space-charge layer*. The latter term originates from the fact that after the recombination of the mobile carriers the immobile positively charged donor ions and negatively charged acceptor ions are left behind in the n^+ and p portions of the depletion layer representing a space charge. The charge distribution is illustrated in Figure 4.1(d). Note that charge neutrality requires that the positive charge Q_+ associated with the donor ions on the n^+ side of the junction must equal the negative charge Q_- associated with the negative acceptor ions on the p side. Because of the high doping concentration in the n^+ region, the charge Q_+ is distributed in a much narrower layer than the charge Q_-, so that most of the depletion region lies in the p-type material, approaching the condition of a one-sided junction. Since the charge density is constant, and, by Maxwell's equations, the divergence of the electrical field equals the charge density over the dielectric constant, an electrical field E is established in the depletion region that decreases linearly from a maximum value at the electrical junction to zero at the outer boundaries of the depletion layer. This is shown in Figure 4.1(f).

In the interior of the n^+- and p-type regions outside the depletion layer, the distance of the band edges from the Fermi level must be the same, as shown in Figure 4.1(a). Therefore, the bands bend in the depletion region, as shown in Figure 4.1(b), forming a barrier to majority carrier injection. The barrier height is given by the difference of the Fermi levels in the n^+- and p-type material and results in the builtin voltage

$$V_d = \frac{\Delta E}{q} \approx \frac{E_g}{q} + \frac{kT}{q} \ln \left[\frac{(N_A - N_D)_p (N_D - N_A)_n}{n_i^2} \right], \tag{4.15}$$

where $(N_A - N_D)_p$ and $(N_D - N_A)_n$ refer to the net carrier concentrations in the p- and n-type regions of the junction, respectively, measured in carriers per cm^3. If an external voltage V is applied such that the p side becomes more positive relative to the n^+ side (forward-bias condition), the applied field opposes the builtin field, thus lowering the barrier to current flow, as shown in Figure 4.1(c). Conversely, if an external voltage of opposite polarity is applied (reverse bias), the barrier height increases, as illustrated in Figure 4.1(e).

The depletion layer width W widens under reverse bias and depends on the bias voltage V according to the relation

$$W = \sqrt{\frac{2\varepsilon\varepsilon_0(|N_A - N_D|_p + |N_D - N_A|_n)(V_d - V)}{q|N_A - N_D|_p|N_D - N_A|_n}} \tag{4.16}$$

where ε is the static dielectric constant of the semiconductor, and $\varepsilon_0 = 8.854 \times 10^{-12}$ F/m is the permittivity of empty space. In cgs units, Equation (4.16) converts into

$$W = \sqrt{\frac{\varepsilon(|N_A - N_D|_p + |N_D - N_A|_n)(V_d - V)}{2\pi q|N_A - N_D|_p|N_D - N_A|_n}}. \tag{4.17}$$

For a one-sided junction Equation (4.16) reduces to

$$W = \sqrt{\frac{2\varepsilon\varepsilon_0(V_d - V)}{q|N_A - N_D|}} \tag{4.18}$$

In the remainder of this book, the quantities $N_A - N_D$ or $N_D - N_A$ are chosen to be positive, referring thus to the net ionized acceptor or donor concentrations in p- or n-type materials. In Equation (4.18), $|N_A - N_D|$ refers to the net carrier concentration in the lightly doped portion of either a p^+n or n^+p junction. Since a change in W is accompanied by a change in the space charge, the depletion layer is associated with a capacitance per unit area

$$C = \frac{dQ}{dV} = \frac{\varepsilon\varepsilon_0}{W} = \sqrt{\frac{\varepsilon\varepsilon_0 q|N_A - N_D|}{2(V_d - V)}} \tag{4.19}$$

Capacitance measurements are routinely achieved with good accuracy so that valuable information on both the builtin voltage and the net carrier concentration can be obtained conveniently from a plot $1/C^2$ versus V.

Figure 4.2 shows this for the example of a heterostructure of highly doped n^+-type CdS grown epitaxially on a lightly doped p-type InP substrate. In view of the large difference in the doping levels, the depletion layer resides overwhelmingly in the InP, so that the device approaches the properties of a one-sided junction. The builtin voltage is in this case 1 V, and the carrier concentration is $2 \times 10^{17}\,\mathrm{cm}^{-3}$. At zero bias the depletion layer width for this device is thus ~ 800 Å, corresponding to a capacitance of $\sim 10\,\mathrm{nF/cm}^2$.

Note that for hetereostuctures, where the electrical junction is formed between n- and p-type layers of different semiconductors, the shape of the energy-band diagram depends on the relative position of the band edges and of the Fermi levels of these semiconductors. The difference in the Fermi levels determines the band bending, as in the case of the homojunction, and is given by the difference in the work functions Φ of the two materials, that is, by the distance of their Fermi levels from the vacuum level. The relative positions of the conduction-band edges is determined by their electron affinities χ. If χ is known for both phases, the valence-band offset follows from the bandgaps E_g. Depending on the band-edge positions, discontinuities appear in the energy-band diagrams at the junction, as shown in Figure 4.3(a)–(d).

The direction of the field **E** in the depletion layer of a pn junction is such

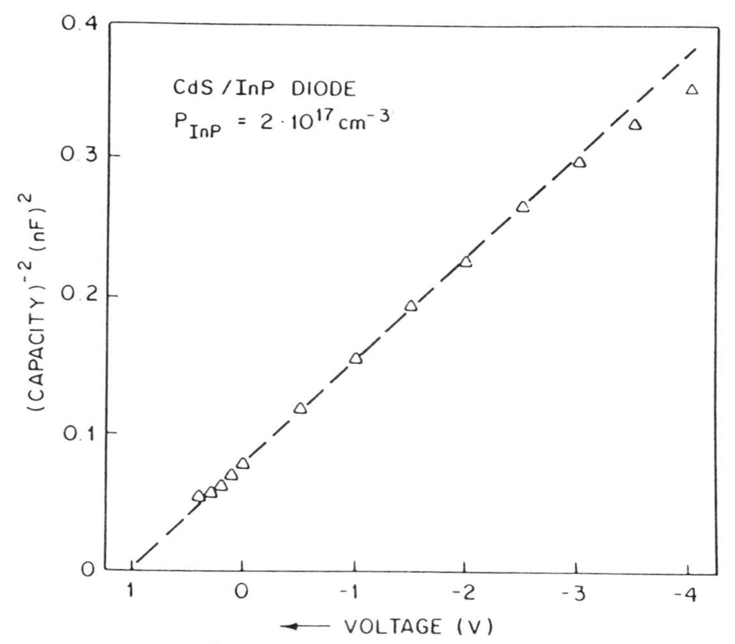

Figure 4.2 Plot of $1/C^2$ versus V for an n-CdS/p-InP diode. After Bettini, Bachmann, and Shay [1]; copyright © 1978, American Institute of Physics, New York.

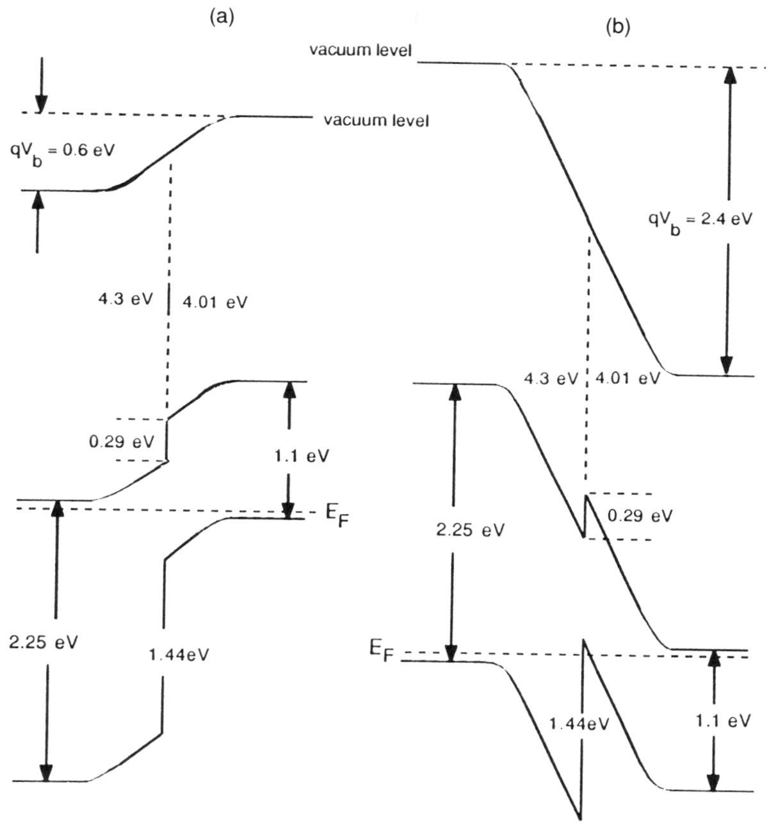

Figure 4.3 Energy-band diagrams for heterojunctions between *n*- and *p*-type portions of silicon ($E_g = 1.1\,\text{eV}$, $\chi = 4.01\,\text{eV}$) and gallium phosphide ($E_g = 2.25\,\text{eV}$, $\chi = 4.3\,\text{eV}$): (a) $E_g(n) > E_g(p)$, $\chi(n) > \chi(p)$; (b) $E_g(n) < E_g(p)$, $\chi(n) < \chi(p)$.

that minority carriers that enter into the depletion region are accelerated across the junction. The force exerted on these carriers is

$$\mathbf{F} = \pm q\mathbf{E} = m^*\mathbf{a} \tag{4.20}$$

where the plus sign applied for holes. Let τ_h and τ_e denote the average time between scattering events for holes and electrons, respectively. Then for average hole velocity $\mathbf{v}_h = 0$ in the absence of a field, in the time interval τ_h the holes are accelerated in the field $\mathbf{E}$ to the average velocity

$$\mathbf{v}_h = \frac{q}{m_h^*}\mathbf{E}\tau_h = \mu_h\mathbf{E} \tag{4.21}$$

where

$$\mu_h = \frac{q\tau_h}{m_h^*} \tag{4.22}$$

is the mobility of holes. This corresponds to a drift current of holes toward the junction

$$\mathbf{I}_{h\,\text{drift}} = qp\mathbf{v}_h = \sigma_h \mathbf{E} \tag{4.23}$$

with the hole conductivity

$$\sigma_h = pq\mu_h. \tag{4.24}$$

Similarly one obtains for the electron drift current

$$\mathbf{I}_{e\,\text{drift}} = -qn\mathbf{E} = \sigma_e \mathbf{E} \tag{4.25}$$

where

$$\sigma_e = nq\mu_e \tag{4.26}$$

is the electron conductivity and

$$\mu_e = \frac{q\tau_e}{m_e^*} \tag{4.27}$$

is the mobility of electrons.

The drift current density due to the attraction of minority carriers from both sides of the junction by the builtin field matches in equilibrium exactly the opposing diffusion current density of the corresponding majority carriers; that is,

$$\mathbf{J}_{h\,\text{diffusion}} = -qD_h \nabla p \tag{4.28}$$

$$\mathbf{J}_{e\,\text{diffusion}} = qD_e \nabla n \tag{4.29}$$

The diffusion constants of the carriers are related to their mobilities by the Einstein relation

$$D = \frac{kT}{q}\,\mu \tag{4.30}$$

The mobility may be measured by the Hall effect, as illustrated in Figure 4.4. Consider the carriers in a slab of a semiconducting material moving under the influence of an electric field $\mathbf{E} \parallel \mathbf{x}$ with the current density

$$\mathbf{J} = \mathbf{J}_e + \mathbf{J}_h = q|N_A - N_D|\mathbf{v}_x = \frac{\mathbf{I}}{wt} \tag{4.31}$$

where $\mathbf{J}_e = qN_D\mathbf{v}_e$ and $\mathbf{J}_h = qN_A\mathbf{v}_h$ are the partial current densities carrier by electrons and holes, respectively. In a magnetic inductance $\mathbf{B} \perp \mathbf{E}$ the carriers experience the Lorentz force $\mathbf{L}_e = -q|\mathbf{v}_e \| \mathbf{B}|$ for electrons and $\mathbf{L}_h = +q|\mathbf{v}_h \| \mathbf{B}|$ for holes. Consequently, a voltage V_H is developed between contacts C1 and C2

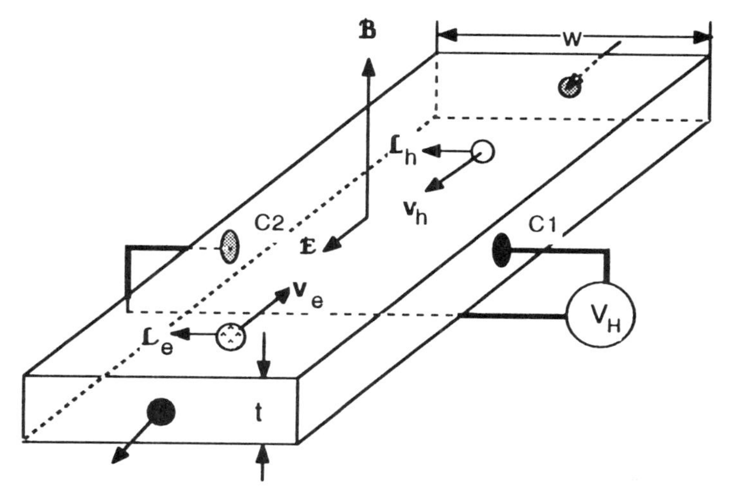

Figure 4.4 Schematic representation of the sample geometry in a Hall-effect measurement.

that is positive for p-type and negative for n-type materials. In steady state the Lorentz force is balanced by the electrical force exerted on the carriers by the electrical field V_H/w; that is

$$q|\mathbf{B}||\mathbf{v}_x| = \frac{|\mathbf{I}||\mathbf{B}|}{wt|N_A - N_D|} = \frac{qV_H}{w} \tag{4.32}$$

so that

$$q|N_A - N_D| = \frac{|\mathbf{I}||\mathbf{B}|}{tV_H} = \frac{1}{R_H} \tag{4.33}$$

where R_H is the Hall coefficient. Since

$$\sigma = |N_A - N_D|q\mu \tag{4.34}$$

the mobility of the majority carriers

$$\mu = \sigma R_H \tag{4.35}$$

can be determined by an independent measurement of the conductivity and the Hall coefficient. The carrier mobility is determined by a variety of scattering mechanisms, for example, impurity scattering and phonon scattering. The mobility observed by a Hall measurement is thus composed of various contributions due to specific scattering mechanisms, that is,

$$\frac{1}{\mu} = \frac{1}{\mu_i} + \frac{1}{\mu_l} + \cdots \tag{4.36}$$

where μ_i and μ_l are the impurity scattering and lattice scattering contributions to the total mobility μ. Since the phonon scattering increases with temperature as $T^{3/2}$ while the impurity scattering decreases as $T^{-3/2}$, in simple cases where these two scattering mechanisms dominate the mobility, the two components may be separated.

Figure 4.5 shows the results of a low-temperature Hall-effect study of Si single crystals grown by the Czochralski technique (see Section 5.3). It demonstrates that the ultimate purity in this case is significantly affected by the choice of the container material, such as synthetic versus natural quartz, because the wall of the SiO_2 crucible is slowly dissolved in the molten silicon that it contains (see Section 5.2). The detection of impurities on the low parts per trillion level is a remarkable achievement that is well beyond current methods of conventional analytical chemistry.

Upon application of a forward bias to a *pn* junction (i.e., the application of a positive potential to the *p*-type side of the junction and of a negative potential to the *n*-type side of the junction), the height of the barrier to current flow is reduced, causing the injection of minority carriers. Note that this injection of

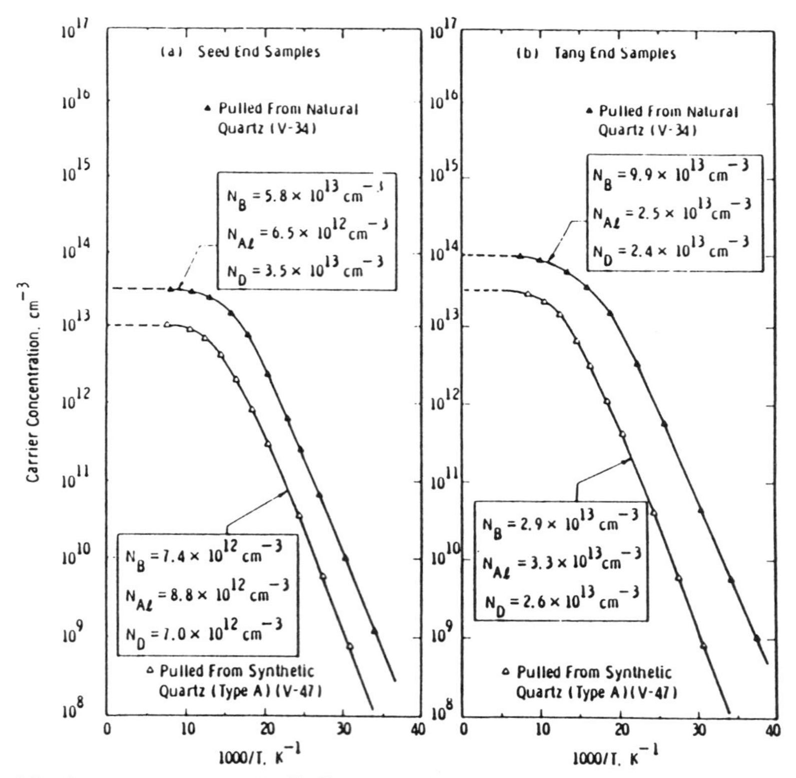

Figure 4.5 Low-temperature Hall-effect data for Czochralski pulled silicon crystals using different sources of the quartz for fabricating the container. After Hobgood et al. [2]; copyright © 1987, United Engineering Trustees, Inc., New York.

minority carriers changes the concentrations of both minority and majority carriers in the vicinity of the junction as compared to their equilibrium values. Consequently the Fermi level splits into two quasi-Fermi levels, one describing the hole distribution and the other describing the electron distribution. In sufficient distance from the junction, that is, at a distance that is large compared to the minority carrier diffusion length, the nonequilibrium electron and hole distributions have decayed and the carrier concentrations are described by a single Fermi level, as illustrated in Figure 4.6 for (a) zero and (b) forward bias conditions.

Figure 4.6(c) shows the general features of the current versus voltage characteristics for a *pn* diode. There are several mechanisms that may contribute to the current, for example, drift and diffusion of carriers, carrier generation/recombination in the space-charge region, and tunneling. Drift of the carriers in the field of the junction increases with increasing forward bias and decreases with increasing reverse bias. The drift current is opposed by diffusion of carriers in the concentration gradients for electrons and holes in the vicinity of the junction. Frequently the $J-V$ characteristics can be expressed as

$$J = J_0[\exp(qV/A_p kT) - 1] \tag{4.37}$$

where A_p is the diode perfection factor [3]. For an ideal diode, where diffusion

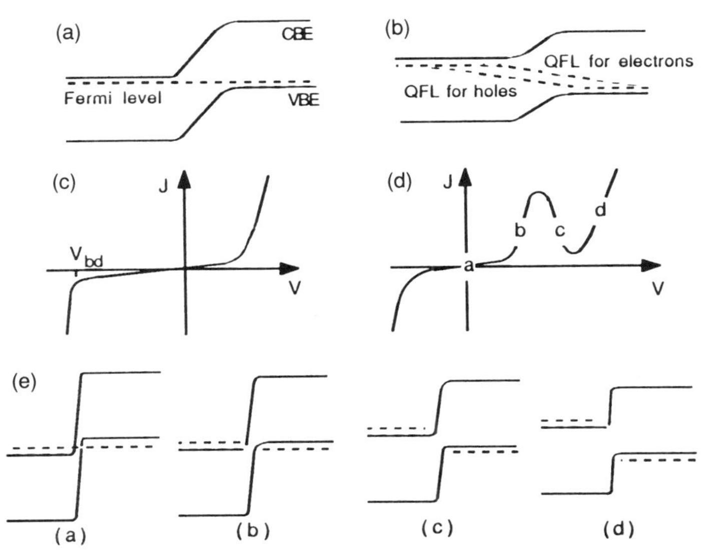

Figure 4.6 Energy band diagrams for a *pn* junction at (a) zero bias and (b) forward bias, respectively. Current-voltage characteristics of (c) a *np* diode, (d) a n^+p^+ tunnel diode. (e) Energy band diagrams for a tunnel diode at (a) zero bias, (b) forward bias permitting majority carrier tunneling, (c) forward bias eliminating majority carrier tunneling, (d) forward bias supporting a large minority carrier injection current.

and drift are the dominant current mechanisms, $A_p = 1$, and the reverse saturation current density is given by

$$J_0 = q\left(\frac{D_e N_D}{L_e} + \frac{D_h N_A}{L_h}\right)\exp\left(\frac{-qV_B}{kT}\right) \tag{4.38}$$

where

$$L_e = (D_e \tau_e)^{1/2} \tag{4.39}$$

$$L_h = (D_h \tau_h)^{1/2} \tag{4.40}$$

are the diffusion lengths of electrons and holes, respectively. If generation/recombination on centers in the space-charge region dominates the current, the forward current density is given by

$$J = \tfrac{1}{2} q W v_{th} \sigma c_t n_i \exp\left(\frac{qV}{2kT}\right) \tag{4.41}$$

so that $A_p = 2$. Figure 4.7 shows the forward log J–V characteristics of the CdS/InP heterodiode. An analysis of the slope results in $A_p \approx 2$, so that the current is dominated by generation/recombination in the space-charge region. Note that upon reverse bias, electrons and holes that enter into the depletion region, either by minority carrier diffusion or tunneling of majority carriers, can gain in the field of the junction sufficient energy to promote impace ionization. Also, under strong reverse bias, tunneling of majority carriers becomes possible, so that the reverse characteristic generally does not saturate at $-J_0$, but exhibits a strongly increasing current at the breakdown voltage V_b. Under idealized conditions of impact ionization,

$$V_b \approx 60\left(\frac{E_g}{1.1}\right)^{3/2}\left(\frac{N_i}{10^{16}}\right)^{-3/4} \tag{4.42}$$

where N_i refers to the impurity concentration [3]. Since tunneling is not a thermally activated process, in the case of a tunneling-dominated current mechanism, the J–V characteristics has a much weaker temperature dependence than in the cases of generation/recombination and drift plus diffusion-dominated current mechanisms. Tunnel diodes employing heavily doped *pn* junctions provide for efficient carrier transport across the junction and exhibit a negative resistance region, as illustrated in Figure 4.6(d). The mechanism that leads to the negative resistance region in the forward J–V characteristics of a tunnel diode is illustrated in Figure 4.6(e), which shows the energy-band diagrams of the tunnel junction for various forward bias conditions. The degenerate doping on both sides of the tunnel junction results according to Equation (4.17) in a very thin depletion region that brings the majority carrier

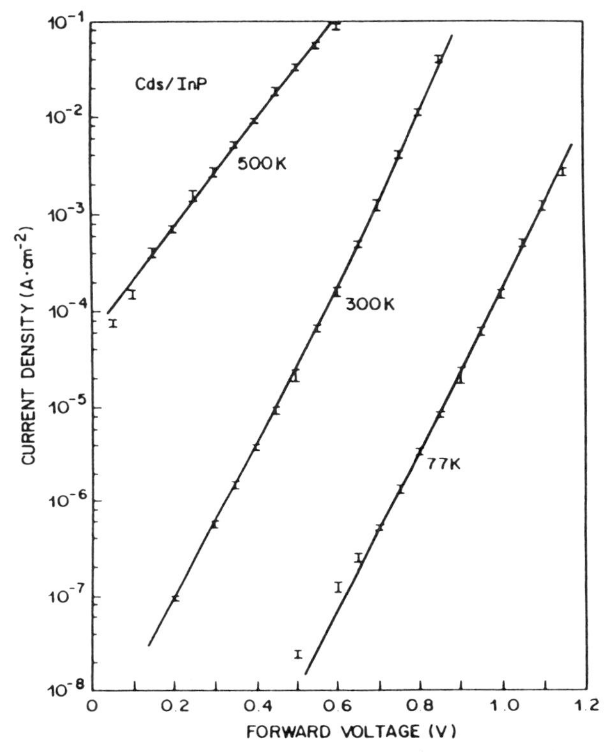

Figure 4.7 Current–voltage characteristics for an n^+-CdS/p-InP heterostructure. After Bettini et al. [1]; copyright © 1978, The American Physical Society, New York.

band edges into close proximity. At sufficiently small forward bias efficient majority carrier tunneling from filled states in the n-type material into empty states in the p-type material results in a rising forward current until the CBE in the n-type crosses the top of the valence band of the p-type portion of the diode. Then tunneling is no longer possible and the current decreases with increasing voltage until the diminishing barrier for minority electron and hole injection takes over and the current rises again, as shown in Figure 4.6(d). Tunnel diodes are used as interconnects in monolithic two-terminal cascade solar cells [4], [5], where the connection of two diodes of different bandgaps optimizes the power conversion efficiency since the high-energy photons absorbed in the wider-gap front cell results in less thermalization and the low-energy photons caught in the lower-gap material result in lower transmission losses than encountered with a medium-bandgap single junction cell. The tunnel diode between the two cells is needed to assure efficient current transport across the interface of the two back-to-back connected diodes. Also, tunnel diodes have been utilized to provide positive feedback in microwave circuits (see Sections 4.3 and 4.4 for a discussion of resonance tunneling and other microwave devices).

4.2 Bipolar Transistors

Large changes in the current can be achieved by a small control current in bipolar *pnp* or *npn* three-terminal devices. Figures 4.8(a–d) show a schematic representation of a *pnp* transistor, the minority carrier distribution in the base, the associated energy-band diagram in equilibrium, and the energy-band diagram under forward bias of the emitter–base junction and reverse bias of the base–collector junction. Under these bias conditions, holes are injected into the base, raising their concentration over the equilibrium value p_{n0}, which results in hole transport toward the collector, governed by the differential equation

$$\frac{\partial^2(p_n - p_{n0})}{\partial x^2} - \frac{p_n - p_{n0}}{L_h^2} = 0 \tag{4.43}$$

which has solutions of the form

$$p_n(x) = p_{n0} + C_1 \exp\left(\frac{x}{L_h}\right) + C_2 \exp\left(-\frac{x}{L_h}\right) \tag{4.44}$$

where C_1 and C_2 are integration constants and p_{n0} is the equilibrium minority

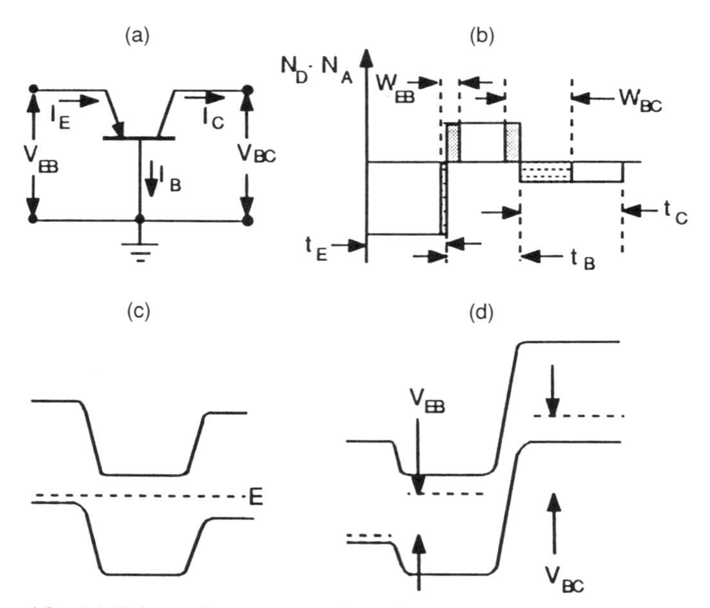

Figure 4.8 (a) Schematic representation of a common base bipolar transistor circuit:; (b) doping profile of a *pnp* transistor; energy-band diagrams at equilibrium (c) and (d) under the conditions of hole injection into the base.

carrier concentration in the base. At the emitter–base junction

$$p_n(0) = p_{n0} \exp\left(\frac{qV_{EB}}{kT}\right) \gg p_{n0} \tag{4.45}$$

because $qV_{EB} \gg kT$, and at the base–collector junction

$$p_n(t_B) \approx p_{n0} \tag{4.46}$$

The current density at the emitter–base junction consists of the injected hole current and a contribution related to electron injection into the emitter

$$J_E = J_{hE} + J_{eE} \tag{4.47}$$

If J_{eE} is small compared to J_{hE}

$$J_E \approx \frac{qD_h}{L_h} p_n(0) \operatorname{cotanh}\left(\frac{t_B}{L_h}\right) \tag{4.48}$$

the base–collector current density is

$$J_C \approx \frac{qD_h}{L_h} p_n(0) \operatorname{cosech}\left(\frac{t_B}{L_h}\right) \tag{4.49}$$

and the base current density is

$$J_B = J_E - J_C \approx \frac{qD_h}{L_h} p_n(0) \tanh\left(\frac{t_B}{2L_h}\right) \tag{4.50}$$

For a narrow base of width $t_B \ll L_h$, Equation (4.50) simplifies to

$$J_B \approx \frac{qD_h}{L_h} p_n(0) \frac{t_B}{2L_h} = \frac{qt_B}{2\tau_h} p_n(0) \tag{4.51}$$

which is very small compared to $J_E \approx J_C \approx qD_h p_n(0)/L_h$. Thus for a very narrow base the common base gain is

$$\alpha_{BT} = \frac{J_C}{J_E} \approx 1 \tag{4.52}$$

and the common emitter gain is

$$\beta_{BT} = \frac{J_C}{J_B} = \frac{\alpha_{BT}}{1 - \alpha_{BT}} \gg 1 \tag{4.53}$$

In the design of homojunction bipolar transistors, a compromise must be made in the choice of the doping levels in the emitter and base regions.

On the one hand, the base doping should be small as compared to the emitter doping to assure a high injection efficiency

$$\gamma_{\mathrm{BT}} = \frac{J_{h\mathrm{E}}}{J_{h\mathrm{E}} + J_{e\mathrm{E}}} \approx 1 \tag{4.54}$$

On the other hand, a low base doping level results in a modulation of the effective base width by the voltage-dependent part of the depletion region in the base and sets a lower limit to the base thickness and thereby an upper limit to the switching speed. Figure 4.9 shows the frequency response of the common emitter and common base gains of a bipolar transistor. The 3 dB points on

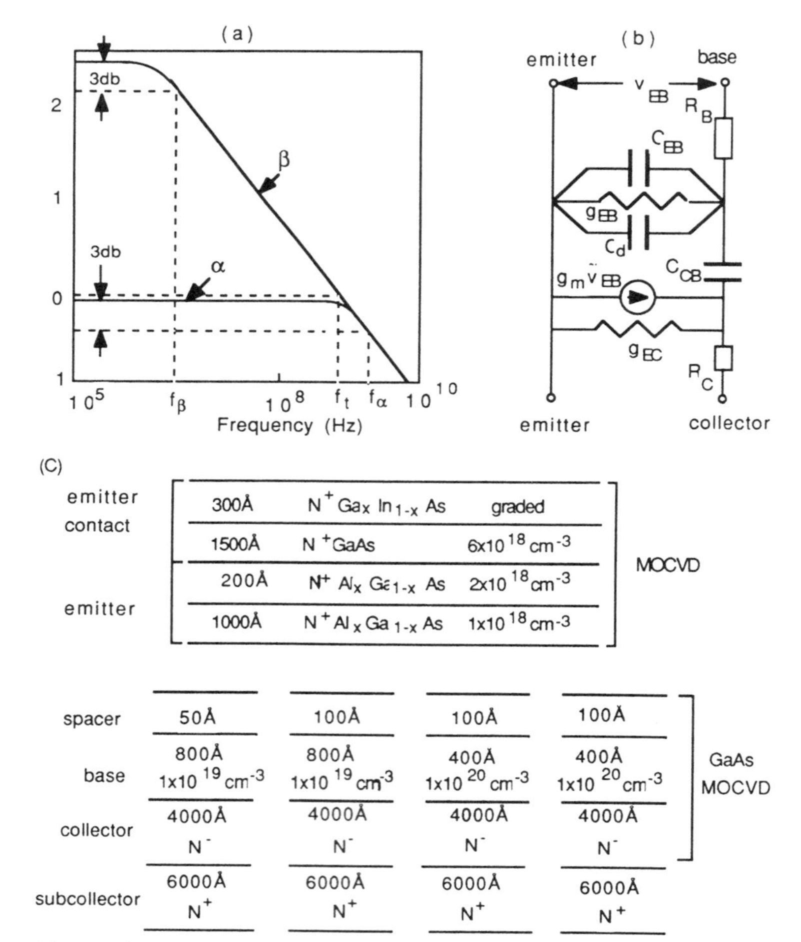

Figure 4.9 (a) Gain–frequency relation and (b) high-frequency equivalent circuit of a bipolar transistor. (c) Schematic representation of a cross section of an $Al_xGa_{1-x}As$/GaAs HBT. After Hobson et al. [7].

these two curves define the common emitter cutoff frequency f_β and the common base cutoff frequency f_α, respectively. An alternative commonly used figure of merit is the unity gain cutoff frequency f_T, which is slightly below the value of f_α. In addition to the cutoff frequency, the transconductance g_m of the transistor

$$g_m = \tilde{i}_c/\tilde{v}_{EB} \tag{4.55}$$

is an important figure of merit since it specifies the relation between the output collector current $\tilde{i}_c$ and the ac voltage $\tilde{v}_{EB}$ at the input terminals. Silicon bipolar transistors with 0.5 μm base aperture achieve at present a gate delay of 21.5 ps/gate at a switching current of 0.32 mA [6].

A significant improvement of the frequency response is achieved by heterojunction base transistors (HBTs), using $Al_xGa_{1-x}As/GaAs$ heterostructures, as illustrated in Figure 4.9(c). The utilization of the wide gap $Al_xGa_{1-x}As$ emitter increases the barrier for the injection of minority carriers from the base into the emitter relative to the barrier for minority carrier injection from the emitter into the base. Thus much higher base doping levels are permitted than for the homojunction bipolar transistor, achieving small base resistance and high injection efficiency [6].

Figure 4.10(a) shows the small-signal equivalent circuit of an *npn*-$InP/Ga_xIn_{1-x}As$ HBT representing the present state of the art [7]. The values of the various elements of this circuit are listed in Table 4.1 along with information concerning the associated time constants and the gain and cutoff frequency. Figure 4.10(b) shows the frequency dependence of the measured and modeled gain and maximum stable power gain (MSG) of the device resulting in a measured unity gain cutoff frequency $f_T = 165$ GHz and a maximum oscillation frequency $f_{max} = 100$ GHz. The emitter–collector delay time

$$\tau_{ec} = R_E C_{je} + (R_E + R_{EE} + R_C)(C_{jc} + C_{FB}) + (\tau_B + \tau_{SCC}) \tag{4.56}$$

for this transistor is 0.97 ps. It is composed of the emitter and collector charging times $\tau_e = R_E C_{je}$ and $\tau_c = (R_E + R_{EE} + R_C)(C_{jc} + C_{FB})$ and the transit times τ_B and τ_{SCC} in the base and collector depletion regions, respectively. The

Table 4.1 Values for the equivalent circuit elements and associated time constants, common-emitter current gain, and cutoff frequencies for the InP/GaInAs HBT of Figure 4.11 (After Chen et al. [8]; copyright © 1989, The Institute of Electrical and Electronics Engineers, New York)

R_{BB}	1.12 Ω	C_{FB}	18.0 fF	τ_c	0.187 ps
R_{EE}	6.19 Ω	C_{jc}	1.0 fF	τ_{ec}	0.965 ps
R_{FB}	11500 Ω	C_{je}	200 fF	f_i	0.318 THz
R_C	2.26 Ω	β	44.5	f_β	3.9 GHz
R_B	122.5 Ω	τ_F	0.5 ps	f_T	165 GHz
R_E	1.39 Ω	τ_e	0.278 ps	f_{max}	100 GHz

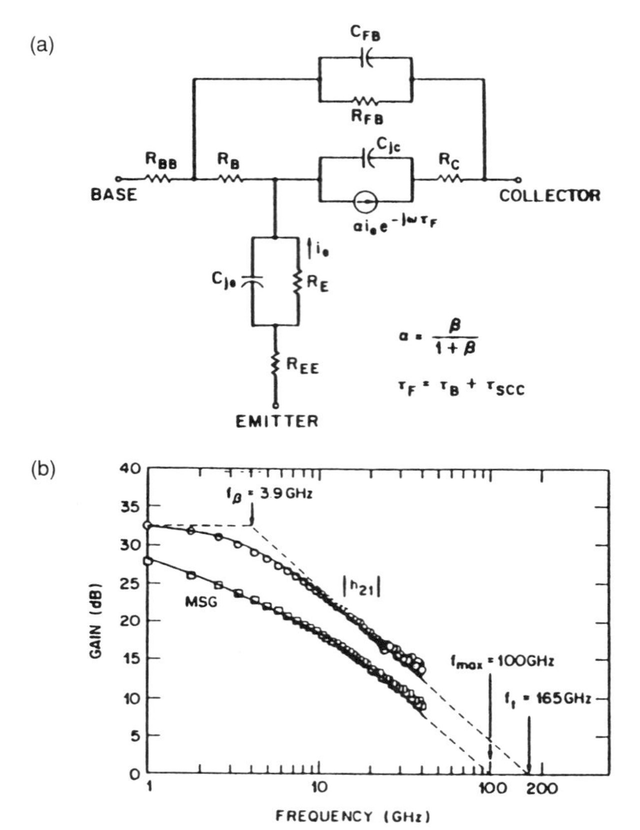

Figure 4.10 (a) Small signal equivalent circuit; (b) Frequency dependence of the measured (open symbols) and modeled (full lines) current gain $|h_{21}|$ and maximum stable power gain (MSG) of a *npn*-InP/$Ga_xIn_{1-x}As$ HBT; (c) Output waveform of a 17-stage NTL-ring oscillator, but with $f_T = 115\,GHz$. After Chen et al. [8]; copyright © (1989). The Institute of Electrical and Electronics Engineers, New York.

power–delay product of ICs is most conveniently evaluated by measuring the output waveform and power consumption of a ring oscillator circuit, which consists of an odd number of inverters connected in series. The output waveform of a 17-stage nonthreshold logic (NTL) ring oscillator using *npn*-InP/$Ga_xIn_{1-x}As$ HBTs with $f_T = 115\,GHz$ and a supply voltage $V_{CC} = 1.4\,V$ exhibits a propagation delay of 14.6 ps and 5.4 mW power consumption per gate. For an emitter-coupled logic (ECL) circuit, a switching delay of 8.8 ps is estimated from the measured equivalent circuit parameters shown in Table 4.1. A delay of 1.9 ps was reported for AlGaAs/GaAs double heterostructure HBTs (DHBTs), albeit at 44 mW power consumption per gate [9]. DHBTs provide complete symmetry of the collector and emitter that are thus interchangeable. This permits exceedingly simple integration schemes in planar structures, as illustrated for the example of the input stage of ECL in Figure 4.11(a) [10].

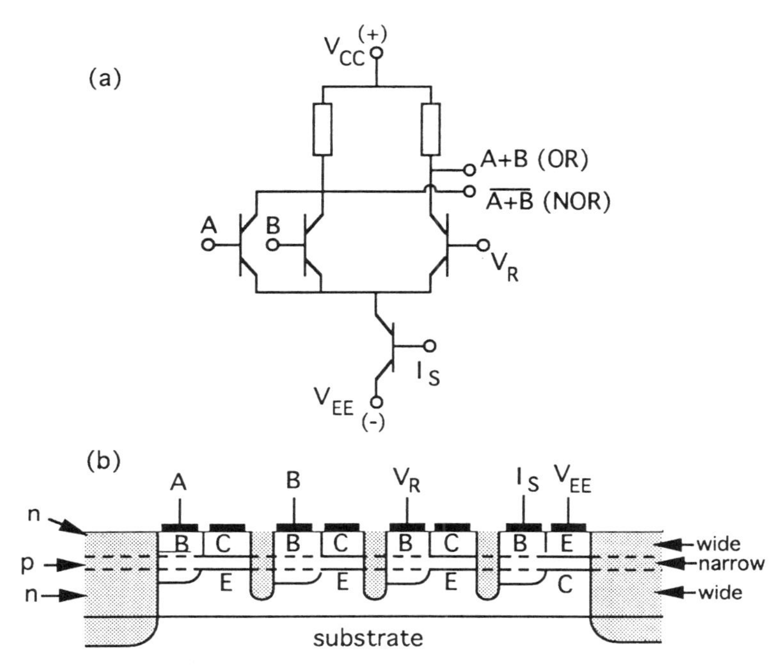

Figure 4.11 (a) Circuit diagram and (b) double heterostructure HBT implementation of the input stage of ECL. After Kroemer [10].

The vertical arrangement of the base region between the emitter and the collector permits to take full advantage of high-resolution heteroepitaxy in minimizing the width of the base. In view of the mature state of silicon technology, considerable interest exists in the utilization of Si_xGe_{1-x} alloys for Si_xGe_{1-x}/Si HBTs [11] and for bipolar inversion channel field-effect transistors (BICFETs) [12]. However, at present, the performance of these devices is inferior to the previously described III–V alloy devices.

4.3 MIS Capacitors and Field-Effect Transistors

The spreading of the depletion layer upon reverse bias of a *pn* junction is utilized for controlling the current flow in a junction field-effect transistor (JFET). Its function is based on the control of the width of the channel between a source and drain contact by the bias on the gate contact, as illustrated in Figure 4.12. If the source is at ground potential, and the drain contact is biased positively with respect to ground, electrons flow from the source to the drain. Since the potential drop across the channel affects the potential under the gate, under the conditions of current flow, the depletion layer spreads out toward

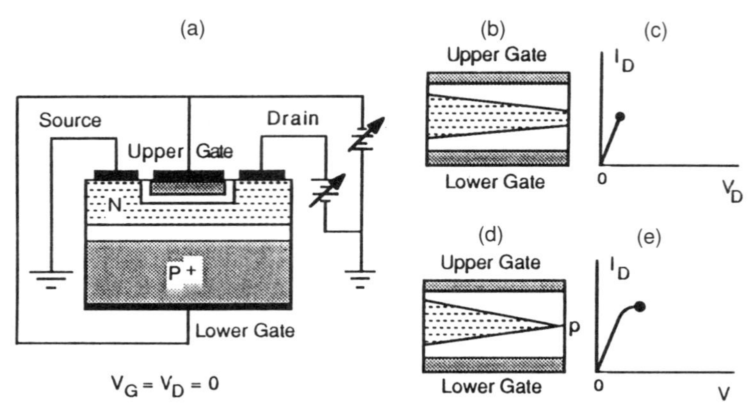

Figure 4.12 (a) Schematic representation of a JFET and its biasing conditions, (a) depletion region under the gate contact (b) and I–V curve, (c) below pinchoff, and depletion region under the gate contact (d) and I–V curve (e) at the pinchoff voltage.

the drain. At low drain voltage, the current increases linearly, but at higher voltages the channel resistance increases due to the increasing constriction of the channel. This leads to a decreasing slope of the I_D–V_D characteristics until at the pinchoff voltage the depletion layer closes the channel at the drain contact. Further increase of the source-to-drain voltage merely pushes the point of closure toward the source and the current saturates. A negative gate bias lowers both the saturation current and the pinchoff voltage. Note that, analogous to the HBT, heterostructure-gate JFETs with a large band-offset at the heterojunction have the advantage of minimizing the increase in the injection current from the gate into the channel with increasing gate bias. Thus larger gate bias can be used without g_m compression.

The MOSFET shown in Figure 1.7 has a similar J–V characteristic as the JFET. It utilizes the excellent properties of the thermal oxide for gate isolation, thus achieving switching by exceedingly small control currents at very high input impedance and low power consumption. Since the function of MOSFETs depends critically on the properties of the metal/insulator/semiconductor (MIS) heterostructure at the gate, we discuss first the properties of MIS junctions and the utilization of MIS capacitors in the context of storage and then address MOSFET technology.

Figures 4.13(a–c) show the energy-band diagrams of MIS junctions under different bias conditions. Choosing a p-type semiconductor with a work function $\phi_s > \phi_M$, the junction is in equilibrium in depletion, as shown in Figure 4.13(b). Forward bias to a voltage $V_{FB} = \phi_M - \phi_s$ achieves flat band conditions. Under even larger forward bias majority carriers are accumulated at the semiconductor–dielectric interface, as shown in Figure 4.13(a). Conversely, at large reverse bias, a condition may be reached where the minority carrier band is closer to the Fermi level than the majority carrier band. In this case, an inversion layer is formed in the semiconductor under the insulator, which is typically 10–100 Å thick. This is shown in Figure 4.13(c). In strong inversion

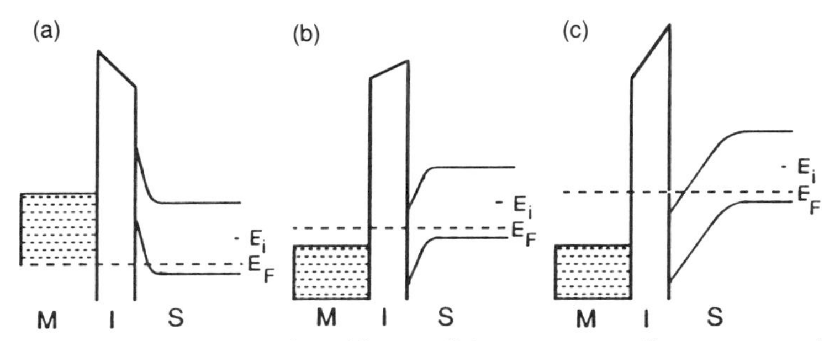

Figure 4.13 MIS junction under various bias conditions corresponding to accumulation (a), depletion (b), and inversion (c).

the concentration of electrons in the inversion layer exceeds by far the concentration of holes.

Under accumulation conditions, the depletion layer in the semiconductor is collapsed; that is, $C_D \rightarrow \infty$. The capacitance of the junction is thus the same as the capacitance $C_i = \varepsilon_i \varepsilon_0 / t_i$ of a plate capacitor consisting of the insulator film between two metal plates, where ε_i and t_i are the dielectric constant and the thickness of the insulator film. In depletion the oxide capacitance is in series with the depletion layer capacitance, so that

$$\frac{1}{C} = \frac{1}{C_i} + \frac{1}{C_D}, \tag{4.57}$$

that is, the capacitance is smaller than in accumulation. In strong inversion the value of the capacitance depends on the frequency. At low frequencies, where the generation/recombination of minority carriers can follow the signal (quasi-static or ac measurements at frequencies < 100 Hz), the differential capacitance is dominated by the charging and discharging of the inversion layer, that is, approaches C_i.

At high frequencies the capacitance–voltage characteristics of an ideal MIS capacitor under accumulation and depletion are identical to the low-frequency characteristics since the charging and discharging associated with the ac signal is governed by a redistribution of majority carriers at the fringes of the depletion layer. The majority carrier response time is of the order of ε_s/σ; that is, for typical conductivities in the $\sigma = 1\,\Omega^{-1}\,\text{cm}^{-1}$ range it is of the order of 10^{-12} s. However, under strong inversion the redistribution of the carriers in the inversion layer is governed by the minority carrier response time, which is many orders of magnitude larger, 0.01–1 s. Therefore, the inversion charge cannot follow the ac signal, and the total capacitance remains at its minimum value since under the influence of the screening by the charge in the inversion layer the depletion layer width does not change very much with increasing bias.

In reality the impedance of MIS devices is modified by the presence of interface trap states that are charged and discharged as the ac signal varies in time. This adds a series combination of a capacitive component C_t and resistive component R_t parallel to the depletion layer capacitance. At low frequency the capacitance is

$$\frac{1}{C_{lf}} \approx \frac{1}{C_i} + \frac{1}{C_s + C_t} \tag{4.58}$$

and the high-frequency C_{hf} capacitance is of the form of Equation (4.57) because the charging/discharging of the interface traps cannot follow the signal. The combination of Equations (4.57) and (4.58) thus allows the determination of C_t according to

$$C_t = \frac{C_{lf} C_i}{C_i - C_{lf}} - \frac{C_{hf} C_i}{C_i - C_{hf}} = qD_t \tag{4.59}$$

where D_t is the trap density in the energy interval $[E, E + dE]$.

Figure 4.14 shows the combined high- and low-frequency $C-V$ characteristics of a Si-MOS capacitor. The density of interface trap levels per unit area per electron volt resulting from an analysis of the data is $2.8 \times 10^{10}\,\text{cm}^{-2}\,\text{eV}^{-1}$. Because the interface trap concentration for Si/thermal oxide interfaces depends on the surface orientation, that is, is an order of magnitude higher for $\{111\}$ as compared to $\{100\}$, the latter orientation is preferred for MOS-type devices and circuits. Recently the difference between $D_t(001)$ and $D_t(111)$ has been narrowed by advances in surface cleaning/conditioning.

Alternatively, the density of interface states may be derived from an analysis

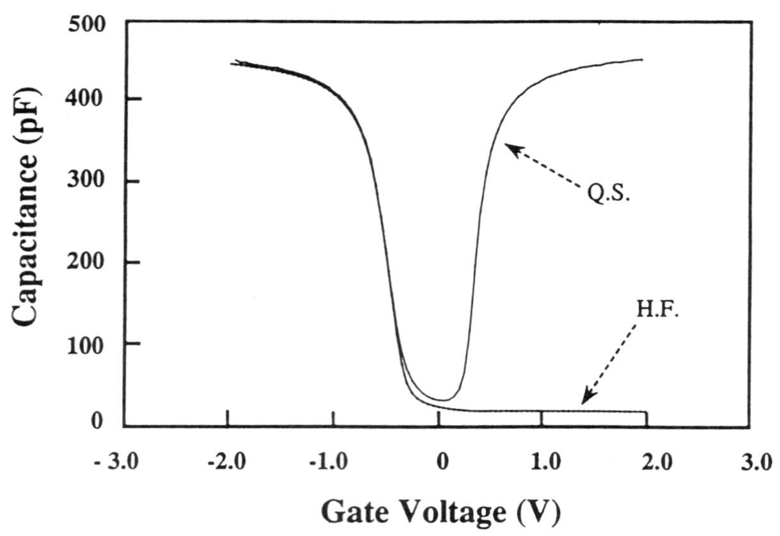

Figure 4.14 $C-V$ characteristics of a Si-MOS capacitor formed by oxide deposition at 200°C followed by in-situ annealing at 400°C for 5 min under UHV conditions. After Lucovsky et al. [13]; copyright © 1990, American Institute of Physics, New York.

of the shape of the high-frequency $C-V$ curve, which is stretched out because charges added to the capacitor are distributed between the surface states and the depletion layer, so that a larger voltage span is needed to drive the device from depletion into accumulation. However, since this analysis requires the knowledge of the undistorted $C-V$ characteristics for the capacitor without interface states (i.e., on a set of assumptions in its calculation), the combination of C_{HF} and C_{LF} data is more reliable.

In addition to interface states, a fixed charge due to excess Si ions may exist in the SiO_2 layer of Si-MOS capacitors, and additional trapped charge may be generated in the course of processing due to radiation damage of the insulator. Also, if the oxide is contaminated by alkali metal ions, a mobile charge may be added that drifts in an uncontrolled fashion and is thus detrimental to the reliability of the device. Generally the addition of charges in the oxide affects the flatband voltage, that is, shifts the $C-V$ curve parallel to the voltage axis. Thermal aging tests under positive and negative bias distinguish between these three mechanisms of charging of the oxide. The fixed charge is not affected by aging treatments and consequently does not cause any shifts in the position of the $C-V$ curve. The annealing of radiation damage shifts in the position of the $C-V$ characteristics toward the position without trapped charge, which is independent of the sign of the bias during aging. Conversely, mobile charges, which are redistributed in different ways under positive and negative bias during thermal treatments, lead to shifts of the $C-V$ curve after aging depending on the sign of the bias. Capacitance measurements, supplemented by conductance measurements, thus offer valuable insights into the properties of MIS junctions [14].

MIS devices that compete with digital transistors in intermediate access time memory applications are charge-coupled devices (CCDs). They are of considerable commercial interest in the context of compact camcorders, high-definition television, and medical imaging systems. The function of CCDs is based on the injection and removal of packets of charge into and out of potential wells established under a set of closely spaced electrodes that are separated from the semiconductor by a thin dielectric film. Filled wells represent the "ones" and empty wells represent the "zeros" of binary coded data. Figure 4.15(a) shows a cross section of a junction-isolated three-phase, two-bit, N-channel CCD, adding an input diode (ID) and gate (IG) and an output diode (OD) and gate (OG) for the injection and detection of the charge packets. Figure 4.15(b) shows the surface potential and charge distribution under the electrodes of the CCD. At time t_1 all electrodes are at a positive bias, repelling majority carriers. Also, a large enough positive bias is applied to ID and OD to prevent inversion under IG, keeping the entire device region under the electrodes in depletion. The voltage $V_{aa'}$ on the contact pair a, a' is larger than $V_{bb'}$ and $V_{cc'}$, and the voltage V_{IG} on IG is such that $V_{aa'} > V_{IG} > V_{bb'} = V_{cc'}$, establishing the surface potential distribution shown in the row labeled t_1 in Figure 4.15(b). At time t_2 the voltage V_{ID} at ID has been lowered and the voltages at IG and contacts a, a' have been raised so that

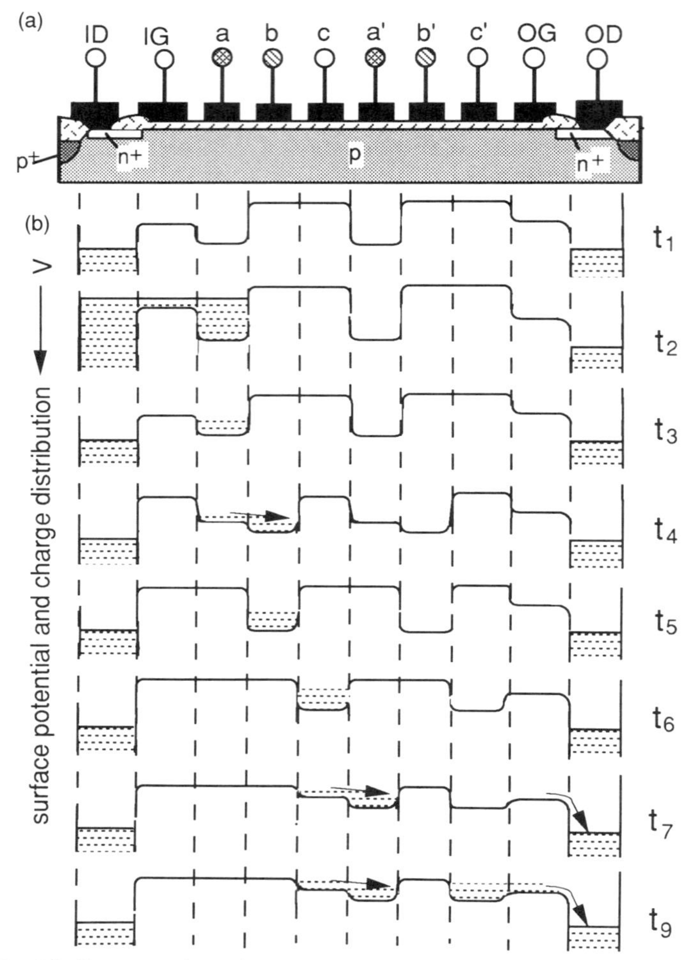

Figure 4.15 (a) Cross section through a three-phase, two-bit, *n*-channel CCD. (b) Input and output signals and clock waveforms for the operation of the CCD. (c) Schematic representation of the potential and charge distributions. After C.-K. Kim [15].

$V_{aa'} > V_{IG} > V_{ID} > V_{bb'} = V_{cc'}$. Electrons residing in the n^+ region of ID under this condition are injected into the region of lower surface potential filling the potential well under contact *a*. At time $t_3 V_{ID}$ has been returned to its original value. Electrons residing under the gate contact and excess electrons under contact *a* thus spill back into ID, leaving a well-defined charge packet under contact *a*. At time t_4, the clock voltage pulses applied to contacts a, a' and b, b', respectively, establish a lower surface potential under *b* and *b'* as compared to *a* and *a'* so that the charge present in the original wells under electrodes *a* and *a'* is transferred into the newly established wells under electrodes *b* and *b'*. At time t_5, this process has been completed. At time t_6 the clock pulses routed to b,b' and c,c' accomplish the transfer of the charge packets into the newly established potential wells under contacts c and c'. At time t_7, the content of the well under contact c is transferred into the newly established well under contact a', and the content of the well under contact c' is pushed out into OD, reading the first of the two bits stored in the device, which is a zero. The next clock cycle starts from the charge distribution at t_8, differing from the initial condition in that the well under contact a' is filled, so that a negative charge packet is read out into OD at the completion of the second cycle at time t_9.

In order to assess which materials properties are critical for building such a circuit, we note that the storage time afforded by a CCD is limited by the filling of the empty wells by internal charge generation and diffusion. It depends thus on the generation rate at traps and surface states in the well region as well as on the rate of diffusion of carriers from the surrounding device regions into the well. Contamination of the silicon by Au, Ni, and Co has been shown to affect the dark current of CCD imagers at concentrations as low as $1 \times 10^9\ \mathrm{cm}^{-3}$ [16]. The trapping of charge in bulk traps and surface states also affects the efficiency of charge transfer during the clock cycle. Large mobility, small width of the MIS stripes, and small capacitance are desirable features to maximize the transfer efficiency. A tradeoff must be made between minimizing the insulator thickness to decrease the effects of fringe fields, which adversely affect the control of the surface potential, and minimizing C_{ox}, which decreases with increasing thickness of the insulator. In particular, the density of slow surface states must be made as small as possible. Surface states that are filled with the well, and emit the trapped charge only slowly when the well is being emptied, continue to emit when the potential no longer provides for charge transfer refilling the supposedly empty well.

Alternatively inexpensive memory cells may be set up using MOSFET technology, which permits more rapid changes in the stored bit pattern than possible with serial access memory circuits such as CCDs. Figure 4.16 shows the circuit diagrams of (a) a CMOS static random access memory (SRAM) cell, and (b) a dynamic random access memory (DRAM) cell. A static RAM provides for permanent storage of information in the bistable states of flip-flop circuits as long as electrical power is provided to the circuit. In case of the DRAM cell the bit value is stored in a capacitor by the presence or absence of charge resulting in a high or low voltage state. The leakage of charge out of the capacitor requires repeated recharging to refresh the memory. In a computer many memory cells are addressably connected to input and output lines. The addressable memory is combined with an arithmetic logic unit (ALU) and I/O devices under the control of a central processing unit (CPU). The CPU contains read-only memories (ROMs) with permanently stored control and microinstruction information, decoders, and logic circuits that enable the timing and control of the interactions. A discussion of these circuits is outside the scope of this book.

An important aspect in determining future materials and processing choices is the dynamics of the transport under the gate contact. Figure 4.17(a) shows the equivalent circuit of an FET, where R_{ds} is the intrinsic source-to-drain

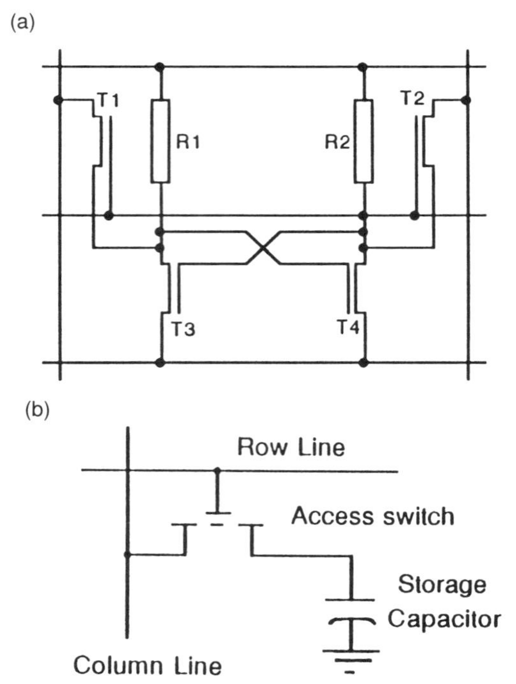

Figure 4.16 Circuit diagrams of (a) SRAM and (b) DRAM cells.

resistance and R_S and R_D are the source and drain resistances, respectively. Since at a short circuit between source and drain, some of the current associated with g_m flows through R_{ds}, the measured short-circuit gain h_{21} is reduced by the factor $\gamma = (R_S + R_D + R_{ds})/R_{ds}$. Also, only the fraction V_{DS}/γ of the applied voltage on the source and drain terminals contributes to the field across the channel $E_{ch} = V_{DS}/\gamma L$. Since for short-channel FETs the transit time of carriers may become comparable to or smaller than the energy relaxation time, carriers injected into the channel at the source with low field mobility μ are accelerated in the field and are collected without establishing steady-state transport conditions. In the limit of very small channel length, ballistic transport is approached, and the effective velocity of the carriers $v_{eff} = 2\pi\gamma L f_T$ [17] is no longer limited by the saturation velocity v_s. A measurement of h_{21} as a function of frequency establishes f_T, which is related to the effective channel mobility by the equation

$$\mu_{ch} = \frac{v_{eff}}{E_{ch}} = \frac{2\pi f_T (\gamma L)^2}{V_{DS}} \tag{4.60}$$

Therefore, a plot of f_T versus the gate length L permits the evaluation of γ and μ_{ch}. Figures 4.17(c) and (d) show plots of $f_T(L)$ and $\mu(L)$ for two sets of E-mode and D-mode $Al_xGa_{1-x}As/GaAs$ heterostructure-insulated-gate field-effect transistors (HIGFETs). The D-mode HIGFETS have lower channel

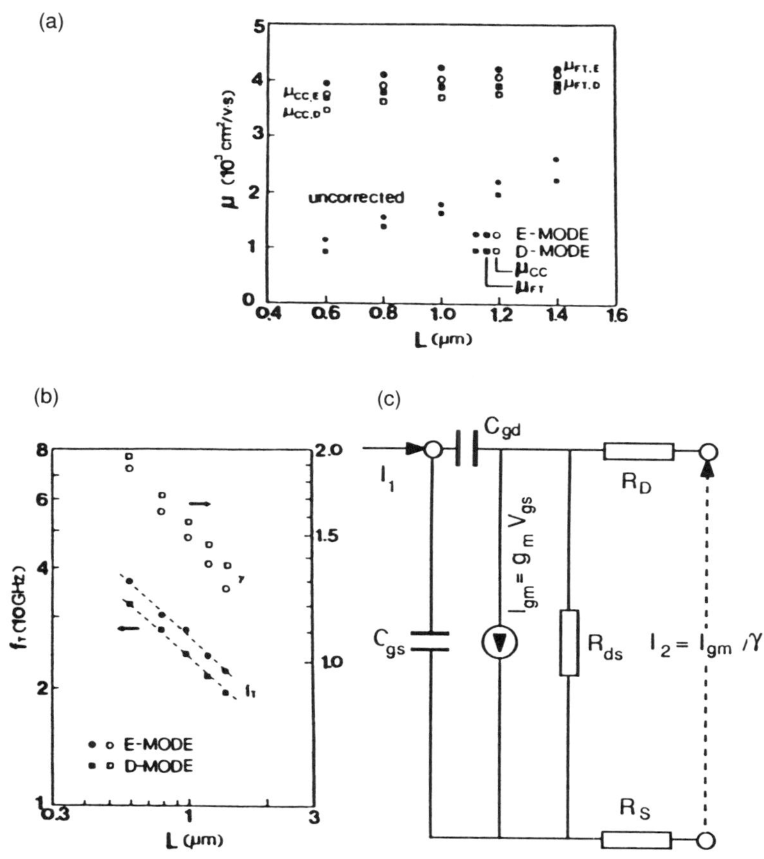

Figure 4.17 (a) Channel mobility μ_{ch} versus channel length L, (b) cutoff frequency f_T versus channel length L, and (c) equivalent circuit of an FET. After Sun et al. [17]; copyright © 1990, The Institute of Electrical and Electronics Engineers, New York.

mobilities because of the implantation of scattering impurities into the channel. An independent evaluation of the channel mobility by capacitance and transconductance measurements on the same sets of transistors gives similar results [labeled μ_{CC} in Figure 4.17(a)] as the f_T method. At $L \geqslant 0.5\,\mu$m, μ_{ch} decreases only weakly with decreasing channel length. The cutoff frequency can thus be engineered by down-scaling, provided that the supply voltage is also scaled because of a steep decrease of μ_{ch} with increasing field at high values ($\geqslant 10^4$ V/cm). Figure 4.18 shows the SEM image of an ultrashort channel MOSFET, employing the design of Figure 1.7, and the output waveform of a 21-stage ring oscillator, operating at 77 K and $V_{DD} = 1.7$ V. The transconductance of this MOSFET is 940 μS/μm at a gate length of 70 nm and a gate delay of 13.1 ps/stage [7] of Chapter 1.

Figure 4.19 shows an evaluation of various effects that limit the feature size of DRAM memory, such as source-drain punch-through, increased subthres-

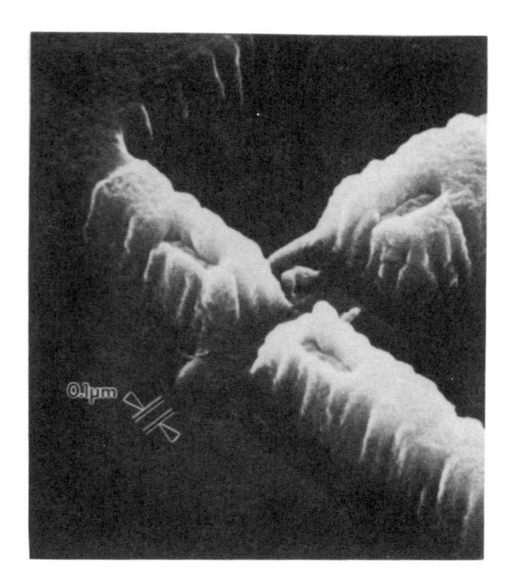

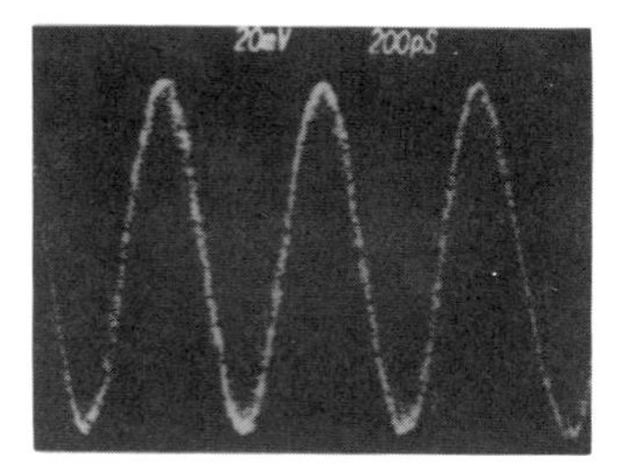

Figure 4.18 (a) SEM image of an ultrashort-channel Si-MOSFET. (b) Output waveform of a 21-stage ring oscillator circuit based on this transistor design. After Kern [7] of Chapter 1; copyright © 1990, Springer Verlag, Berlin.

hold current, the threshold-voltage shift due to hot carrier injection into the gate oxide, soft errors due to radiation damage by high-energy particles, the retention time due to leakage of charge, and the noise margin of the sense amplifier. They all scale to some extent with the supply voltage, in particular, the hot carrier effects that are becoming limiting upon reduction of the feature size at constant voltage which increases the field in the channel region above a critical value. Further discussions of the materials processing problems

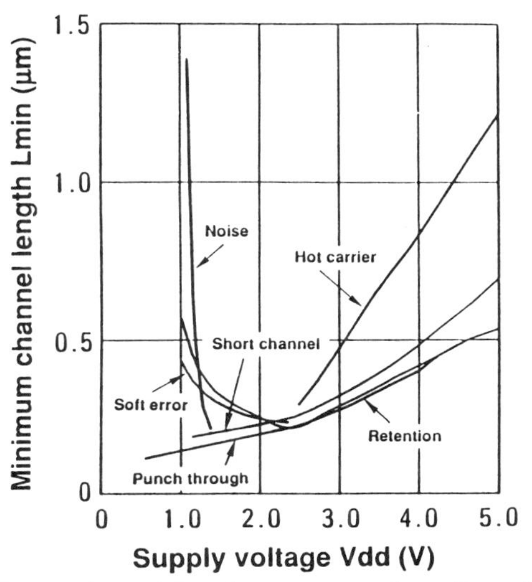

Figure 4.19 Minimum channel length versus supply voltage for MOSFETs used as transfer gates of DRAMs. After Sugano [18]; copyright © 1993, Japanese J. Appl Phys, Tokyo.

associated with these changes in the circuit design and information concerning the strategies that are developed at present to solve these problems are provided in Chapters 7–9.

4.4 A CMOS Processing Sequence

In order to give an overview of the processing steps of the manufacturing of CMOS circuits, we consider here the well-established twin-tub process, using the CMOS inverter and SRAM circuits as examples. The circuit diagrams and schematic representations of cross sections through these circuits are shown in Figures 1.8(a), 4.16(a), 4.20(a) and (b). A summary of the process is shown in Table 4.2 (see Refs. [17] and [19] for details). It starts with a chemomechanically polished *n*-type silicon wafer of usually {100} orientation (see Chapter 5). In order to condition the wafer surface for the epitaxy step 2, a cleaning procedure is carefully executed, including two solvent cleaning cycles (SC1 and SC2) to remove organic matter, and a wet chemical etching step in buffered HF solution to remove residual oxide and metal impurities, and to generate a hydrogen-terminated Si(100)-1 × 1 surface (see Chapter 7). The subsequent epitaxial step has the purpose of providing a reproducible highly perfect, high-purity n^--Si layer in which the circuits are built.

Step 3 establishes a layer of SiO_2 on the surface that is used as a mask for the subsequent definition of the *n*- and *p*-tubs by photolithography (see Section 7.1). There are altogether 10 lithographic pattern definition steps in this

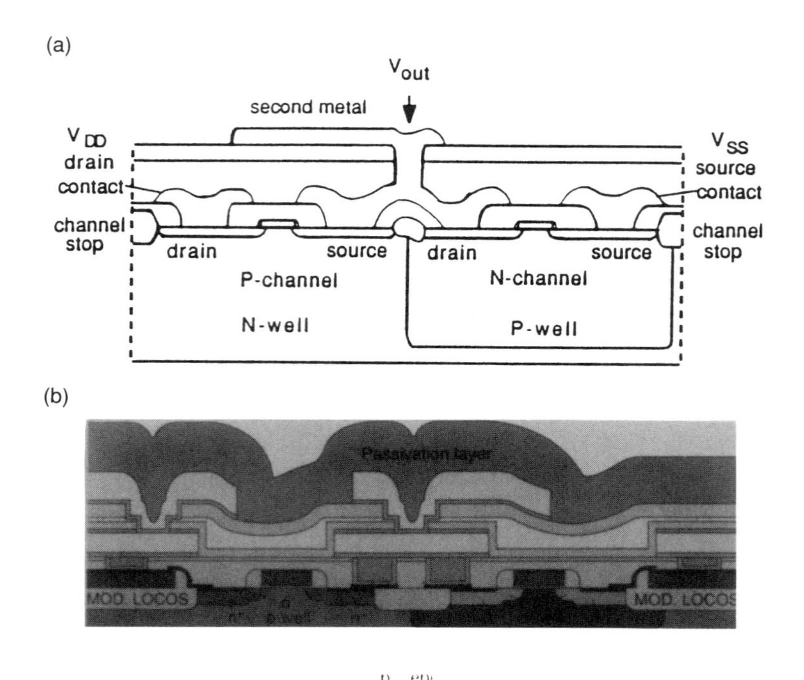

Figure 4.20 Schematic representation of cross sections through CMOS (a) inverter and (b) SRAM circuits. (a) After Maly [18]. (b) After Kramer [19]; copyright © 1989, The Electrochemical Society, Pennington, NJ.

sequence. They consist of spinning a film of photoresist onto the SiO_2-covered wafer, drying the resist layer, exposing selected areas of the surface to light through a mask and development. During development the resist is removed from certain areas of the wafer surface, which allows the opening of windows in selected parts of the SiO_2 layer by etching. Next the boron acceptors that dope the *p*-well region are implanted into the Si areas not covered by the SiO_2/photoresist films (see Section 8.2). The photoresist is stripped after this step. In the following annealing step 6 the implanted B is driven deeper into the Si, removing at the same time the radiation damage to the Si lattice created in the ion implantation step. Also a layer of thermal oxide is grown as part of this annealing procedure over the entire wafer surface. Then the wafer is coated with photoresist, which is exposed through a second mask and development to define the areas where the *n*-tubs are to be established. Arsenic is implanted to provide the *n*-type dopant for the *n*-tub, and the photoresist is removed after the implantation step. A thin film of thermal oxide is grown and at the same time the arsenic is driven into the Si and the radiation damage introduced in step 8 is annealed out.

In step 10 Si_3N_4 is deposited by a vapor deposition method discussed in

Table 4.2 A CMOS processing sequence

1. Wafer cleaning	→ 2. Silicon epilayer growth	→ 3. Thermal oxide growth
→ 4. Definition of *p* well	→ 5. Boron implantation	→ 6. Wafer annealing
→ 7. Definition of *n* well	→ 8. Arsenic implantation	→ 9. Thermal oxide growth
→ 10. Si_3N_4 deposition	→ 11. Tub boundary definition	→ 12. Boron implantation
→ 13. Field oxide growth	→ 14. Si_3N_4 + oxide etching	→ 15. Gate oxide growth
→ 16. Gate metal deposition	→ 17. Gate contact definition	→ 18. Gate contact fabrication
→ 19. Definition of n^+ regions	→ 20. Phosphorus implantation	→ 21. SiO_2 deposition
→ 22. Gate side wall definition	→ 23. Gate side wall etching	→ 24. Oxide growth
→ 25. Window definition	→ 26. Etching of windows	→ 27. Arsenic implantation
→ 28. Definition of windows	→ 29. Etching of windows	→ 30. Boron implantation
→ 31. SiO_2 deposition	→ 32. Contact window definition	→ 33. Etching of windows
→ 34. Aluminum deposition	→ 35. Definition of first metal	→ 36. Etching of first metal
→ 37. Polyimide deposition	→ 38. Planarizing etch	→ 39. SiO_2 deposition
→ 40. Via definition	→ 41. Second metal deposition	→ 42. Definition of second metal
→ 43. Second metal etching	→ 44. *p*-Glass deposition	→ 45. Laser scribing
→ 46. Dicing into chips	→ 47. Wire bonding	→ 48. Packaging

more detail in Section 6.3. It is then patterned by a third photolithographic step to expose the boundaries of the tubs. Boron implantation into this boundary region establishes a high *p*-type dopant concentration on the outside perimeters of the *p*-tub, thus preventing the inversion of the surface. It stops the *n*-channel formed under inversion conditions in the *p*-well. Thick ridges of field oxide (FOX) are grown in a subsequent oxidation step on the areas of the wafer not covered by the Si_3N_4 that passivate the borders of the tubs. Since SiO_2 has a lower density than the Si, a swelling accompanies the oxide growth, which renders the initially flat surface of the wafer uneven. Each selective area oxidation/oxide removal step enhances the surface relief. In order to relieve the strain the oxidation must be carried out at sufficiently high temperature to allow viscous flow in the SiO_2. The FOX growth process is the last high-temperature treatment of the wafer. Therefore, it establishes the final extent of the *n*- and *p*-well penetration into the n^- epilayer. A problem that must be prevented is the latchup of the gate and drain in one tub and the source in a neighboring tub, forming parasitic *npn* and *pnp* bipolar transistors. It is reduced or totally eliminated by etching trenches around the perimeters of the tubs and filling them with SiO_2 or by the replacement of the Si substrate by an insulator (see Chapter 8). In step 14 the Si_3N_4 and remaining thin oxide regions are removed by etching to produce a clean Si surface for the subsequent critical gate oxide growth.

A thin film (250–500 Å for VLSI circuits) of gate oxide (GOX) is grown under carefully controlled conditions, and the gate contact metal degenerately doped poly-Si, metal silicide) is deposited. Since the gate oxide growth requires only a short time, and the metal deposition occurs at low temperature, the dopant distribution is not affected significantly by this step. Then the gate metallization pattern is defined by photolithography, and all excess gate metal/oxide is removed by etching. This fourth photolithographic step defines to a large extent the channel length since the gate is used as a mask for the self-aligned implantation of the source and drain regions.

After masking most of the *n*-well and the outer portions of the *p*-well by photoresist, a shallow implant of *p* is applied that defines the outer edges of the source and drain regions of the NMOS transistors. Simultaneously the channel-stop part of the *n*-tubs is established. In order to fabricate the gate side wall spacers, a film of SiO_2 is deposited by a low-temperature process and is patterned and etched, leaving a thin film of SiO_2 at the perimeter of the gate. In step 24 a film of oxide is grown and covered by photoresist, which is patterned to provide windows in the source and drain regions of the NMOS devices. Arsenic is implanted through these windows to decrease the sheet resistance in the source and drain regions of the NMOS devices. The previously established gate side wall spacers assure that the edges of the source and drain regions remain lightly doped. In steps 28–30 the p^+ source and drain regions of the PMOS transistors are established.

Then SiO_2 is deposited by a low-temperature CVD method and is patterned by photolithography to establish windows for the contacting of the source and drain regions. Subsequently (step 34) a film of aluminum is deposited and patterned to form the first metal contacts and interconnects. Then another thick film of SiO_2 is deposited. In view of the many localized oxidation, deposition, and etching steps, at the last stages of processing, the surface of the IC assumes a rather complex topography [e.g., Figure 4.20(b)]. Therefore, a thick film of photoresist (e.g., polyimide) is deposited that is etched by a dry etching method (see Section 7) with a similar rate as SiO_2, establishing a planarized surface. This surface is further smoothed by the deposition of another layer of oxide. In step 40 vias in the oxide are defined by photolithography and are opened by etching to establish connecting paths to the first metal. Aluminum is deposited and patterned to establish the second metal interconnect lines of the circuit. As a final step of the IC processing, a layer of *p*-glass is deposited to seal the entire circuit. Laser scribing, dicing into chips, wire bonding, and packaging of the chip in a ceramics/polymer enclosure are added to manufacture ICs in a mechanically protected package that provides for optimum heat dissipation. The package must also protect the circuit with regard to moisture and other sources of corrosive environmental attack.

4.5 Microwave Devices and Circuits

Microwave devices and circuits find important applications in communications, materials processing, and consumer products. There are two alternative ways in which active microwave devices are combined with passive circuit elements:

1. Hybrid circuits, where separately fabricated discrete microwave devices are attached to custom-made microwave matching circuits.
2. Monolithic microwave integrated circuits (MMICs), where the active and passive components are integrated on the same chip produced by microelectronics technology.

The development of MMICs is driven by the need for high microwave power at high frequencies in microwave communications. In particular, satellite communications requires miniaturized high-power microwave circuits. MOS transistors offer advantages in high-power applications because they are majority carrier devices with a negative temperature coefficient, that is, decreasing conductance with increasing current. MOS transistors are thus more reliable than bipolar transistors that have a positive temperature coefficient so that the current at hot spots increases, leading to thermal runaway conditions. Figure 4.21 shows a schematic representation of a conventional vertical DMOS power transistor. It consists of many n^+p junctions in parallel, feeding through short n-channels into a large-area backside drain, which optimizes the power distribution and dissipation of heat. Current projections are that in the year 2000, the maximum frequencies of commercial small-signal and 1 W power amplifiers employing single-transistor designs will reach the 100 and 50 GHz ranges, respectively.

Examples of conventional discrete microwave devices are the transferred electron devices (TEDs), the impact avalanche transit time (IMPATT) diode, the trapped plasma avalanche triggered transit (TRAPATT) diode, the barrier injection transit time (BARITT) device, and the tunnel diode. TEDs are based on the scattering of hot electrons into secondary conduction-band minima near the boundaries of the first BZ of, for example, GaAs and InP. The effective masses associated with these secondary conduction-band minima are considerably larger than for carriers residing in the lowest conduction-band minimum at the zone center. Since, at high fields, an increasing number of hot electrons exist that slow down upon transfer into the heavy-mass conduction-band minimum by scattering, a negative slope results in the velocity-field characteristics above a threshold voltage.

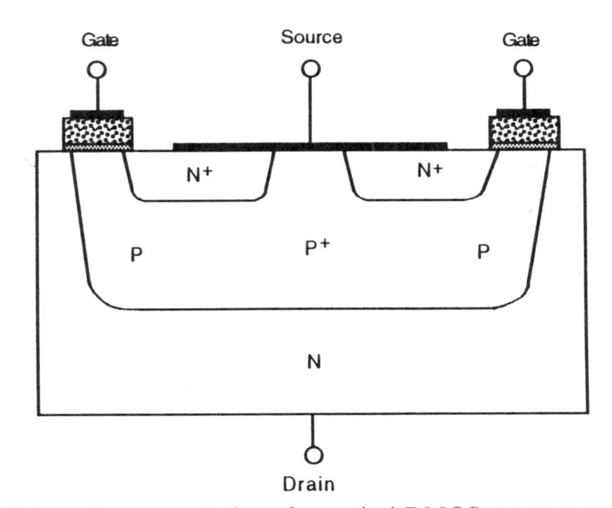

Figure 4.21 Schematic representation of a vertical DMOS power transistor.

A schematic representation of the velocity-field characteristics is shown in Figure 4.22(a). The negative resistance region provides a mechanism for positive feedback and consequently instability with regard to spontaneous microwave oscillations. Typical peak velocities and threshold fields are $v_p = 2.2 \times 10^7$ cm/s, $E_t = 3.2$ keV, and $v_p = 2.5 \times 10^7$ cm/s, $E_t = 10.5$ keV for GaAs and InP, respectively. Figure 4.22(b) shows the conversion efficiency and CW power output of a high-quality n^+nn^+ – InP TED operating at 0.7 A dc current at 4.8 V at a fundamental frequency of ~90 GHz with an output power of 50–70 mW [20]. This diode delivers 7 and 0.2 mW output power in the second- and third-harmonic mode operation at 180 and 272 GHz, respectively [21].

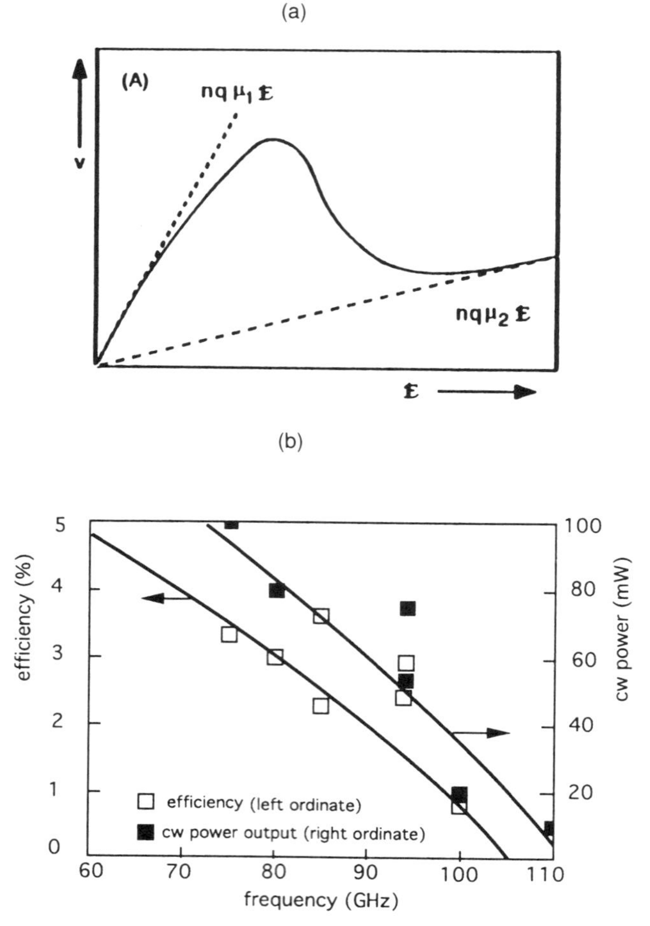

Figure 4.22 (a) Schematic representation of the velocity-field characteristics, and (b) efficiency versus frequency of an InP TED. (b) After diForte-Poisson et al. [20].

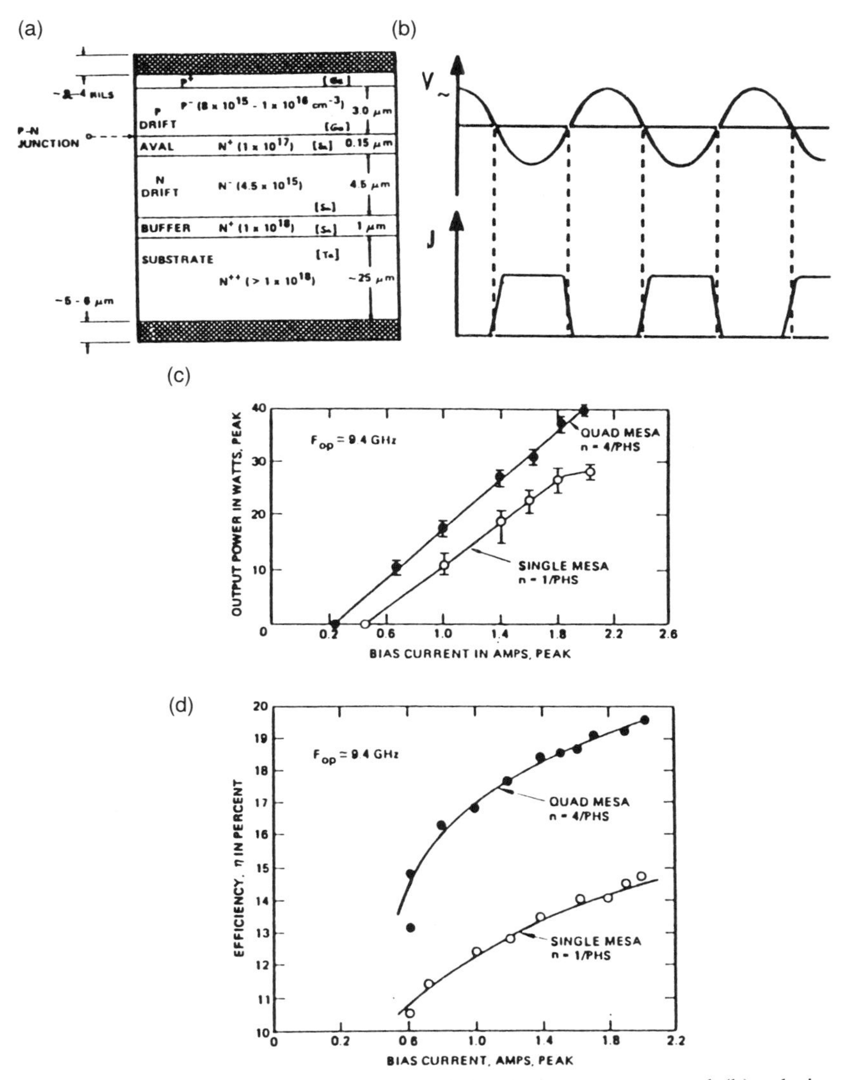

Figure 4.23 (a) Schematic representation of the device structure and (b) relationship between the rf voltage across an IMPATT diode and its current output. (c) Output power and (d) efficiency versus bias current. (c), (d) After P. K. Vasudev [22]; copyright © 1984, The Institute of Electrical and Electronics Engineers, New York.

Single-drift IMPATT diodes consist of a p^+nn^+ structure operated under sufficiently reverse bias that avalanche multiplication occurs. A variety of modifications in the doping profile across the device, improving its dc-to-rf power conversion efficiency and output power, have been made, for example, the addition of an almost intrinsic drift region in the Read diode, the p^+pnn^+ double drift design, and the lo–hi–lo structure. All these devices have in common that the transit time of the electronic charge package traveling through the drift region produces a phase lag of 180° between the rf voltage across the diode and the output current, as shown in Figure 4.23(b). Mounted inside an appropriately designed microwave cavity and biased to the operating dc voltage, the IMPATT diode is thus driven spontaneously by electrical noise into microwave oscillations. Figures 4.23(c) and (d) show the output power and efficiency of dc-to-microwave power conversion as a function of the bias current for a specific device design. Because of its low cost, high power output, and reliabilty, the IMPATT diode is widely employed in microwave communications. A disadvantage of the device is the statistical nature of the

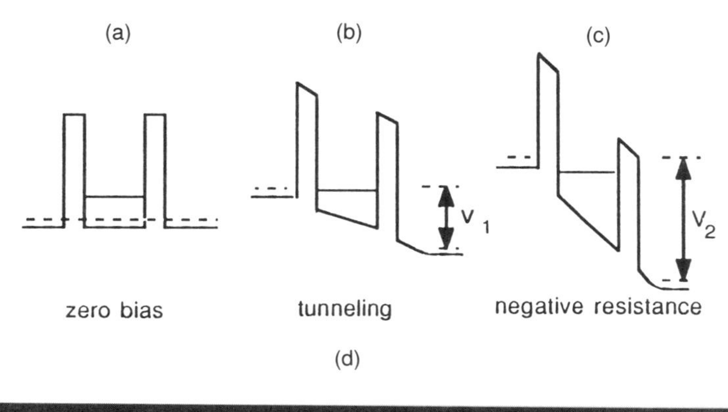

(d)

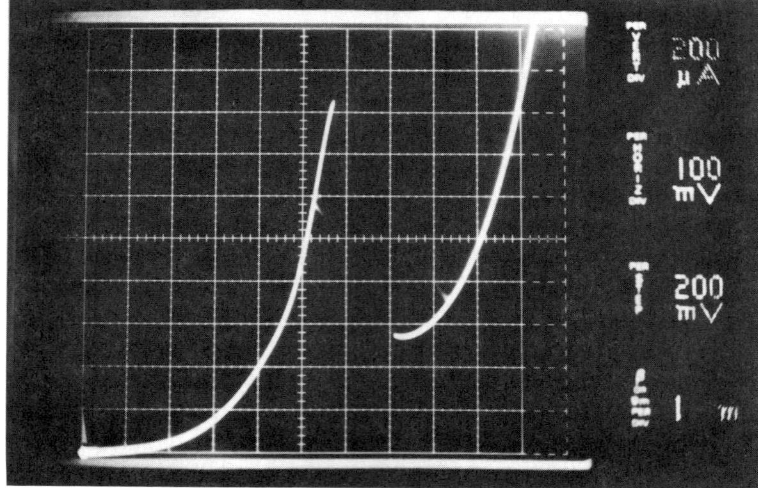

Figure 4.24 (a)–(c) Band diagram of a resonant tunneling diode at different bias conditions. (d) Current–voltage characteristics of a typical resonant tunneling diode. (d) After Watanabe et al. [25]; copyright © 1992, Institute of Electrical and Electronics Engineers, New York, NY.

avalanche process, generating noise in the microwave signal. The TRAPATT diode is similar in construction to the IMPATT diode, but operates by a different principle. It is more efficient and finds application in phased-array radar systems. The BARRITT device is a p^+np^+ structure and competes with TEDs in specialty applications, such as security systems and short-range communications.

Tunnel diodes are heavily doped p^+n^+ junctions that are characterized by a very narrow depletion layer, bringing the conduction band of the n^+ region into sufficiently close proximity of the valence-band edge of the p^+ region that majority tunneling can ensue. This provides a mechanism for a negative resistance region in the J–V characteristics and is utilized for low-power microwave generation. Recently resonant tunneling through quantum wells has been utilized to realize microwave generation at 200 GHz [22] and detection at frequencies $\leqslant$2.5 THz [23]. The principle of operation of a resonant tunneling diode (RTD) is illustrated in Figure 4.24(a)–(c). Figure 4.24(b) shows the current–voltage characteristics of an RTD. Tunneling of carriers through the barriers into the quantum-well states becomes possible at bias voltage V_1, where resonance is established between the energy positions of the quantum-well states and of filled electronic states in the conduction band of the highly doped GaAs source layer. Thus the current increases in the range $0 < V_1$ with increasing bias voltage. However, upon further lowering of the energy position of the quantum-well state below the conduction band edge of the source, the current decreases, and at bias V_2 tunneling through quantum-well states is no longer possible. Therefore, the current must decrease with increasing voltage upon exceeding a threshold $V_1 < V_{th} < V_2$, corresponding to a region of negative resistance [26]. At large bias $V \gg V_2$, the current increases due to hot-carrier transport across and through the very thin triangular part of the biased barriers. If more than one quantum-well state is available, multiple peaks are observed in the J–V characteristics of the RTD at well-defined threshold bias for resonance. RTDs in conjunction with advanced transistor designs based on confined heterostructures open new avenues to circuit design that are discussed in more detail in the next section.

4.6 Confined Heterostructures

A key parameter in the discussion of confined heterostructures is the offset between the valence-band edges of the constituent semiconductors, which is not easily accessed by theory. Empirical knowledge suggests that the valence-band offsets between two materials A and B can be represented as the difference between two energies $E(A) - E(B)$, and a number of attempts have been made to relate these energies to specific models [27]–[33]. The differences between the results of such calculations that are based on different models are at present of the order of 10–20%.

Table 4.3 shows a few examples for selected semiconductor combinations and models. The dielectric midgap energy (DME) model [26] relates the valence-band offset to the Penn gap energy $\hbar\omega_G$, that is, the static dielectric constant ε and the lattice parameter a_0, according to

$$\varepsilon = 1 + \left(\frac{\omega_p}{\omega_G}\right)^2 \tag{4.60}$$

where

$$\omega_p = \frac{128\pi^2 q^2}{m a_0^3} \tag{4.61}$$

is the plasma frequency. Although the results obtained for this model and the calculations of Ref. [28] agree within the above error limits for many heterostructures, in some important cases they do not. Comparisons with the experimental data do not provide for a clearcut preference. Note that, with few exceptions, the experimental data are also subject to considerable error. Therefore, detailed studies on specific well-characterized confined heterostructures are essential for determining reliable energy-band diagrams and for testing of competing theoretical models. Such data are currently becoming available. According to the valence-band offsets, there are two types of superlattices: type I, where the potential wells for electrons and holes are in the same material, and type II, where the confinement states in the valence band and in the conduction band reside in different materials. This is illustrated in Figure 4.25.

For thick quantum-well heterostructures the confined states can be adequately described by the effective-mass approximation. Taking $E_g(A) < E_g(B)$ and the z axis of a Cartesian coordinate system as the confinement direction, the electronic wavefunctions are then represented by a sum over Bloch waves multiplied by an envelope function $\chi_n(\vec{r})$ [34], that is, at the zone center,

$$\Psi = \sum_n \chi_n^{A,B}(\vec{r}) u_{n0}^{AB}(\vec{r}) \tag{4.62}$$

Table 4.3 Valence-band offsets ΔE_V (eV) for selected materials combinations:

DME model [27] (data of Tersoff [28])

Material	Exp.	DME	Tersoff	Material	Exp.	DME	Tersoff
GaP/Si	0.8	0.41	0.45	GaAs/InAs	0.17	0.37	0.20
Si/GaAs	—	0.01	−0.14	ZnSe/GaAs	0.96	1.04	1.20
AlAs/Ge	0.95	0.89	0.87	InAs/GaSb	0.46	0.20	0.43
GaAs/Ge	0.56	0.47	0.32	CdTe/InSb	0.87	0.89	0.84
ZnSe/Ge	1.52	1.51	1.52	CdTe/*a*-Sn	1.0	1.22	—
AlAs/GaAs	0.42/0.55	0.43	0.55	CdTe/HgTe	0.35	0.35	0.51

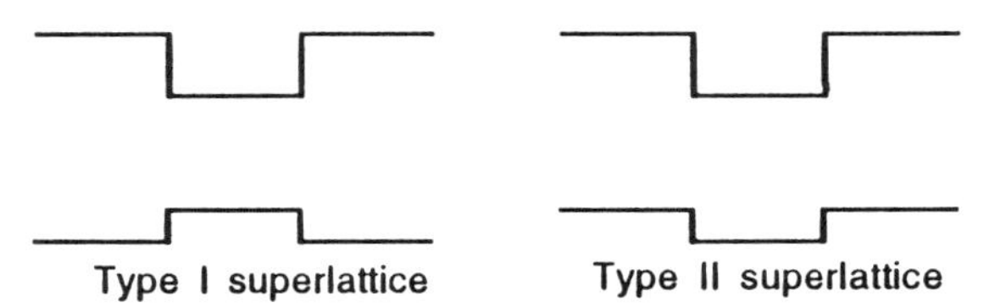

Figure 4.25 Schematic representation of the conduction- and valence-band edge modulation in type I and type II MQW heterostructures.

The $\chi_n(z)$ are solutions to the eigenvalue equation

$$\left(-\frac{\hbar^2}{2m_e^*(z)} \frac{\partial^2}{\partial z^2} + V_c(z) \right) \chi_n(z) = E_{nz}(z) \tag{4.63}$$

where $V_c(z)$ and E_{nz} are the position of the conduction-band edge and the confinement energy eigenvalues, respectively.

For a square well of finite depth, inside the well (i.e., at $|z| \leqslant L_z/2$), $\chi_n(z)$ is either a cosine or sine function, depending on n being odd or even. The energy eigenvalues are given by

$$E_{nz} = E_0 + \frac{\hbar^2 |\hat{\mathbf{k}}|^2}{2m_{eA}^*}, \tag{4.64}$$

where $\hat{\mathbf{k}}$ is the transverse wave vector E_0 is the conduction-band energy at the bottom of the well, which is frequently chosen to be zero. Outside the well $\chi_n(z)$ decays exponentially, and at the interfaces, the boundary conditions

$$\chi_n^A(\pm L_z/2) = \chi_n^B(\pm L_z/2) \tag{4.65}$$

and

$$\frac{1}{m_{eA}^*} \frac{\partial \chi_n^A(\pm L_z/2)}{\partial z} = \frac{1}{m_{eB}^*} \frac{\partial \chi_n^B(\pm L_z/2)}{\partial z} \tag{4.66}$$

must be satisfied, which determines the quantized wave vectors and energies, as discussed for free electrons in an one-dimensional potential well in Chapter 2.

In view of the quantization of E_{nz}, the electronic motion in the z direction is restricted. However, a continuous band of energy eigenvalues

$$E_n = E_{nz} + \frac{\hbar^2}{2m_{eA}^*} (k_x^2 + k_y^2) \tag{4.67}$$

exists, so that the motion of the electrons parallel to the interfaces of the quantum well is not restricted.

The modulation of the valence-band edges is in the case of GaAs-$Al_xGa_{1-x}As$ quantum wells substantially smaller than the conduction-band modulation. Nevertheless, the perturbation term added to the Hamiltonian due to the valence-band-edge modulation has significant effects on the dispersion and lifts the degeneracy of the $J = \frac{3}{2}$ and the $J = \frac{1}{2}$ bands at Γ because they have different effective masses. [35].

For very thin quantum wells, for example, in the case of an $(AlAs)_i$-$(GaAs)_i$ MQW heterostructure with $i < 10$, the effective-mass approximation is not valid. SCF pseudopotential calculations for such structures show that for $i > 2$ the lowest conduction-band state C_1 is located in the AlAs, while the highest valence-band state V_1 is located in the GaAs [36]. This is illustrated in Figure 4.26(top) for a $(GaAs)_4$-$(AlAs)_4$ superlattice. Figure 4.26(bottom) shows the charge-density plot for the first four conduction-band states C1–C4. C1 and C4 are associated with the zone center, and C3 and C4 are associated with the M point of the superlattice BZ. For $i > 10$ the above-discussed type I behavior is observed.

Also, important deviations of the density of states of quantum-well heterostructures from the energy dependence of the density of states of unconfined semiconductors exist, which is illustrated in Figure 4.26. For confinement in one dimension the density of states is

$$N(E) = \sum_n \frac{d_n m_e^*}{\pi \hbar^2} H(E - E_{nz}) \tag{4.68}$$

where d_n is the valley degeneracy and $H(E - E_{nz})$ is the Heaviside step function, adding at each eigenvalue of the confinement energy the constant contribution $d_n m_e^* / \pi \hbar^2$ to the conduction-band density of states. If only the lowest sub-band is occupied, the sheet concentration of the 2D electron gas in the well is

$$N_s = \frac{g_n m_e^*}{\pi \hbar^2} (E_F - E_0) \tag{4.69}$$

In modulation-doped heterostructures, the spillage of carriers from the doped confinement layers into the wells results in a high electron concentration in the wells having exceptional mobility [37].

Since the electron affinities χ_e and bandgaps of $Al_xGa_{1-x}As$ and GaAs provide for a negative spike in the conduction bands at the n^+-$Al_xGa_{1-x}As$/p^--GaAs interface, an approximately triangular quantum well is formed without need for the fabrication of a double heterostructure [38]. This quantum well is filled with electrons from the highly doped $Al_xGa_{1-x}As$. Depending on the relative positions of E_F and the first sub-band, MODFETs utilizing this effect may be either normally on or normally off so that both D- and E-mode devices may be built. The first ICs employing HEMTs have been

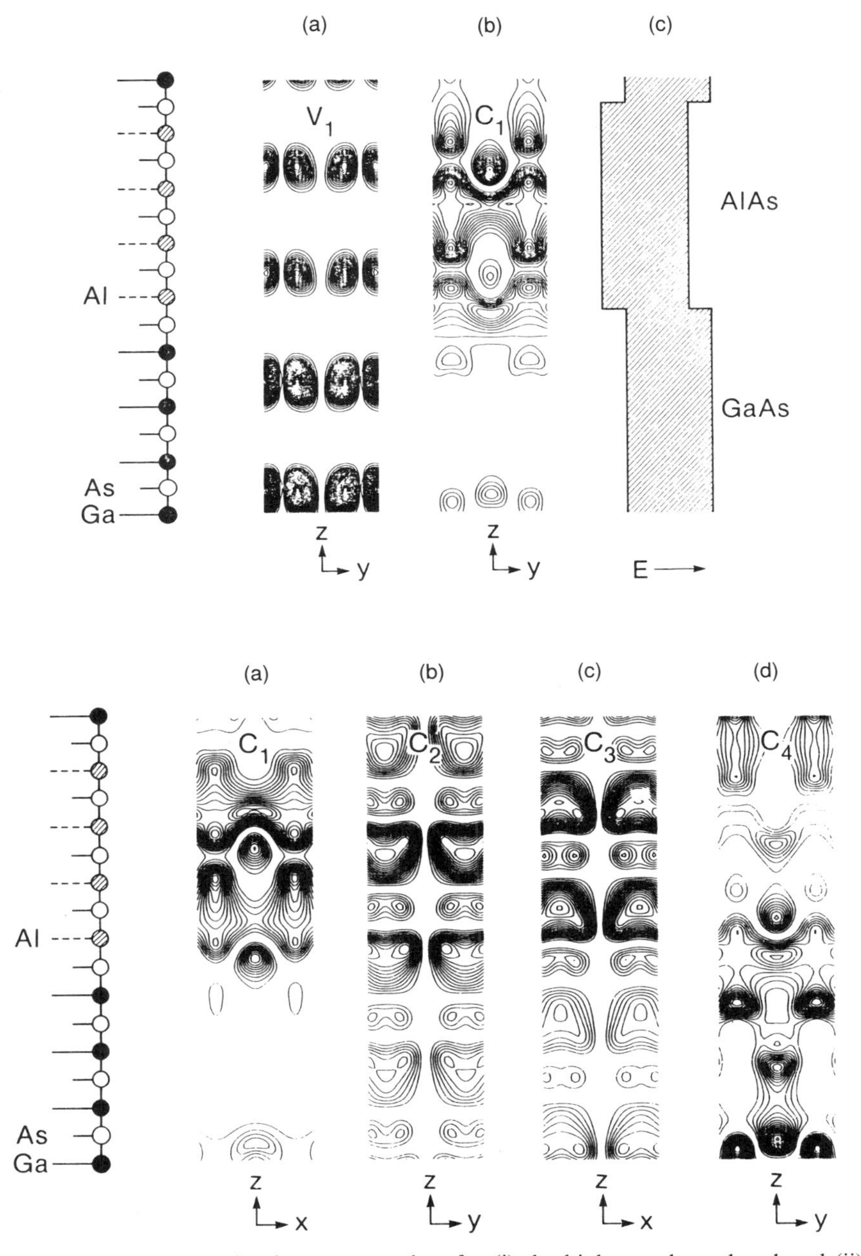

Figure 4.26 (a) Charge-density contour plots for (i) the highest valence band and (ii) the lowest conduction-band states of a $(GaAs)_4$-$(AlAs)_4$ superlattice; (iii) schematic representation of band-edge alignments. (b) Charge-density contour plots for the first four conduction-band states. The atomic plane positons in the [100] direction are shown on the extreme left. Contour spacings 2×10^{-4} a.u. After Batra et al. [36]; copyright © 1987, American Institute of Physics, New York.

based on this technology, and early on they achieved power-delay products in the 50–100 fJ range. For example, the internal logic delay for a divide-by-two circuit was 22 and 36 ps/gate at a power dissipation of 2.8 and 2.9 mW/gate for operation at 77 and 300 K, respectively.

Figure 4.27 shows self-consistent modeling results for the conduction bands at the heterojunction $Al_xGa_{1-x}As$/GaAs for two gate voltages [40]. Note that the parabolic well formed in the $Al_xGa_{1-x}As$ under certain bias conditions results in quantized states on the wide-gap side of the junction. At low bias, the $Al_xGa_{1-x}As$ is depleted, but at larger bias, a significant number of electrons reside in the $Al_xGa_{1-x}As$, filling either sub-band states or neutralizing donor states. An analysis of the gate capacitance $q(dN_e/dV_G)$ as a function of the gate bias V reveals contributions from the free electrons and electrons residing in the three lowest sub-band channels, respectively. Electrons residing in the lowest sub-band make the largest contribution, but the contributions from the other sub-bands are significant.

Figures 4.28(a) and (b) show the mobility as a function of temperature for a high quality $Al_{0.38}Ga_{0.62}As/GaAs/Al_{0.38}Ga_{0.62}As$ double heterostructure with a sheet electron concentration of 3.9×10^{11} cm^{-3} and the spread in f_t for HEMTs made by a 0.5 μm recessed gate process, respectively. Although $Al_xGa_{1-x}As$/GaAs HEMTs were the first useful confined heterostructure devices, fundamental materials properties favor more complex heterostructures (e.g., $Ga_{0.47}In_{0.53}As/Al_{0.48}In_{0.52}As$ on InP or pseudomorphic $Ga_{0.85}In_{0.15}As/Al_{0.15}Ga_{0.85}As$ on GaAs).

Ultrashort-gate-length pseudomorphic HEMTS with $f_t > 200$ GHz and $f_{max} \geqslant 350$ GHz have been reported [42], but are outside the scope of present

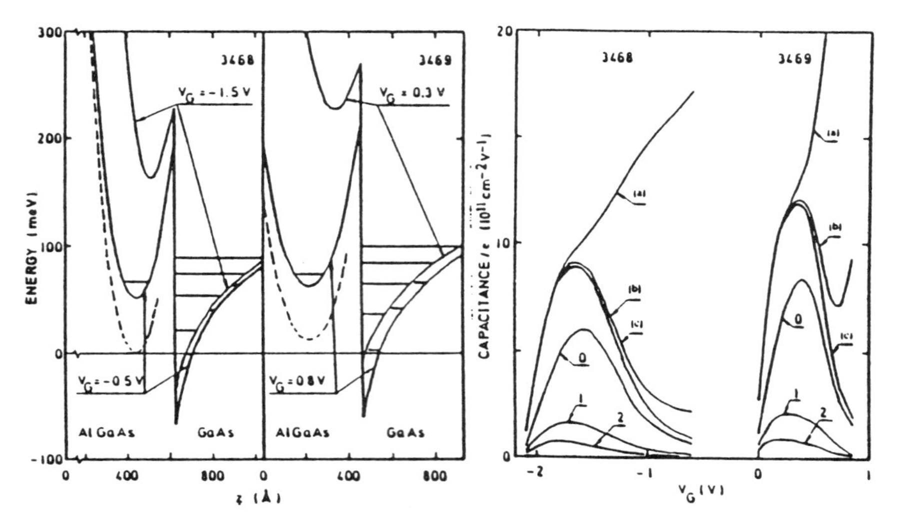

Figure 4.27 Sub-bands in the triangular quantum well in the conduction band at an AlGaAs/GaAs Heterojunction. After Vinter [39]; copyright © 1984, American Institute of Physics, New York, NY.

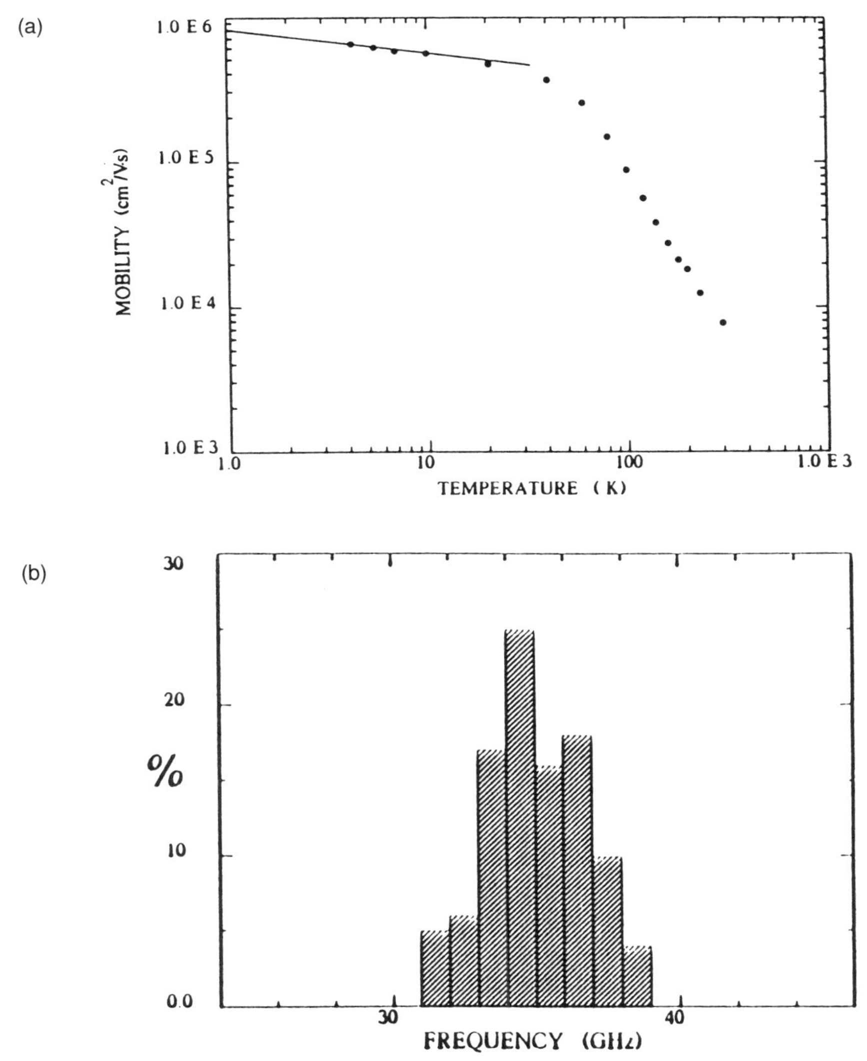

Figure 4.28 (a) Mobility versus temperature for a high quality $Al_{0.38}Ga_{0.62}As/GaAs$ double-heterostructure grown by metalorganic chemical vapor deposition and (b) spread in f_t for HEMTs with 0.5 μm gate length. After Frijlink et al. [39]; copyright © 1987, North Holland Publishing Company, Amsterdam.

uv lithography, which is limited to $\sim$0.25 μm resolution. Current $Al_{0.22}Ga_{0.78}$-$As/Ga_{0.80}In_{0.20}As$ MODFETs that have 0.22–0.25 metallurgical gate length corresponding to 700 nm channel length achieve cutoff frequencies $f_T = 100$ GHz and $f_{max} = 305$ GHz [41].

InGaAs/InAlAs RTDs have been integrated with InGaAs/InAlAs HEMTs in a monolithic structure for implementation of single-transistor SRAM cells.

(a)

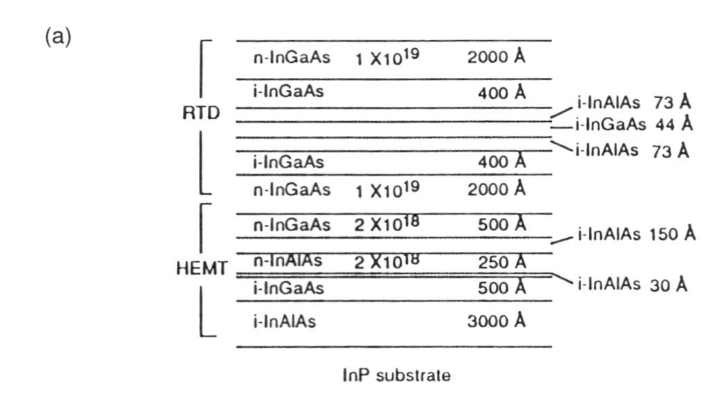

(b)

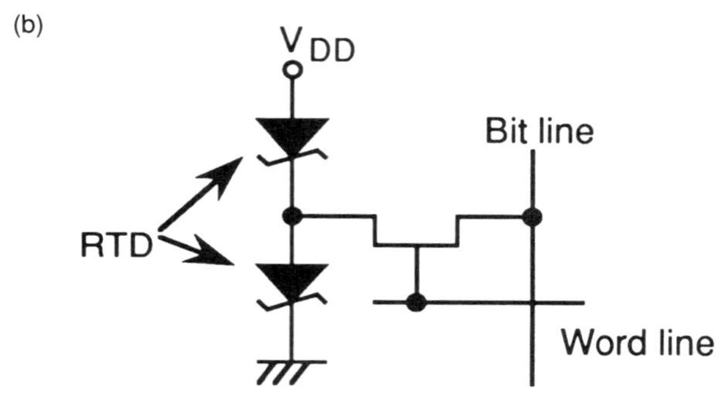

Figure 4.29 (a) Schematic representation of a cross section and (b) current-voltage characteristics of a monolithic RTD/HEMT heterostructure. After Watanabe et al. [25]; copyright © 1992, The Institute of Electrical and Electronics Engineers, New York.

Figures 4.29(a) and (b) show the circuit diagram and a schematic representation of a cross section through such a monolithic RTD/HEMT heterostructure.

Based on the concept of resonant tunneling, also a novel family of bipolar transistors has been introduced that permits a considerable reduction of the complexity of circuits as compared to conventional transistor designs [43], [44]. However, although operation with high gain has been demonstrated for resonant tunneling bipolar transistors (RTBTs), their development is currently still in an experimental stage. Figures 4.30(a) and (b) show a schematic representation of a cross section of an RTBT and its J–V characteristics [45]. Because there are more than one quantum-well states in the well, multiple peaks are observed in the J–V characteristics corresponding to the successive passage of the quantum-well states through resonance with the reservoir of electrons in the emitter. A review of novel circuit designs employing RTDs and RTBTs is provided in Ref. [46].

Through recent advances in patterning and semiconductor processing technologies, both two-dimensionally confined heterostructures (quantum wires) and three-dimensionally confined heterostructures (quantum boxes or

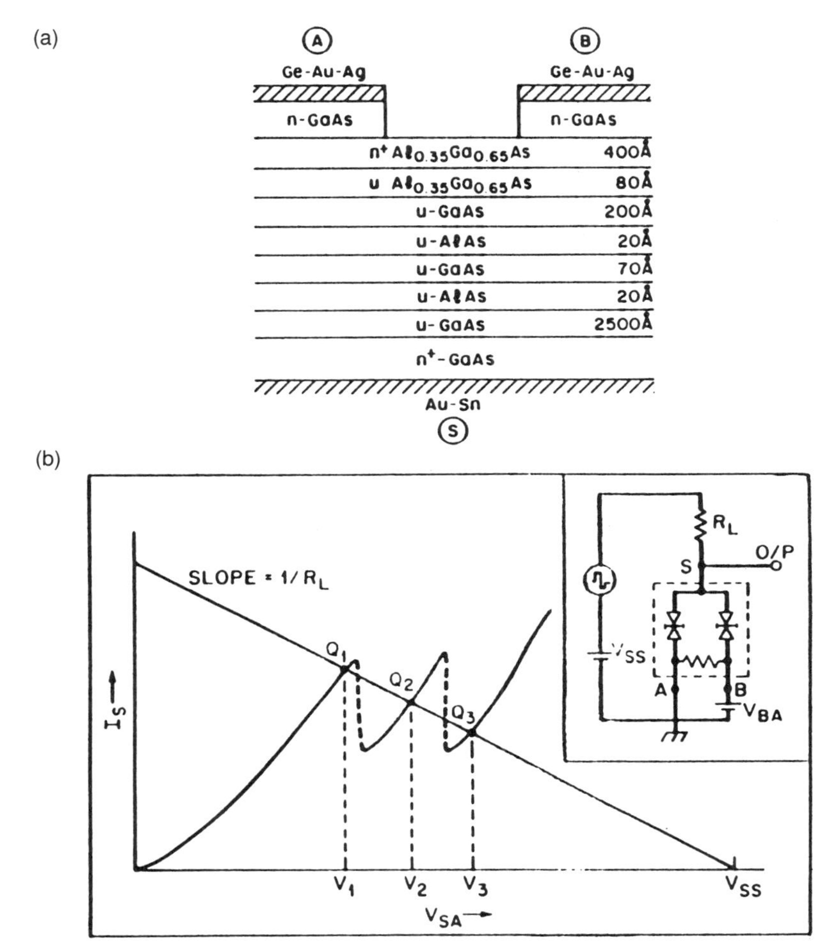

Figure 4.30 (a) Schematic representation of a cross section and (b) current–voltage characteristics of an RTBT. After Sen et al. [44]; copyright © 1987, The Institute of Electrical and Electronics Engineers, New York.

dots) have been realized. For quantum-wire heterostructures, Equation (4.67) is replaced by

$$E_{n,p}(k_y) = E_{n,p} + \frac{\hbar^2 k_y^2}{2m^*} \tag{4.70}$$

In two directions, chosen to coincide with the x and z axes of a Cartesian coordinate system, the energy is quantized with quantum numbers n and p. In the y direction, parabolic bands exist, and the carriers are highly mobile.

Lateral quantization has been reported recently for a variety of single-channel and multichannel FETs, which is reviewed in Ref. [47]. Quantum-wire heterostructures have been made by the growth of fractional layer superlattices on tilted substrates [48], [49], and by a combination of electron beam writing, etching, and epitaxial growth [50], [51], which is discussed in more detail in Chapters 6–8. Electron wave interference and mobility modulation effects in such devices are of considerable interest in the context of both fundamental studies in solid-state physics and access to novel device designs [52]. Although dry etching techniques (see Chapter 7), under unfavorable circumstances, produce side-wall depletion regions in excess of 100 nm and significantly degrade the electron transport in thus-defined quantum wires [53], $Ga_xIn_{1-x}As/InP$ quantum wires have been made successfully by a combination of high-resolution electron beam writing and gentle reactive ion beam etching of lattice-matched $Ga_{0.47}In_{0.53}As/InP$ heterostructures [54]. Thus far such III–V quantum wire heterostructures have been utilized exclusively in fundmental studies of solid state physics.

Figure 4.31(a) and (b) show an image of an 80-nm-wide quantum wire produced in this fashion and the longitudinal magnetoresistance curves for this wire and another wire of 310 nm width, respectively. The longitudinal magnetoresistance is measured in a geometry where a magnetic field **H** is applied parallel to the current density **J**. If the direction of **H** is taken to be the x direction, the electron motion is unaltered by the field in the x direction, but is forced onto a circular path in the yz plane. Assuming free-electron behavior, the energy is quantized according to

$$E = (n_L + \tfrac{1}{2})\hbar\omega_c + gm_B\mathbf{H}m_s + (\hbar^2/2m)k_x^2 \tag{4.71}$$

where $g \approx 2$ and ω_c denotes the cyclotron frequency, the frequency of the circular motion of the conduction electrons about **H**. The quantum number n_L defines a set of Landau levels (one-dimensional sub-bands) separated by the energy $\hbar\omega_c$, each being split into two levels according to the values of the spin quantum number $m_s = \pm\frac{1}{2}$. It can be shown [55] that the position of the Landau levels with regard to the Fermi-level changes with increasing magnetic field in an oscillatory fashion with period

$$\Delta(H^{-1}) = \frac{q\hbar}{mcE_F}. \tag{4.72}$$

A maximum occurs in the magnetoresistance at a field where the diameter of the cyclotron orbit equals the diameter of the quantum wire. It is pronounced for the small-diameter wire, but is still discernible for the quantum wire of 310 nm width at ~0.5 T.

Quantum-wire rings have been prepared by wet etching from a modulation-doped $Al_xGa_{1-x}As/GaAs$ heteroepitaxial layer with a carrier density in the 2D electron gas at the interface of $2 \times 10^{11}\,\text{cm}^{-2}$ and mobility of 970,000 $\text{cm}^2\,\text{V}^{-1}\,\text{s}^{-1}$ [56]. The motion of electrons through the loops and straight quantum-wire sections, connecting them on each side to ohmic contacts, is

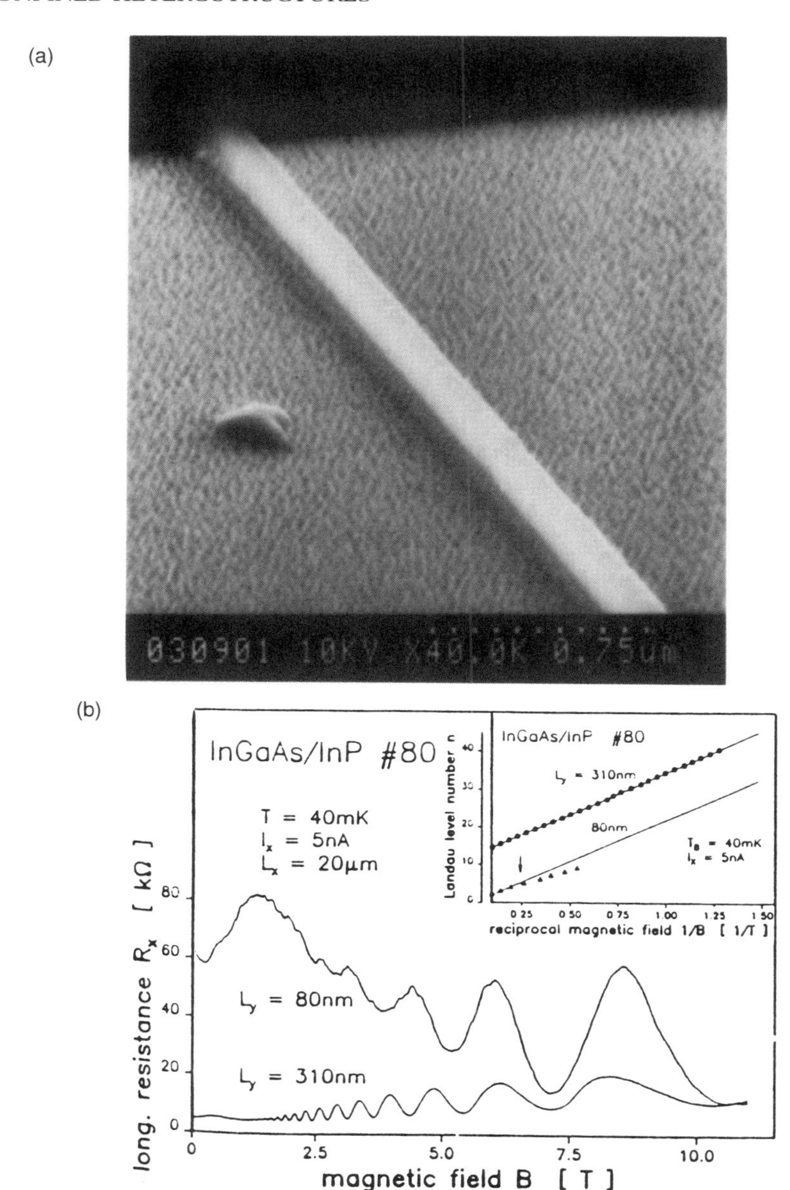

Figure 4.31 (a) SEM image of a GaInAs quantum wire of 80 nm width on an InP substrate. (b) Longitudinal magnetoresistance as a function of the magnetic induction (J‖X). After Mensching et al. [54]; copyright © 1990, American Institute of Physics, New York.

quantized in $2q^2/h$. The conductances of these wire sections and rings were varied by a bias voltage on a gate covering the active area. In a magnetic induction threading the rings, characteristic conductance oscillations occur.

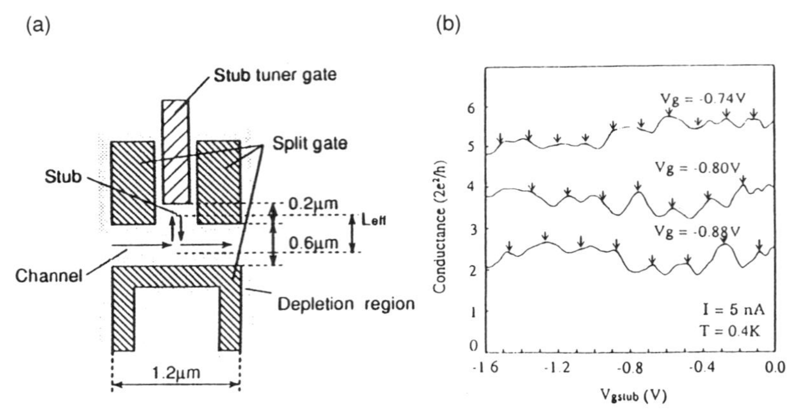

Figure 4.32 (a) Schematic representation of a three-terminal quantum interference device. (b) Conductance of the device as a function of the stub tuning voltage for three values of the split-gate voltage as a parameter. After Aihara, Yamamoto, and Mizutani [57]; copyright © 1992, Institute of Electrical and Electronic Engineers, New York.

They are due to the Aharonov–Bohm effect, the interference of the electronic wavefunctions in the various branches of the multiple-loop structure. Recently three-terminal quantum-wire devices have been built that permit the control of the conductivity by a gate voltage. An example is presented in Figure 4.32(a), which shows a device employing a quantum wire with a stub in a split-gate structure [57]. The control of the conductance modulation due to quantum interference by the gate voltage is illustrated in Figure 4.32(b).

Imposing a periodic potential on a 2D electron gas permits the formation of a lateral surface superlattice (LSSL) with a bias-dependent relative position of the Fermi level to the potential barriers. At low bias, the region in between dots is essentially depleted. Increasing the bias introduces carriers in these regions, that is, increases the coupling [58]. Thus quantum dots can be formed under appropriate bias conditions.

Isolated quantum dots in silicon have been created by severing a voltage-induced inversion channel by the action of an upper-gate grid [59]. Heavily doped poly-silicon are used for making the dual-gate structure shown in Figure 4.33(a). The lower and upper gates are isolated by 10- and 100-nm-thick SiO_2 layers, respectively. An inversion channel beneath the isolation layer is formed under the condition of high lower-gate voltage and small upper-gate voltage, as illustrated in Figure 4.33(b). Applying a large negative upper-gate voltage at a constant moderate lower-gate voltage lowers the Fermi energy, so that the quantum wire associated with the inversion channel is severed into a series of quantum dots, as shown in Figure 4.33(c). Figure 4.34 shows transconductance oscillations versus the lower-gate voltage of the device for various upper-gate voltages as a parameter. Under conditions where quantum dots of $\sim 5000\,\text{nm}^2$ area are formed, the period of these oscillations is $\Delta V_{LG} = 16$–

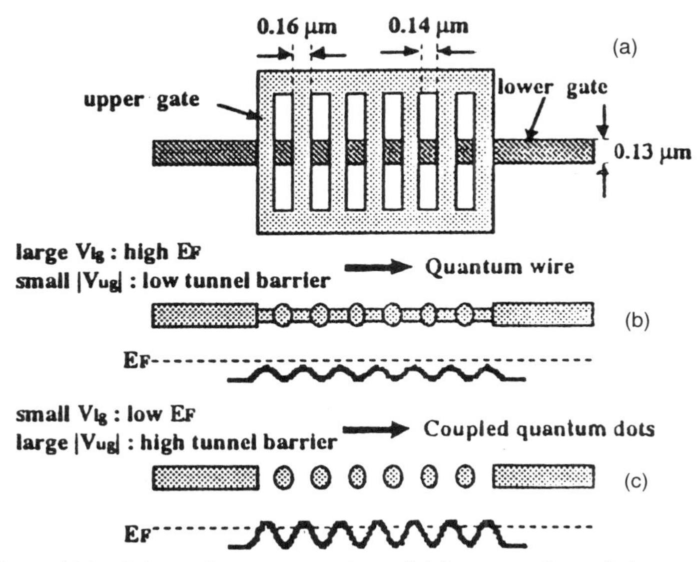

Figure 4.33 Schematic representation of (a) a top view of the upper gate and the quantum wire. (b), (c) Relative positions of the Fermi level to the 1D periodic potential barriers provided by the bias on the large-area lower gate and the serrated upper gate, respectively. After Matsuoka et al. [59]; copyright © 1992, Institute of Electrical and Electronics Engineers, New York.

17 meV. A theoretical analysis of charge transport between quantum dots [60] gives

$$\Delta V_{\mathrm{LG}} = \frac{q}{C_{\mathrm{G}}}\left(1 + \Delta E\,\frac{2(C_{\mathrm{QD}} + C_{\mathrm{G}})}{q^2}\right) \tag{4.73}$$

where ΔE is the distance between quantized energy eigevalues in the quantum dot, and C_{QD} and C_{G} refer to the capacitances associated with the quantum dots and the gates between dots, respectively. For the device under discussion, $C_{\mathrm{QD}} = 3 \times 10^{-17}\,\mathrm{F}$ and $C_{\mathrm{G}} = 1.6 \times 10^{-17}\,\mathrm{F}$, respectively, corresponding to $q^2/2(C_{\mathrm{QD}} + C_{\mathrm{G}}) \approx 1.6\,\mathrm{meV}$ and $\Delta V_{\mathrm{LG}} \approx 10\,\mathrm{meV}$ for single-electron transport between quantum dots.

Single electron transistors (SETs) that are based on the filling and emptying of quantum dot states by single electron switching events would be associated with extremely low power dissipation, but are at present in a very tentative conceptional stage far removed from practical applications [61]. Nevertheless, the interest in quantum dots is sufficient at present to sustain a continuing search for fabrication methods that improve the control of their properties. For example, buried quantum-dot heterostructures have been fabricated from $\mathrm{InP/Ga_{0.47}In_{0.53}As}$ by a combination of lithographic pattern definition with MOCVD growth [62]. They have dimensions $> 100\,\mathrm{nm}$, but are likely to reach dimensions that create strong confinement effects within the near future and are important in the context of future studies.

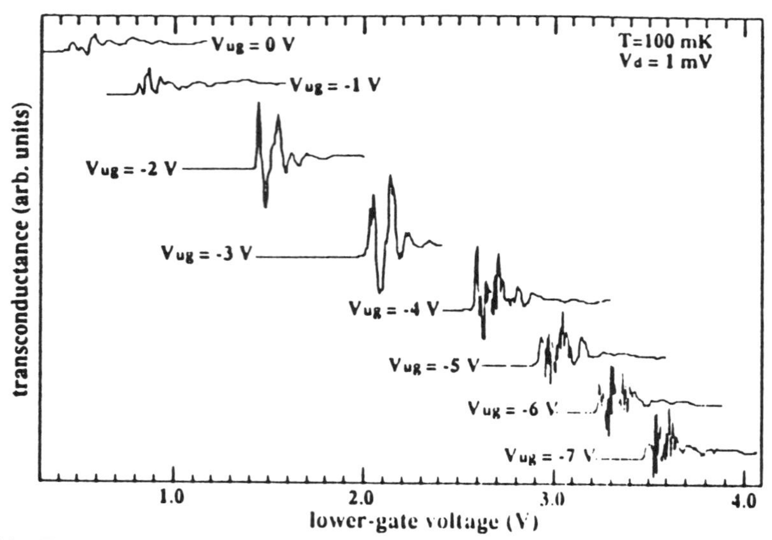

Figure 4.34 Transconductance versus lower-gate voltage for the device shown in Figure 4.31. After Matsuoka et al. [59]; copyright © 1992, Institute of Electrical and Electronics Engineers, New York.

References

1. M. Bettini, K. J. Bachmann, and J. L. Shay, *J. Appl. Phys.* **49**, 865 (1978).
2. H. M. Hobgood, T. T. Braggins, S. McGuigan, D. L. Barrett, and R. N. Thomas, Control of Impurities and Defects in Czochralski Growth of Silicon and GaAs, in *Processing of Electronic Materials*, C. G. Law, Jr., and R. Pollard, eds., The American Institute of Chemical Engineers, New York, 1987.
3. S. M. Sze, *Physics of Semiconductor Devices*, John Wiley & Sons, New York, 1969.
4. K. J. Bachmann, Materials Aspects of Solar Cells, in *Current Topics of Materials Science*, North Holland Publishers, Ltd., Amsterdam, 1979.
5. K. J. Bachmann and H. J. Lewereuz, Solar Cells, in *Encyclopedia of Advanced Electronic Materials*, D. Bloos, R. J. Brook, M. C. Flemings, S. Mahajan and R. W. Cahn, ed., Pergamon Press, Oxford, 1994, p. 2563.
6. S. Nakamura, T. Toyofuku, M. Sueda, K. Hasegawa, I. Kato, and T. Tadada, *IEDM Tech. Dig. 92*, Institute of Electrical and Electronics Engineers, New York, 1992, p. 445.
7. W. S. Hobson, F. Ren, C. R. Abernathy, S. J. Pearton, T. R. Fullowan, J. Lothian, A. S. Jordan, and L. M. Lunardi, *IEEE, Electron Dev. Lett.* **11**, 241 (1990).
8. Y.-K. Chen, R. N. Nottenburg, M. B. Panish, R. A. Hamm, and D. A. Humphrey, *IEEE Electron Dev. Lett.* **10**, 2267 (1989).
9. T. Ishibashi, *IEDM Tech. Dig. 88*, Institute of Electrical and Electronics Engineers, New York, 1988, p. 826.
10. H. Kroemer, *Proc. IEEE* **70**, 13 (1982).
11. J. B. Boos, W. Kruppa, and B. Molnar, *IEEE Electron Dev. Lett.* **10**, 79 (1989).

12. M. E. Mierzwinski, J. D. Plummer, E. T. Croke, S. S. Iyer, and M. J. Harrel, *IEDM Tech. Dig. 92*, Institute of Electrical and Electronics Engineers, New York, 1992, p. 773.

13. G. Lucovsky, S. S. Kim and J. T. Fitch, *J. Vacuum Sci. Technol.* **B8**, 822 (1990).

14. E. H. Nicollian and J. R. Brews, *MOS* (*Metal Oxide Semiconductor*) *Physics and Technology*, John Wiley & Sons, New York, 1982.

15. C.-K. Kim, in *Charge-Coupled Devices and Systems*, M. J. Howes and D. V. Morgan, eds., John Wiley & Sons, Chichester, 1979.

16. W. McColgin, J. Lavine, J. Kyan, D. Nichols, and C. Stancampiano, *IEDM Tech. Dig. 92*, Institute of Electrical and Electronics Engineers, New York, 1992, p. 113.

17. C. C. Sun, J. M. Su, A. Hagley, R. Surridge, and A. Spring Thorpe, *IEEE Electron Dev. Lett.* **11**, 382 (1990).

18. T. Sugano, *Jpn. J. Appl. Phys.* **32**, 261 (1993).

19. R. P. Kramer, in *ULSI Science and Technology*, C. M. Osburn and J. M. Andrews, eds., The Electrochemical Society, Pennington, NJ, 1989, p. 27.

20. M. A. diForte-Poisson, C. Brylinski, G. Colomer, D. Osselin, S. Hersee, J. P. Duchemiu, F. Azan, D. Lechevallier and J. Lacombe, *Electron. Lett.* **20**, 1061 (1984).

21. A. Rydberg, *IEEE Electron Dev. Lett.* **11**, 439 (1990).

22. P. K. Vasudev, *IEEE Trans. Electron Dev.* **ED-31**, 1044 (1984).

23. T. C. L. G. Sollner, W. D. Goodhue, P. E. Tannenwald, C. D. Parker, and D. D. Peck, *Appl. Phys. Lett.* **43**, 588 (1983).

24. E. R. Brown, W. D. Godhue, and T. C. L. G. Sollner, *J. Appl. Phys.* **64**, 1519 (1988).

25. Y. Watanabe, Y. Nakasha, K. Imanishi, and M. Takikawa, *IEDM Tech. Dig. 92*, Institute of Electrical and Electronics Engineers, New York, 1992, p. 475.

26. L. L. Chang, L. Esaki, and R. Tsu, *Appl. Phys. Lett.* **24**, 593 (1974).

27. M. Cardona and N. E. Christensen, *J. Vac. Sci. Technol B* **5**, 1285 (1987).

28. J. Tersoff, *Phys. Rev. B* **30**, 4874 (1984).

29. I. Lefevre, M. Lanoo, C. Priester, G. Allan, and C. Delerue, *Phys. Rev. B* **36**, 1336 (1987).

30. V. Heine, *Phys. Rev. A* **138**, 1689 (1965).

31. M. Jaros, *Phys. Rev. B* **37**, 7112 (1988).

32. N. E. Christensen, *Phys. Rev. B* **37**, 4528 (1988).

33. C. G. Van de Walle and R. M. Martin, *Phys. Rev. B* **35**, 8154 (1987).

34. G. Bastard, *Phys. Rev. B* **24**, 5693 (1981); **25**, 7584 (1982).

35. Y. C. Chang and J. N. Schulman, *Phys. Rev. B* **31**, 2069 (1985).

36. I. P. Batra, S. Ciraci, and J. S. Nelson, *J. Vac. Sci. Technol. B* **5**, 1300 (1987).

37. R. Dingle, H. L. Stormer, A. C. Gossard, and W. Wiegman, *Appl. Phys. Lett.* **33**, 665 (1978).

38. (a) N. T. Linh, in *Semiconductors and Semimetals*, Vol. 24, R. Dingle, ed., Academic Press, Inc., San Diego, CA, 1987, pp. 203–47. (b) M. Abe, T. Mimura, K. Nishiuchi, A. Shibatomi, M. Kobayashi, and T. Mitsugi, in *Semiconductors and Semimetals*, Vol. 24, R. Dingle, ed., Academic Press, Inc., San Diego, CA, 1987, pp. 249–78.

39. B. Vinter, *Appl. Phys. Lett.* **44**, 307 (1984).

40. P. M. Frijlink, J. L. Nicolas, and P. Suchet, *J. Crystal Growth* **107**, 166 (1991).

41. G. M. Metze, T. T. Lee, J. F. Bass, P. L. Laux, H. C. Carlson, and A. B. Cornfeld, *IEEE Electron Dev. Lett.* **11**, 493 (1990).

42. P. C. Chao, *IEEE Trans. Electron Dev.* **36**, 461 (1981).

43. F. Capasso and R. Kiehl, *J. Appl. Phys.* **58**, 1366 (1985).

44. S. Sen, F. Capasso, A. Y. Cho, and D. L. Sivco, *IEEE Trans. Electron Dev.* **ED-34**, 2185 (1987).

45. S. Sen, F. Capasso, A. Y. Cho, and D. L. Sivco, *IEDM Tech. Dig. 88*, Institute of Electrical and Electronics Engineers, New York, 1988, p. 834.

46. F. Beltram, F. Capasso, and S. Sen, Photonic and Electronic Devices Based on Artificially Structured Semiconductors, in *Electronic Materials*, J. R. Chelicowsky and A. Franciosi, eds., Springer Verlag, Berlin, 1993, p. 233.

47. W. Hansen, *Festkörperprobleme* **28**, 121 (1988).

48. K. Tsubaki, T. Honda, H. Saito, and T. Fukui, *Appl. Phys. Lett.* **58**, 376 (1991).

49. T. Fukui and H. Saito, *J. Vac. Sci. Technol. B* **6**, 1373 (1988).

50. B. E. Maile, A. Forchel, R. Germann, and A. Menschig, *J. Vac. Sci. Technol. B* **6**, 2308 (1988).

51. K. Ismail, S. Washburn, and K. Y. Lee, *Appl. Phys. Lett.* **59**, 1998 (1991).

52. H. Sakaki, *Jpn. J. Appl. Phys.* **21**, L381 (1982).

53. G. Timp, A. M. Chang, P. Mankievich, R. Behringer, J. E. Cunningham, T. Y. Chang, and R. E. Howard, *Phys. Rev. Lett.* **59**, 732 (1987).

54. A. Mensching, B. Roos, R. Germann, A. Forchel, K. Pressel, W. Heuring, and D. Grutzmacher, *J. Vac. Sci. Technol. B* **8**, 1353 (1990).

55. G. Burns, *Solid State Physics*, Academic Press, Inc., Orlando, FL, 1985.

56. S. Washburn, *Am. J. Phys.* **57**, 1069 (1989).

57. K. Aihara, M. Yamamoto, and T. Mizutani, *IEDM Tech. Dig. 92*, Institute of Electrical and Electronics Engineers, New York, 1992, p. 491.

58. K. Y. Lee, D. P. Kern, K. Ismail, R. J. Haug, T. P. Smith III, W. T. Masselink, and J. M. Hong, *J. Vac. Sci. Technol. B* **8**, 1366 (1990).

59. H. Matsuoka, T. Ichiguchi, T. Yoshimura, and E. Takeda, *IEDM Tech. Dig. 92*, Institute of Electrical and Electronics Engineers, New York, 1992, p. 781.

60. C. W. J. Beenakker, *Phys. Rev. B* **44**, 1646 (1991).

61. K. K. Likharev, *IBM J. Res. Dev.* **32**, 144 (1988).

62. Y. D. Galeuchet, H. Rothuizen, and P. Roentgen, *Appl. Phys. Lett.* **58**, 2423 (1991).

CHAPTER

5

Crystal Growth

5.1 Materials Purification and Synthesis

In view of the detrimental effects of compensating and trapping impurities on the electrical transport in semiconductors, stringent requirements are placed on their concentrations in the semiconductor-grade starting materials for the fabrication of microelectronic circuits. The vigilance with regard to causes of contamination starts with the selection of the mineral ores and chemical process sources from which the semiconductor-grade (SG) materials are produced. It continues through all synthesis, crystal growth, wafering, polishing, etching, epitaxy, lithography, dielectric deposition, metallization, and packaging steps, including the selection of high-purity container materials, solvents and etchants for processing, and maintenance of exceptional levels of cleanliness in the laboratory and manufacturing environments. Generally the conventional methods of chemical analysis are not sensitive enough to monitor routinely the maintenance of purity in microelectronics processing. Therefore, the ultimate test of a high-quality processing line is given by the reliability of the finished product. Nevertheless, an effort in determining the elemental distribution of impurities in the ores and as far as possible into the purification cycle is desirable because the control of impurities in the source materials usually makes an important contribution to the quality of the final products of bulk crystal growth and epitaxy.

The fabrication of SG silicon starts from quartzite gravel, which contains Al, B, Fe, and P at a level of $\sim 1\%$. In the currently predominant production process the quartzite is reduced in a submerged carbon arc furnace according

Table 5.1 Typical impurity concentrations in metallurgical-grade silicon (after McCormick [1])

Element	Concentration (ppma)	Element	Concentration (ppma)
Aluminum	1000–4350	Manganese	50–120
Boron	40–60	Molybdenum	< 20
Calcium	245–500	Nickel	10–105
Chromium	50–200	Phosphorus	20–50
Copper	15–45	Titanium	140–300
Iron	1550–6500	Vanadium	50–250
Magnesium	10–50	Zirconium	20

to the reaction

$$SiO_2(l) + 2C(s) \rightleftarrows Si(l) + 2CO(g) \tag{5.1}$$

which proceeds at approximately 2000°C at a considerable expense of energy, approximately 690 kJ/mol Si. In excess of 4×10^5 metric tons of this product are generated annually. Since it is used primarily in the aluminum and steel industry, it is called metallurgical grade (MG) silicon. Table 5.1 presents an analysis of the typical concentrations of impurities in MG Si [1]. The level of residual impurities in MG Si is many orders of magnitude higher than can be tolerated in microelectronics.

In order to purify it to SG quality the MG Si is crushed and converted in a fluidized bed process into chlorosilane compounds according to the reaction

$$Si(s) + 3HCl(g) \rightleftarrows SiHCl_3(g) + H_2(g) \tag{5.2}$$

The boiling points (°C) of silane and its chlorination products are SiH_4, −112.3; SiH_3Cl, −30.4; SiH_2Cl_2, 8.3; $SiHCl_3$, 31.5; and $SiCl_4$, 57.6, so that they are conveniently separated by fractioning distillation. Only the trichlorosilane fraction is used to produce semiconductor-grade Si by the Siemens process to be described. The remaining fractions of high-purity lower chlorinated silanes and $SiCl_4$ are primarily used for Si epitaxy, which is discussed in more detail in Section 6.1.

Figure 5.1(a) shows a schematic representation of a Siemens reactor containing under a bell jar an inverted U-shaped assembly of slim rods of ultrapure polycrystalline silicon. The trichlorosilane and hydrogen are injected into the bell jar from below and react on the hot surface of the resistively heated slim rods according to the reaction

$$SiHCl_3(g) + H_2(g) \rightleftarrows Si(s) + 3HCl \tag{5.3}$$

which is the reverse of reaction (5.2). Silicon is added to the rods at a rate of typically 1 mm per hour, forming columnar grains that grow perpendicular to the rod axis. Figures 5.1(b) and (c) show a section of such a rod after approximately 1 day of deposition, revealing the contour of the original slim

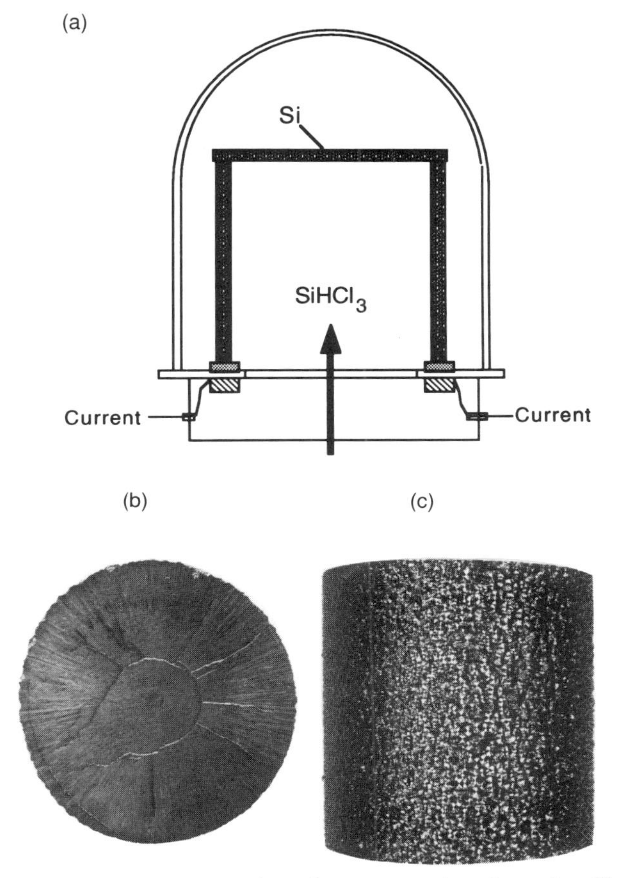

Figure 5.1 (a) Schematic representation of a cross section through a Siemens reactor. (b) Cross section and (c) side view of a polycrystalline silicon rod.

rod and the grain structure of the vapor-deposited layer. Typically the process lasts for several days.

The shift in the reaction from forming $SiHCl_3$ from Si at low temperature to forming Si from $SiHCl_3$ at high temperature is based on the temperature dependence

$$\frac{\partial \ln K_p}{\partial T} = \frac{\Delta H_r}{RT^2} \tag{5.4}$$

of the equilibrium constant based on Equation (5.3)

$$K_p = \frac{p_{SiHCl_3} p_{H_2}}{p_{HCl}^3} \tag{5.5}$$

with $\Delta H_r < 0$, causing the partial pressure of trichlorosilane in equilibrium

Table 5.2 Typical impurity concentrations in semiconductor-grade silicon made by the Siemens process (after Baraclough [2])

Element	Concentration (ppba)	Element	Concentration (ppba)
Arsenic	< 0.001	Gold	< 0.00001
Antimony	< 0.001	Iron	0.1–1.0
Boron	⩽ 0.1	Nickel	0.1–0.5
Carbon	100–1000	Oxygen	100–400
Chromium	< 0.01	Phosphorus	⩽ 0.3
Cobalt	0.001	Silver	0.001
Copper	0.1	Zinc	< 0.1

with solid silicon to decrease with increasing temperature. The Siemens process represented by Equation (5.3) proceeds typically at 1100°C, while the reverse fluidized bed process (5.2) is carried out at ~300°C. A typical analysis of SG silicon is shown in Table 5.2.

An alternative source for the production of ultrapure silicon is SiH_4, which is thermally decomposed to form SG Si. It is made by the following processes that are described in more detail in Ref. [3]

1. The reaction of $SiCl_4$ with lithium hydride,

$$SiCl_4 + 4LiH \rightarrow SiH_4 + 4LiCl \quad (5.6)$$

which provides large batches of SiH_4 (~50 kg) at relatively low cost (~$100/kg), but at a considerable expense of electrical energy.

2. The reaction of $SiCl_4$ with lithium alanate in tetrahydrofuran,

$$SiCl_4 + LiAlH_4 \rightarrow SiH_4 + LiCl + AlCl_3 \quad (5.7)$$

which is less energy consuming than the lithium hydride process.

3. The reaction of Mg_2Si with ammonium chloride in liquid ammonia

$$Mg_2Si + NH_4Cl \rightarrow SiH_4 + 2MgCl_2 + 5NH_3 \quad (5.8)$$

Since diborane is complexed in liquid ammonia to form a $H_3B \leftarrow NH_3$ adduct, the thermal decomposition of silane made in this reaction produces polysilicon with a boron content of <20 ppt [4]. However, reaction (5.8) may result in phosphine contamination of the silane at ⩽10 ppm, requiring subsequent purification.

4. Hydrogenation of $SiCl_4$ and disproportioning distillation of the resulting chlorosilanes,

$$Si(98\%) + 3SiCl_4 + 2H_2 \rightarrow 4SiHCl_3 \quad (5.9)$$

$$2SiHCl_3 \rightarrow SiH_2Cl_2 + SiCl_4 \quad (5.10)$$

$$3SiH_2Cl_2 \rightarrow SiH_3Cl + 2SiHCl_3 \quad (5.11)$$

$$2SiH_3Cl \rightarrow SiH_4 + SiH_2Cl_2 \quad (5.12)$$

which has been developed into a continuous process at Union Carbide Company [5].

In order to convert the silane thus produced into an even purer product, catalytic purification and ion-exchange processes are used after the synthesis steps. For some time the decomposition of silane on powder Si seed material [6] in a fluidized bed process has been pursued by the manufacturers of large-diameter Si substrate crystals in the context of continuous Czochralski pulling. In view of the large surface area for deposition, this method offers high conversion rates of SiH_4 into SG Si pellets at relatively low energy expense and competitive purity as compared to the Siemens process [7]. However, polysilicon in rod form is needed for the growth of ultrapure Si crystals by the float zoning technique. In this technique the segregation of impurities upon passing a molten zone through the polysilicon rod is utilized for purification and crystal growth, which is discussed in more detail in Section 5.3. For special applications (e.g., in the context of solar energy conversion by photovoltaic silicon devices), very large quantities must be made available at relatively low cost. This has led to the development of novel methods of purification starting with SiO_2 that utilize a metallurgical approach [8].

Segregation phenomena in directional solidification processes are the consequence of a separation of the solidus and liquidus in the phase diagram of the multicomponent system being processed. For simplicity we consider here a simple binary alloy system composed of two components A and B, which are soluble in each other in the solid and liquid states. In an open system exchanging with its environment both energy and matter, in the absence of charged species and surface effects, the total differential of the Gibbs free energy is

$$dG = V\,dp - S\,dT + \sum_{i=1}^{k} \mu_i d\,n_i \tag{5.13}$$

where

$$V = \sum_{i=1}^{k} x_i V_i \tag{5.14}$$

$$S = \sum_{i=1}^{k} x_i S_i \tag{5.15}$$

are the molar volume and entropy.

$$V_i = (\partial V/\partial n_i)_{p,\,T,\,nj \neq i} \tag{5.16}$$

$$S_i = (\partial V/\partial n_i)_{p,\,T,\,nj \neq j} \tag{5.17}$$

$$\mu_i = \left(\frac{\partial G}{\partial n_i}\right)_{p,\,T,\,nj \neq i} = \mu_{i_0} + RT \ln a_i \tag{5.18}$$

are the partial molar volume, the partial molar entropy, and the chemical potential of component i. We distinguish between a mechanical mixture where all partial thermodynamic quantities are equal to those of the pure components and solutions where the partial thermodynamic quantities are generally different from those of the pure components. Since in equilibrium between altogether p distinct phases $A, B, \ldots$ the chemical potential of a particular constituent is equal in all phases and so are the pressure and the temperature, there exist $(p-1)(k+2)$ relations between the $p[(k-1)+2]$ variables that are needed to describe the system completely. Consequently the degree of freedom f in choosing variables is specified by

$$f = k + 2 - p \tag{5.19}$$

which is the Gibbs phase rule. For the two-component system A–B Equation (5.19) predicts that the two-phase equilibria are characterized by two degrees of freedom. For the equilibria between solid and liquid phases the equality of the chemical potentials of each component in the solid and liquid phases results in

$$RT \ln \frac{a_i^{\mathrm{s}}}{a_i^{\mathrm{l}}} = \mu_i^{0\mathrm{l}} - \mu_i^{0\mathrm{s}} = \Delta G_{i\mathrm{F}}^0 = \Delta H_{i\mathrm{F}}^0 - T\,\Delta S_{i\mathrm{F}}^0 \tag{5.20}$$

where $\Delta G_{i\mathrm{F}}^0$ is the free energy of fusion of the pure component i at temperature T. In particular at the melting point of pure component i $\Delta G_{i\mathrm{F}}^0 = 0$ so that $\Delta S_{i\mathrm{F}}^0 = \Delta H_{i\mathrm{F}}^0 / T_{i\mathrm{F}}$. Neglecting the temperature dependence of the enthalpy and entropy of fusion, we get for ideal binary solutions, where the activities equal the molar fractions, two equations in two unknowns, x_B^{s} and x_B^{l}

$$\ln \frac{x_B^{\mathrm{s}}}{x_B^{\mathrm{l}}} = \frac{\Delta H_{B\mathrm{F}}^0}{R}\left(\frac{1}{T} - \frac{1}{T_{B\mathrm{F}}}\right) \tag{5.21}$$

$$\ln \frac{1 - x_B^{\mathrm{s}}}{1 - x_B^{\mathrm{l}}} = \frac{\Delta H_{A\mathrm{F}}^0}{R}\left(\frac{1}{T} - \frac{1}{T_{A\mathrm{F}}}\right) \tag{5.22}$$

In accord with Equations (5.21) and (5.22), knowledge of the melting points and enthalpies of fusion for the pure components suffices to calculate the compositions x_B^{s} and x_B^{l} of the coexisting phases at a given T, assuming usually atmospheric total pressure. The curve $T(x_i^{\mathrm{l}})$ and $T(x_i^{\mathrm{s}})$ are liquidus and solidus lines of the system.

Figure 5.2 shows the $T - x$ diagram of the Si–Ge system and as a dashed line the result of a calculation using the ideal solution model. Obviously the model is not sufficiently sophisticated to predict the solidus–liquidus relations accurately. However, at the endpoints the model comes close to the experimental data. Also, at the trace impurity level, there exists generally a range of solid solubility and the solidus and liquidus curves can be linearized so that the segregation of impurity under the conditions of directional solidification can

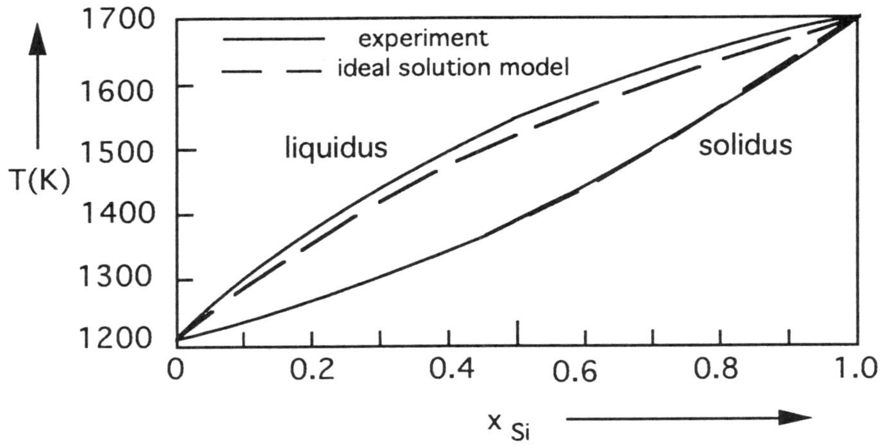

Figure 5.2 Phase diagram of the system Si,Ge. After Thurmond [9]; copyright © 1953, The American Chemical Society, Washington, DC.

be discussed on the basis of Figures 5.3(a) and (b). For example, Ge as an impurity in Si corresponds to Figure 5.3(a), and Si as an impurity in Ge corresponds to Figure 5.3(b). For systems corresponding to Figure 5.3(a), the equilibrium distribution coefficient

$$k_0 = \frac{c_{s0}}{c_{l0}} < 1 \tag{5.23}$$

while for systems corresponding to Figure 5.3(b) $k_0 > 1$.

Consider the directional solidification of a melt contained in a boat proceeding from left to right, as illustrated in Figure 5.3(c) for the conditions of gradient freezing where a boat or crucible containing the melt is kept in a stationary position and the furnace is cooled slowly to move the position of the melting point from the starting to the trailing end of the solidifying ingot. Impurities with $k_{i0} < 1$ will be rejected at the solid–liquid interface so that the impurity concentration in the first to freeze part of the ingot is smaller than the average initial concentration $\bar{c}$. Conversely, for impurities with $k_{i0} > 1$, which are preferentially incorporated into the solid, the first to freeze part of the ingot will be less pure than the last to freeze part. Under idealized conditions, that is, a perfectly flat solid–liquid interface and perfect mixing of the melt, the concentration in the liquid is given by the differential equation

$$d \ln c_l = \frac{1 - k_0}{1 - g}\, dg \tag{5.24}$$

where g is the solidified fraction of the melt, $0 \leqslant g \leqslant 1$. With the initial condition $c_l = \bar{c}$ for $g = 0$ the solution of Equation (5.18) is

$$c_l(g) = \bar{c}k_0(1 - g)^{(k_0 - 1)} \tag{5.25}$$

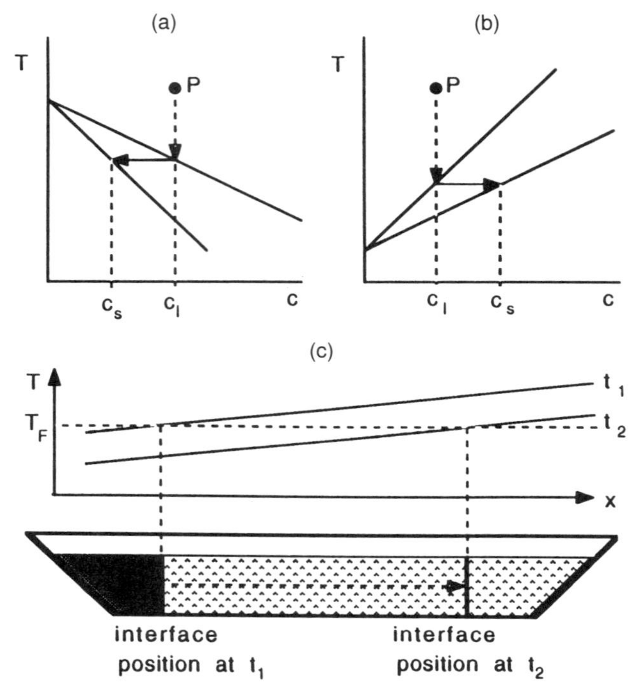

Figure 5.3 Phase relations for a very dilute binary alloy with (a) $k_0 < 1$ and (b) $k_0 > 1$. (c) Schematic representations of the normal freezing of a melt in a boat.

Relaxing the condition of perfect mixing, the solute rejected ($k < 1$) or depleted ($k_0 > 1$) must be removed or replenished by diffusive transport with diffusion constant D across the diffusion layer of characteristic width δ. In steady state, the liquid concentration c_0 at the interface will be such that the rate of generation or consumption of the impurity per unit area at the interface matches the diffusion flux $D(\partial c/\partial x)_{x=0}$. The distribution of the impurity occurs with the distribution coefficient $k_0 = c_s/c_0$. However, in terms of the bulk concentration c_b in the liquid, an effective distribution coefficient

$$k_e = \frac{c_s}{c_b} \neq k_0 \tag{5.26}$$

is defined that replaces k_0 in Equation (5.19) (see Figure 5.4). For a melt solidifying at constant rate v_g, the effective distribution coefficient k_e is related to the equilibrium distribution coefficient k_0 as

$$k_e = \left[1 + \left(\frac{1}{k_0} - 1\right)\exp\left(-\frac{v_g \delta}{D_l}\right)\right]^{-1} \tag{5.27}$$

favoring vigorous stirring and a small growth rate for maximizing the segregation in fractioning crystallization [10]. The directional solidification process

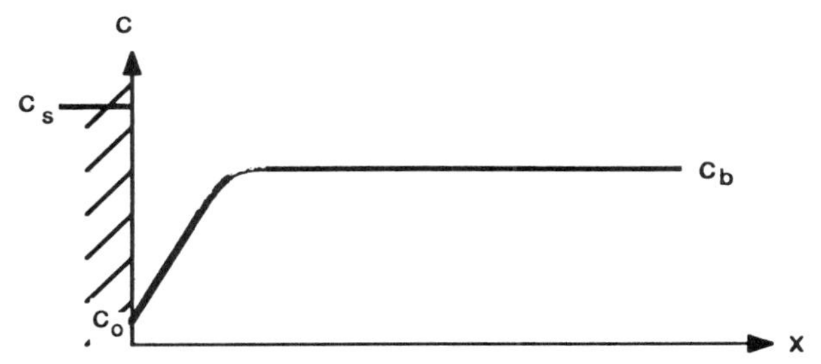

Figure 5.4 (a) Schematic representation of the concentration profile in the vicinity of the solid–liquid interface and the relation of the effective and equilibrium distribution coefficients.

thus provides a means for the separation of impurities, particularly if it is carried out repetitively in a fashion that does not eradicate the impurity distributions of previous melting/solidifying cycles.

Figure 5.5 shows a schematic cross section through a commercial Al recrystallization vessel. Electrorefined 99.99+ Al is molten and is allowed to crystallize on the top surface of the melt. The crystals have higher density than the melt and settle to the bottom. After 70% of the total charge is solidified, the top liquid layer is removed, and the solidified fraction is partially remelted from the top surface down. The second liquid is then removed leaving a 6N pure product. The purification obtained in the process is better than expected from Equation (5.25), presumably because of cyclic remelting and refreezing in the bottom-heated recrystallization vessel with incomplete mixing of the melt [11].

Ultrapure silicon and germanium can be reproduced by the zone melting technique, which was invented in 1956 by W. G. Pfann [12]. It is illustrated in Figure 5.6(a). Consider a molten zone of length l that is established in a solid

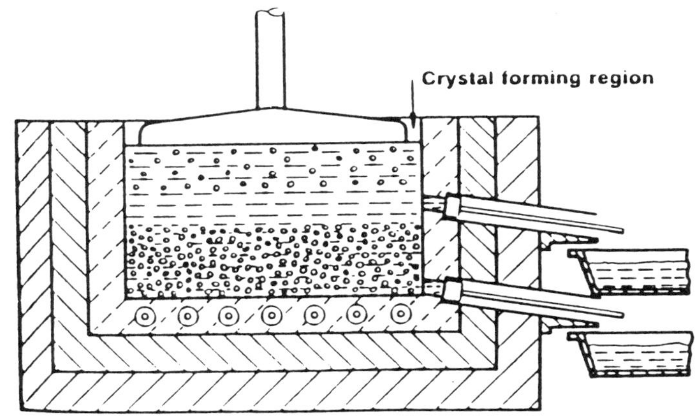

Figure 5.5 Schematic representation of the recrystallization of Al. After Dawless et al. [11]; copyright © 1988, North Holland Publishing Company, Amsterdam.

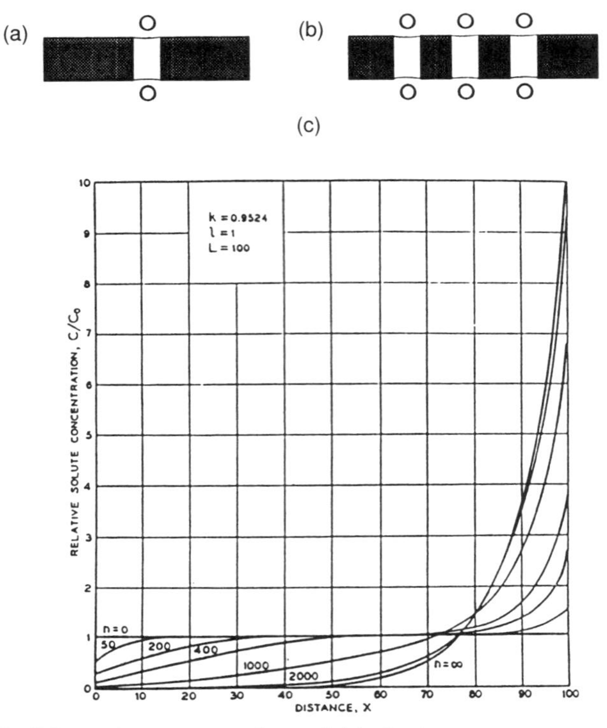

Figure 5.6 Schematic representations of (a) single and (b) multiple zone melting (c) Normalized impurity concentration $c(x)/\bar{c}$ as a function of the normalized distance x/L for various numbers of zone passes as parameter. (c) After Pfann [12]; copyright © 1966, John Wiley & Sons, Inc., New York.

ingot of initially uniform concentrations of impurities $\bar{c}_i$ and total length L. If the zone is made to move from left to right, for $k_0 < 1$ and perfect mixing of the melt, the frozen part on the left trailing edge of the zone is purer than the material that melts in on the right-side leading edge of the zone. Consequently the solid to the left of the zone is purer than the material on its right until the concentration in the molten zone becomes $\bar{c}_i/k_{ei}$. At this point the freezing and melting material have the same concentration, and the impurity level attains a constant value with the exception of the last zone width of the ingot, where the impurity profile corresponds to a normal freezing distribution. The analytical form of the impurity distribution after a single zone pass is

$$c(x) = \bar{c}_i\left[1 - (1 - k_{ie}) \exp\left(- \frac{k_{ie} x}{l}\right)\right] \tag{5.28}$$

If after completion of the first pass the process is repeated, the zone established at $x = 0$ is purer than in the first cycle, and upon motion of the zone even purer material is generated at the first to freeze end of the ingot. The distributions for an arbitrary number of zone passes can be conveniently generated numerically, and a plot $c(x)/\bar{c}$ as a function of x/L for $L/l = 100$ and $k = 0.9524$ is shown for various numbers of zone passes in Figure 5.6(c).

Aluminum and gallium are produced from bauxite in either the wet Bayer process, which relies on the amphoteric behavior of these two elements, or by a dry process, as illustrated in Table 5.3. Based on the substantial separation of the normal potentials of Al and Ga, the latter is separated from the Bayer liquor by electrodeposition at Hg electrodes. A typical analysis of the 96–98% pure Ga thus obtained is shown in column two of Table 5.4. It is purified to 99.99% (4N) purity by acid and base leaching processes, leaving Al, Fe, Mg, Pb, Si, Sn, Ti, and Zn at or above the ppm level, as indicated by the entries in column 3 of Table 5.4. The aluminum is purified to approximately the same level of impurities by molten salt electrolysis. A prepurified Al_2O_3 that is the starting product for the electrochemical processes is obtained by the precipita-

Table 5.3 The fabrication of Ga and Al from bauxite ore

Bauxite world reserves, 4–5 × 10^{10} tons; world production, 8.5 × 10^7 tons (1979). Composition: 40–60% Al_2O_3, 1–25% SiO_2, 1–15% Fe_2O_3, 0.01% Ga_2O_3, 0.05–0.2% other impurities (F, P_2O_5, V_2O_5, trace metals)

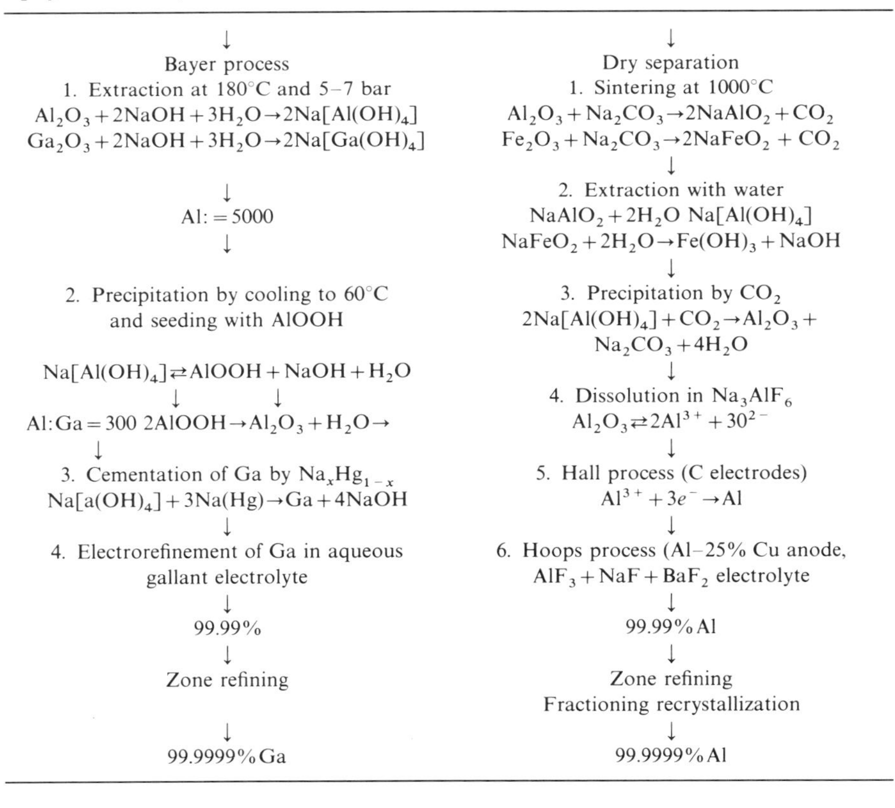

↓ Bayer process	↓ Dry separation
1. Extraction at 180°C and 5–7 bar $Al_2O_3 + 2NaOH + 3H_2O \rightarrow 2Na[Al(OH)_4]$ $Ga_2O_3 + 2NaOH + 3H_2O \rightarrow 2Na[Ga(OH)_4]$	1. Sintering at 1000°C $Al_2O_3 + Na_2CO_3 \rightarrow 2NaAlO_2 + CO_2$ $Fe_2O_3 + Na_2CO_3 \rightarrow 2NaFeO_2 + CO_2$
↓ Al: = 5000 ↓	↓ 2. Extraction with water $NaAlO_2 + 2H_2O$ $Na[Al(OH)_4]$ $NaFeO_2 + 2H_2O \rightarrow Fe(OH)_3 + NaOH$
2. Precipitation by cooling to 60°C and seeding with AlOOH	↓ 3. Precipitation by CO_2 $2Na[Al(OH)_4] + CO_2 \rightarrow Al_2O_3 + Na_2CO_3 + 4H_2O$
$Na[Al(OH)_4] \rightleftarrows AlOOH + NaOH + H_2O$ ↓ ↓ Al:Ga = 300 $2AlOOH \rightarrow Al_2O_3 + H_2O \rightarrow$	↓ 4. Dissolution in Na_3AlF_6 $Al_2O_3 \rightleftarrows 2Al^{3+} + 3O^{2-}$
↓ 3. Cementation of Ga by Na_xHg_{1-x} $Na[a(OH)_4] + 3Na(Hg) \rightarrow Ga + 4NaOH$	↓ 5. Hall process (C electrodes) $Al^{3+} + 3e^- \rightarrow Al$
↓ 4. Electrorefinement of Ga in aqueous gallant electrolyte	↓ 6. Hoops process (Al–25% Cu anode, $AlF_3 + NaF + BaF_2$ electrolyte
↓ 99.99%	↓ 99.99% Al
↓ Zone refining	↓ Zone refining Fractioning recrystallization
↓ 99.9999% Ga	↓ 99.9999% Al

Table 5.4 Typical impurity levels (ppmw) at various stages in the purification of Ga (after Papp and Solymar [13])

Element	Bayer process	Acid/base leaching	500 zone passes
Aluminum	100–1000	7	<1
Calcium	10–100	ND	ND
Copper	100–1000	2	<1
Iron	100–1000	7	<1
Lead	< 2000	30	ND
Magnesium	10–100	1	ND
Mercury	10–100	ND	ND
Nickel	10–100	ND	ND
Silicon	10–100	~1	ND
Tin	10–100	~1	ND
Titanium	10–100	1	<1
Zinc	30,000	~1	ND

tion of the $Al(OH)_3$ from the Bayer liquor, followed by calcination or is produced by dry processing of the ore in cases where the bauxite is contaminated by large concentrations of SiO_2. The Al_2O_3 is dissolved in molten cryolite and is reduced electrochemically in the well-known Hall process. The 99.9+ % pure Al thus produced is either purified by halide transport according to

$$Al(l) + \tfrac{1}{2}X_2(g) \rightleftarrows AlX(g) \tag{5.29}$$

or is alloyed with 25% copper for further electrorefinement in the three-layer Hoops process. In this process a layer of a fluoride molten salt electrolyte of lower density than the Cu alloy, but larger density than pure aluminum, is located in between the heavy alloy anode and the cathode. A liquid pool of aluminum is formed on the cathode that floats on the molten salt electrolyte and is easily separated and removed without interrupting the process. Both Al and Ga can be purified by electrorefinement in complex organometallic electrolytes [14], which will be discussed in more detail in the context of the purification of indium and of the organometallic source materials for organometallic chemical vapor epitaxy of the III–V compounds and their alloys. Alternatively the ultrapurification of Al to a 6N pure product may be achieved by recrystallization and/or zone melting.

Since for most of the residual impurities in Ga $k_0 \approx 1$, many passes are required to achieve a significant purification effect. However, in view of the favorably low melting temperature, 29.8°C, this can be accomplished with minimum reintroduction of impurities from the container walls into the metal. An exceedingly simple implementation of the zone refining of Ga is shown in Figure 5.7. The gallium is contained in a plastic tube wrapped around a rotating cylinder, which is immersed in a cooling bath except for the top portion. A heater wire strung across the top part of the Ga-containing plastic

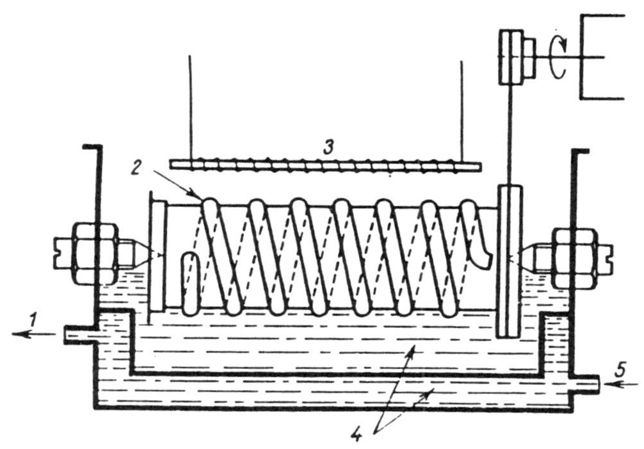

Figure 5.7 Arrangement for the zone melting of Ga: (2) Plastic tube filled with gallium (3) heater, (1), (4), and (5) cooling bath. After Papp and Solymar [13]; copyright © 1960, Hungarian Academy of Sciences, Budapest.

coil establishes a series of molten zones that pass upon rotation of the drum by one helical segment per revolution.

For Al, the container is a more problematic source of impurities However, the distribution coefficients encountered with regard to most impurities are favorable, which may be understood on the basis of the phase diagrams shown in Figure 5.8. Most metals form with Al either monotectic or eutectic dilute alloy systems as illustrated for a system A–B in Figure 5.8. Therefore, their distribution coefficients are <1, and zone refining is effective in their simultaneous removal [15]. Only a few transition metals, for example, Mn, Mo, V, W, and Zr, form peritectic alloy systems with Al and are thus enriched at the first to freeze part of zone-melted Al. Cu, Si, and Fe, which have distribution coefficients between 0.15 and 1, and are the most difficult impurities to separate. Fortunately, electrorefinement in complex aluminumalkyl electrolyte is a very efficient method of purification with regard to these impurities.

The purification of indium represents a particular challenge since it occurs jointly with many other elements as a minor constituent in galena and sphalerite [16]. It is enriched in the ZnO flumes and the slag generated in the PbS blast furnaces as well as in the lead residues of the ZnS roasting process. There are several important differences in the compositions of galena and sphalerite. For example, Fe and Te are found primarily in the galena but not at high conentration in the sphalerite. Conversely, Co, Ga, Ge, and Hg are commonly found in the sphalerite samples, but not at significant concentration in the galena. Therefore, the trace metal composition depends very much on the mineral sources, which even within one class of minerals show significant variations depending on the location of the deposit. Changes in results of the synthesis of a particular compound semiconductor may thus ensue if a supplier changes the mineral source even if the purification procedure is carefully

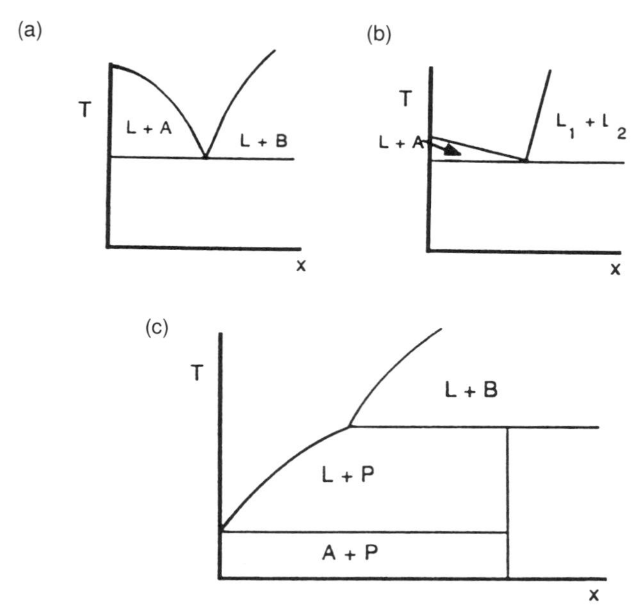

Figure 5.8 Types of phase diagrams of dilute alloys of aluminum with the following impurities: (a) eutectic systems; (b) monotectic systems; and (c) peritectic systems.

standardized. Since indium in sphalerite is only infrequently observed at levels exceeding 0.01% while Cd, Cu, Mn, and Pb are fairly frequently found at higher levels, the separation process is tedious and starts with a series of chemical separation processes based on differences in the solubilities of the various constituents in acids and bases, precipitation of the heavy metals by H_2S, cementation of In by Al or Zn, and electrorefinement. The crude metal produced from the lead residues of the ZnS roasting process contains typically 0.8% Pb, 0.5% Zn, 0.5% Sn, 0.01–0.05% Fe, plus Ag, Al, Ca, Mg, Ni, Sb, Si, and Tl at lower concentrations. Further purification employing a combination of zone refining, vacuum baking, and chemical separations allows the fabrication of a 5N+ pure In metal containing Al, Cu, Fe, Mg, Pb, and Sn as primary impurities. Zone refining is not very efficient to purify In, but reduces the residual concentrations of Ag, Cu, and Ni, which have the distribution coefficients 0.07, 0.06, and 0.01, respectively. Also, it permits the reduction of the Te level, which is difficult to remove by vacuum baking [17].

An alternative purification method for indium is its electrorefinement in organometallic electrolytes [14]. Two purification principles are combined: (1) The separation of certain groups of elements due to differences in their ability to form volatile organometallic compounds. (2) The selective plating and anodic dissolution of the elements based on the differences in their standard potentials. For example, considering the formation of alkyl compounds, the group IIIA to VIIA transition metals and the lanthanides are separated from the group IB, IIB, and IA–VIIA elements in the synthesis step since only the latter form stable alkyl compounds. The thermally very unstable group IB alkyl compounds and the nonvolatile alkyl compounds of the group IA and IIA elements, with the exception of BeR_3, are subsequently separated from the volatile and relatively stable IIIB–VIB by fractioning distillation. The remaining elements that form stable alkyl compounds are separated in the electrorefinement process at constant potential in group IIIA alkyl compound based complex electrolytes.

Equilibrium at an inert metal electrode (e.g., a Pt wire) in a solution containing a redox couple is established by the exchange of electrons (represented in reaction equations by the symbol e^-) across the metal–solution interface according to the electrode reaction

$$C_{\text{red}} \rightleftarrows C_{\text{ox}} + ze^- \tag{5.30}$$

where C_{red} and C_{ox} represent the reduced and oxidized components, of the redox couple, respectively. For example, in the redox reaction

$$[\text{Fe(CN)}_6]^{4-} \rightleftarrows [\text{Fe(CN)}_6]^{3-} + e^- \tag{5.31}$$

the ferricyanate ion $[Fe(CN)_6]^{3-}$ and the ferrocyanate ion $[Fe(CN)_6]^{4-}$ correspond to the oxidized and reduced components and $z = 1$. In the redox reaction

$$\text{HAsO}_2 + 2\text{H}_2\text{O} \rightleftarrows \text{H}_3\text{AsO}_4 + 2\text{H}^+ + 2e^- \tag{5.32}$$

$HAsO_2$ and H_3AsO_4 correspond to reduced and oxidized components and $z = 2$. The Galvani potential drop at the metal electrolyte interface cannot be measured directly, but can be compared to the Galvani potential difference between the electrolyte and a suitable reference electrode, for example, the hydrogen electrode under standard conditions (i.e., at 25°C and 1 atm pressure), forming the electrochemical cell:

|Pt| redox electrolyte| hydrogen ion containing electrolyte|hydrogen|Pt|

If the two half-cells employ two different metal electrodes, the measured cell voltage contains the contact potential between metals 1 and 2. The Galvani potential distribution for such a cell is shown in Figure 5.9.

The electrode reaction for the hydrogen electrode, that is, a Pt wire in contact with hydrogen gas in a hydronium ion containing electrolyte, is

$$\text{H}_2\text{O} + \tfrac{1}{2}\text{H}_2 \rightleftarrows \text{H}_3\text{O}^+ + e^- \tag{5.33}$$

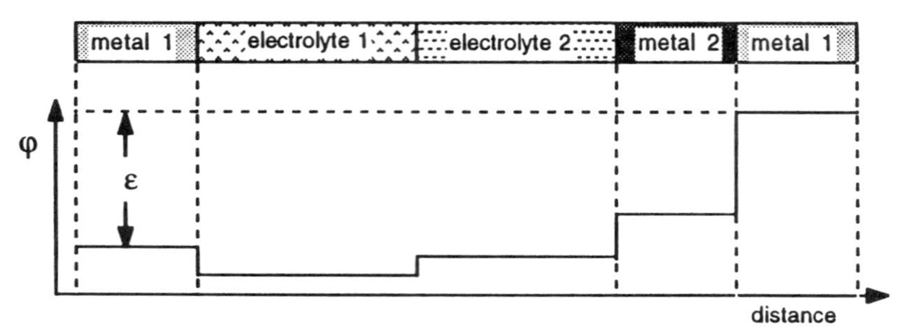

Figure 5.9 Galvani potential distribution in an electrochemical cell.

which upon subtraction from Equation (5.30) results in the cell reaction

$$C_{\text{red}} + z\text{H}_3\text{O}^+ \rightleftarrows C_{\text{ox}} + \frac{z}{2}\,\text{H}_2 + z\text{H}_2\text{O} \tag{5.34}$$

If the comparison is made to a standard hydrogen electrode (1 atm hydrogen pressure and unity hydrogen ion activity), the cell potential depends on the activities of C_{red} and C_{ox} in the electrolyte according to the Nernst equation

$$\varepsilon_0 = \frac{\Delta G}{zF} = \varepsilon^0 + \frac{RT}{zF}\ln\frac{a_{\text{ox}}}{a_{\text{red}}} \tag{5.35}$$

In Equation (5.35), $\Delta G = \Sigma v_i \mu_i$ is the free energy of the cell reaction (5.34), taking the coefficients $v_i > 0$ and <0 for the reaction products and reactants on the right and left sides of the cell reaction, respectively. In this definition of the cell voltage in terms of ΔG, the assumption is made that the positive direction of the current is from left to right. For metal/metal ion electrodes, the charge transfer at the interface is not carried by electrons, but by metal ions of charge $+zq$, and the half-cell reaction is represented by

$$M \rightleftarrows M^{z+} + ze^- \tag{5.36}$$

for example, for the cadmium/cadmium ion electrode

$$\text{Cd} \rightleftarrows \text{Cd}^{2+} + 2e^- \tag{5.37}$$

that is, $z = 2$.

Clarification is needed with regard to the relation between the electrochemical potential scale relative to the standard hydrogen electrode and the energy scale relative to the vacuum level that is used in defining band-edge energies. Based on a thermodynamic evaluation for the Ag/Ag^+ electrode, the

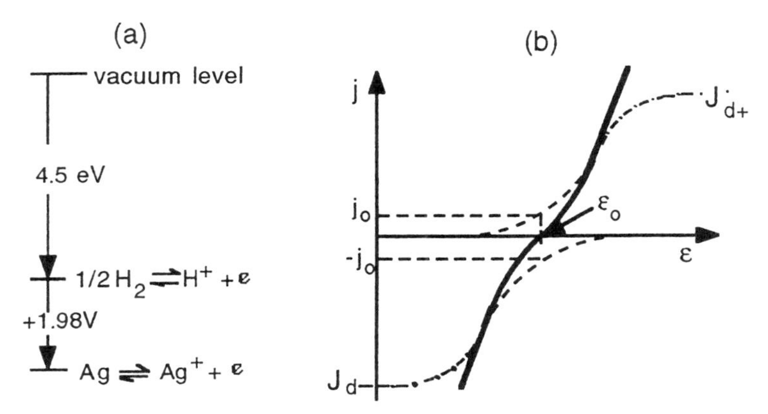

Figure 5.10 (a) Standard electrode potentials plotted versus the vacuum level. (b) Schematic representation of the J–ε characteristics according to the Butler–Volmer equation for $\alpha = 0.65$.

absolute potential of the hydrogen electrode under standard conditions has been determine and is located at approximately -4.5 V relative to the vacuum level [18]. A subsequent experimental evaluation resulted in the value -4.73 V [19]. The definition given of ε^0 corresponds to a downward shift from the position of the standard hydrogen electrode for positive values of ε^0, as illustrated in Figure 5.10(a).

If the only barrier to electrical transport is the activation energy for the chare-transfer reaction at the eletrode, the dependence of the current density J on the potential ε is given by the Butler–Volmer equation

$$J = J_0\left[\exp\left(\frac{\alpha zF}{RT}\eta\right) - \exp\left(-\frac{(1-\alpha)zF}{RT}\eta\right)\right] \tag{5.38}$$

The deviation of the electrode potential ε upon current flow from the equilibrium potential ε_0 is called the *overvoltage* $\eta \equiv \varepsilon - \varepsilon_0$. In the absence of other kinetic barriers, it represents a measure of the barrier to charge transfer. The charge-transfer coefficient $0 \leqslant \alpha \leqslant 1$ is a measure of the symmetry of this barrier [18]. At $\varepsilon > \varepsilon_0 (\eta > 0)$, anodic oxidation with $J > 0$ results. For a metal/metal ion electrode, this corresponds to dissolution. At $\varepsilon < \varepsilon_0 (\eta < 0)$, cathodic reduction with $J < 0$ results. For a metal/metal ion electrode this corresponds to plating. Figure 5.10(b) shows a plot of J versus ε according to Equation (5.38), where the anodic and cathodic current densities J_+ and J_-, respectively, are represented by dashed lines and the total current density is represented by a full line for $\alpha = 0.5$. Note that at the equilibrium potential, the anodic and cathodic current densities are the same, resulting in zero external current. However, the equilibrium is dynamic; that is, in equilibrium the redox and metal ion electrode reactions occur in both directions with exchange current density $J_0 = J_+(\varepsilon_0) = |J_-(\varepsilon_0)| \neq 0$.

Generally, the rate of diffusion of ions in the solution is also limited, causing deviations from equilibrium in the concentrations of C_{ox} and C_{red} at the electrode surface. Also, homogeneous or heterogeneous chemical reactions may precede or succeed the charge-transfer step, resulting in changes in the concentrations of the species participating in the charge-transfer reaction. Since this causes deviations of the electrode potential from the equilibrium value, in addition to the charge-transfer overvoltage, more complex current–voltage characteristics than shown in Figure 5.10(b) result. This is indicated in Figure 5.10(b) by the dash-dotted curves, which show saturation of the current at large η. A diffusion and/or reaction-limited current density J_d is expected as a result of the limit to the concentration gradient at the interface when one of the components needed for sustaining the electrode reaction is depleted to zero concentration. Since the value of J_d depends on the concentration of the critical component in the bulk of the solution, a measurement of J_d provides for quantitative analysis.

In an electrorefinement process, the absence of dissolution of all metals with equilibrium potentials greater than the potential applied for the anodic dissolution of the metal to be purified causes them to accumulate as an insoluble residue in the anode compartment. Also the elements, with equilibrium potentials that are smaller than the potential applied in the cathodic deposition of the metal to be purified, do not plate out, thus resulting in an overall purification effect. In order to carry out electrorefinement at constant potential, a three-electrode arrangement is used, employing an electronic potentiostat to match the electrode potential, measured against a suitable reference electrode, to a chosen set-point value by appropriate regulation of the current density on the working electrode. This is shown schematically in Figure 5.11(a). In cyclic voltammetry, the set-point voltage is cycled between chosen limits, and the current density versus potential curve is recorded. As the potential sweeps through the equilibrium potential of a particular metal/metal ion electrode in the cathodic direction, plating coupled to a depletion of the electrolyte of these metal ions at the electrode surface results in a reduction peak. Further cathodic peaks associated with the discharge of other metal ions in the same solution may result at other characteristic potential values. This is illustrated in Figure 5.11(b) for In and Al plating from an electrolyte with the composition: $C_6H_5CH_2(CH_3)_3NF$:$Al(C_2H_5)$:$In(C_2H_5)_3$:$C_6H_5CH_3 = 1{:}2.4{:}x{:}3$. The peak height is proportional to the concentration; that is, the three curves shown in Figure 5.11 correspond to three different triethylindium concentrations in the electrolyte. Note that the cathodic reduction waves, corresponding to the deposition of In and Al, are separated by ~ 0.8 V. This permits the efficient separation of In and Al in this electrolyte under potentiostatic control of the electrorefinement process [20].

The aluminum alkyl compounds used as the major constituent of the electrolyte in the electrorefinement cycle are made on a large scale by the chemical industry. The Ziegler process invented in 1955 proceeds by the direct reaction of hydrogen and isobutene at 120–150°C and 100–200 atm with aluminum metal to tri-*i*-butylaluminum [21]. This compound is subsequently

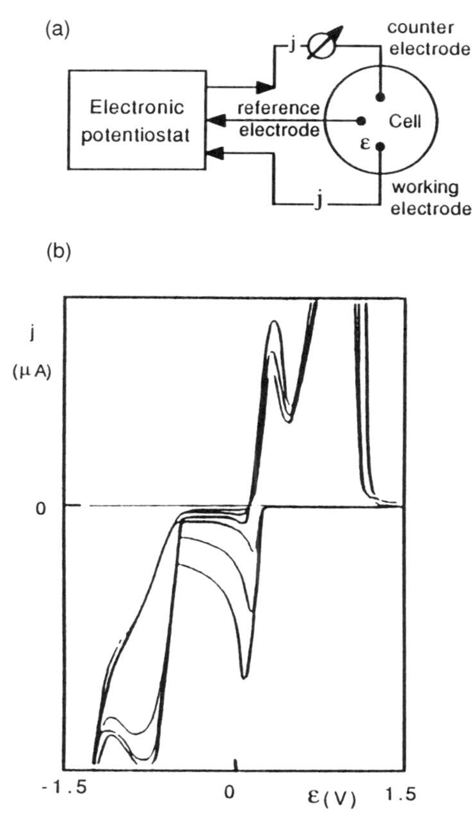

Figure 5.11 (a) Schematic representation of a three-electrode electrochemical cell under potentiostatic control. (b) Cyclic voltammogram for a complex organometallic electrolyte composed of trimethylbenzylammonium fluoride + triethylaluminum + triethylindium + toluene. After Bachmann and Su [20]; copyright © 1987, United Engineering Trustees, Inc., NY.

converted into triethylaluminum in a replacement reaction with ethylene recycling the liberated *i*-butene back into the reaction with additional aluminum metal. The overall reaction is thus

$$Al + \tfrac{3}{2}H_2 + 3C_2H_4 \rightarrow Al(C_2H_5)_3 \tag{5.39}$$

Organometallic compounds of B, Pb, P, Sn, and Zn are produced from AlR_3 by the reaction

$$nAlR_3 + MeCl_n \rightarrow MeR_n + nAlR_2Cl \tag{5.40}$$

The syntheses of the alkyl compounds that are used in the OMCVD of compound semiconductors can be achieved on a laboratory scale by a large variety of methods and can be purified by distillation. Excellent comprehensive reviews of this field exist

[22]. Particularly important syntheses are based on Grignard reactions, such as

$$6R\text{Br(ether solution)} + 2\text{In(s)} + 3\text{Mg(s)} \rightarrow 2\text{In}R_3\text{(ether solution)} + 3\text{MgBr}_2 \quad (5.41)$$

Friedel–Crafts-type reactions in the presence of aluminum chloride catalyst, for example,

$$AsCl_3 + 2C_5H_{12} + 2C_6H_6 \rightarrow C_6H_5AsH_2 + C_6H_5C_5H_{11} + 2HCl \quad (5.42)$$

and exchange reactions of the metal halides with Li, Na, or Hg organic compounds. Because of the extreme toxicity of mercuryalkyl compounds, particularly stringent precautions must be taken when working with this material. The group V halides can be purified to very high purity by distillation. Also, the group III halides can be made in very high purity and are well suited for zone refining.

Electronic-grade organometallic compounds generally are synthesized from the purest starting materials available. Large batches of indium and gallium alkyl are conveniently prepared by direct electrolysis of Grignard solutions (RMgI) in ether (di-isopentylether, anisole) [23]. The Grignard reagent can be prepared from alkylhalide and magnesium metal in the electrolysis cell with the Ga pool or In metal electrodes in place. During electrolysis magnesium is deposited on the cathode, but is chemically dissolved if excess alkylhalide is provided. Thus the ether adduct of the group III alkyl compound is formed in the overall reaction

$$3R\text{MgI} + 3R\text{I} + 2M + 2R'_2\text{O} \rightarrow 2R_3M{:}\text{O}R'_2 + 3\text{MgI}_2 \quad (5.43)$$

where M stands for Ga or In, R is an alkyl (e.g., CH_3), and R' is an alkyl or aryl (e.g., C_6H_5). The group III alkyl compounds are thermally liberated from the ether adducts in a subsequent distillation step. Their further purification by physical separation methods is possible, for example, by fractioning distillation or sublimation. Major impurities in organometallic compounds are traces of organic contaminants. These are primarily incompletely utilized reactants (e.g., alkyl halides) and solvents, of which ethers are particularly detrimental since they contain oxygen, which may interfere in subsequent OMCVD processes [24]. Although distillation of the neat group III alkyl compounds removes paritial oxidation products (e.g., $[In(CH_3)_2]_2O$ from $In(CH_3)_3$ [25]), as well as some of the residual impurities, it is not very effective in removing volatile OM impurities, such as silanes and group IIB alkyl compounds. Recrystallization of adduct compounds with nonvolatile Lewis bases, in particular 1,2-bis-diphenylphospine-ethane, followed by thermal decomposition of the purified adduct at 100–150°C, has been found more efficient in achieving ultrapure products [26]. Figure 5.12 shows the inductively coupled plasma emission (ICP) signals for Zn and Si in $In(CH_3)_3$ before and after purification by this

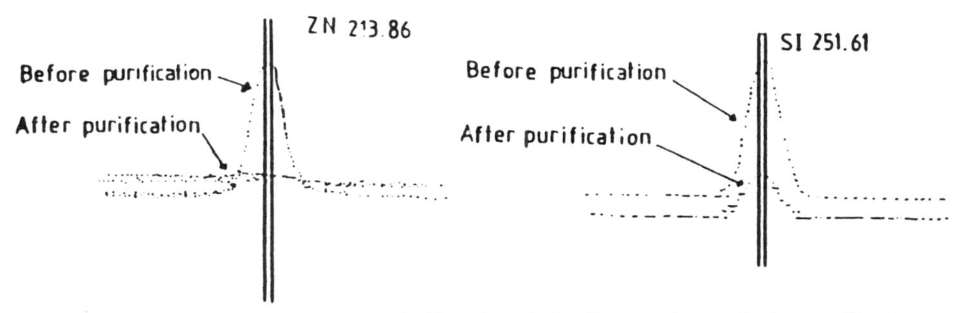

Figure 5.12 ICP signals for Zn and Si in trimethyl indium before and after purification by the diphos-adduct method. After Jones et al. [28]; copyright © 1986, Elsevier Science Publishers B.V., Amsterdam.

adduct method. On-site purification has been essential for certain dopant source materials, such as biscyclopentadienyl magnesium [27], since they decompose slowly during storage upon exposure to light.

Phosphorus is produced worldwide at a rate of ~1 million tons per annum from phosphate rock by carbon reduction. The phosphorus vapor is condensed and is further purified by fractioning distillation. The product obtained is the white metastable modification of phosphorus that is pyrophoric and not suitable for direct use in the synthesis of compound semiconductors. It is converted into the violet nonpyrophoric modification by heating for prolonged periods of time. Ideally violet phosphorus crystallizes in the monoclinic structure and has a well-defined sublimation pressure versus temperature behavior, permitting its use for the control of the phosphorus partial pressure during compound synthesis. However, in practice the 6N pure product, which is commercially available, is not completely crystallized; that is, it represents a complicated cross-linked structure with vapor pressure characteristics in between those of the white and the monoclinic violet modifications. Since this material transforms upon heating toward the violet modification with a concomitant change in the vapor pressure characteristics, reliable pressure control via the two-phase equilibrium can only be expected for completely transformed violet phosphorus. For temperature distributions above the critical temperature of phosphorus, the desired vapor pressure may be established by sealing an appropriate amount of excess phosphorus in the ampule. In this case, it is important to recognize that the equilibrium

$$P_4 \rightleftarrows 2P_2 \tag{5.44}$$

results for nearly ideal behavior (i.e., at low vapor density) in a pressure versus temperature relation

$$p = (1 + \alpha)\frac{n}{v}RT \tag{5.45}$$

where α is the degree of dissociation of P_4 into P_2 molecules. At high vapor density the ideal gas behavior is no longer observed. However, in the pressure range required for the synthesis of III–V compounds, the p–T relation is very well described by the van der Waals equation

$$\left(p + \frac{a}{V^2}\right)(V - b) = RT \tag{5.46}$$

where $a \approx 30.1$ liter^2 atm/mole and $b \approx 0.109$ liter/mol [29]. Figure 5.13 shows the pV–p diagram for phosphorus. In the temperature range $800 \leqslant T \leqslant 1100$ K the second virial coefficient is given by

$$B = b - \frac{a}{RT^x} \tag{5.47}$$

where $1.03 \leqslant x \leqslant 1.06$.

Arsenic is made primarily by roasting of arsenopyrite according to the reaction

$$2FeAsS(s) \rightarrow 2FeS(s) + As_2(g) \tag{5.48}$$

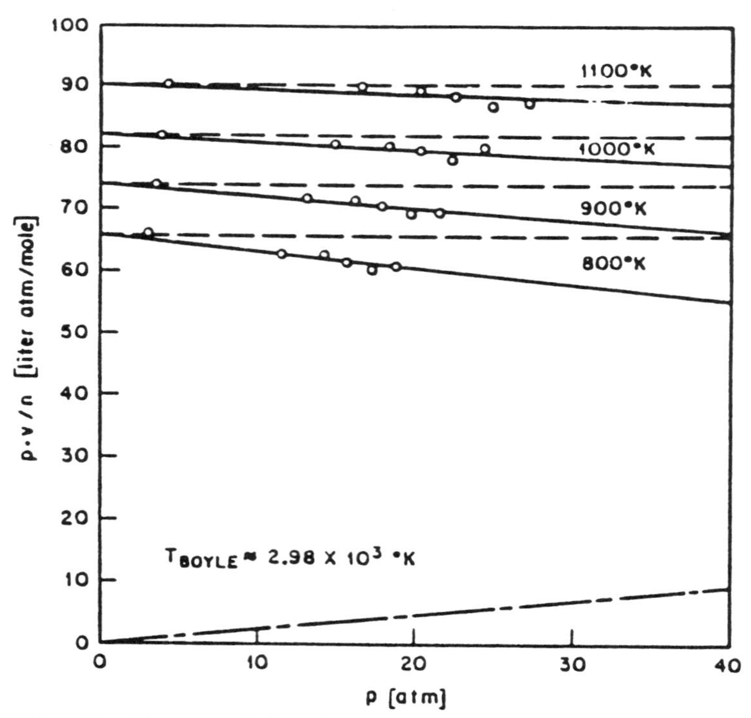

Figure 5.13 pV–p diagram of phosphorus. Dashed lines: $pV = RT$. Solid lines: least-squares fit of experimental data to $pV = RT + Bp$. Dash-dotted line: Boyle curve. After Bachmann and Buehler [29]; copyright © 1974, The Electrochemical Society, Pennington, NJ.

The As thus obtained is dissolved in lead and is then sublimed off the alloy, which binds sulfur more strongly than arsenic. Table 5.5 shows the concentrations of impurities in a typical batch of arsenopyrite. Note that Se, which exists in the ore as a minor constituent, is particularly tenacious in remaining in the arsenic throughout the purification cycles, so that in the ultrapure As it is a major contributor to the residual impurity composition. The concentrations of impurities in nominally 99.9999% (6N) pure As and P are in the ppb range, corresponding to typically 10^{14}–10^{16} cm^{-3}.

The group V hydrides are made from the elements either directly by reaction with hydrogen or via the zinc or calcium compounds, which are hydrolyzed or reacted with sulfuric acid. On a laboratory scale also the reaction of arsenic trichloride with lithium hydride and the decomposition of phosphonium iodide in KOH result in high-purity AsH_3 and PH_3, respectively. Both PCl_3 and $AsCl_3$ are purified to exceptional purity by distillation and are essential materials for the halide transport of III–V compounds and alloys.

Table 5.6 shows the concentrations of impurities in commercial high-purity arsine. Comparison with Table 5.5 shows that the majority of the impurities are already present in the arsenopyrite. However, the relative ratios are quite different, reflecting the different efficiencies of purification processes for different components. For example, the Se concentration in the arsine normalized to the Cu concentration is far higher than in the arsenopyrite, reflecting the very similar properties of Se and As and the difficulty of separating H_2Se and AsH_3. Se is a shallow donor in GaAs. Since arsine and phosphine undergo wall reactions with the stainless steel cylinder in which they are contained [30], contaminating the gases with water, on-site purification is required. This is usually accomplished by inserting molecular sieve traps into the source lines and by scrubbing with Al–Ga–In eutectic. Similar point-of-use purification is applied to most gases used in semiconductor processing, that is, hot Ti and Cu for the gettering of noble gases and nitrogen, respectively, and a variety of molecular sieve filters for trapping residual water vapor and organic contaminants.

The analytical evaluation of the trace impurities in semiconductors is most

Table 5.5 Impurity concentrations in arsenopyrite (ppm) (after Fleischer [16])

Impurity	Concentration	Impurity	Concentration
Ag	90	Ni	< 3000
Au	8	Pb	50
Co	30,000	Pt	0.4
Cu	200	Re	50
Ge	30	Se	50
Mn	3000	V	300
Mo	60	Zn	400

Table 5.6 Impurity concentrations (ppm) in electronic-grade AsH_3 (after Kroll [30])

Impurity	Concentration	Impurity	Concentration	Impurity	Concentration
Cu	0.04–0.51	Na	0.009–0.50	Se	9.78–16.27
Fe	2.71–3.77	Pb	0.53–1.53	Si	2.02–3.41
Ge	2.00–3.07	Sb	3.90–6.89	Zn	3.50–9.98

sensitively and accurately done by electrical measurements, such as Hall effect measurements for the shallow impurities and DLTS for deep traps, which make it possible to detect certain impurities at the parts per trillion level. However, the above-discussed starting materials for compound semiconductor synthesis and crystal growth include metals and insulators. Therefore, other less sensitive methods of evaluation must be employed. They fall into 5 categories: (1) methods of analysis based on electronic transitions in free atoms or molecules; (2) neutron activation analysis (NAA); (3) mass spectrometric analysis; (4) methods of analysis based on vibrational and rotational spectroscopies; (5) electrochemical methods. These methods are often complementary, and a combination of different techniques is thus desirable for pinpointing the concentrations of specific impurities. For example, NAA is highly sensitive for the detection of In, Dy, Rh, and Re, which can be detected in the low ppba range. However, it is quite insensitive to Ca and Fe, which can only be detected at a level >1 ppma. On the other hand, Ca and Mg can be detected at sub-ppba level by inductively coupled plasma (ICP) emission or absorption spectroscopy, while Re and Pd can be detected only at >100 ppba level by this method.

For both optical and mass spectrometric measurements on solid samples, appropriate vaporization techniques must be developed. Dissolution and spray evaporation in a flame or a plasma torch is a possible option, but dilutes the sample and is a source of systematic errors. Spark source mass spectrometric (SSMS) analysis, which in the early phase of trace metal analysis in the context of electronics applications was the workhorse of the industry, is at present being increasingly replaced by other methods.

Figure 5.14 shows a schematic representation of the apparatus used for sputter-induced resonance ionization spectroscopy (SIRIS) [31]. The sample surface is sputtered by a pulsed microbeam of Ar^+, and selected atoms are ionized by resonance ionization by a tuned dye laser beam. The ions are energy analyzed and detected in either a time of flight or magnetic sector mass spectrometer. Sputtering typically 5×10^8 atoms per ion pulse and achieving a yield of 0.3% in the ions detected per atom sputtered result with single-ion-counting techniques in a sensitivity in the low-ppb range. For specific cases (e.g., ^{56}Fe in Si), detection limits in the mid-ppt range have been achieved.

A method that is widely used in the semiconductor industry for depth profiling is secondary ion mass spectroscopy (SIMS). In SIMS the sample surface is sputtered by an oxygen or cesium ion beam that scans the surface of

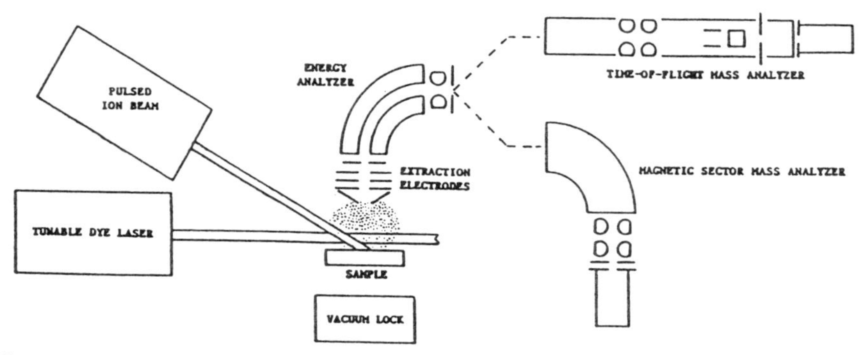

Figure 5.14 Schematic representation of the apparatus for SIRIS analysis. After Parks et al. [31]; copyright © 1988, North Holland Publishing Company, Amsterdam.

the sample, forming in time a shallow flat crater. Secondary ions of the sputtered surface layer are extracted into a double focusing mass spectrometer for analysis. Only a small area in the center of the sputtered crater is used for the analysis, providing excellent depth resolution. Matrix effects are important in SIMS and require careful calibrations for quantitative analysis. Also, the difficulty of separating trace elements with similar mass/charge ratio as the major constituents degrades the detection limit.

For example, the detection limit for Zn and Cu in GaAs by SIMS is poor (see Table 5.7), but a considerable improvement is possible in such cases by a combination of SIMS with chemical separation techniques [32]. In the case of GaAs, after dissolution of a typical 200 mg sample, anion exchange on Dowex 2×8 resin successfully separates Zn from Ga, as shown in Figure 5.15. Also, Mn and Cr are separated at the nanogram level from As by liquid chromatography on a $25\,\text{cm} \times 1\,\text{cm}^2$ column, and the separation of the Cu from the Ga is possible by liquid extraction methods, utilizing the difference in their

Table 5.7 Detection limits and precision of direct SIMS analysis and chemical SIMS analysis of GaAs (after Tanaka and Kurosawa [32])

Element	Analytical Method	Precision (RSD%)[a]	Detection Limit (ppb)[a]	Precision (RSD%)[b]	Detection Limit (ppb)[b]
Cr	Isotope[c]	10	7	5	3
Mn	Internal[d]	10	5	5–10	0.5
Cu	Isotope	10	350	5	5
Zn	Isotope	10	5000	5	5
Al	Internal	10	1	5–10	50

[a]Direct SIMS.
[b]Chemical SIMS.
[c]Isotope dilution method.
[d]Internal standard method.

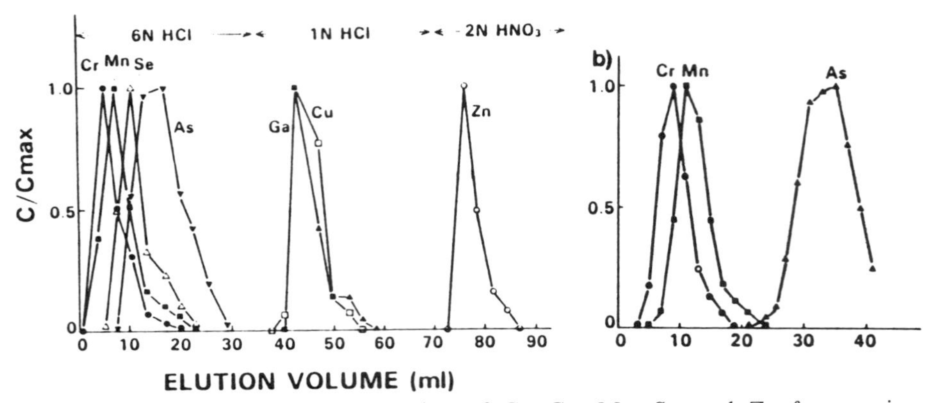

Figure 5.15 Anion exchange separation of Cr, Cu, Mn, Se, and Zn from major constituents Ga and As. After Tanaka and Kurosawa [32]; copyright © 1986, The Electrochemical Society, Pennington, NJ.

solubilities in two immiscible liquids. After evaporation of the solvents from droplets of the separated liquid fractions on a thoroughly cleaned silicon wafer, thin films of these fractions left behind were analyzed by SIMS, achieving the improved detection limits and precisions listed in Table 5.7.

Also, purely chemical and electrochemical methods can be combined to achieve a precision of 10^{-4}–10^{-5} in the determination of major-phase composition. For example, the deviations from stoichiometry of nominally undoped semi-insulating GaAs has been determined by completely converting a precisely weighed sample of the semiconductor into ions of well-defined charge state (i.e., Ga^{3+} and As^{5+}), using oxidative digestion in nitric acid in a pressurized autoclave. Then the Ga^{3+} ions are complexed with a well-known amount of ethylenediamine-tetra-acetic acid (EDTA). Knowing the weight of the GaAs sample, the weight of the EDTA is chosen to exceed slightly the amount required for complete complexation under the assumption of an exactly stoichiometric atom ratio for the GaAs. The precise value of this excess is determined by back-titration with Hg^{2+} ions formed by anodic oxidation on a mercury-plated gold electrode. The charge expended in the anodic generation process is precisely measured. The endpoint of the back-titration reaction is determined using a three-electrode potentiostatic circuit, as illustrated in Figure 5.11(a) for keeping the working electrode at the appropriate constant potential for the formation of the Hg^{2+}–EDTA complex. The current density changes steeply when the excess EDTA is used up, providing for accurate endpoint determination. Similarly, the As^{5+} ions are first chemically reduced to the As^{3+} state and are then titrated by anodically produced iodine on a Pt electrode, again using amperometric endpoint determination at constant potential. For a recent review of electroanalytical methods of trace impurity determinations and stoichiometry evalutions providing more detailed information, the reader is referred to Ref. [33].

5.2 Equilibrium Conditions and Kinetic Limitations of Growth

In microelectronics the uniformity of the properties of individual circuit elements separated by large distances on the semiconductor wafer is essential for the overall performance and reliability of the IC. Therefore, the circuits are fabricated on single crystal substrates that are characterized by the perfect preservation of the translational symmetry of the lattice through their entire volume under ideal conditions. In principle, substrate crystals can be grown from a variety of nutrients, including solution growth, recrystallization, and vapor transport. However, the slow rate of growth from dilute nutrients permits the use of solution growth and vapor transport only in the case of specialized applications of materials with a high markup in value (e.g., radiation detector materials). Also, the formation of defects in the crystal volume swept by the grain boundaries during the recrystallization of semiconductors (see Figure 3.12) is generally incompatible with the requirements of circuit manufacturing. Therefore, crystal growth from the melt is preferred. Important aspects of the growth of large single crystals are:

1. Seeding, that is, the control of the nucleation of secondary grains and of the crystal orientation;
2. The control of the density and distribution of extended defects;
3. The control of the distribution of residual strain;
4. The control of the concentration and distribution of point defects;
5. The control of the stoichiometry of compound semiconductors.

Since the bonding of the atoms in the surface layer of a solid generally differs from the bonding in the bulk, the free enthalpy of a solid is composed of a volume and a surface term

$$G = G_{\text{volume}} + G_{\text{surface}} = G_{\text{volume}} + \sum \sigma_{hkl} O_{hkl} \tag{5.49}$$

where the σ_{hkl} are surface free energy values that depend on the surface orientation (hkl). O_{hkl} denotes the area of the surface of orientation (hkl). The surface energy term increases proportionally to the surface area and thus is proportional to the square of the radius r of the particle, while the volume free energy increase is proportional to r^3. For infinitely extended phases $(r \to \infty)$, that is, crystals of macroscopic dimensions, the ratio of the surface to the volume contribution approaches zero so that the contribution of the surface to the free energy is negligible. The phase rule derived in Section 5.1 without consideration of surface contributions to the free energy thus applies to infinitely extended phases. However, in the nucleation of a new phase, where the fraction of atoms residing in surface positions is substantial, surface effects must be expected to play an important role.

This is illustrated in Figure 5.16 for the simple case of a one-component system. The application of the phase rule to such a system, for example, Si,

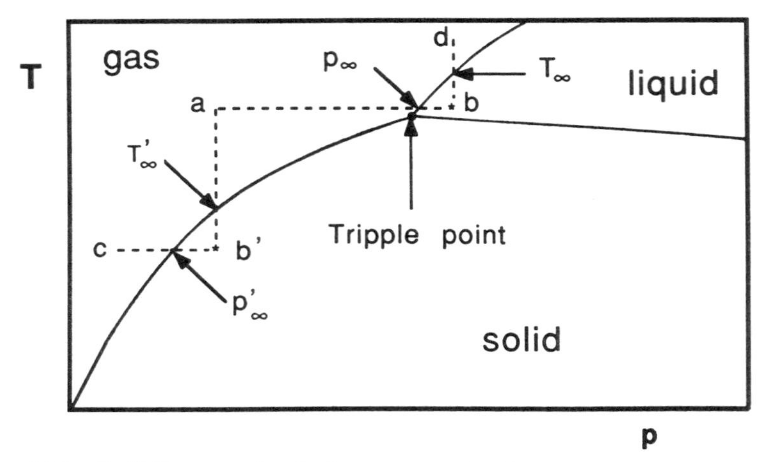

Figure 5.16 P–T diagram of a single-component system.

results in $f = 3 - p$. The two-phase equilibria are in this case characterized by one degree of freedom, so that for any chosen value of the temperature, the pressure is fixed. The single-phase fields in which the temperature and pressure can be chosen independently are thus separated from each other by curves $p(T)$ representing equilibrium between two phases and the equilibrium between three phases; for example, solid (s), liquid (l), and vapor (g), is characterized by an invariant point (triple point).

For the formation of a droplet of a liquid in a vapor volume that is slowly compressed at constant temperature along the path $a \rightarrow b$ in Figure 5.16, the phase rule predicts that once the two-phase line is reached at point p_∞, the liquid phase should exist in equilibrium with the vapor phase. In practice the compression must be continued beyond p_∞ to point b, that is, to a pressure $p_r > p_\infty$ to observe the nucleation of a second phase. The quotient $(p_r - p_\infty)/p_\infty$ is called the *supersaturation ratio*, which must be provided for nucleation to occur. Note that point b could be reached along an alternative path $d \rightarrow b$ by cooling the vapor at constant pressure from an initial temperature T_i at point d to the final temperature T_f at point b, which is lower than the temperature T_∞ at which two infinitely extended phases would be expected to exist in equilibrium; that is, the supersaturated state corresponds to a supercooled state, both requiring the crossing of the boundary between the single-phase fields defined for infinitely extended phases in equilibrium [34]. An analogous behavior is observed for the nucleation of a solid particle from the vapor phase along the path $a \rightarrow b'$ or $c \rightarrow b'$.

In the case of an isotropic surface free energy σ, that is, for the nucleation of droplets in a vapor phase or of small gas bubbles in a liquid phase, the minimum surface area at constant volume, that is, the minimum in the surface contribution to the free energy, is attained for spherical shape. Consider the following circular process: (1) Transport of a small number dn of molecules corresponding to the infinitesimal volume $dv = V\,dn$ from the infinitely extended phase to a cluster of radius r of the same material. The work performed in this process is $(\mu_r - \mu_\infty)dn$, where μ_r and μ_∞ refer to the chemical potentials of the cluster and the infinitely extended phase. (2) Removal of the same number of atoms and transfer back to the infinitely extended phase, reducing the surface of the cluster by the area dO corresponding to the work $-\sigma\,dO$. If the process is conducted reversibly, the sum of the work performed must be zero, that is,

$$(\mu_r - \mu_\infty)\,dn = \sigma\,dO \tag{5.50}$$

so that with $dO = 8\pi r\,dr$ and $dv = 4\pi r^2\,dr$

$$\mu_r - \mu_\infty = \sigma\,\frac{dO}{dv} = \frac{2\sigma V}{r} \tag{5.51}$$

For the formation of a gas bubble in a liquid at moderate pressure, we may relate the chemical potential to the vapor pressure by the relation

$$(\partial\mu/\partial p)_{T,\mathrm{ni}} = V_\mathrm{m} = RT/p,$$

so that

$$\mu_r - \mu_\infty \approx RT\int_{p_\infty}^{p_r} d\ln p = RT\ln\frac{p_r}{p_\infty} = \frac{2\sigma V_\mathrm{m}}{r} \tag{5.52}$$

which is Thomson's equation.

For a crystal where the surface free energy is anisotropic [i.e., depends on the orientation (hkl) of the crystal faces], a similar circular process can be carried out, replacing the radius in the isotropic case by the central distances d_{hkl} of the crystal faces as indicated for the cross section through a crystal in Figure 5.17. It results in the Gibbs–Thomson equation

$$\mu_{d_{hkl}} - \mu_\infty = RT\ln\frac{p_{d_{hkl}}}{p_\infty} = \frac{2\sigma_{hkl}V_\mathrm{m}}{d_{hkl}} = C' \tag{5.53}$$

Since at equilibrium the vapor pressure over all faces appearing on the equilibrium shape of the crystals must be the same, at constant temperature Equation (5.53) equates to a constant C'. Dividing by $2V_\mathrm{m}$, which is also a constant quantity, results in Wulff's law

$$\frac{\sigma_{hkl}}{d_{hkl}} = C_\mathrm{W} \tag{5.54}$$

stating that in equilibrium the central distances to the crystal facets that bound the crystal must be proportional to the associated surface free energy values, all with the same proportionality constant $1/C_\mathrm{W}$. A mathematical proof for this law is provided by von Laue in Ref. [35].

In order to construct the equilibrium shape of a crystal nucleus, we need the surface free energy as function of the crystallographic orientation. An estimate

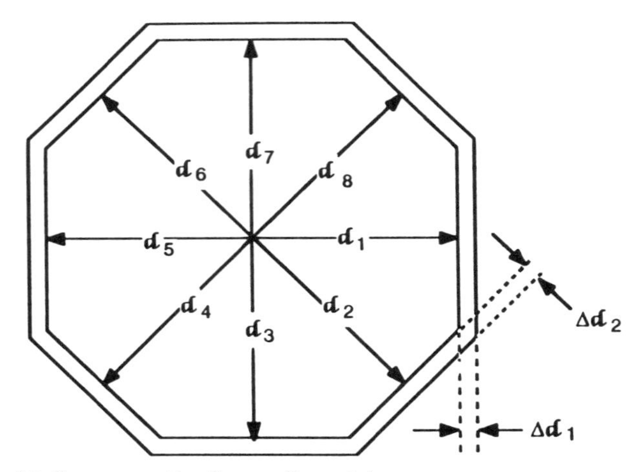

Figure 5.17 Uniform growth of a small crystal.

of this relation may be obtained on the basis of a simple model introduced by W. Kossel, which evaluates the bonding energy of the atoms or molecules of which the crystal is composed by a count of their first-, and if higher accuracy is desired, second- and higher-order nearest neighbors. Figure 5.18 illustrates this for a simple cubic structure, representing the building blocks of the crystal as cubes. Let the bond energy for nearest-neighbor bonds be denoted by φ. Then in a model that neglects all but nearest-neighbor bonding, it suffices to count the number of nearest neighbors of an atom to estimate the bonding energies for the various positions on the surface and inside the crystal. The labeling in Figure 5.18 is based on a nearest-neighbor count, so that the atoms in positions 1, 2, 3, 4, and 5 are bonded to the crystal by φ, 2φ, 3φ, 4φ, and 5φ, respectively. An atom in the interior of the crystal is bonded to 6 nearest neighbors, that is, has a bonding energy 6φ. Therefore, it requires an energy 6φ to bring an atom from the interior of the crystal into the vapor phase. However, since in this process a vacancy is created that can annihilate without change to the surface configuration at position 3, the average bonding energy per atom is for the infinitely extended crystal 3φ. In thermal equilibrium, the probability for addition and removal of an atom from kink position 3 must be the same. Since the atoms impinging on the surface and bonding to the crystal in positions 2 and 1 are not as strongly bonded as in position 3, they are more likely to re-evaporate than to stay on the surface. On the other hand, atoms incorporated in positions 4 and 5 are more strongly bonded than the atoms in position 3; that is, they are more likely to stay attached to the crystal than to re-enter the vapor. Consequently position 3 is the site where a crystal building block reaching the surface becomes incorporated into the lattice.

For estimating the surface free energy, we note that in separating the simple cubic lattice on (001) breaks 1 nearest-neighbor bond and generates two

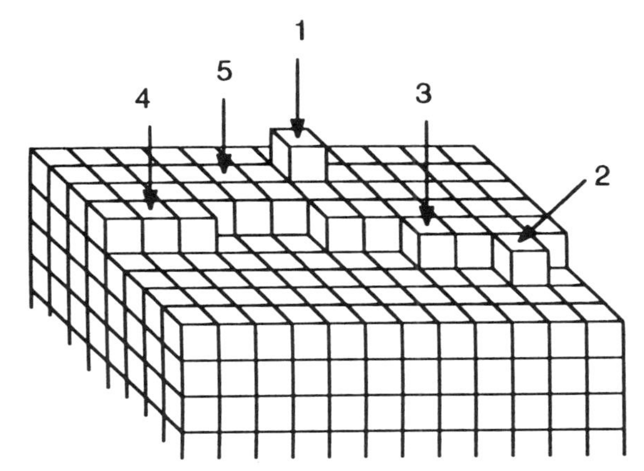

Figure 5.18 Kossel model of a solid crystallizing in the simple cubic structure.

surfaces of area a_0^2 per atom, where a_0 is the lattice constant, which, in this case, equals the nearest-neighbor distance. Consequently the surface free energy on (001) is $\sigma_{001} = \varphi/2a_0^2$. An inclined surface that is composed of (001) steps results in both (001) and (010) surface components, as illustrated in Figure 5.19(a). For an average surface face inclination with regard to (001) by an

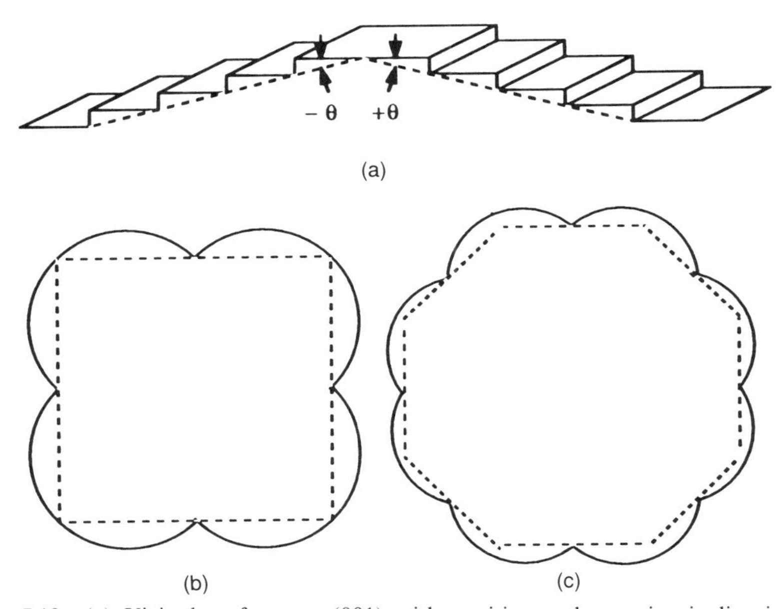

Figure 5.19 (a) Vicinal surfaces to (001) with positive and negative inclinations θ. Polar plots of the surface free energies and traces of the crystal facets expected to appear on the equilibrium shape. (b) Considering nearest-neighbor contributions only; (c) Considering nearest- and second-nearest-neighbor contributions to the bonding.

angle θ about the x axis, the projected (001) and (010) surface components per unit area are given by $|\cos\theta|$ and $|\sin\theta|$, respectively. Consequently the surface free energy per unit area is given by

$$\epsilon(\theta) \cong \epsilon_{(001)}|\cos\theta| + \epsilon_{(010)}|\sin\theta| \tag{5.55}$$

that is, $\sigma(\theta) = \sigma_{001}(|\cos\theta| + |\sin\theta|)$, assuming that $\sigma_{001} = \sigma_{010}$. Since $|\cos\theta| + |\sin\theta| \geqslant 1$, a polar plot is thus characterized by cusps on the $\{100\}$ faces, as shown in Figure 5.19(b) on a relative scale. Because the slope of $\sigma(\theta)$ is discontinuous at the position of the cusps, the corresponding surfaces are called *singular faces* of the crystal. They have relatively low surface free energies and are usually the faces that appear on the stationary growth shape in largest dimensions. Figure 5.19(c) shows a plot including second-nearest-neighbor contributions to the bonding. Using Wulff's law, the traces of the crystal facets that appear on the equilibrium shape are drawn in both figures. The construction of the equilibrium shape proceeds as follows: Vectors proportional to $\sigma(\theta)$ are drawn from a central point, and planes perpendicular to these vectors are constructed that intersect in a closed figure of minimum volume, representing the equilibrium crystal shape. An absolute value of σ_{001} may be estimated from the enthalpy of sublimation, which for the simple cubic lattice in the nearest-neighbor approximation equals $3N_A\varphi$, where N_A is Avogadro's number. For other lattices analogous evaluations can be made with little effort, and the reader is encouraged to work out a few examples as an exercise. For example, in an fcc lattice one obtains for the (111) surface $\sigma = \varphi/\sqrt{3}a_0^2$ and 6φ for the bonding energy in the kink site. Taking a typical value for the heat of sublimation and lattice constant of fcc metals results in a value for the surface free energy of the order of 10^3 erg/cm^2, which is in agreement with the list of experimental data shown in Table 5.8. Also, the free energy values scale approximately with the enthalpies of evaporation, which are by one to two orders of magnitude smaller for ionic materials and molecular crystals, respectively, corresponding to surface free energies of typically hundreds and tens of erg/cm^2 for these materials.

The connection of Kossel's atomistic model to the Thomson–Gibbs equation is made by considering the average energy per atom for the removal or addition of a net plane to a crystal of finite dimensions. Let n denote the number of atoms on the edge of the crystal, as illustrated in Figure 5.20 and consider the average energy per atom for removal of the top net plane. In a net plane of n^2 atoms there are $(n-1)(n-1)$ atoms bonded with energy 3φ, $2(n-1)$ atoms bonded with energy 2φ, and one atom bonded with energy φ. Consequently the average energy per atom is

$$\bar{\varphi} = 3\varphi - \frac{2}{n}\varphi = \varphi_{1/2} - \frac{2}{n}\varphi \tag{5.56a}$$

which is identical to the chemical potential of the crystal. For $n \to \infty$ the average energy per atom becomes 3φ, in accord with the previously assessed

Table 5.8 Surface free energies σ (erg/cm^2) and their linear temperature coefficients $\partial\sigma/\partial T$ (erg/K cm^2) for the surfaces of selected materials (after Kern [36])

Material	(*hkl*)	σ	$\partial\sigma/\partial T$	Material	(*hkl*)	σ	$\partial\sigma/\partial T$
A	(100)	32	—	Cr	(110)	2775	0.12
Xe	(100)	63	—	Cr	(100)	3644	0.16
K	(100)	150	—	W	(110)	~3000	—
Cu	(111)	2554	0.18	α-Fe	(110)	3032	0.12
Ca	(100)	2932	0.13	α-Fe	(100)	4010	0.17
Ag	(111)	1693	0.14	Ni	(111)	3246	0.19
Ag	(100)	1944	0.08	Ni	(100)	3720	0.14
Au	(100)	2218	0.14	Pt	(111)	3294	0.16
Au	(100)	2547	0.10	Pt	(100)	3781	0.15
Mg	(0001)	739	0.11	NaF	(100)	306	—
Zn	(0001)	909	0.16	NaCl	(100)	170	—
Cd	(0001)	624	0.13	KCl	(100)	152	—
Al	(111)	1692	0.15	KBr	(100)	137	—
Al	(100)	1941	0.19	KI	(100)	112	—
Diamond	(111)	5650	—	KI	(100)	112	—
Si	(111)	1240	—	LiF	(100)	347	—
Ge	(111)	1100	—	CaF_2	(111)	450	—
Pb	(111)	774	0.09	MgO	(100)	1200	—
Pb	(100)	892	0.07	$CaCO_3$	$(10\bar{1}0)$	230	—

value for the infinitely extended crystal. The symbol $\varphi_{1/2}$ is chosen for the bonding energy in the kink site because it is bonded with half the energy of an atom in the interior of the crystal ("Halbkristall" position). Because the average energy for the addition or removal of an atom is identical with the definition of the chemical potential,

$$\mu_r - \mu_\infty = -(\bar{\varphi} - \varphi_{1/2}) = \frac{2}{n}\,\varphi \tag{5.56b}$$

which is the Thomson–Gibbs equation in the language of the Kossel model.

Since the surface of a particle increases with $n^{2/3}$ while its volume increases in proportion to n, as illustrated in Figure 5.21, the free energy fluctuation for its generation is composed of a volume term $\Delta G_v = -n\,\Delta\mu < 0$ and a surface term $\Delta G_s = a\sigma n^{2/3} > 0$,

$$\Delta G(n) = -n\,\Delta\mu + a\sigma n^{2/3} \tag{5.57}$$

where $\Delta\mu$ is the supersaturation and $an^{2/3}$ is the surface area of the particles assuming isotropic surface free energy σ. At small radii the $n^{2/3}$ term exceeds the linear term in n, which becomes dominant, however, at large radii. There

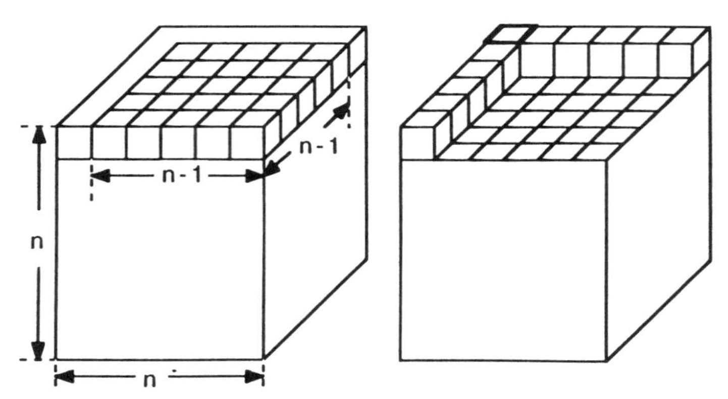

Figure 5.20 Kossel model for the evaluation of the average bonding energy per atom for the top plane of atoms on a crystal of dimensions na_0.

thus exists a maximum in the curve $\Delta G(n)$, which is easily determined from the condition $\partial \Delta G/\partial n = 0$ for a critical number of atoms in the nucleus

$$n_c = \left(\frac{2a\sigma}{3\,\Delta\mu}\right)^3 \tag{5.58}$$

corresponding to a critical dimension r_c. Particles of subcritical size are more likely to disappear than to grow because it takes energy to increase n. On the other hand, once the size of the particle becomes larger than r_c, it is more likely to grow than to disappear because any addition of atoms lowers the free energy. Consequently the particle of size r_c is called the *critical nucleus* and the free energy fluctuation

$$\Delta G_c = \frac{4}{27}\frac{a^3\sigma^3}{\Delta\mu^2} \tag{5.59}$$

required for its generation is called the *Gibbs free energy of nucleation.* Note that these expressions hold only in the small supersaturation regime, where n_c is sufficiently large to vary quasicontinuously with $\Delta\mu$.

In practice homogeneous nucleation is rarely observed because it requires generally a larger undercooling than heterogeneous nucleation on some sort of substrate, which may be a dust particle, the container wall, or the surface of a seed crystal. It can be shown [37] that the steady-state nucleation rate on a substrate surface is given by

$$I = \omega_{n+} N_0 \Gamma_Z \exp\left(-\frac{4}{27}\frac{a^3\sigma^3}{kT\,\Delta\mu^2}\,\phi(\theta)\right) \tag{5.60}$$

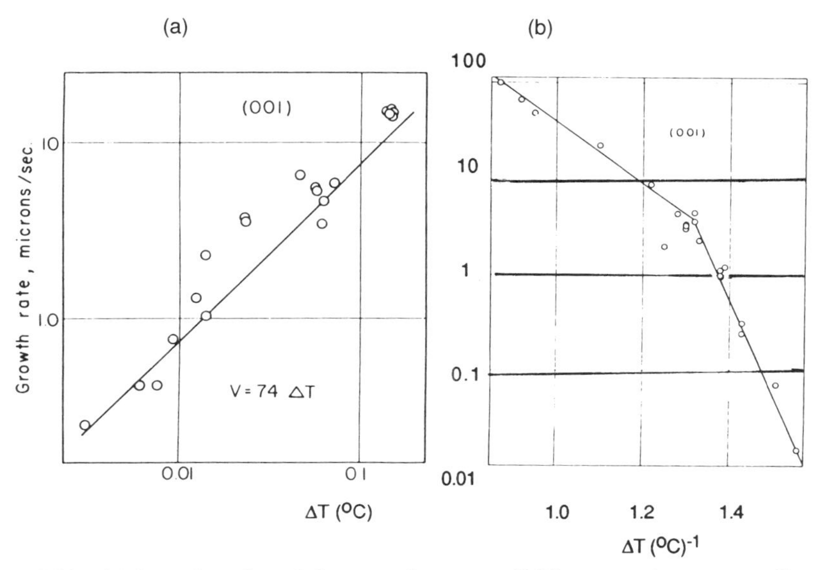

Figure 5.21 (a) Log–log plot of the growth rate on (001) versus the supercooling for a dislocated gallium crystal. (b) Plot of the logarithm of the growth rate on (001) versus $1/\Delta T$ for a perfect gallium crystal. After Pennington et al. [42]; copyright © 1970, Pergamon Press, Oxford.

where N_0 and ω_{n+} are the number of adsorption sites and frequency of adding an atom to the nucleus of size n and Γ_Z is the Zeldovich factor [38]. The function $\phi(\theta)$ accounts for the shape of the nucleus. Further details concerning the conditions of nucleation and growth of a crystal on a foreign substrate will be considered in Chapter 6.

Note that as the formation of three-dimensional nuclei in the formation of a new phase from a supersaturated nutrient requires a fluctuation of the free energy to overcome the nucleation barrier, so, according to Volmer's classical theory of crystal growth [34], does the addition of a new layer of atoms on a singular facet of a crystal require the formation of a two-dimensional nucleus. The reason for this requirement is that the average bonding energy per atom on the edges of a two-dimensional island of finite size is smaller than that for the infinitely extended phase. For example, a square nucleus of n atoms on the (001) face of a simple cubic crystal in the nearest-neighbor approximation has average energy $3\phi - \phi/\sqrt{n}$ per edge atom. Since the length of the edges on the two-dimensional island increases as $\sqrt{n}$, the free energy fluctuation required for the formation of a square two-dimensional island of n atoms is

$$\Delta G(n) = -n\,\Delta\mu + 4a_0\sqrt{n} \tag{5.61}$$

If, α is the free energy per unit length of the step, the maximum in $\Delta G(n)$

corresponds to a critical size of the 2D nucleus

$$n_c = \left(\frac{2a_0\alpha}{\Delta\mu}\right)^2 \tag{5.62}$$

$$\Delta G_c = \frac{4a_0^2\alpha^2}{\Delta\mu} \tag{5.63}$$

so that

$$I = \omega_{n+} N_0 \Gamma_Z \exp\left(-\frac{4a_0^2\alpha^2}{kT\,\Delta\mu}\right) \tag{5.64}$$

The phenomenon of two-dimensional nucleation has been observed under the conditions of electrocrystallization on perfect silver crystal surfaces, which block at subcritical cathodic overvoltage the flow of a Faradaic current. Upon application of a short overvoltage pulse exceeding the nucleation barrier, a Faradaic current flows at the subcritical overvoltage for a duration that corresponds to the charge for the deposition of exactly one monolayer [39].

For crystal growth from the melt, the growth rate is given by

$$v_g = \frac{aD}{\delta^2} f_s \left[1 - \exp\left(-\frac{\Delta\mu}{kT}\right)\right] \tag{5.65}$$

where η_m is the viscosity of the melt and ΔT is its supercooling [40]. Whether the crystal is bound by facets as assumed in Volmer's theory or possesses a surface that is rough on an atomic scale and grows by the random addition of atoms to this surface depends on the value of the parameter

$$\alpha_J = \frac{L}{kT_m} g_E \tag{5.66}$$

where L_m is the latent heat of melting, T_m is the melting temperature, and g_E is the fraction of the total energy stored in a plane parallel to the crystal surface [41]. For $\alpha_J > 2$, facetted growth is observed, which is generally the case for organic crystals, but only in rare cases for metals.

Figure 5.21 shows the growth rate as a function of supercooling on the (001) facet of dislocated (a) and perfect (b) crystals of gallium growing in a glass tube with the facet filling the tube completely. The growth rate is of the form

$$v_g = u_g \exp\left(-\frac{C_g}{\Delta T}\right) \tag{5.67a}$$

Table 5.9 Average edge energies of laterally moving steps on various faces on perfect Ga crystals (after Pennington et al. [42])

Face	Step height h (Å)	Edge energy α (10^{-8} ergs/cm)	α/h (ergs/cm^2)
(111)	2.92	50	17.2
(001)	3.83	40	10.5
(211)	1.95	12.5	6.5
(010)	2.25	7.1	3.2
(120)	2.00	5.3	2.7

where u_g and C_g are system-specific constants. An analysis of the velocity of lateral step motion, assuming nucleation at the periphery of the facet, results in the edge free energies per unit length shown for different crystal facets in Table 5.9.

The rate of crystal growth from the melt on facets containing self-propagating steps is of the form

$$v_g = \frac{h_{st} v_{st} \Delta\mu}{19\Omega\alpha_{st}} \cong C\,\Delta T^q \qquad (5.67b)$$

where h_{st}, v_{st}, and α_{st} are the height, velocity, and free energy of the spiral steps, $\Omega = a^3$ is the volume of a crystal building block in the lattice and $1 \leqslant q \leqslant 2$, depending on the overlap of the steps during growth [44]. The growth rate of

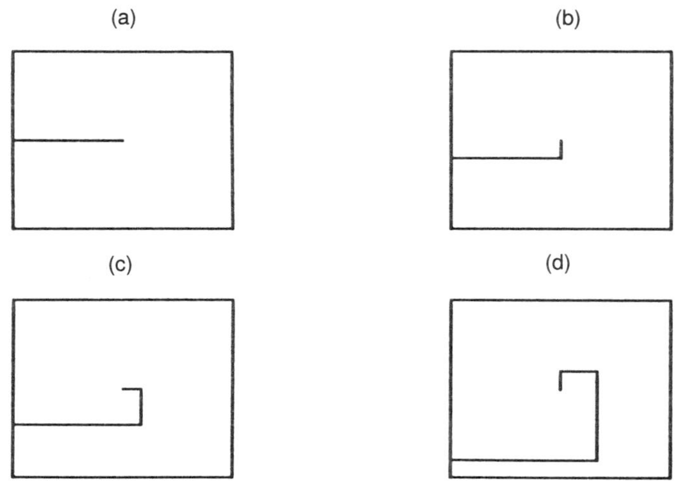

Figure 5.22 Schematic representation of a growth spiral about the emergence point of a dislocation.

the dislocated Ga crystal shown in Figure 5.21 corresponds to $q \approx 1$.

Figure 5.22 shows schematically the formation of a self-propagating system of steps at the emergence point of a screw dislocation (a) in the initial stage and (b)–(d) at later stages of growth [43]. Addition of atoms to the initially straight step advances the step in the direction of its normal until the edge created by this advance exceeds the dimensions of the critical 2D nucleus, which in Figure 5.22 is assumed to have the shape of a square; that is, the free edge energy α_{hk} has the angular dependence shown in Figure 5.20 for a 2D nucleus formed on (100). In this case the minimum edge free energy about the emergence point of the dislocation exists for a growth spiral bound by the 10, 01, $\bar{1}0$, and $0\bar{1}$ edges. If the edge free energy is isotropic, a growth spiral with rounded edges develops that has at its apex the radius r_c of the critical nucleus.

A similar nucleation barrier as observed in the case of growth must be overcome during etching of the crystal surface in an undersaturated phase with the difference that the crystal edges may provide a line of attack. The emergence points of dislocations that usually accommodate impurities in the core region provide for additional locations of attack [45]. Figure 5.23 shows rounded spiral steps of monolayer height formed during the evaporation on the (100) surface of a NaCl crystal about the emergence points of dislocations of the Burgers vector $(a/2)[110]$, where $a = 5.62$ Å [46]. In addition, there are dislocations with Burgers vector $a[100]$ that, in the case of

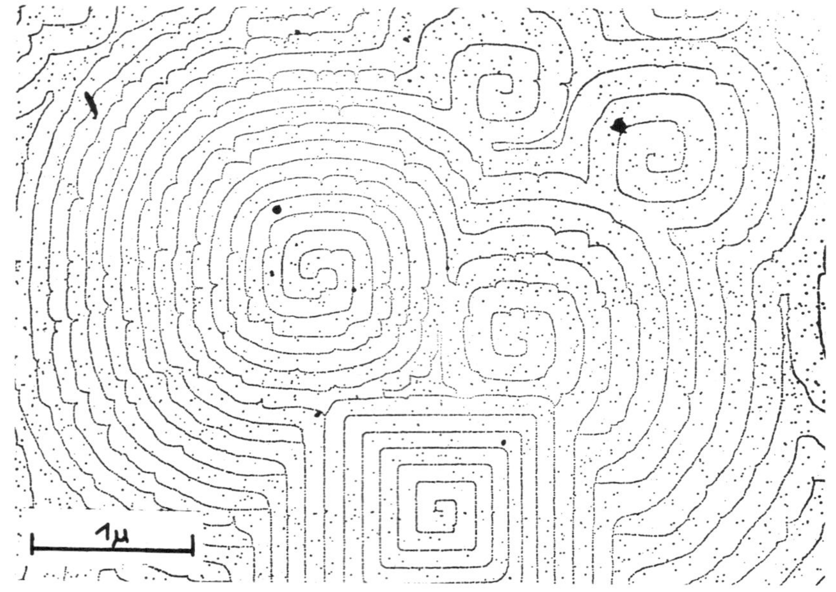

Figure 5.23 Spiral steps produced at the emergence points of dislocations during the evaporation of NaCl. After Keller [46]; copyright © 1975, North Holland Publishing Company, Amsterdam.

the double spiral in the left part of the figure, is dissociated into to partial dislocations according to

$$\mathbf{b} = a[100] = \frac{a}{2}[100] + \frac{a}{2}[1\bar{1}0] \tag{5.68}$$

so that the height of the branches of the spiral is also 2.86 Å. However, the square spiral developed about the emergence point of an undissociated dislocation with Burgers vector $a[100]$ has double-layer height, as revealed by the interactions of the monolayer steps with the steps of this spiral.

Although the emergence points of dislocations thus play an important role in crystal growth and etching, the significance of the screw dislocation mechanism has been overrated in the past, particularly in the context of semiconductors characterized by strong directional bonds, where surface bonding effects (both reconstruction and bonding to adsorbates) dominate the etching and growth behavior. Nevertheless, the kinetics of the growth of dislocated and perfect semiconductor crystals shows marked differences (see Section 5.3). For example, the supercooling sustained on the facets of perfect Si crystals under the conditions of Czochralski pulling is typically a few degrees, while for dislocated crystals it is a few tenths of a degree [47], [48].

For large supersaturation, that is, small size of the critical nucleus, continuum theories are insufficient to calculate the required energy fluctuation for nucleation. HF calculations have been performed for silicon clusters of up to 10 atoms in size [49]. They show considerable deviations of the coordination numbers and bond angles from the bulk sp^3 geometry. The clusters Si_3 and Si_4 assume planar forms, such as triangle and rhombus shapes of C_{2v} and D_{2h} symmetry, respectively. The clusters with more than four silicon atoms have more compact geometries and lower energies than microcrystalline arrangements of atoms. They are shown in Figure 5.24. The edge-capped trigonal bipyramid and bicapped tetrahedron geometries of C_{2v} symmetry for Si_6 have nearly identical energy so that the ground-state geometry of this cluster is at present not identified. The bicapped octahedral form of the Si_8 cluster undergoes in the ^1E_g state a Jahn–Teller distortion into C_{2h} symmetry, lowering the energy so that in this case the distorted singlet state represents the ground state. Similarly, for Si_9 the ground state has C_s symmetry derived by a Jahn–Teller distortion from the tricapped octahedron of D_{3h} symmetry shown in Figure 5.24 [50]. The lowest-energy configuration for the Si_{10} cluster in the ground state is a tetracapped octahedron of T_d symmetry.

Table 5.10 shows the results of total energy calculations by the HF calculations with and without M4 corrections, which refer to a fourth-order Møller–Plesset perturbation calculation of electron correlation effects. They are large, and the best match to experimental total binding energies is provided by the scaled M4 calculations shown in column 4 [49]. Figure 5.25 shows the MOs of the particularly stable clusters Si_4, Si_6, Si_7, and Si_{10}. The lower MOs have predominantly s and the highest MOs have predominantly p character.

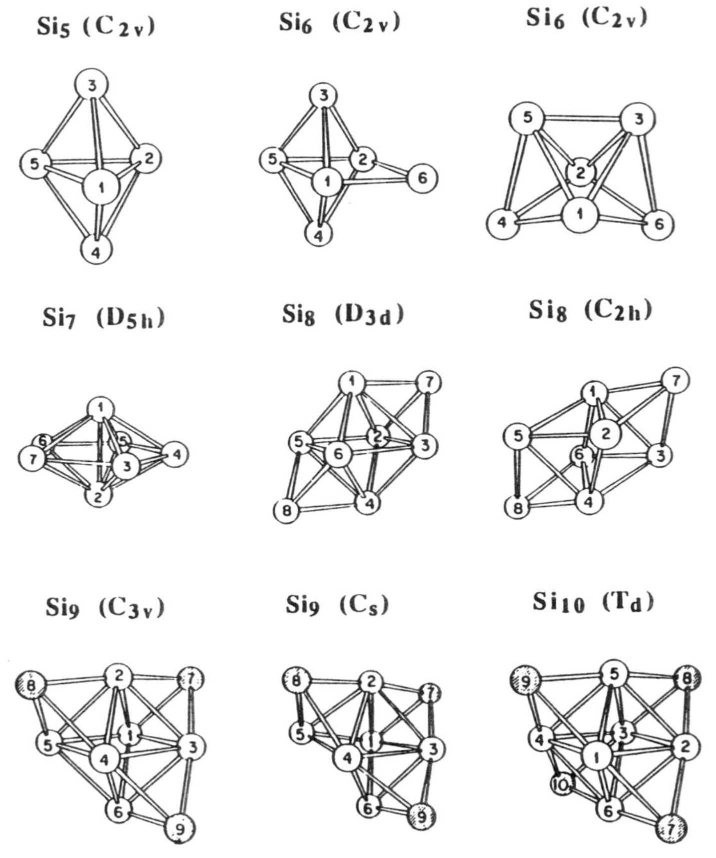

Figure 5.24 Lowest-energy configurations of atoms in silicon clusters. After Raghavachari and Rohlfing [50]; copyright © 1988, American Institute of Physics, New York.

Although mixing occurs, it is far from the sp^3 hybridization of bulk crystalline silicon. The highest occupied MO (HOMO) and the lowest unoccupied MO (LUMO) both have principally p character. They are in the cluster range of Figure 5.25 separated by gaps of 1–1.5 eV, which is similar to the bandgap of bulk silicon, even though the cluster geometries are quite different.

Note that, where strong directional bonds are formed between the atoms of a cluster, the minimum free energy may not be attained for a three-dimensonal cluster at all. For example, carbon clusters nucleating in a supersaturated carbon vapor made by laser evaporation of a carbon film, for fewer than 12 atoms, prefer the shape of a chain. The bending of the chain into rings and even more so into 3D structures, at this size, increase the energy [51]. For more than 12 atoms the formation of rings becomes energetically favorable, since the opportunity for multiple bonding in conjunction with resonance stabilization

Table 5.10 Results of quantum-chemical calculations for silicon clusters containing up to 10 atoms[a] (after Raghavachari [49])

Cluster	Total binding energy (eV)			Binding energy per atom (eV)		
Si_2	1.51	2.64	3.17	0.76	1.32	1.58
Si_3	3.04	6.42	7.70	1.01	2.14	2.57
Si_4	6.04	10.71	12.85	1.51	2.68	3.21
Si_5	7.42	13.92	16.70	1.48	2.78	3.34
Si_6	10.13	18.26	21.91	1.69	3.04	3.65
Si_7	11.69	21.9[b]	26.3[b]	1.67	3.13[b]	3.76
Si_{10}	18.11	32.6[b]	39.2[b]	1.81	3.26[b]	3.92[b]

[a]Columns 2 and 5: HF calculation; columns 3 and 6: MP4 calculation; columns 4 and 7: scaled MP4 calculation.
[b]Estimated values (see [9] for more accurate calculations).

overcompensates the bond strain. A three-dimensional structure is formed only for carbon clusters containing more than 27 atoms, with a particularly stable configuration at 60 atoms [52].

The C_{60} cluster (buckminsterfullerene) is capable of accepting/shedding reversibly two electrons. The buckminsterfullerene anion forms saltlike complexes with divalent metal ions (e.g., Ni^{2+}) that can be crystallized and are thus suitable for diffraction studies [53]. The electronic structure and stability of clusters, which bridge the gap between small molecules and colloidal particles, are of considerable interest in the context of both fundamental questions of

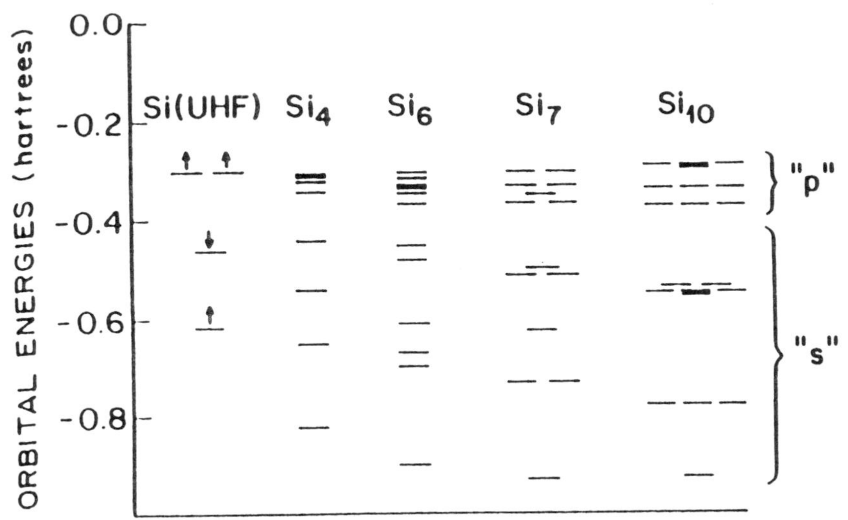

Figure 5.25 Molecular orbitals of the silicon clusters Si_4, Si_6, Si_7, and Si_{10}. After Raghavachari and Rohlfing [50]; copyright © 1988, American Institute of Physics, New York.

chemistry and the future engineering of quantum dots. Unfortunately, HF calculations are too expensive to guide this development, but local-density-functional calculations have been explored with excellent success in the context of both fullerenes and large aromatic ring systems, such as corannulene.

Directional bonding effects also must be considered under the conditions of

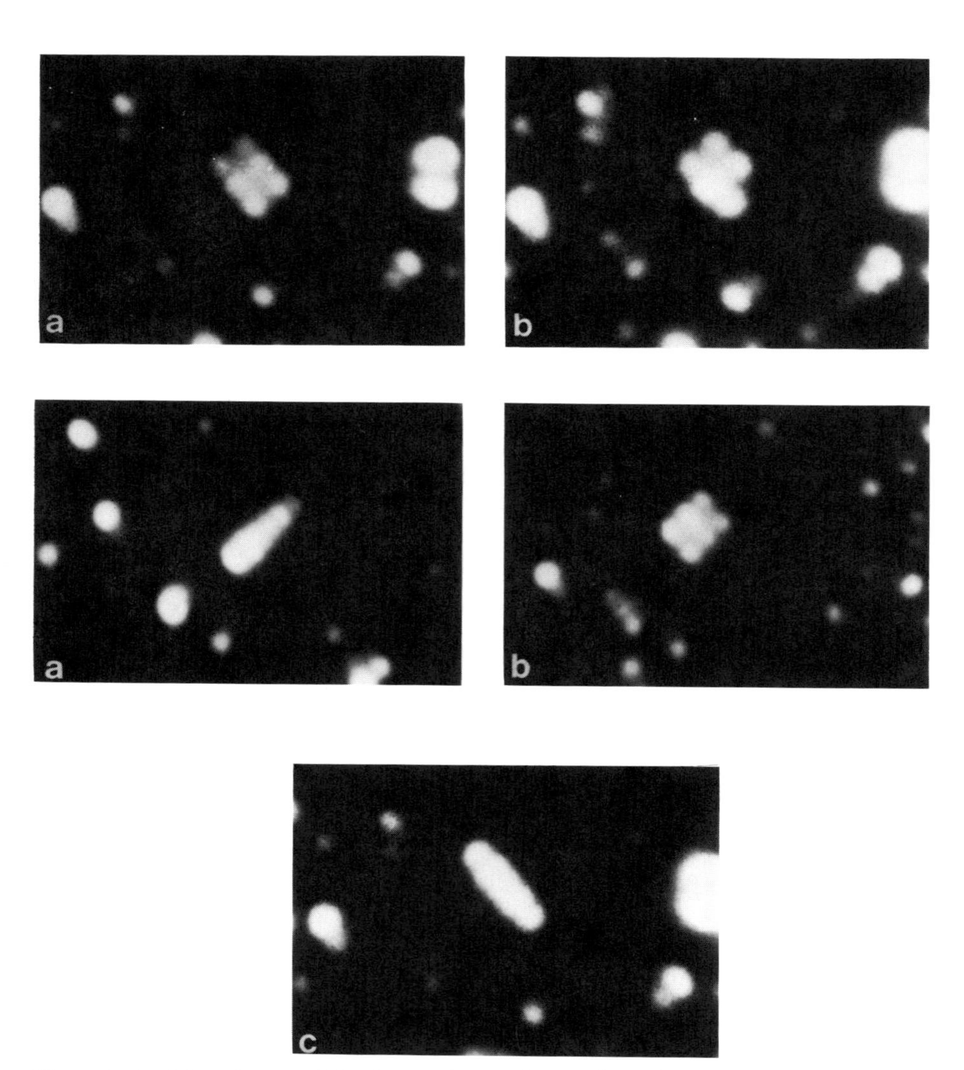

Figure 5.26 The direct FIM observations of the clustering Ir atoms under the conditions of field evaporation on a close-packed facet of an Ir single crystal. After Schoelbel and Kellogg [54]; copyright © 1988, American Institute of Physics, New York.

crystal growth and etching/evaporation on perfect facets, which does not necessarily proceed via two-dimensional nucleation. For example, Figure 5.26 shows the clustering of Ir atoms on a close-packed facet of an FIM emitter made from Ir metal [54] under the conditions of field evaporation. Figure 5.26(a) shows a 2D island of 6 Ir atoms. After field evaporation of one atom, the 2D cluster shown in Figure 5.26(b) becomes unstable and unravels into the stable linear chain configuration of Figure 5.26(c). On the other hand, upon addition of a sixth atom [Figure 5.26(d)], the linear chain becomes unstable and rearranges into a two-dimensional island, as shown in Figure 5.26(e). Thus there is a threshold for two-dimensional island formation versus linear chain formation in the initial stages of nucleation and growth due to chemical bonding forces, adding a chemical aspect to crystal growth theory that has been largely ignored in the past. Chemical reactions on the surface of the nuclei can significantly enhance the nucleation rate [55] and plays an important role in both technological processes, for example, the nucleation of SiO_2 in the slag gas effluent of high-temperature coal combustion [56] and scientific studies, such as condensation processes in supersonic beams [57], which may become useful in the context of the formation of microcrystalline materials.

In addition, the knowledge of the real structure of the surface must be incorporated into models of crystal growth. Since substantial changes in the surface structure often occur upon the adsorption of foreign atoms, not only the reconstruction of the surface as an initial condition, but also the dynamics of changes in the reconstruction during heteroepitaxial growth, must be incorporated into the modeling. For example, reconstruction of the steps on the Si(001)-2 $\times$ 1 surface is revealed by STM [59]. On vicinal (001) surfaces tilted about $[1\bar{1}0]$ steps are observed for both Si and Ge [58], [60]. Figure 5.27 shows the STM image of double steps on Si(001) tilted by 4° about $[1\bar{1}0]$. Kinks exist in steps 3 and 5. Note that where kinks occur the double step breaks up into single steps. The formation of double steps on silicon and their integrity are of utmost importance in the context of the heteroepitaxial growth of III–V compounds and alloys on silicon substrates since single steps result in antiphase domain boundaries [61] in the III–V epilayer that are detrimental to the performance of devices.

5.3 The Fabrication of Substrate Crystals

The most important method for the growth of silicon and germanium crystals is the Czochralski method, which was initially developed for the growth of metal crystals in 1918 [62]. Czochralski growth is initiated by dipping a seed crystal of a chosen crystallographic orientation onto the surface of a melt contained in the heated crucible. The seed is then slowly withdrawn, solidifying the adherent column of the melt in steady state at an average rate given by the sum of the pulling rate and the rate of lowering the solid–liquid interface due to the depletion of the melt in the crucible. The latter may be eliminated either

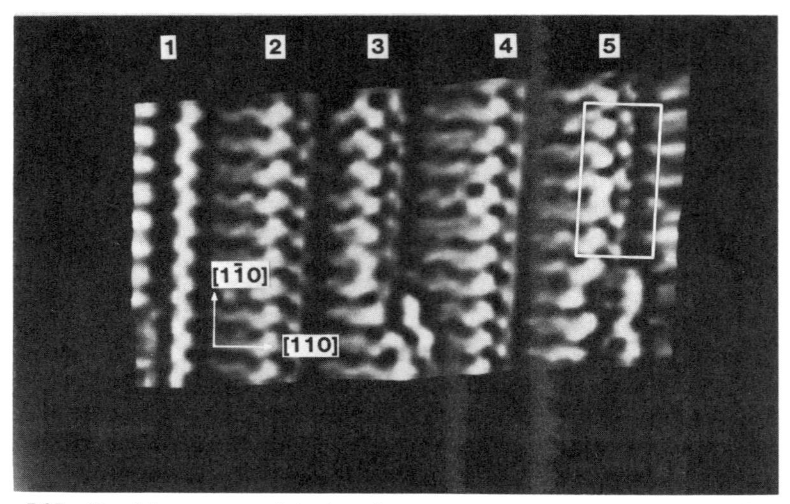

Figure 5.27 Double steps on vicinal surface of Si(001) tilted by 4° about [1$\bar{1}$0]. After Griffith et al. [58]; copyright © 1988, American Vacuum Society, New York.

by a crucible lift mechanism or by melt replenishment. The control of the desired crystal diameter is achieved by closely controlling the heat balance at the solid–liquid interface. The diameter remains constant if the heat flow from the hotter parts of the melt to the interface plus the latent heat released upon crystallization balances the heat conducted away from the interface through the seed holder plus the heat irradiated off the surface of the crystal and seed. The diameter shrinks if the heat input to the interface is larger than the drainage of heat to the crystal surface and the seed holder. Conversely, if the heat flow from the melt into the interface is smaller than the radiative heat loss off the surface and the conductive heat loss through the seed, the diameter increases. This is illustrated in Figure 5.28, which shows various stages of the growth of a metal crystal by the Czochralski technique: (a) the necking stage and (b) the constant-diameter growth stage. The formation of a thin crystal neck at the beginning of the growth procedure has the purpose of eliminating the propagation of defects from the seed into the growing crystal. The decrease of the crystal diameter toward the end of the run has the purpose of reducing the formation of defects due to thermal shock upon separation of the crystal and the melt.

The Czochralski pulling of silicon was first reported by Thiel and Buehler in 1952 [64]. In conventional Czochralski pulling of silicon, fused silica crucibles of high purity are used (compare Section 3.2) that are completely empty at the end of the run. In view of its long duration, the automation of the process is of considerable interest in all commercial silicon pullers. This is usually accomplished by reflecting a light beam on the meniscus of the melt

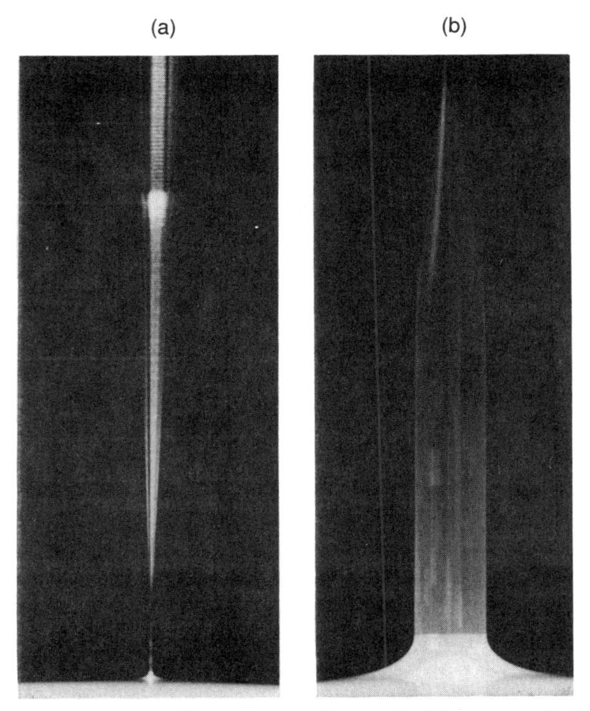

Figure 5.28 Czochralski pulling of a metal crystal with automatic diameter control: (a) necking; (b) constant-diameter growth; (c) separation stage. After Bachmann et al. [63]; copyright © 1970, North Holland Publishing Company, Amsterdam.

column that adheres to the growing crystal and is supported by surface tension, as illustrated in Figure 5.29 [65].

Note that this sensing of the meniscus shape is not only important for diameter control, but is essential for initiating and monitoring dislocation-free growth. Since the dislocations in the silicon crystals are aligned along specific directions, such as [110] and [211], they can be grown out in the neck portion if the axis is chosen to be inclined to these directions (e.g., $\langle 111 \rangle$ or $\langle 100 \rangle$) [66]. As in the growth of metal crystals, first stable growth at the seed diameter is established. Whether dislocations formed in the growing crystal multiply or grow out depends on the balance of heat flow and morphological features at the solid–liquid interface. (A comprehensive analysis of the conditions of Czochralski pulling of silicon is provided in Ref. [67].) At a small pulling rate, where the heat balance is dominated by conduction from the melt into the crystal, a convex interface is formed that has a central facet, as shown in Figure 5.30(a). Note that there are additional inward-pointing inclined $\{111\}$ facets (e.g., at an angle of 70.53° for pulling in the $\langle 111 \rangle$ direction). Dislocations that are formed due to growth errors and thermal stress caused by the temperature gradient across these facets extend upon growth inward

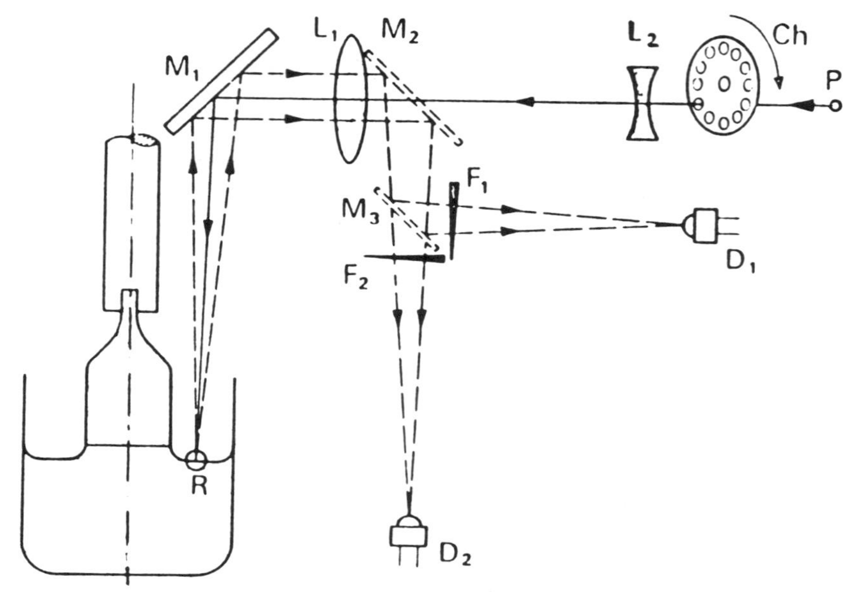

Figure 5.29 Control of Czochralski pulling by top viewing of the reflecting surface R of the melt adhering to the crystal. P, light source; Ch, Chopper; L1, L2 lenses; M1–M3, mirrors; F1, F2, filters; D1, D2, detectors sensing changes in the curvature of R. After van Dijk et al. [65]; copyright © 1974, *Acta electronica*, Limeil Brevannes.

and multiply. Therefore, dislocation-free growth is not possible under this condition.

However, upon acceleration of the pulling rate at otherwise constant conditions, the increasing latent heat liberated per unit time at the solid–liquid interface changes its shape. First a concave interface region forms in the center of the crystal. In the case of a ⟨111⟩ pulling direction, a ring-shaped (111) facet separates the concave and convex portions of the interface. Upon further increasing the pulling rate, the convex portion of the interface shrinks and disappears at a critical pulling rate, as illustrated in Figure 5.30(b), (c). Since the inward-pointing inclined peripheral {111} facets are eliminated at this point, dislocation-free growth commences.

The onset of dislocation-free growth is accompanied by the phenomenon of lockon, that is, a sudden increase in the crystal diameter at constant growth conditions. The cause for this phenomenon is evident from the balance of surface tension forces at the perimeter of the crystal, which is shown in Figure 5.30(b) and (c) for pulling in the ⟨111⟩ direction. Due to the much smaller surface free energy of the (111) facet as compared to other interface orientations (see Table 3.3), a temporary imbalance in the surface tension forces ensues once the (111) ring facet reaches the outer perimeter of the crystal. This imbalane self-adjusts by a change in the slope of the meniscus, which is accompanied by an increase of the diameter to a new condition of stable

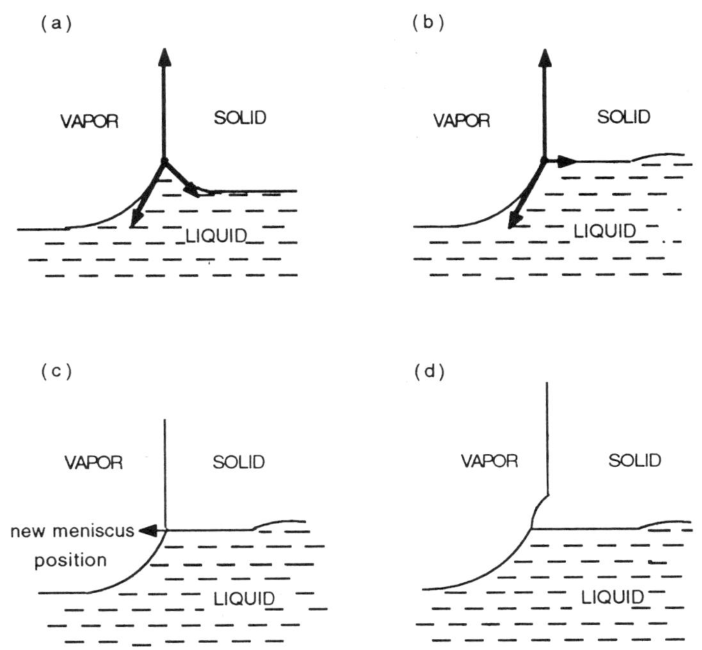

Figure 5.30 Schematic representation of the changes in the solid–liquid interface shape encountered at various stages of initiating dislocation-free growth of silicon crystals by Czochralski pulling in the ⟨111⟩ direction: (a) Convex interface dominated by heat transport from the melt perimeter toward the solid–liquid interface; (b) transient stage of imbalance between the surface tension forces upon expansion of the (111) ring-facet to the outer crystal perimeter; (c) meniscus change associated with this imbalance and lockon; (d) conditions of constant diameter growth just after lockon. After Wilkes [67].

growth, explaining the lockon phenomenon. Note that once dislocation-free growth is established, the diameter can be increased to the desired value without loss of the peripheral facet, so that dislocation-free growth continues. Since multiplication of dislocations (e.g., by the Frank–Read mechanism) inside the crystal is no longer possible, dislocation-free growth is relatively easy to maintain. Currently dislocation-free silicon crystals are produced routinely at a rate of $\sim 10^4$ tons per year worldwide [67].

Figure 5.31 shows a photograph of the top part of a 150 mm diameter Si crystal of (001) orientation revealing narrow facets that continue for dislocation-free crystals all the way down on the outer surface of the crystal. Although 150 mm diameter is currently a standard size for IC production, production lines based on 200 mm diameter wafers are already in use, and even larger wafer diameters are expected to come into use in the future.

Since this requires the handling of very large volumes of silicon, continuous replenishment of the melt by the addition of polycrystalline Si has been explored. Pelletized polysilicon produced by silane decomposition is particu-

Figure 5.31 Top part of a Czochralski-pulled silicon crystal of 150 mm diameter. Courtesy Dr. Kenneth E. Benson, AT&T Bell Laboratories, Allentown, PA.

larly suited for melt replenishment because it can be fed into the crucible at a well-controlled rate [68]. If the dopant concentration in the added feed material is the same as the dopant concentration in the frozen Si (i.e., $k_e c_1$), the melt concentration and, consequently, the concentration in the growing crystal can be kept constant over extended periods of crystal growth. The uniformity of the material produced in this way is further improved by maintaining a constant position of the interface, that is, steady-state convective mixing controlled to achieve constant melt composition for all impurities. The electrical properties of the continuously pulled crystal are not strictly independent of the axial position. Also, the lifetime of crucibles under conditions of continuous pulling is a problem. Therefore, the current crystal production is based on the conventional process.

The limited life of the silica crucibles used to contain the silicon melt is due to both devitrification and chemical interactions that contaminate the melt and consequently the crystal. Fused quartz reacts with molten Si according to

$$SiO_2(s) + Si(l) \rightleftarrows 2SiO(g) \tag{5.69}$$

so that silicon monoxide bubbles are formed at the crucible wall. Also, the melt becomes saturated with oxygen, leading to the incorporation of oxygen into the crystal. Oxygen is located in the Si lattice at interstitial positions at the midpoints between the Si–Si bonds [69]. Its distribution coefficient is 0.7. However, under the conditions of conventional Czochralski pulling, the oxygen concentration decreases from the first to freeze to the last to freeze ends of the crystal due to the combination of a constant rate of oxygen loss at the melt surface with a decreasing generation rate at the silica–melt interface as the melt surface is lowered inside the crucible upon pulling. Under the conditions of continuous growth, where the melt surface-to-volume ratio is kept constant, the steady-state oxygen concentration at the interface can be kept constant over an extended period of time, thus resulting in a nearly constant axial oxygen concentration profile.

The saturation concentration of oxygen in solid Si decreases with decreasing temperature according to the relation [70]

$$[\mathrm{O}] = 9 \times 10^{22} \left(\frac{\text{atoms}}{\text{cm}^3}\right) \exp\left(-\frac{1.52(\text{eV})}{kT}\right) \tag{5.70}$$

Therefore, upon cooling, the crystal becomes supersaturated with regard to the oxygen. If the supersaturation exceeds a critical value, small SiO_2 particles are nucleated and grow by diffusion of oxygen from the surrounding supersaturated region to the precipitate. Since the density of SiO_2 differs substantially from the density of Si, a considerable stress is generated in the vicinity of the precipitates, which can produce other extended defects. Also, it favors the segregation of impurities that do not fit into the unstrained Si lattice in the vicinity of the precipitates. This may be utilized to advantage for the gettering of impurities by the deliberate generation of precipitates of SiO_2 in the volume just below the active device regions of the wafer (see Section 8.1).

Another important method of silicon bulk crystal growth is float zoning (FZ) [71]. The float zoning of silicon was invented in 1952 and employs the motion of a molten zone to convert an initially polycrystalline rod into a single crystal [72]. It produces the purest bulk Si single crystals known to date. The method does not involve any container, but rather relies on the support of the molten zone by surface tension [73]. Growth is initiated from a thin seed crystal that is attached to a pendant drop generated by rf heating on a polycrystalline Si rod. This inverted growth direction as compared to Czochralski pulling is predicated by considerations of the stability of the molten zone. After seeding the crystal is allowed to grow out to the desired diameter, which usually is wider than both the opening in the rf coil and the feed rod diameter, striking a fine balance in the shape of the liquid surface that must be maintained throughout the process. A substantial stabilizing electrodynamic force component is provided by the use of rf heating at 2–3 MHz frequency using a flat coil. Both the crystal and the feed rod rotate in opposite directions to improve the thermal geometry and are translated with independently controlled rates to allow the stretching or squeezing of the zone as required.

Crystals of 125–150 mm diameter weighing up to 40 kg are routinely produced [74], [75] by this method, representing a significant accomplishment of crystal growth engineering. Dopants can be brought into the molten zone either by the use of doped polysilicon or by a vapor-phase equilibrium at its surface. The latter doping mode allows for maintaining a constant average doping level. However, the incorporation of dopants is nevertheless nonuniform, since the FZ process necessarily proceeds with a nonplanar interface morphology if the heat input is made from the outside. In the growth of dielectric crystals, where purity is a less critical issue, resistively heated metal ribbons containing one or several holes for passage of the melt may be employed and achieve planar solid–liquid interfaces. However, in the growth of reactive materials with metallic melt properties (e.g., Si), this method is not suitable. Since under the conditions of float zoning the heat input is by eddy currents induced in a very thin skin at the outer surface of the molten zone, steep temperature gradients exist that establish vigorous convective mixing. Time variations in the convective rolls modulate both the microscopic growth rate and the diffusion layer thickness at the crystal–melt interface. The variations in the dopant concentration caused by these fluctuations in the growth conditions follow the shape of the interface, so that wafers cut perpendicular to the axis show swirl-like radial variations in the doping and in the concentrations of other residual point defects.

These doping inhomogeneities can be largely avoided by the use of transmutation doping [76], that is, by the growth of nominally undoped ultrapure crystals that are subsequently exposed to a highly uniform flux of thermal neutrons. The conversion of the stable isotope ^{30}Si, which exists naturally at 3.05% concentration, into the unstable isotope ^{31}Si according to the nuclear reaction

$$^{30}\mathrm{Si}(n,\gamma)\,^{31}\mathrm{Si} \rightarrow {}^{31}\mathrm{P} \tag{5.71}$$

followed by decay with a half-life of 2.6 h into stable isotope ^{31}P [77], dopes the silicon n type. The secondary reaction

$$^{31}\mathrm{P}(n,\gamma)\,^{32}\mathrm{P} \rightarrow {}^{32}\mathrm{S} \tag{5.72}$$

produces ^{32}S, but at quantities that do not significantly affect the properties. However, the longer half-life of ^{32}P of 14.22 d requires a cooling period of a few days after irradiation. The uniform distribution of phosphorus provides for uniform breakdown behavior and excellent reliability of neutron transmutation-doped (NTD) Si devices. This is important in the context of high-power device applications [78]. However, due to the limitations to the maximum diameter and to the maximum doping level ($10^{16}\,cm^{-3}$) of NTD silicon, as well as the comparatively high cost, float zoning does not compete with the Czochralski method in the fabrication of substrates for the production of circuits in low-power applications, which represent by far the largest segment of the present solid-state electronics market.

For the fabrication of stochiometric III–V compound substrate crystals, detailed knowledge of the phase relations is essential. Figure 5.32 shows the p–T–x phase diagram of the In,P system as an example. Liquidus compositions on both the In-rich and In-deficient sides of the phase diagram are in isothermal equilibrium with the only compound in the system InP. It melts congruently at 1062°C. The total phosphorus pressure over the stoichiometric liquid at this temperature is 27.5 atm. At dilute P concentrations ($x < 0.3$) the liquidus of the In,P system behaves nearly as an ideal solution. However, in general III,V systems deviate from ideal behavior.

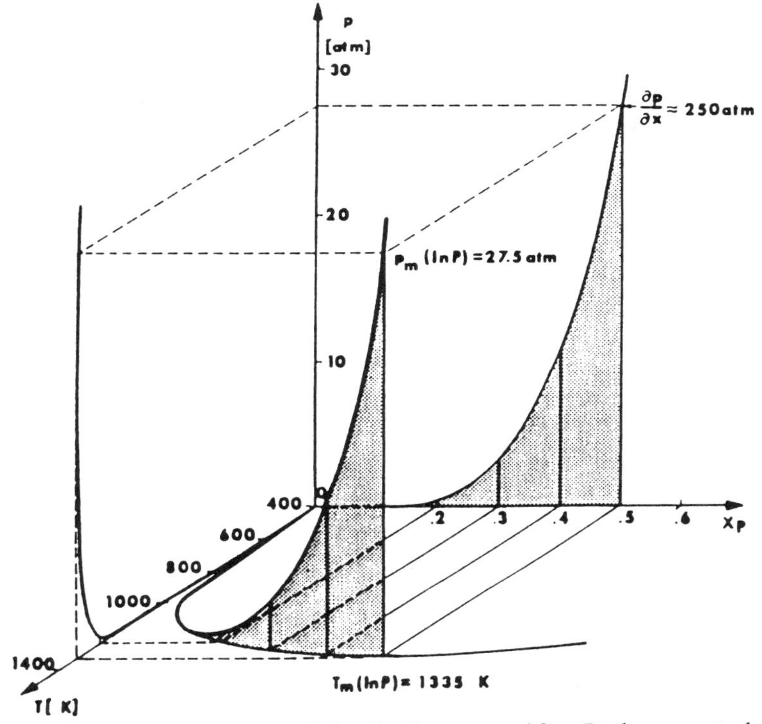

Figure 5.32 The p–T–x diagram of the In–P system. After Bachmann et al. [79]; copyright © 1974, The Metallurgical Society of AIME, New York.

For nonideal solutions the partial excess free energy of mixing of component C_i in the mixture is defined as

$$\Delta G_{mi}^{xs} = \Delta G_{mi} - \Delta G_{mi}^{ideal} = RT \ln \gamma_i \tag{5.73}$$

which follows from $\Delta G_{mi} = \mu_i - \mu_i^0 = RT \ln a_i = RT \ln(\gamma_i x_i) = RT \ln x_i + RT \ln \gamma_i$. The total excess free energy of mixing for one mole of a mixture consisting of k components is thus

$$\Delta G_m^{xs} = \sum_{i=1}^{k} x_i \Delta G_{mi}^{xs} = RT \sum_{i=1}^{k} x_i \ln \gamma_i \tag{5.74}$$

reflecting the nature and strength of the interactions between the constituents of the mixture.

For a binary system A, B the crystal AB is at the liquidus temperature T_L in equilibrium with a liquid of composition $x_A = 1 - x_B$. For one gram atom solution

$$\begin{aligned} G_{liquid}(T_L) &= \tfrac{1}{2}(G_{A0} + RT_L \ln \gamma_A x_A + G_{B0} + RT_L \ln \gamma_B x_B) \\ &= G_{solid}(T_L) \end{aligned} \tag{5.75}$$

A supercooled stoichiometric liquid of composition $x_A = x_B = 0.5$ has at the same temperature the free energy

$$G^{\text{sl}}(T_{\text{L}}) = \tfrac{1}{2}(G_{A0} + RT_{\text{L}} \ln \tfrac{1}{2}\gamma_A^{\text{sl}} + G_{B0} + RT_{\text{L}} \ln \tfrac{1}{2}\gamma_B^{\text{sl}}) \tag{5.76}$$

so that

$$\Delta G = G^{\text{sl}}(T_{\text{L}}) - G_{\text{solid}}(T_{\text{L}}) = \frac{RT}{2}\left(\ln \frac{\gamma_A^{\text{sl}}\gamma_B^{\text{sl}}}{\gamma_A \gamma_B} + \ln \frac{1}{4x(1-x)}\right) \tag{5.77}$$

In the case of III–V systems, such as in the case of the In–P system, where the interactions between the constituents are not very strong, the high-temperature behavior governing the conditions of melt growth is reasonably well described by the regular solution model. In this model random mixing, but nonzero enthalpy of mixing, is assumed. Consequently

$$\Delta S_{\text{m}}^{\text{xs}} = \Delta S_{\text{m}}^{\text{real}} - \Delta S_{\text{m}}^{\text{ideal}} = 0 \tag{5.78}$$

A simple statistical model considering the random placement of atoms A and B in a lattice of w nearest neighbors results in an excess enthalpy of mixing $\Delta H_{\text{m}}^{\text{xs}} = \Delta H_{\text{m}}^{\text{real}}$ that is symmetric to $x = 0.5$,

$$\Delta H_{\text{m}} = \Omega x(1 - x) \tag{5.79}$$

and the interaction parameter Ω is related to the enthalpies H_{AA}, H_{BB}, and H_{AB} for the bonds between atoms A–A, B–B, and A–B, respectively, according to

$$\Omega = Nw\left(H_{AB} - \frac{H_{AA} + H_{BB}}{2}\right) \tag{5.80}$$

It can be shown that the activity coefficients of the components A and B of a binary system are related to the interaction parameter according to

$$\ln \gamma_i = \frac{(1 - x_i)^2 \Omega}{RT} \tag{5.81}$$

Therefore, the term in the activity coefficients in Equation (5.77) may be rewritten in terms of the interaction parameter

$$\ln \frac{\gamma_{\text{A}}^{\text{sl}}\gamma_B^{\text{sl}}}{\gamma_A \gamma_B} = -\frac{2\Omega}{RT_{\text{L}}}(x - 0.5)^2 \tag{5.82}$$

Also, the free energy difference between a supercooled stoichiometric liquid and the stoichiometric solid may be calculated by summing the free energy changes in the following process of: (1) Heating the crystal from T_{L} to the melting point T_{F} of the compund AB; (2) melting AB at this temperature; (3) cooling the liquid AB back to the temperature T_{L}. Since the second process proceeds at no change in the free energy and $\partial G/\partial T = -S$, the free energy difference between a supercooled stoichiometric liquid and the stoichiometric solid is the sum of two integrals

$$\Delta G = -\int_{T_L}^{T_F} S^{ss}(T)\,dT - \int_{T_F}^{T_L} S^{sl}(T)\,dT \tag{5.83}$$

where S^{ss} and S^{sl} are the entropies of the stoichiometric solid and the stoichiometric liquid, respectively. Let $\Delta S_F = \Delta H_F/T_m$ denote the entropy of fusion of the compound AB. Then, neglecting terms in ΔC_p, Equation (5.78) reduces to

$$\Delta G = \Delta S_F(T_F - T_L) \tag{5.84}$$

Combining Equation (5.84) with Equations (5.77) and (5.82) results in

$$\ln\frac{1}{4x(1-x)} - \frac{2\Omega}{RT_L}(x-0.5)^2 - \frac{2\,\Delta H_F}{R}\left(\frac{1}{T_L} - \frac{1}{T_F}\right) = 0 \tag{5.85}$$

which uniquely relates the liquidus composition x to the liquidus temperature if the melting point and the enthalpy of fusion of the compound AB and the interaction parameter Ω are known [80]. For an ideal solution, Equation (5.85) reduces with $\Omega = 0$ to

$$\ln\frac{1}{4x(1-x)} = \frac{2\,\Delta H_F}{R}\left(\frac{1}{T_l} - \frac{1}{T_m}\right) \tag{5.86}$$

so that Ω can be expressed in terms of the deviation ΔT of the measured liquidus from the ideal liquidus line [5.81].

Figure 5.33 shows the liquidus composition as a function of the reciprocal temperature for selected of III–V systems utilizing the simple solution model, where $\Omega = a + bT$ for the calculated curves shown as full lines [82]. This model implies the existence of a small excess entropy of mixing term so that random mixing is no longer a good approximation.

Note that there are for each temperature two liquidus points tying into the solid of nominal composition $x = 0.5$. Since these two liquidus compositions correspond to different chemical potentials, the representation of the solidus by a line requires closer scrutiny; that is, the tie points on the solidus line from the metal-rich and metal-deficient sides of the phase diagram have different compositions. The extent of the homogeneity range defined in this way is small for all III–V compounds. It can be calculated by considering the temperature-dependent point-defect equilibria in the solid, which requires a selection of the point defects incorporated into the model and the identification of the thermodynamic functions describing the equilibria. Only very few experimental investigations exist [83], [84], which is unfortunate because the thermodynamic calculations are subject to considerable error, but understandable because of the substantial effort required for obtaining meaningful experimental data.

For example, initially a maximum melting point on the Ga-rich side of the phase diagram was predicted for the Ga–As system [85]. However, more recent experimental and theoretical results favor a location of the maximum

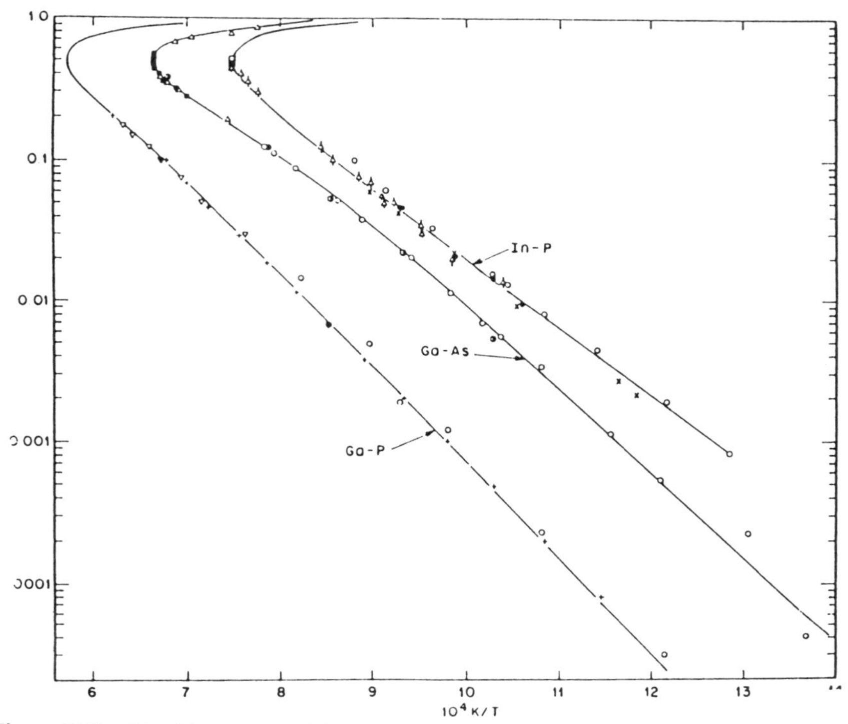

Figure 5.33 Liquidus composition versus reciprocal temperature for the Ga–P, Ga–As, and In–P systems. After Panish [82]; copyright © 1974, North Holland Publishing Company, Amsterdam.

melting point at an As-rich composition [86]. This is demonstrated in Figure 5.34, which shows (a) the results of analyses of the head and tail sections of GaAs crystals and (b) the $T - X$ diagram for the Ga–As system concluded from these analytical results. The first-to-freeze portion (head: analytical result shown as an open circle) of the crystal pulled from a Ga-rich melt at 1.045 Ga:As ratio and even its last-to-freeze portion (tail: analytical result represented by a closed circle) are As-rich, indicating an As-rich maximum melting point and a relatively flat solidus sloping downward toward Ga-rich compositions. The two crystals grown from As-rich melt compositions are both As-rich. However, while the crystal grown from a lightly As-rich melt corresponding to Ga:As = 0.968 has increasing As concentration from head to tail, the crystal grown from a melt of larger As concentration corresponding to Ga:As = 0.943 has decreasing As concentration from head to tail. This indicates a turning point in the solidus, as illustrated in Figure 5.34(b).

In the application of Czochralski pulling to compound III–V compounds, it is necessary either completely to encase the crystal puller in an inert enclosure that is held above the condensation temperature of the group V element or to seal the melt by an inert liquid membrane (e.g., B_2O_3). The

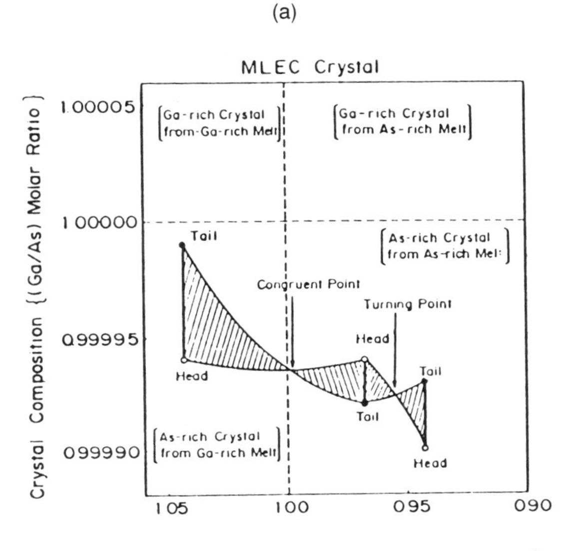

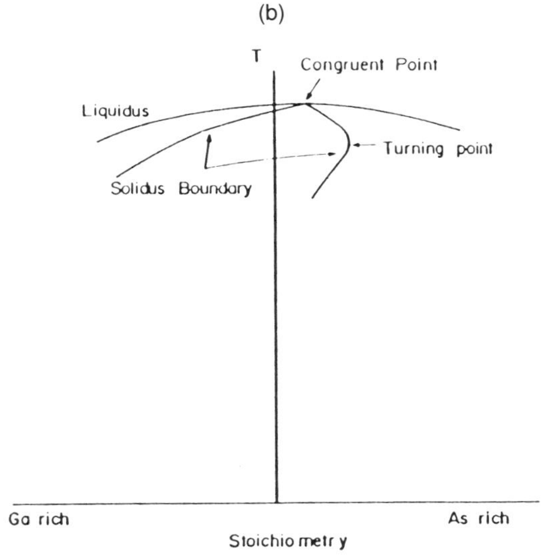

Figure 5.34 *T–x* phase diagrams of the Ga,As system. After Terashima et al. [86]; copyright © 1986, North Holland Publishing Company, Amsterdam.

crystals analyzed in Figure 5.29 were grown by the latter method, which is preferred for production. It is called liquid encapsulated Czochralski (LEC) pulling and was first described by Metz, Miller, and Mazelski in 1962 [87]. The escape of gas bubbles formed from the volatile constituent of the melt (e.g., As_2 in the case of GaAs) is prevented by pressurization with an inert gas such as nitrogen on the outside of the crucible. This allows the use of water-cooled stainless steel enclosures and O-ring sealed moving parts in the construction of the puller, contributing to both the convenience and safety of the operation.

Figure 5.35(a) shows a schematic cross section of an LEC puller that allows viewing of the meniscus from atop similar to the Czochralski pulling of silicon [88]. Figure 5.35(b) shows a semi-insulating (SI) crystal of InP grown by this method. Alternatively, a variety of methods (e.g., the weighing method, x-ray imaging, and pulling out of a coracle) have been employed successfully for controlling the crystal diameter under the conditions of LEC pulling [89], [90]. The thermal conditions in LEC pulling are not favorable for achieving low dislocation density. Due to the strong convective heat transport from the crystal surface to the water-cooled steel walls of the vessel by the dense nitrogen atmosphere, steep radial temperature gradients result in the crystal. Assuming that the dislocation density is proportional to the excess shear-stress sum,

$$\Sigma_e = \sum_{i=1}^{p} |\Sigma_i(\mathbf{r}, \theta)| - \Sigma_c \tag{5.87}$$

one can calculate the dislocation density at a particular location $(\mathbf{r}, \theta)$ on a cross section of the crystal [91]–[93]. The $\Sigma_i(\mathbf{r}, \theta)$ are the measured resolved shear-strss components, and Σ_c is th critical resolved shear-stress [terms $|\Sigma_i(\mathbf{r}, \theta)| - \Sigma_c \leqslant 0$ make no contribution to Σ_e]. The resulting theoretically predicted distribution of dislocations is in approximate agreement with the measured dislocation density distribution for crystals grown by the Czochrals method and LEC pulling. More accurate quantitative evaluations must incorporate the effects of activation of more than one slip system and of interactions of the dislocations that tend to tangle, particularly at large dislocation densities. Generally, dislocations are undesirable since they may propagate during the growth of epitaxial layers into the active device region, where threading dislocations alter locally the reverse breakdown characteristics and the dark current. In particular, nonhomogeneous defect distributions are always a problem, and there exists a considerable interest in controlling the density and distribution of defects. Dislocation-free GaAs crystals of small diameter have been produced by LEC pulling from dislocated seeds in nominally undoped material [94]. Doping to high levels with either electrically active shallow impurities [95] or electrically nonactive isoelectronic impurities, such as In or Al doping of GaAs or G doping of InP, hardens the material and allows the growth of larger-diameter dislocation-free crystals [96].

Figures 5.36(a) and (b) show an x-ray topograph and a double-crystal spectrometer rocking curve, respectively, of wafers cut from a heavily doped InP crystal, revealing nearly perfect behavior. Of course, heavy doping is only suitable for applications requiring low-resistivity substrates, as is the case in certain optoelectronic application. The hardening of the material by the incorporation of isoelectronic impurities is more generally applicable, but the same changes in the mechanical properties that lead to the reduction of the

(a)

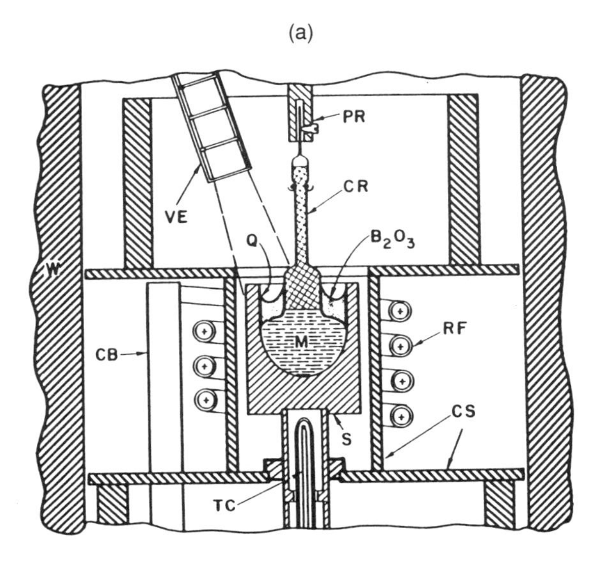

(b)

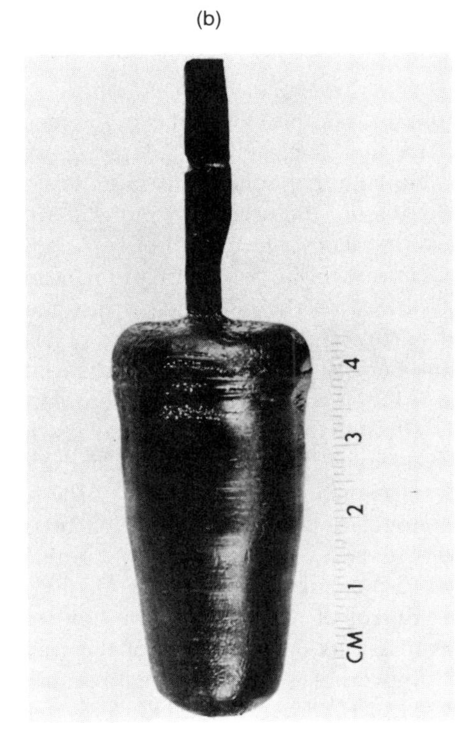

Figure 5.35 (a) Schematic representation of a cross section through an LEC puller. (b) LEC-pulled semi-insulating InP crystal. After Bachmann, Buehler, and Strnad [88]; copyright © 1975, The Minerals, Metals and Materials Society, Warrendale, PA.

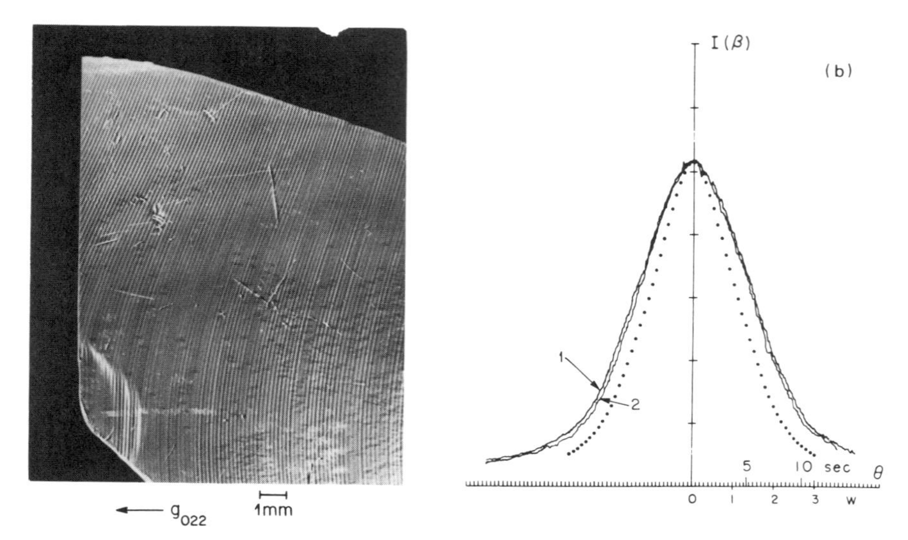

Figure 5.36 X-ray topograph and rocking curves of highly doped InP crystals. (a) Zn doped to $6 \times 10^{18}\,cm^{-3}$; (b) outer curve: Se doped to $2 \times 10^{19}\,cm^{-3}$; middle curve: Zn doped to $6 \times 10^{18}\,cm^{-3}$; dotted curve calculated for ideal InP crystal. After Matsui, Watanabe, and Seki [95]; copyright © 1979, North Holland Publishing Company, Amsterdam.

dislocation density cause undesirable side effects in the machining and handling of the crystal wafers, so that considerable interest exists in controlling the dislocation density by controlling the thermal strain during growth and cooling of the crystals, thus eliminating the need for solution hardening.

Figure 5.37(a) shows a perfect single crystal of GaAs grown by enlarging the thickness of the B_2O_3 layer without penetrating the encapsulant, avoiding the cooling of the crystal surface by the dense gas atmosphere [97]. In order to prevent stress generation during the cooling of the crystal to room temperature, thick layers of the encapsulant on the crystal surface are not desirable, since they cause strain and in extreme cases severe surface damage on the outside of the crystal. Alternatively thermal geometries that reduce the radial gradient by the supply of extra heat by after-heaters may be applied, which reduces viscosity and consequently the thickness of the adhering B_2O_3 layer. However, surface decomposition is a problem, particularly in the case of phosphides. This problem has been successfully controlled by providing a low-density phosphorus atmosphere above the B_2O_3 layer [98].

In addition to the control of dislocations, the control of the density and distribution of the point defects, in particular of the traps in semi-insulating material, is extremely important for achieving high reliability in the subsequent processing of GaAs ICs. For IC processing, SI GaAs is required, which may be fabricated by the addition of impurities (e.g., Cr) to the melt or by the control of the growth conditions to favor the formation of EL2 traps (see Chapters 3 and 8), thus trapping the excess electrons in the nominally undoped material.

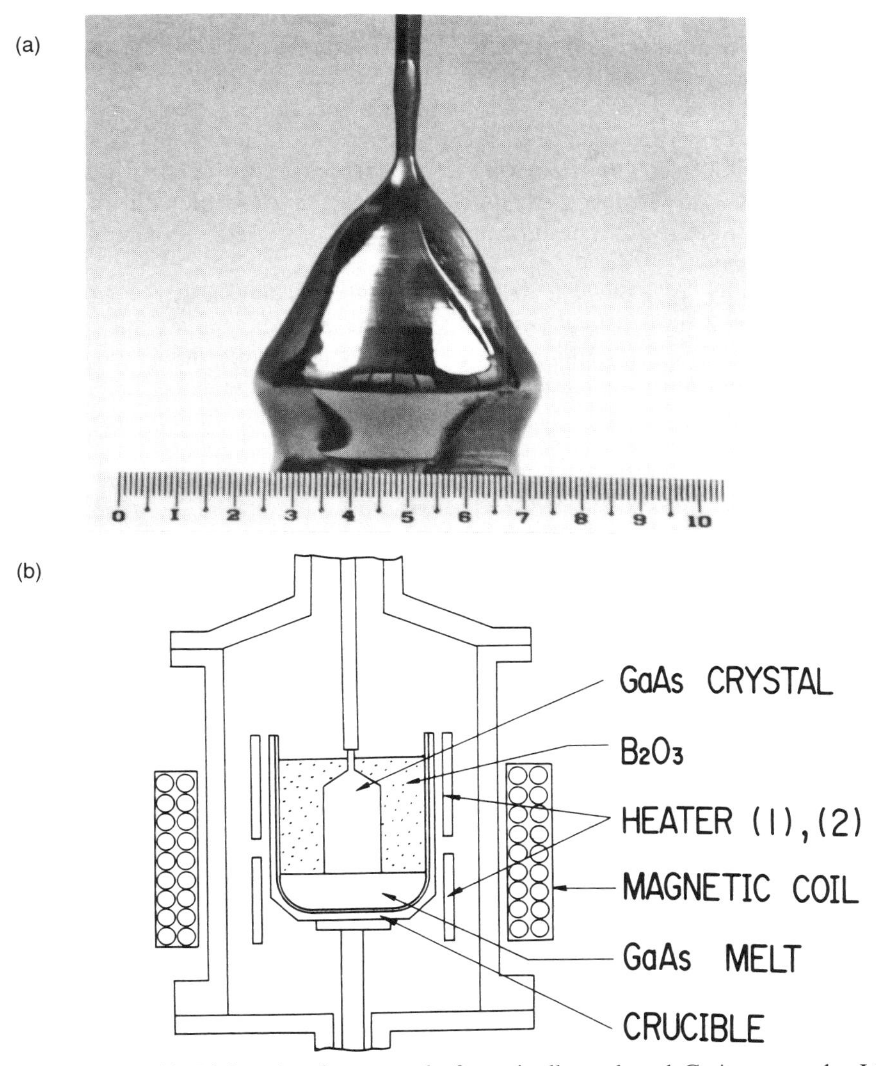

Figure 5.37 (a) Dislocation-free crystal of nominally undoped GaAs grown by LEC pulling in the arrangement shown in (b). After Kohda, Yamada, Nakanishi, Kobayashi, and Hoshikawa [97]; copyright © 1985, North Holland Publishing Company, Amsterdam.

In addition to the strong effects of the control of the stoichiometry on the trap density, the concentrations of residual impurities clearly affect the density and distribution of the EL2 trap in GaAs. In fact, originally the achievement of semi-insulating properties in nominally undoped GaAs was attributed to residual oxygen, but it was later shown that the trap concentration exceeds the oxygen concentration so that no direct correlation exists. The distribution of the EL2 trap in wafers of nominally undoped and In-doped GaAs has been measured by the absorption of infrared radiation tuned to transitions from the midgap trap levels to the conduction band [99]. Stress plays a role in the distribution of these trap states, both by affecting the point-defect distribution as well as via the distribution of dislocations. Dislocations may act as sources or sinks of point defects participating in the formation of the complexes that constitute the traps. However, the most important factor in the control of the trap concentration and distribution is the control of the stoichiometry of the melt and the growth rate.

Since the uniformity of the distribution of defects is significantly affected by the fluctuations in both the growth rate and the boundary layer thickness in the melt, the control of the uniformity of the electrical properties of the crystal requires, in addition to the control of the distribution of the stress, the control of the microscopic growth rate and diffusion conditions [100], [101]. For low-frequency fluctuations (i.e., fluctuations at a time scale $t > 2\pi\delta^2/D$) the change in the solute concentration Δc_s in the crystal upon a change of the growth rate Δv_g can be calculated from the steady-state relation between c_s and v_g using Equation (5.27):

$$\frac{\Delta c}{c} = \frac{(1 - k_0)(v_g\delta/D)\exp(-v_g\delta/D)}{k_0 + (1 - k_0)\exp(-v_g\delta/D)} \frac{\Delta v_g}{v_g} \tag{5.88}$$

For a given value of k_0, the relative variation of the concentration inside the crystal $\Delta c/c$ normalized to the fluctuation in the growth rate $\Delta v_g/v_g$ has a maximum at an intermediate value of $v_g\delta/D$. In order to minimize the effect of a growth rate fluctuation, it is thus necessary either to choose $v_g\delta/D \ll 1$ so that

$$\frac{\Delta c}{c} = (1 - k_0)\frac{v_g\delta}{D}\frac{\Delta v_g}{v_g} \tag{5.89}$$

or $v_g\delta/D \gg 1$, which is, however, limited by the onset of constitutional supercooling. For high-frequency growth rate fluctuations, that is, at a time scale $t \ll \delta^2/D$,

$$\frac{\Delta c/c}{\Delta v_g/v_g} = (1 - k_0)\frac{v_g\delta/D}{(2\omega\delta^2/D)^{1/2}} \tag{5.90}$$

Significant modulations of the diffusion layer thickness δ by perturbations of the convective flow occur at low frequencies. The cutoff frequency above which the diffusion layer is unaffected by growth rate fluctuations is

$$\omega_1 = \frac{\nu_v}{\delta_m(\delta_m - \delta)} \tag{5.91}$$

where ν_v is the kinematic viscosity of the melt and δ_m is the momentum boundary layer thickness to which the diffusion layer thickness is related as

$$\delta = \delta_m(D/\nu_v)^{1/3} \tag{5.92}$$

[102]. Taking typical values for semiconductor melt growth results in a cutoff frequency of the order of 1 Hz.

Under the conditions of Czochralski pulling growth rate variations may have a number of causes [101]:

1. Thermal convection that modulates the thermal transport conditions at the interface, that is, the dissipation of the latent heat upon freezing, and consequently the growth rate of the crystal;
2. Lack of coincidence of the axis of crystal rotation and the thermal symmetry axis of the melt, causing the off-axis areas of the solid–liquid interface to rotate between hotter and colder parts of the melt, leading to periodic growth rate variations;
3. Variations in the effective pulling speed, which may be related to nonuniformities in the mechanical part of the motion mechanism as well as to vibrations;
4. Solutal convection that modulates the solute concentration at the interface.

A quantitative postgrowth assessment of the growth rate modulations occurring during crystal growth is possible by providing in the course of the solidification process deliberate markers (i.e., controlled growth rate modulations) at regular intervals in time. This is accomplished either by periodic generation of a mechanical vibration or, more elegantly, by the provision of current pulses, which, due to the Peltier effect, momentarily provide additional cooling at the solid–liquid interface. The heat flow at the interface matches in steady state the latent heat generated upon freezing, that is, in the absence of a current pulse

$$\kappa_{ts} \nabla T_s - \kappa_{tl} \nabla T_l = \rho H_F v_g(0) \tag{5.93}$$

and in the presence of a current pulse

$$\kappa_{ts} \nabla T_s - \kappa_{tl} \nabla T_l = \rho H_F v_g(J) + pJ \tag{5.94}$$

where κ_t and p are the thermal conductivity and the Peltier coefficient and the subscripts s and l refer to the solid and the liquid at the interface. Assuming that, for short current pulses, the temperature gradient at the interface does not change

$$v_g(J) = v_g(0) - \frac{pJ}{\rho H_F} \tag{5.95}$$

or

$$p = \frac{\rho H_F [v_g(0) - v_g(J)]}{J} \tag{5.96}$$

Therefore, a measurement of the growth rate during current flow and in the absence of a current provides a means for determining the Peltier coefficient of a material. The change in the composition of the solid associated with a change in the growth rate during the presence of a current pulse may be revealed by etching, since the local dissolution rate of the crystal for suitably selected etchants depends on the solute concentration (see Section 7.]. Thus contrast can be observed in an optical microscope on etched axial cross sections of crystals that reveals the interface shape at the time of the application of the current pulse. Also, the measurement of the length of crystal added between two current pulses of known distance in time determines the average growth rate in this time interval.

Figure 5.38 shows a block diagram of the electronics needed for implemented current-controlled growth by means of pulse generator A and buffer C1 and interface demarcation by shorter superimposed pulses provided by pulse generator B and buffer C2, so that the growth rate can be measured for high and low output of pulse generator A. For InSb, the method results in $p = -0.098$ V, so that Peltier cooling at a current density $10 \leqslant J \leqslant 100$ A/cm^2 dissipates approximately 1–10 W/cm^2. Figure 5.39 shows the interface demarcation by short current pulses and the growth-rate variation by longer superimposed pulses for a crystal of Te-doped InSb. Note that the striations produced by the current pulses are all parallel, which is characteristic of the growth on a low-index facet, while off-facet striations usually are curved [104].

Figure 5.40 (top) shows an optical micrograph of an etched axial cross section of a Ga-doped germanium crystal that is schematically presented in the bottom part of the figure. The stained cross section reveals striations and demarkations by 7 A/cm^2 current pulses of 30 ms duration in 0.5 s intervals that are visible as a set of sharp dark lines in regular intervals. They permit the analysis of the microscopic growth rate [curve (a) and right-side scale in the

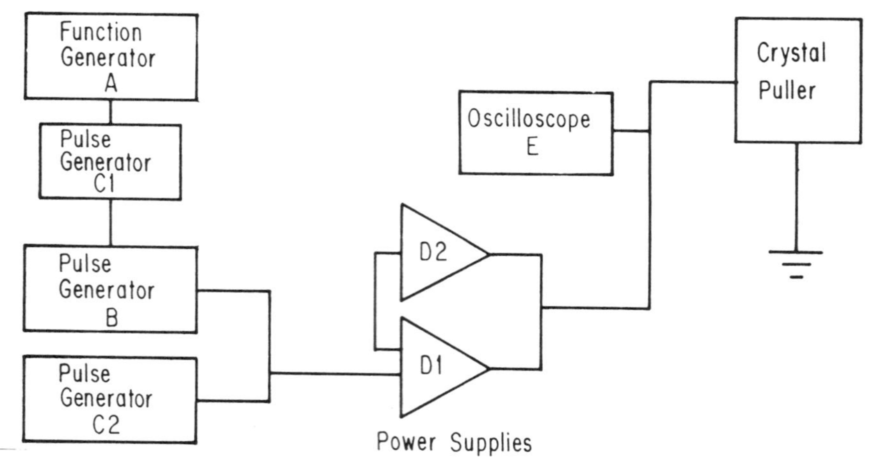

Figure 5.38 (a) Block diagram of the electronics utilized for a superposition of long current pulses for control of the growth rate and short current pulses for interface demarcation by the Peltier effect. (a) After Wargo and Witt [103]; copyright © 1984, North Holland Publishing Company, Amsterdam.

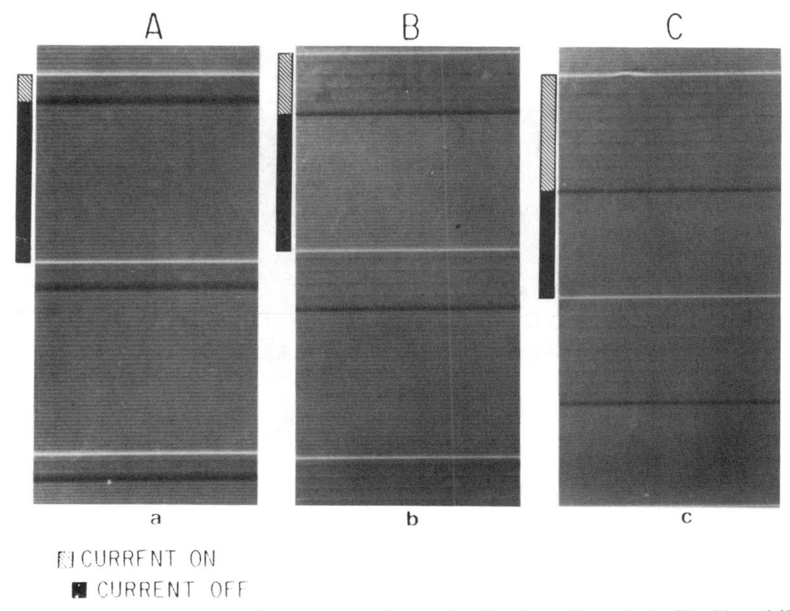

Figure 5.39 (a–c) Stained axial cross sections of a Te-doped crystal of InSb exhibiting growth-rate variations induced by long current pulses of different durations, as illustrated by shading on the margins, and interface markers created by short current pulses. After Wargo and Witt [103]; copyright © 1984, North Holland Publishing Company, Amsterdam.

middle part of Figure 5.40]. Curve (b) shows the carrier concentration obtained from spreading resistance measurements at positions that are visible by small light markings in the lower part of the stained cross section. In this particular experiment, the carrier concentration follows approximately the microscopic growth rate variation. In addition to striations due to random variations in the microscopic growth rate, periodic growth rate modulations (rotational striations) appear as broad white striae. If substantial radial temperature gradients exist in the melt, even cycling between growth and remelting may occur. Materials with high melting points are especially prone to this type of behavior. The average rate of solidification during the growth periods is then in all parts faster than the pulling speed.

Note that nonrotational striations are closely related to natural convection, which is the consequence of temperature gradients and concentration gradients in a liquid and at its surface. Since these gradients are associated with density gradients and surface tension gradients, they generate a buoyancy force in the volume and a shear stress at the surface causing fluid flow. Both natural convection and forced convection due to externally imposed stirring actions (e.g., stirring of the melt by crystal and/or crucible rotations) affect the diffusion

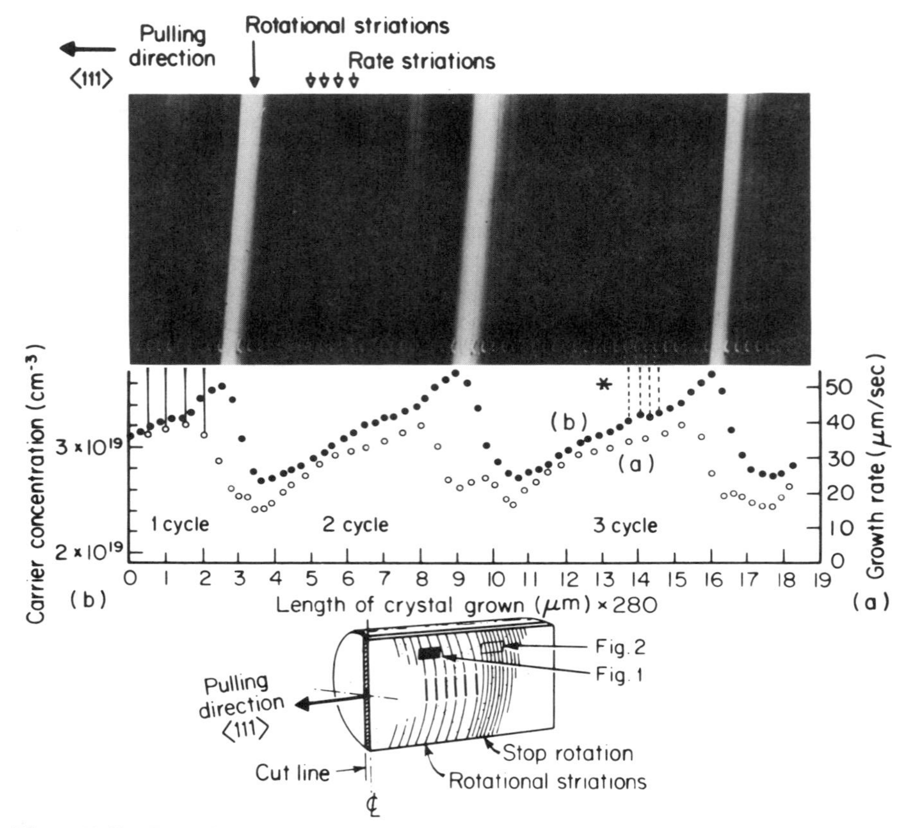

Figure 5.40 Rotational and nonrotational striations in a Czochralski pulled crystal of Ga-doped Ge (top), microscopic growth rate (a) and carrier concentration (b) as a function of the position along the pulling direction (middle), and schematic representation of the crystal and the axial cut position (bottom). After Witt, Lichtensteiger, and Gatos [105]; copyright © 1973, The Electrochemical Society, Pennington, NJ.

layer thickness and thereby the solute flux at the interface. In addition, convection affects the heat transport at the interface and consequently the growth rate so that both heat transport (thermal convection) and solute transport (solutal convection) contribute to the local variations in the solute incorporation into the solid. A gauge for the relative importance of convective heat and solute transport as compared to diffusive heat and solute transport is given by the thermal and solutal Peclet numbers

$$\mathrm{Pe_t} = \frac{v_c l_c}{\kappa_t} \tag{5.97}$$

$$\mathrm{Pe_s} = \frac{v_c l_c}{D} \tag{5.98}$$

where v_c and l_c are characteristic velocity and length scales. Boundary layers form and determine the mode of transport for large values of the Peclet numbers or the Reynolds number

$$\mathrm{Re} = \frac{v_c l_c}{\nu_v} \tag{5.99}$$

The onset of natural steady-state convection in an initially quiescent liquid layer that is heated from below is governed by a critical value of the Rayleigh number

$$\mathrm{Ra} = \frac{\alpha_t g h \nabla T_v}{\kappa_t \nu_v} \tag{5.100}$$

where α_t is the thermal expansion coefficient of the liquid, g is the gravitational constant, h is the height of the melt, and ∇T_v is the vertical temperature gradient. The value of the critical Rayleigh number Ra_c depends on the boundary conditions and on the flow phenomena to which it applies, usually being the onset of turbulence. However, critical Rayleigh numbers also may be defined for other flow phenomena, e.g., the spontaneous break up of a bottom heated liquid into hexagonal cells with fluid circulation up in the cell walls and down in their centers. Localized heating on the bottom, resulting in deviations from a strictly vertical temperature gradient, induces flows even at small values of the driving force. Oscillatory instabilities ensue for Grashof numbers

$$\mathrm{Gr} = \frac{\rho \alpha_t g l^3 \Delta T}{\nu_v^2} \gtrsim 10^9 \tag{5.101}$$

The Prandtl number

$$\mathrm{Pr} = \frac{\nu_v}{\kappa_t} \tag{5.102}$$

and Schmid number

$$\mathrm{Sc} = \frac{\nu_v}{D} \tag{5.103}$$

measure the relative importance of, respectively, viscous transport of heat versus thermal conduction and viscous transport of solute versus diffusion in the liquid. Redefining the Rayleigh number as the product of the Grashof and Prandtl numbers, several modes of convection-driven fluid flow may be defined in terms of ranges in the Rayleigh number of the system under study.

Figure 5.41 shows a schematic representation of the modification of the effective distribution coefficient governing the axial segregation (top figure) and the radial segregation Δc (middle figure) in these regimes of fluid flow. Under the conditions of crystal growth from the melt, turbulence is usually avoided, but fluctuations in the convective mixing generally exist. Attempts have been made to diminish fluid flow instabilities caused by the fluctuations in natural

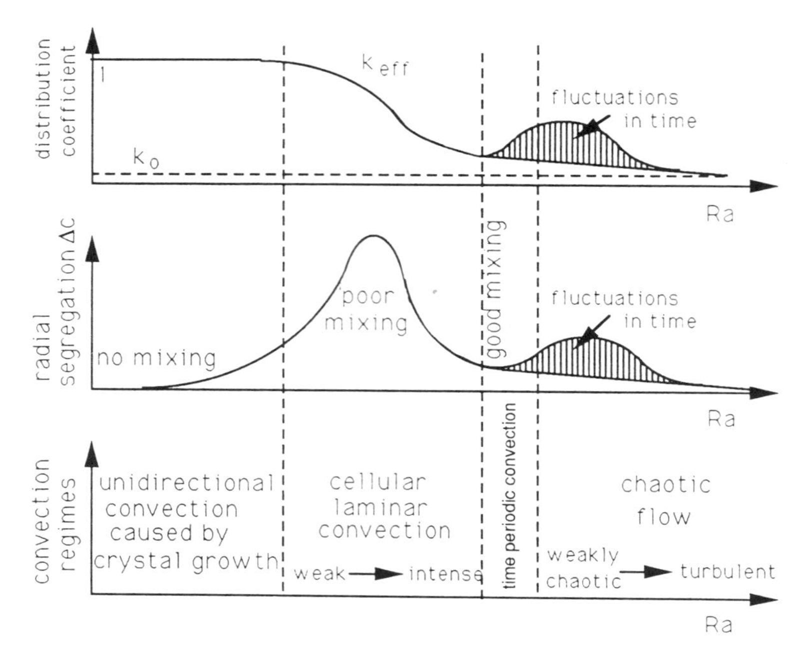

Figure 5.41 Regimes of fluid flow defined by the Rayleigh number of the system and the characteristic of radial and axial segregation in these regimes.

convection by combinations of crucible and seed rotations. Figure 5.42 shows a schematic representation of the flow patterns developed under the conditions of Czochralski pulling in the melt for various combinations of seed and crucible rotations. Complex flow patterns develop in all combinations except for fast synchronous rotation of the seed and the crucible in the same direction.

Accurate descriptions of fluid flow under the conditions of crystal growth may be established by solving the Navier–Stokes equation with appropriate bondary conditions. In the Boussinesq approximation, that is, assuming that the thermal and solutal expansion coefficients $\alpha_t = (1/\rho)(\partial\rho/\partial T)_{p,c}$ and $\alpha_c = (1/\rho)(\partial\rho/\partial c)_{p,T}$ are constants, they take the form

$$\rho\frac{\partial \mathbf{v}}{\partial t} + \mathbf{v}\cdot\nabla\mathbf{v} = -\nabla\mathbf{p} + \nu_N\nabla^2\mathbf{v} + \rho\mathbf{g}[1 - \alpha_t(T - T_r) + \alpha_c(c - c_r)] + \mathbf{F}_b \tag{5.104}$$

$$\nabla\cdot\mathbf{v} = 0 \tag{5.105}$$

$$\frac{\partial T}{\partial t} + \mathbf{v}\cdot\nabla T = \kappa_t\nabla^2 T \tag{5.106}$$

$$\frac{\partial c}{\partial t} + \mathbf{v}\cdot\nabla c = D\,\nabla^2 c \tag{5.107}$$

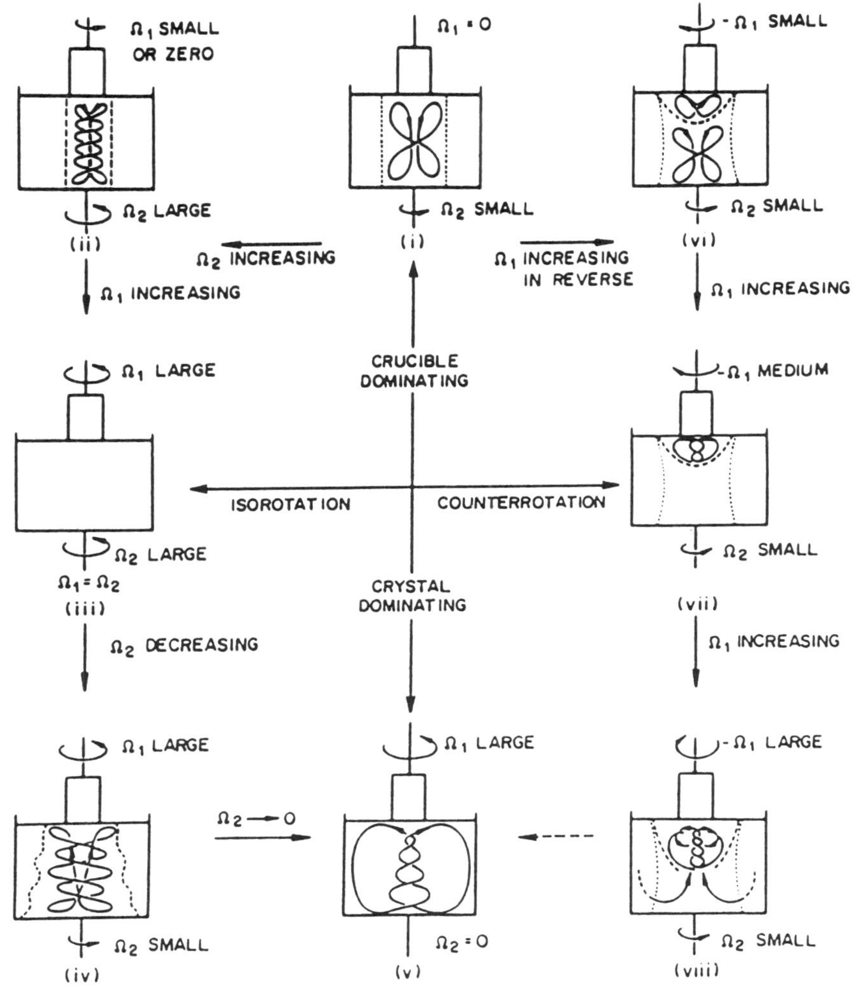

Figure 5.42 Schematic representation of typical flow patterns developed under the conditions of Czochralski pulling for different combinations of seed and crucible rotation. After Carruthers and Witt [101]; copyright © 1975, North Holland Publishing Company, Amsterdam.

[107]. In Equation (5.104)–(5.107), **v** and **p** are velocity and pressure field vectors, **g** is the gravity vector, and $\mathbf{F}_b$ is a body force that may act on the system.

Surface tension effects must be added to the consideration of buoyancy-driven convection in the bulk. In fact, they are often dominant, as, for example, in the original investigations of Benard instabilities in bottom-heated liquid layers, where the observed breakdown into hexagonal cells was caused by surface tension gradients rather than buoyancy-driven convection. Surface effects are particularly important under the conditions of Czochralski pulling and float zoning, where the shape and temperature distribution on the free surface of the liquid critically affects the control of the crystal diameter and the stability of the molten zone. The maximum length of a molten zone that is stable under the conditions of vertical float zoning of a rod of diameter d_r is

$$\frac{l_{max}}{d_r} = \frac{\pi}{(1 + \mathrm{Bo})^{1/2}} \tag{5.108}$$

where the Bond number

$$\mathrm{Bo} = \frac{\Delta\rho\, g d_r^2}{\sigma_s} \tag{5.109}$$

At zero gravity Bo = 0 and $l_{max} = \pi d_r$, while on earth Bo $\gg 1$ so that

$$l_{max} = \pi \left(\frac{\sigma_s}{\Delta\rho\, g}\right)^{1/2} \tag{5.110}$$

where $\Delta\rho$ is the difference between the density of the molten zone and the surrounding medium [106]. Rotation of the zone to improve its thermal symmetry reduces l_{max} from the value predicted by Equation (5.108), which applies to a nonrotating zone to

$$\frac{l_{max}}{d} = \frac{\pi}{(1 + \mathrm{Bo})^{1/2}(1 + \mathrm{Bo}_r)^{1/2}} \tag{5.111}$$

with the rotational Bond number

$$\mathrm{Bo}_r = \frac{\Delta\rho\, r^2\omega^2}{\sigma_s} \tag{5.112}$$

[107].

Since the temperature and solute concentration on the surface of the melt is a function of position, the surface tension is expressed as

$$\sigma_s = \sigma_{sr}\left(1 + \frac{\partial\sigma_s}{\partial T}(T - T_r) + \frac{\partial\sigma_s}{\partial c}(c - c_r)\right) \tag{5.113}$$

where σ_{sr}, c_r, and T_r are reference values of the surface tension, concentration, and temperature. The driving force for fluid motion associated with the shear stress imposed by gradients in the surface tension at the melt surface is characterized in relation to the viscous damping of the flow by the Marangoni number

$$\mathrm{Ma} = \frac{\sigma(\partial\sigma/\partial T)h^2\,\Delta T}{\rho v_v^2} \tag{5.114}$$

Since the effects of surface tension gradient driven convection and buoyancy-driven convection in the bulk mutually reinforce, the condition for the onset of natural convection may be written in the form

$$\frac{\mathrm{Ra}}{\mathrm{Ra_c}} + \frac{\mathrm{Ma}}{\mathrm{Ma_c}} = 1 \tag{5.115}$$

where $\mathrm{Ma_c}$ is the critical Marangoni number. Figure 5.43 shows the results of simulations of convective flow patterns in vertical molten zones under various conditions of rotation of the solid rods suspending this zone.

Since simulations based on numerical solutions to the Navier–Stokes equation with appropriate boundary conditions and incorporation of surface forces show that boundary-layer modulations cannot be suppressed successfully by externally imposed stirring actions, a considerable effort has been made in recent years to modify the conditions of crystal growth by either removing some of the driving forces (e.g., by experimentation in a microgravity environment) or by utilizing the body force in Equation (5.104). For example, in the case of Czochralski pulling of a material, for which the melt is electrically conductive with conductivity σ_e, in an axial magnetic field the Lorentz force

$$\mathbf{F}_b = \sigma_e(\mathbf{v} \times \mathbf{B} \times \mathbf{B}) \tag{5.116}$$

is added [108]. Simulations under the assumption of axially symmetric flow for a system with Re ≈ 500 shows that the radial flow velocity component under

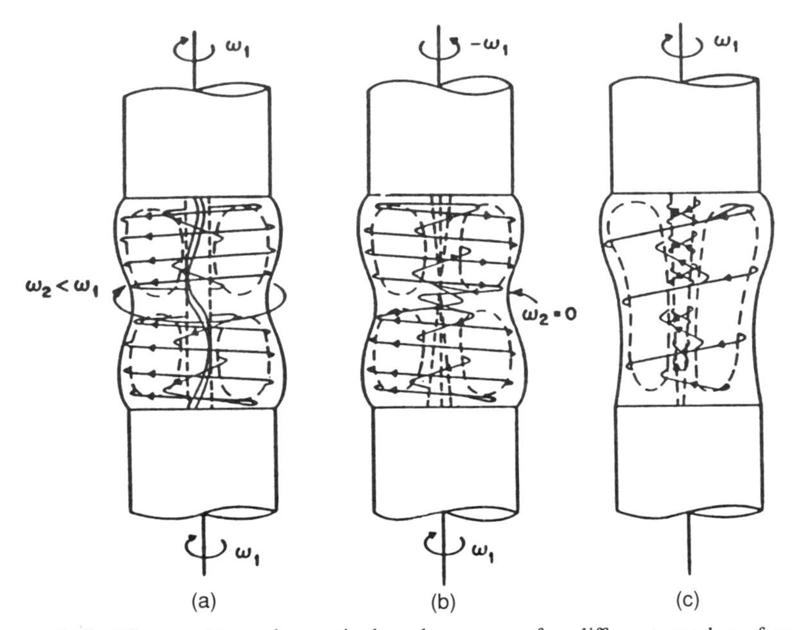

Figure 5.43 Flow patterns in vertical molten zones for different modes of rod rotations. After Carruthers [102]; copyright © 1975, Plenum Press, New York.

the rotating crystal scales inversely with the magnetic interaction parameter

$$N = \frac{\sigma_e \mu_m^2 H_0^2}{\rho \omega} \tag{5.117}$$

where H_0 is the magnetic field and ω is the angular velocity of the crystal rotation [109]. The pumping action of the rotating crystal leading to upward axial flow under the crystal and downward flow beyond the crystal periphery is thus substantially restricted by the magnetic field; that is, it is confined to the immediate vicinity of the crystal. Also, convective flow in the melt is suppressed by the presence of an axially symmetric field, which increases the critical Raleigh number [107]. The ratio of the Lorentz force and the viscous force is characterized by the Hartmann number

$$\mathrm{Ha} = \frac{|\mathbf{B}| l_c \sigma_e^{1/2}}{\rho^{1/2} \nu_v^{1/2}} \tag{5.118}$$

The critical Rayleigh number for a rectangular cavity heated from below increases for low aspect ratios h/w as

$$\mathrm{Ra}_c = \left(\frac{\pi}{2} \frac{h}{w} \mathrm{Ha}\right)^2 \tag{5.119}$$

[110].

X-ray topography of axial sections of a Si crystal pulled in different axial magnetic fields reveals the suppression of striations at relatively small fields (500–1000 Oe) and a concomitant drastic improvement of the resistivity spread in the material pulled in the presence of a moderate field [111]. Since the suppression of convective mixing of the melt reduces the transport rate for oxygen in the Si melt, its concentration in the crystal can be controlled by the field, which affects both the carrier concentration and lifetime [112].

Figure 5.44(a) shows a plot of the positron annihilation spectroscopy signal, which is proportional to the oxygen concentration (full line) and the carrier lifetime (dashed curve) for Si crystals as a function of the magnetic field employed in the pulling process. Figure 5.44(b) shows the axial microsegregation impurity profiles obtained by infrared absorption measurements for crystals of Si pulled with (bottom) and without (top) magnetic field [113]. The bottom figure shows that the magnetic field is effective in eliminating the uncontrolled modulations of the rotational striation pattern by boundary layer modulations. However, a comparison of the top and bottom figures also reveals that while the oxygen concentration is decreased by the presence of a magnetic field, the carbon concentration is increased. As demonstrated by Figure 5.44(a), the life time increases with increasing field, which is clearly a positive effect, but does not reveal the full consequences of the changes in the point defect concentrations by pulling in a magnetic field to the subsequent

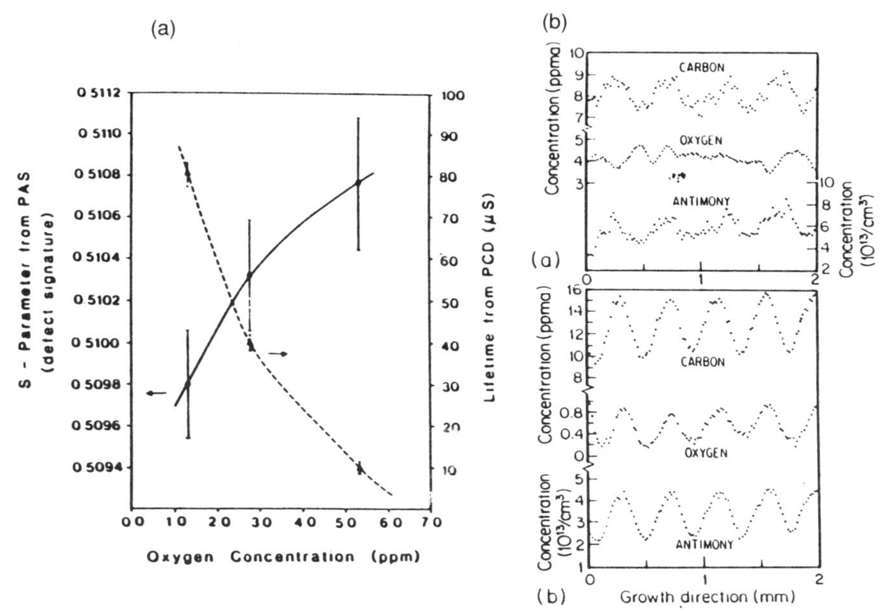

Figure 5.44 (a) The effects of an axial magnetic field during the Czochralski pulling of silicon on the carrier lifetime and (b) the microsegregation profiles of impurities. After (a) Rohatgi [112] and (b) Yao and Witt [113]; copyrights (a) © 1986, IEEE, New York, and (b) © 1987, North Holland Publishing Company, Amsterdam.

processing of circuits. For LEC growth of GaAs the application of a magnetic field has been shown to reduce the concentration of mid-gap levels substantially and to result in a stronger dependence of the EL2 concentration on the Ga:As ratio in the melt than under the conditions of conventional LEC growth [114], which may be related to the reduction of the transport of segregating components to the interface and of the residual strain in the lattice.

A sensitive measurement of weak long-range deformations of a crystal lattice is made possible by the analysis of Pendellösung fringes that appear in the x-ray topographs of nearly perfect crystals. They are due to the interference of the x-ray waves inside the crystal, causing the energy flow to swap back and forth between the forward and back diffracted beam directions [115]. Fringes appear in the diffracted beams upon break-up of the wave field at the exit surface of the crystal. This is illustrated for a perfect crystal lattice in Figure 5.45(a) and (b). An example of experimentally observed Pendellösung fringes in an x-ray topograph is given in Figure 5.42(a). In the case of a weak strain gradient in the crystal, the trajectories of the x-rays become bent as illustrated in Figure 5.45(c). Thus, in the presence of a strain gradient, the Pendellösung fringes have different spacings than in the perfect crystal case. If the strain gradient is inhomogeneous also, the shape of the Pendellösung fringes changes [116].

As an example, we consider the Pendellösung fringes observed in the γ-ray

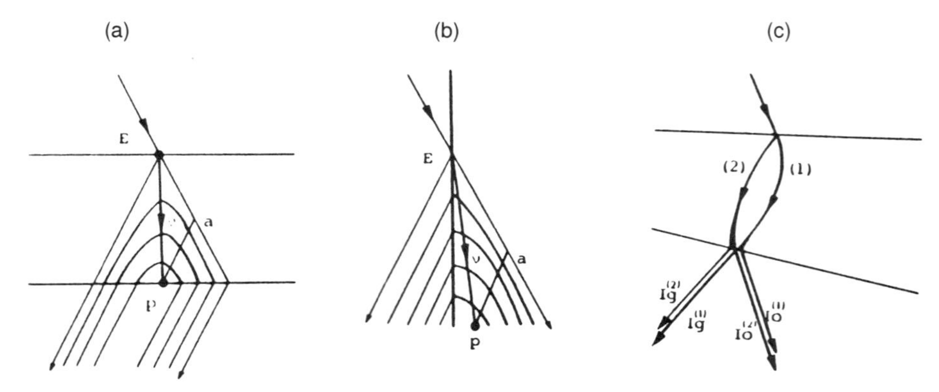

Figure 5.45 Schematic representations of the conditions for the formation of Pendellösung fringes upon diffraction on a perfect crystal (a) Laue case, (b) Bragg case, and (c) on a strained crystal. After Kato [117]; copyright © 1975, North Holland Publishing Company, Amsterdam.

spectra for wafers cut from silicon crystals, that were grown by Czochralski pulling in the absence and presence of a magnetic field, respectively. Figure 5.46 shows plots of the integrated reflected power R_H for diffraction of γ-rays of 0.0392 Å wavelength on such wafers, versus the abscissa (A) for growth (top) without and (bottom) with an axial magnetic field.

The parameter

$$A = \frac{r_e |F'_H| \lambda t_c \exp(-W)}{V_{cell}\sqrt{\gamma_0 \gamma_H}} \tag{5.120}$$

where γ_0 and γ_H are the direction cosines. For comparison, the function

$$R_{H\,dyn} \approx \frac{r_e |F'_H| \lambda^2 \exp(-W)}{2V_{cell}\sqrt{\gamma_0/\gamma_H}\sin 2\Theta_B} \times \exp\left(-\frac{\mu_0 t_c}{\cos\Theta_B}\right) \int_0^{2A} J_0(x)\,dx \tag{5.121}$$

predicted by the dynamical theory of x-ray diffraction is shown. $J_0(x)$ is the zero-order Bessel function of the first kind [118]. Note that for the crystal grown in the absence of a magnetic field, the fringe spacing is shorter (8 fringes experimentally instead of 7 fringes for the simulated perfect crystal case) and that the oscillations in the experimental R_H are located on a slope, indicating a significant residual strain in the crystal. For the crystal grown in a magnetic field, the fringe spacings match the perfect crystal simulation and the slope is very small, so that, for the chosen diffraction conditions, minimal residual strain is observed.

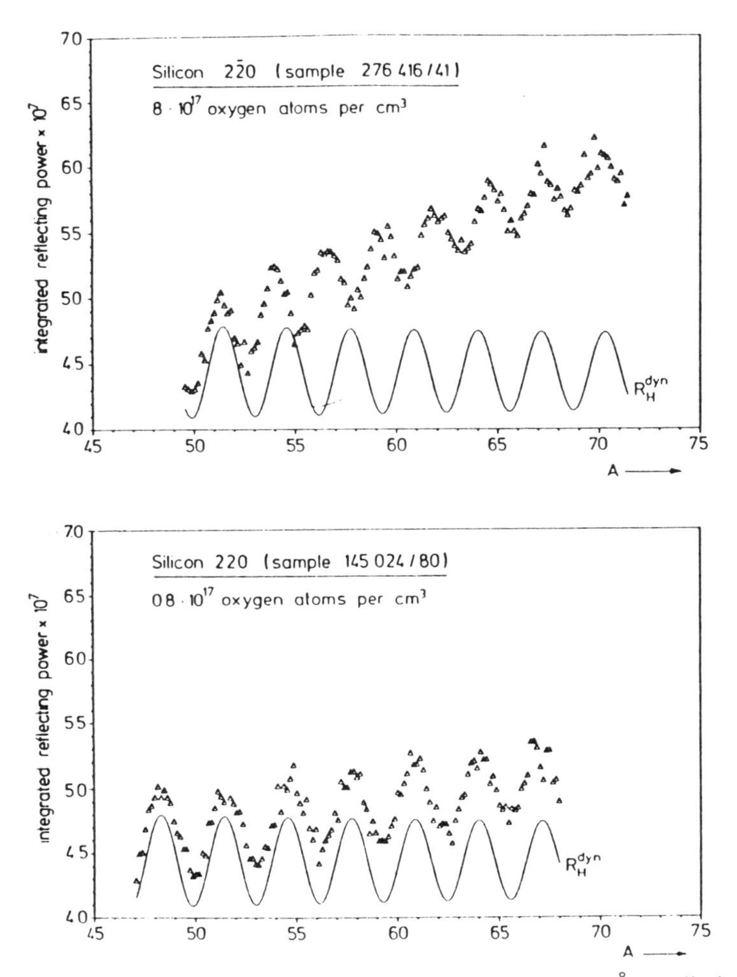

Figure 5.46 Pendellösung intensity beats measured with 0.0392 Å γ radiation on wafers of Si crystals grown (top) in the absence and (bottom) in the presence of a magnetic field. The full lines are the results of simulations of $R_H(A)$ for a perfect Si lattice. After Schneider et al. [118]; copyright © 1989, North Holland Publishing Company, Amsterdam.

In addition to Czochralski pulling, gradient freezing is a particularly attractive method of substrate crystal production for the III–V compounds because of its safety features and minimum need for operator intervention. Figure 5.47 shows a schematic representation of a cross section through a high-pressure apparatus for the synthesis of III–V compounds from the elements and vertical gradient freezing of the melt. Vertical gradient freezing (VGF) has been developed to considerable perfection and is used today to produce large single crystals of III–V compounds. A melt is formed inside a pyrolytic BN capsule that is heated by a resistance heating element. A reservoir

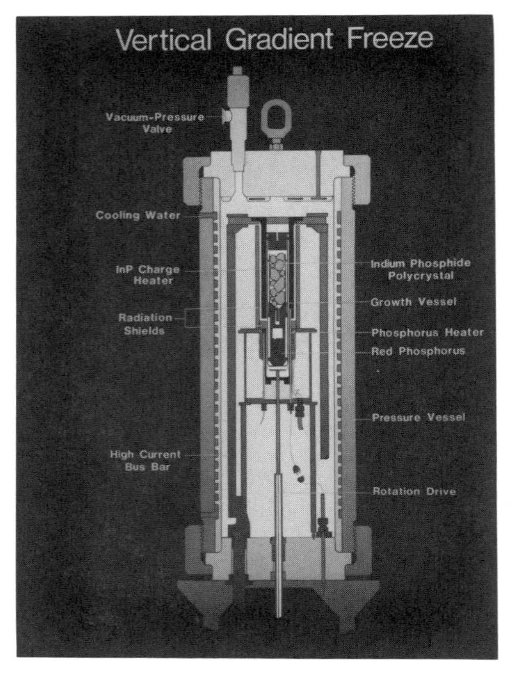

Figure 5.47 (a) Schematic representation of the apparatus for the synthesis of compound semiconductors and crystal growth by the VGF method. After Gault et al. [119]; copyright © 1986, North Holland Publishing Company, Amsterdam.

of the volatile group V component is provided in the bottom portion of the BN capsule, which is sealed by a BN plug. The melt is contained in a pyrolytic BN crucible located inside the BN capsule. It is separated from the group V charge by a BN spacer containing a series of slots that permit communication of the vapor atmosphere above the melt and the group V charge in the reservoir. Seeding can be initiated from a seed contained in a seed pocket in the bottom of the crucible. Note that the apparatus can serve for both synthesis of the compound semiconductor from the elements and crystal growth.

This is illustrated for the VGF growth of a GaP crystal in Figure 5.48(a). The reservoir is filled with red phosphorus, and a GaP seed and presynthesized GaP are loaded into the crucible. The charge is initially heated to a set point that establishes a temperature distribution, placing the melting temperature at the seed–melt junction and keeping the volume above this position at a higher temperature. Then the temperature is slowly lowered, advancing the solid–liquid interface upward. Note that a relatively steep temperature gradient is

(a)

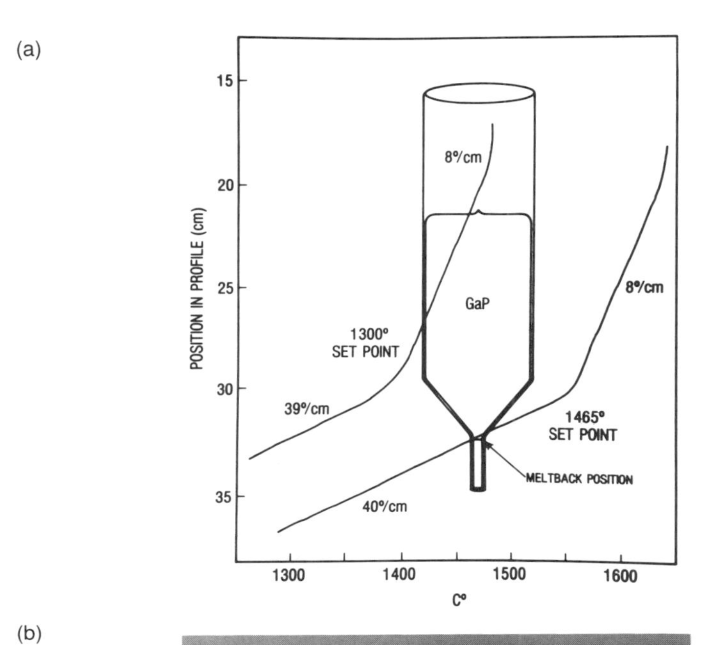

(b)

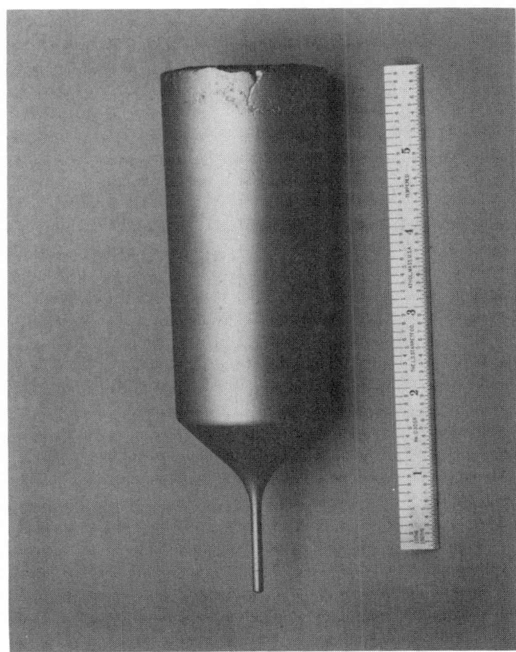

Figure 5.48 (a) Schematic representation of the temperature distribution in the crucible under the conditions of VGF growth of GaP. (b) GaP crystal grown by the VGF method. After Gault, Monberg, and Clemans [119]; copyright © 1986, North Holland Publishing Company, Amsterdam.

used during the initial seeding procedure and that in the later stages of VGF growth the temperature gradient is kept at 8°C/cm. Figure 5.48(b) shows a photograph of a GaP crystal grown by VGF.

Figure 5.49 shows stained cross sections of the seed/shoulder regions of InP crystals grown by VGF. The striations revealed by the stain decorate the shape of the interface, which can be made perfectly flat by slight adjustments in the heat extraction conditions (compare the bowing of the striations of the top and bottom parts of Figure 5.49). In this regard VGF growth has a significant advantage over Czochralski pulling, where maintaining a perfectly flat interface is much more difficult to achieve.

For specialized applications in optoelectronics the Bridgman method has also been applied for the production of substrate crystals, especially GaAs. It accomplishes the directional solidification of the melt by moving the furnace relative to the crucible or boat, causing directional solidification. A fused silica cloth is usually employed as a liner to minimize nucleation of competing grains on the walls of the boat and stresses in the solidified ingot due to binding to the crucible walls. The method is particularly appropriate for the fabrication of Si-doped substrates for optoelectronics applications, where heavily doped crystals are used. For the fabrication of FET circuits, more stringent requirements exist on the residual concentration of shallow impurities. They can be achieved in carefully controlled crystal growth experiments utilizing pyrolytic boron nitride ware. However, at present LEC pulling and VGF growth are the preferred methods of III-V compound crystal growth.

Similar to the results of Czochralski pulling and float zoning, magnetic fields also effect the results of vertical Bridgman growth and gradient freezing. This has been analyzed on the basis of numerical solutions to the Boussinesq equations [20]. For gallium-doped germanium as a model system, the following characteristic parameters are valid for a particular set v_c, l_c: $\mathrm{Re} = 500$, $\mathrm{Ra} = 3 \times 10^8$, $\mathrm{Pe} = 1.6 \times 10^{-2}$, $\mathrm{Pr} = 6.0 \times 10^{-3}$, $\mathrm{Sc} = 5.4$, and $\mathrm{Ha} = 10^3$ at $|\mathbf{B}| = 2.4\,\mathrm{kG}$. Figure 5.50 shows the conversion of the flow fields in the melt above the solid–liquid interface under the conditions of the vertical Bridgman method from cellular convection at small values of the Hartmann number to unidirectional flow at high magnetic fields.

Figure 5.51 shows the radial segregation of the dopant Δc in percent as a function of the Rayleigh number for selected values of Ha as the parameter. At $\mathrm{Ra} = 1 \times 10^7$, cellular convection results in good mixing in the absence of a magnetic field that is decreased in the presence of a weak magnetic field. Thus the radial nonuniformity of doping is enhanced and goes through a maximum for intermediate fields. At high magnetic fields the flow is suppressed completely, resulting in the absence of radial segregation and diffusion-controlled axial segregation. Clearly a judicious choice must be made in the selection of the field and other process parameters to optimize the control of the homogeneity of the crystal.

An electric field affects the critical Rayleigh number of moderately conducting systems both through the temperature dependence of σ_e and of the dielecric constant ε. For insulators, the electrical force is scaled to the viscous

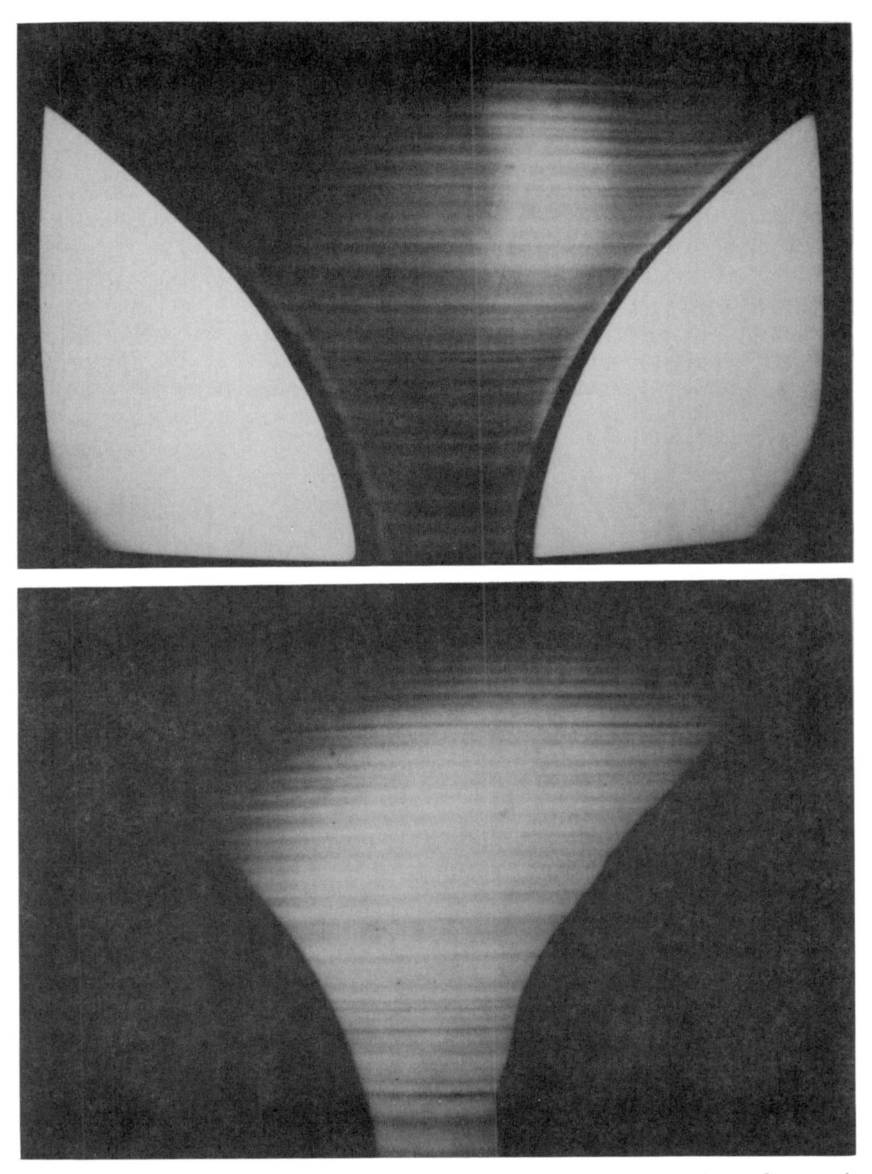

Figure 5.49 Stained axial section of an InP crystal grown by the VGF method revealing striations. After Gault, Monberg, and Clemans [119]; copyright © 1986, North Holland Publishing Company, Amsterdam.

force by the Senftleben number.

$$\mathrm{Se} = \frac{l^2 E^2 \nabla T(\partial \varepsilon / \partial T)}{\rho v_v^2} \tag{5.122}$$

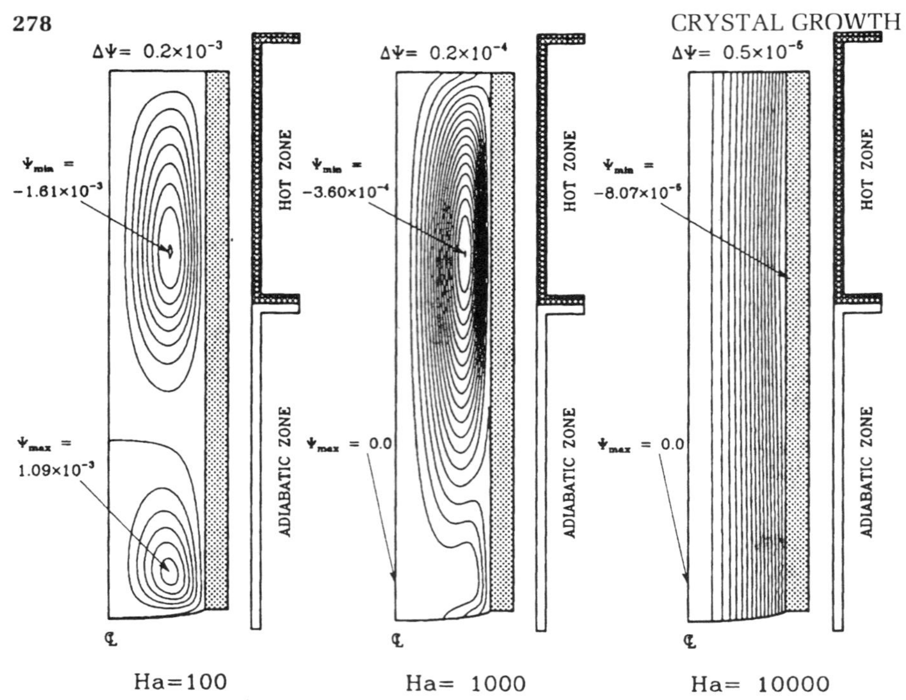

Figure 5.50 (a) Flow fields for vertical Bridgman growth with $Ra = 1 \times 10^7$ at different values of Ha. After Kim, Adornato, and Brown [120]; copyright © 1988, North Holland Publishing Company, Amsterdam.

The local variation of ε due to the temperature gradient ∇T is always destabilizing and leads to electrostrictive convection, which increases in the presence of a field the transport of heat from hot wires in gases over the rate expected for natural convection [122].

An interesting option for stabilization of flow is given by centrifugal fields, restricting the flow to two dimensions [123]. For large Taylor numbers

$$\mathrm{Ta} = \left(\frac{2\omega l^2}{\nu_v}\right)^2 \tag{5.123}$$

cellular convective flow is eliminated, that is, the critical Rayleigh number increases with increasing Taylor number [121]. Because the realization of directional solidification under the conditions of centrifugation is difficult, it plays at present no significant role in the fabrication of substrate crystals of silicon and of the III–V compounds. However, it may become of interest in the context of solution growth of substrate crystals of advanced materials for special applications that cover the cost. As melt and solution grow, also the growth of crystals from the vapor phase may be adversely affected by fluid dynamics. These phenomena are discussed in more detail in chapter 6.

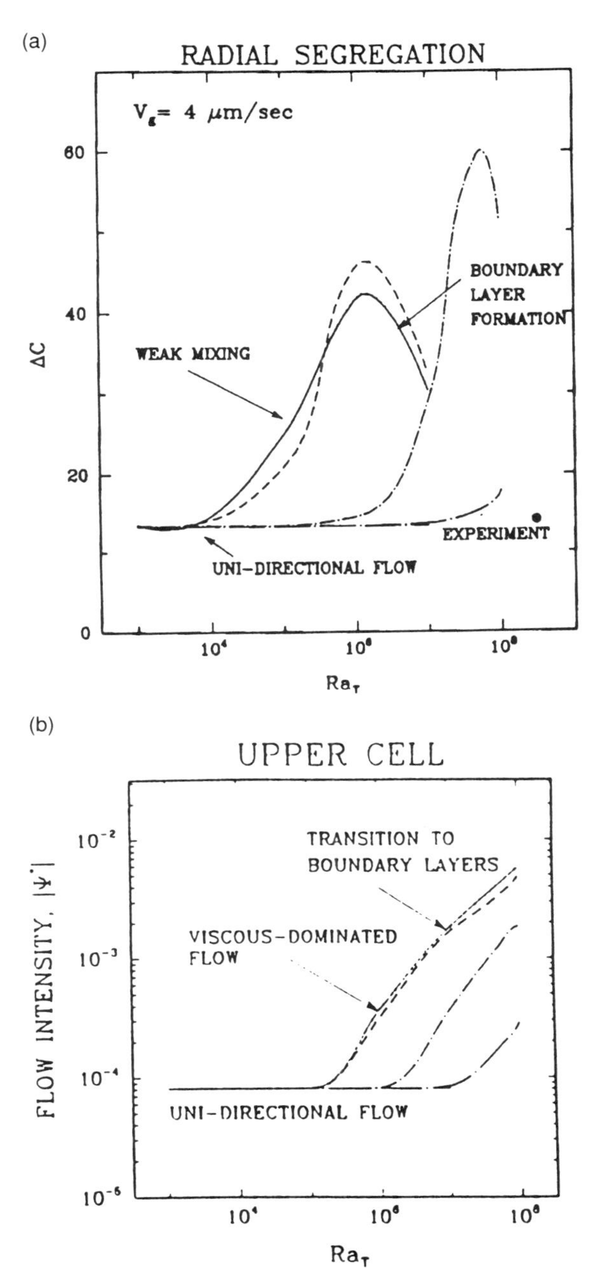

Figure 5.51 (a) Family of curves percentage radial segregation as a function of Ra for different values of Ha as the parameters: ——Ha = 0, --- Ha = 1×10^2, —·— Ha = 5×10^3, (b) Flow intensity as a function of the thermal Rayleigh number. After Kim, Adornato, and Brown [120]; copyright © 1988, North Holland Publishing Company, Amsterdam.

References

1. J. R. McCormick, *Conf. Rec. 14th IEEE Photovolt. Specialists Conf.*, San Diego, CA, 1980, p. 298.
2. K. G. Baraclough, in *The Chemistry of the Semiconductor Industry*, S. J. Moss and A. Ledwith, eds., Blackie & Sons, Ltd., Glasgow, England, 1987.
3. P. A. Taylor, *J. Crystal Growth* **89**, 28 (1988).
4. A. Yusa, Y. Yatsurugi, and T. Takaishi, *J. Electrochem. Soc.* **122**, 1700 (1975).
5. P. Taylor, *Solid State Technol.* **30**, 53 (1987).
6. G. Hsu, N. Rohatgi, and J. Houseman, *AlChE J.* **33**, 784 (1987).
7. S. K. Iya, R. N. Flagella, and F. S. DiPaolo, *J. Electrochem. Soc.* **129**, 1531 (1982).
8. J. Dietl, D. Helmreich, and E. Sirtl, *Conf. Rec. 19th IEEE Photovolt. Specialists Conf.*, New Orleans, LA, 1987, p. 345.
9. C. Thurmond, *J. Phys. Chem.* **57**, 827 (1953).
10. J. A. Burton, R. C. Prim. and W. P. Slichter, *J. Chem. Phys.* **21**, 1987 (1953).
11. R. K. Dawless, R. L. Troup, D. L. Meier, and A. Rohatgi, *J. Crystal Growth* **89**, 68 (1988).
12. W. G. Pfann, *Zone Melting*, John Wiley & Sons, New York, 1966.
13. E. Papp and K. Solymar, *Acta Chim. Hung.* **24**, 451 (1960).
14. R. Dötzer, *Chem. Ing. Techn.* **36**, 616 (1964).
15. V. N. Vigdorovich, *Purification of Metals and Semiconductors by Crystallization*, Freund Publishing House, Ltd., Tel Aviv, 1971.
16. M. Fleischer, in *Economic Geology*, 50th Aniv. Vol., A. M. Bateman, ed., The Economic Geology Publishing Company, Lancaster, PA, 1955, p. 970.
17. K. F. Hulme and J. B. Mullin, *Solid State Electronics* **5**, 211 (1962).
18. W. Lohmann, *Ber. Bunsen Ges. Phys. Chem.* **10**, 87 and 428 (1966).
19. R. A. L. Van den Berghe, F. Cardon, and W. P. Gomez, *Ber. Bunsen Ges. Phys. Chem.* **78**, 331 (1974).
20. K. J. Bachmann and M. S. Su, in *Processing of Electronic Materials*, C. Law and G. R. Pollard, eds., The American Institute of Chemical Engineers, New York, 1987, p. 415.
21. K. Ziegler, *Experientia*, Suppl. II, 274 (1955); K. Ziegler, R. Koster, and H. Lehmkuhl, *Z. Angew. Chem.* **67**, 213 and 424 (1955).
22. Comprehensive Organometallic Chemistry, G. Wilkinson, G. A. Stone, and E. W. Abel, eds., Pergamon Press, Oxford, 1982.
23. A. C. Jones, A. K. Holliday, D. J. Cole-Hamilton, M. M. Ahmad, and N. D. Gerrard, *J. Crystal Growth* **68**, 1 (1984).
24. J. I. Davies, R. C. Goodfellow, and J. O. Williams, *J. Crystal Growth* **68**, 10 (1984).
25. J. M. Olson, S. R. Kurtz, and A. E. Kibbler, *J. Crystal Growth* **89**, 131 (1988).
26. A. H. Moore, M. D. Scott, J. I. Davies, D. C. Bradley, M. M. Factor, and H. Chudzynska, *J. Crystal Growth* **77**, 19 (1986).

27. J. D. Cape, L. M. Fraas, P. S. McLeod, and L. D. Partain, *Research on Multigap Solar Cells, Chevron Research Company, Ann. Subcontr. Rep. 1984–1985*, SERI/STR 211–2825 DE86004422, 1986.

28. A. C. Jones, P. R. Jacobs, R. Caferty, M. D. Scott, A. H. Moore, and P. Wright, *J. Crystal Growth* **77**, 47 (1986).

29. K. J. Bachmann and E. Buehler, *J. Electrochem. Soc.* **121**, 835 (1974).

30. W. Kroll, *Solid State Technology* **27**, 220 (1984).

31. J. E. Parks, M. T. Spaar, and P. J. Cressman, *J. Crystal Growth* **89**, 4 (1988).

32. T. Tanaka and S. Kurosawa, *J. Electrochem. Soc.* **133**, 416 (1986).

33. M. H. Yang, M. L. Lee, and H. L. Hwang, in *Non-Stoichiometry in Semiconductors*, K. J. Bachmann, H.-L. Hwang, and C. Schwab, eds., Elsevier Science Publishers B. V., Amsterdam, 1992.

34. M. Volmer, *Die Kinetik der Phasendildung*, Steinkopff-Verlag, Dresden and Leipzig, 1939.

35. M. von Laue, *Z. Kristallogr.* **105**, 124 (1943).

36. E. Kern, in *Current Topics in Materials Science,* Vol. 3, E. Kaldis, ed., North Holland Publishing Company, Amsterdam, 1979, p. 131.

37. S. Stoyanov, in *Current Topics in Materials Science*, Vol. 3, E. Kaldis, ed., North Holland Publishing Company, Amsterdam, 1979, p. 421.

38. Zeldovich, *Zh. Eksper. i Theor. Fiz.* **12** (1942).

39. E. Budevski, T. Vitanov, and W. Bostanov, *Phys. Stat. Sol.* **8**, 369 (1965); E. Budevski, W. Bostanov, T. Vitanov, Z. Stoinov, A. Krotzewa, and R. Kaischew, *ibid.* **13**, 577 (1966).

40. F. Frenkel, *Phys. Z. Sowietunion* (1932).

41. K. A. Jackson, in *Prog. in Solid State Chem.*, Vol. 4, H. Reis, ed., Pergamon Press, Oxford, 1967, p. 53.

42. P. R. Pennington, S. F. Ravitz, and G. Abbaschian, *Acta Met.* **18**, 943 (1970).

43. W. K. Burton, N. Cabrera, and F. C. Frank, *Phil. Trans. Roy. Soc. (London)* **243**, 299 (1951).

44. A. A. Chernov, *Modern Crystallography III—Crystal Growth*, Springer Verlag, Berlin, 1984.

45. N. Cabrera, in *The Art and Science of Growing Crystals*, J. J. Gilman, ed., John Wiley & Sons, Inc., New York, 1963, p. 1.

46. K. W. Keller, in *Crystal Growth and Characterization*, R. Ueda and J. B. Mullin, eds., North Holland Publishing Company, Amsterdam, 1975, p. 361.

47. V. V. Voronkov, *Kristallographiya* **15**, 1120 (1970).

48. T. Abe, *J. Crystal Growth* **24**/**25**, 463 (1974).

49. K. Raghavachari, *J. Chem. Phys.* **83**, 5668 (1982); **84**, 5672 (1986).

50. K. Raghavachari and C. M. Rohlfing, *J. Chem. Phys.* **89**, 2219 (1988).

51. J. Bernholc and J. C. Phillips, *J. Chem. Phys.* **85**, 3258 (1986).

52. E. Blaisten-Barogas and D. Levesque, *Phys. Rev. B* **34**, 3910 (1986).

53. B. Feuston, R. K. Kalia, and P. Vashita, *Phys. Rev. B* **37**, 6297 (1988).

54. P. R. Schwoebel and G. L. Kellogg, *Phys. Rev. Lett.* **61**, 578 (1988).

55. J. L. Katz and M. D. Donohue, *Adv. Chem. Phys.* **40**, 137 (1979).

56. J. L. Katz and M. D. Donohue, *J. Colloid and Interface Sci.* **85**, 267 (1982).

57. P. P. Wegener, *Acta Mech.* **21**, 65 (1975).

58. J. E. Griffith, J. A. Kubby, P. E. Wierenga, R. S. Becker, and J. S. Vickers, *J. Vac. Sci. Technol. A* **6**, 493 (1988).

59. R. J. Hammers, R. M. Tromp, and J. E. Demuth, *Phys. Rev. B* **34**, 5343 (1986).

60. T. Berghaus, A. Brodde, H. Neddermeyer, and S. Tosch, *J. Vac. Sci. Technol. A* **6**, 478 (1988).

61. H. Kroemer, *J. Crystal Growth* (1987).

62. J. Czochralski, *Z. Phys. Chem.* **92**, 219 (1918).

63. K. J. Bachmann, K. H. Kirsch, and K. J. Vetter, *J. Crystal Growth* **7**, 290 (1970).

64. G. K. Teal and E. Buehler, *Phys. Rev.* **87**, 190 (1952).

65. J. A. van Dijk, J. Goorisen, U. Gross, R. Kersten, and J. Pistorius, *Acta Electronica* **17**, 45 (1974).

66. G. Ziegler, *Z. Naturforschg.* **16a**, 219 (1961).

67. John Wilkes, in *Proc. of the International Summer School of Crystal Growth*, Palm Springs, CA, 1992, in preparation.

68. R. E. Lorenzini, A. Iwata, and K. Lorenz, U.S. Patent No. 4,036,595 (1977).

69. W. Kaiser, P. H. Keck, and C. F. Lange, *Phys. Rev.* **101**, 1264 (1956).

70. J. C. Mickelsen, Jr., in *Oxygen, Carbon, Hydrogen and Nitrogen in Crystalline Silicon*, J. C. Mickelsen, Jr., S. J. Pearson, J. W. Corbett, and S. J. Pennycook, eds., Materials Research Society, Pittsburg, PA, 1986, p. 16.

71. P. H. Keck and M. J. Golay, *Phys. Rev.* **89**, 1297 (1953).

72. H. C. Theurer, *U.S. Patent* 3,060, 123 (1962), filed December 17, 1952.

73. W. Heywang, *Z. Naturforschg.* **11a**, 238 (1956).

74. W. Keller and A. Muhlbauer, *Floating-Zone Silicon*, Marcel Dekker, New York (1981).

75. T. Abe, in *VLSI Electronics Microstructure Science*, G. N. Einspruch and H. R. Huff, eds., Academic Press, New York, 1985, p. 3.

76. J. W. Cleland, K. Lark-Horowitz, and J. C. Pigg, *Phys. Rev.* **78**, 814 (1950).

77. M. Tanenbaum and A. D. Mills, *J. Electrochem. Soc.* **108**, 171 (1961).

78. W. E. Haas and M. S. Schnöller, *J. Electron Mater.* **5**, 57 (1976); *IEEE Trans. Electr. Dev.* **ED-23**, 803 (1976).

79. K. J. Bachmann, E. Buehler, and J. H. Wernick, *J. Electron. Mater.* **3**, 279 (1974).

80. L. J. Vieland, *Acta Met.* **11**, 137 (1963).

81. W. Schottky and R. Bever, *Acta Met.* **6**, 320 (1958).

82. M. B. Panish, *J. Crystal Growth* **27**, 6 (1974); M. Ilegems, M. B. Panish, and J. R. Arthur, *J. Chem. Thermodyn.* **6**, 157 (1974).

83. A. S. Jordan, A. R. von Neida, R. Caruso, and C. K. Kim, *J. Electrochem. Soc.* **121**, 153 (1974).

84. *Non-Stoichiometry in Semiconductors*, K. J. Bachmann, H.-L. Hwang, and C. Schwab, eds.,

Elsevier Science Publishers BV, Amsterdam, 1992.

85. R. M. Logan and D. T. J. Hurle, *J. Chem. Phys. Solids* **32**, 1739 (1971).

86. K. Terashima, J. Nishio, A. Okada, S. Washiuka, and M. Watanabe, *J. Crystal Growth* **79**, 463 (1986).

87. E. P. A. Metz, R. C. Miller, and R. Mazelski, *J. Appl. Phys.* **33**, 2016 (1962).

88. K. J. Bachmann, E. Buehler, and R. Strnad, *J. Electron Mater.* **4**, 741 (1975).

89. W. Bardsley, G. W. Green, C. H. Holliday, and D. T. J. Hurle, *J. Crystal Growth* **16**, 277 (1972).

90. H. D. Pruett and S. Y. Lien, *J. Electrochem. Soc.* **121**, 822 (1974).

91. A. S. Jordan, A. R. Von Neida, R. Caruso, and J. W. Nielsen, *J. Appl. Phys.* **52**, 3331 (1981).

92. A. S. Jordan, *J. Crystal Growth* **49**, 631 (1980); *ibid.* **71**, 559 (1985).

93. A. S. Jordan, A. R. Von Neida, and R. Caruso, *J. Crystal Growth* **76**, 243 (1986).

94. B. C. Grabmaier and J. G. Grabmaier, *J. Crystal Growth* **13/14**, 635 (1971).

95. J. Matsui, H. Watanabe, and Y. Seki, *J. Crystal Growth* **46**, 563 (1979).

96. G. Jacob, M. Duseaux, J. P. Farges, M. M. B. van den Boom, and P. J. Roksnoer, *J. Crystal Growth* **61**, 417 (1983).

97. H. Kohda, K. Yamada, H. Nakanishi, T. Kobayashi, and K. Hoshikawa, *J. Crystal Growth* **71**, 813 (1985).

98. Tatsumi, in *Non-Stoichiometry in Semiconductors*, K. J. Bachmann, H.-L. Hwang, and C. Schwab, eds., Elsevier Science Publishers BV, Amsterdam, 1992.

99. P. Dobrilla and J. S. Blakemore, *J. Appl. Phys.* **58**, 204 and 208 (1985).

100. D. T. J. Hurle and E. Jakeman, *J. Crystal Growth* **5**, 227 (1969).

101. J. R. Carruthers and A. F. Witt, in *Crystal Growth and Characterization*, Proc. ICCG Spring School Japan 1974, R. Ueda and J. B. Mullin, eds., North-Holland Publishing Company, Amsterdam, 1975, p. 107.

102. J. R. Carruthers, in *Treatise of Solid State Chemistry*, N. B. Hannay, ed., Plenum Press, New York, 1975, Vol. 5.

103. M. J. Wargo and A. F. Witt, *J. Crystal Growth* **66**, 289 (1984).

104. P. J. Roksnoer, M. M. B. van Rijbroek-van den Boom, *J. Crystal Growth* **66**, 317 (1984).

105. A. F. Witt, M. Lichtensteiger, and H. C. Gatos, *J. Electrochem. Soc.* **120**, 1119 (1973).

106. J. R. Carruthers and M. Grasso, *J. Appl. Phys.* **43**, 436 (1972).

107. L. M. Hocking, *Mathematika* **7**, 1 (1960).

108. S. Chandrasekhar, *Phil. Mag.* **43**, 501 (1952).

109. N. Riley, *J. Crystal Growth* **85**, 417 (1987).

110. U.R. Kurzweg, *Int. J. Heat Mass Transfer* **8**, 35 (1965).

111. K. Hoshikawa, *J. Appl. Phys.* **21**, L545 (1982).

112. A. Rohatgi, *Conf. Rec. 19th IEEE Photovolt. Spec. Conf.*, New Orleans, 1986, p. 1500.

113. K. H. Yao and A. F. Witt, *J. Crystal Growth* **880**, 453 (1987).

114. K. Terashima, A. Yahata, and T. Fukuda, *J. Appl. Phys.* **59**, 982 (1986).

115. N. Kato and A. R. Lang, *Acta Cryst.* **12**, 787 (1959).

116. N. Kato, *J. Phys. Soc. Japan* **18**, 1785 (1963); **19**, 67 (1964).

117. N. Kato, in *Crystal Growth and Characterization, Proc. ICCG Spring School Japan 1974*, R. Ueda and J. B. Mullin, eds., North Holland Publishing Company, Amsterdam, 1975, p. 279.

118. J. R. Schneider, H. A. Graf, O. Goncalves, W. von Ammon, P. Stallhofer, and H. Walizki, *J. Crystal Growth* **80**, 225 (1987).

119. W. Gault, E. M. Monberg, and J. E. Clemans, *J. Crystal Growth* **74**, 491 (1986).

120. D. H. Kim, P. M. Adornato, and R. A. Brown, *J. Crystal Growth* **89**, 339 (1988).

121. S. Chandrasekhar and D. D. Elbert, *Proc. Roy. Soc. A* **231**, 198 (1955).

122. P. S. Lykoudis and C. P. Yu, *Int. J. Heat Mass Transfer* **6**, 853 (1963).

123. S. Chandrasekhar, *Hydrodynamic and Hydromagnetic Stability*, Clarendon Press, Oxford, 1961.

CHAPTER

6

Epitaxy and Dielectric Deposition

6.1 Chemical Vapor Deposition at Atmospheric Pressure

In this Section we consider chemical vapor deposition at atmospheric pressure (APCVD), starting with a discussion of the thermodynamic basis and the kinetics of heterogeneous reactions. Next we consider the interactions of specific reactants of silicon APCVD processes with the reconstructed silicon surface. The discussion of APCVD is completed by a review of the modeling of the coupled gas-flow dynamics and chemical kinetics of APCVD of Si and selected III–V compounds.

After grinding an orientation flat of usually (110) orientation parallel to the axis of the single crystal boules produced by the processes described in Chapter 5, wafers of controlled crystallographic orientation are cut, chamfered at the outer edges, lapped and polished by chemomechanical polishing methods. Although high-quality single crystal wafers have been used for the direct fabrication of FET circuits, in modern processing, epitaxial steps are added to improve the quality of the material in the active device regions, and to produce heterostructures that are essential for the realization of certain device designs, such as HBTs. In particular, in the processing of silicon ICs, CVD is being used for the growth of Si epilayers, the deposition of nominally undoped and *P*-doped SiO_2, Si_3N_4, and BN layers, the fabrication of the polysilicon gate contact, and selective metallization.

In applications where the epitaxial step is a front end process, for example, in the twin-tub CMOS processing sequence discussed in Chapter 4, high temperatures can be tolerated. This permits the utilization of the hydrogen reduction of silicon tetrachloride or trichlorosilane, which proceed in the ranges between 1150 to 1250°C and 1100 to 1200°C, respectively, according to the net reactions

$$SiCl_4 + 2H_2 \rightleftarrows Si + 4HCl \tag{6.1}$$

$$SiHCl_3 + H_2 \rightleftarrows Si + 3HCl \tag{6.2}$$

Usually this CVD process is carried out at atmospheric pressure (APCVD), producing highly perfect epilayers. Autodoping can become a problem in high-temperature epitaxial processes. It is caused by impurities that enter into the gas stream by etching of the less pure substrate by the vapor phase in the reverse reactions of Eqs. (6.1) and (6.2). They are carried in the vapor flow to downstream regions of epitaxial growth where they are incorporated into the epitaxial film. Also, at the high T employed in Si APCVD, impurities may diffuse from the substrate into the epitaxial film. This leads to a graded doping profile at the epilayer/substrate interface. In addition, it results in the spreading of buried doped regions created prior to the epitaxial growth by diffusion or ion implantation. This is illustrated in Figure 6.1 for an As-doped region in a silicon wafer, which is subsequently overgrown by an epitaxial Si layer.

Autodoping and interdiffusion are reduced by lowering the processing temperature, which is made possible by the replacement of trichlorosilane and silicon tetrachloride by other source materials, such as silane or dichlorosilane. The optimum range of epitaxial growth is 1000–1100°C for SiH_2Cl_2 and 950–1050°C for SiH_4. Attention to cleanliness is especially important at these lower growth temperatures. Particulate matter that is either present on the surface due to improper cleaning or is generated in the process may lead to the

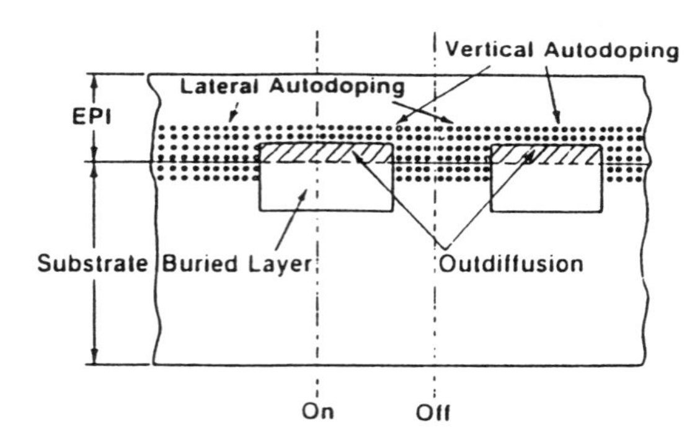

Figure 6.1 Autodoping at the Si epilayer–substrate interface. Vertical cross section through a buried layer structure. After Srinivasan [1]; copyright © 1984, North Holland Publishing Company, Amsterdam.

formation of stacking faults. Stacking faults in the epilayer that are decorated by impurities are electrically active and degrade the device performance. They increase significantly the dark current and thus degrade the properties of the diodes [2].

For CVD processes that are based on the temperature dependence of the equilibrium constant of reversible gas-phase reactions, the composition of the gas phase in the reactor as a function of temperature can be calculated on the basis of thermodynamic principles. Consider a general gas-phase reaction

$$v_1 C_1(\mathrm{g}) + v_2 C_2(\mathrm{g}) + v_3 C_3(\mathrm{g}) + \cdots \rightleftharpoons v_n C_n(\mathrm{s}) + v_{n+1}\ C_{n+1}(\mathrm{g}) + \cdots +\ n_N C_N(\mathrm{g}) \tag{6.3}$$

where the v_i are stoichiometric numbers. By convention $v_i < 0$ for $i < n$ and $v_i > 0$ for $i \geqslant n$. The C_i represent the various constituents of the vapor phase participating in the reaction. At constant pressure and temperature the equilibrium condition is

$$\Delta G = \left(\frac{\partial G}{\partial \lambda_{\mathrm{r}}}\right) = \sum v_i \mu_i = 0, \qquad i = 1, 2, \ldots, N \tag{6.4}$$

where λ_{r} is the extent of the reaction that changes by 1 for the progression of the reaction by one formula unit from left to right. For gas-phase reactions

$$\sum v_i \mu_i = -RT \ln K_{p*} = \Delta H - T\,\Delta S \tag{6.5}$$

where

$$K_{p*} = \prod [p_i^*]^{v_i} \tag{6.6}$$

In the atmospheric to subatmospheric pressure range of CVD processing the fugacities p_i^* of the constituent gas-phase components C_i are close to the partial pressures p_i. Therefore, K_{p*} may be replaced in thermochemical calculations for these processes by K_p. The temperature dependence of K_p is given by van't Hoff's reaction isobar

$$\frac{\partial \ln K_p}{\partial T} = \frac{\Delta H}{RT^2} \tag{6.7}$$

where the enthalpy of the reaction is defined as

$$\Delta H = \sum v_i H_i \tag{6.8}$$

Therefore, in the case of an exothermic reaction, that is $\Delta H < 0$, K_p decreases with increasing T, causing the reaction to shift to the left. The partial pressures of the reactants in equilibrium with the solid and the reaction products thus increase with increasing T. For endothermic reactions, that is, $\Delta H > 0$, K_p increases with increasing temperature. This shifts the equilibrium with increasing temperature to the right. In a closed system, the transport by an endothermic reaction proceeds thus from a cold source to a hot substrate. This is, for example, utilized in the equalization of the diameter of an electrically heated tungsten wire in a chlorine atmosphere, where tungsten is removed at the larger than average diameter segments of the wire and is deposited at the hotter smaller than average diameter segments by the transport reaction

$$WCl_6(g) \rightleftarrows W(s) + 3Cl_2(g) \tag{6.9}$$

Similarly reaction (6.9) prevents the condensation of tungsten on the colder surface of the glass bulb [3].

The value of $K_p(T)$ can be calculated by integration of Eq. (6.6) if the temperature dependence of $\Delta H(T)$ is known. The latter can be calculated according to

$$\Delta H(T) = \Delta H_{T=0} + \int_0^T \Delta C_p(T)\, dT \tag{6.10}$$

where

$$\Delta C_p(T) = \Sigma \nu_i C_{pi}(T) \tag{6.11}$$

is a function of temperature. In the case of gas-phase reactions, the heat capacities C_{pi} of the constituents usually can be represented as power series in T,

$$C_{pi} = a_i + b_i T + c_i T^2 + \cdots \tag{6.12}$$

Consequently,

$$\Delta H_i(T) = \Delta H^0_{T=0} + \sum \nu_i \left(a_i T + \frac{b_i}{2} T^2 + \frac{c_i}{3} T^3 + \cdots \right) \tag{6.13}$$

so that

$$\frac{d \ln K_p}{dT} = \frac{\Delta H_{T=0}}{RT^2} + \frac{\Sigma \nu_i}{R} \left(\frac{a_i}{T} + \frac{b_i}{2} + \frac{c_i T}{3} + \cdots \right) \tag{6.14}$$

$$\ln K_p = -\frac{\Delta H_{T=0}}{RT} + \frac{\Sigma \nu_i}{R} \left(a_i \ln T + \frac{b_i T}{2} + \frac{c_i T^2}{6} + \cdots \right) + c' \tag{6.15}$$

There exist a number of thermochemical tables that list the standard enthalpies and entropies as well as the heat capacities and their temperature coefficients for a large variety of materials [4]. Given the input vapor flow rates (i.e., the initial concentrations of the reactants and the law of the conservation of mass), it is thus often possible to determine the equilibrium concentrations of the products and reactants as a function of temperature on the basis of the tabulated thermochemical data. The recent extension of CVD techniques to the deposition of intermetallic compounds, Si/metal composites, and a variety of dielectrics requires the consideration of multicomponent systems of increasing complexity.

Figure 6.2 shows the calculated composition of the gas phase in the hydrogen reduction of $SiCl_4$–VCl_4 mixtures used in the fabrication of vanadium silicide films by CVD [5]. The endpoint at zero VCl_4 concentration characterizes the vapor composition in Si epitaxy using $SiCl_4$ as the source material. Note that in addition to the reactants and products of the net reaction (6.1), a variety of intermediate products are present that participate in the reaction. Of course, kinetic limitations always cause a difference between the vapor composition at the surface of the substrate and the bulk vapor atmosphere, so that thermodynamics does not provide the information necessary to predict the growth rate and the segregation behavior. However, it is always helpful to determine by thermochemical calculations which species are expected to be present at what concentration in equilibrium. Thus experimentally determined deviations from equilibrium establish the conditions of supersaturation or undersaturation under which the growth or etching of the solid proceeds. This provides the quantitative basis for the interpretation of the deposition process in terms of kinetic models that characterize the rate-determining steps in the complex combination of parallel and sequential reactions.

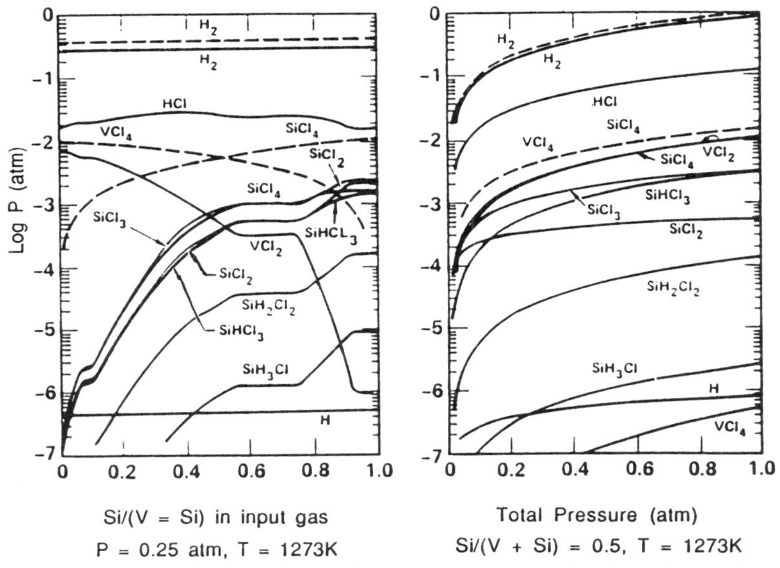

Figure 6.2 Calculated composition of the vapor phase in the deposition of vanadium silicide by the hydrogen reduction of mixtures of $SiCl_4$ and VCl_4; --- reactants, —— equilibrium; $H/H+Cl = 50.95$; total flow rate 1406 sccm. After Wang and Spear [5]; copyright © 1984, The Electrochemical Society, Pennington, NJ.

The kinetics of CVD reactions generally involves the following steps:

1. Homogeneous vapor-phase reactions generating the reactants that participate in the surface reaction.
2. Transport of these species in the vapor phase to the location of the substrate and across the diffusion layer to the surface.
3. Adsorption of some or all of the reactants at the surface.
4. Surface reaction and incorporation into the growing film.
5. Desorption of the reaction products that are not incorporated into the film.
6. Transport across the diffusion layer in the vapor phase and possible homogeneous follow-up reactions of these waste products.

The deposition process is thus usually complex, and the overall rate law relating the growth rate to the concentrations of the input reactants and exhaused waste products does not reveal the mechanism. However, for practical reasons it is useful to determine the rate equation, which for some reactions may be expressed as

$$v_r = \prod p[C_i]^{\lambda_i} \tag{6.16}$$

where the λ_i are the reaction orders with regard to the species C_i in the reaction and k_r is the rate constant. Its temperature dependence is often satisfactorily described by the two-parameter equation

$$k_r(T) = A_r T^{\beta} \exp(-E_a/kT) \tag{6.17}$$

where A_r and β are constants. If the pre-exponential factor is a weak function of T, the temperature dependence of the growth rate is represented by the simpler exponential Arrhenius equation.

Figure 6.3 shows the growth rate v_g as a function of temperature for silicon epitaxy using $SiCl_4$, $SiHCl_3$, SiH_2Cl_2, and SiH_4 as source materials. Two temperature domains are observed:

1. A high-temperature regime where the surface kinetics is fast and the growth rate is controlled by the mass transport.
2. A low-temperature regime where the deposition rate is controlled by the surface kinetics.

Surface-controlled reactions are ubiquitous in chemistry and have been investigated thoroughly in the context of heterogeneous catalysis. The concept and models developed in these investigations are an excellent basis for similar studies concerning surface-controlled CVD processes. For ideal adsorption of a single molecule of C_1 without dissociation, the rate of desorption is proportional to the fraction θ of occupied surface sites and the rate of adsorption is proportional to $[C_1](1 - \theta)$, where $[C_1]$ is the concentration of C_1 in the gas phase at the solid–gas interface. At constant temperature, the coverage is thus described by the Langmuir isotherm

$$\theta = \frac{K[C_1]}{1 + K[C_1]} \tag{6.18}$$

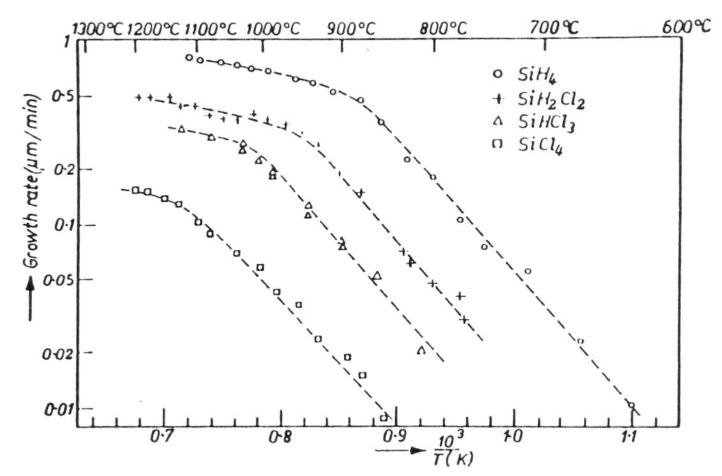

Figure 6.3 Growth rate versus $1/T$ for Si epitaxy from $SiCl_4$, $SiHCl_3$, SiH_2Cl_2, and SiH_4. After Eversteyn [6]; copyright © 1974, Philips Research Laboratories, Eindhoven.

where K is the quotient formed of the rate constants of the adsorption and desorption processes [7]. If C_1 dissociates upon adsorption into two equal fragments, the adsorption isotherm is of the form

$$\theta = \frac{\sqrt{K}\sqrt{[C_1]}}{1 + \sqrt{K}\sqrt{[C_1]}} \tag{6.19}$$

and for competitive adsorption of two vapor species C_1 and C_2 competing for the same surface sites

$$\theta_1 = \frac{K_1[C_1]}{1 + K_1[C_1] + K_2[C_2]} \tag{6.20}$$

$$\theta_2 = \frac{K_2[C_2]}{1 + K_1[C_1] + K_2[C_2]} \tag{6.21}$$

Both dissociative and competitive adsorption are frequently encountered in CVD processes. The surface reaction rates are closely related to the coverages of the reactants. In unimolecular reactions it is directly proportional to θ, so that

$$v_r = k_r \frac{K[C_1]}{1 + K[C_1]} \tag{6.22}$$

An example of such a reaction is the thermal decomposition of PH_3 on a glass surface [8]. The rate law Eq. (6.22) predicts saturation of v_r for large coverages; that is, the reaction becomes of zero order at large concentration of the reactant. At moderate temperatures, the pyrolysis of PH_3 at a glass surface is inefficient. However, it is catalyzed on a III–V surface so that, under the conditions of thermally activated APCVD, epitaxial growth is sustained at $T > 500°C$. The catalytic activity of a surface corresponds to a lowering of the energy barrier for the reaction by adsorption.

In case of poisoning of the surface by an inhibitor C_p the reaction rate is of the form

$$v_r = k_r \frac{K_1[C_1]}{1 + K_1[C_1] + K_p[C_p]}, \tag{6.23}$$

That is, due to the competitive adsorption of C_p, according to Eq. (6.23), the coverage by C_1 may be substantially reduced, leading to a diminished rate. Bimolecular surface reactions proceed by reaction of an adsorbed species C_1 with another species C_2 that is either also adsorbed on the surface or resides in the gas phase. Since these two mechanisms were first investigated by Langmuir [9], Hinshelwood [10], and Rideal [11] they are called the Langmuir–Hinshelwood and Langmuir–Rideal mechanisms, respectively. The associated rate laws are of the form

$$v_r = k_r \cdot \theta_1 \theta_2 = k_r \cdot \frac{K_1 K_2 [C_1][C_2]}{(1 + K_1[C_1] + K_2[C_2])^2} \tag{6.24}$$

$$v_r = k_r \cdot \theta_1 [C_2] = k_r \cdot \frac{K_1[C_1][C_2]}{1 + K_1[C_1] + K_2[C_2]} \tag{6.25}$$

if the two species participating in the bimolecular surface reaction bind preferably to different surface sites, the rate law (6.24) is replaced by

$$v_r = k_r \cdot \frac{K_1 K_2 [C_1][C_2]}{(1 + K_1[C_1])(1 + K_2[C_2])} \tag{6.23}$$

The configuration of the adsorbed molecules on the surface of the solid, including changes in the surface reconstruction due to the interactions of the adsorbate with the solid, may be determined by the minimization of the total energy in quantum-chemical calculations. The change of this minimal energy as a function of the adsorbate position during the course of a surface reaction defines the activation energy and is thus a valuable supplement to experimental investigations of the deposition mechanism. Since in the case of chemisorption the bonding is highly localized, only a relatively small number of neighboring atoms of the solid need to be considered. Although such cluster calculations are an adequate simplification, considerable care must be taken in the definition of the cluster termination to obtain realistic results. A method that has been applied with particular success in such calculations is the scattered-wave $X\alpha$ method introduced by Slater in 1952 [12]. In this method, the space considered in the calculation is partitioned into atomic, interatomic, and extramolecular regions, in which the potential is spherically and volume averaged. The initial charge-density distribution is calculated by superimposing atomic charge densities. The HF approximation to exchange is replaced by setting the exchange correlation proportional to the $\frac{1}{3}$ power of the electronic charge density. The Schrödinger equation is solved separately for each of the partitioned regions. At their interfaces the resulting wavefunctions are joined with continuous first derivatives as in cellular methods of band-structure calculations [13]. Solving the Schrödinger equation with the molecular potential thus established results in a yet improved potential and so on until self-consistency in the energy is achieved. The fast convergence of the secular equations stimulated applications of the scattered-wave $X\alpha$ method to a number of problems of catalysis and chemical vapor deposition and etching processes. For example, based on $X\alpha$ calculations for metal clusters, the suggestion has been made that the activity of surface steps in catalytical processes [14] relates to their electronic structure. Thus the preferred performance of vicinal cuts of substrates in CVD applications may not be simply explained by the provision of kink sites for the incorporation of atoms, but may affect the kinetics of the deposition process in a more subtle way, as they are preferable sites for chemical surface reactions. Recent studies of surface reactions on semiconductors revealed a significant role of the domain boundaries in the reconstructed surfaces as active sites. Therefore, the changes in the surface reconstruction must also be incorporated into the analysis of surface reactions.

For example, the interaction of atomic hydrogen with the Si(111)-7×7 surface has been investigated by STM/STP. This investigation revealed two reactions channels: (1) binding of the hydrogen to existing dangling bonds on the adatoms in the 7×7 surface unit cell and (2) desorption of the hydrogen-saturated top surface atoms and binding to newly created dangling bonds in the stacking fault layer surface atoms [15]. At low coverage, the 7×7 unit cell structure is maintained, but some of the adatoms in the structue disappear at low-voltage imaging conditions and reappear, in part, at higher voltage. This is explained by the saturation of the dangling bonds on the adatoms in the 7×7 unit cell and is corroborated by LEED studies [16]. Further insights can be obtained from the change in the $J-V$ curves $(\partial I/\partial V)/(I/V)$ as a function of the sample bias. The structure in the plot $V\partial I/I\partial V$ versus energy associated with filled and empty adatom states at 0.4 and 0.5 eV, respectively, vanishes at low hydrogen exposure of the surface. Upon large exposure of the surface to atomic hydrogen, the boundaries of the 7×7 structure are still maintained, but a sublattice of 1×1 structure is formed that corresponds to the stacking fault layer beneath the adatom surface layer [see Figure 3.55(b)], and a bandgap of 1.2 eV opens up. This implies removal of the adatom layer during the reaction through channel number 2 and binding of hydrogen to the 43 dangling bonds per 7×7 unit cell on the Si surface atoms in the stacking fault layer positions, which is corroborated by the mass spectrometric observation of SiH_3 and SiH_4 upon H exposure of the silicon surface.

Also, the adsorption of disilane gas on the Si(111)-7×7 and (100)-2×1 surfaces has been investigated by vibrational spectroscopies and STM/STL [18] and [19], respectively. Figure 6.4 shows the vibrational spectra for a Si(111)-7×7 surface exposed to various doses of Si_2H_6 ranging from 20 to 1000

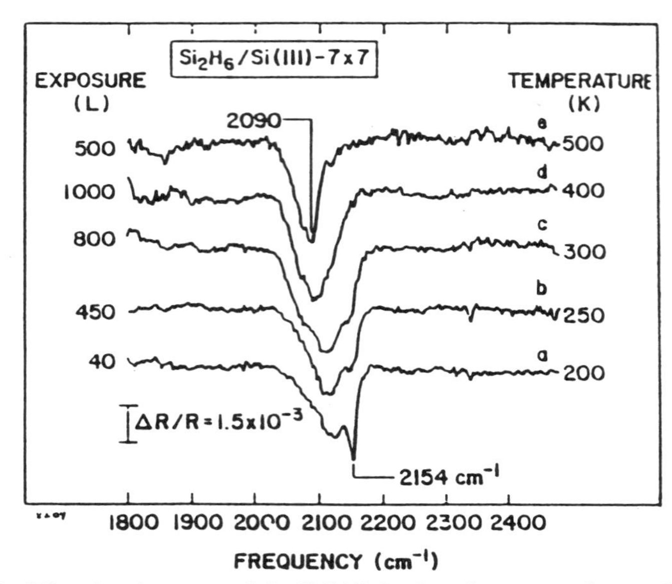

Figure 6.4 Vibrational spectra of the Si(111)-7 × 7 surface exposed to various doses of Si_2H_6 listed on the left side of curves a–e at various temperatures listed on the right side of the same family of curves. After Uram and Jansson [18]; copyright © 1991, Elsevier Science Publishers BV, Amsterdam.

Langmuir at temperatures ranging from 120 to 500 K. In this temperature range the adsorption process is dissociative according to reaction (6.27a) and results at low temperatures in a dominant sharp feature at 2154 cm^{-1} and a shoulder at 2130 cm^{-1}, corresponding to the symmetric and asymmetric stretching vibrations of SiH_3, respectively. The most likely symmetry of the SiH_3 fragments bonded to the silicon surface is C_{3v}, as expected for binding to the adatoms in the 7 × 7 unit cell. Since the uptake of SiH_3 decreases with increasing temperature, the sticking coefficient for this species decreases as in a chemisorption process with negative activation energy, which is −1.9 kcal/mole [18]. Consequently the features at 2154 and 2130 cm^{-1} in Figure 6.4 are reduced at temperatures <200 K. At 500 K the spectrum is dominated by a peak at 2090 cm^{-1} associated with the silicon monohydride bond. Experiments with polarized light reveal a dipole moment parallel to the surface, that is, a tilt of the Si–H bond with respect to the surface normal. The silyl species on the Si surface at $T < 250$ K enter into surface reactions of the types

$$Si_2H_6 + 2S^* \rightarrow 2SiH_3(ad) \quad (6.27a)$$

$$SiH_3(ad) + S^* \rightarrow SiH_2(ad) + H(ad) \quad (6.27b)$$

$$SiH_2(ad) + S^* \rightarrow SiH(ad) + H(ad) \quad (6.27c)$$

$$SiH_3(ad) + SiH(ad) \rightarrow 2SiH_2(ad) \quad (6.27d)$$

where S^* refers to an empty surface site. Although at higher temperatures most of the surface is covered by monohydride, SiH_3 and SiH_2 species still remain in adatom positions near defects.

At high exposure and low temperature, multilayer adsorption of molecular Si_2H_6 is observed on both the Si(111)-7 × 7 and Si(100)-2 × 1 surfaces. Figure 6.5(a) and (b) show the STM images of a Si(100)-2 × 1 surfaces under clean conditions and after saturation exposure to Si_2H_6. At elevated temperature, that is, $T \geqslant 670\,K$, the silyl groups formed in the dissociative interactions of Si_2H_6 with the surface lose hydrogen according to reactions (6.27a)–(6.27d). Upon the desorption of the hydrogen, this leads to the formation of silicon monohydride dimer islands, as shown in Figure 6.6(a). Upon further exposure to disilane, epitaxial growth ensues, as illustrated in Figure 6.6(b) (compare with the discussion of Si MBE in Section 6.4).

The modeling of the coupled gas-flow dynamics of APCVD requires solving the Navier–Stokes equations as discussed in Chapter 5, but with different boundary conditions that account for the different geometry and physico-chemical parameters under the conditions of vapor transport and the coupling the chemical kinetics to the gas-flow dynamics. The geometry of various types of silicon CVD reactors is illustrated in Figures 6.7(a)–(c), which show schematic representations of a tubular reactor, a barrel reactor, and a pancake reactor. The uniformity of the deposition in these reactors is typically ±5–10%. In all cases the walls are cold and the backside of the wafers are in contact with a hot susceptor made for example from SiC-coated graphite.

Figure 6.5 STM images of a Si(100)-2 × 1 surface (a) clean, (b) after saturation with Si_2H_6 at low temperature. After Boland [19]; copyright © 1991, The American Physical Society, New York.

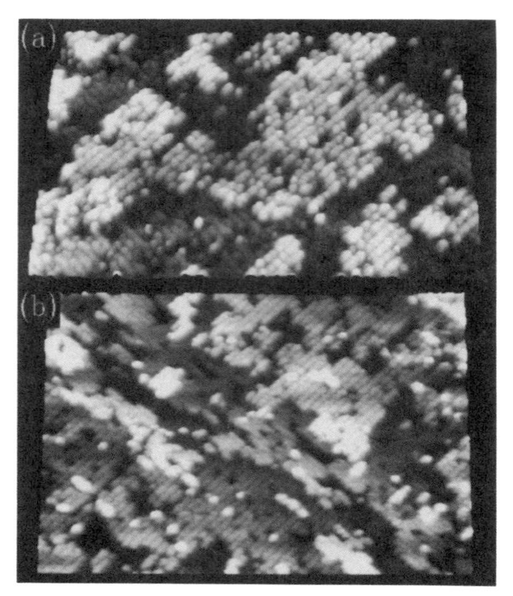

Figure 6.6 (a) Epitaxial growth on a hydrogen-terminated surface exposed to 2.4×10^{-5} torr · s Si_2H_6 at 690 K; (b) epitaxial overgrowth formed upon an additional exposure to 4.8×10^{-5} torr · s Si_2H_6 at 690 K followed by a 300 s postanneal at 650 K. After Boland [19]; copyright © 1991, American Physical Society, New York.

Barrel reactors are currently preferred to pancake reactors because of better control of the growth conditions. Laminar flow is desirable in all cases of APCVD since turbulence in the vapor stream severely degrades the uniformity of the deposition process. This imposes certain restrictions on the flow velocity that differ for vertical and horizontal reactors. For laminar viscous flow in a tubular reactor with horizontal susceptor, the diffusion layer thickness is related to the Reynolds number as

$$\delta = \sqrt{\frac{d}{\mathrm{Re}} x} \tag{6.28}$$

where x is the distance on the suspector parallel to the axis and d is the diameter of the reactor. The increase of the diffusion layer thickness with increasing distance along the substrate in a horizontal reactor implies different growth rates at the leading and trailing edges of the wafers. Constricting the cross section gradually toward the trailing edge [compare Figure 6.7(a) and (b)] evens out the diffusion layer thickness by increasing the average flow velocity $\bar{v}_f$. However, in order to maintain laminar flow, the Reynolds number must be restricted to values $\mathrm{Re} < \mathrm{Re}_c$ ($\mathrm{Re}_c \approx 2.3 \times 10^3$ if hydrogen is used as the carrier gas).

In order to discuss the modeling of the coupling of the gas-flow dynamics

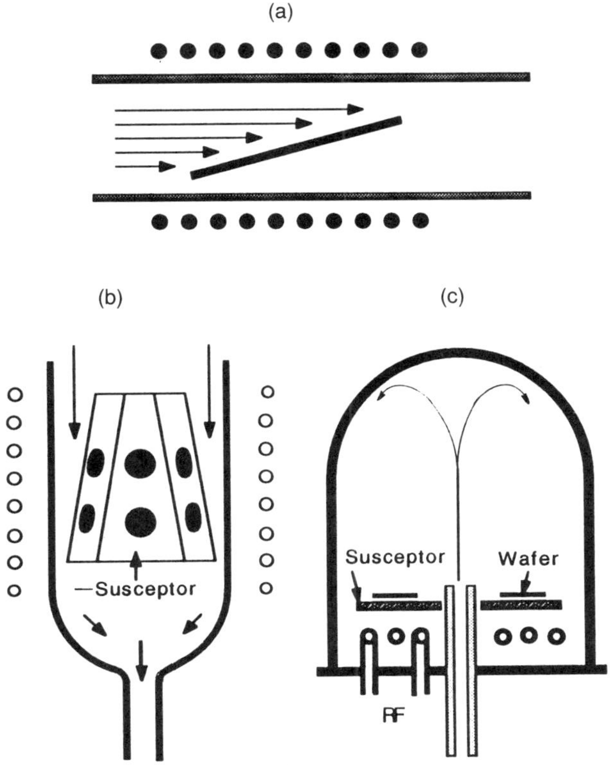

Figure 6.7 Various reactor designs used in Si epitaxy. (a) Tubular reactor, (b) barrel reactor, (c) pancake reactor.

to the chemical kinetics of CVD reactions, we consider the deposition of Si by silane pyrolysis as an illustrative example. Simplified models of APCVD in a channel [20] and in a vertical reactor [21], respectively, have been developed, including Soret diffusion, that is, the tendency of heavier molecules to accumulate in the regions of space at relatively low temperature. Figure 6.8(a) shows a schematic representation of the vertical reactor injecting the reactant and carrier gas mixture through a porous surface on top of a rotating substrate. Although the simplifications used to make this analysis one dimensional exclude a realistic evaluation of the limitations to the uniformity of the process in the peripheral region of the wafer, it gives some insights into the complexity of this seemingly simple CVD process. Table 6.1 shows the reactions considered in the modeling of the coupled flow dynamics and chemical kinetics of silane pyrolysis on a silicon surface [17]. This list is by no means complete. However, some of the possible reactions do not make important contributions to the growth rate and can be neglected. The major contributions from various species are shown in Figure 6.8(b) and (c) for the vertical reactor geometry of Figure 6.8(a). They demonstrate the significant role of the carrier gas. In a H_2

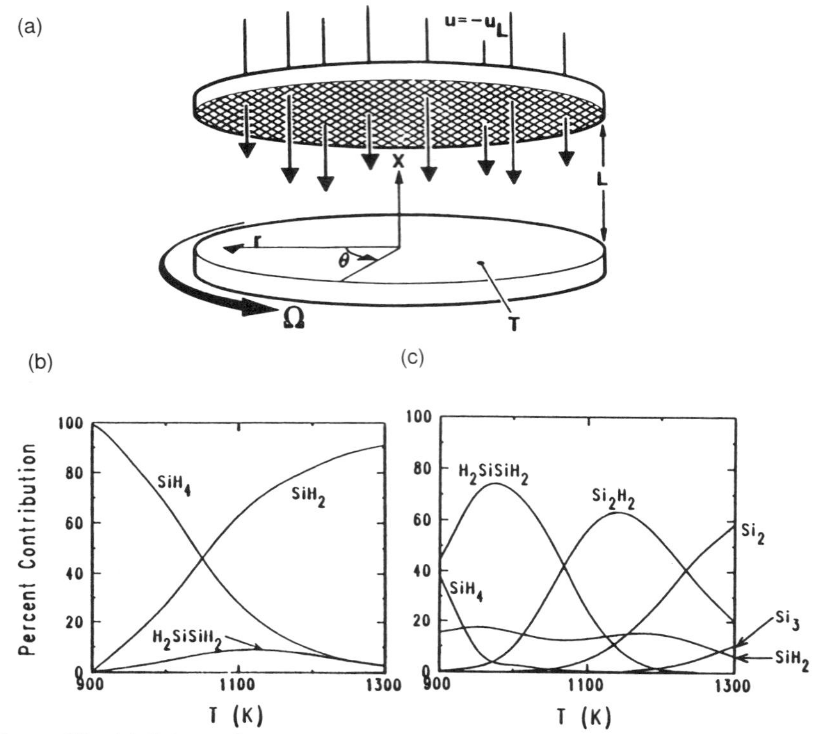

Figure 6.8 (a) Schematic representation of the flow conditions in the modeled vertical reactor. (b) and (c) Contributions of various vapor species to the growth rate of the silicon film as a function of the substrate temperature. After Coltrin, Kee, and Evans [21]; copyright © 1989, The Electrochemical Society, Pennington, NJ.

carrier, silylene formed in reaction R1 of Table 6.1 makes by far the greatest contribution to the growth rate at the preferred growth temperature. However, in an inert carrier, the growth rate is determined by the formation of Si_2H_4 and of Si_2H_2 depending on the temperature. Thermodynamic calculations of the equilibrium concentrations of Si_x ($x = 2, 3$) suggest that in the temperature range of Si epitaxy from silane such clusters should be present in insignificant concentrations. Yet there is experimental evidence for the presence of Si_2 above the substrate surface at substantial concentrations under the conditions of silane decomposition in an inert gas carrier. This observation demonstrates the substantial excursions from equilibrium in CVD processes.

In the presence of destabilizing temperature gradients, the flow becomes modulated by thermal convection, further complicating the analysis of the deposition process. A three-dimensional model incorporating thermal convection effects has been developed in Ref. [22] for a rectangular horizontal reactor. Restricting the discussion to H_2 as a carrier gas, only reactions R1 and R3 of Table 6.1 are considered in this model. The relative importance of the surface and mass transport kinetics is defined more precisely by the surface Damköhler

Table 6.1 Reactions considered in the modeling of Si epitaxy by silane pyrolysis under the condition of APCVD (after Coltrin, Kee, and Miller [17])

Reaction number	Reaction	Pre-exponential factor[a]	E_a (kcal/mole)
R1	$SiH_4 \rightarrow SiH_2 + H_2$	5.00E12	52.2
R2	$SiH_4 \rightarrow SiH_3 + H$	3.69E15	93
R3	$SiH_4 + SiH_2 \rightarrow Si_2H_6$	5.01E12	1.29
R4	$Si_2H_4 + H_2 \rightarrow SiH_4 + SiH_2$	6.22E16	2
R5	$SiH_4 + H \rightarrow SiH_3 + H_2$	1.04E14	2.5
R6	$SiH_4 + SiH_3 \rightarrow Si_2H_5 + H_2$	1.77E12	4.4
R7	$SiH_4 + SiH \rightarrow SiH_3 + SiH_2$	1.38E12	11.2
R8	$SiH_4 + SiH \rightarrow Si_2H_5$	2.93E12	2
R9	$SiH_4 + Si \rightarrow 2SiH_2$	9.31E12	2
R10	$Si + H_2 \rightarrow SiH_2$	1.15E14	2
R11	$SiH_2 + SiH \rightarrow Si_2H_3$	1.26E13	2
R12	$SiH_2 + Si \rightarrow Si_2H_2$	7.24E12	2
R13	$SiH_2 + Si_3 \rightarrow Si_2H_2 + Si_2$	1.43E11	18.8
R14	$Si_2H_2 + H_2 \rightarrow Si_2H_4$	2.45E14	2
R15	$Si_2H_4 + H_2 \rightarrow Si_2H_6$	9.31E12	2
R16	$SiH + H_2 \rightarrow SiH_3$	3.45E13	2
R17	$Si_2 + H_2 \rightarrow Si_2H_2$	1.54E13	2
R18	$Si_2H_3 + H_2 \rightarrow Si_2H_5$	2.96E13	2
R19	$Si_2H_2 + H \rightarrow Si_2H_3$	8.63E14	2
R14	$Si_3 + Si \rightarrow 2Si_2$	2.06E12	24.1

[a]The unit of A_r depends on the reaction order and is given in terms of mols, cm^3, and s.

numbers for the participating gas-phase species

$$Da_i = \gamma_i \sqrt{(kT/2\pi m_i)}\, \frac{D_i(T)}{h_r} \tag{6.29}$$

where D_i is the diffusion constant of component C_i in the gas phase above the susceptor, γ_i is its surface reaction probability, and h is the height of the susceptor. Table 6.2 shows the calculated effective Damköhler numbers, including the surface reactions of Si_2H_6 and SiH_2. Also the concentrations of SiH_2 and Si_2H_6 and the contributions of each species to the growth rate are listed for a SiH_4 partial pressure of 0.76 Torr and a hydrogen partial pressure of 759.24 Torr. At $T > 1200$ K, $D_a > 1$ and the reaction becomes mass transport limited.

Figure 6.9(a) shows the development of lateral flow in the horizontal rectangular CVD reactor, which is assumed to have adiabatic side walls with a susceptor heated to 1373 K at the bottom and a top surface temperature of 300 K. The effective Rayleigh number of this system at an average temperature of 815.5 K is 227, which is well below the critical value of 2056 for convective instabilities. Figures 6.9(b) and (c) show growth rate distributions for a

Table 6.2 Effective Damköhler numbers, silene and disilane concentrations, and their contributions to the growth rate (after Moffat and Jensen [22])

T (K)	Da[a]	$p(SiH_2)$[b] (Torr)	$p(Si_2H_6)$[b] (Torr)	$v_g(Si/SiH_2)$ (μm/min)	$v_g(Si/Si_2H_6)$ (μm/min)	$v_g(SiH_4)$ (μm/min)
800	0.007	9.9×10^{-9}	2.37×10^{-4}	2.6×10^{-6}	4.0×10^{-4}	8.5×10^{-4}
900	0.029	4.99×10^{-7}	3.16×10^{-4}	1.1×10^{-3}	2.6×10^{-3}	3×10^{-3}
1000	0.10	1.16×10^{-5}	4.17×10^{-4}	2.0×10^{-3}	1.1×10^{-2}	8×10^{-3}
1100	0.32	1.52×10^{-4}	5.51×10^{-4}	2.3×10^{-2}	3.1×10^{-2}	1.79×10^{-2}
1200	1.1	1.24×10^{-3}	6.13×10^{-4}	1.67×10^{-1}	6.8×10^{-2}	3.48×10^{-2}
1300	4.4	7.62×10^{-3}	7.31×10^{-4}	9.39×10^{-1}	9.9×10^{-2}	6.11×10^{-2}
1400	14.91	3.55×10^{-2}	8.47×10^{-4}	3.73	1.08×10^{-1}	9.87×10^{-2}

[a]Effective Damköhler numbers including the contributions of the SiH_2 and Si_2H_6 reactions.
[b]For $p(SiH_4) = 0.76$ Torr and $p(H_2) = 759.24$ Torr.

horizontal position of the susceptor and a tilted position of the susceptor by 5°, which significantly improves the uniformity of the deposition process.

As in silicon CVD, ultrapure epitaxial layers are obtained by the utilization of chloride transport at atmospheric pressure for the III-V compounds and their alloys. Two options exist in the choice of the group V source, which may be either a halide [23] or a hydride [24]. In the halide process, hydrogen is bubbled through liquid group V halide (e.g., arsenic trichloride), held at a well-defined constant temperature to control their partial pressures. The $AsCl_3$-saturated H_2 stream is then diluted by additional hydrogen and is injected into a tubular hot wall reactor with three temperature zones. In the first zone it reacts with hydrogen to form arsenic vapor and HCl gas according to

$$4AsCl_3(g) + 6H_2(g) \rightleftarrows 2As_2(g) + 12HCl(g) \tag{6.30}$$

$$2As_2(g) \rightleftarrows As_4(g) \tag{6.31}$$

$$4PCl_3(g) + 6H_2(g) \rightleftarrows 2P_2(g) + 12HCl(g) \tag{6.32}$$

$$2P_2(g) \rightleftarrows P_4(g) \tag{6.33}$$

which are swept by the hydrogen carrier stream into zone 2. Located in this zone is a boat filled with the liquid group III metal. In the initial phase of the experiment, the liquid metal saturates to the concentration in equilibrium with the solid III–V compound, e.g., GaAs or InP, at the chosen liquidus temperature. After saturation is reached, a crust of the compound forms on the surface of the Ga or In melt. Complete coverage of the solution by the III-V compound is important for the control of the epitaxial process. For example, the GaAs crust reacts with the HCl in the gas stream to form the volatile gallium monochloride and arsenic vapor, that is,

$$GaAs(s) + HCl(g) \rightleftarrows GaCl(g) + \tfrac{1}{2}As_2(g) + \tfrac{1}{2}H_2(g) \tag{6.34}$$

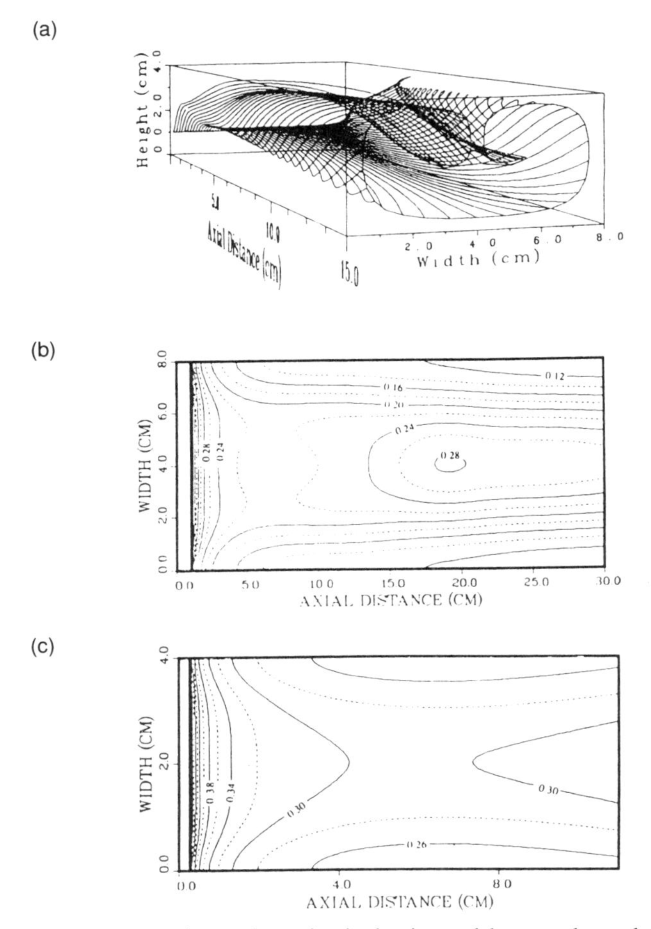

Figure 6.9 Representative trajectories in horizontal bottom-heated rectangular reactors of (a) 4 cm × 8 cm cross section. (b) Growth rate distribution on the susceptor surface for silane decomposition at $T = 1373$ K at a flow velocity of 17.54 cm/s of an injected gas mixture of 0.76 Torr SiH_4 + 759.24 Torr H_2. (c) Susceptor tilted 5° off the horizontal position at the same conditions as in (b). After Moffatt and Jensen [22]; copyright © 1988, The Electrochemical Society, Pennington, NJ.

At the colder substrate the equilibrium (33) is shifted to the left, establishing the supersaturation required for epitaxial growth. Note that the reverse reaction is not complete. Consequently, substantial concentrations of all constituents are retained in the vapor phase, which condense at the cold downstream end of the reactor and must be removed periodically.

Figure 6.10 shows the dependence of the growth rate of GaAs on the crystallographic orientation of the substrate under the conditions of halide CVD for a constant source temperature of 750°C. Both the thermally activated behavior and the strong dependence of the growth rate on the surface

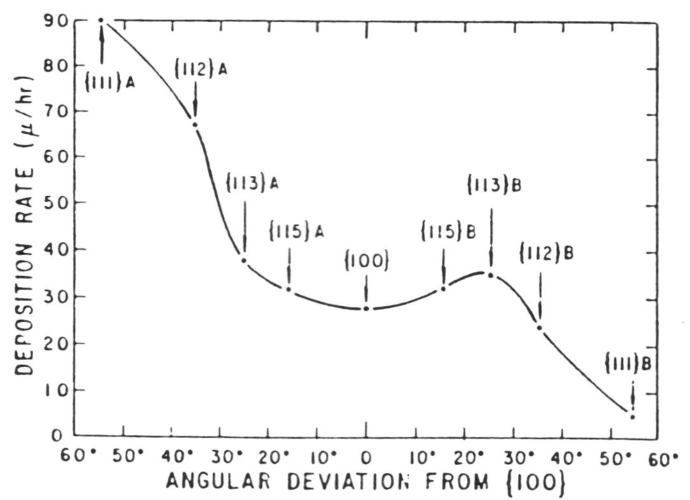

Figure 6.10 The growth rate of GaAs as a function of the substrate orientation under the conditions of the halide process. The source temperature is 750°C. After Shaw [25]; copyright © 1968, The Electrochemical Society, Pennington, NJ.

orientation show that the reaction is surface controlled. The highest and lowest growth rates are obtained for the (111)*A* and (111)*B* faces, respectively [25]. Further details of the surface kinetics are revealed by the quantitative study of reference [26].

Figure 6.11 shows the growth rate dependence of the GaAs on the partial pressures of GaCl. Note that although the data are consistent with the Langmuir–Hinshelwood mechanism, they do not prove that the adsorbed species are GaCl, As_2 and As_4. Thus, further surface analysis methods must be employed to clarify the reaction mechanism. Similar observations as for GaAs have been reported for the halide APCVD of InP [27].

Figure 6.12 shows the dependence of the growth rate of InP for a given supersaturation for a variety of surface orientations. Note that the saturation and falloff at the higher substrate temperatures are due to the decrease in the deposition rate caused by the diminishing supersaturation, which vanishes at 750°C substrate temperature. As in the case of GaAs epitaxy, the growth rate is strongly dependent on the surface orientation, that is, the growth kinetics is surface controlled. At low V/III ratio in the vapor phase, nominally undoped InP grown by the halide method is n-type. In the range $0.1 \leqslant \mathrm{V/III} \leqslant 0.3$, the electron concentration drops precipitously, causing type conversion at $\mathrm{V/III} \approx 0.33$ [28]. By carefully controlling the V:III ratio and the purity of the source materials, an electron mobility of $121{,}050\,\mathrm{cm^2\,V^{-1}\,s^{-1}}$ at 77 K has been achieved for nominally undoped InP films [29].

GaAs epilayers have been made with a total ionized impurity concentration of $<10^{14}\,\mathrm{cm^{-3}}$ and an electron mobility $>2 \times 10^5\,\mathrm{cm^2\,V^{-1}\,s^{-1}}$ at 77 K [30]. Si and Zn have been found to be the primary donor and acceptor impurities

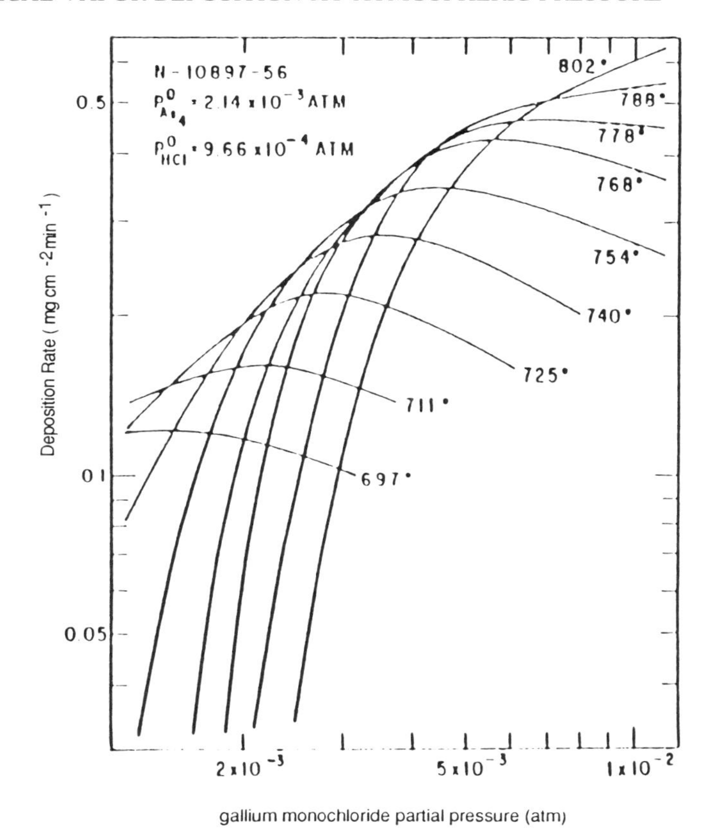

Figure 6.11 Growth rate of GaAs versus the partial pressures of GaCl under the conditions of the halide process. After Shaw [26]; copyright © 1970, The Electrochemical Society, Pennington, NJ.

in GaAs grown by the halide process, with hydrogen being the carrier gas. The source of Si is the interaction of the HCl-containing vapor atmosphere with the fused silica enclosure. A significant reduction in the Si levels is achieved by the use of nitrogen as a carrier gas, resulting in electron mobilities of $250{,}000\,cm^2\,V^{-1}\,s^{-1}$ at 77 K, with sulfur and carbon being the primary donor and acceptor impurities [31]. Unfortunately the same approach does not work in the case of InP, which is more limited by source purification problems than GaAs deposition.

In the hydride process the group V hydride is mixed into the vapor downstream of the position of the boat with the source metal. The group III monochloride is thus formed separately by reaction of HCl gas with the pure metal, permitting greater flexibility in the control of the gas composition at the location of the substrate. HCl gas is available in high purity, but the possible phosphine and particularly arsine contamination by wall reactions is of concern and requires point of use purification of the hydrides. GaAs epilayers

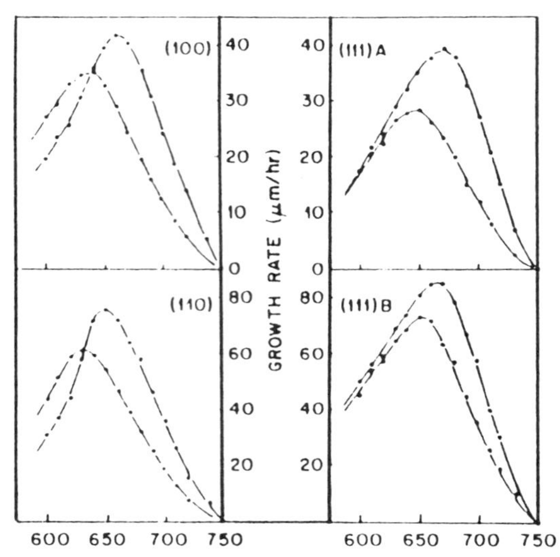

Figure 6.12 Growth rate vs. temperature plots for InP films grown by halide CVD. After Mizuno [27]; copyright © 1975, Japanese J. Appl. Phys., Tokyo.

grown by the hydride process have achieved an electron mobility of 201,000 $cm^2\ V^{-1}\ s^{-1}$ [32]. For InP the hydride process results in an electron mobility of 137,000 $cm^2\ V^{-1}\ s^{-1}$ at 77 K [33].

6.2 Low-Pressure and Plasma-Enhanced Vapor Deposition

Since the Reynolds number varies linearly and the Rayleigh number varies quadratically with the vapor density, substantial advantages in the control of the gas flow conditions are achieved by reducing the pressure from 760 to $\sim$1 Torr. Figure 6.13 shows a representation of a typical LPCVD reactor for multiwafer processing and an enlarged view of the wafer holder that is made from fused silica. Under the conditions of low-pressure CVD (LPCVD), the linear flow velocity is generally higher than under the conditions of APCVD. However, in terms of the mass flow, this increase in the velocity is overcompensated by the reduction in density. The mean free path λ of the gas molecules in LPCVD remains well below the characteristic dimension of the flow tube, so that viscous flow. conditions exist. Typically, the Reynolds number is reduced from $10 \leqslant Re \leqslant 1000$ for APCVD to $Re \approx 1$ for LPCVD. Since thermal convection is no longer a problem, LPCVD provides for better uniformity than APCVD, that is, roughly $\pm 1\%$ in thickness.

An important advantage of LPCVD is the reduction of the minimum temperature at which epitaxial growth is achieved [36]. Close to this limit the growth is surface reaction controlled, with hydrogen acting as an inhibitor.

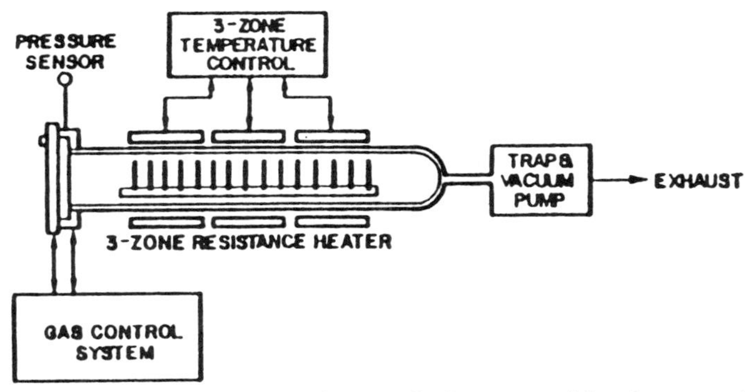

Figure 6.13 Schematic representation of an LPCVD reactor. After Jensen and Graves [34]; copyright © 1983, The Electrochemical Society, Pennington, NJ,

Below 1200 K, LPCVD results in the growth of polycrystalline silicon. Patterns that are defined by photolithography prior to epitaxial growth are transferred by LPCVD with smaller distortions than generated under the conditions of APCVD. For a given layer thickness this difference in the distortions between APCVD and LPCVD increases with decreasing deposition temperature. The price for the advantage of LPCVD is a substantially lower deposition rate than is achieved under the conditions of APCVD. Therefore, at a given growth rate and layer thickness, the distortions generated by LPCVD are actually larger than those generated by APCVD [35].

Because deposition at low temperature minimizes the thermal effects on underlying device regions, LPCVD is a preferred method for the deposition of the poly-Si gate contacts and related poly-Si composite metallization patterns. The lower limit for the growth of polycrystalline Si by LPCVD is 800 K. Near this lower temperature limit the grain size is $\sim 0.1-1.0\,\mu$m. Below 800 K the transition from microcrystalline to amorphous silicon films occurs. Hydrogenated amorphous silicon (α-Si:H) that can be doped to either n- or p-type conduction has received considerable attention in the context of thin film solar cells because of its relatively low cost and the simplicity of the fabrication process. Because of the rapid deterioration of the minority carrier life time with increasing dopant concentration, most a-Si:H solar cells employ nearly intrinsic material to maximize the depletion layer width at a rectifying junction, e.g. a thin film Schottky barrier, and use dopants only for the formation of an ohmic contact (see Chapter 8). Thus, a large fraction of the photogenerated electron-hole pairs are formed within the built-in field of the junction, assuring a reasonable collection efficiency. A disadvantage of α-Si:H solar cells is their degradation under illumination (Stabler–Wronski effect), which is reduced in properly engineered microcrystalline silicon solar cells. Also, α-Si:H and microcrystalline Si are currently under study in the context of thin film transistors (TFTs) with potential applications in liquid crystal displays.

Since under the conditions of LPCVD of poly-Si the adsorption of SiH_2

competes with H_2 for the same surface sites, a rate law of the form of Eq. (6.25) has been proposed [34], assuming that the growth rate is dominated by the decomposition of the adsorbed silene molecules into Si and H_2. As a consequence of the inhibiting function of hydrogen, larger growth rates are observed using nitrogen instead of hydrogen as the carrier gas.

Also, the addition of dopants alters the growth kinetics significantly. For example, PH_3 and AsH_3 depress the growth by blocking active surface sites for the adsorption of Si-containing vapor molecules and cause a substantial nonuniformity in the dopant distribution, that is, impair the growth of doped epilayers. An important advantage of LPCVD is that autodoping by As or P is significantly reduced [37]. Boron doping substantially increases the growth rate as compared to the growth of nominally undoped films.

Further advances have been made in recent years towards even lower processing temperature for epitaxial growth by the use of plasma-enhanced CVD. Plasmas are collections of electrons and ions and are established by the initiation of an electrical breakdown between two electrodes at a pressure of 0.1–1 torr. Although there are a variety of implementations of plasma CVD, the simplest and frequently chosen geometry in plasma deposition and etching employs parallel electrodes, one of which carries the substrate wafers. The voltage required for discharge depends on the product of the distance d between the electrodes and density ρ of the gas, the charged-particle energy distribution E/ρ, and the ratio of the frequency to the mean free path ω/λ. While initially during the breakdown process the field is uniform in the gap region, upon establishing the plasma, a nonuniform space-charge distribution results due to the vastly different mobilities of the electrons and ions. The voltage drops over sheath regions near the electrodes in a manner that assures equal rates for electron and ion loss to the electrode surfaces and chamber walls; that is, in an electropositive gas, the electrons are repelled, and ions are accelerated to achieve a well-defined ambipolar diffusion rate. In between the sheath regions is the glow region of constant E/ρ, where radiative transitions in electron-impact-excited gas molecules and atoms result in the emission of light at characteristic frequencies that provide information about the plasma composition. The ionic content of glow discharge plasmas is dilute, and the ion temperature is moderate because of efficient energy exchange with neutral gas molecules. For the electrons the mass difference with regard to the neutral gas molecules does not provide for efficient energy exchange. Therefore, the electron energies in plasmas peak in the range between 10 and 100 eV. The average energy of the plasma, $3kT_P/2$, corresponds typically to electron temperatures 10^4 K. The uniqueness of plasma processing thus lies in the generation of energetic precursors for chemical reactions that are not available under the conditions of ordinary CVD.

Figure 6.14 shows the calculated distribution of monomer and dimer radicals between parallel electrodes in a symmetric rf discharge in silane at 0.25 torr and a power deposition of 0.25 W/m [38]. Plasma-enhanced epitaxial growth of silicon is achieved at temperatures between 750 and 800°C. Oxygen

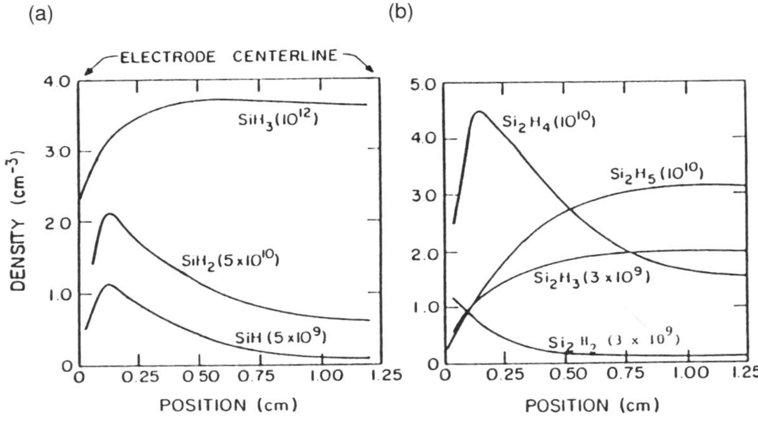

Figure 6.14 Radical density distribution between the electrodes in a symmetric rf discharge sustained at 0.25 torr silane at a power deposition of 0.25 W/cm: (a) monomer radicals, (b) dimer radicals. After Kushner [38]; copyright © 1988, The American Institute of Physics, New York.

and carbon are the primary impurities in the epilayers that exhibit optimum electrical properties for growth at 800°C. At 750°C the free carrier concentration exceeds 10^{15} cm^{-3}, and the carbon level is as high as 5×10^{17} cm^{-3} [39]. Figure 6.15 shows the low-field I_D versus V_G characteristic of Si PMOS transistors fabricated in the PECVD epilayers. The hole channel mobility calculated from the slope of this plot is 213 cm^2 V^{-1} s^{-1}, which is comparable to the 218 cm^2 V^{-1} s^{-1} channel mobility for devices fabricated in a bulk Si

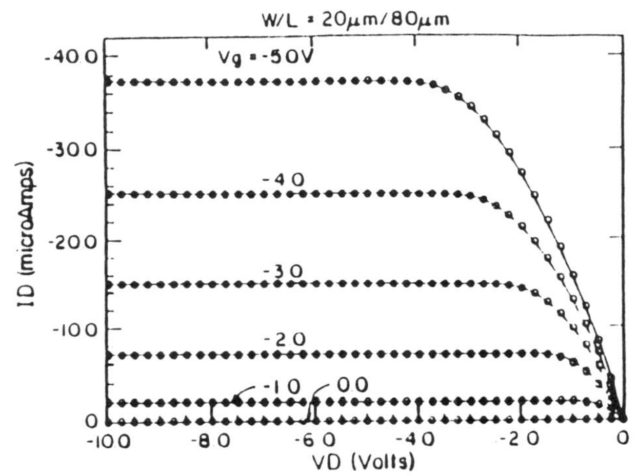

Figure 6.15 (a) I–V characteristics of a PMOS transistor fabricated in a silicon epilayer grown by PECVD at 800°C. (b) Plot of I_D versus V_G, the control device refers to bulk silicon. After Burger and Reif [40]; copyright © 1987, The Electrochemical Society, Pennington, NJ.

control wafer [40]. Vertical autodoping effects are virtually absent in epilayers grown over heavily As-, B-, or P-doped silicon substrates.

Remote plasma-enhanced chemical vapor deposition (RPECVD) has achieved the homoepitaxial growth of silicon epilayers at even lower temperatures, $150 \leqslant T_s \leqslant 350°C$ [41a]. The plasma excitation region is, in this case, separated from the growth region. Thus the energy transfer occurs between long-lived excited species extracted from the plasma and gas molecules injected downstream from the plasma region through multiple gas-phase collisions. In the course of these collisions highly reactive precursors are generated that gate the surface reaction pathway. The primary type of defect present in the epilayers is dislocation loops that are formed during the growth process and that have a strong temperature dependence in both size and density. At low rf power density, dislocation densities below $10^6\,cm^{-2}$ have been observed at $T_s = 150°C$. However, there exists a pronounced growth rate dependence on the power density, so that a tradeoff must be made between growth rate and defect density [41b]. The dependence of the growth rate on the partial pressure of SiH_4 molecules in the growth chamber is weak. This permits the conclusion that the arrival rate of molecular SiH_4 on the surface of the silicon substrate is not a controlling factor of the growth kinetics; that is, the rate of the surface reaction is controlled by the incident flux of growth assisting charged and energetic neutral particles emanating from the plasma, the concentration of which depends only weakly on the silane partial pressure. That charged particles are involved in the growth kinetics is implied by a strong substrate bias dependence of the rate [41b]. An advantage of RPECVD is its utility for in situ surface cleaning by a short hydrogen plasma burn that removes carbon and oxygen from the silicon surface [42]. However, care must be taken to avoid excessive damage and plasma-generated contamination of the silicon surface. Therefore, alternative plasma cleaning techniques have been developed that rely on a mild oxidation of the Si surface followed by thermal desorption in a rapid thermal annealing step [43].

An important attribute of the deposition of dielectrics by either PECVD or RPECVD onto silicon is the substantially lower processing temperature than possible under the conditions of thermal oxidation of silicon. In the limit of very large He dilution of the reactants, the direct method results in high-quality SiO_2/Si interfaces ($N_{ss} \leqslant 3 \times 10^{10}\,cm^{-2}\,eV^{-1}$) [44]. However, in general, the control of the stoichiometry and surface chemistry is less flexible than under the conditions of remote excitation (see Section 8.1).

Figure 6.16 shows a typical RPECVD system. Only the anion-forming gases are subjected to the plasma, while the silane is brought into the deposition chamber through a gas dispersal ring located downststream of the plasma region. The downstream indirect deposition of SiO_2 was first investigated in 1982 [46]. Commercial direct PECVD reactors developed for Si IC processing, using either inductive or capacitative coupling, have existed for more than a decade. They provide for automatic wafer exchange and multiwafer processing. DC, rf and microwave plasmas are being used in the context of PECVD.

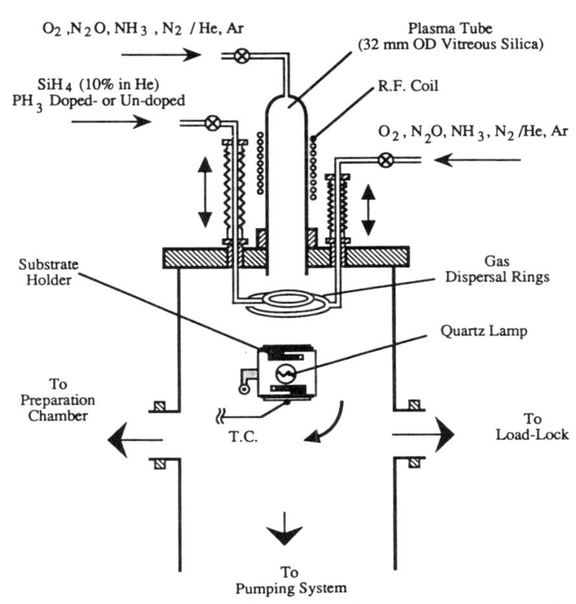

Figure 6.16 Schematic representation of an RPECVD reactor. After Lucovsky, Kim, and Fitch [45]; copyright © 1990, American Institute of Physics, New York.

The current trend is towards increasing utilization of microwave plasma deposition [47].

Also, for dielectrics the temperature of equivalent PECVD processes as compared to LPCVD is significantly reduced. For example, under the conditions of LPCVD, the deposition of SiO_2 by the reaction of SiH_4 and O_2 in a nitrogen or argon carrier and by the pyrolysis of tetraethylorthosilicate (TEOS) requires processing at 900 and 700°C, respectively. Also, Si_3N_4 deposition by reaction of SiH_4 with ammonia requires processing at 750°C, and BN deposition by the reaction of BCl_3 with ammonia proceeds at 700°C. This compares with the plasma-assisted deposition of silicon dioxide by the oxidation of dichlorosilane at 350°C, of silicon nitride by the reaction of silane with ammonia at 300°C, and of boron nitride by the reaction of diborane with ammonia at 250°C. Because of the presence of reactive vapor species in the plasma, which are normally absent under the conditions of LPCVD, the composition of plasma-deposited silicon dioxide, silicon nitride, and boron nitride differs from the compounds deposited by ordinary LPCVD. Depending on the gas composition of the plasma region and on the substrate temperature, varying amounts of nitrogen and hydrogen are incorporated in the SiO_2, and hydrogen is incorporated into both the Si_3N_4 and BN films grown by the PECVD processes described. Consequently the properties of plasma-deposited oxides and nitrides, such as their etch rates, dielectric strength, and density, differ substantially from those of the corresponding stoichiometric compounds produced by CVD.

Valuable insights into the chemical bonding of the residual impurities in the deposited dielectric films, which determine their properties, can be gained from infrared spectroscopy [48]. The intensity of the ir absorption peak at optical frequency ω_{nm} associated with the transitions between vibrational states n and m is proportional to the square of the dipole moment gradient along the relevant normal coordinate. MO calculations establishing these dipole moment gradients and intensities for the most important ir lines of a variety of molecules thus provide the basis for the interpretation of the spectra [49], [50].

Figure 6.17 shows the ir spectra and the vibrational modes associated with the primary features due to Si–O vibrations of films of SiO_2 deposited by RPECVD on Si. The incorporation of hydrogen from the SiH_4 source gas into the oxide can be controlled by the choice of the substrate temperatures and the ratio of the oxygen to helium flow rates. These parameters have been chosen to result in stoichiometric SiO_2 [Figure 6.19(a)], SiO_2 containing O–H groups as revealed by the presence of a broad band at 3640 cm^{-1} associated with the O–H stretching vibration [Figure 6.19(b)], and SiO_2 containing Si–H bonds as revealed by the peaks at 875 and 2260 cm^{-1} corresponding to the Si–H bending and the Si–H stretching vibrations, respectively [Figure 6.19(c)]. Si–H groups in SiO_2 are only observed for dilutions of the oxygen below 1% in He. Compared to stoichiometric SiO_2, the films containing OH groups contain more oxygen and the films incorporating SiH groups contain less oxygen per silicon atom. Therefore, they are sometimes referred to as silicon-deficient and silicon-rich (suboxide) films.

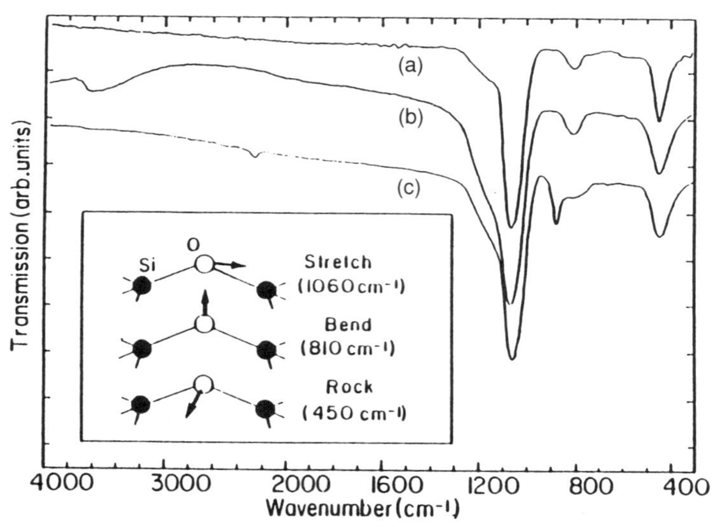

Figure 6.17 Infrared absorption spectra of silicon dioxide growth by RPECVD: (a) stoichiometric SiO_2, (b) silicon-deficient SiO_2, (c) Si-rich SiO_2. Courtesy of D. V. Tsu and G. Lucovsky, Department of Physics, North Carolina State University, Raleigh, NC [50].

In silicon-deficient oxide films, the ir absorption band between 3700 and 3400 cm^{-1} broadens asymmetrically on the low-frequency side with increasing concentration of silanol groups. This has been explained in Ref. [50] by increasing hydrogen bonding between neighboring O–H groups with their increasing concentration in the films. On the other hand, water molecules that are incorporated into the films form bridging hydrogen bonds to neighboring oxygens in the SiO_2 network and are thus distinguishable by a characteristic spectroscopic signature from silanol groups. Also, residual strain, causing deviations of the Si–O–Si bond angle from the ideal angle in fully relaxed thermal oxide, results in a shift of the Si–O–Si vibration to 10–20 cm^{-1} below the peak position for a fully relaxed film. Further experimental corroboration of the predicted locations of the spectral features associated with hydrogen incorporation into dielectric films produced by RPECVD can be obtained by deuterium doping of the films. For example, a shift from 3650 cm^{-1} for the O–H stretching vibration to 2655 cm^{-1} for the O–D stretching vibration is observed upon deuterium doping of silicon–deficient SiO_2 films in accordance with a frequency ratio equal to the square root of the ratio of the reduced masses.

Figures 6.18(a)–(c) shows the ir absorption spectra of H-doped films of silicon nitride grown by RPCVD at different substrate temperatures. Trace (d) corresponds to a deuterium-doped film grown at the same substrate temperature as the film associated with trace (b). Figure 6.19 shows the various modes of vibrations contributing to the spectral features: N–H stretching and bending vibrations at 3340 and 1175 cm^{-1}, respectively, the Si–N stretching vibration between 830 and 885 cm^{-1}, and the Si breathing vibration between 430 and

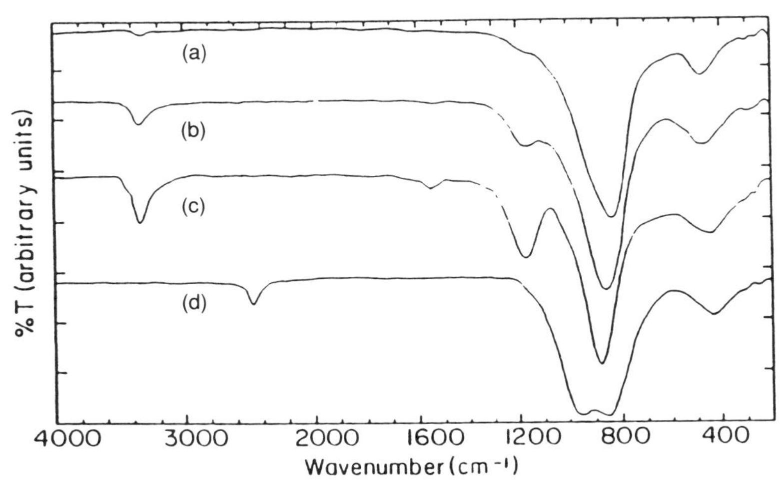

Figure 6.18 Infrared transmission spectra of silicon nitride films grown by RPECVD, utilizing excited NH_3 at (a) 500°C, (b) and (d) 250°C (c) 100°C substrate temperature. Trace (d) refers to a deuterium-doped film. Courtesy of D. V. Tsu and G. Lucovsky, Department of Physics, North Carolina State University, Raleigh, NC [50].

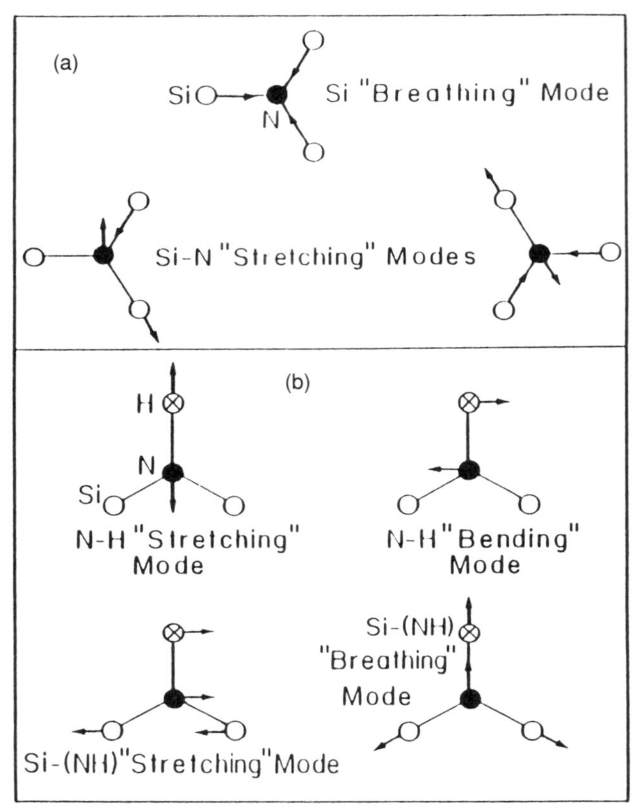

Figure 6.19 Displacements in ir-active vibrations involving Si, N, and H atoms. Courtesy of D. V. Tsu and G. Lucovsky, Department of Physics, North Carolina State University, Raleigh, NC [50].

490 cm^{-1}, depending on the substrate temperature. The peak shifts from 3340 and 1175 cm^{-1} for the N–H vibrations to 2475 and 970 cm^{-1} for the N–D vibrations again correspond to the frequency ratios expected from the square root of the reduced mass ratio.

Significant differences exist in the ir spectra of Si_3H_4 films grown from excited NH_3/He and N_2/He mixtures, respectively. In the latter case, features related to both N–H and Si–H bonds are detected, while in the former case only features related to N–H bonds are observed. The concentration of N–H groups in films grown from excited ammonia depends on the substrate temperature, that is, varies in the range from 0.5×10^{22} to 4.4×10^{22} cm^{-3} for $500 \geqslant T \geqslant 50°C$. The lower temperature limit results thus in a substantial molar fraction of N–H groups in the film, which incorporates structural elements of silicon diimide $Si(NH)_2$. In accord with the predictions of the VSEPR model discussed in Section 2.4, the bond angle in the diimide configuration is in between that of the tetrahedral Si bond angle and the

Si–O–Si bond angle. The latter is wider by the influence of the two lone pairs on the oxygen, which exceeds the valence-shell electron-pair repulsion associated the bonding electrons of the N–H bond. Helium dilution of the NH_3 prior to the plasma excitation improves the stoichiometry of the films, so that Si_3N_4 films can be deposited by RPECVD at a temperature as low as 250°C. In addition to silicon dioxide and silicon nitride, a variety of alternative dielectrics, such as silicon oxynitride and AlN, are currently under development. Some of these materials have attractive physical properties, for example, high thermal conductivity, excellent chemical stability, high resistivity, and thermal expansion coefficients that provide for a better match to Si and GaAs than SiO_2 and Si_3N_4. However, in each case, the optimum processing parameters must be evaluated experimentally under well-controlled conditions regarding the initial cleanliness and reconstruction of the silicon surface, which is still an ongoing research task.

Figure 6.20 shows a plot of the density of interface traps (D_{it}) at the SiO_2/Si interface and values of the flat-band voltage shift for MOS capacitors, employing oxide layers that are deposited by RPCVD, versus the deposition temperature, including a postmetallization anneal of the MOS test structures. The data reveal a process window close to 200°C deposition temperature, where D_{it} values are observed that are close to the interface trap density achieved at thermal oxide/silicon interfaces (see Section 4.3). At growth temperatures above and below 200°C, subcutaneous oxidation of the silicon by diffusion of activated molecular oxygen extracted from the plasma to the oxide/silicon interface (see Section 8.1) and increasing incorporation of OH groups, respectively, degrade the interfacial properties. Note that, although it is appropriate to utilize D_{it} measurements for monitoring the effects of

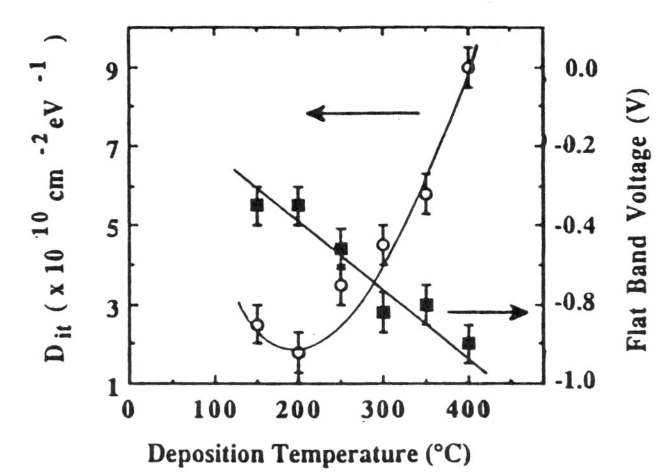

Figure 6.20 Midgap density of interface states and flat-band voltage shifts for MOS capacitors made by RPECVD of the oxide film as a function of the deposition temperature. After Lucovsky, Kim, and Fitch [45]; copyright © 1990, American Institute of Physics, New York.

variations in processing parameters, it is not a sufficient figure of merit for assessing the value of a processing technology in the context of IC manufacturing, which requires measurements of the channel mobility and stress tests for sufficiently large sets of MOSFET devices having a realistic feature size. Initial results of such measurements exist, but do not permit at present a definitive statement concerning the merit of deposited gate oxide layers as opposed to thermal oxide, which is the preferred choice of current IC manufacturing. However, one can predict that advanced ULSI circuits employing gate isolation layers of 30–50 Å thickness will require a change in the process technology favoring deposited oxynitride films.

In the case of III–V MIS structures, a low deposition temperature is even more important than in the case of silicon because of the low thermal decomposition temperatures of compound semiconductors. Also radiation damage is of considerable concern in this context, since the native defects generated in III–V compounds and alloys by the exposure to energetic particles often are electrically active to the extent of causing type conversion close to the surface. RPECVD has been utilized successfully for the fabrication of $Ga_xIn_{1-x}As$ FETs using SiO_2 as the gate dielectric [51]. Alternative dielectrics may become available in time with the development of novel source materials.

At a given temperature, a significant enhancement of the growth rate of compound semiconductors is achieved by PECVD. For example, an enhancement of the growth rate of GaAs by more than an order of magnitude as compared to low pressure MOCVD has been achieved by PECVD at tempera-

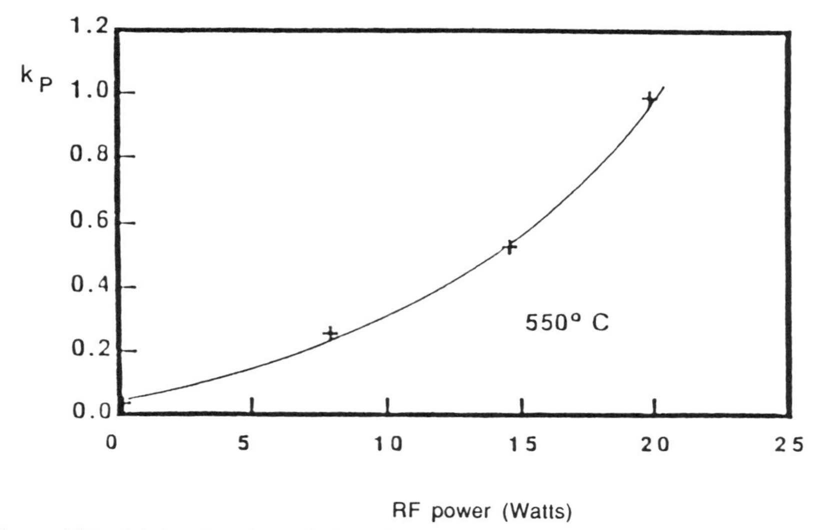

Figure 6.21 Molar fraction of phosphorus in the solid over the molar fraction of phosphorus in the vapor phase for the growth of GaP_yAs_{1-y} by PECVD versus the rf power. After Huelsman and Reif [55]; copyright © 1989, The Minerals, Metals, and Materials Society of Warrendale, PA.

tures between 500 and 600°C [52]. Also, a significant enhancement of the formal distribution coefficient $k_P = x_P(\text{solid})/x_{PH_3}(\text{gas phase})$ for the incorporation of phosphorus into III–V alloys grown by PECVD has been demonstrated, which is illustrated in Figure 6.21 [53]. The value of k_P approaches 1 at sufficiently high rf power. Epitaxial GaP_yAs_{1-y} growth is observed at a substrate temperature as low as 550°C [54].

In view of the enhanced sensitivity of compound semiconductors to radiation damage, RPECVD is preferred to PECVD. Figure 6.22 shows a schematic cross section of an RPECVD system for III–V epitaxy. The group V hydride source vapor is generated in situ by injecting the vapor of a heated elemental group V source into a hydrogen plasma. The group III alkyl source

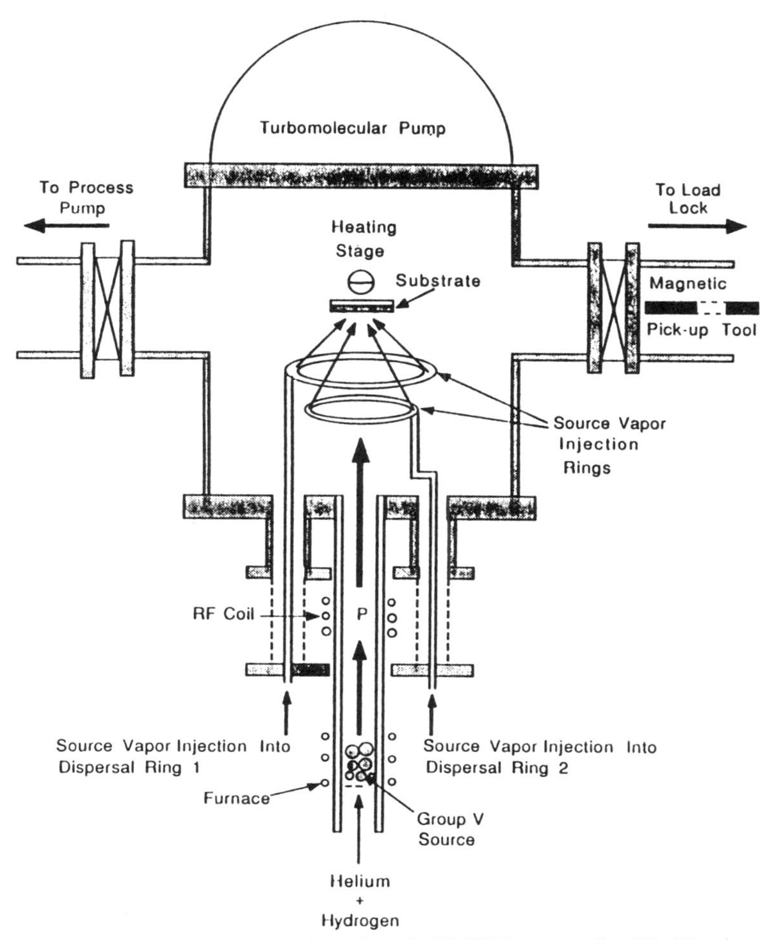

Figure 6.22 Schematic representation of an RPECVD system for III–V epitaxy. After Choi, Lucovsky, and Bachmann [55]; copyright © 1992, Materials Research Society, Pittsburgh, PA.

vapors are injected downstream from the plasma region through perforated dispersal rings. This in situ generation process is generally applicable to a variety of hydride sources, including chalcogens [57]. Since highly reactive fragments of the group V hydrides are formed in the plasma [56], the growth temperature is significantly lower than under the conditions of conventional LPMOCVD. The elimination of the need for storage of large reservoirs of toxic and pyrophoric hydrides is a distinct advantage of RPECVD with regard to the safety of the process, which is self-terminating in the case of a power failure or an alarm condition. However, the possible plasma generation of contaminants and radiation damage of the substrate require careful process control to achieve useful electrical properties.

6.3 Organometallic Chemical Vapor Deposition

The growth of epitaxial films of the III–V compounds by pyrolysis of the group III alkyl compounds in the presence of the group V hydrides was demonstrated first by H. Manasevit in 1968 in an APCVD reactor for GaAs using $Ga(CH_3)_3$ and AsH_3 as source materials [58]. According to the overall reaction

$$\mathrm{III}R_3(\mathrm{g}) + \mathrm{V}R'_3(\mathrm{g}) \rightarrow \mathrm{IIIV}(\mathrm{s}) + 3RR'(\mathrm{g}) \tag{6.35}$$

Because of the excellent surface morphology and uniformity of epitaxial growth provided by organometallic chemical vapor deposition (OMCVD), it has become an important method of compound semiconductor processing. In this section, we discuss the thermodynamic driving force for the deposition process, effects of the flow dynamics and chemical kinetics, and the incorporation of dopants and safety issues, including the selection of alternative source materials. The section is concluded with a review of flow rate modulated OMCVD and atomic layer epitaxy (ALE).

A typical tubular cold wall OMCVD reactor is shown in Figure 6.23. Commercially barrel reactors are also used. The system consists of three major subassemblies:

1. A source vapor conditioning and metering part;
2. The reactor including in situ diagnostics and temperature measurement and control equipment;
3. A pressure regulating and waste treatment part.

Ideally the source vapors are mixed upstream, are pyrolyzed, and react at the surface of the substrate, which is located on an rf- or lamp-heated susceptor. In commercial systems, rotation of the substrate wafer is provided either on a motor-driven pedestal or, more elegantly, by liftoff on a cushion of hydrogen that is generated by hydrogen injection through properly designed channels in the graphite susceptor to effect both separation of the wafer and its rotation.

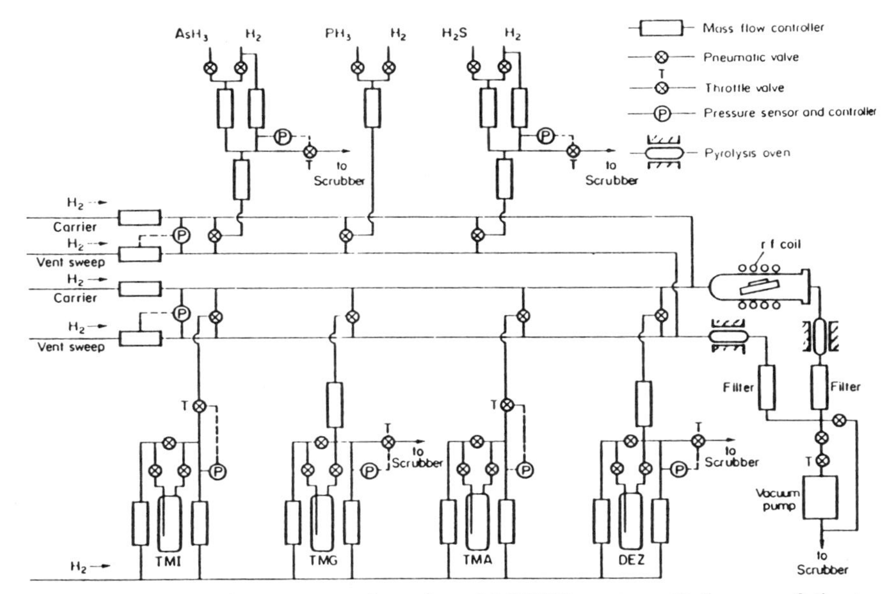

Figure 6.23 Schematic representation of an MOCVD system. Only one of the two parallel sets of organometallic sources is shown. After Bhat and Mahajan [59]; copyright © 1992, Pergamon Press, Oxford.

A typical OMCVD process involves the following steps:

1. Saturation of the carrier gas by the various OM source materials in the bubblers to the partial pressures determined by the bubbler temperature.
2. Dilution of these source vapor streams by excess carrier gas and mixing with the group V hydrides.
3. Flow of the gas mixture to the location of the substrate.
4. Homogeneous gas-phase reactions in the heated gas volume in the vicinity of the susceptor and diffusion of the reaction products to the surface of the substrate.
5. Adsorption of reactants on the surface of the substrate.
6. Surface reaction and incorporation of the film-forming reaction products into the lattice.
7. Desorption of waste products from the surface.
8. Diffusion of waste products across the diffusion layer in the gas phase.
9. Flow of waste products and nonpyrolyzed reactants to the exit port of the OMCVD reactor.
10. Waste treatment to remove toxic and pyrophoric components.
11. Ejection of the remaining gases into the exhaust system.

Figure 6.24 shows schematically the separation of the thermodynamic driving force for the reaction into components that are expended in steps 4–8, which determine the growth rate. As in other CVD processes, the fractions of the total driving force expended in the transport of reactant and waste

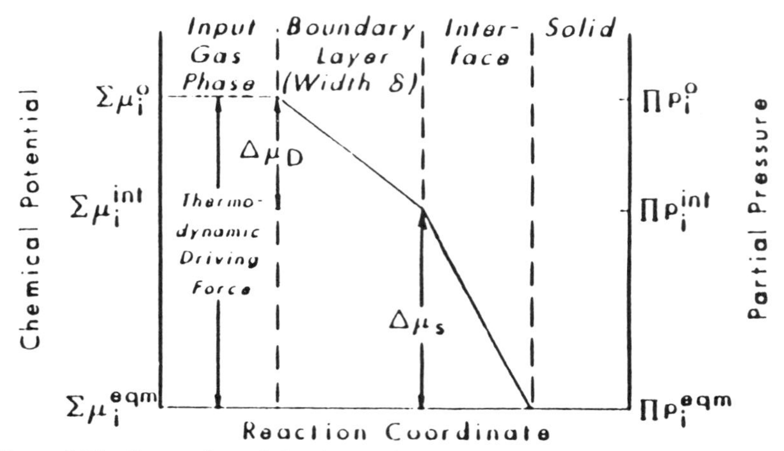

Figure 6.24 Separation of the thermodynamic driving force for OMCVD into components expended to drive the boundary-layer diffusion and surface reaction steps. After Stringfellow [60(a)]; copyright © 1984, North Holland Publishing Company, Amsterdam.

products across the diffusion layer and the surface reaction (i.e., the ratio $\Delta\mu_D:\Delta\mu_s$) depend on the growth temperature.

Figure 6.25 shows the temperature dependence of the growth rate of GaAs using $Ga(CH_3)_3$ and AsH_3 sources. The growth rate is normalized to the $Ga(CH_3)_3$ flow rate to permit comparison of the results of different authors (see [60a] for references). In the region between 550 and 780°C the growth rate is independent of the temperature and surface orientation, that is, is diffusion controlled. The rate-limiting step is in this case the arrival rate of group III metal atoms at the surface. At low temperatures (i.e., at $T \leqslant 550°C$ in this example), the reaction is surface controlled because the surface-catalyzed decomposition reactions become rate limiting.

The condition of both the homogeneous gas-phase reactions and the transport of reactants and waste products depend on the flow dynamics in the reactor. It may be characterized by laser velocimetry, which is based on the measurement of the scattering of light by small particles out of the volume of a gas-phase region, where two laser beams create an interference fringe pattern. The timed detection of the light scattered by a particle out of the bright fringe regions and their known separation provides for a measurement of the particle velocity, so that with three color-separated wavelengths the three components of the gas flow velocity in a particular region in the reactor can be mapped out simultaneously. A comparison of the results with the susceptor at elevated and room temperature, respectively, reveals the dominance of thermal convection under typical growth conditions of GaAs MOCVD [61].

Figure 6.26(a) and (b) shows the measured growth rate variations along the direction of the axis of the flow tube on a GaAs substrate wafer in the MOCVD of GaAs from $Ga(CH)_3$ and AsH_3 in a hydrogen carrier gas flow, using a similar susceptor geometry as shown in Figure 6.23 at reduced pressure

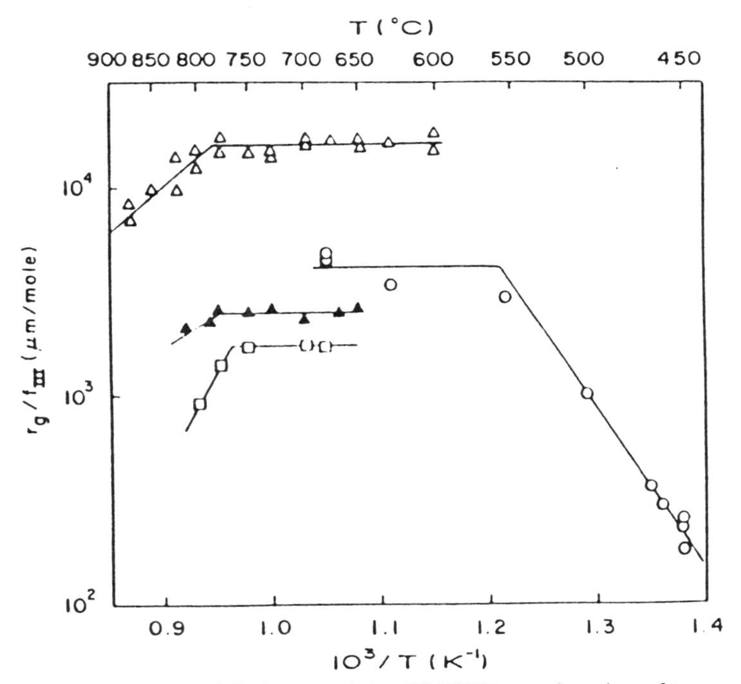

Figure 6.25 The rate of GaAs growth by OMCVD as a function of temperature. After Stringfellow [60(a)]; copyright © 1984, North Holland Publishing Company, Amsterdam.

for two different flow velocities [62]. The measured results are compared with two-dimensional simulations using a commercially available fluid-dynamics modeling code for CVD (FLUENT [63]) and the activation energy for the heterogeneous reaction of trimethyl gallium and arsine on the surface of GaAs of Ref. [64]. Soret diffusion, that is, the tendency for the heavier constituents of the vapor phase to separate out in the cooler parts of the cavity, and homogeneous vapor-phase reactions are not included in this simulation and account for part of the deviations of the experimental and calculated growth rate distribution. The thermal boundary above the susceptor may be evaluated by Raman spectroscopy without interfering with the process using the ratio of the intensity of the Stokes and anti-Stokes lines to determine the temperature [65].

An extensive amount of physico-chemical data exist concerning the thermal decomposition kinetics of organometallic compounds. Unfortunately, these investigations are not representative of the conditions of OMCVD because of the synergistic effect between the decompositions of the group III and group V precursor molecules. Therefore, methods that allow the study of the reaction kinetics of MOCVD in situ are essential. Coherent anti-Stokes Raman scattering (CARS) [67], which converts the radiation associated with the vibrational and rotational molecular transitions by third-order frequency mixing with

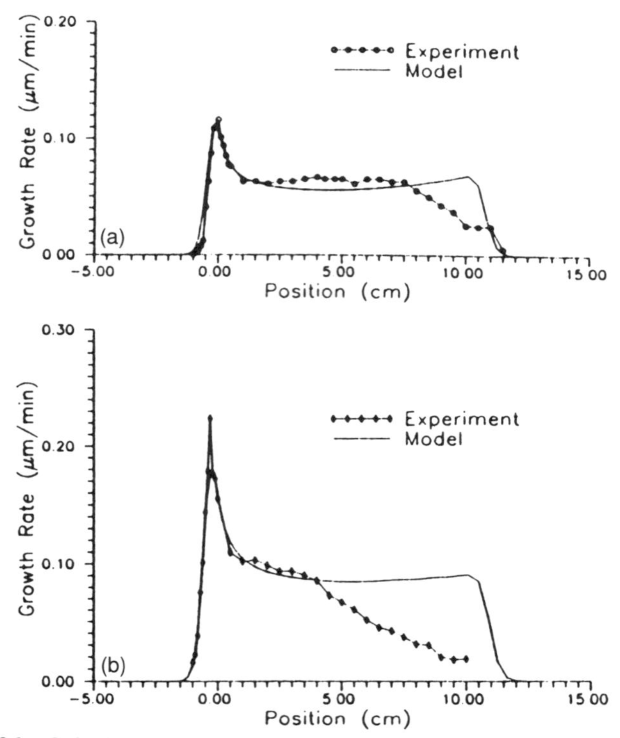

Figure 6.26 Calculated and experimental growth-rate distributions for MOCVD of GaAs at 0.1 atm total pressure and inlet flow velocities: (a) 90 cm/s and (b) 45.3 cm/s. After Black, Clark, Fox, and Jesser [62]; copyright © 1991, North Holland Publishing Company, Amsterdam.

visible laser radiation into a frequency shift in the visible part of the spectrum, has the advantage that holographic gratings and photomultipliers of high resolution and sensitivity can be employed in the measurements. CARS has been applied to the investigation of the kinetics of both group V hydrides and group III alkyl compounds in the context of MOCVD [68], [69]. The CARS spectrum results from transitions between the rotational/vibrational states of particular bonds, which according to the Born–Oppenheimer approximation may be represented as a characteristic set of eigenvalues of energy within the electronic energy well (see Figure 2.6). Since the thermal excitation of rotational/vibrational modes of the molecules with increasing temperature alters the CARS spectrum in addition to providing information on the concentration of specific reactants and waste products, it provides for the contactless measurement of the temperature distribution in the reactor.

Figure 6.27 shows typical CARS spectra of AsH_3 and of the H_2 generated upon AsH_3 decomposition. The analysis of these spectra taken in different environments reveals that on a SiO_2 surface 3/2 moles of hydrogen are produced for each mole of pyrolyzed arsine, that is, the molecules decompose

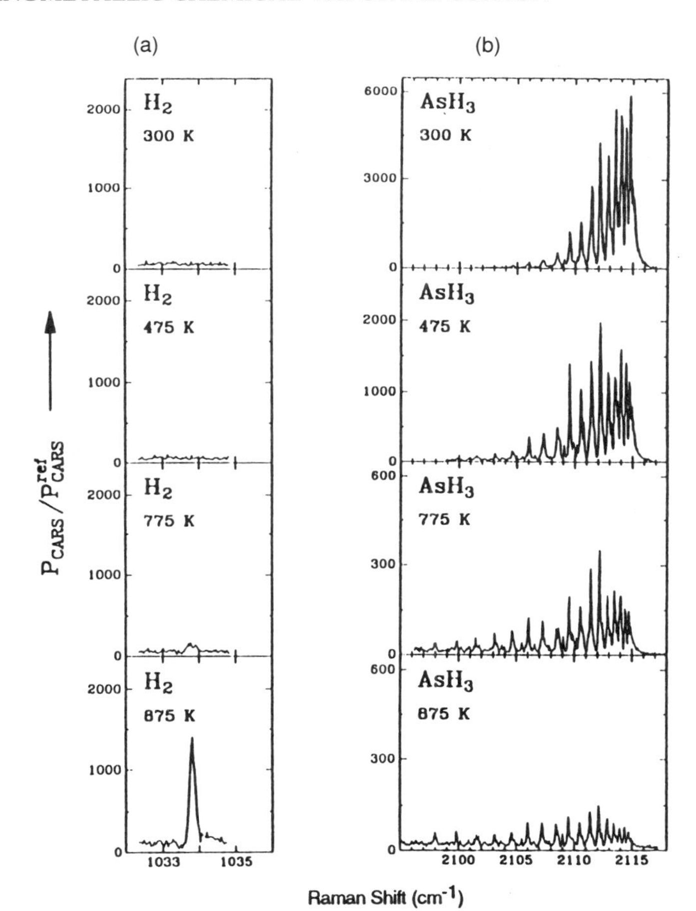

Figure 6.27 CARS spectra of (a) H_2 and (b) AsH_3 as a function of the temperature. (b) After Richter et al. [43]; copyright © 1991, Elsevier Science Publishers B.V., Amsterdam.

completely. On a GaAs surface, the onset of arsine pyrolysis is shifted toward a lower temperature and less than the maximum amount of hydrogen per pyrolyzed arsine molecule is produced, indicating a catalytic enhancement of the reaction rate at only partial dehydrogenation.

Infrared spectroscopic measurements of the concentration of methyl radicals generated by the pyrolysis of $In(CH_3)_3$ at a given substrate temperature as a function of the AsH_3 partial pressure, normalizing the signal to the radical concentration for trimethyl indium pyrolysis in the absence of arsine, reveal a decrease of this normalized concentration with increasing arsine concentration. This suggests that the methyl radicals interact with the arsine molecules according to the reaction

$$AsH_n + H_3C^{\bullet} \rightarrow AsH_{n-1} + CH_4 \tag{6.36}$$

Disproportionation of the methyl radicals according to

$$H_3C^{\cdot} + H_3C^{\cdot} \rightarrow CH_4 + CH_2 \tag{6.37}$$

generates the unstable carbene radical that readily undergoes further dehydrogenation, leading to the incorporation of carbon into the semiconductor lattice. Indeed, for a given flow rate of $Ga(CH_3)_3$, the concentration of carbon acceptors in the GaAs decreases with increasing AsH_3 flow rate, indicating an increased scavenging of methyl radicals by reaction (36). The intentional addition of methane to the gas phase does not contribute to the C acceptor concentration [70].

Figure 6.28(a) shows schematically a cross section of a GaAs test structure grown by the successive deposition of layers using ^{13}C-labeled $As(CH_3)_3$ and AsH_3, respectively, in the OMCVD process. The C concentration cannot be measured on the basis of the ^{12}C and ^{13}C mass spectra. However, as shown in Figure 6.28(b), the SIMS profile at $m/q = 88$ ($As^{13}C^-$) easily reveals carbon incorporated from the $As(CH_3)_3$ source at a mid-$10^{16}\,cm^{-3}$ levels [78]. It is important to note that at the higher growth temperature the C concentration in the epilayer is increased. Also, it has found that reducing the pressure generally increases the C incorporation. The fact that ^{13}C-labeled methane is found to be the primary decomposition product adds further credence to the assumption that the incorporation of carbon into the lattice starts with the formation of $H_3C^{\cdot}$.

Replacing trimethylgallium by triethylgallium reduces the carbon incorporation since the ethyl radicals can disproportionate into ethene and ethane, which are both stable gases. Only slight increases in the C levels are observed upon the intentional addition of acetylene, ethene, and other alkenes to the gas

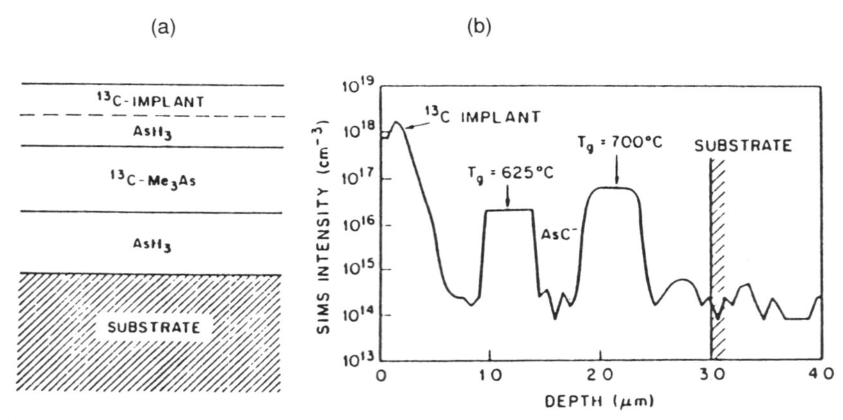

Figure 6.28 (a) Schematic representation of an epitaxial GaAs test structure. (b) SIMS profile of the epilayer. After Lum, Klingert, and Lamont [70]; copyright © 1988, North Holland Publishing Company, Amsterdam.

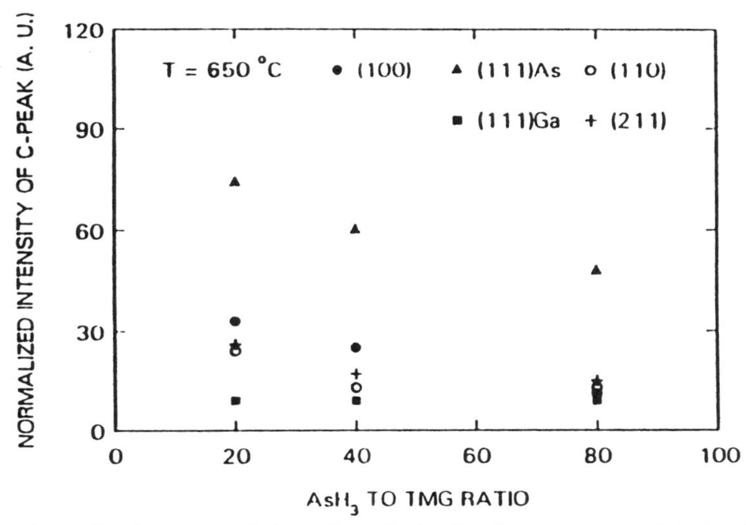

Figure 6.29 Normalized integrated intensity of the luminescence associated with carbon acceptors in GaAs for epilayers grown from trimethylgallium and arsine at $T_s = 650°C$ as a function of the V:III ratio in the gas phase for selected surface orientations of the substrates. After Kuench and Veuhoff [70]; copyright ©1984, North Holland Publishing Company, Amsterdam.

phase; that is, once desorbed from the surface, C_2H_4 is effectively removed [71]. Details of the surface reaction may be deduced from the results of thermal desorption and x-ray photoemission [72], [73]. $Ga(C_2H_5)_3$ is adsorbed at low temperatures (150 K) undissociated and desorbs from the surface of GaAs at 170–220 K. However, not all of the $Ga(C_2H_5)_3$ desorbs; at this low temperature the desorption process is linked to decomposition. Relevant surface reactions in the growth processes involving $Ga(C_2H_5)_3$ are:

$$Ga(C_2H_5)_3(g) + S^* \rightleftarrows Ga(C_2H_5)_3(ads) \tag{6.38a}$$

$$Ga(C_2H_5)_3(ads) + S^* \rightarrow Ga(C_2H_5)_2(ads) + C_2H_5(ads) \tag{6.38b}$$

$$Ga(C_2H_5)_2(ads) \rightarrow Ga(C_2H_5)_2(g) + S^* \tag{6.38c}$$

$$Ga(C_2H_5)_2(ads) + 2S^* \rightarrow Ga(ads) + 2C_2H_5(ads) \tag{6.38d}$$

$$C_2H_5(ads) \rightarrow C_2H_4(g) + H(ads) \tag{6.38e}$$

$$C_2H_5(ads) + H(ads) \rightarrow C_2H_6(g) + 2S^* \tag{6.38f}$$

$$2H(ads) \rightarrow H_2(g) + 2S^* \tag{6.38g}$$

At 620–670 K hydrogen, ethene, and ethane desorb. Above 700 K, the growth process becomes mass transport limited. For lower temperatures, the growth rate becomes limited by the competition between $Ga(C_2H_5)_3$ and the more tightly bonded decomposition products and group V hydride-derived species for active surface sites.

Figure 6.29 shows the integrated intensity of the photoluminescence peak at 1.489 eV, which is associated with carbon acceptors in GaAs, normalized to the integrated intensity of the band-edge luminescence at 1.513 eV as function of the V:III ratio in the vapor phase for various surface orientations of the substrates. The highest and lowest carbon concentrations are observed for the growth of GaAs on the As(111) and Ga(111) faces, respectively. The electrical properties of the epitaxial films are given in Table 6.3. Note that while the reduction of the C-acceptor level is an important goal in the context of many applications, in the fabrication of HBTs and certain other classes of devices, access to high carbon doping is of substantial technological benefit.

Table 6.4 presents information on the vapor pressures and safety data for selected organometallic compounds and hydrides. For example, the group V

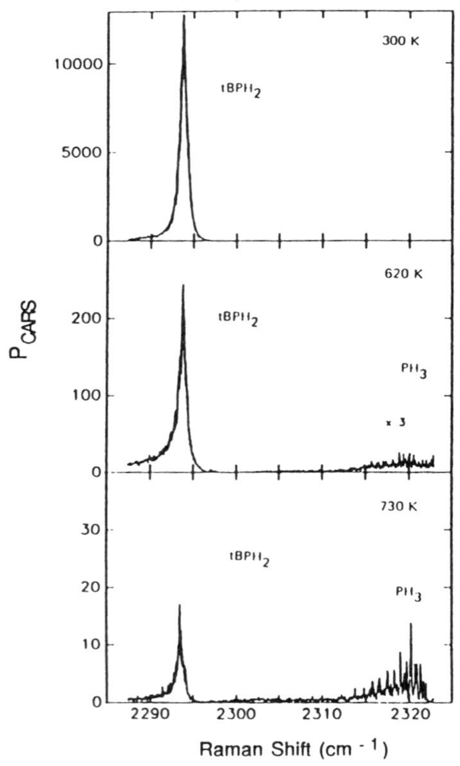

Figure 6.30 CARS spectra of tertiarybutylphosphine at different temperatures. After Richter et al. [43]; copyright © 1991, Elsevier Science Publishers B.V., Amsterdam.

Table 6.3 Electrical properties of GaAs epilayers grown by OMCVD at $AsH_3/Ga(CH_3)_3$ ratio of 80 and 40, respectively, for 78 and 300 Torr reactor pressure using H_2 and He as carrier gases (after Kuench and Veuhoff [70])

Carrier Gas	T_s (°C)	p (Torr)	μ(77 K) ($cm^2 V^{-1} s^{-1}$)	N_D-N_A (cm^{-3})
H_2	650	78	123,000	1.4×10^{14}
He	650	300	56,700	5.5×10^{14}
H_2	650	300	59,900	6.3×10^{14}
He	800	300	9,210	2.1×10^{16}
H_2	800	300	9,960	2.1×10^{16}

hydrides, diborane, hydrogen selenide, and silane have LC-50 (4 h exposure) levels in the ppm range, that is, are dangerous toxins resulting in death at very low concentrations. Reliable monitors that detect these hydrides at sub-TLV (safe 8 h exposure) or sub-TWA (maximum allowed workplace concentration) levels are available. However, the safety data for some of the OM source materials are still incomplete. Note that although phosphine is by itself not pyrophoric at room temperature, diphosphine, which is frequently present as an impurity, renders phosphine self-igniting. Pure PH_3 becomes pyrophoric at 42°C. Also, it should be noted that silane, though being pyrophoric, may form in contact with air a self-protecting bubble that upon rupture may explode with considerable force [76].

Absorption on activated charcoal, oxidation, and reactive scrubbing are the predominantly used methods of waste management. Care must be taken that the oxidative reactions are carried out under well-controlled conditions and that the complete toxic gas content of the effluent of the OMCVD reactor is removed. Monitoring of the gases injected after waste treatment into the stack for toxic components is a mandatory requirement for protecting the environment. The oxidation products of some of the materials listed in Table 6.4 are by themselves dangerous materials that must be collected and disposed of safely [81], adding thus to the cost and danger of the operation. Organometallic group V sources that are liquid at room temperature with subatmospheric vapor pressures and significantly higher LC-50 values (see Table 6.4) are available, but in most cases, do not provide the essential active hydrogen, as, for example, delivered in the pyrolysis of AsH_3. For example, the utilization of diethyl arsine produced GaAs epilayers with $N_D - N_A = 3 \times 10^{14}\,cm^{-3}$ and $\mu_{77K} = 64{,}000\,cm^2\,V^{-1}\,s^{-1}$ [82]. However, its small vapor pressure limits its use in APOMCVD. Tertiary-butyl arsine results in conjuction with triethyl gallium $N_D - N_A = 5 \times 10^{15}\,cm^{-1}$ and $\mu_{77K} = 53{,}000\,cm^2\,V^{-1}\,s^{-1}$ [79]. Also monophenyl arsine is an attractive new source since benzene does not contribute significantly to the carbon uptake by GaAs [83]. However, none of the substituted arsines has achieved the same quality as AsH_3 in the growth of group III arsenides.

Table 6.4 Properties of source materials for OMCVD

Source	Boiling point (°C)	Vapor pressure	Remarks
Et_3Al	194	$10.784-3625/T$	Pyrophoric liquid, $T_m = -58°C$[a]
Me_3Al	126	$7.3147-1534.1/(T\text{-}53)$	Pyrophoric liquid, $T_m = 15.4°C$[a]
Et_3Ga	55,7	$8.224-2222/T$	Pyrophoric liquid, $T_m = -82.3°C$[a]
Me_3Ga	143	$8.07-1703/T$	Pyrophoric liquid, $T_m = -15.8°C$[a]
Et_3In	184	1.2 Torr (44°C)	Pyrophoric liquid, $T_m = -32°C$[a]
Me_3In	133.8	$10.520-3014/T$	Pyrophoric solid, $T_m = +88.4°C$[a]
B_2H_6	−92.5		0.1 ppm TWA[e]
SiH_4	−112.3		Explosion hazard, 0.5 ppm TWA[e]
NH_3	−33.4		50 ppm TWA[e]
H_3As	−62.5	205 psig (20°C)	5–50 ppm LC-50,[b] 0.05 ppm TWA[e]
Et_3As	140	5 Torr (20°C)	1065 ppm LC-50[c]
Me_3As	50	235 Torr (20°C)	2×10^4 ppm LC-50[b]
$t\text{-}BuAs_2$	65	~100 Torr (−10°C)	70 ppm LC-50[c]
$PhAsH_2$	148	2 Torr (20°C), 12 Torr (50°C)	Poison B
H_3P	−87.8		0.1 ppm TWA,[d] $T_m = -133.8°C$
Et_3P	127	$7.86-2000/T$	Pyrophoric liquid, $T_m = -88°C$[a]
Me_3P	37.8	$7.7329-1512/T$	Pyrophoric liquid, $T_m = -85°C$[a]
$t\text{-}BuPH_2$		286 Torr (23°C)	>1000 ppm LC-50[d]
Me_3Sb	80.6	$7.728-1709/T$	Pyrophoric liquid, $T_m = -86.7°C$[a]
Tris	33	115 Torr (−10°C)	≥ 475 ppm LC-50[c]
H_2S	−60.8		10 ppm TWA[e]
H_2Se	−41.3		0.5 ppm TWA[e]
HCl	−85		5 ppm TWA[e]

[a] [77].
[b] [78].
[c] [60b] and [80].
[d] [79].
[e] [76].

Since the group III alkyl compounds are Lewis acids and the group V hydrides or alkyls are Lewis bases, adducts $R_3\text{III}-\text{V}R'_3$ form by a homogeneous gas-phase reaction. In the case of $R = CH_3$ and $R' = H$ polymerization by the elimination reaction

$$R_3\text{III}-\text{V}R'_3 \rightarrow -R_2\text{III}-\text{V}R'_2- + RR'$$
$$-R_2\text{III}-\text{V}R'_2- + (n+1)R_3\text{III}-\text{V}R'_3 \rightarrow R_3\text{III} -\text{V}R'_2\{-R_2\text{III}-\text{V}R'_2-\}_n\,\text{III}R_2-\text{V}R'_3 \tag{6.39}$$

is a problem in certain cases. For example, under the conditions of InP deposition using trimethyl indium and PH_3 in an APCVD reactor, CH_4 elimination readily leads to polymerization. This is a nuisance because of the formation of wall deposits of the polymer upstream of the location of the susceptor. On the other hand, there are certain adducts that do not polymerize in the presence of phosphine and that have been used as source materials in OMCVD processing, such as in the formation of alloys

$$In(CH_3)_3 + P(C_2H_5)_3 \rightarrow In(CH_3)_3{:}P(C_2H_5)_3 \tag{6.40a}$$

$$In(CH_3)_3{:}P(C_2H_5)_3 + yPH_3 + (1-y)AsH_3 \rightarrow InP_yAs_{1-y} + 3CH_4 + P(C_2H_5)_3 \tag{6.40b}$$

Since the performed III–V bonds in the adducts are not very strong, they dissociate at elevated temperature. For example, the dissociation of the $(CH_3)In{-}P(C_2H_5)_3$ adduct is complete at 200°C [74]. A variety of organometallic compounds containing preformed III–V bonds, such as of the type $[R_2M(\mu\text{-}t\text{-}Bu_2X)_2M'R'_2$, where $R = CH_3$, $X =$ P, As, and M, $M' =$ Al, Ga, In, have been introduced [75], but have at present no significant impact on industrial MOCVD processing at atmospheric pressure. However, compounds with preformed III–V bonds may become of value in future in the context of quantum dot heterostructures, and a continuing search for alternative source materials for conventional MOCVD is important because the cost of monitoring and waste disposal and the danger of the currently used source materials seriously affects the efficiency of processing and cost. In some cases this search has already led to significant advances.

For example, tertiary-butyl phosphine (t-butPH_2) is an excellent alternative to phosphine since it has an order of magnitude higher TLV [84]. Figure 6.30 shows the CARS spectra for t-butPH_2 at three temperatures, revealing an onset of decomposition into C_4H_8 and PH_3 below 350°C. The pyrolysis of t-ButPH_2 proceeds at lower temperature than the pyrolysis of PH_3 because of the relatively easy cleavage of the phosphorus bond to the bulky t-butyl group. Of the possible reactions following this bond cleavage, the production of C_4H_8 and PH_3 is only of secondary importance [69]. It has been shown that the $C_3H_9^{\cdot}$ radical is a useful source of reactive hydrogen atoms in statu nascendi [60b], [85]. Since H atoms are known to facilitate the pyrolysis of trimethyl gallium [70], the admixture of azo-t-butane, which decomposes thermally into nitrogen and $C_3H_9^{\cdot}$ radicals, can be used to enhance the growth rate of GaAs from AsH_3 and $Ga(CH_3)_3$ at a given temperature in the surface controlled regime. Radical assisted OMCVD achieves GaAs epitaxy at 390°C [85]. InP epilayers have been made by OMCVD exhibiting a net ionized donor concentration $N_D - N_A = 2 \times 10^{14}\,cm^{-3}$ and an electron mobility $\mu_n(InP) = 147{,}000\,cm^2\,V^{-1}\,s^{-1}$ at 77 K [86].

A number of hydride and metal alkyl doping sources, such as GeH_4, SiH_4, H_2S, and H_2Se and the group II alkyl compounds, are available for doping. Also, cyclopentadienyl complexes of group IIA elements, such as magnesocene, are alternative acceptor sources, and transition metal alkoxides, cyclopentadienyl, carbonyl, and phosphine complexes are suitable doping sources for the fabrication of semi-insulating InP and GaAs epitaxial layers [87], [88]. Since OMCVD provides for very sharp interfaces, in properly designed reactors [89], [90] complex doping profiles may be synthesized by planar doping techniques. Planar doping relies on the provision of sheets of highly doped atomic layers by appropriate timing of dopant precursor pulses and pulses of precursors to compound growth under the conditions of interrupted-cycle OMCVD; that is, an interruption of the group III source flow is followed by injection of a dopant gas pulse and subsequent resumption of the group III source flow. A high dopant concentration is thus established at the interface of the epitaxial III–V layers grown prior to and after the dopant pulse. Figure 6.31 shows a complex doping profile synthesized by a series of planar doping planes inserted during OMCVD into an epitaxial layer of GaAs [89].

In general, flow-rate-modulated growth [91], [92] provides for abrupt step-free interfaces at relatively low temperatures because of the high mobility of the OM molecules on the surface. The rearrangement of the surface positions of the precursors to growth during the waiting period prior to the admission of reaction partners that lock them into a bonded configuration results in smoother interfaces than attained by simultaneous mixing of the reactants on the surface. Therefore, interrupted-cycle OMCVD improves the uniformity of growth, and is thus particularly useful for the fabrication of quantum-well heterostructures [93].

The sequential exposure of the surface of a semiconductor to appropriate precursors that strongly interact with the terminal surface groups or atoms, but only weakly interact among themselves, may result in the self-terminating formation of chemisorbed monolayers, knitting reactively one atomic layer after the other onto a polar surface, irrespective of the precursor dose admitted

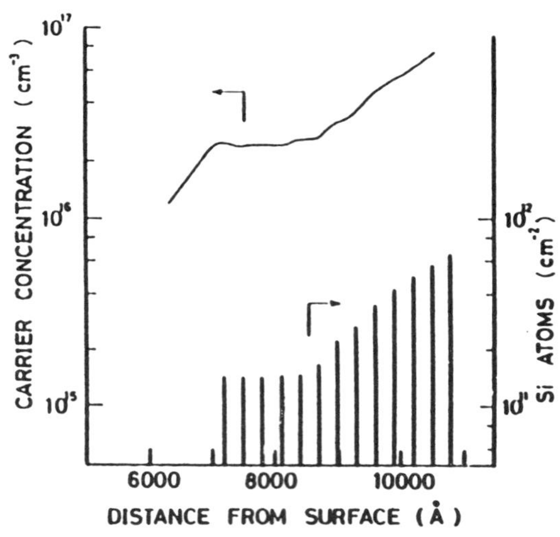

Figure 6.31 Free-carrier profile synthesized by planar Si doping in GaAs. After Ohno, Ikeda, and Hasegawa [91]; copyright © 1984, North Holland Publishing Company, Amsterdam.

in each step to the surface. This process, which provides inherently for excellent step coverage, was suggested in 1974 by Suntola and Antson in the context of II–VI epitaxy [94] and is called *atomic layer epitaxy* (ALE) [96]. For example, on a polar surface of a ZnS wafer, such as the S-terminated (111) face, exposure to a pulse of $ZnCl_2$ at elevated temperature results in the bonding of Zn to the sulfur surface atoms. This surface reaction is self-terminating because Zn–Zn bonds are thermally unstable. Similarly subsequent exposure of the now at least partially Zn-terminated surface to a pulse of H_2S vapor results in bonding of sulfur to the Zn surface atoms. Again the surface reaction is self-terminating because S–S bonds are not stable at the growth temperature (typically 800°C). Therefore, a maximum of one atomic layer is added to the substrate lattice upon injection of each precursor pulse. Note that if a one monolayer per pulse growth mechanism were operative, one gram equivalent of HCl should be released after each precursor pulse (H_2S or $ZnCl_2$). However, experiments show that less than a monolayer is added to the surface after each pulse. The fraction of a monolayer added per pulse, in this case, depends on the growth temperature.

Quantum-chemical calculations [95] reveal that only partial coverage of a clean sulfur-terminated surface may be expected upon exposure to a $ZnCl_2$ pulse. For statistical individual adsorption events, $\frac{1}{3}$ of a monolayer coverage at maximum is expected due to interactions between the adsorbed $ZnCl_2$ moieties, while the formation of $ZnCl_2$ chains on the S-terminated surface permits coverage of half of the surface. The chain mechanism results in the release of one gram equivalent of HCl after every combination of a $ZnCl_2$ with an H_2S pulse, while the statistical addition of $ZnCl_2$ to the surface results in a more complex mechanism. The start of such a cycle is illustrated in Figure 6.32. The negative numbers refer to the calculated energies. Comparing the energy for the surface complex containing both $ZnCl_2$ and H_2S with that of the surface complex obtained upon release of HCl after the first cycle shows that the former is energetically favored. Since the number of atoms in this complex does not permit ring closure and neighboring complexes are too far apart to interact, at least one further $ZnCl_2$ molecule must be added before ring closure under release of HCl becomes possible. Thus a strict ALE mechanism is not operative even though the condition of self-limiting adsorption upon sequential exposure is met.

Nevertheless, the extension of the concept of ALE to III–V compounds has received considerable attention. In implementations of interrupted-cycle OMCVD (ICOMCVD) in the viscous flow regime, the precursors must be separated by the insertion of plugs of a nonreacting carrier gas, which is illustrated for the example of hydrogen pulses in between AsH_3 and $Ga(CH_3)_3$ pulses in Figure 6.33. It requires the prevention of mixing in the vapor phase, which is difficult to accomplish. An elegant alternative, eliminating the need for intermittent plugs of pure carrier gas, is described in [97] and relies on the sequential rotation of the substrate into spatially separated streams of $Ga(CH_3)_3 + H_2$ and $AsH_3 + H_2$. A drawback of this approach is that precise-

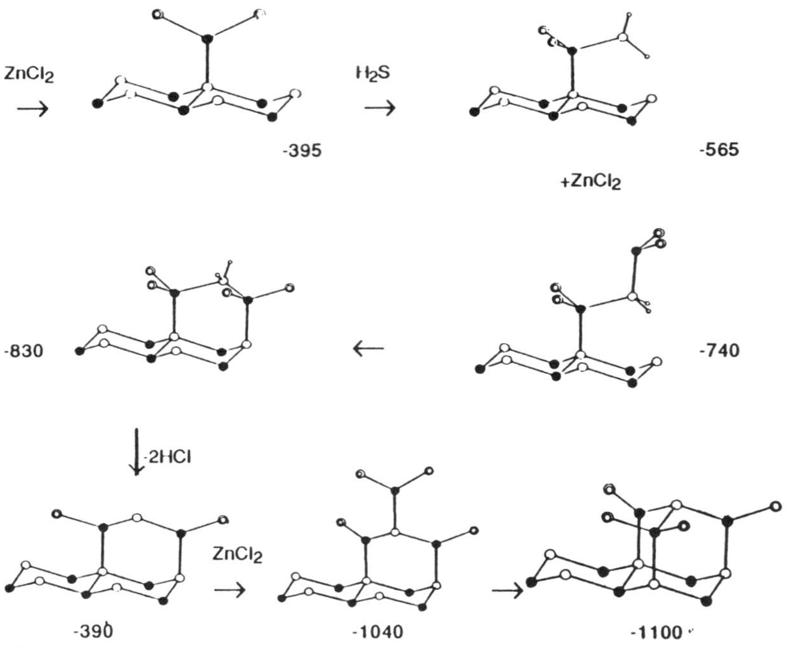

Figure 6.32 (a) 2 cycle–1 layer and (b) 3 cycle–1 layer mechanisms of the reaction of $ZnCl_2$ and H_2S with a hydrogen-terminated S surface of ZnS. (c) Beginning pathway of the 3 cycle–1 layer mechanism. After Pakkanen et al. [95]; copyright © 1987, North Holland Publishing Company, Amsterdam.

ly machined gaps between the substrate wafer and a stationary enclosure must be maintained to achieve effective separation of the two gas streams. An interesting alternative route to ALE has been introduced in the context of silicon deposition [98]. It employs sequential exposures of the surface of a silicon substrate to pulses of disilane and uv light, resulting in the dissociative

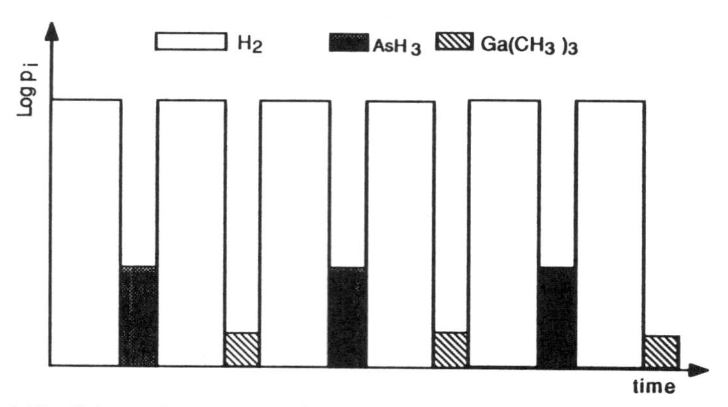

Figure 6.33 Schematic representation of the timed sequence of AsH_3 and $Ga(CH_3)_3$ precursor pulses separated by plugs of hydrogen in ICOMCVD of GaAs.

adsorption of Si_2H_6 followed by the dehydrogenation of the surface by photolysis of the Si–H bonds. Of course, under ballistic transport conditions (i.e., in an ultrahigh vacuum), the mixing in the gas phase is avoided completely and provides thus for advantageous processing conditions in the context of molecular layer epitaxy, which is reviewed in Section 6.4.

6.4 Molecular Beam Epitaxy and Chemical Beam Epitaxy

Molecular beam epitaxy was introduced in the early 1960s [99] and contributed greatly to pioneering developments in solid-state electronics. Figure 6.34(a) shows the layout of a typical MBE system for III–V compounds, which consists of an ultrahigh-vacuum chamber containing Knudsen cells for establishing the required source beams, the substrate heating/manipulation stage, an Auger spectrometer for studies of the surface composition, a mass analyzer, and an electron gun for studies of the surface structure by RHEED. The Knudsen cells are usually loaded with elemental source materials. They are independently heated and have individual shutters to establish the desired effusion fluxes at the surface of the substrate. In the molecular flow region (i.e., for a mean free pass $\bar{l}$ of the atoms or molecules of the source vapors that is very large

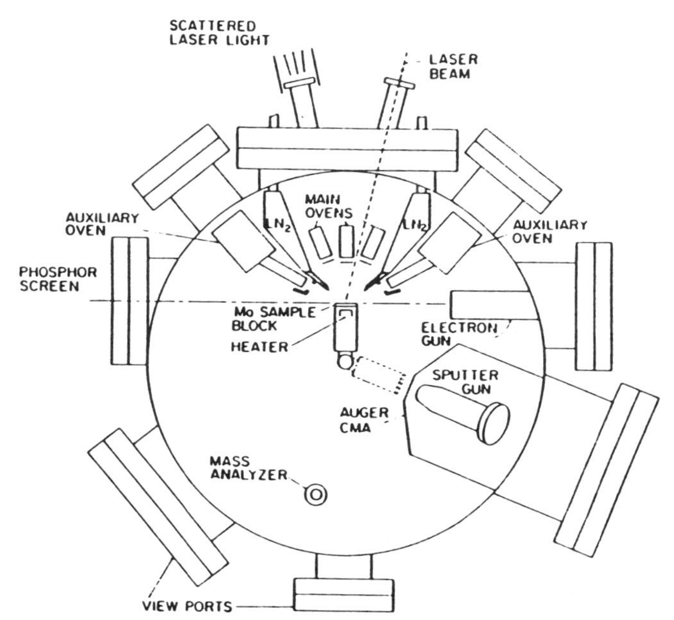

Figure 6.34 Cross section through a conventional MBE system for the growth of compound semiconductor heterostructures. After Miller et al. [100]; copyright © 1977, The Electrochemical Society, Pennington, NJ.

compared to the diameter d_0 of the beam-defining orifice), the flux f_i of the source labeled i out of the orifice is related to the partial pressure p_i inside the cell as

$$f_i = \frac{\pi d_0^2 p_i}{4\sqrt{2\pi m_i kT}} \tag{6.41}$$

where m_i is the mass of the effusing species. However, under the conditions of MBE, the desire to establish high fluxes usually requires relatively large openings in the cells so that their output differs from that predicted by Eq. (6.41). The flux at the substrate location scales as d_0^2/L^2, where L is the distance from the source. Typical conditions for GaAs MBE are: $T_{Ga} \approx 1080°C$ and $T_{As_4} \approx 330°C$, corresponding to fluxes of Ga atoms and As_4 molecules of $\sim 10^{15}\,cm^{-2}\,s^{-1}$. The effective temperature of the source material may be affected by the opening and closing of the shutter. Therefore, the flux must be calibrated for each source at the site of the substrate. This may involve a mass spectrometric evaluation, a measurement with an ion gauge that is temporarily rotated into the substrate position, or an ex situ assessment after completing the growth. The latter is frequently done for the calibration of dopant fluxes. The growth rate is dominated by the group III atom flux since the sticking coefficient is close to 1. Oval defects in the surface of GaAs epilayers grown by MBE, which were a matter of concern in early MBE studies, have been related to the choice of the crucible material for the Ga source. Pyrolytic boron nitride crucibles for containing the Ga source, which have been conditioned by a preceding use as aluminum source containers in MBE, eliminate this problem [102]. Since the group V flux is composed of tetramer and dimer molecules, dissociative reaction with surface sites is required, which is illustrated in Figure 6.35.

A significant improvement of the efficiency of MBE growth is achieved by pyrolysis of AsH_3 and PH_3 in appropriately designed cracker cells, which provide substantially higher As_2/As_4 and P_2/P_4 flux ratios than achieved with conventional elemental sources [103]. Also, gas source MBE facilitates the control of the composition and optimizes the quality of alloy heterostructures in mixed phosphide–arsenide systems. The higher efficiency of the utilization of the source beams with higher dimer to tetramer molecular flux ratio also benefits the maintenance of the MBE chamber, which must be cleaned periodically to remove built-up phosphorus and arsenic deposits.

Figure 6.36(a) shows a high-pressure source, where the hydrides are cracked at $\approx 900°C$ at a pressure of typically several hundred Torr. In this cell a gas-phase equilibrium is established between the dimer and tetramer group V molecules that escape from the high-pressure region through a narrow crack in the confining Al_2O_3 tube. Figure 6.36(b) shows a low-pressure gas source where the PH_3 and AsH_3 molecules are cracked at the surface of a Ta mesh held at 1000°C. The mass spectrometric evaluation of the dimer to tetramer

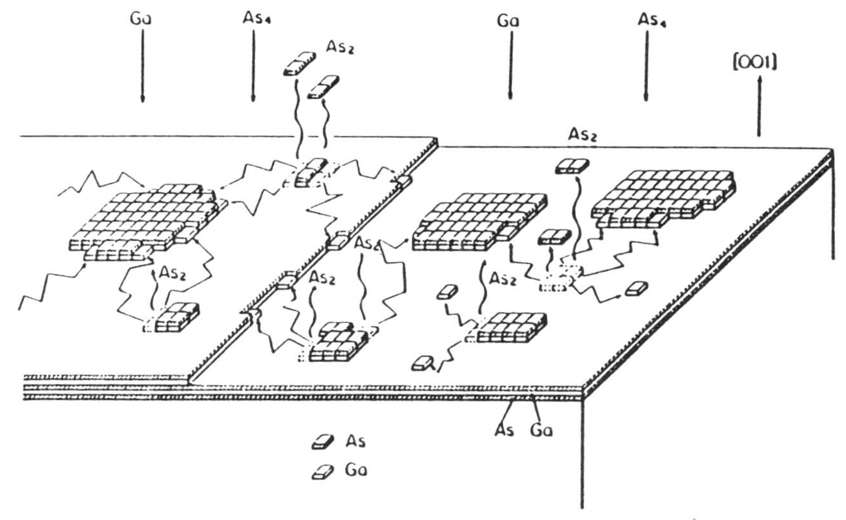

Figure 6.35 Schematic representation of the growth mechanism of GaAs under the conditions of MBE. After Horikoshi, Kawashima, and Yamaguchi [101]; copyright © 1989, The Electrochemical Society, Pennington, NJ.

flux ratios results in peak arsenic ion current ratios $10 < As_2^+/As_4^+ < 130$ and $7 < As_2^+/As_4^+ < 30$ for high- and low-pressure cells, respectively, depending on the flow rate. For phosphorus, the high-pressure cell results typically in $P_2^+/P_4^+ \approx 100$, while the low-pressure cell delivers $20 < P_2^+/P_4^+ < 50$ [03b].

In nominally undoped MBE layers of GaAs carbon and sulfur are the prevalent residual impurities. The former enters into the MBE layers from the residual gas atmosphere in the growth chamber and residual surface contami-

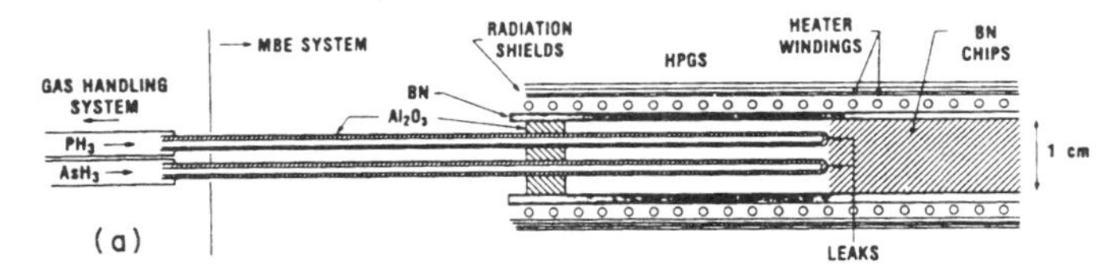

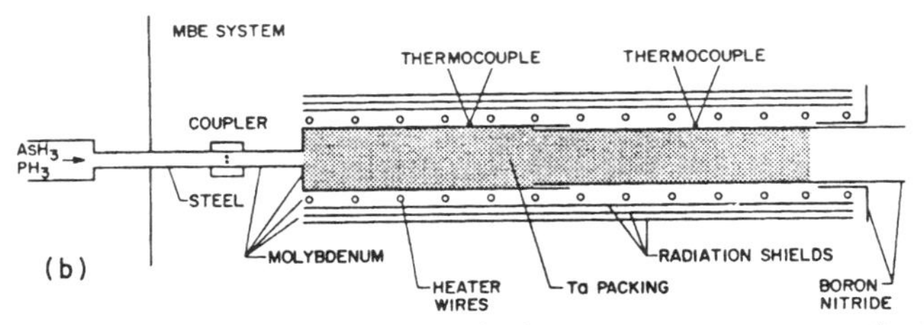

Figure 6.36 (a) High-pressure gas source; (b) low-pressure gas source for MBE. After Panish [103b]; copyright © 1987, North Holland Publishing Company, Amsterdam.

nation, while the latter is a known impurity in As sources. By careful baking and vigilance in the substrate and source preparation, a residual acceptor concentration of $2.4 \times 10^{13}\,cm^{-3}$ and donor concentration of $1.5 \times 10^{14}\,cm^{-3}$, corresponding to a compensation ratio $N_A/N_D = 0.16$ and a Hall mobility of $163{,}000\,cm^2\,V^{-1}\,s^{-1}$ at 77 K, have been achieved in MBE GaAs [104]. Group VI elements evaporated from compound sources, such as PbS, PbSe, and SnTe, are used as n-type dopants. Beryllium is used frequently as a *p*-type dopant. Since C doping achieves both very high acceptor concentration and stable steep acceptor gradients, halocarbons (e.g., CCl_4) have been introduced during MBE growth of GaAs as sources for deliberate C doping [105].

The incorporation of dopants under the conditions of MBE is clearly affected by the native point-defect chemistry. Therefore, a close control of the processing conditions is essential for controlling the incorporation of dopants and the concentration of traps in MBE GaAs and InP. For example, Ge is incorporated into GaAs as an acceptor and a donor, respectively, under gallium- and arsenic-stablized surface conditions corresponding on (001) to a 1×1 and 2×4 surface mesh, respectively [106]. Shallow donors in wide-bandgap III-V compounds, such as $Al_xGa_{1-x}As$ and GaP_yAs_{1-y}, exhibit in addition to the shallow effective-mass state near the CBE, deep states that trap the electrons and consequently limit the doping efficiency. The depth of these *DX* centers depends on the alloy composition and on the chemical nature of the impurity [107], [108].

An important advantage of ballistic beam conditions is that they permit in situ studies of the crystal growth processes by advanced surface analysis methods. For example, low-energy ion scattering (LEIS) has been employed to study the evolution of the atomic structure and composition of the topmost surface layer [109]. In this method, a well-collimated pulsed He ion beam is scattered by the surface atoms, and the backscattered ions are energy analyzed by a time-of-flight (TOF) energy analyzer. Figure 6.37(a) shows a schematic representation of the scattering conditions for an ideally terminated (111) surface of a zincblende structure semiconductor, for example, the InP–*A* surface. If the beam is aligned with the $\langle 110 \rangle$ direction, ions that are backscattered from both the In and P atoms are detected, while for alignment with the $\langle 111 \rangle$ direction only the In atoms are seen. Surface reconstruction will change these conditions so that the angular dependence of the LEIS signal will reveal the surface structure. Figure 6.37(b) shows the LEIS spectrum obtained for indium atoms on a Si(100) surface using 1.6 keV incident He ions. The In and Si peaks are located at 1.395 and 0.915 eV, respectively. Since the TOF of backscattered ions from the substrate surface to the microchannel plate detector depends on their velocity, the ions having lower energy are detected at a higher TOF value.

As an alternative, information concerning the surface composition may be obtained by Auger electron spectroscopy (AES), which is discussed in more detail in Chapter 8. Figures 6.38(a) and (b) show the Auger electron spectra obtained for an InP surface without any conditioning in UHV and after

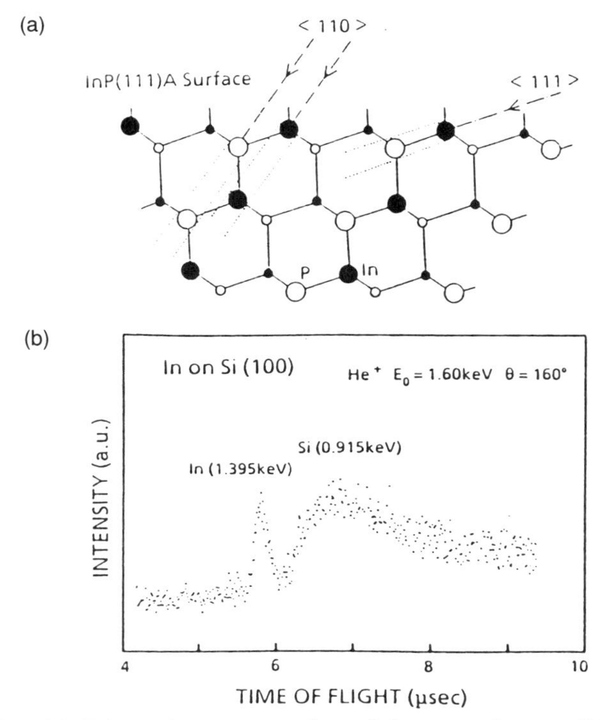

Figure 6.37 (a) Schematic representation of the scattering conditions for an ideally terminated InP surface. (b) LEIS spectrum of indium atoms on a Si(100) surface. After Kubo and Narusawa [109]; copyright © 1990, The American Institute of Physics, New York.

sputtering with 3 keV Ar^+ at 1.5 mA/cm^2 for 10 min, respectively. Oxygen, residual Cl, and C on the untreated surface are removed by the sputtering. In order to remove the sputtering-induced damage, the surface must be annealed, which is done for InP at 365°C. Note the strong effect of the surface cleaning on the amplitude and shape of the P signal, which is a sensitive gauge for surface contamination.

Information concerning the surface structure is conveniently obtained by RHEED, which, in view of the glancing incidence of the electron beam, requires a minimum compromise in the arrangement of the Knudsen cells. Since the intensity of the RHEED signal depends on the surface coverage, that is, is reduced under the conditions of fractional surface coverage and reaches a maximum for complete coverage, the RHEED intensity oscillates upon growth with a period corresponding to the addition of one monolayer. This offers an opportunity for in situ studies of the growth rate as well as the uniformity of the deposition process [110]. Under arsenic-stabilized surface conditions, Ga atoms impinging onto the surface react with arsenic to form islands of GaAs so that abrupt interfaces are difficult to obtain. Improvements in the interface

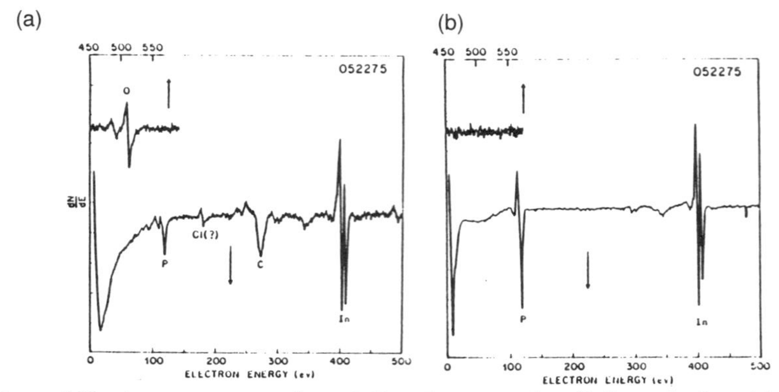

Figure 6.38. Auger spectra of an InP substrate prior (left) and after (right) sputter cleaning. After McFee, Miller, and Bachmann [100]; copyright © 1977, The Electrochemical Society, Penning-ton, NJ.

roughness have been obtained by interrupted MBE growth, allowing the coalescence of surface islands and by migration-enhanced epitaxy (MEE), where the surface is exposed to alternating doses of gallium atoms and arsenic molecules to force monolayer-by-monolayer growth. Excellent control of the interfaces of quantum-well heterostructures is obtained in this manner [127].

Figure 6.39 shows the RHEED specular beam intensity trace for several MEE cycles concerning the growth of GaAs at 580°C. The Ga atom dose is chosen to correspond to exactly one monolayer coverage per cycle, that is, $6.4 \times 10^{14}\,\mathrm{cm}^{-3}$ for GaAs(001). If a different dose per cycle is chosen, islands remaining after the first cycle are filled up in the next cycle and so on until after an appropriate number of cycles a flat interface results. Proof for this self-flattening mechanism is given in Figure 6.39(a), where a dose of 0.8

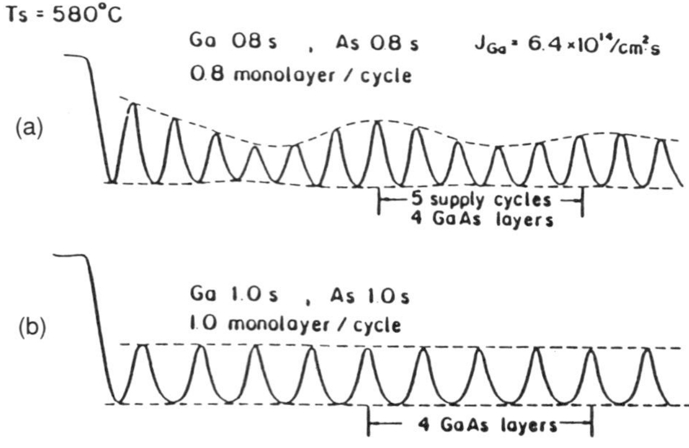

Figure 6.39 Specular beam intensity trace recorded during the growth of GaAs by MEE. After Horikoshi, Kawashima, and Yamaguchi [101]; copyright © 1989, The Electrochemical Society, Pennington, NJ.

monolayer coverage is applied per cycle and a modulation of the specular RHEED intensity trace with a period of 5 cycles is observed.

While at normal growth temperatures the control of the dose of group V molecules is noncritical since any excess is rejected, at low temperature, where As_4 adsorption interferes, a specific dose must be provided to obtain the maximum recovery of the specular RHEED intensity. For GaAs the optimized As_4 dose is half the Ga dose. Figure 6.40 shows the persistence and decay of the RHEED oscillations for optimized and deficient Ga:As_4 flux ratios in the upper and lower traces, respectively. The RHEED pattern, corresponding to the Ga-covered surface at the minima and As-stabilized surface at the maxima in the RHEED intensity oscillations, is shown as an inset into Figure 6.40.

Figure 6.41 shows the RHEED oscillations observed in the growth of GaAs for different widths of the surface area probed by the incident electron beam changing from 13.5 mm in trace (a) to 4.2 mm in trace (d) [111]. The surface-stabilizing arsenic beam is on during the entire duration of the experiments, but the shutter to the Ga cell is opened after 5 s and is closed after 125 s, as indicated by the start of RHEED oscillations and the discontinuity and rise of the signal upon terminating the growth process. A rapid decay of the amplitude of the RHEED oscillations and a beat in the periodic signal are observed. The beat pattern becomes more visible with increasing surface area sampled by the specularly reflected electron beam and is related to the variation of the flux, that is, the growth rate as a function of position on the sample surface [112]. Assuming a linear growth rate variation permits a quantitative interpretation of the beat pattern in terms of the nonuniformity of the growth process, which is ~4% in this particular case. Note that the flux variation depends on the positioning of the effusion cell with regard to the

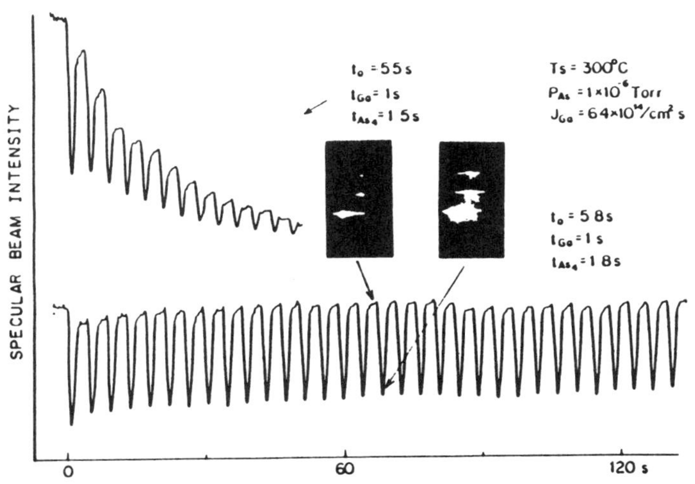

Figure 6.40 RHEED oscillations recorded for constant Ga dose, but two different As_4 doses per cycle during MEE of GaAs. After Horikoshi [101]; copyright © 1989, The Electrochemical Society, Pennington, NJ.

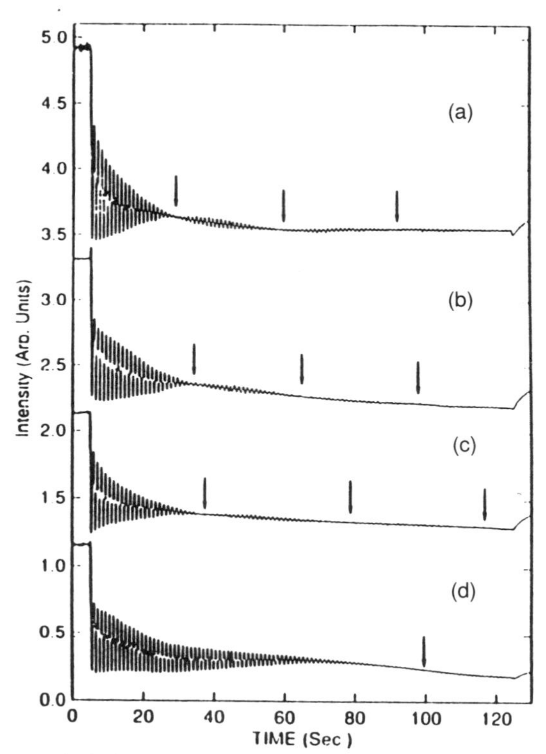

Figure 6.41 Dynamic behavior of the RHEED intensity of the specular diffraction beam as a function of time. Width of the probed area: (a) 13.5 mm, (b) 12.7 mm, (c) 9 mm, and (d) 4.2 mm. Arrows indicate the minima positions in the beat pattern. After Shanabrook, Katzer, and Wagner [112]; copyright © 1991, American Institute of Physics, New York.

substrate. Also the specularly reflected beam measures only the component of this nonuniformity in the direction of $\mathbf{k}_0$, but not perpendicular to this direction. Therefore, knowledge of the alignment of the flux nonuniformity and the specularly reflected beam is important for the interpretation of the results. Using two effusion cells for the same group III element in an appropriate relative position to the substrate allows a correction of the flux variations [113]. Also, the design of the shape of the crucibles containing the source materials inside the Knudsen cell permits improvements in the flux uniformity [114].

In addition to MBE of III–V compounds, there also exists at present a considerable interest in the MBE of group IV elements [115]–[117]. In view of the large latent heat of evaporation of silicon, the conventional Knudsen cell is replaced by electron beam heating for the silicon source. Germanium can be vaporated from conventional source ovens. Dopants are a problem in the context of Si MBE since Al, Ga, In, and Sb, which can be evaporated from conventional cells, have strongly temperature-dependent sticking coefficients. High-temperature Knudsen cells were developed for boron doping, which is well behaved and has become a viable alternative to the group III metals listed. The combination of RHEED with STM studies has revealed details of the epitaxial growth process on both Si and GaAs surfaces.

We consider here the growth on (001) surfaces and vicinal surfaces close to (001) as an example. Figure 6.42(a)–(d) shows schematic representations of two distinct types of surface steps running parallel and perpendicular to the dimer rows along $[\bar{1}10]$ on the Si(001)-2 × 1 surface, respectively. They are labeled by the subscripts A and B, respectively. The steps are identified in Figure 6.42 by dashed lines using the letter S to denote single steps and the letter D to denote double steps [118]. The calculated step formation energies per unit length in eV are $E(S_A) = (0.01 \pm 0.01)/a$, $E(S_B) = (0.15 \pm 0.03)/a$, $E(D_A) = (0.54 \pm 0.1)/a$, and $E(D_B) = (0.05 \pm 0.02)/a$, where a is the lattice

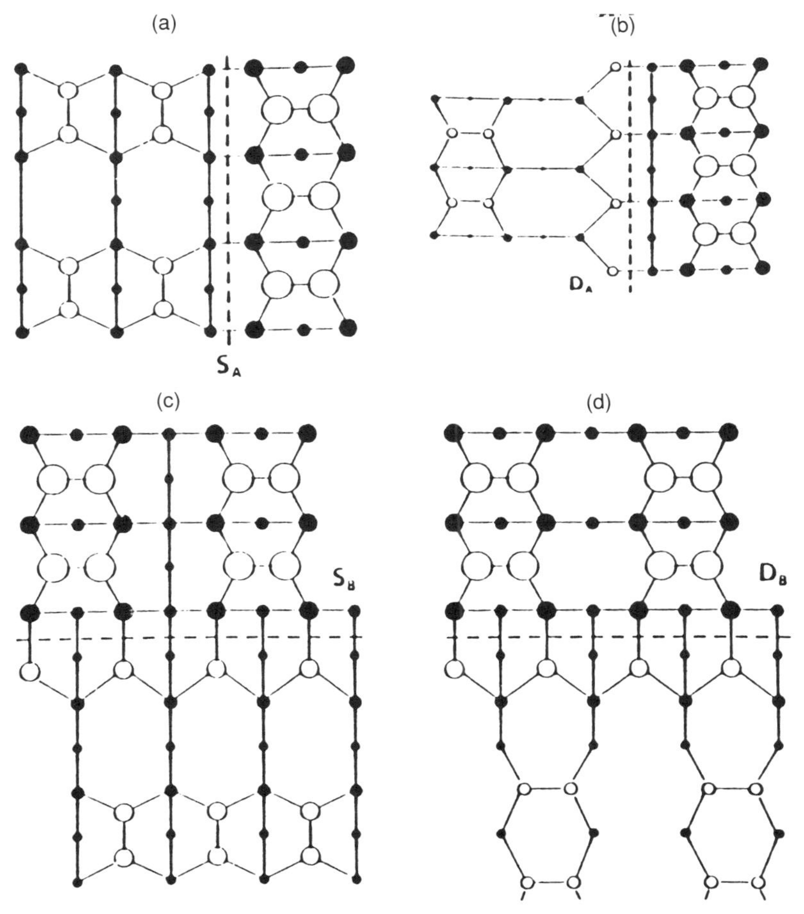

Figure 6.42 Schematic representation of top views of (a) S_A, (b) S_B, (c) D_A, and (d) D_B steps on a Si(001)-2 × 1 surface. After Chadi [118]; copyright © 1987, The American Institute of Physics, New York.

constant of Si and the energies are relative to the fully relaxed Si(001)–2×1 surface and all dimers are asymmetric [118]. For low kink density, the variation of the step energy along a direction at an angle $0 < \Theta < \pi/4$ away from [$\bar{1}$10] and toward [110] is:

$$E = [E(S_A) + E(S_B)](1 + \tan\Theta)\cos\Theta \tag{6.42}$$

$$E = [E(D_A) + E(D_B)\tan\Theta]\cos\Theta \tag{6.43}$$

There exists a critical angle above which single steps become energetically favored to double steps defined by the relation for stability of double layers

$$\tan\Theta < \frac{E(S_A) + E(S_B) - E(D_B)}{E(D_A) - E(S_A) - E(S_B)}, \tag{6.44}$$

that is, $\Theta_{max} = 16°$ for the step formation energies given on p. 339.

Figure 6.43 shows an STM image of a vicinal Si(001) surface that is misoriented by 0.5° in the ⟨110⟩ direction after an exposure to 0.7 ML of Si at 750°C [119]. Growth occurs by both two-dimensional island formation/Ostwald ripening and the addition of atoms on double steps.

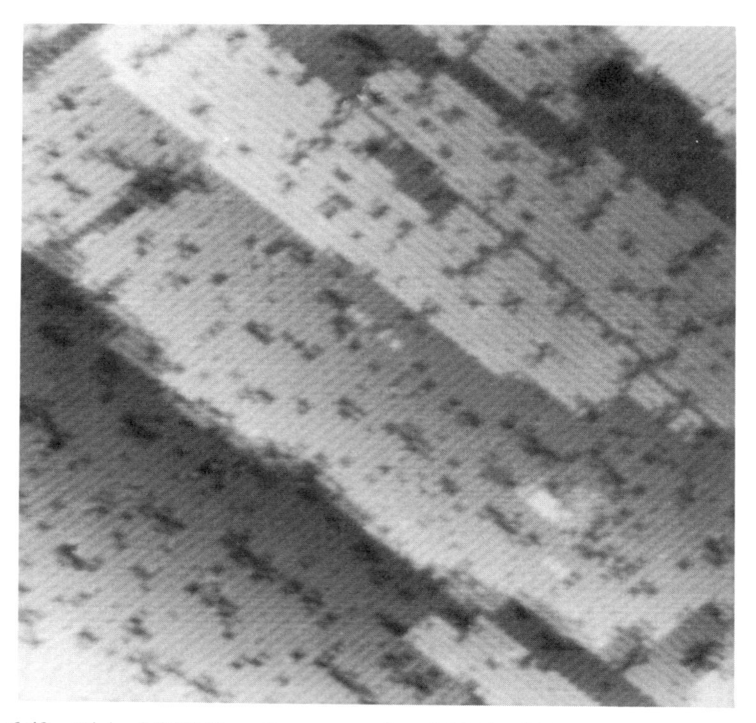

Figure 6.43 Vicinal Si(001) surface exposed to 0.7 ML of Si at 750°C. After Hoeven et al. [119]; copyright © 1990, American Institute of Physics, New York.

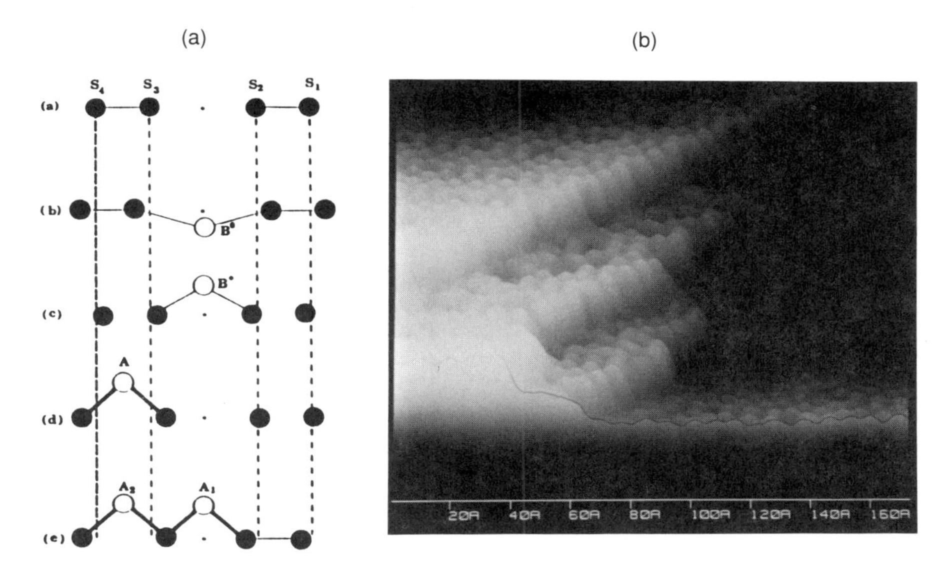

Figure 6.44 Schematic representations of (a) possible positions of ad-atoms on a Si(001) 2×1 surface and (b) STM image of steps on Si(III)-7×7. (a) After Zhang and Metiu [124] and (b) reprinted by permission of Omicron Vakuum Physik GmbH, Taunusslein.

Figure 6.44(a) shows possible positions of ad-atoms on the Si(001)-2×1 surface. Monte Carlo and molecular dynamics calculations [124] show that the dimer atoms bonding to an incoming ad-atom in position A relax into nearly bulk (001) positions. Also, pairs of ad-atoms in positions A_1 and A_2 open the dimer row at the point of contact and result thus in a local unraveling of the 2×1 reconstruction. However, ad-atoms in positions B* and B_0, or dimer atoms that are punched into position B_0 below the surface by an incoming ad-atom, do not undo the existing surface reconstruction. The activation energies for surface diffusion of ad-atoms parallel and perpendicular to the dimer rows are expected to differ, which must be incorporated in the modeling of crystal growth. Also, the attachment energies on steps A and B are different and affect the incorporation of dimers and single ad-atoms at the steps.

Neglecting details of the local changes in the surface reconstruction by the deposition of ad-atoms and tunnelling motion, the energy barrier for the rearrangement of atoms by hopping on the reconstructed surface may be described by the relation

$$E_h = E_s + mE_d + nE_{\parallel} + pE_{\perp} \tag{6.45}$$

where E_s is the activation energy for surface diffusion of adatoms, E_d is the energy of dimer formation, and $E_{\parallel}$ is the interaction energy in directions $\parallel$ to the dimer rows in the same plane. $E_{\perp}$ is the interaction energy in a direction

$\perp$ to the dimer rows between atoms not forming dimers, and the integers $m,p = 0,1$ and $n = 0,1,2$ count numbers of nearest neighbors. A comparison of simulations of MBE growth with the experimental STM images of the Si(001)-2×1 surface results in $E_s = 1.15\,\text{eV}$, $E_d = 0.45\,\text{eV}$, $E_{\parallel} = 0.2\,\text{eV}$, and $E_{\perp} = 0.1\,\text{eV}$ within an error of $\sim \pm 0.1\,\text{eV}$ (see [120] for details).

The contribution to anisotropic growth from incorporation at kinks depends on their density

$$n_{ks} = n_{od} \exp(-E_{ks}/kT) \tag{6.46}$$

where n_{od} is the density of dimer sites per unit length and E_{ks} is the kink energy. Both differ for A and B steps, because $E_{ks}(A) = E_{\parallel}$ and $E_{ks}(B) = E_{\perp}/2$ and $n_{od}(B) = n_{od}(A)/2$. For the values of $E_{\parallel}$ and $E_{\perp}$ given, $n_{ks}(A) < n_{ks}(B)$, so that the propagation of B steps by the incorporation of dimers at kinks is faster than the propagation of A steps. Also, the bonding of dimers to B steps at sites that are not kinks occurs with energy $2E_{\parallel}$, while their bonding to A steps occurs with energy $E_{\perp}$. In the temperature range where a significant difference results in the residence times and capture of additional adatoms at the adsorbed dimers on A and B steps, the latter will propagate faster, explaining in part the observed anisotropy of growth for both terraces and islands that are bound by A and B steps.

Similar observations hold on the Si(111)-7×7 surface, where the amplitude of RHEED oscillations during growth decreases with increasing temperature [121a]. At 700°C, RHEED oscillations cease to exist, and the 7×7 surface reconstruction is maintained during growth up to the edges of steps. Figure 6.44(b) shows an STM image of step edges on a Si(111)-7×7 surface (see also Figure 3.61). At lower temperature (380–600°C), both the 7×7 and a 5×5 structures [121b], [122] exist simultaneously on the surface. An Arrhenius plot of the sizes of these domains, estimated from the length of the streaks in the RHEED pattern, results an energy of formation for these domains of 0.9 eV.

In the MBE growth of GaAs on a GaAs(001) surface under arsenic-rich conditions, both the 2×4 reconstruction and a related $c(2 \times 8)$ structure are observed as illustrated in Figure 3.62. STM studies of MBE growth on vicinal GaAs(001) surfaces tilted by 2° along $\langle 110 \rangle$ and $\langle \bar{1}10 \rangle$ towards the (111)-A and (111)-B surfaces, respectively, reveals that the arsenic dimers in the reconstructed surface are formed along the $\langle \bar{1}10 \rangle$ direction. At a growth temperature of $\sim$600°C, which is a few degrees above the oxide desorption temperature on GaAs, the growth occurs predominantly by step flow; that is, two-dimensional nucleation on the terraces does not contribute to the growth rate. The growth of semiconductor compounds and alloys by MBE on vicinal surfaces offers an opportunity for the construction of quantum wire heterostructures by the restriction of growth to only fractional surface coverage [124]. Since MEE provides for smoother step edges than obtained under the conditions of continuous growth [125], a periodic interruption of the group V flux is advantageous in the growth of fractional-layer superlattices [126].

The perfectly flat interface obtained under the conditions of MBE with interrupted group V flux [127] permits the incorporation of dopant sheets by the appropriate timing of dopant pulses relative to the pulses of group V flux. This process was introduced in 1980 in the context of digital designs of doping profiles, as discussed in Section 6.3 in the context of planar doping under the conditions of OMCVD [128]. It has achieved very high sheet conductance in δ-doped structures, where the deformation of the bands in the vicinity of a highly doped sheet of atoms provides for confinement effects, which may be utilized, for example, for the realization of advanced FET designs [129]. Figure 6.45 shows a schematic representation of the band-edge modulations in a periodic δ-doped GaAs structure with alternating Si- and Be-doped sheets separated by nominally undoped layers. The width of these layers determines the period z_p and thereby the coupling of the V-shaped quantum wells of the sawtooth doping superlattice (SDS). For small period (type-*A* SDS), the wavefunctions strongly overlap, forming bands that are separated by a gap smaller than the bandgap of the host crystal, while for large spacing (type-*B* SDS), the quantum-well states are localized and are separated in energy by larger values than the bandgap of the host crystal. This permits the design of a variety of device structures with potential applications in optoelectronics and microelectronics, which are reviewed in Ref. [129].

A further enhancement in the surface mobility and gains in the flexibility of the precursor and process designs are achieved by replacing the ballistic beams of metal atoms escaping from high-temperature Knudsen cells by molecular beams of volatile organometallic compounds. This method of chemical beam epitaxy (CBE) [130] also has the virtue of permitting the storage and flow control of the organometallic sources outside the UHV chamber. Figure 6.46 shows a schematic representation of a CBE system for growth of compound semiconductors [130]–[134].

Although the control of the carbon background level in the epitaxial GaAs

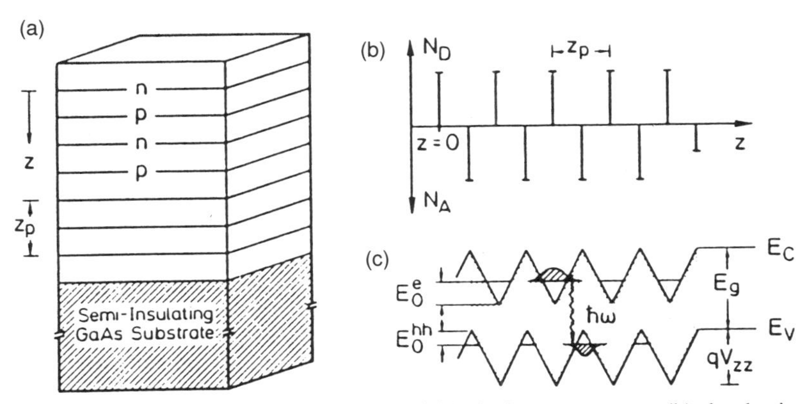

Figure 6.45 Schematic representations of (a) the layer structure, (b) the doping profile, and (c) the band-edge modulation in a GaAs SDS. After Ploog, Hauser, and Fischer [129]; copyright © 1988, Springer Verlag, Berlin.

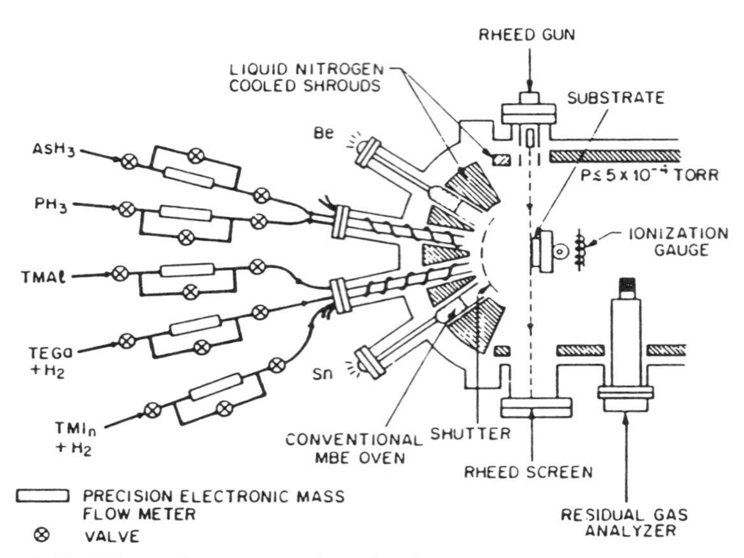

Figure 6.46 Schematic representation of a chemical beam epitaxy system. After Tsang [131]; copyright © 1987, Elsevier Science Publishers B.V., Amsterdam.

films is a matter of concern, high-resolution device-quality heteroepitaxial structures have been demonstrated [135]. Using triethyl gallium and arsine, net acceptor concentrations of $1.2 \times 10^{15}\,\text{cm}^{-3}$ and hole mobility $m_h = 400\,\text{cm}^2/\text{V s}$ have been attained in nominally undoped GaAs [136]. On the other hand, completely ionized C doping has been observed up to $N_A - N_D = 1.5 \times 10^{21}\,\text{cm}^{-3}$ under the conditions of CBE without need for halomethane injection [137]. *n*-type Si doping, using a conventional thermal source for the dopant, has been achieved for both InP [138] and GaAs [139].

Since the ballistic beam conditions of CBE provide for the clean separation of sequentially applied metal alkyl and group V hydride pulses, it is ideally suited for the testing of ALE mechanisms. The residence time of the alkyl radicals on the surface of a GaAs surface exposed to a gallium alkyl flux depends on both their chemical nature and the substrate temperature [140]. A self-limiting adsorption mechanism is only observed under the conditions of interrupted-cycle CBE (ICCBE), when the surface is protected by alkyl groups, that is, at low temperature and for a limited time interval between the group III and group V precursor pulses [141]. Planar doping under the conditions of ICCBE of GaAs using dimethyl cadmium, diethyl zinc, disilane, diethyl selenium, and diethyl tellurium as dopant sources has been demonstrated [142] and is an interesting aspect of this method since the timing of the dopant precursor injection relative to the sequence of trimethyl gallium and arsine pulses determines the incorporation efficiency. CBE lends itself also to light assisted ALE, as discussed in Section 6.3. For example, the reaction of trimethylgallium with an arsenic-stabilized surface of GaAs may be triggered by Ar^+ laser pulses [143]. This may become important for the controlled heteroepitaxy of III–V compounds on a Si surface at low temperatures that prevent interdiffusion and minimize the strain due to the mismatch of the thermal expansion coefficients.

Conventional CBE growth of epitaxial nearly lattice-matched GaP on Si has been achieved at 310°C [144], and Si/GaP/Si double heterostructures have been prepared by a combination of CBE and RPCVD [145]. Since large bandoffsets exist at the Si/GaP interfaces (compare Fig 4.3), GaP in combination with low dielectric constant interlayers may become useful in the context of epitaxial dielectric isolation of silicon and optically interconnected ICs, which are discussed in more detail in Chapter 9.

References

1. G. R. Srinivasan, *J. Crystal Growth* **70**, 201 (1984).
2. C. J. Varker and K. V. Ravi, *J. Appl. Phys.* **45**, 272 (1974).
3. H. Schafer, *Chemical Transport Reactions*, Academic Press, New York, 1964.
4. *JANAF Thermochemical Tables*, 2nd ed., D. R. Stull and H. Prophet, eds., NBS NSRDS-NBS 37, Washington, 1958; supplements to Ref. [] by M. W. Chase, J. L. Curnut, A. T. Hu, H. K. E. Spear, in *Chemical Vapor Deposition 1984*, McD. Robinson, C. H. J. van den Brekel, G. W. Cullen, J. M. Blocher, Jr., and P. Rai-Choudhury, eds., The Electrochemical Society, Pennington, 1984, p. 98.
6. J. F. C. Eversteyn, *Philips Res. Repts.* **29**, 45 (1974).
7. I. Langmuir, *J. Am. Chem. Soc.* **38**, 2221 (1916).
8. R. M. Barrer, *Trans. Faraday Soc.* **32**, 496 (1936).
9. I. Langmuir, *Trans. Faraday Soc.* **17**, 621 (1921).
10. C. N. Hinshelwood, *Kinetics of Chemical Change in Gaseous Systems*, Clarendon Press, Oxford, 1926.
11. E. K. Rideal, *Proc. Cambr. Philos. Soc.* **35**, 130 (1939).
12. J. C. Slater, *Quantum Theory of Molecules and Solids*, Vol. 4, McGraw–Hill, New York, 1974.
13. J. M. Ziman, *Principles of the Theory of Solids*, Cambridge University Press, London, 1964.
14. J. C. Slater and K. H. Johnson, *Physics Today*, 34 (1974).
15. S. J. Tauster, S. C. Chung, R. T. K. Baker, and J. A. Horsley, *Science* **211**, 1121 (1981).
16. K. Mortensen, D. M. Chen, P. J. Bedrossian, J. A. Golovchenko, and F. Besenbacher, *Phys. Rev. B* **43**, 1816 (1991).
17. M. E. Coltrin, R. J. Kee, and J. A. Miller, *J. Electrochem. Soc.* **131**, 425 (1984).
18. K. J. Uram and U. Jansson, *Surface Sci.* **249**, 105 (1991).
19. J. J. Boland, *Phys. Rev. B* **44**, 1383 (1991).
20. M. E. Coltrin, R. J. Kee, and J. A. Miller, *J. Electrochem. Soc.* **133**, 1206 (1985).
21. M. E. Coltrin, R. J. Kee, and Evans, *J. Electrochem. Soc.* **135**, 819 (1989).
22. H. K. Moffatt and K. E. Jensen, *J. Electrochem. Soc.* **135**, 4549 (1988).

23. J. R. Knight, D. Effer, and P. R. Evans, *Solid State Electronics* **8**, 178 (1965).

24. J. Tietjen and J. Amick, *J. Electrochem. Soc.* **113**, 724 (1966).

25. D. W. Shaw, *J. Electrochem. Soc.* **115**, 405 (1968).

26. D. W. Shaw, *J. Electrochem. Soc.* **117**, 683 (1970).

27. O. Mizuno, *Japan. J. Appl.* **14**, 451 (1975).

28. R. C. Clarke, *Inst. Phys. Conf. Ser.* **45**, 19 (1979).

29. K. Fairhurst, D. Lee, D. S. Robertson, H. D. Parfitt, and W. H. E. Wilgoss, *J. Mater. Sci.* **16**, 1031 (1981).

30. C. M. Wolfe, G. E. Stillman, and E. B. Brown, *J. Electrochem. Soc.* **117**, 129 (1970).

31. L.-Y. Lin, Y.-W. Lin, X.-R. Zhong, Y.-Y. Zhang, and H. L. Li, *J. Crystal Growth* **56**, 344 (1982).

32. J. K. Abrokwah, T. N. Peck, B. A. Walterson, G. E. Stillman, T. S. Low, and B. Skromme, *J. Electron. Mater.* **12**, 681 (1983).

33. S. W. Sun, A. P. Constant, C. D. Adams, and B. W. Wessels, *J. Crystal Growth* **64**, 140 (1983).

34. K. F. Jensen and D. B. Graves, *J. Electrochem. Soc.* **130**, 1950 (1983).

35. J. F. Corboy, G. W. Cullen, and R. H. Pagliaro, Jr., *Proc. 9th Int. Conf. on CVD*, Cincinatti, OH, The Electrochemical Society, Pennington, NJ, 1984, p. 434.

36. J. P. Duchemin, M. Bonnet, and F. Kroelch, *J. Electrochem. Soc.* **125**, 637 (1978).

37. W. C. Benzing, *Proc. 9th Int. Conf. on CVD*, Cincinatti, OH, The Electrochemical Society, Pennington, NJ, 1984, p. 373.

38. M. J. Kushner, *J. Appl. Phys.* **63**, 2532 (1988).

39. L. M. Garverick and R. Reif, *J. Electrochem. Soc.* **135**, 2620 (1988).

40. W. R. Burger and R. Reif, in *Processing of Electronic Materials*, C. G. Law and R. Pollard, eds., The Electrochemical Society, Pennington, 1987, p. 153.

41. (a) L. Breaux, B. Anthony, T. Hsu, S. Banerjee, and A. Tasch, *Appl. Phys. Lett.* **55**, 1885 (1989); (b) T. Hsu, B. Anthony, L. Breaux, R. Qian, S. Banerjee, and A. Tasch, *Proc. TSM Ann. Meeting*, February 19, 1990.

42. R. A. Rudder, G. G. Fountain, and R. J. Markunas, *J. Appl. Phys.* **60**, 3590 (1986).

43. W. Richter, P. Kurpas, R. Lückerath, M. Motzkus and R. Waschbüsch, J. Crystal Growth 107, 13 (1991).

44. J. Batey and E. Tierney, *J. Appl. Phys.* **60**, 3136 (1986).

45. G. Lucovsky, S. S. Kim, and J. T. Fitch, *J. Vac. Sci. Technol. B* **8**, 822 (1990).

46. L. G. Meiners, *J. Vac. Sci. Technol.* **21**, 655 (1982).

47. M. R. Wertheimer and M. Moissan, *J. Vac. Sci. Technol. A* **3**, 2643 (1983).

48. D. Stelle, in *Advances in Infrared and Raman Spectroscopy*, R. J. Clark and R. E. Hester, eds., Heyden, London, 1975, Vol. 1, p. 232.

49. J. Sadlej, *Semi-Empirical Methods of Quantum Chemistry*, Polish Scientific Publishers, Warszawa, 1985.

50. D. V. Tsu, *Thesis*, under the direction of G. Lucovsky, North Carolina State University, Raleigh, NC 1989.

51. G. Lucovsky, M. J. Mantini, J. K. Srivastava, and E. E. Irene, *J. Vac. Sci. Technol. B* **5**, 530 (1987).

52. A. D. Huelsman, R. Reif, and C. G. Fronstad, *Appl. Phys. Lett.* **50**, 206 (1987).

53. A. D. Huelsman and R. Reif, *J. Electron. Mater.* **18**, 91 (1989).

54. A. D. Huelsman and R. Reif, *J. Vac. Sci. Technol. A* **7**, 2554 (1989).

55. S. W. Choi, K. J. Bachmann, and G. Lucovsky, *J. Vac. Sci. Technol. B* **10**, 1070 (1992).

56. M. Naitoh, T. Soga, T. Jimbo, and M. Umeno, *J. Crystal Growth* **93**, 52 (1988).

57. T. R. Omstead, A. V. Annapragada, and K. F. Jensen, *Appl. Phys. Lett.* **57**, 2543 (1990).

58. H. M. Manasevit, *Appl. Phys. Lett.* **12**, 156 (1968).

59. R. Bhat and S. Mahajan in *Concise Encyclopedia of Semiconducting Materials and Related Technologies*, S. Mahajan and L. C. Kimerling, Pergamon Press, Ltd., Oxford 1992, p. 349.

60. G. B. Stringfellow (a) *J. Crystal Growth* **68**, 111 (1984); (b) *ibid.* **105**, 260 (1990).

61. E. J. Johnson, P. V. Hyer, P. W. Culotta, L. R. Black, I. O. Clark, and M. L. Timmons, *J. Crystal Growth* **109**, 24 (1991).

62. L. R. Black, I. O. Clark, B. A. Fox, and W. A. Jesser, *J. Crystal Growth* **109**, 241 (1991).

63. M. Z. Sheikholeslami, T. Jasinski, and K. W. Fretz, in *Proc. AIAA 1st Natl. Fluid Dynamics Congr.,* Cincinnati, OH, 1988, p. 1616.

64. M. E. Coltrin and R. J. Lee, *Mater. Res. Soc. Symp. Proc.* **145**, 119 (1989).

65. Y. Monteil, M. P. Berthet, R. Favre, A. Harris, J. Bouix, M. Vaille, and P. Gibart, *J. Crystal Growth* **77**, 172 (1986).

66. M. Mashita, S. Horiguchi, M. Shimitsu, K. Kamon, M. Mihara, and M. Ishii, *J. Crystal Growth* **77**, 194 (1986).

67. S. A. J. Druet and J. -P. E. Taran, *Progr. Quant. Electr.* **7**, 1 (1981).

68. R. Lückerath, P. Tommack, A. Hertling, H. J. Koss, P. Balk, K. F. Jensen, and W. Richter, *J. Crystal Growth* **93**, 151 (1988).

69. R. Lückerath, H. -J. Koss, P. Tommack, W. Richter, and P. Balk, *Mat. Res. Soc. Symp. Proc.* **131**, 91 (1989).

70. T. F. Kuech and E. Veuhoff, *J. Crystal Growth* **68**, 148 (1984).

71. T. F. Kuech, G. J. Scilla, and F. Cardone, *J. Crystal Growth* **93**, 550 (1988).

72. A. J. Murrell, A. T. S. Wee, D. H. Fairbrother, N. K. Singh, J. S. Foord, G. J. Davis, and D. A. Andrews, *J. Crystal Growth* **105**, 199 (1990).

73. T. Maeda, J. Saito, and K. Kondo, *J. Crystal Growth* **105**, 191 (1990).

74. R. Karlicek, J. A. Long, and V. M. Donelly, *J. Crystal Growth* **68**, 123 (1984).

75. (a) A. M. Arif, B. L. Benac, A. H. Cowley, R. L. Geerts, R. A. Jones, K. B. Kidd, J. M. Power, and S. T. Schwab, *J. Chem. Soc. Chem. Commun.* 1543 (1988); (b) A. H. Cowley, B. L. Benac, J. C. Eckert, R. A. Jones, K. B. Kidd, J. Y. Lee, and J. E. Miller, *J. Am. Chem. Soc.* **110**, 628 (1988).

76. E. Luthardt and H. Jürgensen, *SEMICON 1988*, Zürich, Switzerland.

77. M. J. Ludowise, *J. Appl. Phys.* **58**, R31 (1985).

78. R. M. Lum, J. K. Klingert, and M. G. Lamont, *J. Crystal Growth* **89**, 137 (1988).

79. G. T. Muhr, D. A. Bohling, T. R. Omstead, S. Brandon, and K. F. Jensen, *Chemtronics* **4**, 26 (1989).

80. G. B. Stringfellow, *J. Electron. Mater.* **17**, 327 (1988).

81. M. L. Cotton, N. D. Johnson, and K. G. Wheeland, *Removal of Arsine from Process Emission*, The Metallurgical Society of CIM Annual Volume, 1977, p. 205.

82. R. Bhat, M. A. Koza, and B. J. Skromme, *Appl. Phys. Lett.* **50**, 1194 (1987).

83. A. Brauers, O. Kayser, R. Kall, H. Heinecke, and P. Balk, *J. Crystal Growth* **93**, 7 (1988).

84. C. H. Chen, C. A. Larsen, G. B. Stringfellow, D. M. Brown, and A. J. Robertson, *J. Crystal Growth* **77**, 11 (1986).

85. S. H. Li, C. H. Chen, D. H. Jaw, and G. B. Stringfellow, *Appl. Phys. Lett.* **59**, 2124 (1991).

86. M. A. Di Forte-Poisson, C. Brylinski, and J. P. Duchemin, *Appl. Phys. Lett.* **46**, 476 (1985).

87. R. T. Huang, A. Appelbaum, D. Renner, and S. W. Zehr, *Appl. Phys. Lett.* **58**, 170 (1991).

88. M. Akiyama, Y. Kawarada, and K. Kaminishi, *J. Crystal Growth* **68**, 39 (1984).

89. P. M. Frijlink and J. Maluenda, *Jpn. J. Appl. Phys.* **21**, L514 (1982).

90. R. D. Dupuis, *J. Crystal Growth* **55**, 213 (1981).

91. H. Ohno, E. Ikeda, and H. Hasegawa, *J. Crystal Growth* **68**, 15 (1984).

92. B. I. Miller, E. F. Schubert, U. Koren, A. Ourmazd, H. Dayem, and R. I. Capik, *Appl. Phys. Lett.* **49**, 1384 (1986).

93. R. C. Miller, R. D. Dupuis, and P. M. Petroff, *Appl. Phys. Lett.* **44**, 508 (1984).

94. T. Suntola and J. Antson, Finnish Patent No. 52359 (1974); U.S. Patent No. 4,058,430 (1977).

95. T. A. Pakkanen, V. Nevalainen, M. Lindblad, and P. Makkonen, *Surface Science* **188**, 456 (1987).

96. T. Suntola, *Materials Science Reports* **4**, 261 (1989).

97. M. A. Tischler and S. Bedair, *J. Crystal Growth* **77**, 89 (1986).

98. Y. Suda, D. Lubben, T. Motooka, and J. E. Greene, *J. Vac. Sci. Technol. B* **7**, 1171 (1989); *ibid. A* **8**, 61 (1990).

99. J. R. Arthur, *J. Appl. Phys.* **39**, 4032 (1968).

100. J. H. McFee, B. I. Miller, and K. J. Bachmann, *J. Electrochem. Soc.* **124**, 359 (1977).

101. Y. Horikoshi, M. Kawashima, and H. Yamaguchi, in *Heteroepitaxial Approaches in Semiconductors: Lattice Mismatch and its Consequences*, A. T. Macrander and T. J. Drummond, eds., The Electrochemical Society, Pennington, 1989, p. 358.

102. N. Chand, *Appl. Phys. Lett.* **56**, 466 (1990).

103. M. B. Panish (a) *J. Electrochem. Soc.* **127**, 2729 (1980); (b) *J. Crystal Growth* **81**, 249 (1987).

104. E. C. Larkins, E. S. Hellman, D. G. Schlom, J. S. Harris, Jr., M. H. Kim, and G. E. Stillman, *J. Crystal Growth* **81**, 345 (1987).

105. T. J. de Lyon, J. M. Woodall, J. A. Kash, D. T. McInturff, R. J. S. Bates, P. D. Kirchner, and F. Cardone, *J. Vac. Sci. Technol. B* **10**, 846 (1992).

106. A. Y. Cho and I. Hayashi, *J. Appl. Phys.* **42**, 4422 (1971).

107. D. J. Chadi and K. J. Chang, *Phys. Rev. B* **39**, 1063 (1989); *Phys. Rev. B* **39**, 1063 (1989).

108. J. M. Salese, D. K. Maude, M. L. Fille, U. Willke, J. C. Portal, and P. Gibart, in *Non-Stoichiometry in Semiconductors*, K. J. Bachmann, H. -L. Hwang, and C. Schwab, eds., Elsevier Science Publ. B.V., Amsterdam, 1992, p. 229.

109. M. Kubo and T. Narusawa, *J. Vac. Sci. Technol. B* **8**, 697 (1990).

110. J. J. Harris, B. A. Joyce, and P. J. Dobson, *Surf. Sci.* **103**, L90 (1981).

111. J. M. van Hove, P. R. Pukite, and P. I. Cohen, *J. Vac. Sci. Technol. B* **3**, 563 (1985).

112. B. V. Shanabrook, D. S. Katzer, and R. J. Wagner, *Appl. Phys. Lett.* **59**, 1317 (1991).

113. J. P. A. van der Wagt, K. L. Bacher, G. S. Solomon, and J. S. Harris, Jr., *J. Vac. Sci. Technol. B* **10**, 825 (1992).

114. G. C. Ayers and Z. R. Wasilewski, *J. Vac. Sci. Technol. B* **10**, 815 (1992).

115. J. C. Bean, *J. Crystal Growth* **81**, 411 (1987).

116. E. Kasper, H. J. Herzog, and K. Wörner *J. Crystal Growth* **81**, 458 (1987).

117. E. Kasper, *Festkörperprobleme* **27**, 265 (1989).

118. J. Chadi, *Phys. Rev. Lett.* **59**, 1691 (1987).

119. A. J. Hoeven, D. Dijkkamp, E. J. van Loenen, J. M. Lessink, and J. Dieleman, *J. Vac. Sci. Technol. A* **8**, 207 (1990).

120. H. B. Elswijk, A. J. Hoeven, E. J. van Loenen, and D. Dijkamp, *J. Vac. Sci. Technol. B* **9**, 451 (1991).

121. H. Nakahara, and A. Ichimiya (a) *Surface Sci.* **241**, 124 (1991); (b) *J. Crystal Growth* **95**, 472 (1989).

122. M. Ichikawa and T. Doi, in *Reflection High-Energy Electron Diffraction and Reflection Electron Imaging of Surfaces*, P. K. Larsen and P. J. Dobson, eds., Plenum Press, New York, 1988, p. 343; *Appl. Phys. Lett.* **50**, 1141 (1987).

123. M. D. Pashley, K. W. Harberern, and J. M. Gaines (a) *J. Vac. Sci. Technol. B* **9**, 938 (1991); (b) *Appl. Phys. Lett.* **58**, 406 (1991).

124. Z. Zhang and H. Metiu, *Surface Sci.* **245**, 353 (1991).

125. F. Briones, D. Golmayo, L. Gonzalez, and A. Ruiz, *J. Crystal Growth* **81**, 19 (1987).

126. J. M. Gaines, P. M. Petroff, H. Kroemer, R. J. Simes, R. S. Geels, and J. H. English, *J. Vac. Sci. Technol. B* **6**, 1378 (1988).

127. Y. Horikoshi, M. Kawashima, and H. Yamaguchi, *Jpn. J. Appl. Phys.* **25**, L868 (1986).

128. C. E. C. Wood, G. Metze, J. Berry, and L. F. Eastman, *J. Appl. Phys.* **51**, 383 (1980).

129. K. Ploog, M. Hauser, and A. Fischer, *Appl. Phys. A* **45**, 233 (1988).

130. W. T. Tsang, *Appl. Phys. Lett.* **45**, 1234 (1984).

131. W. T. Tsang, *J. Crystal Growth* **81**, 261 (1987).

132. V. J. Mifsud, P. W. Sullivan, and D. Williams, *J. Crystal Growth* **105**, 289 (1990).

133. S. D. Hersee and J. M. Ballingall, *J. Crystal Growth* **105**, 282 (1990).

134. T. B. Joyce, *J. Crystal Growth* **105**, 299 (1990).

135. W. T. Tsang, A. F. J. Levi, and E. G. Burkhardt, *Appl. Phys. Lett.* **53**, 983 (1988).

136. S. Horiguchi, K. Kimura, K. Kamon, M. Mashita, M. Shimazu, M. Mihara, and M. Ishii, *Jpn. J. Appl. Phys.* **25**, L979 (1986).

137. M. Konagai, T. Yamada, T. Akatsuka, S. Nazaki, R. Miyake, K. Saito, T. Fukamachi, E. Tokumitsu, and K. Takahashi, *J. Crystal Growth* **105**, 359 (1990).

138. W. T. Tsang, B. Tell, J. A. Ditzenberger, and A. H. Dayem, *J. Appl. Phys.* **60**, 4182 (1986).

139. M. J. Collum, S. L. Jackson, I. Szafranek, and G. E. Stillman, *J. Crystal Growth* **105**, 316 (1990).

140. M. Yu, U. Memmert, and T. F. Kuench, *Appl. Phys. Lett.* **55**, 1011 (1989).

141. T. H. Chiu, J. E. Cunningham, A. Robertson, Jr., and D. L. Malm, *J. Crystal Growth* **105**, 155 (1990).

142. J. -I. Nishizawa, H. Abe, and T. Kurabayashi, *J. Electrochem. Soc.* **136**, 478 (1989).

143. K. Nagata, Y. Iimura, Y. Aoyagi, and S. Namba, *J. Crystal Growth* **105**, 52 (1990).

144. J. T. Kelliher, N. Dietz, J. Thornton, G. Lucovsky, and K J. Bachmann, *Materials Science and Engineering B* **22**, 97 (1993).

CHAPTER

7

Pattern Definition and Etching

7.1 Photolithographic Pattern Definition

Pattern transfer and doping are essential steps in integrated circuit manufacturing and are usually applied at several stages in the processing sequence. For example, the CMOS process discussed in Section 4.4 requires 11 lithographic steps and 6 ion implantation/diffusion steps. In this section, we discuss photolithography as well as modern pattern definition methods that are based on direct electron beam and ion beam writing and micromachining techniques. Also, wet and dry etching are discussed in the context of both patterning and surface conditioning, including a brief discussion of vacuum microelectronics based on field emission from arrays of semiconductor or metal tips of nanometer radii of curvature.

Conventional photolithographic pattern definition involves the application of a few drops of photoresist solution onto the surface of a substrate. The substrate is held in place by a vacuum chuck and is rotated at high angular velocity to spread the resist uniformly over the entire wafer, as illustrated in Figure 7.1(a). Subsequently it is transferred into a prebake oven to drive out the solvent, thus creating a well-adhering polymer coating. The wafer is then transferred into a mask aligner. There the photomask, which contains the desired pattern in the form of transparent regions in a translucent film on a glass plate, is aligned with regard to previously established features of the evolving circuit. There are three possible configurations of the mask during exposure: (1) Contact printing, where the mask is in contact with the wafer; (2) proximity printing, where a gap is provided between the mask and the wafer;

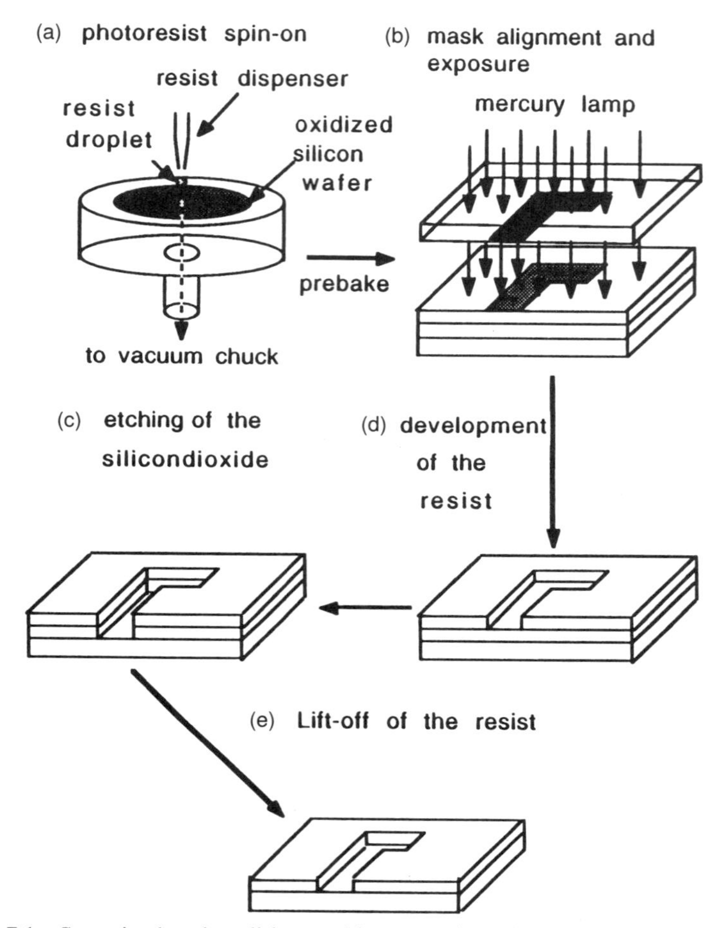

Figure 7.1 Steps in the photolithographic patterning of a semiconductor.

and (3) projection printing, where a demagnifying optics is inserted in between the mask and the wafer. Illumination through the mask, usually accomplished by a mercury lamp, transfers the pattern as a latent image into the resist layer, which is developed in a suitable developer. This developer removes either the exposed or unexposed regions of the resist, depending on whether a positive or negative tone resist material is chosen for the process. After rinsing to remove the developer solution and drying, a postbake is applied to condition the resist remaining on the wafer for the subsequent etching step, which transfers the pattern into the underlying dielectric. After etching, the resist is stripped off, completing the lithographic processing step.

The overall error in the lithographic pattern transfer is composed of the individual errors of the mask-making process, mask alignment, exposure, the swelling and shrinking of the resist during development and postbaking, and the errors in the transfer of the resist pattern into the dielectric upon etching. In the current implementation of the process, each of these sources of error is significant. The resolution of conventional photolithography is restricted by the diffraction limit to 0.5 μm with conventional high-intensity mercury lamps. Excimer lasers are available for achieving a resolution between 0.25 and 0.5 μm, pushing the limits of deep ultraviolet (DUV) lithography in the fabrication of 64 M circuits, as illustrated in Figure 7.2. Typical specifications for a KrF laser source are: $\lambda = 248.4$ nm, $\Delta\lambda$(FWHM) = 0.005–0.007 nm, reduction ratio 5:1, numerical aperature 0.35–0.40, and a field size of 20–21.2 mm diameter [4].

A problem of DUV exposure systems is the difficulty of mask alignment, which must use the same DUV light as applied in processing, because the lens system is corrected for chromatic aberations at only this wavelength, and the pulsed operation of the excimer laser makes the detection of the alignment position difficult. Super-resolution optical lithography enhances the depth of focus (DOF) and edge contrast by annular/quadrupole illumination, phase shifting masks, and high spatial frequency enhancing filters, as illustrated in Figure 7.3(a). Figure 7.3(b) and (c) show the gains by these techniques in the DOF for a given pattern size and in the practical resolution for a given critical contrast. Patterns of 0.2–0.3 μm resolution are achieved with little restrictions to conventional circuit designs that suffice for 64 Mbit memory. For the realization of feature sizes below this limit, that is, 256 Mbit circuits and beyond, alternative ultrahigh-resolution lithography methods must be developed (Section 7.2).

Generally, photoresists are either one- or two-component organic materials that upon exposure change their solubility in a suitable developing solution.

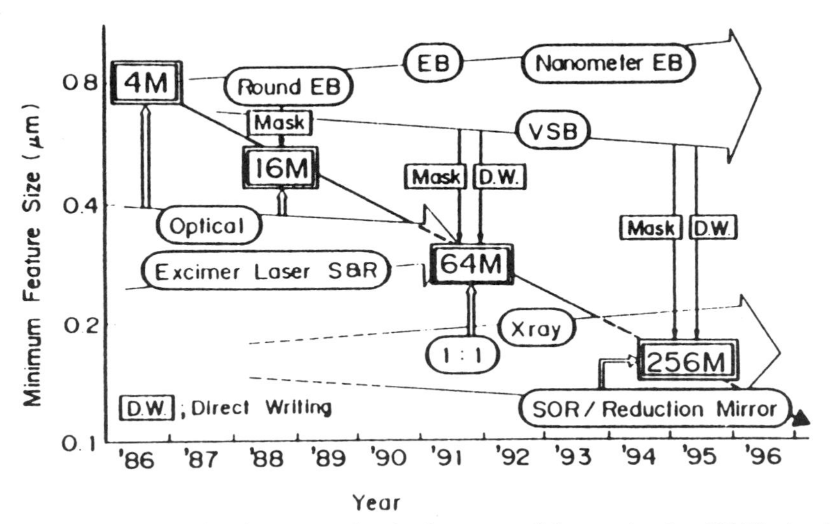

Figure 7.2 Projected development of submicrometer lithography for ULSI circuits. After Horiike et al. [1]; copyright © 1987, Materials Research Society, Pittsburgh, PA.

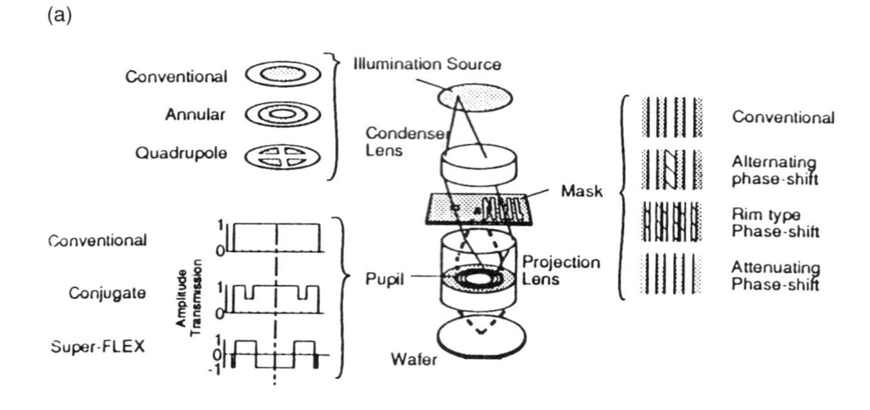

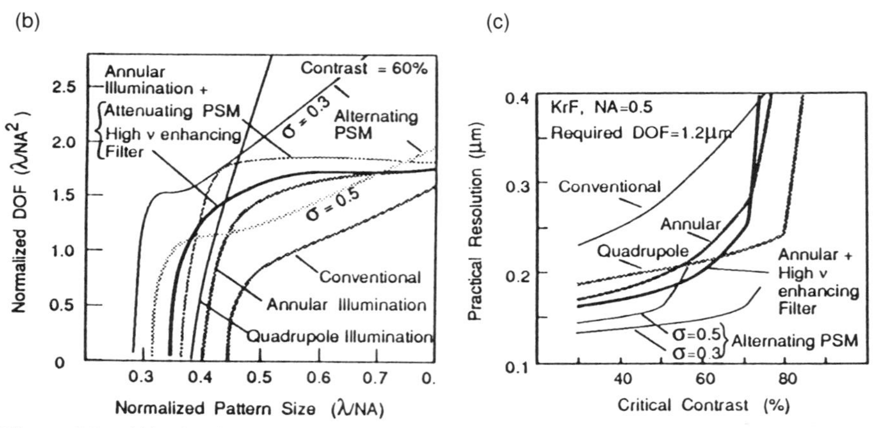

Figure 7.3 (A) Optical elements that are used to improve the resolution of uv lithography, (B) normalized DOF versus the normalized pattern size, and (C) the practical resolution versus the critical contrast for various DOF and edge-contrast enhancing methods. After Fukuda et al. [3]; copyright © 1992, The Institute of Elecrical and Electronics, New York.

The most common positive photoresists are two-component materials consisting of a phenolic polymer (novolac) and a light-sensitive inhibitor, which renders the mixture insoluble in an aqueous base developer. Novolac resins are formed by polymerization of formaldehyde and phenols with various substituents. For example, using *m*-cresol one obtains a resist of the formula

$$\text{HO-C}_6\text{H}_4\text{-CH}_3 + \text{H}_2\text{C}=\text{O} \xrightarrow{\text{H}^+} \left(\text{C}_6\text{H}_2(\text{OH})(\text{CH}_3)\text{-CH}_2\text{-C}_6\text{H}_2(\text{OH})(\text{CH}_3)\text{-CH}_2 \right)_n \tag{7.1}$$

Novolacs typically have molecular weights of 2000–12000 Dalton. Both the novolac and the inhibitor, which is a substituted diazonaphthoquinone, are soluble in organic solvent, forming a viscous solution for spin coating. The exposure to light does not affect the novolac component, but initiates photodecomposition of the diazo compound, which adsorbs in the wavelength region where intense mercury lines exist. Figure 7.4(a) shows the absorbance of a cresylic acid novolac film and of a typical diazonaphthoquinone sensitizer as an example.

The photodecomposition of α-diazoketones is initiated by the excitation of the molecule and elimination of the diazo nitrogen upon absorption of a photon. The excited singlet ketocarbene radical thus generated relaxes to the ground state [5], which remains vibrationally excited and undergoes a Wolff rearrangement of bonds [6] into a ketene with oxirene as a possible energetic transition state. This is illustrated in Figure 7.4 on the basis of an extended Hückel MO calculation of the enthalpies of the intermediates in the photo- and thermal decomposition for the simple example of 1-diazo-2-ethanone. The initial product of the thermal decomposition, formylmethylene and its zwitterion, can either decompose via the oxirene or the bridged intermediate. The latter has significantly lower enthalpy and constitutes an orbital symmetry allowed transition state in the formation of the ketene so that the photo- and thermal decomposition pathways differ. In the case of 1-oxo-2-diazonaphthoquinone, the ketene reacts with residual water in the resist to form indene carboxylic acid according to the abbreviated scheme

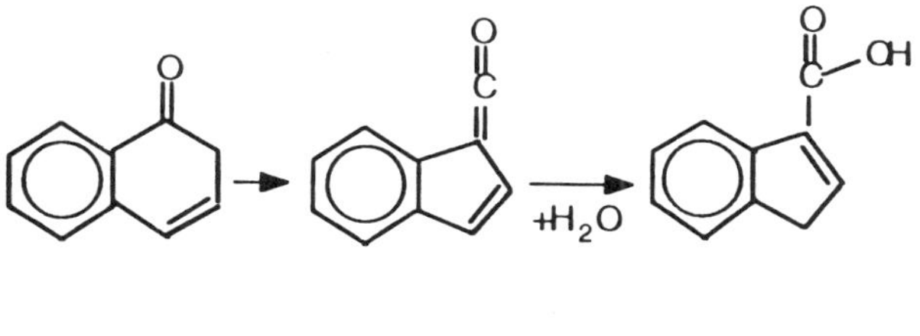

$+H_2O$ (7.2)

Because of the acidic carboxyl group, this end product dissolves readily in aqueous base, rendering the exposed resist areas soluble. The positive novolac resist can be converted into a negative tone resist by doping with monazoline, which upon exposure and baking decarboxylates the indene carboxylic acid according to the reaction

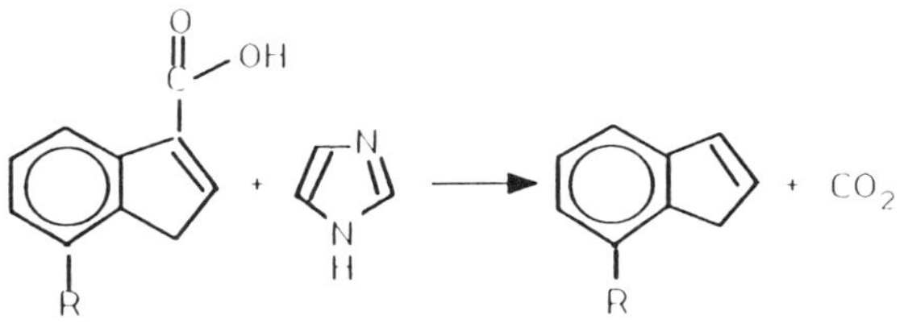

$+ CO_2$ (7.3)

(a)

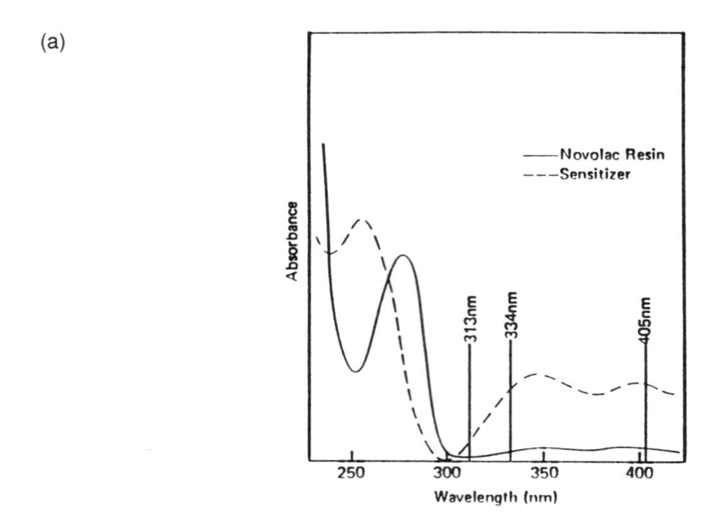

(b)

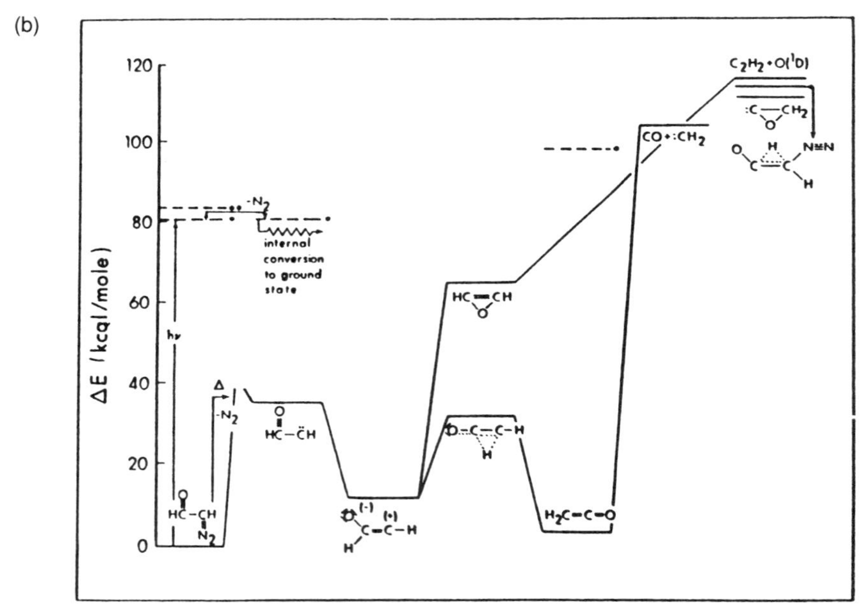

Figure 7.4 (a) Absorption spectra of a cresylic novolac and of the 1-oxo-2-diazo-naphthoquinone-5-arylsulfonate inhibitor. (b) Enthalpy diagram for the thermal and photodecomposition, respectively, of diazoethanone. (a) After Willson [2]; copyright © 1983 and (b) after Czizmadia et al. [5]; copyright © 1973, The American Chemical Society, Washington, DC.

rendering the exposed regions insoluble. In a subsequent flood exposure without subsequent baking, the previously unexposed region is rendered soluble and is developed as usual, creating a negative tone image. If the baking step after the first exposure is avoided, the same photoresist can be used to develop a positive tone pattern. The more common negative tone photoresists are based on the cross linking of a base polymer component, which may be a cyclized polyisoprene rubber or a substituted polystyrene, which are soluble in organic solvents. The cross linking is accomplished by the photodecomposition of a bisazide component of the resist.

A specific problem of photolithography is the fact that the absorption of light in the upper portion of the resist decreases its intensity toward the resist–semiconductor interface. This causes a tapering of the resist edges, that is, a wider opening on the top of the resist layer than on its bottom for trenches developed after exposure in a positive photoresist. As a consequence of this phenomenon, errors in the linewidth result in the patterning of the resist on a nonplanar surface, which is illustrated in Figure 7.5(a) and represents a serious limitation in the later stages of IC fabrication. In addition, multiple reflections and interference of the radiation at the resist–air and resist–dielectric–silicon or metal interfaces generate an intensity modulation in the film that depends on the film thickness, as illustrated in Figure 7.5(b). For the generally complex composite structure of an IC, combining materials with difference refractive indices and cross sections, complicated interference patterns modulate the exposure level in the resist, so that upon developing unexpected modulations in the line patterns evolve [7].

This problem is solved by the use of multilayer resists, which were introduced in the late 1970s. The principle is illustrated in Figure 7.6 for a two-layer resist. A thick planarizing film of organic material is covered by a thin film of a resist that is photosensitive. Upon exposure and development this thin film forms a transportable conformal mask (TCM) on the underlying thick film,

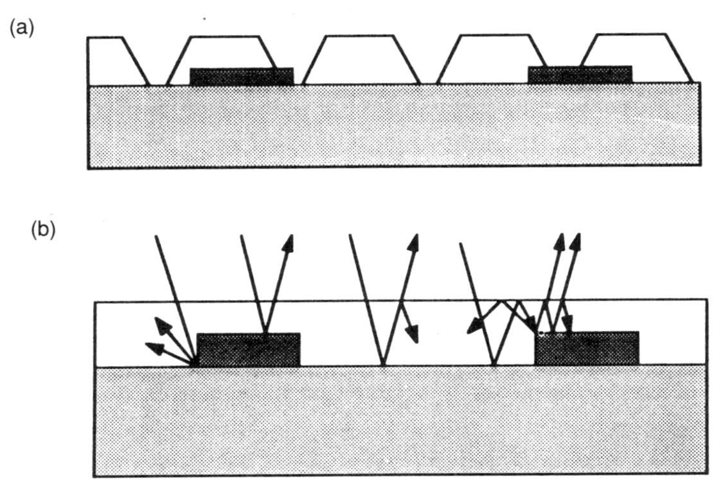

Figure 7.5 (a) Linewidth errors due to tapering in the pattern transfer by a positive photoresist on a nonplanar surface. (b) Multiple reflections in a resist.

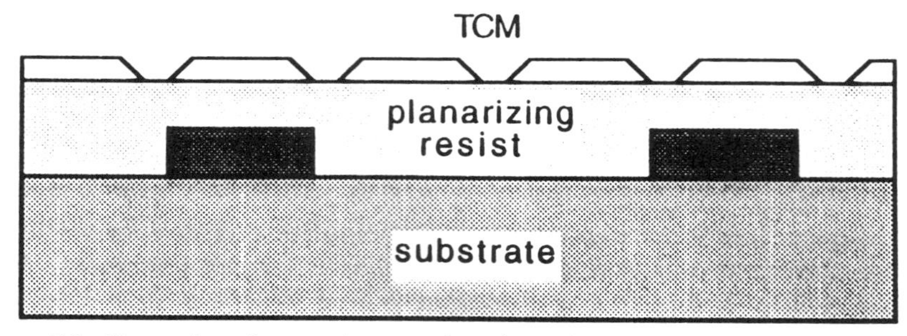

Figure 7.6 Patterning of a nonplanar surface through a transportable conformal mask.

which can be plasma etched to develop a resist pattern. Such a pattern is largely decoupled from surface relief effects because the nontransparent top resist effectively shields the surface of the wafer so that no multiple reflections occur. Both inorganic and organic resists have been used as top layers. For example, a few thousand angstrom thick film of $GeSe_2$ glass may be deposited by evaporation onto a planarizing organic polymer film [8]. The $GeSe_2$ film is photosensitized by the reaction with a complex silver ion containing solution according to the reaction

$$\begin{aligned} &GeSe_2 + 4K[Ag(CN)_2] + 8KOH \\ &\quad \rightarrow 2Ag_2Se + K_4GeO_4 + 8KCN + 4H_2O \end{aligned} \tag{7.4}$$

uv exposure generates an insoluble Ag emulsion in the exposed areas. A welcome edge-sharpening effect occurs due to diffusion of Ag^+ from the unexposed region to the edges of the exposed region, resulting in excellent contrast and submicrometer resolution, which is further improved by the very thin vertical dimension of the inorganic resist. The unexposed film is dissolved in a suitable solution, such as Kl/Kl_3 forming the TCM. The underlying organic material is removed by an oxygen plasma etch. Since the $GeSe_2$ film is not transparent to uv light, multiple reflection effects are absent. In addition, since the film is transparent to longer-wavelength light, mask alignment through the film is possible at 500 nm.

7.2 Ultrahigh-Resolution Patterning

X-ray lithography, introduced in 1972 by Spears and Smith [9], is one of the options for carrying mass production of ICs to the deep submicrometer resolution required for advanced ULSI (see Figure 7.2). Major obstacles to the realization of this technology are:

1. The lack of low-cost factory-size x-ray sources;
2. Difficulties in mask making;
3. Difficulties in mask alignment;
4. The need for resists that efficiently absorb x rays;
5. Radiation damage upon x-ray exposure.

Since the cross section for x-ray absorption generally decreases steeply with decreasing wavelength, soft x-rays are preferred for x-ray lithography. Thus the common monochromatic x-rays used for diffraction studies are not suitable for this purpose. Conventional electron-impact x-ray sources achieve too low a power density at the wafer plane, typically 10^{-4} W/cm^2, which needs to be increased for efficient throughput. Focused plasma beams of excited noble gas atoms and focused laser beams can produce upon impact onto metal targets high-intensity x-ray pulses in the appropriate wavelength regime, but are handicapped by a relatively low repetition rate. Synchrotron radiation pro-

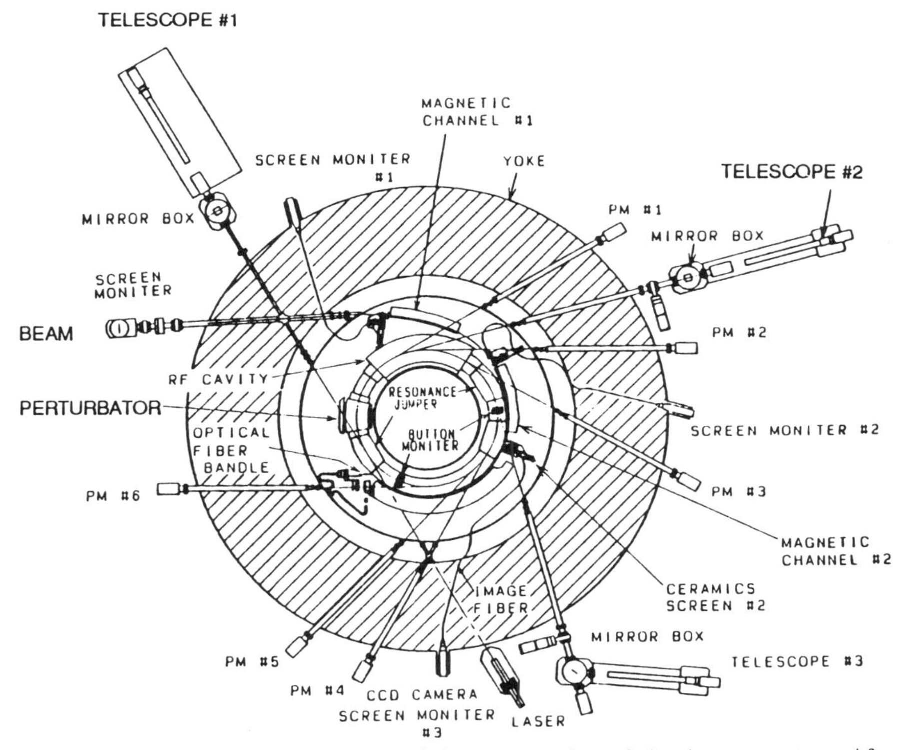

Figure 7.7 Schematic representation of the storage ring of the Aurora system. After Yamada [7]; copyright © 1990, The American Institute of Physics, New York.

vides for 3 orders of magnitude higher power density than electron-impact x-ray sources, but require downscaling to serve in an industrial production environment. Compact x-ray sources of a few meters diameter are currently available [7], [8]. Figure 7.7 shows a schematic representation of such a compact x-ray source. The outer diameter and height of the iron yoke are 3 and 2.2 m, respectively. The elliptical beam for this particular source has major and minor axis dimensions of 1.2 mm × 0.14 mm at 650 MeV.

In proximity printing, where the mask is brought close to the wafer for a 1:1 image transfer, the x-ray wavelength is $4 \leqslant \lambda \leqslant 10\,\text{Å}$, and the resolution is

$$r_{\text{proximity}} = k_1\sqrt{\lambda d_g/2} \tag{7.5}$$

where d_g is the gap between the mask and the wafer and k_1 is a constant. Figures 7.8 and 7.9 show a schematic representation of a particular implementation of x-ray proximity printing [8] and a plot of the resolution versus the DOF

$$\Delta_f = \pm\frac{\lambda}{2(\text{NA})^2} \tag{7.6}$$

where NA is the numerical aperture. A resolution of $\sim 0.05\,\mu$m should be obtainable with a viable value for the depth of focus. In order to implement circuit manufacturing at 0.1 μm resolution, the alignment of the mask must be done with an accuracy of 100 Å. This accuracy has already been demonstrated

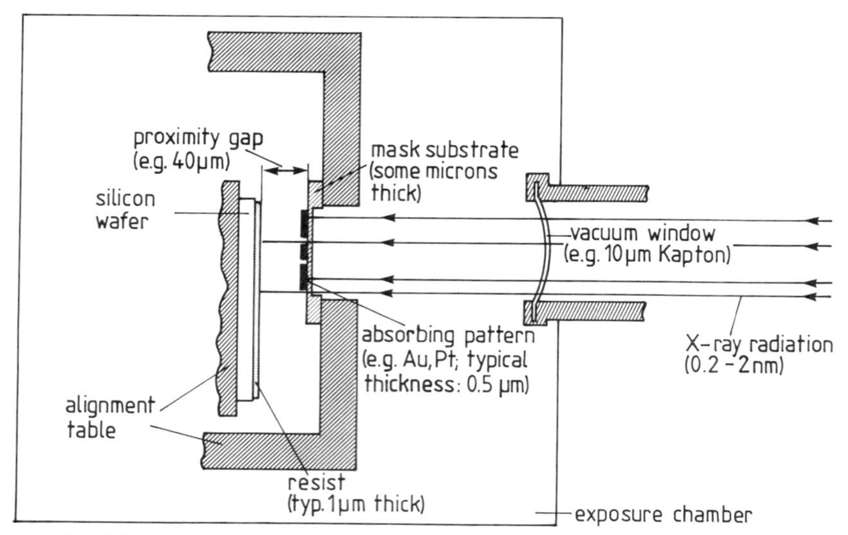

Figure 7.8 Schematic representation of an arrangement for x-ray proximity printing. After Heuberger [8]; copyright © 1988, American Institute of Physics, New York.

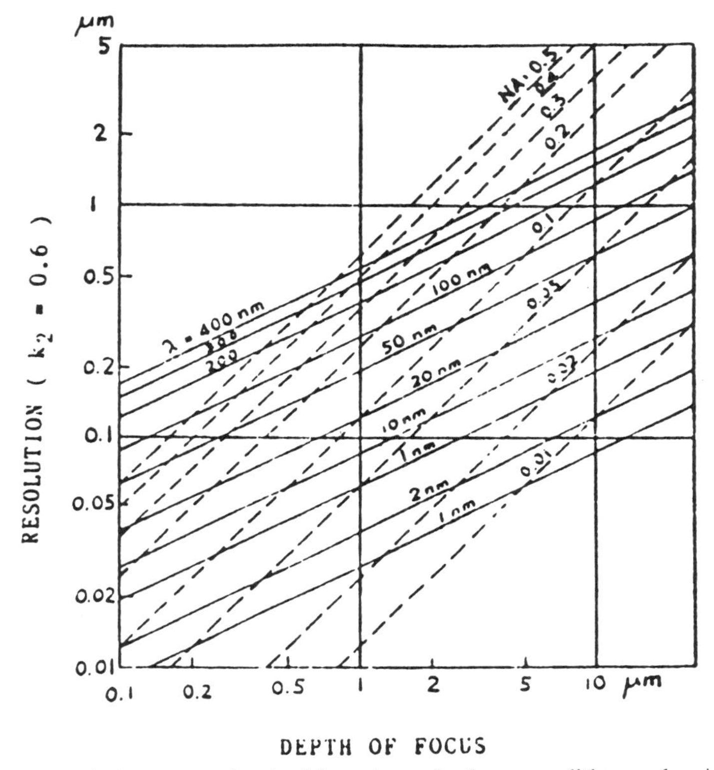

Figure 7.9 Resolution versus depth of focus in projection x-ray lithography. After Hoh [10]; copyright © 1989, The Electrochemical Society, Pennington, NJ.

[11]. In projection printing the resolution is

$$r_{\text{projection}} = k_2 \frac{\lambda}{\text{NA}} \tag{7.7}$$

where $k_2 \approx 0.61$. Therefore, the resolution decreases in proximity printing with λ sublinearly, while in the case of projection printing it decreases linearly. This provides an incentive for the development of projection printing.

Figure 7.10(a) shows a schematic representation of a projection printing system using a Schwarzschild objective with Mo/Si multilayer coatings that achieve a reflectance of 40% at 14 nm wavelength. Figure 7.10(b) shows a set of 50 nm wide lines projection printed in 60 nm of PPMA after wet developing, using 14 nm radiation provided by the VUV storage ring at the National Synchrotron Light Source at the Brookhaven National Laboratory. The aperture used in this experiment corresponding to NA $\approx$ 0.16 is not optimized with regard to aberrations, but provides the resolution required for the printing of 50 nm lines [12].

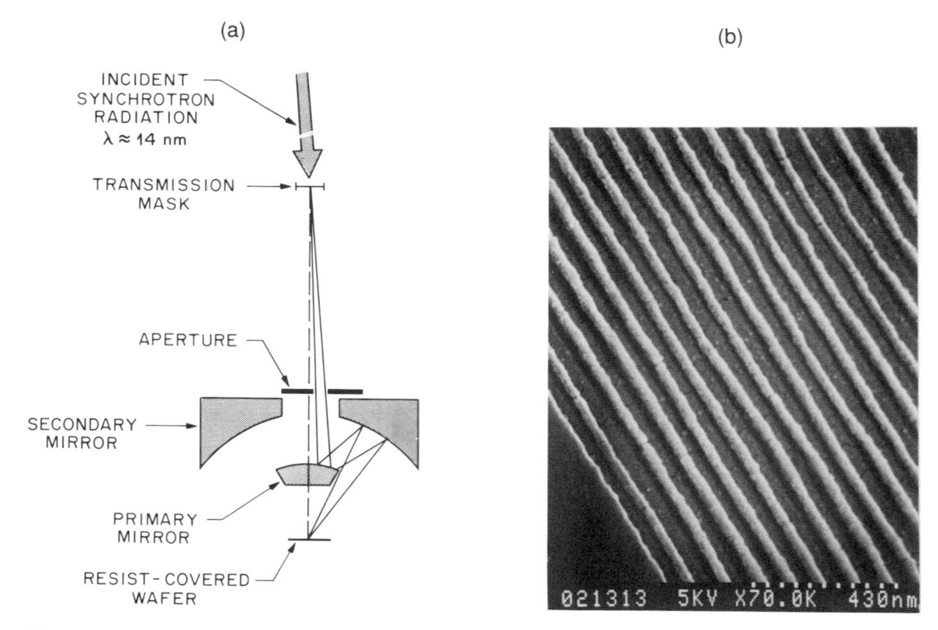

Figure 7.10 (a) Schematic representation of a system for x-ray projection printing. (b) SEM image of 50 nm wide lines in PPMA. After Bjorkholm et al [12]; copyright © 1990, American Vacuum Society, New York.

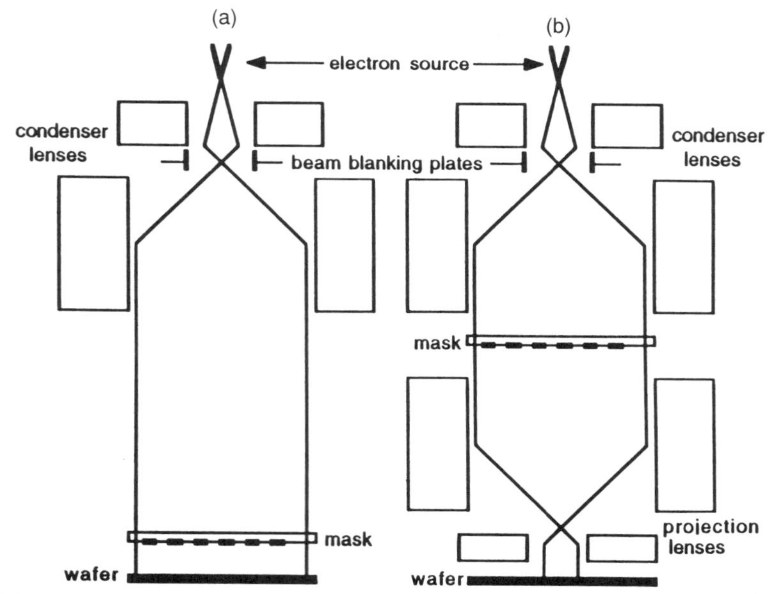

Figure 7.11 Schematic representations of proximity and projection printing by broad electron beams.

In the pattern generation by broad electron-beam exposures, either proximity or projection printing can be implemented with well-established electron optics, as illustrated in Figure 7.11. Since the energy of the radiation used in both electron-beam and x-ray exposures exceeds the bond energy of a variety of polymer molecules, chain scission reactions can be initiated by the exposure without the presence of sensitizers. A versatile group of monocomponent resists for electron and x-ray lithography is thus based on polymethyl-methacrylate (PMMA), which is formed from the monomer in a radical-induced polymerization reaction:

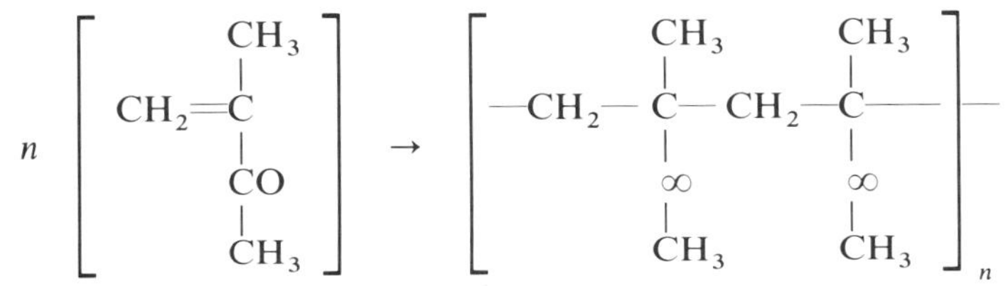

$$n\left[\begin{array}{c} CH_3 \\ | \\ CH_2{=}C \\ | \\ CO \\ | \\ CH_3 \end{array}\right] \rightarrow \left[\begin{array}{c} \quad CH_3 \qquad\quad CH_3 \\ \quad | \qquad\qquad\quad | \\ -CH_2-C-CH_2-C- \\ \quad | \qquad\qquad\quad | \\ \quad CO \qquad\quad\; CO \\ \quad | \qquad\qquad\quad | \\ \quad CH_3 \qquad\quad CH_3 \end{array}\right]_n \qquad (7.8)$$

The chain scission upon exposure to energetic radiation proceeds as

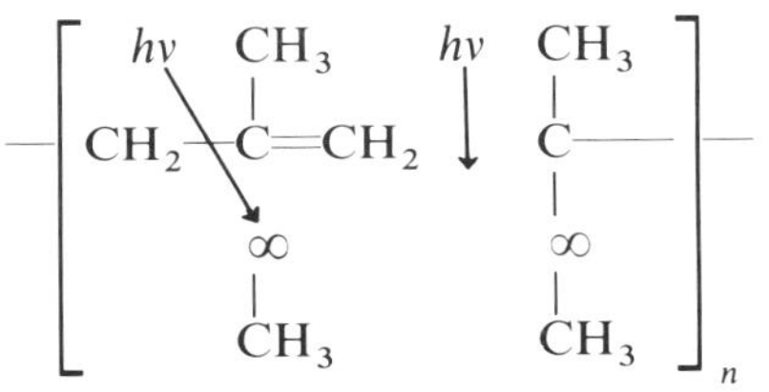

$$-\left[\begin{array}{c} h\nu \quad CH_3 \qquad h\nu \quad CH_3 \\ \qquad | \qquad\qquad\quad | \\ CH_2 \searrow C{=}CH_2 \downarrow C- \\ \qquad | \qquad\qquad\quad | \\ \qquad CO \qquad\quad\; CO \\ \qquad | \qquad\qquad\quad | \\ \qquad CH_3 \qquad\quad CH_3 \end{array}\right]_n- \qquad (7.9)$$

Although PMMA provides excellent resolution, well into the 100 Å range, its sensitivity to electron exposure ($\sim 10^3$ mJ/cm^2) exceeds by at least an order of magnitude the required value for ULSI processing [8]. Significant progress can be made by copolymerization. However, since PMMA and its derivatives have poor etching stability, other resists are being pursued at present.

Novolac-based resists with and without chemical amplification have attracted attention in this context. They employ an inhibitor that is destroyed in the exposed volume by the action of an acid which is formed in the primary photochemical decomposition of a sensitizer [13]. Thus a sensitivity $\leqslant 100\,\mathrm{mJ\,cm^{-2}}$ and excellent resolution have been achieved with a resist that retains the well-established etching stability of conventional novolacs. For example, EPTR (epoxy novolac-type resist), a glycidyl ether novolac resin sensitized with triphenylsulfonium hexafluoroantimonate, achieves high resolution in both electron and x-ray printing with relatively low-contrast (2.5:1) masks in the x-ray exposures [14]. Propylene glycol methyl ether acetate (PGMEA) is a suitable developer for this negative resist that exhibits minimum swelling upon wet development.

Since x-ray exposure modifies the diffusivities of organometallic molecules (e.g., trimethyltin chloride) or transition metal halides (e.g., $TiCl_4$), treatments of the resist after exposure with such materials result in significantly different concentrations in the film decorating the pattern. In chain scission-type resists the diffusivity is enhanced in the exposed regions. Upon dry etching with oxygen, the metal oxide formed in these regions inhibits etching and results thus in a negative tone image. Positive images employing such a builtin etch mask can be obtained by the use of cross-linking resists that have diminished diffusivities in the exposed regions.

Focused-electron-beam writing achieves excellent resolution, but low throughput limits its utility for the manufacturing of high-density ULSI circuits. The total pattern definition time t_P is

$$t_P = N_{es}\left(\frac{S}{J} + t_s\right) + t_0 \tag{7.10}$$

where N_{es}, S_r, J, and t_s are the number of exposure shots, the resist sensitivity, the current density, and the settling time for electrostatic deflection per shot, respectively. The overhead time t_0 sums over all other contributions to t_P, such as the time required for evacuation, calibration, stage movement, etc. N_{es} values for the gate layer of 64 and 256 Mb DRAMS, respectively, are 3×10^8 and 1×10^9 shots per chip, corresponding to exposure times of tens to hundreds of seconds [15]. Although direct electron-beam writing by a single electron beam thus is an inadequate tool for the manufacturing of ULSI circuits, it has become a valuable tool for fabricating confined heterostructures [16] and is indispensible as a means for fabricating the master masks for both x-ray and broad electron-beam printing. In computer-aided design, the pattern is generated on the videoscreen of a computer and is then stored in digital form. The digital data are utilized to drive a pattern generator, which controls the position of the electron or ion beam on the mask.

A schematic representation of the trilevel processing sequence of an x-ray mask is shown in Figure 7.12. The procedure starts with the patterning of the electron-beam resist on an intermediate layer of Si_3N_4 by electron-beam exposure. The patterned resist serves as a mask for the pattern transfer into the Si_3N_4 by reactive ion etching (RIE) in a CHF_3/O_2 plasma (see Section 7.3). The patterned Si_3N_4 is then used for masking a 2–3 μm thick bottom resist layer. After opening lines and windows in this layer by RIE in an O_2 plasma, the passivation layer is removed in these opened regions by a subsequent RIE in a CHF_3/O_2 plasma followed by electroplating of gold that acts as the x-ray absorber. This method produces 0.3 μm features with an aspect ratio of 5 and a loss of linewidth of 50 nm per edge. Generally the metal will exert in the as-deposited state a stress onto the membrane, which is reduced subsequently by annealing at ~80°C. Alternatively, direct patterning of heavy metal films

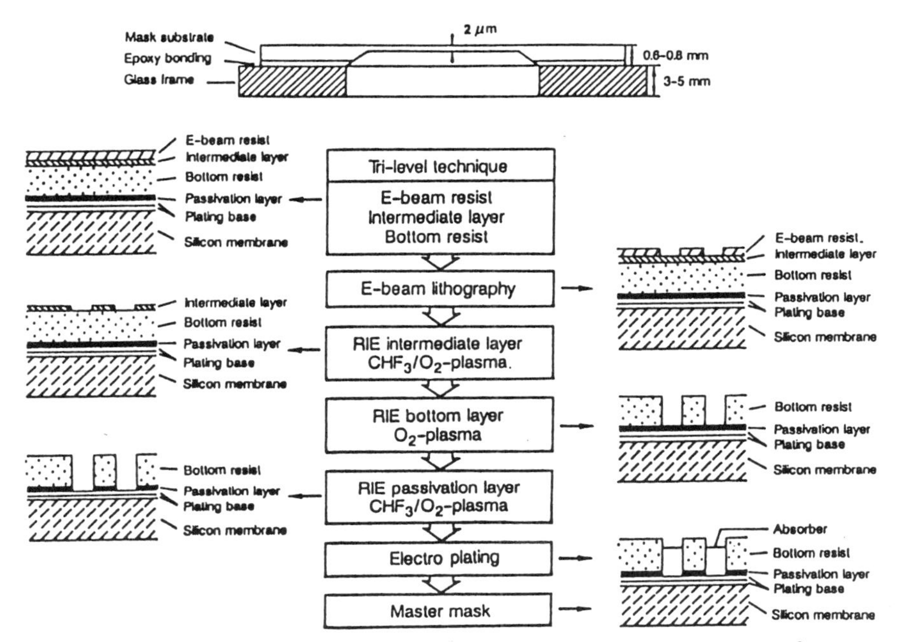

Figure 7.12 Schematic representation of the trilevel processing sequence of an x-ray mask for ULSI circuit manufacturing. After Petzold [17]; copyright © 1988, Gordon and Breach Science Publishers, London.

with stress relief by ion bombardment has been utilized [18]. In contrast to photolithography, the resolution of the pattern-definition process by electrons and ions is not wavelength limited. However, limitations to the resolution exist due to proximity effects, which will be explained in the following. The use of thin single resist and thin metal films to generate a low-contrast mask that is utilized to produce a high-contrast replica mask achieves 60 nm features at a contrast of 1.7–10 [19].

A schematic representation of a cross section of a finished x-ray mask is shown in Figure 7.13. It must employ high contrast, which is achieved by the utilization of light element foils, for example, Si_3N_4 and Si in combination with heavy element absorbing films made, for example, from Au, Ta, or W. Also SiC and BN have been used in the fabrication of x-ray masks. BN has excellent transparency for high-energy radiation, but, for epitaxial films containing hydrogen, suffers from radiation-induced distortions [20]. The same effect also occurs for CVD-generated Si_3N_4 membranes. The linear distortion of the pattern due to absorber stress σ_a can be estimated by the equation

$$\frac{\Delta x}{x} = \frac{\sigma_a \zeta}{B_a(1-\zeta)} \frac{t_a}{t_m B_m} \tag{7.11}$$

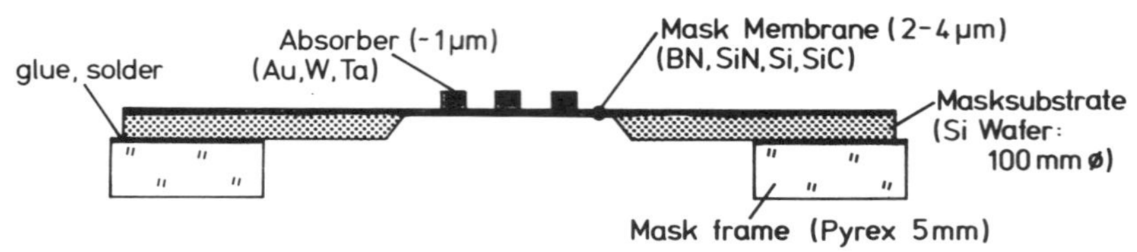

Figure 7.13 Schematic representations of cross sections through an x-ray mask. After Petzold [17]; copyright © 1988, Gordon and Breach Science Publishers, London.

where t_a, t_m, and B_a, B_m refer to the thicknesses and Young's moduli, respectively, of the absorber and the supporting membrane [21]. For a given distortion tolerance, absorber stress, and Young's modulus of the absorber, the parameter t_a/t_m, which determines the contrast, scales thus with B_m, requiring membrane materials with high Young's modulus. Within the group of materials listed, the Young's moduli range from 1×10^{12} (BN) to 3.5×10^{12} dyne cm^{-2} (SiC), favoring SiC in terms of mask distortion resistance and contrast. In terms of mask engineering, a straight Si technology is the most advanced and serves well at relatively low x-ray energies. Mask blanks are produced by anisotropic etching (see Section 7.3) of nominally undoped Si wafers covered by a typically 2–3 μm thick epilayer of heavily B-doped Si, which acts as an etch stop. In order to compensate for the stress created by the heavy p doping that substitutes a large number of Si atoms by smaller B atoms, the epilayer is also doped with Ge atoms, whose size compensates for the B atoms, but which do not affect the conductivity [17], [22].

Figure 7.14(a) shows the results of Monte Carlo simulations of the backscattering of electrons, demonstrating the substantial widening of the exposed area of the resist as compared to the incident electron beam. Nevertheless both electron- and ion-beam writing can produce the submicrometer patterns required for the fabrication of masks for ULSI applications. Since ion-beam exposure results in substantially smaller proximity effects than electron-beam exposure, both broad-beam lithography and focused-ion-beam writing are of considerable interest in the context of microstructure fabrication [23].

In addition to lithography by broad ion beam sources, focused-ion sources are of considerable interest in the context of x-ray mask repair, that is, of the removal of opaque defects from areas on an x-ray mask that are intended to be transparent. The complementary task of repairing transparent defects in the opaque areas of the mask requires localized deposition of metals, which may be achieved by laser-assisted selected-area OMCVD followed by the removal of excess metal by focused-ion-beam milling [17]. Figure 7.15 shows a schematic representation of ion milling. Both highly focused proton sources and liquid metal ion sources (LMISs) are available. The function of the LMIS is based on the flow of a liquid metal from a reservoir to the apex of a sharp needle of a conducting solid that is wetted by the liquid metal. In a sufficiently

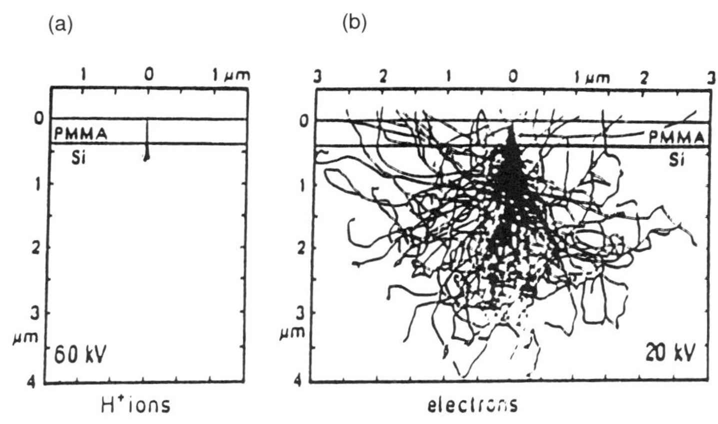

Figure 7.14 Proximity effects due to backscattered ions (a) and electrons (b) for a resist-covered semiconductor. After (a) Saitoh et al. [23] and (b) Kyser and Viswanathan [24]; copyrights: (a) © 1988, and (b) © 1975, The American Institute of Physics, New York.

high electric field, a narrow protrusion forms at the tip of the Taylor cone that approximately describes the liquid shape under static conditions [25]. Field evaporation of metal atoms from this sharp protrusion and ionization establishes a high flux of ions that are focused and scanned by the two-lens focusing optics. Liquid Ga and liquid Au sources are particularly convenient because of the excellent wetting properties of these metals on tungsten and were utilized first for microlithography in 1979 [26]. A spot size of ~60 nm at 25 keV beam

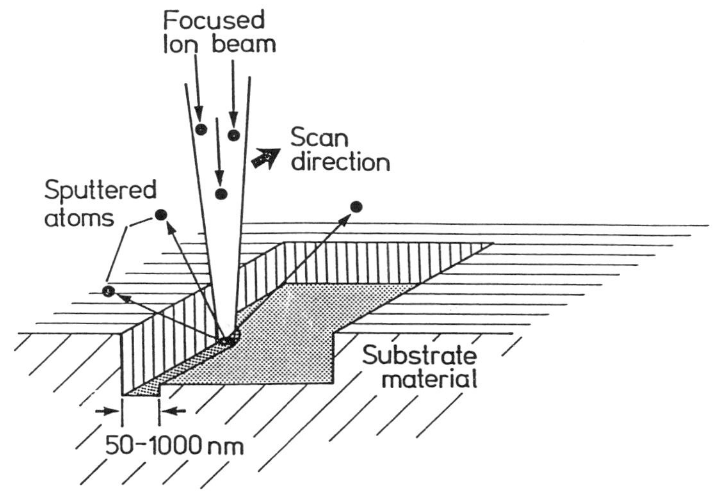

Figure 7.15 Schematic representation of a focused-ion-beam milling. After Petzold [17]; copyright © 1989, Gordon and Breach Science Publishers, London.

energy is readily available, corresponding to a current density of $\sim 3.5\,A/cm^2$ at the target and a total beam current of $\sim 0.1\,nA$. Focused-ion-beam sources have high brightness and thus achieve large cutting rates in micromachining applications [27]. In mask repair, the focused-ion source serves a dual purpose: (1) The removal of the opaque defect by ion milling; and (2) display of the repair area prior to and after the work to locate the defect and monitor its removal. Secondary electrons emitted from the ion-beam scanned surface are utilized to create the image.

Figure 7.16(a) shows a schematic representation of proximity effects caused by the redeposition of ion-milled metal on neighboring walls. Figure 7.16(b) shows an image of a cutout section of a gold layer that is intended to have

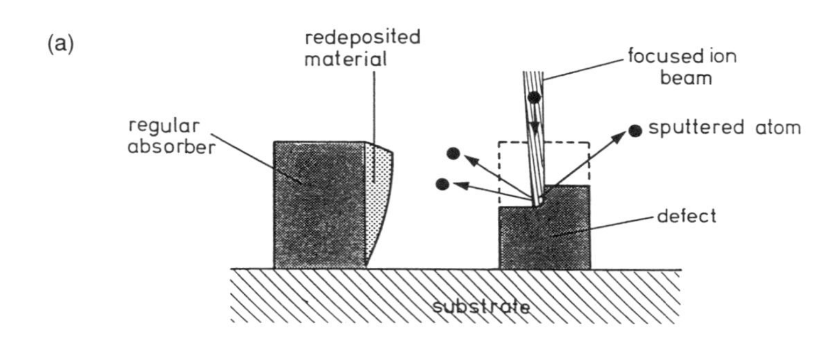

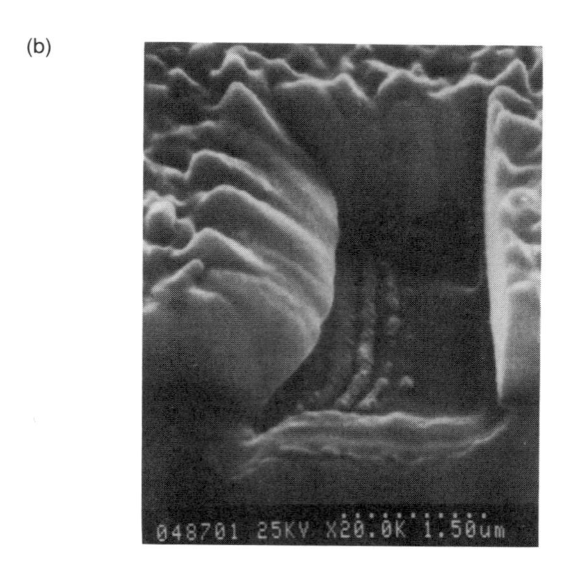

Figure 7.16 (a) Schematic representation of proximity effects in focused-ion-beam milling. (b) Milling of a rectangular slot into the edge of a gold film. After Petzold [17]; copyright © 1989, Gordon and Breach Science Publishers, New York.

rectangular shape, but is distorted due to proximity effects. Stepwise removal schemes can greatly reduce redeposition effects, and rectangular features of micrometer dimensions can be cut into metal films with commercial focused-ion-beam sources. Also, STM liquid metal ion sources have been constructed for laboratory use and have achieved the cutting of 1 μm deep holes of 2000 Å diameter in GaAs and Si [28]. In addition to mask repair and

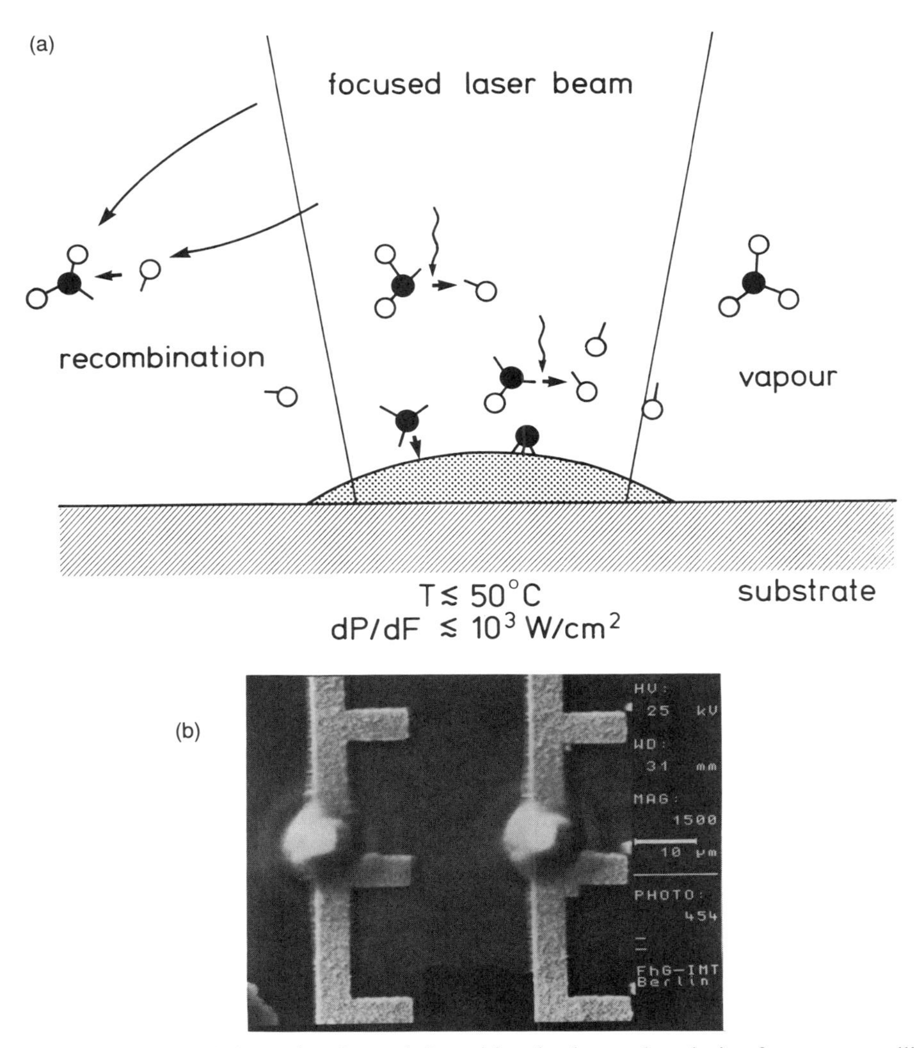

Figure 7.17 (a) Schematic of metal deposition by laser photolysis of organometallic precursor molecules. (b) SEM image of a selected area on a Si x-ray mask repaired by photolytic Sn deposition. After Petzold [17]; copyright © 1989, Gordon and Breach Science Publishers, New York.

lithography, focused-ion sources may also become useful for selected-area ion implantation. Several alloy sources have been developed for this purpose [29].

A further alternative in mask repair is the photolytic decomposition of organometallic precursors by uv laser pulses, as illustrated in Figure 7.17(a). Figure 7.17(b) shows a Sn deposit produced on a transparent defect in an x-ray mask by uv laser photolysis of $Sn(CH_4)$ at 20–50 mbar pressure using frequency-doubled Ar^+ laser radiation at 257 nm. The repaired section is transferred into photoresist upon exposure to synchrotron soft-x-ray radiation with excellent contrast. Also ion-beam exposure can be utilized for direct writing of metals lines in a thin film of a suitable organometallic precursor, such as gold hexafluoropentanedionate [30], [31].

Electron-beam writing in the STM permits pattern definition at sub-100 nm resolution since its low-energy operation eliminates to a large extent the proximity effects that limit the resolution of direct electron-beam writing at tens of keV energy. Monte Carlo simulation of the effect of the backscattered electrons on the linewidth achievable by lithography with the STM suggests an exposure at 10% of the primary beam intensity over several beam diameters [32]. The limitations to this ultrahigh-resolution electron-beam writing method are illustrated in Figure 7.18. However, STM-written lines on the 10 nm scale have been produced already, and further advances may be expected in the near future.

Figure 7.19 shows SEM images of single lines written at STM tip voltages, from left to right: −50, −35, and −25 V, corresponding to linewidths of 90, 42,

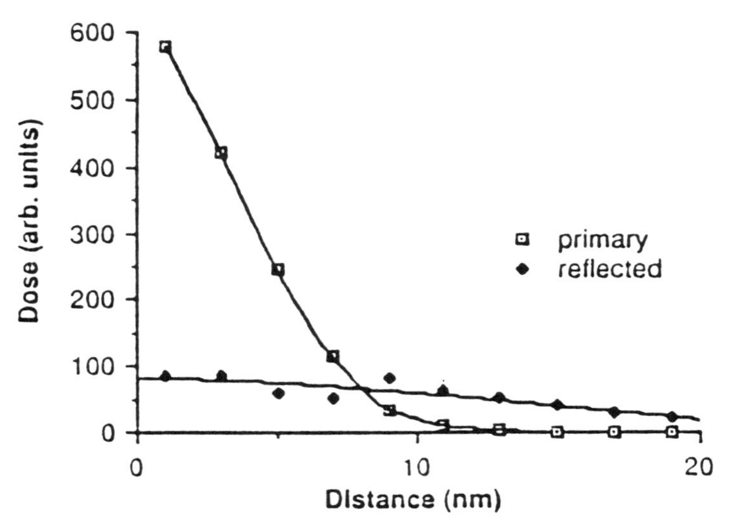

Figure 7.18 The distribution of primary and backscattered intensity for electron-beam writing with the STM. Tip diameter 20 nm, gap 6 nm. After McCord and Pease [33]; copyright © 1987, North Holland Publishing Company, Amsterdam.

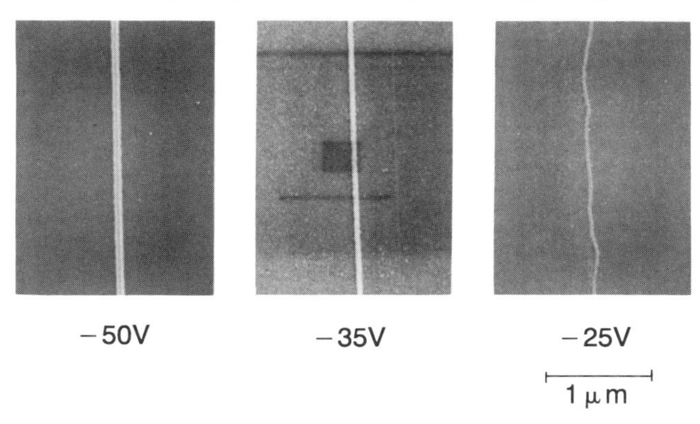

Figure 7.19 Single lines written by an STM tip at different bias voltages. After Dobisz and Marian [38]; copyright © 1991, American Institute of Physics, New York.

and 27 nm, respectively. Furthermore, the direct deposition of Au atom clusters from an STM tip [34], the localized desorption of adsorbates from the surface of silicon [35], and the transport of atom clusters by means of the STM tip have been demonstrated, achieving at present feature sizes of a few nanometers [36]. Therefore, the capabality of manipulating features on the surface of selected materials with nanometer resolution is currently at hand, making the engineering of surface features with atomic resolution a realistic goal of future research [37].

7.3 Wet Etching

The etching processes applied in microelectronic processing employ either liquid (wet etching) or gaseous (dry etching) etchants. Wet etching is essential for the following tasks:

1. The removal of the work-damaged surface layer on machined substrate wafers;
2. The delineation of junctions, grain boundaries, and individual defects in semiconductor wafers;
3. The conditioning of surfaces for epitaxy and other processing steps;
4. The development of anisotropic features along specific crystallographic planes for the fabrication of trench capacitors and a variety of micromachined silicon parts, such as single tips for STM and AFM as well as field-emitter arrays for vacuum microelectronics.

Selected etchants for Si, III–V, and II–VI compounds are listed in Table 7.1. Etchants for selected dielectrics on silicon are listed in Table 7.2 [50]. For a fast removal of work damage, HF/HNO_3 is a preferred etchant for silicon and is used for ingot etching after the grinding of the orientation flats on the silicon crystal boules. The nitric acid acts as an oxidant, and the hydrofluoric acid dissolves the oxidation product according to the reactions

$$Si + 4HNO_3 \rightarrow SiO_2 + 4NO_2 + 2H_2O \tag{7.12}$$

$$SiO_2 + 6HF \rightarrow H_2SiF_6 + 2H_2O \tag{7.13}$$

After cutting and lapping, the edges of the wafers are rounded off to avoid buildup of a crown on their periphery during photoresist coating and epitaxy, which would interfere with the pattern transfer. A caustic etch is used to remove residual work damage after this machining step. It proceeds by the formation of silicate:

$$Si + 2OH^- + H_2O \rightarrow SiO_3^{2-} + 2H_2 \tag{7.14}$$

Table 7.1 Selected etchants for semiconductors

Material	Etchant	Remarks
Silicon	$HF{:}5M\,CrO_3{:}H_2O = 1{:}1{:}1$ [39]	Defects on (111), 3.5 μm/min
Silicon	$HF{:}0.15M\,K_2Cr_2O_7 = 2{:}1$ [40]	Defects on (100), 1.5 μm/min,
Silicon	$HF{:}HNO_3{:}5M\,CrO_3{:}CH_3COOH{:}H_2O = 2{:}1{:}1{:}2{:}2 + 1\,g\,Cu(NO_3)_2/120\,ml$[41]	Defects on various surfaces, 1.0 μm/min
Silicon	$HF{:}HNO_3{:}CH_3COOH = 2{:}5{:}3$ [42]	Fast polishing etch
Silicon	50 g KOH:200 g *n*-propanol:800 g H_2O [43]	Anisotropic etch
GaAs	$H_2SO_4{:}H_2O_2{:}H_2O = 5{:}1{:}1$ [44]	Polishing etch
	Fused KOH at 300°C [45]	Good dislocation etch
	1 ml HF, 2 ml H_2O, 8 mg $AnNO_3$, 1g CrO_3 [45]	*A–B* etch
GaP	as GaAs, $HF{:}CH_3COOH$:saturated $KMnO_4$ [45]	Stain for revealing striations
InP	$HNO_3{:}HCl{:}H_2O = 6{:}1{:}1$ [46]	Dislocations on (111), 60°C
InSb	$HNO_3{:}CH_3COOH{:}HF5{:}3{:}3$ [47]	CP4 polishing etch
	0.2*N* solution of $FeCl_3$ in HCl [48]	Pitting etch
	$1HF{:}1CH_3COOH{:}12N\,KMnO_4$ [49]	Stain for striations
CdS	$HNO_3{:}CH_3COOH{:}H_2O = 6{:}6{:}1$ [45]	Dislocation etch pits, Distinguishes *A*, *B* faces

Table 7.2 Wet etchants for selected dielectrics (After Kern and Schnabel [7.50])

Material	HF–H_2O	HF–NH_4F	$H_3PO_4^a$	$H_2SO_4^a$	30%NaOH[a]
SiO_2	H[b]	H	L[b]	0	L
Al_2O_3	H	H	M[b]	0	0
Ta_2O_5	M	L			
TiO_2	H	M	L	L	
PSG	H	H	L	0	L
Si_3N_3	L	L	M	0	0
SiN_xH	H	M	H	0	0
$SiO_xN_yH_z$	H	M	L	0	0

[a]Hot.
[b]H = high, M = intermediate, L = low etching rates.

In the subsequent polishing of semiconductor wafers, a combination of chemical and mechanical removal is used. For example, in Syton polishing, a suspension of colloidal SiO_2 in an alkaline solution is applied. Dilute solutions of bromine in methanol are suitable for chemically polishing of a variety of compound semiconductors [51]. Wafers that are shipped by the manufacturers with polished surfaces require a careful cleaning procedure prior to processing to remove contamination of the surface by metal ions, organic matter, and particulates [52], [53]. Clean silicon surfaces thus produced oxidize spontaneously in air which can be corrected by a buffered HF dip.

The removal of the suface oxide of silicon by HF or NH_4F is an important processing step, since the hydrogen termination of the surface obtained as a result of the etching protects it from oxidation and, with care, permits the loading of wafers into processing chambers and tubes in a well-defined ultraclean initial state. LEED studies of Si(111) surfaces reveal that for both HF and NH_4F etching a 1×1 surface mesh is obtained. Figure 7.20 shows the STM images of HF- and buffered NH_4F/HF-etched silicon (111) surfaces. While the as-HF-etched surface is atomically rough [54], the buffered NH_4F/HF-etched surface is atomically flat, with terraces extending over thousands of angstroms [55]. This pH dependence of the surface is reflected in the density of surface states for MOS capacitors prepared by remote plasma deposition of silicon dioxide. HF-etched surfaces at the optimum pH value followed by plasma oxidation and an 800°C anneal results similar densities of mid-gap interface states for both {100} and {111} as obtained for thermally oxidized Si{100}. Figure 7.21 shows the channel mobility vs. field characteristics of MOSFETs prepared with different gate oxides. The peak mobility obtained for MOSFETs made by a combination of plasma preoxidation and rapid thermal oxide deposition followed by an 800°C anneal approaches at present the value

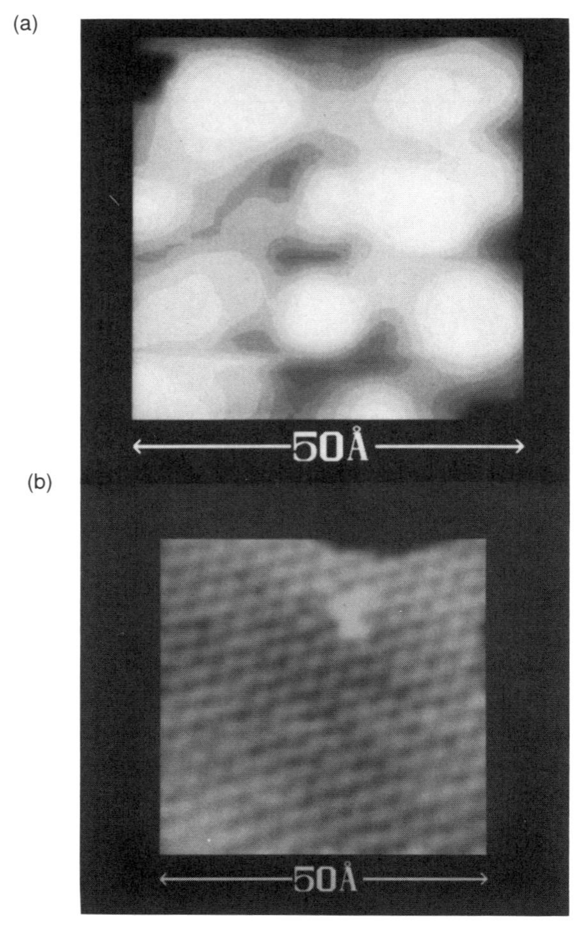

Figure 7.20 STM images of H-terminated Si(111)-1 × 1 surfaces produced by (a) HF and (b) NH_4F etching. After Higashi et al. [55]; copyright © 1991, American Institute of Physics, New York.

obtained by the conventional thermal process, which yields still the best high field characteristics.

By diluting pure HF solutions, the removal rate of residual native oxide can be made comparable to the etching rate of thermal and TEOS oxide layers so that cleaning in critical window areas does not significantly alter the thickness of other oxide structures on the chip [57]. The etching rate decreases with increasing dilution. Using the contact angle of 0.5 μl droplets of H_2O as an indicator, the etching time in 0.5 and 0.1% HF for complete removal of the native oxide is ~60 and 260 s, respectively. A deionized water rinse after etching results in a moderate decrease of the contact angle.

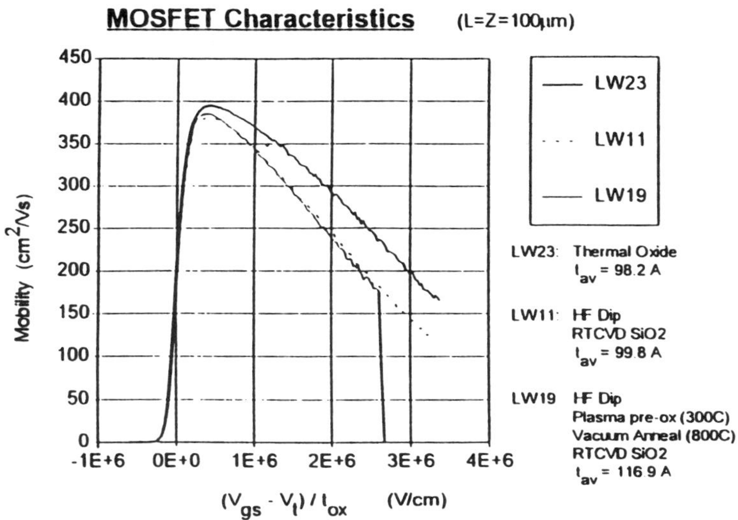

Figure 7.21 Channel mobility of silicon MOSFETs prepared with three different gate oxide fabrication methods vs. $V_{gs} - V_t$, L = Z = 100 μm. Courtesy of Gerald Lucovsky North Carolina State University, Raleigh, North Carolina [56].

Figure 7.22 shows p-polarized attenuated-total-reflection (ATR) spectra of the silicon–hydrogen stretching vibrations on a Si(111) surface, assigning the absorption features to the following surface configurations: a coupled –SiH on a double step (M), a –SiH terminating dangling bond into the [111] direction (M'), a coupled –SiH on a (100) step (M''), a coupled –SiH_2 on a (100) step (D), and a –SiH terminating dangling bond in the [111] direction (T) [58]. Symmetric and asymmetric modes are denoted by the subscripts SS and AS. Immersion in water at room temperature after 1.5% HF etching for 30 s removes silicon hydrides at steps and converts them into silicon monohydride. It also removes fluorine, changing from initially 12% to 5% ML after 5 s immersion. A similar passivation upon HF etching is also observed for the Si(001) surface [59].

In advanced silicon processing tools, wet HF etching after the usual RCA cleaning procedure has been combined with in situ HF vapor etching after loading into the multichamber reactor [60]. Figure 7.23 shows the residual surface contamination by various metal impurities after the RCA cleaning procedure, after RCA cleaning followed by wet HF etching, and after RCA cleaning/wet HF etching followed by in situ vapor etching, respectively. In some cases, significant improvements result after the wet HF etching step. However, the in situ vapor etch seems not to be effective. Nevertheless, improvements in the yield and electrical performance of MOS devices are obtained, as shown in Figure 7.24, which is presumably due to improved removal of residual surface oxide and better control of the initial phase of oxidation.

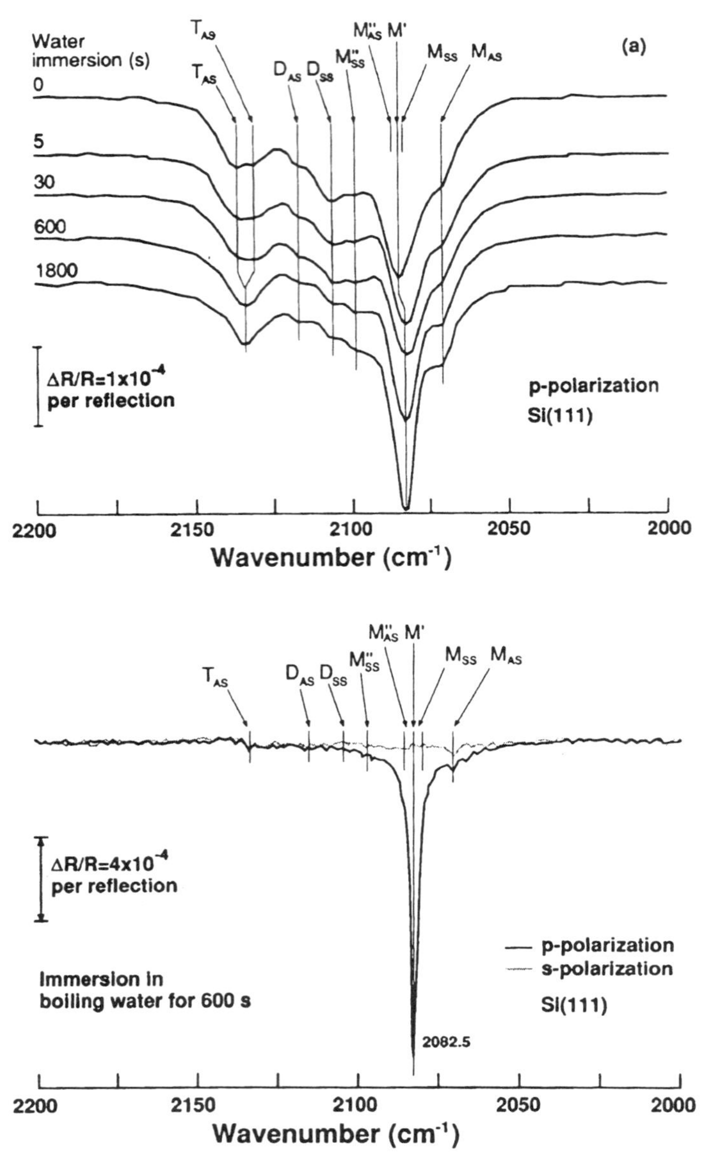

Figure 7.22 Polarized ATR spectra of Si(111)-1 × 1 after HF etching and immersion into water (a) at room temperature, (b) at 100°C. After Watanabe, Nakayama, and Ito [58]; copyright © 1991, American Institute of Physics, New York.

For GaAs and InP, etching in HSO_4:H_2O_2:H_2O in the volume ratios 5:1:1 and 2:1:1, respectively, followed by a deoxidizing etch in HF–ethanol inside a dry box, provides for an oxide-free surface of good morphology [61]. Instead of the deoxidizing etching step, the native oxide can be desorbed by rapid

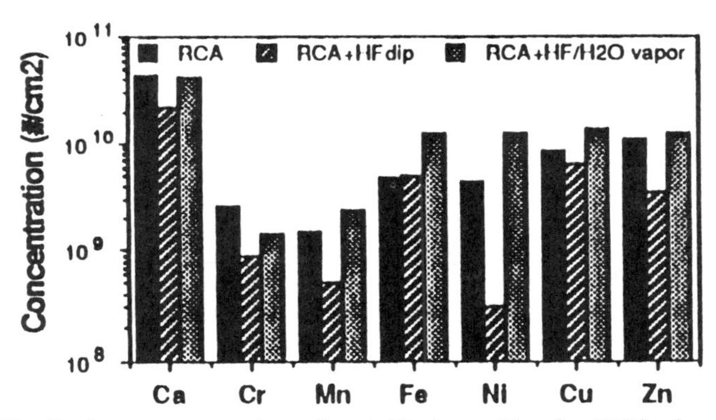

Figure 7.23 Surface concentration of metallic impurities for RCA cleaned Si wafers after various surface cleaning steps. After Werkhoven et al. [60]; copyright © 1992, Institute of Electrical and Electronics Engineers, New York.

thermal annealing in t-$ButAsH_2$ and t-$ButPH_2$ under the conditions of LPMOCVD at 50 mTorr partial pressure [62] or under a stabilizing P_2 flux of $\sim 5 \times 10^{14}\,cm^{-2}\,s^{-1}$ at $T \leqslant 490°C$ or a stabilizing As_4 flux of $4 \times 10^{13}\,cm^{-2}\,s^{-1}$ at 520°C under the conditions of MBE or CBE [63].

The kinetics of chemical dissolution of a semiconductor can be either surface or diffusion controlled. Since under diffusion-controlled conditions the etching rate does not depend on local differences of the surface structure, polishing etches usually operate in the diffusion-controlled regime. Figure 7.25 shows the etching rate as a function of the reciprocal temperature for silicon in an oxidizing etch. For low oxidant concentration and high concentration of complexing fluorine ions, the reaction is surface controlled over the entire temperature range investigated, but for a high oxidant concentration and low

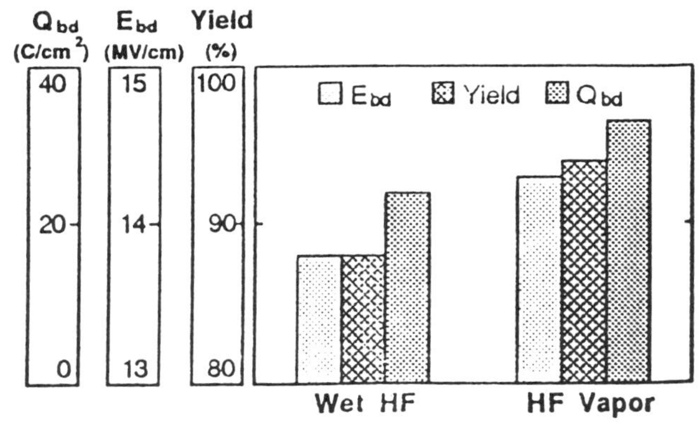

Figure 7.24 Yield of 100 MOS capacitors per wafer surviving 12 MV/cm field strength (E_{bd}) and 80% probability value of a 0.1 A/cm^2 current stress (Q_{bd}) test. After Werkhoven et al. [60]; copyright © 1992, Institute of Electrical and Electronics Engineers, New York.

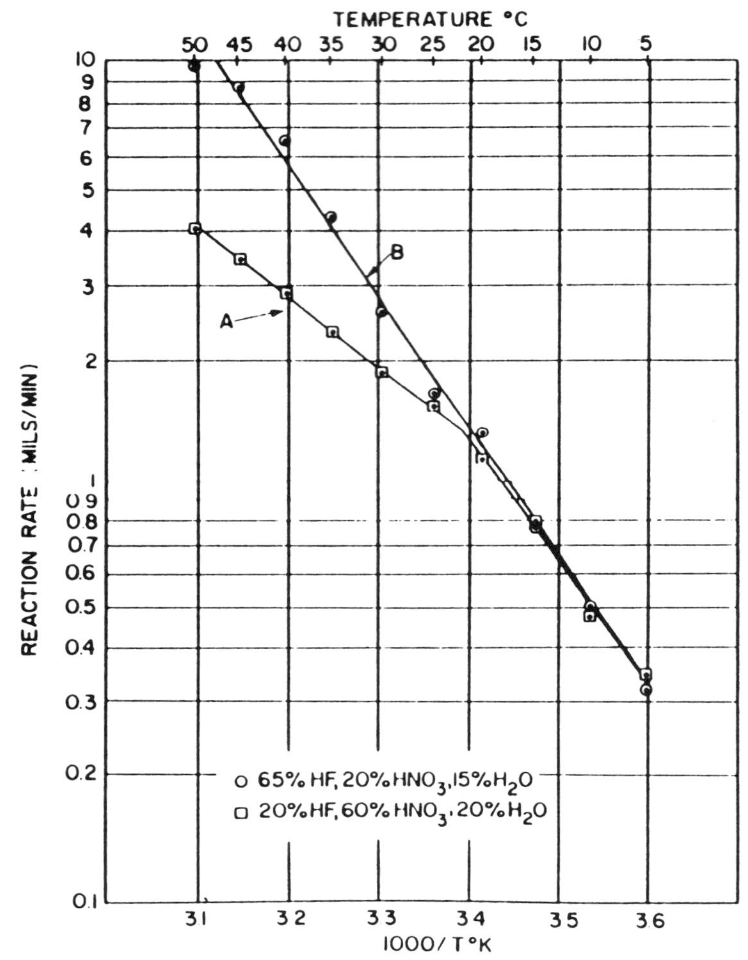

Figure 7.25 Etching rate of silicon versus reciprocal temperature for two etchant compositions. After Schwartz and Robbins [63]; copyright © 1961, The Electrochemical Society, Pennington, NJ.

concentration of complexing fluorine ions, the reaction becomes diffusion limited at a threshold temperature. This is indicated by the break in the curve that occurs because diffusion in the etchant solution has a lower activation energy than the surface reaction.

Surface-controlled reactions depend critically on the local bonding and proceed in the vicinity of the emergence points of dislocation at the surface with faster rates than on the surrounding perfect crystal surface since the dislocation core is frequently decorated by impurities and strained bonding. Thus etch pits develop. The slope of the side faces of these pits depends on the ratio of the vertical component of the etch rate at the dislocation core to the

(a)

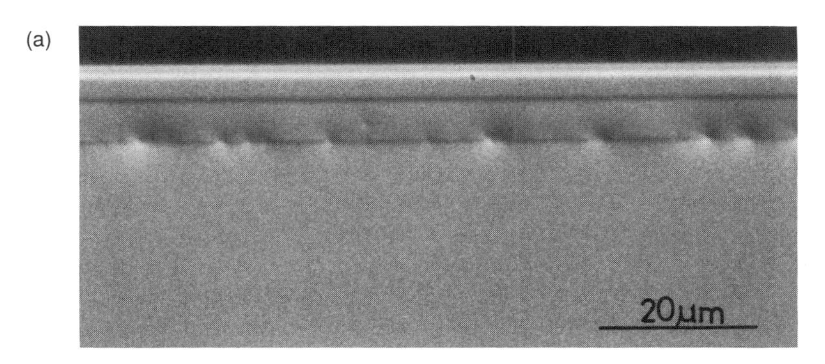

(b)

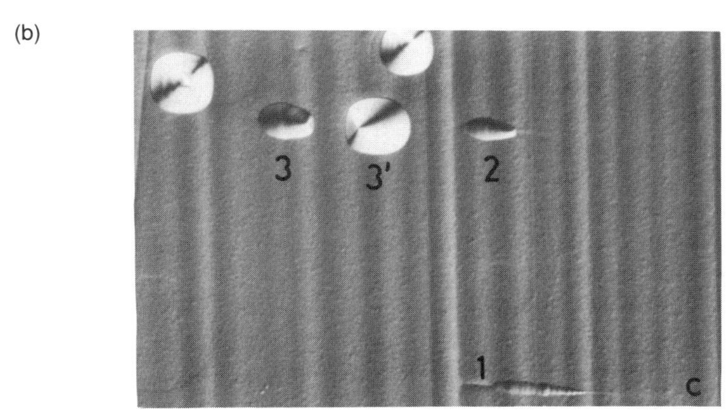

Figure 7.26 Photochemical etching features revealing (a) emergence points of dislocations and striations on a wafer of InP(001) and (b) a three-layer epitaxial structure on an InP substrate for a cleaved (110) cross section. After Weyher and Giling [64]; copyright © 1985, American Institute of Physics, New York.

horizontal component of the etch rate. If the edge free energy is highly anisotropic, crystallographic features develop, for example, triangular pits on {111} (see Fig. 3.43). However, this is not always the case, as, for example, for the dislocation etch pit obtained with the Secco etch [34] on the (001) surface of a silicon crystal or the shallow photoetched pits at emergence points of dislocations on InP(001) as shown in Figure 7.26 [60].

Photochemical etching in diluted Sirtl etch ($HF-CrO_3-H_2O$) utilizes photogenerated minority holes in *n*-type semiconductors in bond-breaking surface reactions that delineate dislocations in III–V compounds and alloys and also reveal the location of *pn* junctions and striations for extremely small etch depth (i.e., typically 50 nm). It must be emphasized that etch pits are a reliable means of determining dislocation densities only if a 1:1 correspondence has been established for the particular etch by another means (e.g., x-ray topography).

Staining etches reveal slight differences on the doping level of semiconduc-

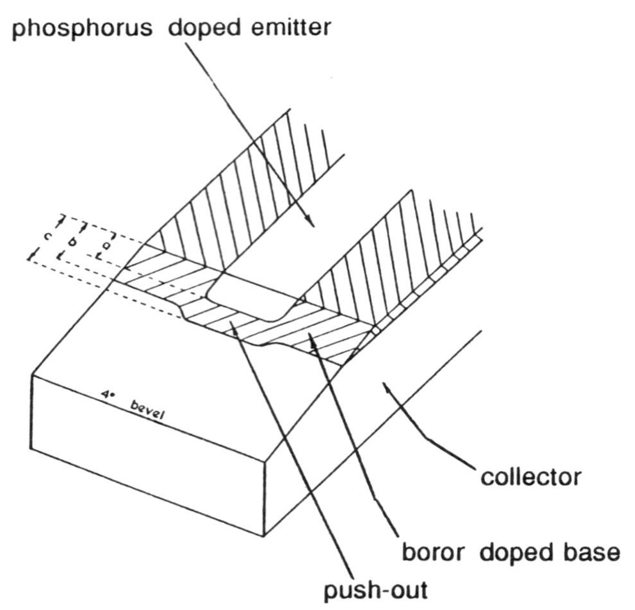

Figure 7.27 Schematic representation of a 4° angle lap to enhance vertical features of an *npn* bipolar transistor structure. After Lee [65]; copyright © 1974, Philips Research Laboratories, Eindhoven.

tors. In addition to decorating striations in bulk crystals, they are useful in delineating of *pn* junctions by etching of angle-lapped cross sections that amplify the vertical dimensions by a reciprocal sine factor, as illustrated in Figure 7.27. Figure 7.27 shows the emitter push effect in an *npn* bipolar transistor for a stained angle-lapped cross section [59]. The anomalous diffusion behavior that is responsible for the faster diffusion of the boron-doped base below the phosphorus-doped emitter is explained in Chapter 8.

Orientation-dependent etching of silicon in KOH etchant proceeds along the $\langle 100\rangle$ and $\langle 110\rangle$ directions with much faster rates than along the $\langle 111\rangle$ direction. The etching rate v_e depends on the KOH concentration and peaks at ~16–20%. Figure 7.28 shows the etching rate for 44% KOH solution at unobstructed faces in the angular ranges between (011), (111), and (100). At a concentration of 44 wt %, $v_{e\langle 110\rangle}:v_{e\langle 100\rangle}:v_{e\langle 111\rangle} = 600:300:1$ [66]. This anisotropy in the etching rate is widely utilized for the fabrication of x-ray reflectors and gratings in the micromachining of field emitters for vacuum electronics applications, mechanical and chemical sensors, trench capacitors, microporous silicon, and vertical multijunction solar cells. Although the etching rate is strongly anisotropic, circular semispherical mirrors can be etched into Si(001) surfaces by KOH etching [68]. Note that the etching rate on Si(110) by aqueous KOH solutions is 100–1000 times faster for SiO_2. This is illustrated in Figure 7.29, which shows the temperature dependence of the etching rate for Si(110) and SiO_2. The large etching rate difference has been utilized in the

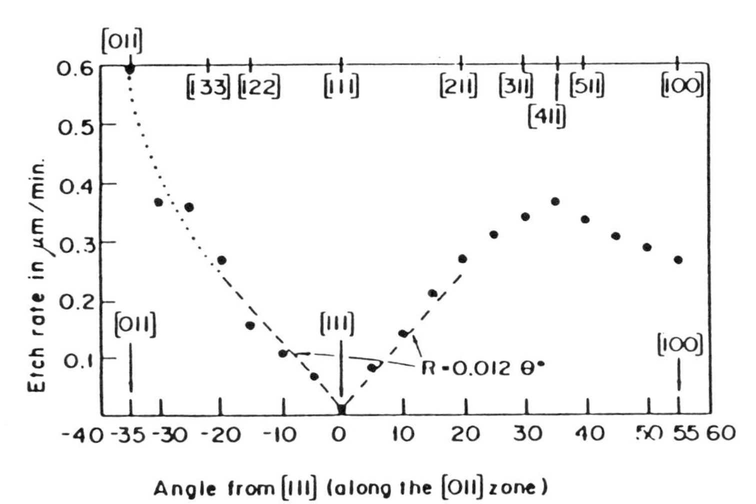

Figure 7.28 Etching rate in a 40% aqueous solution of KOH at 62°C for unobstructed faces of Si in the angular ranges {110} → {111} and {111} → {100}. After Kendall [67]; reproduced with permission, from the Annual Reviews of Materials Science, Vol. 9, copyright © 1979, Annual Reviews, Inc., Palto Alto, CA.

fabrication of microporous materials and has spurred a variety of ideas concerning commercial applications [69].

Recently micromachining of Si has become of particular interest in the context of AFM. In the construction of a Si sensor tip on a silicon cantilever attached to a silicon handling block, a force constant of $0.1 \leqslant C_S \leqslant 100$ N/m and resonance frequency $f_R \geqslant 100$ kHz are required for high sensitivity and bandwidth of operation [70]. A rectangular silicon cantilever beam of $3 \times 20 \times 100$ mm dimensions satisfies this condition with $C_S = 20$ N/m and $f_R = 400$ kHz. It is fabricated by first defining the cantilever width by photolithography and RIE. Then the silicon below the ridge thus formed is removed by an anisotropic etch, relying on its appropriate crystallographic orientation. Either aqueous ethylene diamine/procatechol or KOH solutions are suitable for this step. The tip is defined by providing a circular mask that is undercut by etching until a cone with a sharp apex is produced and the mask is removed. The integral cantilever–stylus assembly is then mounted on a silicon frame by thin silicon brackets, completing the sensor fabrication.

In applications where the shaped structure needs to be integrated with microelectronic circuits and the metal ion contamination by the KOH etching solution is a problem, the dopant-selective etching by HF can be utilized to underetch the cantilever beams [72]. Further enhancements in the control of etching and in the range of products are achieved by electrochemical etching. Of particular interest are:

1. The formation of black silicon upon anodic polarization of an *n*-type silicon

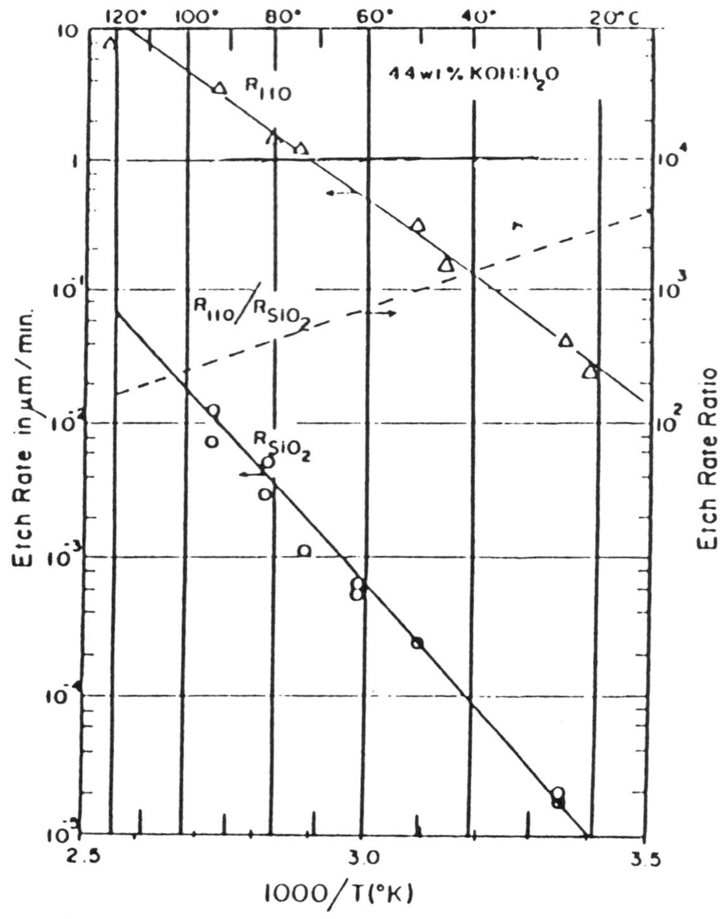

Figure 7.29 Arrhenius plot of the etching rates on Si(110) and SiO_2 by a 44 wt % aqueous KOH solution. After Kendall [67]; reproduced with permission, from the Annual Reviews of Materials Science, Vol. 9 copyright © 1979, Annual Reviews, Inc., Palo Alto, CA.

electrode under illumination with photons of above-bandgap energy in a dilute HF electrolyte [73];

2. The formation of porous silicon upon anodic treatments of *p*-type silicon under illumination with photons of above-bandgap energy in a dilute HF electrolyte [74], [75];
3. The formation of trenches in backside-illuminated *n*-type silicon wafers upon front-side anodization in dilute HF electrolyte [71].

An explanation for the formation of the black-surface structure and high aspect ratio holes/trenches in front- and backside-illuminated *n*-type silicon is given in Figure 7.30. We recall from Section 5.1 that a depletion layer is formed at

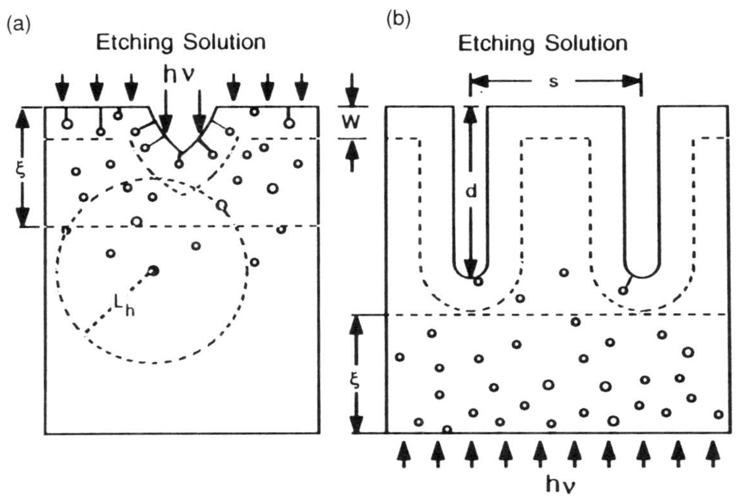

Figure 7.30 The formation of (A) black silicon and (B) high aspect ratio trenches in front and backside illuminated N-silicon, respectively, contacted on the front face by a photoetchant solution. O-Symbol for a photogenerated hole.

contacts between properly chosen electrolytes and semiconductors. For moderately doped semiconductors the depletion layer $W \approx 100\,\text{nm}$, which is smaller than the minority hole diffusion length L_h and the absorption length ξ of the incident light (depending on the energy $h\nu > E_g$). Within the depletion region the hole motion follows the field lines, while outside this region the hole diffusion follows a random path. The probability of a hole reaching the depletion region from the outside parts of the semiconductor is proportional to the distance of its photogeneration point to the depletion layer fringes divided by the diffusion length. Therefore, a majority of holes is collected at the bottoms of the surface depressions, where they enter into bond-breaking reactions at the solid/liquid interface if the quasi-Fermi level for holes is more negative than the potential $\varepsilon = \Delta G_r/zF$ measured from the vacuum level. Other charge-transfer reactions scavenging holes at the semiconductor/electrolyte interface decrease the rate of this process and must be minimized. This governs the choice of the electrolyte. Because of the strong decrease of the density of the photogenerated minority carriers from the entrance surface into the interior, upon front illumination, the surface depressions widen toward the entrance surface, while nearly vertical side faces are maintained upon backside illumination [71].

The emission of visible light from photoexcited porous silicon has found widespread interest. However, since porous silicon is a rather complex material, incorporating crystalline and amorphous regions of varying size and surface properties, this effect is currently not completely understood. In the opinion of the author, the expected problems with the reliability of electroluminescent devices favor alternative approaches to monolithic optical interconnect and sensor functions on Si chips.

7.4 Dry Etching

In dry etching processes, the species that attack the surface of the semiconductor are either neutral gas molecules that generate volatile reaction products (vapor etching) or energetic ions that remove surface atoms either by momentum transfer only (sputtering) or ions that assist the chemical attack by neutrals (ion-assisted etching) or directly chemically react with the surface atoms to form volatile molecules (reactive ion etching). In addition, electron-assisted etching and photon-assisted etching processes are being used in microelectronic processing. Etching by sputtering is nonspecific since the sputtering yields of different materials varies only within a factor of ~ 3. Also, the sputtered vapor may condense in the vicinity of the eroded areas, leading to uncontrolled modifications of the surface relief, which is generally an

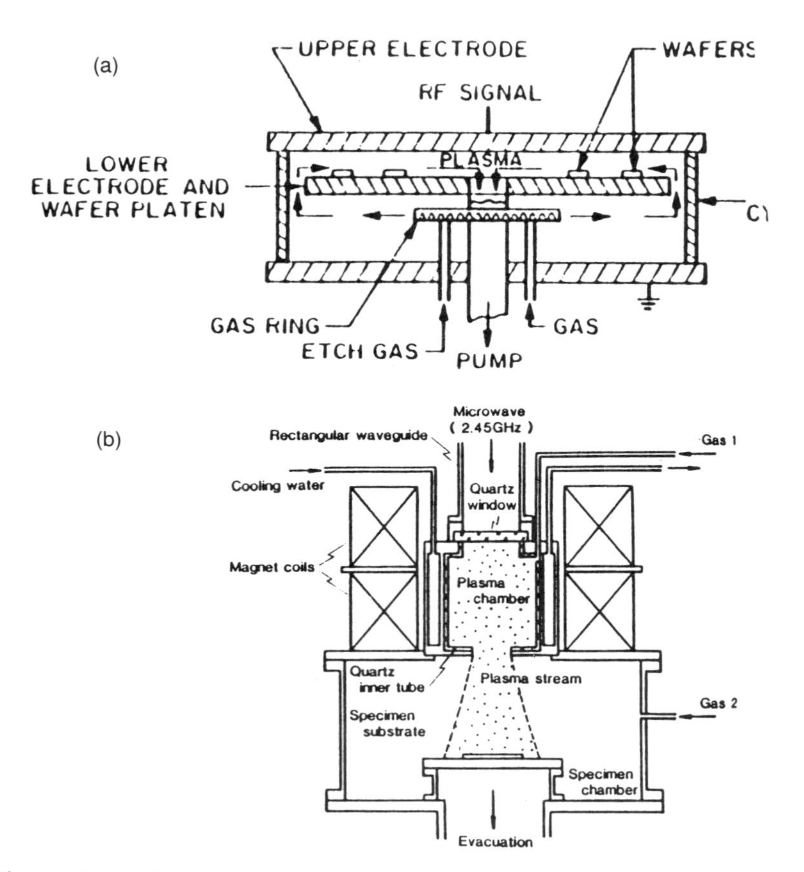

Figure 7.31 Schematic representations of (a) a Reinberg-style rf plasma etcher and (b) an ECR etcher. After (a) Flamm and Mucha [76] and (b) Ohno, Oda, Takahashi, and Matsuo [77]; copyright (b) © 1987, American Institute of Physics, New York.

undesirable feature. Etching processes that generate volatile molecules avoid this problem.

A frequently employed experimental plasma etching arrangement is illustrated in Figure 7.31(a). The pressure in such rf plasma etchers that operate at 13.6 MHz is typically 0.01–1 Torr, and the density of charged particles in the resulting glow discharge plasmas is small (10^9–$10^{12}\,cm^{-3}$) as compared to the neutrals (10^{15}–$10^{17}\,cm^{-3}$). Most of the collisions in the vapor phase generate excited neutrals and radicals. Since the recombination velocity of charged particles in the plasma is fast as compared with the gas-phase chemical reactions, the energetic neutrals may be utilized for reactive etching downstream of the plasma without bombardment of the surface with charged particles, similar to the conditions of RPECVD. This is utilized in photoresist stripping and other mild etching processes where anisotropy is not an issue. There are a variety of source gases with and without additives and materials that can be etched by them under the conditions of plasma etching listed in Table 7.3.

Electron-cyclotron-resonance (ECR) etchers operate at substantially lower pressure, typically 10^{-5}–10^{-4} Torr, where the charged particle recombination is slow as compared to gas transport processes and the mean free path is several tens of centimeters. Consequently highly directional low-energy ions, typically of 20 eV energy, can be extracted from such plasmas by means of a biased grid or magnetic field for ion etching with minimum ion damage of the

Table 7.3 Examples of plasma etching gases and radicals (after Ibotson and Flamm [78])

Etching species	Source gas	Additive	Additive effect	Materials etched	Etching mechanism	Selective over
F	CF_4	O_2	Oxidant	SiO_2	Ion-energetic	GaAs
	SF_6	O_2				InP
	NF_3	Ar	Diluent			Resist
CH_x?	CH_4	H_2	Radical scavenger	GaAs	Ion-energetic (?)	SiO_2
			Group V etchant	InP		Al
			Oxide etchant	$Ga_xIn_{1-x}P_yAs_{1-y}$		
H	H_2	None		GaAs, InP	Chemical	Si, SiO_2
				$GaAsO_x$		Metals
Cl	Cl_2	None		GaAs	Chemical	SiO_2
		None		GaAs, InP	Ion-energetic	SiO_2
		H_2O, O_2	Inhibitor	GaAs, InP	Chemical	SiO_2
		$C_2F_2Cl_2$	Inhibitor		Ion–inhibitor	$Al_xGa_{1-x}As$
		BCl_3	Inhibitor	Al, GaAs, $Al_xGa_{1-x}As$	‖	Resist
$SiCl_4$	Oxide Etchant			SiO_2		
		CCl_4	Inhibitor			
Br	Br_2	He	Stabilizes plasma	GaAs	Chemical	SiO_2
I	I_2	None		GaAs, InP	Chemical	SiO_2

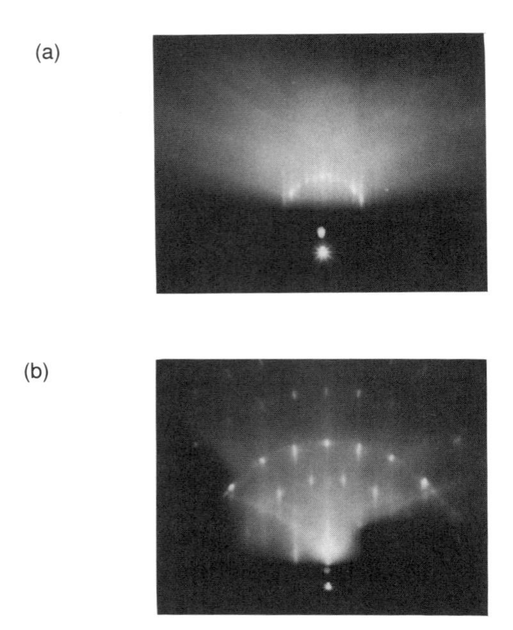

Figure 7.32 RHEED pattern observed after 1 h etching in a 20 eV hydrogen ECR plasma beam at 10° incidence on GaAs(001) at 450°C: (a) [110] azimuth and (b) [$\bar{1}$10] azimuth. After Suemune, Kunitsugu, Kan, and Yamanishi [79]; copyright © 1989, American Institute of Physics, New York.

semiconductor at a relatively small flux of reactive neutrals. However, the gain in the control of damage in ECR etching is paid for by a lower removal rate, which may be welcome in research tasks, but is an impediment in industrial processing. A typical ECR etching arrangement is shown schematically in Figure 7.31(b).

An important parameter of ECR plasmas etching is the angle of incidence. Figure 7.32 shows RHEED patterns of the GaAs(100) surface after 1 h etching in an ECR hydrogen plasma beam at 10° incidence. Under this condition, momentum transfer parallel to the surface enhances significantly the surface migration and consequently the flattening of the surface. Thus perfectly flat GaAs(100)-2 × 8 surfaces are obtained, which is not the case for 90° incidence under otherwise similar conditions [79]. Note that charging of the dielectric mask surface areas alters the angle of incidence, leading to deviations of the etched profiles from the intended shape. Figure 7.33 shows the bowing in the ion trajectories upon (a) positive and (b) negative charging of the mask surface and the associated distortions when large open spaces exist close to the areas, where smaller features are to be etched. The use of conducting masks eliminates this problem, which may, however, reappear if by patterning the conducting mask is partitioned into isolated islands.

In rf plasma etching the control of the ion energy and impingement angles is less direct, and the control of the anisotropy thus relies frequently on a

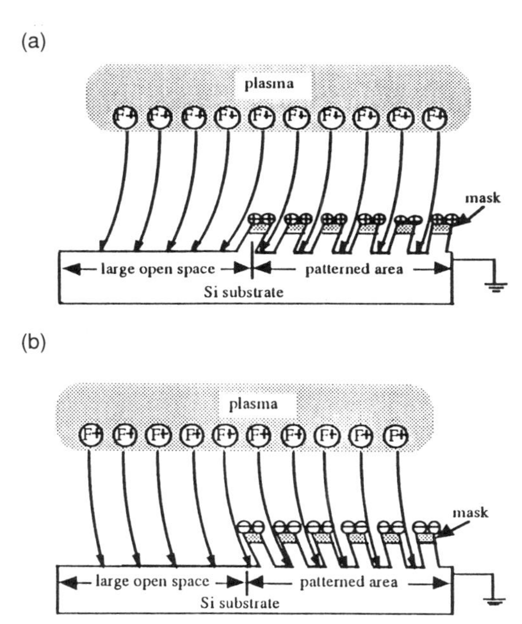

Figure 7.33 Etching profile distortions by (a) positive and (b) negative charging of dielectric mask areas on a silicon surface. After Murakawa, Fang, and McVittie [81]; copyright © 1992, The Institute of Electrical and Electronics Engineers, New York.

judicious choice of the composition of the processing vapor. For example, the undercut in the plasma etching of Si by Cl_2 can be controlled by the addition of C_2F_6. The dissociative electron collision of this molecule generates CF_3, which is absorbed at the walls during etching. The etching is accomplished by chlorine atoms formed in the primary dissociative electron collisions of the Cl_2 molecules. Recombination of the chemically active chlorine atoms with CF_3 to CF_3Cl slows down the etching. Desorption of the CF_3 from the walls and the dissociation of CF_3Cl by ion collisions thus provides the mechanism for controlling the anisotropy of this etching process [80]. Although Cl atoms are effective for etching both silicon and III–V compounds, certain materials (e.g., Al), are difficult to etch and require special tricks, as, for example, inert ion-assisted etching. Small amounts of water added to the etchant vapor provide for a reliable etch stop in the etching of $GaAs/Al_xGa_{1-x}As$ [76].

Halocarbons (e.g., CF_4) are preferred etchant vapors because they can be handled far more conveniently than the halogens themselves and their hydrides. Upon fragmentation in the plasma, they tend to form oligomers of the unsaturated fragments. Although this is a welcome effect in side-wall-protected ion-assisted etching, it is clearly a nuisance in etching processes that aim at high rate because the formation of free halogen atoms often controls the etching mechanism and is suppressed by the oligomerization of the unsaturated fragments. Admixture of oxygen up to an optimum concentration helps in this case since the added oxidant will preferably react with the unsaturated fragments of the halocarbons. A substantial amount of work has been devoted to the investigation of the reaction kinetics of reactive ion etching, combining mass spectrometry, spectral analysis of the radiation emitted from the plasma [80], and surface analysis techniques, in particular XPS. For example, the emission lines of F appear as a prominent feature in the emission spectra of a CF_4/O_2 plasma. They may be utilized for determining the fluorine concentration, which depends critically on the oxygen concentration. For a more comprehensive review of the emission spectra of plasma processes, the reader is referred to Ref. [80].

Silicon etching by fluorcarbon plasmas proceeds via a branching mechanism that is discussed in more detail in Ref [76]. XPS shows that the formation of Si–F bonds are not restricted to the surface, but continues to a depth of ~ 10 atomic layers into the silicon [82]. It is significant that, in contrast to fluorine atoms, fluorine molecules form only a monolayer of fluorosilicon species at the

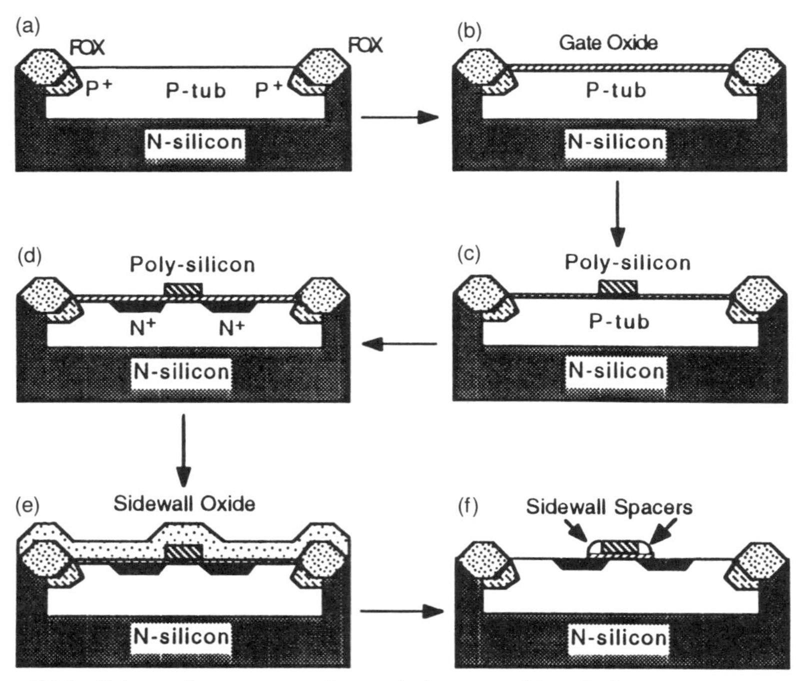

Figure 7.34 Schematic representations of the gate sidewall fabrication process: (a) Junction-isolated p-tub in an n-type silicon epilayer with periferal surface passivation by field oxide and an ion implanted p^+ channel-stop ring to counteract latch-up; (b) after gate oxidation; (c) after polysilicon gate metal deposition and paterning; (d) after self-aligned shallow n^+ implants to define the n-channel region under the gate; (e) after sidewall oxide deposition; (f) after sidewall spacer fabrication and gate oxide patterning by dry etching. A deeper n^+ implant following step (f) that is self-aligned with regard to the sidewall spacer edges defines the source and drain contact regions of the N-channel transistor.

surface. The stability of the surface in the presence of a flux of reactive F atoms, the transport in the fluorinated film, and the reaction mechanism at the film–silicon interface are at present not completely understood. However, since Si carries in SiF_4 a net positive charge, while it is covalently bonded in the silicon lattice, evidently a charge transfer occurs during the reaction that is oxidative with regard to the silicon atoms and reductive with regard to the fluorine atoms, which have a strong oxidation potential and supply the driving force for the reaction.

The dry etching of SiO_2 is important for pattern transfer, as, for example, in the fabrication of the gate side-wall spacer (step 23 in Table 4.3). Figure 7.34 illustrates this processing step, which affects the critical edge region of the doping profiles under the source and drain of MOSFETs and the channel width. High selectivity of the dry etching process is essential in this case and is accomplished by fluorine-deficient fragments of fluorocarbon compounds (e.g., the CF_3 and CF_2 moieties), which attack SiO_2 at a much higher rate than Si. The goal of SiO_2 plasma etching is thus to maximize the fraction of unsaturated fragments relative to the fluorine atom concentration in the plasma. Excellent results have been obtained with mixtures of fluorocarbon compounds and hydrogen-containing compounds, such as C_2F_6/CHF_3 mixtures [83]. Gas mixtures are also used for the etching of Si_3N_4, (e.g., with CF_4/O_2). The etching of polymer coatings by an oxygen plasma has been discussed already in the context of photolithography and proceeds via hydrogen abstraction followed by autoxidation.

An important issue in the mechanism of both plasma and reactive ion beam etching is the interplay between ion bombardment effects and chemical attack. This interplay is more straight forward under the conditions of ion-assisted etching, where the ions enhance the interactions of the neutrals with the surface in two different ways:

1. A weakening of the bonding of the surface atoms by the ion bombardment so that the reaction rate with the simultaneously impinging neutrals is enhanced. In such an energy-driven etching mechanism the neutrals by themselves generally do not attack the surface at a significant rate.
2. The ions remove an inhibiting film that is formed upon the spontaneous interaction of the neutrals with the surface. The growth of a protective film on the side walls of the surface depression generated by this mechanism is exploited to suppress undercutting of the surface mask edges.

Since the damage and the sputtering rate both depend on the ion energy, the control of the etching anisotropy involves the control of the neutral and ion fluxes and of the ion energy.

In inert-ion-assisted etching, where a jet of neutrals of a reactive gas impinges simultaneously with an inert ion beam on the surface, these control parameters are separable [84]. Figure 7.35(a) shows a schematic representation

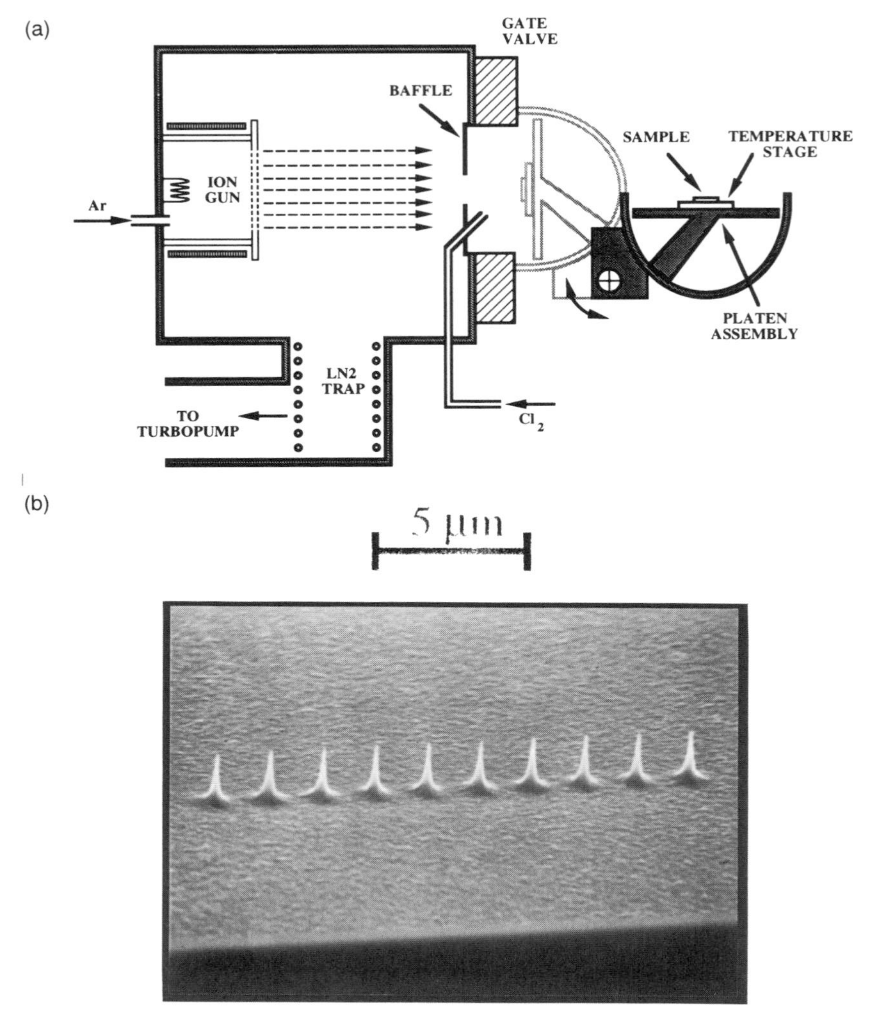

Figure 7.35 (a) Schematic representation of an apparatus for inert-ion-assisted etching and (b) array of GaAs field emitters made by Ar^+-assisted chlorine etching of GaAs. After Grande, Johnson, and Tang [85]; copyright © 1990, American Institute of Physics, New York.

of a typical experimental arrangement providing for inert-ion bombardment from a Kaufman ion source and chemical etching in chlorine gas. Figure 4.35(b) shows a linear array of GaAs field emitters with tip radii $\leqslant 50$ nm in the as-machined state [85]. Also slots of <50 nm width have been etched by ion-assisted chlorine etching into GaAs and Si using argon [84] and gallium [86] ions, respectively, at normal incidence.

The fabrication of field-emitter arrays nested in a set of gate electrodes is of interest for vacuum microelectronics, replacing transistors by triodelike devices that conduct entirely on the basis of ballistic electron transport in vacuum. Vacuum microelectronics may make an impact in the context of microwave generation and amplication, flat panel dislays, and multiple electron sources for SEMs and related instruments. For example, addressably operated gated electron emitter arrays may serve as intense sources for parallel ultrahigh-resolution electron-beam writing. Attractive properties of gated electron emitter arrays are their inherent radiation hardness, their relative insensitivity to temperature changes as compared to silicon transistors, the ease of implementing focusing electron lenses as part of the circuit, and the high predicted cutoff frequency, which is of the order of 100 GHz–1 THz [89]. Figure 7.38 shows a typical arrangement for the testing of field emitter devices are characterized by small leakage currents [87]. However, the generally larger voltage ($\geqslant$100V) required for the operation of gated field-emitter arrays as compared to transistor circuits is clearly a disadvantage. Also, their fabrication with reproducible and uniform properties is not a trivial task. Clearly alternative processes and materials must be developed in the context of flat panel displays, which requires the processing of large areas and includes the development of efficient phosphors. On a laboratory scale, valuable insights into electron stimulated emission can be gained with presently available technology.

An attractive processing route for the fabrication of silicon field emitters is shown in Figure 7.36. Anisotropic etching that undercuts a set of circular SiO_2

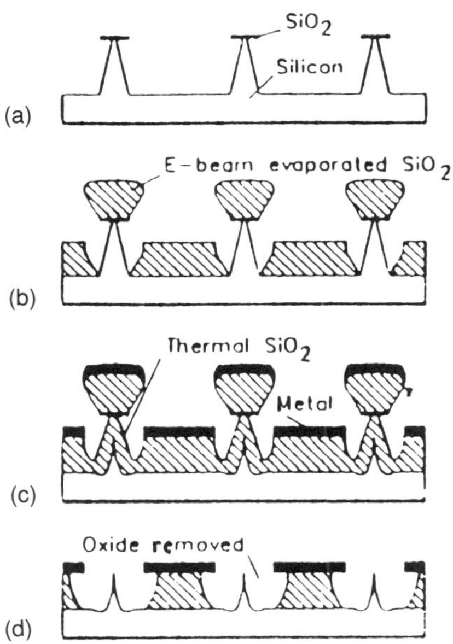

Figure 7.36 Schematic representation of a processing sequence for the fabrication of a gated field-emitter array. After McGruer et al. [87]; copyright © 1991, The Institute of Electrical and Electronics Engineers, New York.

[87] or Si_3N_4 [90] masks permits the fabrication of pyramidal field-emitter arrays. However, although the micromachining of silicon is in an advanced stage of development, silicon field emitters are subject to blunting and chemical degradation in air. This motivates research on the fabrication of transition metal field-emitter arrays and silicon field emitters coated by more durable thin films (e.g., SiC [94]). For example, the replacement reaction

$$2WF_6(g) + 3Si(s) \rightarrow 3SiF_4(g) + 2W(s) \tag{7.15}$$

has been utilized to convert silicon tips into metal tips [91]. Figure 7.37(a) and (b) shows a schematic representation of a cross section through a gated field-emitter array and an SEM micrograph, respectively, revealing the nesting of individual emitters in openings in the gate metal that is isolated from the base by a layer of SiO_2. Large arrays at $\sim 2\,\mu m$ spacing between individual emitters have been produced [88].

They are appropriately shaped from the point of view of heat dissipation, but are not ideal in their field-emission characteristics, which is optimized by an Eiffel tower geometry. This nearly ideal emitter shape is realized most closely by silicon whiskers that are sharpened by sequential oxidation and etching [92]. Silicon whiskers are grown by the vapor–liquid–solid mechanism [93], which is based on the orders of magnitude higher rate of silicon incorporation by CVD on the surface of a silicon–gold alloy droplet at the

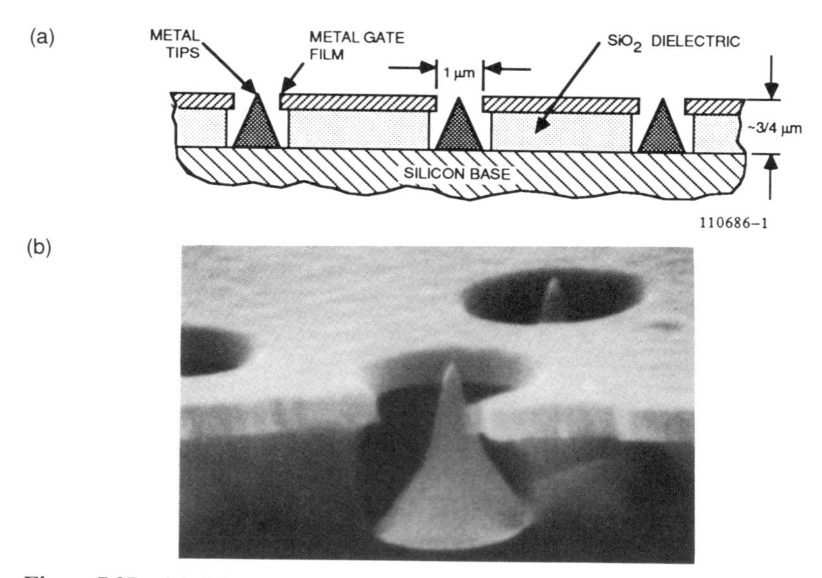

Figure 7.37 (a) Schematic cross section through a gated field-emitter array. (b) SEM image of such an array. After Spindt [88]; copyright © 1992, Institute of Electrical and Electronics Engineers, New York.

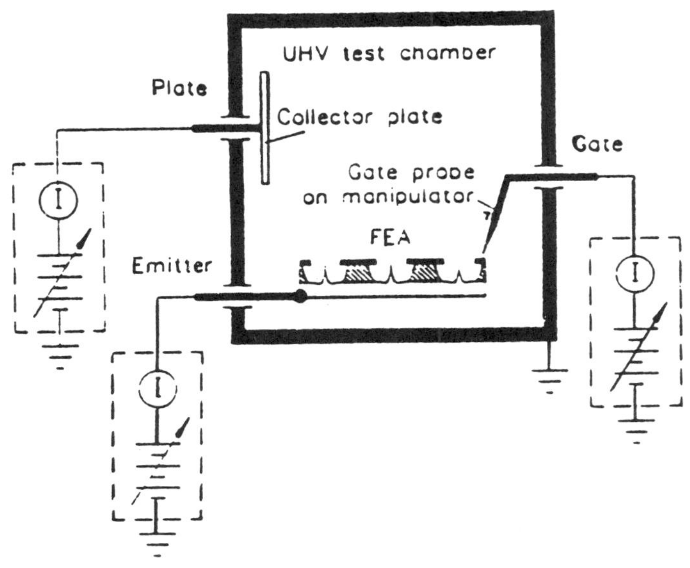

Figure 7.38 Schematic representation of a gated field emitter array test circuit. After McGruer et al. [87]; copyright © 1991, The Institute of Electrical and Electronics Engineers, New York.

whisker tip and diffusion to the solid–liquid interface as compared to the deposition rate on the solid silicon whisker side walls. Note that, though less desirable in shape, the micromachined pyramids [90] can also be sharpened by oxidation to 10 nm radius of curvature and below. Their use provides for advantages with regard to the reproducibility of the spacings and heights of the field emitters. An alternative mechanism of tip sharpening by interactions with the imaging gas under the conditions of FIM has been reported for soft metal tips [95].

As a supplement to focused ion beam etching, electron-assisted etching is capable of maskless pattern transfer with submicrometer resolution [96]. The momentum transfer to the surface atoms is negligible in this case, and the enhancement of the etching rate is primarily due to chemical changes on the surface. Thus latent images can be generated by electron-beam writing and subsequently developed by chemical etching, avoiding the need for integration of the electron-beam writing tool and etching chamber [97]. Figure 7.39 shows the experimental arrangement for showered electron-assisted etching and a set of 0.3 μm lines etched by this method into GaAs [98].

Maskless etching with submicrometer resolution is also possible by photon enhancement of the rate of attack [99], which is based on a combination of minority carrier generation, photolysis and activation of vapor precursors to etching, and surface heating. The latter is only important in laser-assisted etching at relatively high power density. Figure 7.40 shows the maximum local temperature versus laser power for photon-assisted etching GaAs, InP, and

(a)

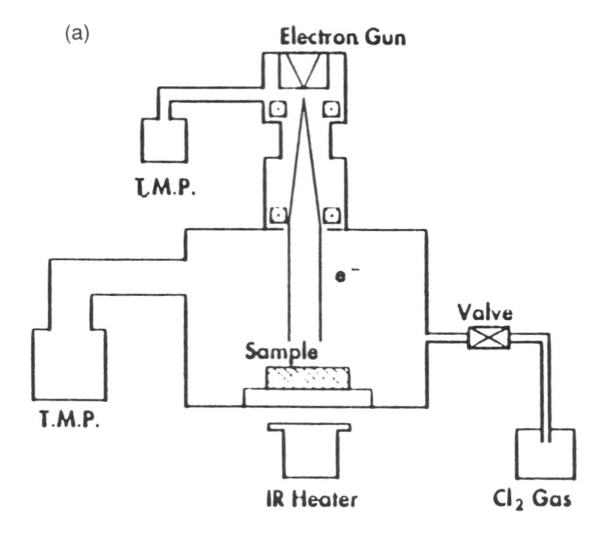

(b)

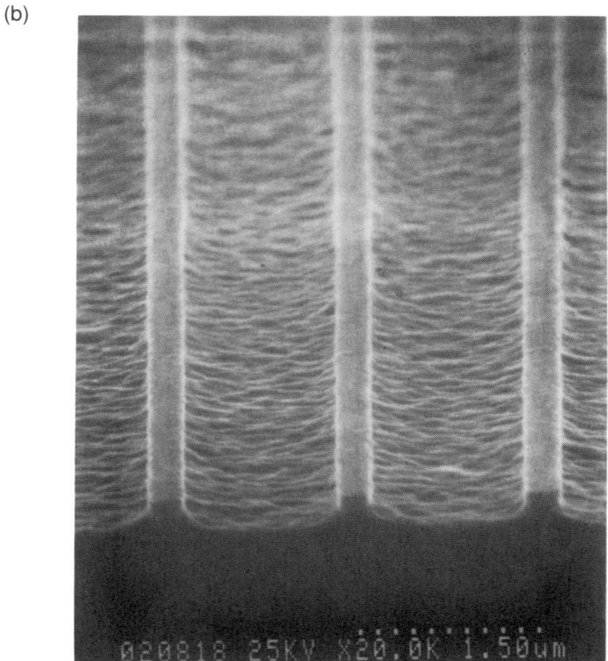

Figure 7.39 (a) Schematic cross section of the apparatus for showered electron-beam-assisted etching. (b) SEM image of 0.3 μm wide lines in GaAs. After Matsui and Watanabe [98]; copyright © 1989, The Electrochemical Society, Pennington, NJ.

InSb with 514.5 nm Ar^+ laser radiation focused to 13.2 μm ($1/e$ intensity) [100] [101]. Thermochemical etching in CCl_4 becomes dominant at the power densities and temperatures indicated by broken lines. Photoluminescence scans

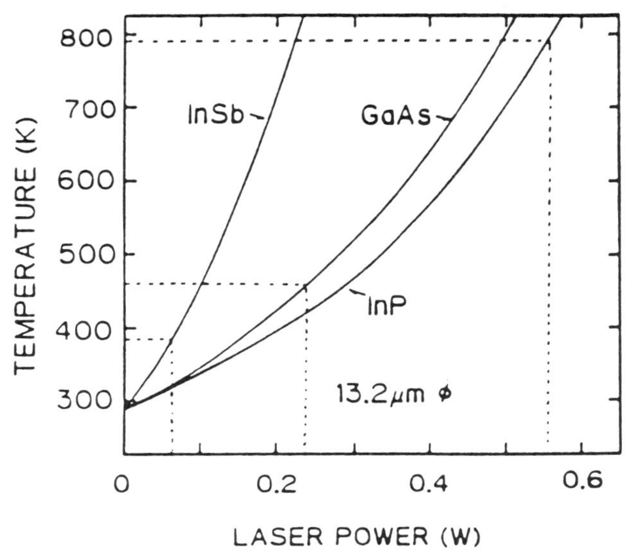

Figure 7.40 Maximum local temperature versus laser power for GaAs, InP, and InSb. After Takai et al. [100]; copyright © 1988 Springer Verlag, Berlin.

of the etched and unetched parts of a GaAs surface reveal damage at a laser power density in the thermal etching regime, which is probably related to changes in the stoichiometry, leading to enhanced recombination. Arrhenius plots of the etch rates result in activation energies $E_a(\text{GaAs}) = 0.168\,\text{eV}$, $E_a(\text{InP}) = 0.156\,\text{eV}$, and $E_a(\text{InSb}) = 0.161\,\text{eV}$, which is significantly lower than the activation energies obtained under the conditions of plasma etching [102].

The rate of photochemical etching is generally wavelength dependent both because of the dependence of the adsorption length of the light in the semiconductor on the photon energy and because of specific thresholds for precursor activation in the gas phase. For example, GaAs cannot be etched in HBr at the 248 nm KrF laser wavelength, but is etched upon illumination with 193 nm ArF radiation [103]. Figure 7.41 shows the etch-rate dependence of Si by Cl_2 on the laser power for selected wavelengths at a chlorine pressure of 300 mbar. The rate for a given power depends strongly on the wavelength and is zero for 647.1 nm Kr^+ laser radiation [104]. The rate of photon-assisted etching of semiconductors also depends on the doping level, which entails both the effects of doping on the local chemical bonding and minority-carrier-assisted etching, which depends on the doping-dependent surface field that accelerates the majority carriers toward the surface [105].

Since the capacitance values used in IC designs are more or less fixed, the size of capacitors cannot be scaled in accord with the scaling of the feature size of transistors. This presents a challenge in the design and construction of ULSI circuits, in particular, large DRAM memory that relies on a high packing density of storage cells. Several approaches have been used to address this

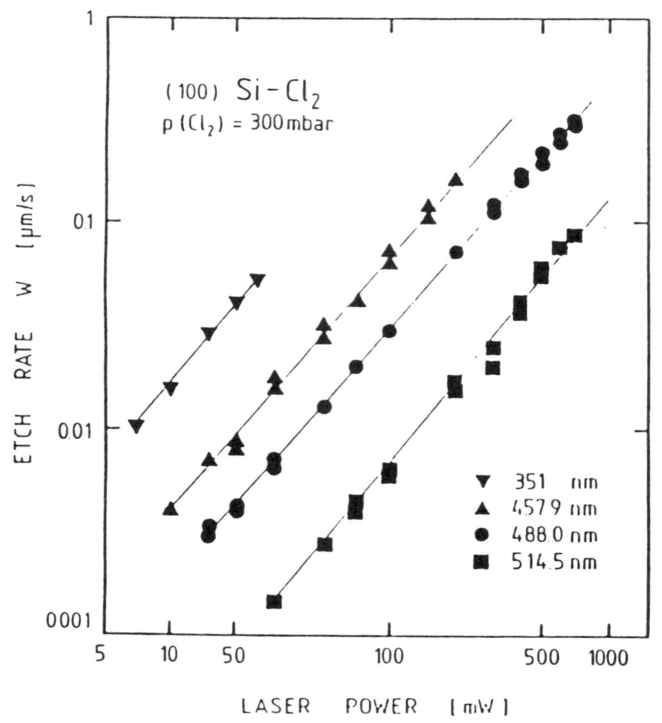

Figure 7.41 Etch rate versus laser power for the photon-assisted etching of Si in Cl_2 at selected wavelengths. After Mogyorósi et al. [104]; copyright © 1988, Springer Verlag, Berlin.

problem that rely critically on the highly directional etching processes discussed here. For example, trench capacitors are fabricated in vertical trenches etched into the silicon, which are coated by a thin dielectric film and then filled by the electrode metal. They reduce the real estate occupied by the capacitor on the silicon surface [107], but cannot satisfy the needs beyond 256 Mbit DRAMs. Innovative improvements utilize hemispherical grained (HSG) silicon electrodes in the construction of cylindrical capacitors with a factor 1.9 gain in the capacitance, as compared to straight cylindrical electrode capacitors of otherwise the same construction with little penalty in the leakage current. Figure 7.42 presents (a) successive stages in the manufacturing of the HSG-Si electrode and (b) the capacitance–voltage characteristics of stacked electrode, cylindrical electrode, and HSG cylindrical electrode capacitors employing a silicon oxide/silicon nitride (ON) dielectric [106]. The HSG-processing sequence starts with the patterning and growth of the polysilicon contact stem through openings in a SiO_2 isolation layer, followed by the deposition of a phosphorus-doped amorphous silicon layer and boron-phosphorus-silicate glass (BPSG). The P-doped α-Si/BPSG double layer is then patterned into plugs that are coated on the outside by further P-doped α-Si, which, after

(a)

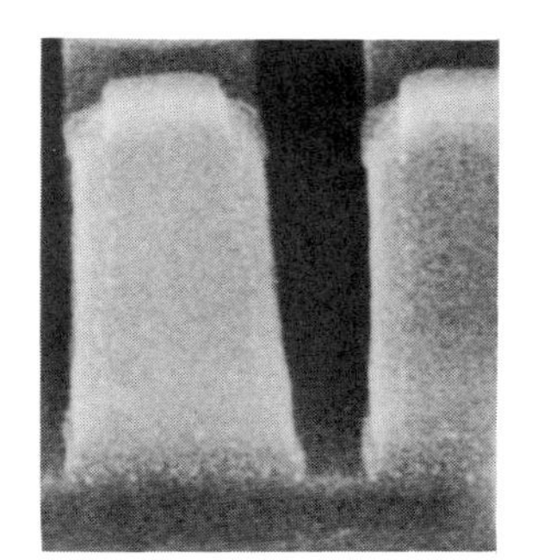

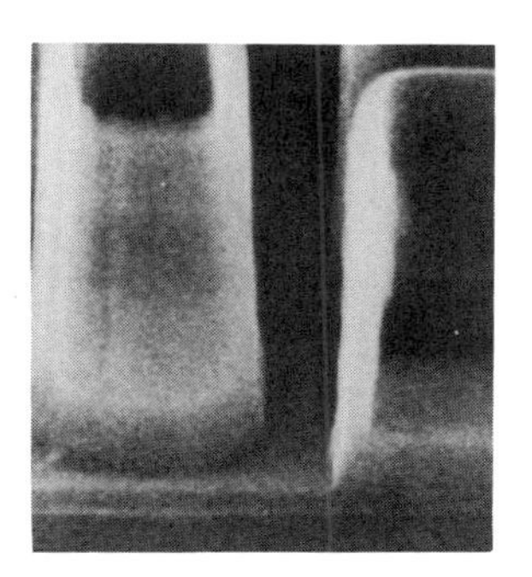

(b)

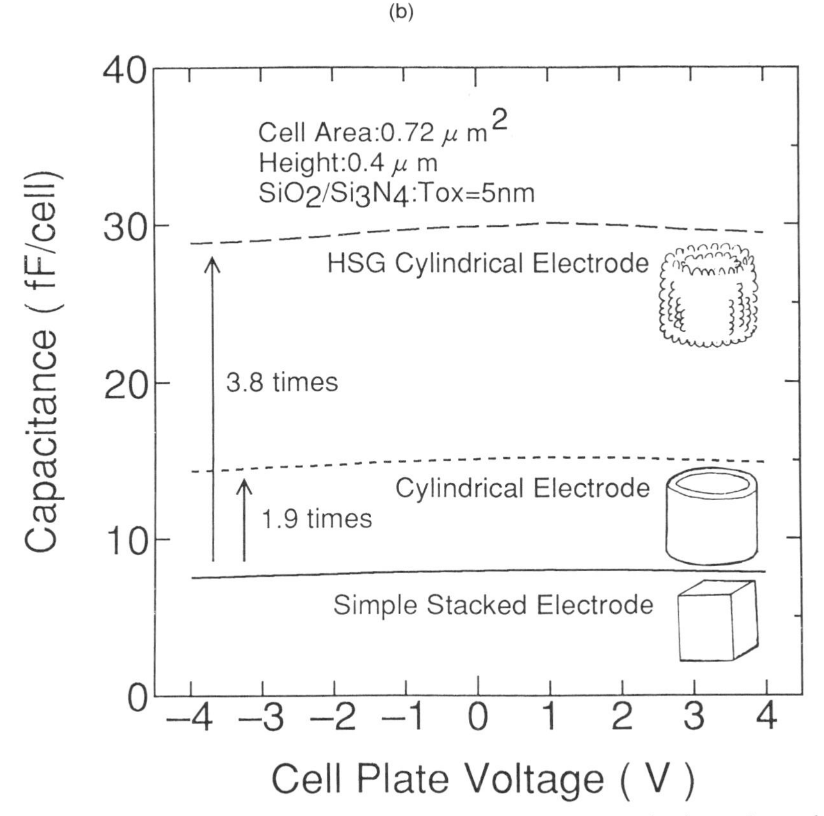

Figure 7.42 (a) Stages in the manufacturing of the HSG-Si electrode and (b) capacitance–voltage characteristics of stacked electrode, cylindrical electrode, and HSG-Si electrode capacitors of same cell area and dielectric composition and thickness. after Watanabe et al. [106]; copyright © 1992, Institute of Electrical and Electronics Engineers, New York.

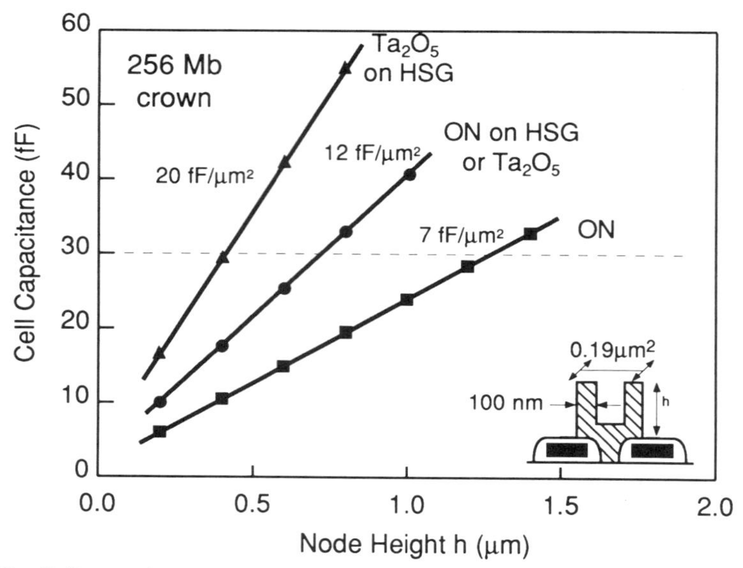

Figure 7.43 Cell capacitance versus node height for various dielectrics and electrode configurations in the construction of capacitors for 256 Mbit DRAMs. After Fazan et al. [108]; copyright © 1992, Institute of Electrical and Electronics Engineers, New York.

selective etching of the BPSG, forms hollow cylinders that are contacted on the bottom by a polysilicon stem. An HSG silicon coating is deposited onto the inside and outside walls of the cylinders by LPCVD, working within a very narrow process window just at the border between amorphous and microcrystalline silicon deposition. Vacuum annealing of the as-deposited HSG silicon electrode permits further optimization of the properties of the capacitors that are finished by the deposition of a polysilicon cell plate overgrowing both the inside and outside of the cylinder walls.

Further improvements may become accessible by the use of higher-dielectric-constant materials. Ferroelectrics represent the ultimate choice, but must be incorporated into silicon processing, which is difficult because of a lack of suitable electrode materials. Tantalum oxide on HSG silicon electrodes shows some promise, as illustrated in Figure 7.43, which shows a comparison of the cell capacitance as a function of the node height of plain ON, ON/HSG-Si, and Ta_2O_5/HSG-Si capacitors [108].

Figure 7.44 shows (a) a schematic representation of a cross section through two W/TiN/Ta_2O_5/HSG-Si capacitors and (b) an XTEM image of HSG-Si/Ta_2O_5/cell plate interfaces. Using bonded silicon-on-insulator for the construction of the transistors, the cell capacitors can be buried beneath the thin silicon film, allowing for large area and depth with concomitant gains in the capacitance. For example, 25 fF capacitors should be a realistic goal with ONO insulator technology and a cell size of 0.18 μm^2 for 1 Gbit DRAMs [109].

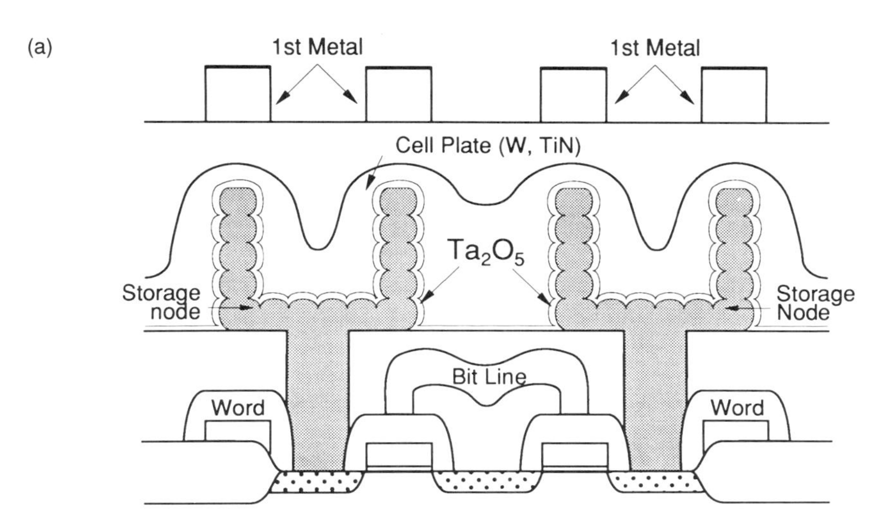

Figure 7.44 (a) Schematic representation and (b) TEM image of cross sections through W/TiN/Ta_2O_5/HSG-Si storage nodes. After Fazan et al. [108]; copyright © 1992, The Institute of Electrical and Electronics Engineers, New York.

References

1. Y. Horiike, R. Yoshikawa, H. Okano, M. Nakase, H. Komano, and T. Takigawa, in *Science and Technology of Microfabrication*, R. E. Howard, E. L. Hu. S. Namba, and S. W. Pang, eds. *MRS Symp. Proc.* **76**, 35 (1987).

2. C. G. Willson, in *Introduction to Microlithography*, L. F. Thompson, C. G. Willson and M. J. Bowden, eds., ACS Symp. Ser. 219, The American Chemical Society, Washington, DC, 1983, p. 87.

3. H. Fukuda, R. Yamanaka, T. Terasawa, T. Hama, T. Tawa, and S. Okazaki, *IEDM Techn. Dig.*, San Francisco, 1992, p. 49.

4. J. H. Bennewitz, The Electrochemical Society Spring Meeting, Los Angeles, CA, May 7–12, 1989, Extended Abstracts, p. 222.

5. I. G. Csizmadia, H. E. Gunning, R. K. Gosavi, and O. P. Strausz, *JACS* **95**, 133 (1973).

6. L. Wolff, *Justus Liebigs Ann. Chem.* **325**, 129 (1902).

7. H. Yamada, *J. Vac. Sci. Technol. B* **8**, 1628 (1990).

8. A. Heuberger, *J. Vac. Sci. Tech. B* **6**, 107 (1988).

9. D. L. Spears and H. I. Smith, *Solid State Technol.* **15**, 21 (1972).

10. K. Hoh, The Electrochemical Society Spring Meeting, Los Angeles, CA, May 7–12, 1989, Extended Abstracts, p. 224.

11. J. Ito, T. Kanayama, N. Atoda, and K. Hoh, *J. Vac. Sci. Technol. B* **6**, 409 (1988).

12. J. E. Bjorkholm, J. Bokor, L. Eichner, R. R. Freeman, J. Gregus, T. E. Jewell, W. M. Mansfield, A. A. MacDowell, E. L. Raab, W. T. Silfvast, L. H. Szeto, D. M. Tennant, W. K. Waskiewisz, D. L. White, D. L. Windt, O. R. Wood II, and J. H. Bruning, *J. Vac. Sci. Technol. B* **8**, 1509 (1990).

13. K. F. Dossel, H. L. Huber, and H. Oertl, *Microelectronic Engineering* **5**, 97 (1986).

14. K. G. Chiong, S. Wind, and D. Seeger, *J. Vac. Sci. Technol. B* **8**, 1447 (1990).

15. Y. Nakayama, S. Okazaki, N. Saitou, and H. Wakabayashi, *J. Vac. Sci. Technol. B* **8**, 1836 (1990).

16. K. Furuya and Y. Suematsu, *JEOL News* **29**, 10 (1991).

17. H. -C. Petzold, X-Ray Lithography, in *Applications of Synchrotron Radiation*, Gordon and Breach Publishers, Inc., 1988.

18. T. Kanayama, M. Sugawara, and J. Itoh, *J. Vac. Sci. Technol. B* **6**, 174 (1988).

19. V. White, J. Wallace, F. Cerina, Y. Vladimirski, Y. Su, and J. Maldono, *J. Vac. Sci. Technol. B* **8**, 1595 (1990).

20. W. A. Johnson et al., *J. Vac. Sci. Technol. B* **5**, 257 (1987).

21. H. Betz et al., *Microelectronic Engineering* **5**, 41 (1986).

22. H. J. Herzog, L. Csepregi, and H. Seidel, *J. Electrochem. Soc.* **131**, 2969 (1984).

23. K. Saitoh, H. Onoda, H. Morimoto, T. Katayama, Y. Watanabe, and T. Kato, *J. Vac. Sci. Technol. B* **6**, 1032 (1988).

24. D. Kyser and N. S. Viswanathan, *J. Vac. Sci. Technol.* **12**, 1305 (1975).

25. D. R. Kingham and L. W. Swanson, *Appl. Phys. A* **34**, 123 (1984).

26. R. L. Seliger, J. W. Ward, V. Wang, and R. L. Kubena, *Appl. Phys. Lett.* **34**, 310 (1979).

27. J. Puretz, R. K. DeFreez, R. A. Elliot, and J. Orloff, *Electron. Lett.* **32**, 700 (1986).

28. E. A. Bell, K. Rao, and L. W. Swanson, *J. Vac. Sci. Technol. B* **6**, 306 (1988).

29. M. J. Bozak, L. W. Swanson, and A. E. Bell, *J. Mater. Sci.* **22**, 2421 (1987).

30. G. M. Shed, H. Lezec, A. D. Dubner, and J. Mengailis, *Appl. Phys. Lett.* **49**, 1584 (1986).

31. Y. Ohmura, T. Shiokawa, K. Toyoda, and S. Namba, *Appl. Phys. Lett.* **51**, 1501 (1987).

32. L. F. Thompson and M. J. Bowden, in *Introduction to Microlithography*, L. F. Thompson, C. G. Wilson, and M. J. Bowden, eds., ACS Symp. Ser. 219, Washington, DC, 1983, p. 15.

33. M. A. McCord and R. F. W. Pease, *Surface Sci.* **181**, 278 (1987).

34. H. M. Mamin, P. H. Guether, and D. Rugar, *Phys. Rev. Lett.* **65**, 2418 (1990).

35. L. -W. Lyo and Ph. Avouris, *J. Chem. Phys.* **93**, 4479 (1990).

36. I. -W, Lyo and Ph. Avouris, *Science*, 253, 173 (1991).

37. Nanotribology, J. F. Belak, ed., *MRS Bulletin* **18**, 15 (1993).

38. E. A. Dobisz and C. R. K. Marian, *Appl. Phys. Lett.* **58**, 2526 (1991).

39. E. Sirtl and A. Adler, *Z. Metallkd.* **52**, 529 (1961).

40. F. Secco D'Aragona, *J. Electrochem. Soc.* **119**, 948 (1972).

41. M. W. Jenkins, *J. Electrochem. Soc.* **124**, 757 (1977).

42. S. K. Ghandi, *VLSI Fabrication Principles*, John Wiley, New York, 1983, p. 477.

43. K. E. Bean, *IEEE Trans. Electron. Dev.* **ED-25**, 1185 (1978).

44. J. Massies and J. P. Contour, *J. Appl. Phys.* **58**, 806 (1985).

45. D. C. Miller and G. A. Rozgonyi, in *Handbook of Semiconductors*, T. S. Moss, ed., S. P. Keller, vol. ed., North Holland Publishing Company, Amsterdam, 1980, p. 217.

46. K. J. Bachmann, in *Ann. Rev. Mater. Sci.* R. A. Huggins, R. H. Bube, and D. A. Vermilyea, eds., Vol. 11, Annual Reviews, Inc., Palo Alto, 1980.

47. P. J. Holmes, *Proc. Inst. Electr. Eng. B* **106**, 861 (1959).

48. C. S. Fuller and H. W. Allison, *J. Electrochem. Soc.* **109**, 880 (1962).

49. A. F. Witt, *J. Electrochem. Soc.* **114**, 298 (1967).

50. W. Kern and G. L. Schnabel, in *The Chemistry of the Semiconductor Industry*, Blackie & Sons, Ltd., Glasgow, 1987, p. 225.

51. B. H. Chin and K. L. Barlow, *J. Electrochem. Soc.* **135**, 3120 (1988).

52. W. Kern and D. A. Puotinen, *RCA Review* **31**, 187 (1970).

53. M. Liehr, in *Integrated Processing for Micro and Optoelectronics*, Y. I. Nissim, A. Katz, and G. W. Rubloff, eds., Elsevier Science Publishers B.V., Amsterdam, 1993.

54. Y. J. Chabal, G. S. Higashi, K. Raghavachari, and V. A. Burrows, *J. Vac. Sci. Technol. A* **7**, 2104 (1989).

55. G. S. Higashi, R. S. Becker, Y. J. Chabal, and A. J. Becker, *Appl. Phys. Lett.* **58**, 1656 (1991).

56. T. Yasuda, D. R. Lee, C. H. Bjorkman, Y. Ma, and G. Lucovsky, unpublished results.

57. S. Verhaverbeke, M. Iovacci, P. Mertens, M. Meuris, M. Heyns, R. Scheutelkamp, K. Maex, J. Alay, W. Vandervorst, R. DeBlank, M. Kubota, and A. Philipossian, *IEDM Techn. Dig.*, San Francisco, 1992, p. 637.

58. S. Watanabe, N. Nakayama, and T. Ito, *Appl. Phys. Lett.* **59**, 1458 (1991).

59. N. Yabumoto, K. Saito, M. Morita, and T. Ohmi, *Appl. Phys. Lett.* **55**, 562 (1989); *ibid.* **56**, 457 (1990).

60. C. Werkhoven, E. Granneman, M. Hendriks, R. DeBlank, S. Verhaverbeke, P. Mertens, M. Meuris, W. Vandervorst, M. Heijns, and A. Philipossian, *IEDM Techn. Dig.*, San Francisco, 1992, p. 633.

61. A. Saletes, F. Turco, J. Massies, and J. P. Contour, *J. Electrochem. Soc.* **135**, 504 (1988).

62. A. Katz, A. Feingold, S. J. Pearton, C. R. Abernathy, M. Geva, and K. S. Jones, *J. Vac. Sci. Technol. B* **9**, 2466 (1991).

63. R. Averbeck, H. Riechert, H. Schlotterer, and G. Weinmann, *Appl. Phys. Lett.* **59**, 1732 (1991).

64. J. L. Weyher and L. J. Giling, *J. Appl. Phys.* **58**, 219 (1985).

65. D. B. Lee, *Philips Res. Repts. Suppl.* (1974), p. 1.

66. J. B. Price, in *Semiconductor Silicon*, H. R. Huff and R. R. Burgess, eds., The Electrochem. Soc., Princeton, NJ, 1973, p. 339.

67. D. L. Kendall, in *Ann. Rev. Mater. Sci.*, R. A. Huggins, ed., Vol. 9, Ann. Rev. Inc., Palo Alto, CA, 1979, p. 373.

68. D. L. Kendall, G. R. de Guel, S. Guel-Sandoval, E. J. Garcia, and T. A. Allen, *Appl. Phys. Lett.* **52**, 836 (1988).

69. D. L. Kendall and G. R. de Guel, in *Micromachining and Micropackaging of Transducers*, C. D. Fung, P. W. Cheung, W. H. Ko, and W. D. Fleming, eds., Elsevier Science Publishers B.V., Amsterdam, 1985, p. 107.

70. O. Wolter, Th. Bayer, and J. Greschner, *J. Vac. Sci. Technol. B* **9**, 1353 (1991).

71. H. Foll, *Appl. Phys. A* **53**, 8 (1991).

72. C. J. M. Ejkel, J. Branebjerg, M. Elwenspoek, and F. C. M. Van De Pol, *IEEE Electron Dev. Lett.* **11**, 588 (1990).

73. V. Lehmann and H. Foll. *J. Electrochem. Soc.* **137**, 2556 (1990).

74. A. Prasad, S. Balakrishnan, S. K. Jain, and G. C. Jain, *J. Electrochem. Soc.* **129**, 596 (1982).

75. Z. -Z. Tu, *J. Vac. Sci. Technol. B* **6**, 1530 (1988).

76. D. L. Flamm and J. A. Mucha, in *The Chemistry of the Semiconductor Industry*, S. J. Moss and A. Ledwith, eds., Blackie, Glasgow and London, 1987, p. 343.

77. T. Ohno, M. Oda, C. Takahashi, and S. Matsuo, *J. Vac. Sci. Technol. B* **4**, 696 (1986).

78. D. E. Ibotson and D. L. Flamm, *Solid State Technology* (1988), p. 105, Penn Well Publishing Company, Tulsa.

79. I. Suemune, K. Kunitsugu, Y. Kan, and M. Yamanishi, *Appl. Phys. Lett.* **55**, 760 (1989).

80. D. L. Flamm and D. L. Donelly, *Plasma Chem. and Plasma Proc.* **1**, 317 (1981).

81. S. Murakawa, S. Fang, and J. P. McVittie, *IEDM Techn. Dig.*, San Francisco, 1992, p. 57.

82. (a) F. R. McFeely, J. F. Morar, N. D. Shinn, G. Landgren, and F. J. Himpsel, *Phys. Rev. B* **30**, 764 (1984); (b) F. R. McFeely, J. F. Morar, and F. J. Himpsel, *Surface Sci.* **165**, 277 (1986).

83. S. Matsuo, *J. Vac. Sci. Technol.* **17**, 775 (1980).

84. M. W. Geis, G. A. Lincoln, N. Efremov, and W. J. Piacentini, *J. Vac. Sci. Technol.* **19**, 1390 (1981).

85. W. J. Grande, J. E. Johnson, and C. L. Tang, *J. Vac. Sci. Technol. B* **8**, 1075 (1990).

86. K. Gamo and S. Namba, *J. Vac. Sci. Technol. B* **8**, 1927 (1990).

87. N. E. McGruer, K. Warner, P. Singhal, J. J. Gu, and C. Chan, *IEEE Trans. Electron Dev.* **38**, 2389 (1991).

88. C. Spindt, C. Holland, A. Rosengreen, and I. Brodie, *IEDM Techn. Dig.*, San Francisco, 1992, p. 363.

89. T. Utsumi, *IEEE Trans. Electron Dev.* **38**, 2276 (1991).

90. G. J. Campisi and H. F. Gray, in *Science and Technology of Micofabrication*, R. E. Howard, E. L. Hu, S. Namba, and S. W. Pang, eds., MRS Symp. Proc. **76**, 67 (1987).

91. R. B. Marcus, T. S. Ravi, T. Gmitter, H. H. Busta, J. T. Niccum, K. K. Chin, and D. Liu, *IEEE Trans. Electron Dev.* **38**, 2289 (1991).

92. T. S. Ravi, R. B. Marcus, and D. Liu, *J. Vac. Sci. Technol. B* **9**, 2733 (1991).

93. R. S. Wagner and W. C. Ellis, *Appl. Phys. Lett.* **4**, 89 (1964).

94. J. Liu, U. T. Son, A. N. Stepanova, K. N. Christensen, G. J. Wojak, E. I. Givargizov, K. J. Bachmann and J. J. Hren, *J. Vac. Sci. Technol.* **B12**, 717 (1994).

95. S. Nakamura, T. Hashizume, and T. Sakurai, *J. de Physique* **47**, C2-431 (1986).

96. S. Matsui, T. Ichihashi, and M. Mito, *J. Vac. Sci. Technol. B* **7**, 1182 (1989).

97. E. M. Clausen, Jr., J. P. Harbinson, C. C. Chang, H. G. Craighead, and L. T. Florez, *J. Vac. Sci. Technol. B* **8**, 1830 (1990).

98. S. Matsui and H. Watanabe, *Appl. Phys. Lett.* **59**, 2284 (1991).

99. M. Takai, J. Tokuda, H. Nakai, K. Gamo, and S. Namba, *Jpn. J. Appl. Phys.* **22**, L757 (1983); *ibid.* **23**, L852 (1984).

100. M. Takai, J. Tsuchimoto, J. Tokuda, H. Nakai, K. Gamo, and S. Namba, *Appl. Phys. A* **45**, 305 (1988).

101. M. Takai, H. Nakai, S. Nakashima, J. Tuchimoto, K. Gamo, and S. Namba, *Jpn. J. Appl. Phys.* **24**, L705 (1985).

102. Y. Yuba, K. Gamo, X. G. He, Y. S. Zhang, and S. Namba, *Jpn. J. Appl. Phys.* **22**, 1211 (1983).

103. P. D. Brewer, D. McClure, and R. M. Osgood, Jr., *Appl. Phys. Lett.* **47**, 3104 (1985).

104. P. Mogyorósi, K. Piglmayer, R. Kullmer, and D. Bäuerle, *Appl. Phys. A* **45**, 293 (1988).

105. C. I. H. Ashby, *Appl. Phys. Lett.* **46**, 752 (1985).

106. H. Watanabe, T. Tasumi, S. Ohnishi, I. Homa, and T. Kikkawa, *IEDM Techn. Dig.*, San Francisco, 1992, p. 259.

107. R. Sinclair, K. B. Kim, O. Shippou, and H. Iwasaki, *J. Electrochem. Soc.* **136**, 511 (1989).

108. P. Fazan and V. Mathews, *IEDM Techn. Dig.*, San Francisco, 1992, p. 263.

109. T. Nishihara, N. Ikeda, H. Aozawa, Y. Miyazawa, and A. Ochiai, *IEDM Techn. Dig.*, San Francisco, 1992, p. 803.

CHAPTER

8

Oxidation, Doping, and Metallization

8.1 Thermal Oxidation and Dielectric Isolation of Silicon

In this chapter we complete the review of IC processing steps by a discussion of (1) the thermal oxidation of silicon in the context of gate isolation and the isolation of individual IC components from each other; (2) the doping of semiconductors by ion implantation and diffusion; (3) special problems in the use of compound semiconductors for IC manufacturing related to their native oxides; and (4) the metallization of semiconductors in the context of Schottky barriers, ohmic contacts, and interconnects. As part of this review, a discussion of several characterization methods is added, such as high-resolution electron energy loss spectroscopy (HREELS), Rutherford backscattering spectroscopy (RBS), secondary ion mass spectrometry (SIMS) and deep-level transient spectroscopy (DLTS), Auger electron spectroscopy (AES), extended x-ray absorption fine structure spectroscopy (EXAFS), x-ray/ultraviolet photoemission spectroscopies (XPS/UPS), and ballistic electron emission microscopy/spectroscopy (BEEM/BEES).

As discussed in Section 5.2, the preferred method of gate oxide fabrication is the thermal oxidation of silicon. The oxidation of silicon by both dry oxygen and water vapor occurs at the silicon–oxide interface. The overall oxidation process consists of four steps:

1. Diffusive transport of oxygen across the diffusion layer in the vapor phase adjacent to the silicon oxide–vapor interface;
2. Incorporation of oxygen at the outer surface into the silicon oxide film;

3. Diffusive transport across the silicon oxide film to its interface with the silicon lattice;
4. Reaction of oxygen with silicon at this inner interface.

The driving force for steps 1–3 is the depletion of oxygen at the inner interface by the oxygen-consuming reaction. In steady state, the fluxes of oxygen $\mathbf{J}_1$, $\mathbf{J}_2$, $\mathbf{J}_3$, and $\mathbf{J}_4$ associated with the four steps listed are equal. In general, the steady-state diffusive flux of a component C_j caused by a gradient ∇c_j of its concentration c_j is given by Fick's first law

$$\mathbf{J}_{\mathrm{d}}(C_j) = D_j \nabla c_j \tag{8.1}$$

where the diffusion coefficient is a function of composition and temperature. In the Deal–Grove model [1] for thermal oxide growth on silicon, the diffusion constant of oxygen D_{s} in the silicon oxide film is assumed to be a constant, and the diffusion problem is treated in one dimension as illustrated in Figure 8.1. In view of the small thickness of the film as compared to the wafer diameter, this is reasonable. Therefore,

$$\mathbf{J}_1 = D_{\mathrm{v}}(\mathrm{O}_2) \frac{\partial c}{\partial x}\bigg|_{x=0} \approx D_{\mathrm{v}}(\mathrm{O}_2) \frac{c_{\mathrm{v}} - c_{\mathrm{s}}}{\delta} = \mathbf{J}_2 = k_{\mathrm{o}}(c_{\mathrm{s}} - c_{\mathrm{o}}) \tag{8.2}$$

$$\mathbf{J}_3 = D_{\mathrm{v}}(\mathrm{O}_2) \frac{\partial c}{\partial x}\bigg|_{x=d_{\mathrm{ox}}} \approx D_{\mathrm{s}}(\mathrm{O}_2) \frac{c_{\mathrm{o}} - c_{\mathrm{i}}}{d_{\mathrm{ox}}} = \mathbf{J}_4 = k_{\mathrm{i}} c_{\mathrm{i}} \tag{8.3}$$

where k_{o} and k_{i} are rate constants and c_{v} and c_{o} are the oxygen concentrations in the vapor phase and in the oxide, respectively, at the outer interface, and c_{i} is the oxygen concentration at the inner interface. Note that the consumption of oxygen at the inner interface lowers both c_{o} and c_{i} below the equilibrium concentration of oxygen in silicon dioxide c_{ox}, which has at 1000°C the values

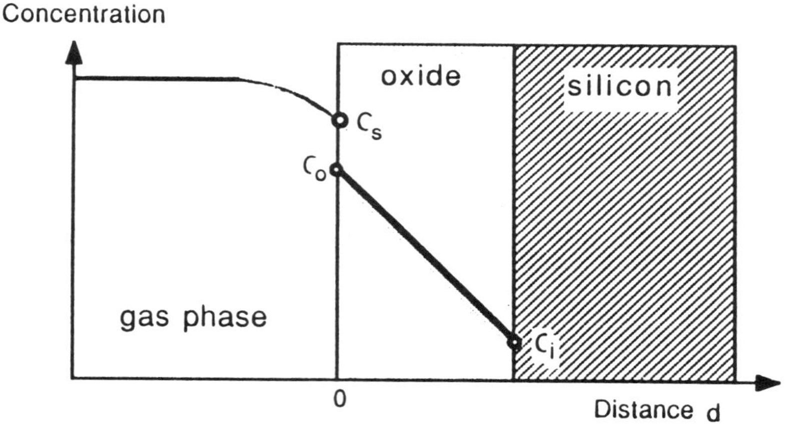

Figure 8.1 Schematic representation of the oxidant distribution in the oxide and vapor phases during silicon oxidation.

$c_{ox} \approx 5.2 \times 10^{16}\,\text{cm}^{-3}$ and $c_{ox} \approx 3 \times 10^{19}\,\text{cm}^{-3}$ for dry and steam oxidation, respectively. Solving for c_o results in

$$c_o = c_i\left(1 + \frac{k_s d_{ox}}{D_s}\right) \tag{8.4}$$

with

$$c_i = \frac{c_{ox}}{1 + \dfrac{k_s}{k_v} + \dfrac{k_s d_{ox}}{D_s}} \tag{8.5}$$

and the growth rate for the oxide film

$$|\mathbf{v}_g| = |\mathbf{J}_4| = \frac{k_s c_{ox}}{1 + \dfrac{k_s}{k_v} + \dfrac{k_s d_{ox}}{D_s}} = \rho_{ox}\frac{d(d_{ox})}{dt} \tag{8.6}$$

where ρ_{ox} is the formal number density of oxidant molecules existing in the oxide film, that is, $2.2 \times 10^{22}\,\text{cm}^{-3}\,O_2$ and $4.4 \times 10^{22}\,\text{cm}^{-3}\,H_2O$ for dry and wet oxidation, respectively. Solving for $d_{ox}(t)$ with the initial condition $d_{ox}(t = 0) = d_i$ results in

$$d_{ox}^2 + Ad_{ox} = B(t + \tau_i) \tag{8.7}$$

so that

$$d_{ox} = \frac{A}{2}\left[\left(1 + \frac{4B(t + \tau_i)}{A^2}\right)^{1/2} - 1\right] \tag{8.8}$$

where $A = 2D_s(k_s + k_v)/k_s k_v$ and $B = 2D_s c_{ox}/\rho_{ox}$ and $\tau_i = (d_i^2 + Ad_i)/B$ are characteristic constants for the oxidation process. For long oxidation: $d_{ox} \approx \sqrt{Bt}$; that is, the film thickness depends only on D_s, but not on the rate constants. On the other hand, for short oxidation $[(t + \tau_i) \ll A^2/4B]$: $d_{ox} \approx B(t + \tau_i)/A$, so that the film thickness depends only on k_s and k_v, but not on D_s. Therefore, the growth process changes from being initially reaction controlled into a diffusion-controlled mode. Note that in the limit $D_s \to 0$, $c_i \to 0$, and $c_o = c_{ox}$, so that under diffusion control the oxidant concentration at the oxide–silicon interface tends to zero, and the maximum concentration gradient across the oxide develops. On the other hand, under reaction-controlled conditions (i.e., $k_s d_{ox} \ll D_s$), $c_i \approx c_o = c_{ox}/(1 + k_s/k_v)$. A recent investigation [2] of the silicon oxidation data given in [3] reveals that the rate constant B/A of the "linear" rate law depends on a log-linear fashion on d_{ox}.

Because the number density of Si atoms is substantially smaller in SiO_2 than in Si, the oxidation of silicon is associated with a volume change by a factor of 2.2. In the initial stages of silicon oxidation, interstitial silicon atoms are formed at the oxide–semiconductor interface to accommodate this change in number density. In the later stages, the motion of the interface requires a supply of vacancies from and the diffusion of Si interstitials into the underlying

Si lattice. Under unfavorable conditions, extrinsic stacking faults may be formed [4]. The increased leakage current at electrical junctions associated with oxidation-induced stacking faults (OSFs) degrades the storage time of MOS capacitors. Low-temperature oxidation limits the formation of OSFs [5]. It is achieved, for example, by plasma oxidation [6]. Also, HCl or Cl_2 addition to dry oxygen is helpful since it increases the oxidation rate [7] and converts impurities in volatile chlorides that are removed in the oxidation process, reducing the concentration of traps and increasing the breakdown field of the oxide. However, at temperatures below 950°C, the lack of viscous flow and stress relaxation becomes a problem.

Also, in the fabrication of trench capacitors for ULSI circuits, high oxidation temperatures are required to achieve optimum electrical properties. This poses a problem since the thermal budget in ULSI processing is extremely restricted. Under the conditions of single-wafer ULSI processing, the thermal budget is minimized by the use of either plasma-enhanced low-temperature reactions or by rapid thermal processing (RTP). A schematic representation of the apparatus for rapid thermal oxidation (RTO) is shown in Figure 8.2(a). The wafer is heated radiatively by a set of tungsten–halogen lamps through a fused silica bell jar. Since it presents only a small thermal mass, rapid changes in the wafer temperature are possible. This is illustrated in Figure 8.2(b), which shows heating and cooling transients of 15 s and a total processing time of ~1 min. Of course, a short cycle time is equally important in the context of throughput, which must be high to make single-wafer processing a viable approach to manufacturing. Ultimately, single-wafer processing must be conducted as an integrated process entailing all manufacturing steps. Today this goal is only partly realized; traditional processing steps and rapid thermal processing are mixed, as illustrated in Table 8.1.

Figures 8.3(a) and (b) show the oxide thickness versus oxidation time data for the RTO of planar silicon surfaces and the temperature dependence of the linear and parabolic rate constants derived from these data. Their activation energies are $E_a(B/A) = 1.98\,\text{eV}$ and $E_a(B) = 1.42\,\text{eV}$. Figures 8.3(c) and (d) show plots of the leakage current versus applied voltage and time to 50% cumulative failure versus stress current density for trench capacitors made by RTO. The data for high-temperature oxidation are clearly superior. Also the density of interface states decreases strongly with increasing RTO temperature.

In general, the thermal oxidation results in a density of midgap interface states of the order of $10^{11}\,\text{cm}^{-2}\,\text{eV}^{-1}$, which, for Si(100), is improved by a subsequent anneal in forming gas to $\sim 10^{10}\,\text{cm}^{-2}\,\text{eV}^{-1}$. The passivation of the silicon–oxide interface in this anneal is due to the reaction of residual dangling-bond states with hydrogen according to

$$\text{Si}\cdot + H_2 \rightleftarrows \text{Si–H} + \text{H} \tag{8.9}$$

with an activation energy of 1.66 eV [9]. The depassivation reaction

$$\text{Si–H} \rightleftarrows \text{Si} + \text{H} \tag{8.10}$$

(a)

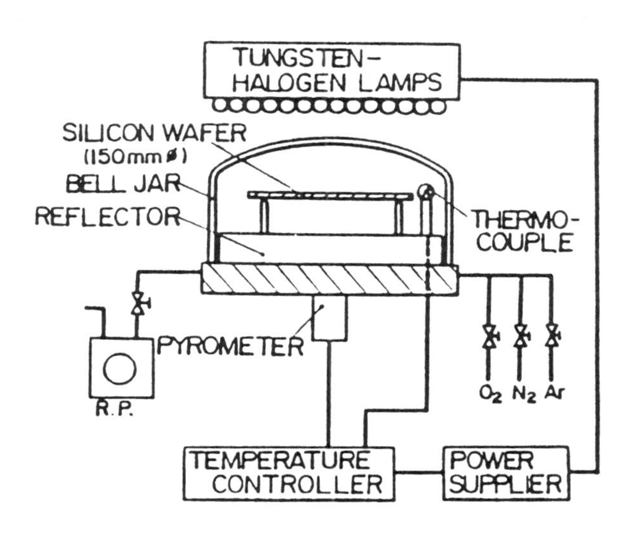

(b)

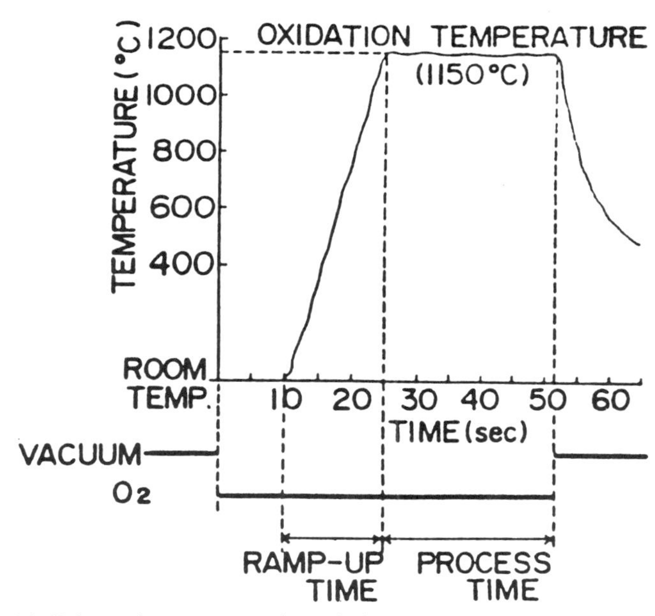

Figure 8.2 (a) Schematic representation of the RTO apparatus. (b) A typical RTO cycle. After Miyai et al. [F2]; copyright © 1988, The Electrochemical Society, Pennington, NJ.

Table 8.1 MOS capacitors fabrication sequence including RTO (after Miyai et al. [12]; copyright © 1988, The Electrochemical Society, Pennington, NJ).

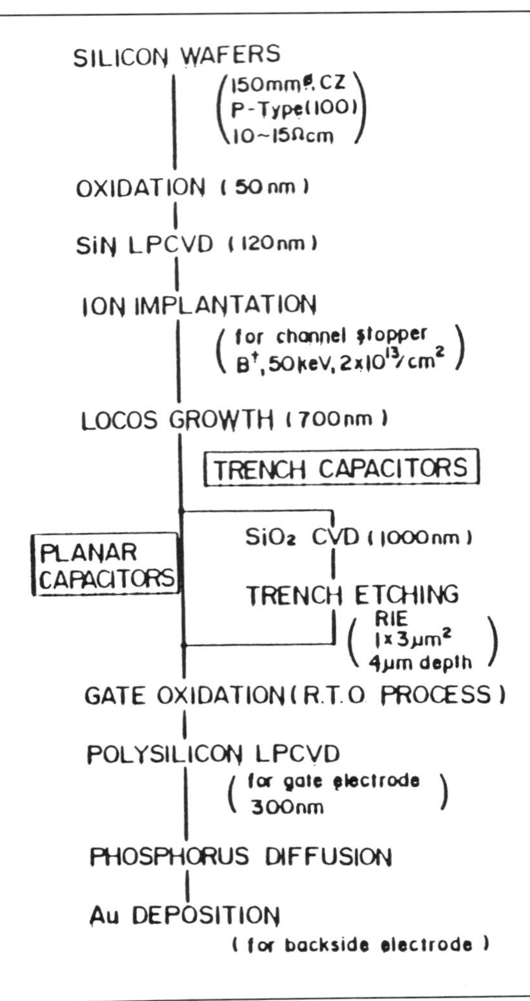

occurs with an activation energy of 2.5 eV [11], [12]. Another important aspect of the post-oxidation annealing is stress-relief and the minimization of the density of defects in the gate oxide that affects the electric breakdown characteristics. The latter becomes increasingly relevant for the reliability of ICs with decreasing gate oxide thickness. As illustrated in Table 8.2, considerable improvements are made by the introduction of stacked oxide growth, where thermal preoxidation establishing an optimum $Si\text{-}SiO_2$ interface, is combined with a relatively low thermal budget CVD oxide deposition [13]. The yield Y shown in Table 8.2 is defined as the percentage of sites passed for

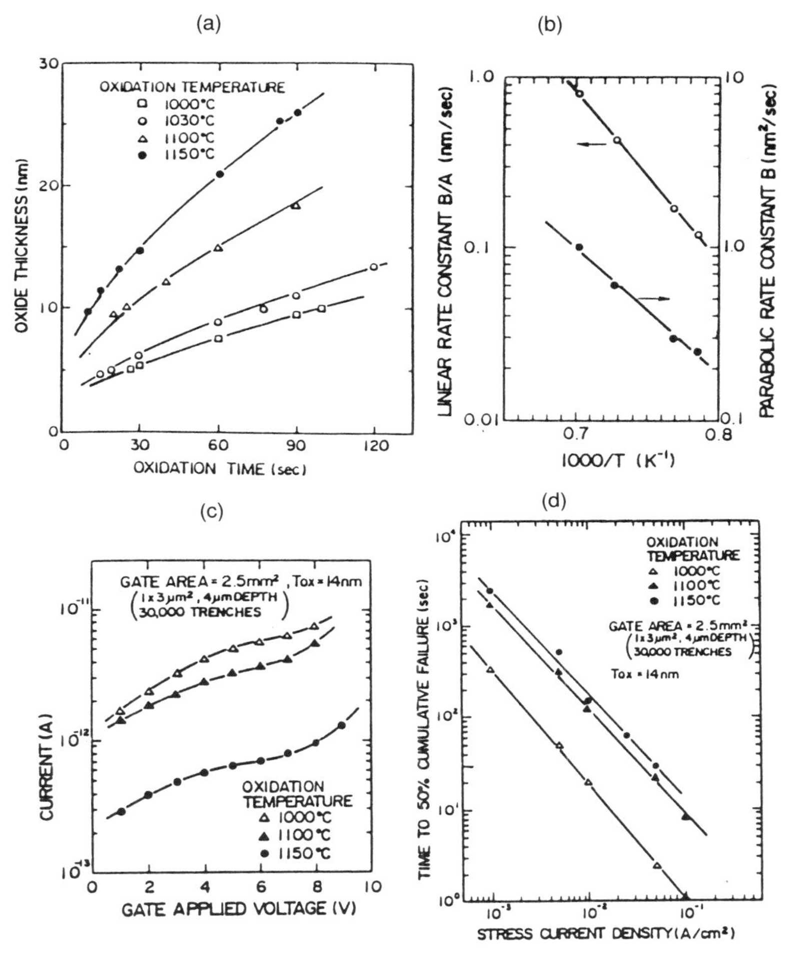

Figure 8.3 (a) Oxide thickness versus oxidation time and (b) Arrhenius plots of the linear and parabolic rate constants for the RTO of planar Si surfaces; (c) leakage current versus voltage and (d) time to 50% cumulative failure as a function of the stress density for trench capacitors made by RTO. After Miyai et al. [n2]; copyright © 1988, The Electrochemical Society, Pennington, NJ.

a chosen breakdown criterium (e.g., break down voltage V_b at 1 μA leakage current) for area A. It is related to the defect density D_0 as $Y = \exp(-AD_0)$. In view of the small values of D_0 for high quality gate oxides, a very large number of capacitors must be evaluated to assure a high level of confidence in the data. Primary causes for degraded interfacial oxide properties are:

1. Residual impurities and native oxide on the silicon surface;
2. Microroughness of the silicon surface;

Table 8.2 Dielectric Yield and Defect Density in Gate Oxide Films (after Roy and Sinha [13]).

Technology	Gate Oxide Process	Dielectric Yield (%)	$D_0(cm^{-2})$
1.25 μm	250 Å conventional SiO_2	90	0.35
(1 Mb DRAM)	250 Å stacked SiO_2	98	0.07
1.25 μm	210 Å conventional SiO_2	90	0.35
	210 Å stacked SiO_2	98	0.10
0.6 μm	125 Å conventional SiO_2	74	1.00
	1.25 Å stacked SiO_2	94	0.20
0.6 μm	100 Å conventional SiO_2	68	1.28
2	100 Å stacked SiO_2	92	0.28

3. Uncontrolled growth of the first few monolayers of the oxide during heating to the oxidation temperature.

Therefore, the preparation of the first monolayer of oxide is closely linked to the cleaning/HF etching procedure discussed in Chapter 7, and particular attention must be given to the critical phase in the heating cycle when the surface-terminating hydrogen atoms are desorbed and replaced by either high-quality or defective oxide, depending on the care invested in controlling the process.

Figure 8.4 shows the heating cycles used (a) in conventional silicon oxidation and (b) in a combination of controlled preoxidation at 300°C with conventional oxide growth at 900°C. The latter procedure seals the silicon surface by a monolayer of oxide in an ultraclean environment (e.g., <5 ppb moisture in the argon blanket during heating) during the critical phase in the heating procedure. Significant improvements are obtained by controlled pre-oxidation in both the distribution of the dielectric breakdown voltage of $Al/SiO_2/Si(100)$ capacitors and the threshold shift of NMOSFETs. This is illustrated in Figure 8.5(a) and (b), which compares the results of preoxide growth and conventional oxidation, which is associated with the uncontrolled formation of 1.4 nm interfacial oxide during loading into the oxidation tube and heating.

A better understanding of the early stages of silicon oxidation, which are not well represented by the Deal–Grove model, is obtained by spectroscopic studies, such as electron energy loss spectroscopy (EELS), in conjunction with advanced imaging methods like HREM and STM. EELs is based on the scattering of electrons on the surface of a solid, conserving energy and momentum,

$$E(\mathbf{k}_s) = E_o(\mathbf{k}) - \hbar\omega \tag{8.11}$$

$$\mathbf{k}_{s\parallel} = \mathbf{k}_{\parallel} - \mathbf{q}_{\parallel} \pm \mathbf{G}_{hk} \tag{8.12}$$

(a)

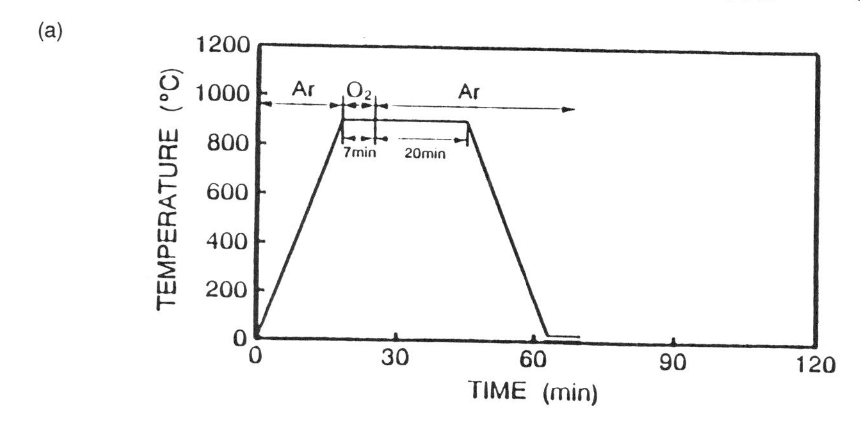

(b)

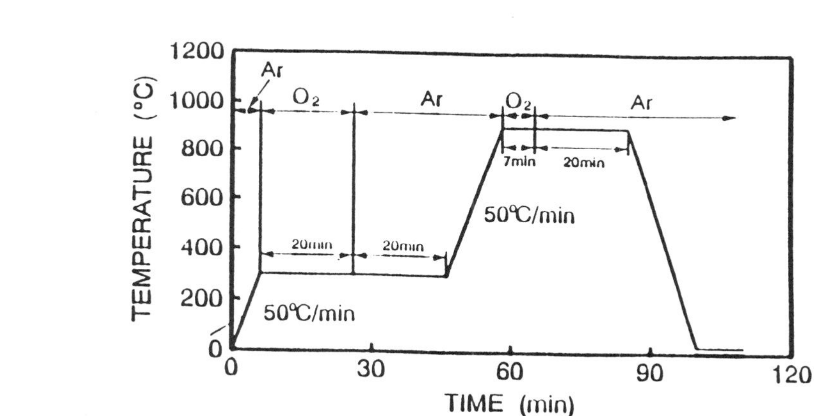

Figure 8.4 (a) Temperature profiles of ultraclean oxidation (a) without and (b) with preoxide growth. After Makihara et al. [10]; copyright © 1993, J. Appl. Phys., Tokyo.

where $\hbar\omega$ is the energy loss of the electron upon scattering by the surface and $\mathbf{k}$ and $\mathbf{k}_s$ are the wave vectors of the incoming and scattered electrons. The vector $\mathbf{q}_{\parallel} \pm \mathbf{G}_{hk}$ is a vector in the two-dimensional reciprocal lattice of the surface, and $\mathbf{k}_{s\parallel}$ and $\mathbf{k}_{\parallel}$ are the components of $\mathbf{k}$ and $\mathbf{k}_s$ parallel to the surface. The energy loss of the electron imparted in the scattering event onto the surface may be due to excitation of optical surface phonons, of surface plasmons, of surface electronic transitions, and of vibrational modes of molecules that are adsorbed at the surface. Thus for reproducible results the method requires UHV conditions and well-controlled surface preparation. Since the vibrational energies of adsorbed molecules are of the order of 10–500 meV, highly monochromatic ($\Delta\lambda/\lambda \sim 10^{-4}$) electrons must be used that have $10 \leqslant E_o \leqslant 100\,\text{eV}$ primary energy for high-resolution evaluations of the surface chemistry and structure.

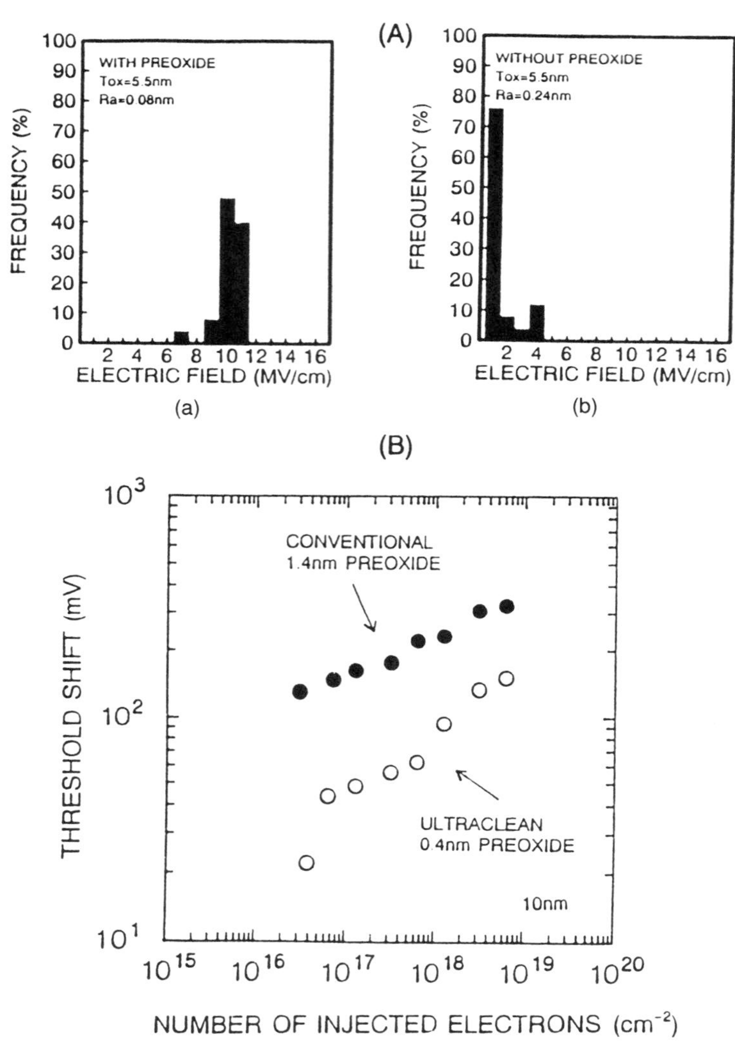

Figure 8.5 (A) Histograms of the breakdown voltage of Al/SiO_2/Si(100) capacitors (i) with and (ii) without preoxide growth. (B) Comparison of the threshold shift as a function of the number of injected electrons for NMOSFETs prepared with and without preoxide growth. After Makihara et al. [10]; copyright © 1993, Japanese J. Appl. Phys., Tokyo.

For example the hydrogen saturated silicon (111) 7×7 surface exhibits energy loss peaks at 79 meV and 259 meV, corresponding to the bending mode and stretching mode of the Si-H bond, and the scissor mode of the SiH_2 surface configuration results in a peak at 110 meV. The investigations of other reconstructions of the (111) surface and different surface orientations of silicon result in very similar loss spectra indicating that the bonding of the hydrogen finds similar sites on all surfaces. The quenching of specific features in the loss spectra can be utilized in studies of the desorption from specific surface sites and has been combined successfully with TDS in the evaluation of the desorption kinetics of hydrogen and other gases from the surface of silicon. Furthermore, in the case of the (100) Si surface, the density of sites for SiH_2 formation correlate with the density of surface steps. Therefore, hydrogen titration, using the EELS signal at 110 meV as indicator, can be utilized for the evaluation of the step density.

Figure 8.6 shows the energy loss spectra of an oxygen-exposed cleaved Si(111) surface. The loss associated with the surface phonon at 56 meV for the 2×1 reconstruction of the clean surface persists to a coverage of 0.2. Two additional features that correspond to different bonding configurations are observed at 94 and 130 meV upon absorption of oxygen, which is nondissociative. A third feature at 175 meV appears as a shoulder at the highest coverage and is related to the O–O stretching vibration, which is excited to small intensity because of the strong coupling of the two oxygen atoms in the absorbed quasimolecule.

Information on the initial stages of oxidation has been obtained from

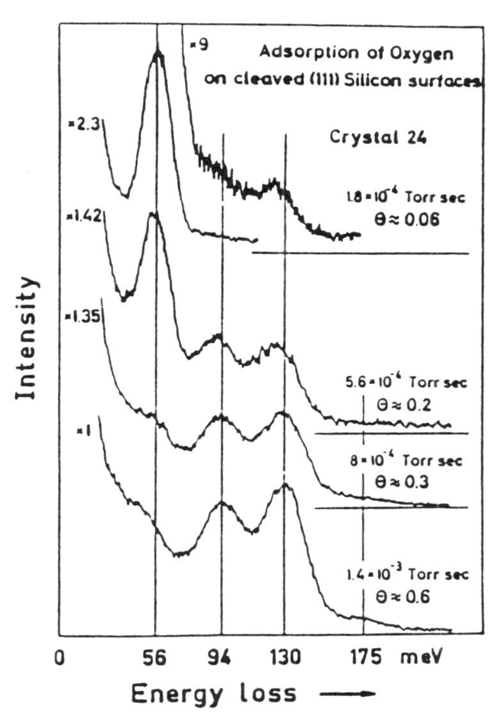

Figure 8.6 Electron energy loss spectra of an as cleaved Si(111) surface exposed to various coverages by oxygen. After Ibach et al. [14]; copyright © 1973, North-Holland Publishing Company, Amsterdam.

excitations of silicon core levels at higher energies than exhibited in Figure 8.6 [18]. Figure 8.7 shows the electron energy loss spectra due to Si $2p$ core-level excitation for (a) a clean silicon (100)-2 × 1 surface, (b) and (c) at a coverage of 0.6 and 1.0 monolayers of oxygen, respectively, and (d) coverage of the silicon by amorphous SiO_2. For comparison, the energy loss spectrum of a clean silicon surface, which was made amorphous by Ar^+ bombardment, is dotted into curve (a). The loss peaks at 104.4 and 106.5 eV are attributed to the initial formation of Si–O bonds. EELS [17] shows that the bonding of oxygen to two silicon atoms is the prevalent bonding configuration in the very early stage of oxidation.

Additional information may be obtained by STM/STP and HREM investigations of the silicon–oxide interface. STM studies of the Si(111)-7 × 7 surface

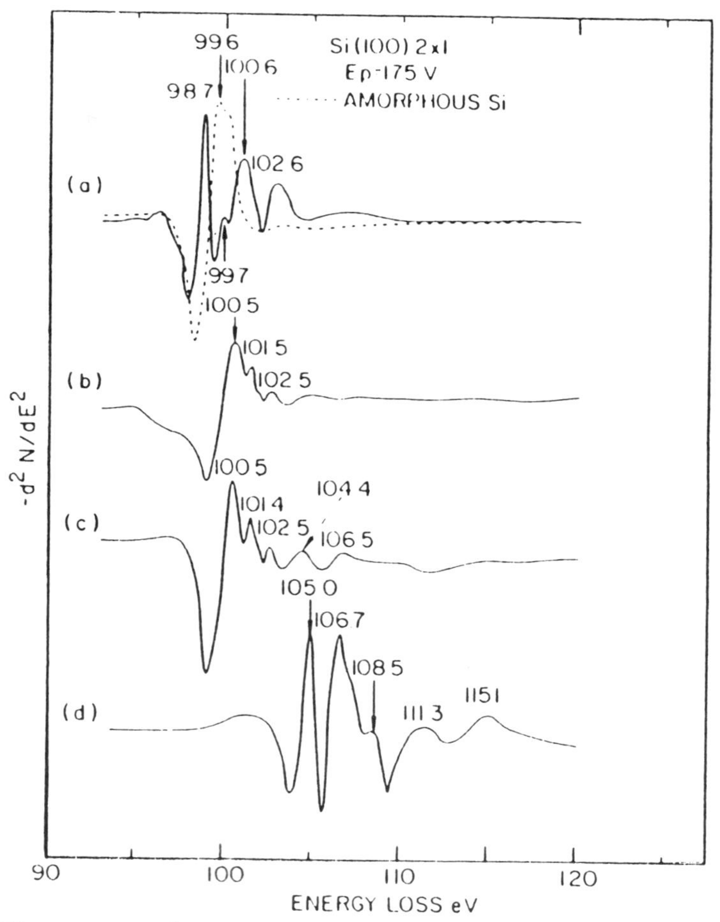

Figure 8.7 Electron energy loss spectra of the silicon (100)-2 × 1 surfaces. After Koma and Ludeke [15]; copyright © 1975, American Institute of Physics, New York.

at low oxygen dose have been reported in Ref. [16]. At 0.1 L dose (1 L = 10^{-6} torr s) certain adatom positions of the DAS model are attacked. The $J-V$ characteristics taken at these sites are consistent with the bonding configuration of Figure 8.8(b). At the higher exposure further oxidation results in the disappearance of adatoms in the topographical image, which corresponds to the bonding configuration of Figure 8.8(c). Note that the oxidation selects adatom positions preferably from the faulted half of the 7 × 7 unit cell and corner rather than center adatom positions; that is, the reactivity of the solid surface is intimately linked to its reconstruction. In fact, STM investigations on the oxidation of Ge surfaces show that the attack proceeds preferably from disordered boundaries between 2 × 8 domains of different orientations rather than surface steps [19]. Therefore, not only the kinetics of crystal growth, but also the kinetics of corrosion and oxidation processes, requires reevaluation, focusing onto the bonding of surface atoms rather than idealized geometrical surface features.

The oxidation of the Si(001)-2 × 1 surface has been investigated by STM, using larger exposures (15 L) under conditions where the loss of surface oxide by SiO sublimation is negligible and oxide growth commences (1.33 × 10^{-4} Pa oxygen pressure at 600°C). Two different oxidation products are found: (1) dark patches that are unstable at 600°C (i.e., behave similar to residual native oxide patches formed at room temperature) and (2) persistent oxidation features (sequences of dots) that are thought of as an initial form of the thermal oxide. Figure 8.9(a) shows a model of this initial oxide containing –O–Si–O– units that are similar to the bonding in β-cristobalite. HREM images for the [110] and [1$\bar{1}$0] projections of the silicon–silicon dioxide interface reveal crystalline order in the oxide over a thickness of ~5 Å at least on a local scale, albeit being consistent with the tridymite phase of SiO_2, as illustrated in Figure 8.9(b) [20]. Although there exists thus at present no agreement on the specific structure of the interfacial oxide, there is mounting evidence for significant differences in the lattice spacing and bonding at the Si/SiO_2 interface, which may explain the observed inhomogeneities in the oxidation process [22].

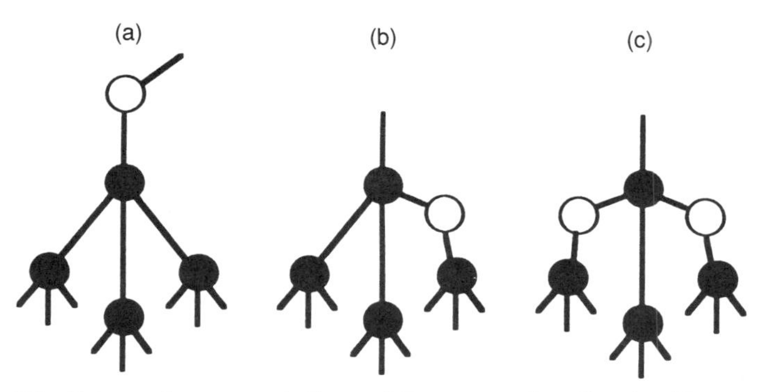

Figure 8.8 Schematic representations of the oxygen bonding to silicon surface atoms in adatom positions of the Si(111)-7 × 7 surface unit cell. After Pelz and Koch [16].

Despite the advances in ULSI processing discussed here and the increasing interest in dielectric isolation, a substantial part of today's IC processing refers to conventional VLSI technology. Therefore, the passivation of the boundaries betwen the N-and P-channel transistors of CMOS circuits by local oxidation of silicon (LOCOS) [23] remains an important step that is illustrated in Figure 8.10. A typical field oxidation run with steam at 1 atm for 5 h generates a FOX layer of 9000 Å, adding a 100 Å thick GOX pad under 1000 Å Si_3N_4. During the oxidation process more than 50% of the channel-stop dopant is removed from the silicon under the FOX. The same thickness of the FOX and GOX are obtained by steam oxidation at 10 atm at 850°C for 1.5 h, removing only 20% of the channel-stop dopant. However, the increase in c_{ox} under the conditions of high pressure oxidation moves the growth rate from the parabolic into the linear regime, increasing the bird's beak structure at the fringes of the FOX due to the increased lateral oxide growth [21].

Figure 8.11 shows the bird's beak structure extending under the oxidation stack for (a) conventional LOCOS processing (step in Table 4.3) and (b) modified LOCOS processing using oxynitride instead of oxide for stress relief under the silicon nitride pads. The replacement of the conventional oxide layer under the Si_3N_4 pad (step 9 in Table 4.3) by an oxynitride layer reduces the

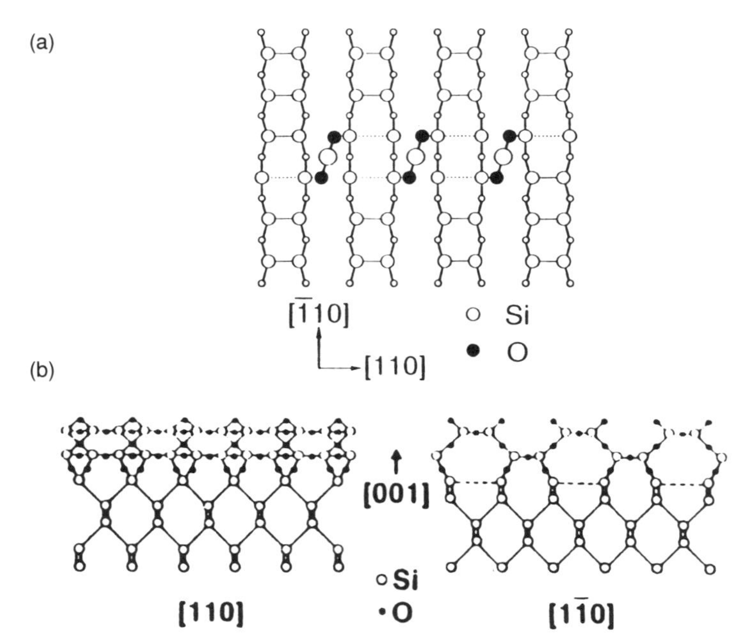

Figure 8.9 (a) Schematic representations of the oxygen bonding to silicon surface atoms on the Si(001)-2 × 1 surface. After Udagawa, Niwa, and Sumita [18]; copyright © 1993, Japanese Physical Society, Tokyo. (b) Interfacial atomic arrangement at the Si/SiO_2 interface deduced from lattice imaging. Dashed lines indicate possible dimerization of unsaturated bonds. After Ourmazd, Taylor, and Rentschler [20]; copyright © 1987, American Institute of Physics, New York.

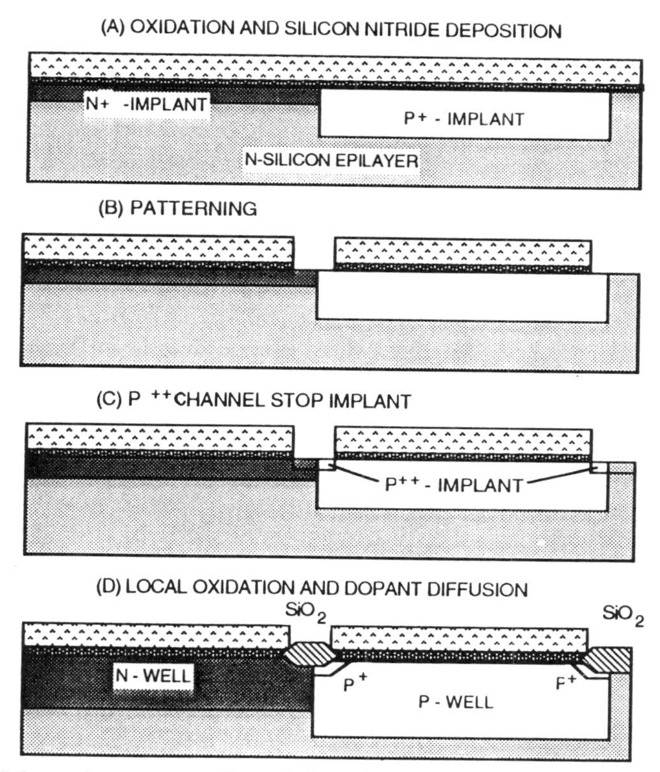

Figure 8.10 Schematic representation of the LOCOS processing step in a twin-tub CMOS processing sequence (a) prior and (b) after oxidation.

extent of the bird's beak due to the retardation of the oxidation of silicon along the periphery of the oxynitride pad [19] of Chapter 4, [25]. Figure 8.11(c) shows a top view of the FOX ridges and associated bird's beak fringes for part of the SRAM cell shown in Figure 4.20(b). Note that there are reductions and extensions of the bird's beak in the vicinity of outside and inside corners of the oxide mask, as indicated by arrows a and c in Figure 8.11(c), which requires modifications in the fabrication process [19] of Chapter 4.

A detailed investigation of the effects of the gate bird's beak (GBB) on the performance and reliability of submicrometer NMOSFETs [14] reveals a significant degradation of the transconductance, the subthreshold swing and the threshold voltage. A comparison of the threshold voltage of *n*- and *p*-MOS transistors of CMOS circuits built with LOCOS and nitrided-oxide LOCOS (NOLOCOS) technologies is presented in Figure 8.12 [26]. The threshold variations ΔV_{th} for *n*- and *p*-channel MOSFETs, made by the NOLOCOS process and having an effective channel width of 1.2 and 1 μm, are within 40 and 120 mV, respectively. The use of polysilicon-buffer-recessed LOCOS technology for the isolation of 256 Mbit DRAM cells has achieved a cell size of

(a)

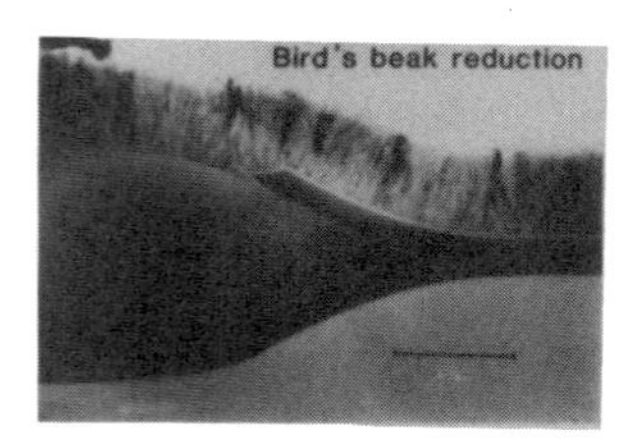

(b)

(c)

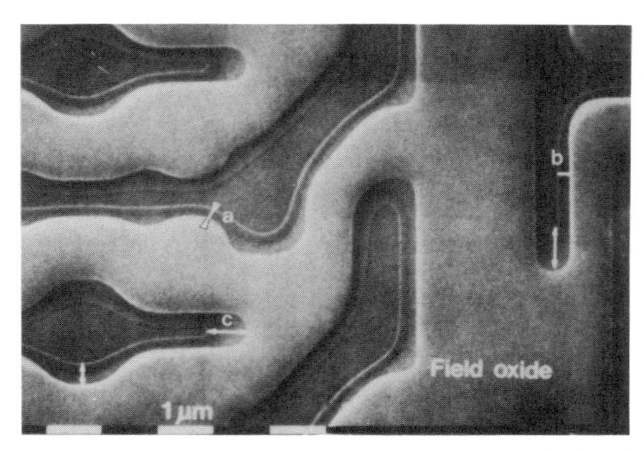

Figure 8.11 (a) Bird's beak formation under the conditions of (a) LOCOS and (b) modified LOCOS processing. (c) Top view of part of the SRAM cell of Figure 4.20(b). After Kramer [19] of Chapter 4; copyright © 1989, The Electrochemical Society, Pennington, NJ.

$0.6 \times 1.2\,\mu m^2$ for double cylindrical capacitors with 5.3 nm thick ONO isolation at a capacitance of 25 fF for 0.55 μm height (compare Section 7.4) [27].

As an alternative to LOCOS technology, trench insulation has been introduced by Tamaki et al. [29]. This technology employs refilling of deep

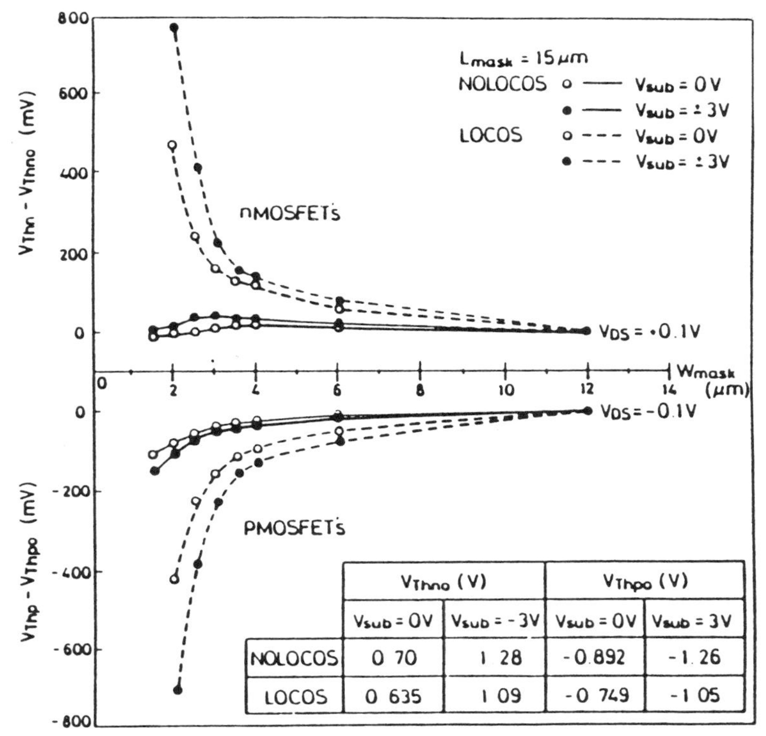

	VThno (V)		VThpo (V)	
	Vsub = 0V	Vsub = -3V	Vsub = 0V	Vsub = 3V
NOLOCOS	0 70	1 28	-0 892	-1 26
LOCOS	0 635	1 09	-0 749	-1 05

Figure 8.12 Narrow width effect on the threshold voltage of CMOS transistors employing NOLOCOS and LOCOS isolation. After Tsai, Yu, and Wu [26]; copyright © 1989, The Institute of Electrical and Electronics Engineers, New York.

trenches with Si_3N_4/high-temperature oxide/polysilicon followed by oxidation of the poly-Si, as illustrated in Figure 8.13(a). Trenches of $\leqslant 0.75\,\mu m$ width have been filled without void formation and are very well suited for isolation of CMOS transistors with excellent suppression of latchup. Trench isolation provides for lower leakage and higher breakdown voltages. Figure 8.13(b) shows a comparison of the threshold voltage shifts as a function of the effective channel width for narrow-channel NMOSFETS fabricated by LOCOS and trench isolation technologies, respectively. At submicrometer channel width, the trench isolation approach has the advantage.

Even more attractive in the context of ULSI is the dielectric isolation of thin films of single crystalline silicon by methods that permit one to tailor the silicon thickness to the feature size of the circuit. This may be accomplished by:

1. The epitaxial growth of thin silicon films on suitable single crystal dielectric substrates, such as sapphire or magnesium–aluminum spinel;
2. Epitaxial lateral overgrowth (ELO) of amorphous dielectric layers on a

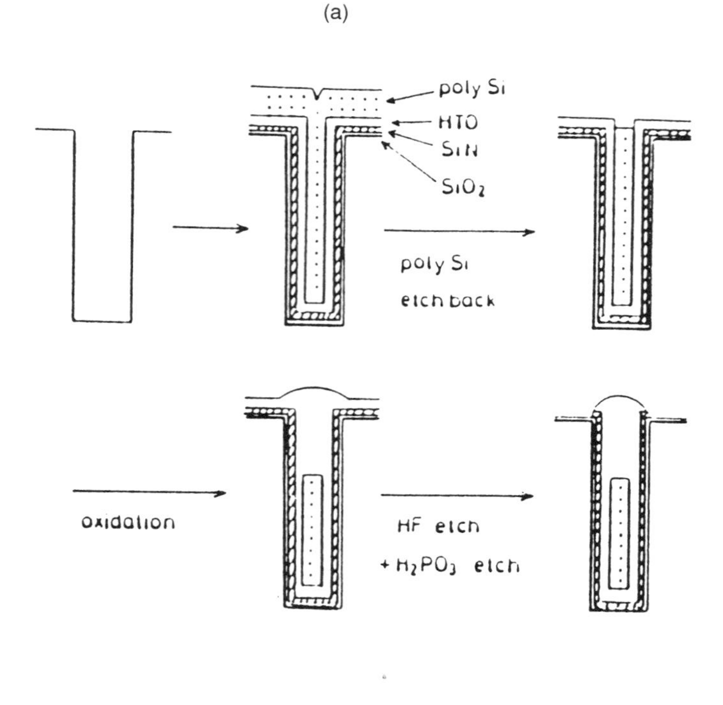

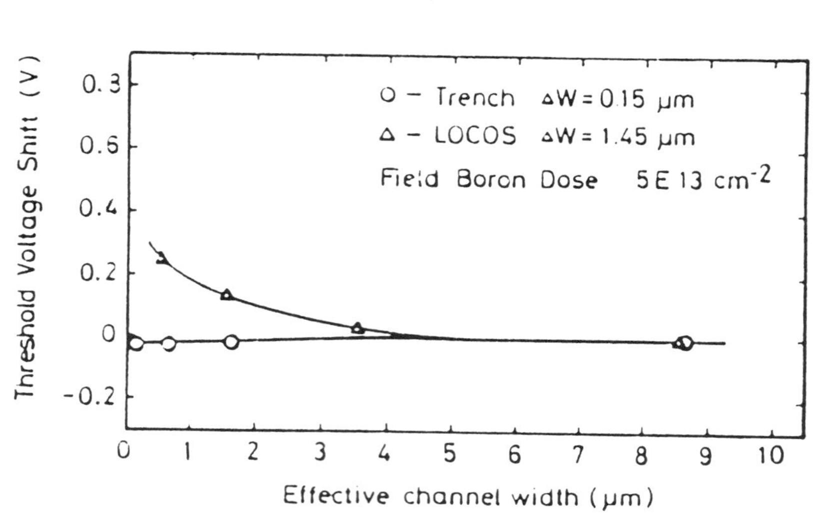

Figure 8.13 (a) Process flow in the implantation of trench isolation. (b) A comparison of the threshold voltage shift as a function of the effective channel width for conventional LOCOS and trench isolation. After Onishi et al. [28]; copyright © 1989, The Electrochemical Society, Pennington, NJ.

single crystalline silicon substrate wafer by silicon epilayers seeded through openings in the dielectric layer;
3. The generation of subcutaneous layers of amorphous dielectrics underneath a thin crystalline silicon film.

For example, porous silicon retains in its cell walls [30] the crystalline structure of the underlying *p*-silicon wafer [31] and can be overgrown epitaxially by MBE [32]. Also, porous silicon provides for a very large surface area to weight ratio [33]. Consequently, an oxidation front spreads rapidly beneath *n*-type silicon layers grown on *p*-type porous silicon, thus establishing a dielectric isolation layer. Details of the interaction of porous silicon with water vapor have been elucidated by Fourier transform infrared spectroscopy [34]. At room temperature, H_2O vapor adsorbs on the surface of porous silicon dissociatively, forming Si–OH and Si–H bonds. Figure 8.14(a) and (b) shows the annealing temperature dependence of the integrated infrared absorbance of the Si–H, Si–OH, SiO–H, and Si–O–Si stretching vibrations at 2090, 775, 3680, and 980 cm^{-1} and of the Si–D, SiO–D, Si–OD, and Si–O–Si stretching vibrations at 1513, 2707, 835, and 980 cm^{-2} for H_2O and D_2O saturated porous silicon surfaces, respectively. Upon annealing above 650 K, the Si–OH bonds are progressively replaced by Si–O–Si bonds and additional Si–H bonds that decrease, however, in density due to desorption of H_2. The generation of a buried oxide layer by ion implantation (SIMOX process) is discussed in Section 8.2.

The epitaxial lateral overgrowth of silicon by CVD nucleating through vias established by lithography in a SiO_2 film on a silicon wafer is illustrated in Figure 8.15. The challenge in this work is to prevent the nucleation of misoriented grains on the silicon dioxide surface and to avoid defect formation in the silicon film due to strain generated upon cooling of the epitaxial structure due to the difference in the thermal expansion coefficients for silicon and SiO_2. The former goal is achieved by the combination of growth and etching cycles. During the etching phases of the cycles, misoriented grains are removed preferentially from the SiO_2 surface so that after a number of cycles the wafer is coated by a single crystal film with few inclusions of misoriented silicon grains [35]. The formation of defects is critically dependent on the orientation of the edges of the silicon dioxide films defining the vias with regard to the silicon lattice. If the edge is along one of the ⟨110⟩ directions on an (100)-oriented wafer, the resolved stress generated upon cooling of the overgrown structure on the {111} slip planes can easily generate dislocations that move in the stress field by glide and multiply, leading to a high density of defects. On the other hand, orienting the edges along ⟨100⟩ directions reduces both the nucleation rate and the mobility of the dislocations that are not contained in a slip plane and consequently move by climb, reducing their multiplication [36]. High-quality ELO films exhibit carrier lifetimes and mobilities comparable to bulk silicon and are thus suitable for the manufacturing of MOS and bipolar transistor circuits [37]. A disadvantage of the

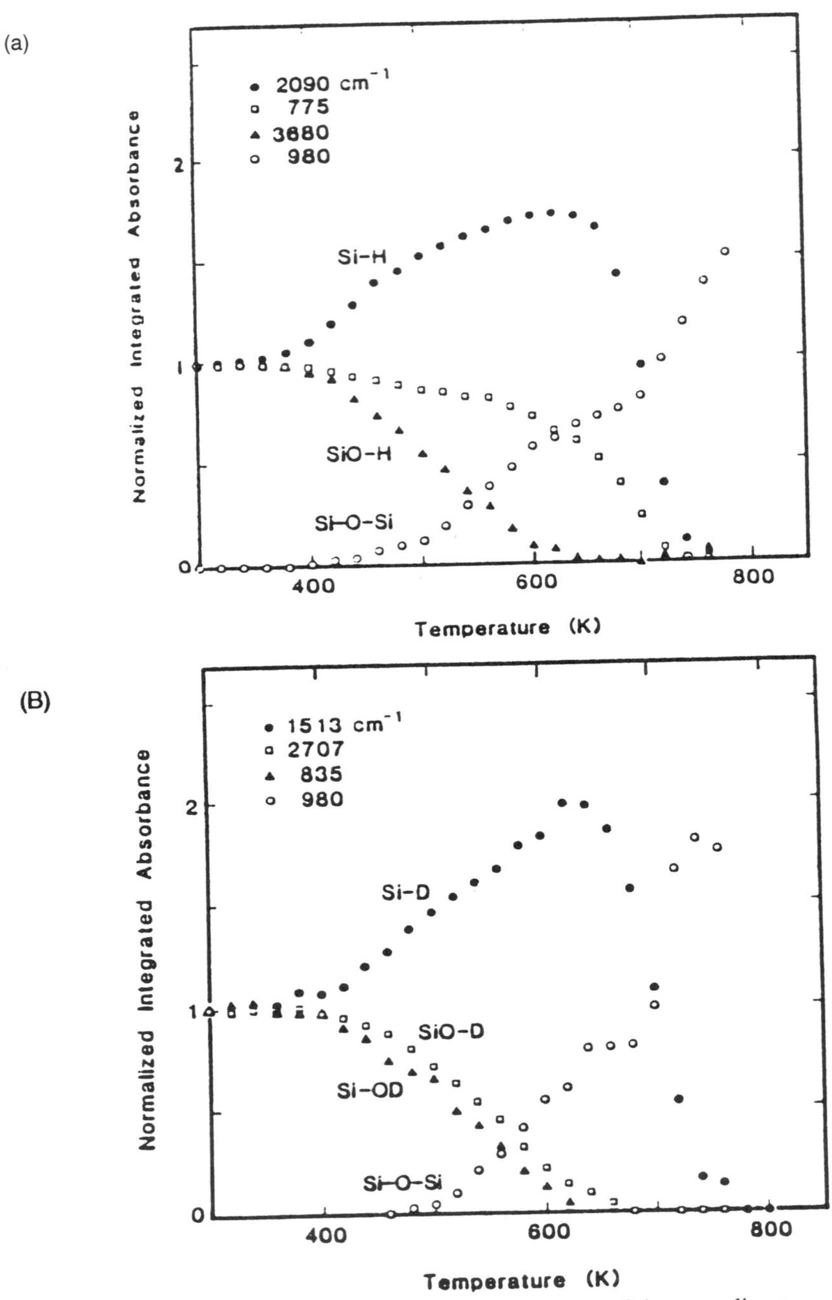

Figure 8.14 Integrated infrared absorbance as a function of the annealing temperature for porous silicon surfaces after saturation by dissociative adsorption of (a) H_2O and (b) D_2O vapor at room temperature. After Gupta et al. [34]; copyright © 1991, Elsevier Science Publishers B.V., Amsterdam.

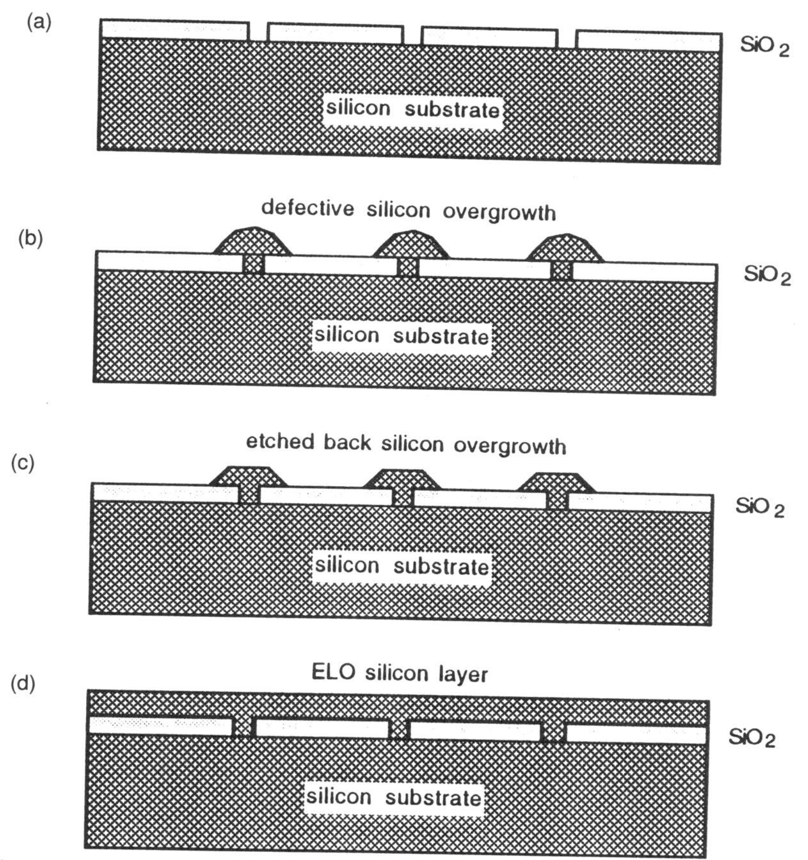

Figure 8.15 Epitaxial overgrowth of an SiO_2 film on a silicon wafer by CVD of silicon. (a) Structured SiO_2 film with openings for seeding. (b) Nucleation and growth. (c) After etching phase. (d) After complete overgrowth.

method is that relatively thick films, typically of the order of micrometers are formed while high-speed CMOS circuits are preferably fabricated in 0.2–0.5 μm thick films. Therefore, a thinning step must be included in the IC manufacturing process with this type of dielectric isolation.

Alternatively dielectric isolation has been achieved by silicon on sapphire (SOS), involving a substantial mismatch in the lattice parameters, and by epitaxial silicon deposition on layers of epitaxial dielectrics, such as $MgAl_2O_4$ [38], [39], [40] and CaF_2 [39], [40]. Figure 8.16(b) shows the lattice mismatch to silicon for a variety of semiconductors and the group IIA fluorides. While CaF_2 is nearly lattice-matched to Si, $MgAl_2O_4$ is not. However, $MgAl_2O_4$ provides for a large density of coincidence sites to Si on the (001) interface.

Figure 8.16(a) shows a schematic representation of a CVD system for chloride transport of Mg and Al in a nitrogen carrier to the mixing zone, where

(a)

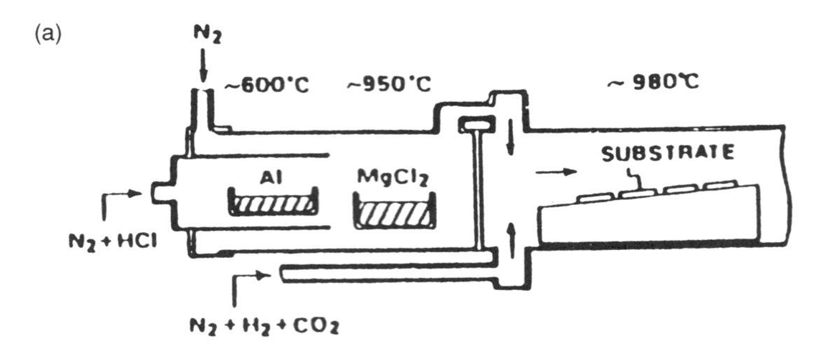

(b)

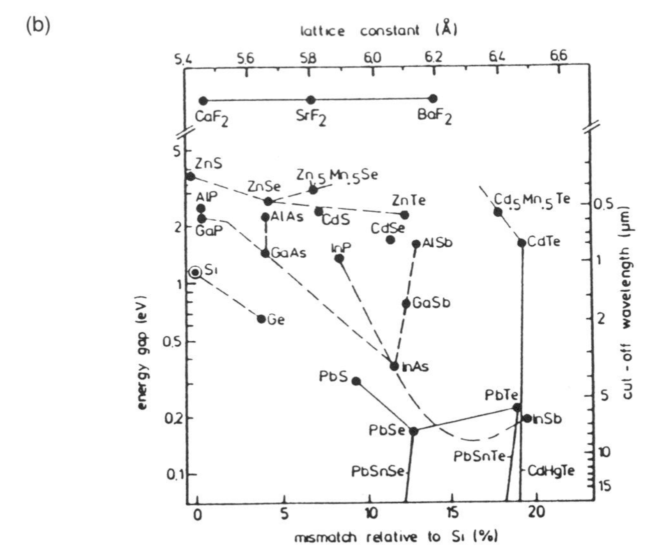

Figure 8.16 (a) Schematic representation of a CVD system for the growth of epitaxial spinel layers on silicon. (b) Lattice constants/mismatch to Si versus bandgaps. (a) After Hokari et al. [38]; copyright © 1985, Institute and Electronics Engineers, New York. (b) After Zogg et al. [39]; copyright © 1989, The Electrochemical Society, Pennington, NJ.

a mixture of nitrogen, hydrogen, and CO_2 is added to the vapor stream at flow rates of 56 l/min N_2, 20 l/min H_2, and 100–500 cm^3/min H_2. The transport rates are 0.3–0.8 g/h Al and 6–12 g/h $MgCl_2$. The oxygen supplied by the CO_2 provides for the deposition of an epitaxial layer of magnesium-aluminum spinel on a set of silicon substrate wafers located downstream from the mixing zone at a temperature in the range 850–1050°C. Si on $MgAl_2O_4$ MOSFETs achieve with a 3 μm thick Si epilayer a channel mobility of 440 $cm^2 V^{-1} s^{-1}$. This compares to 550 and 370 $cm^2 V^{-1} s^{-1}$ for the same FET using a bulk silicon wafer and silicon on sapphire, respectively. The perfection of the spinel improves with increasing layer thickness, as revealed by rocking curve measurements [41]. However, the surface roughness of the spinel film also increases with increasing layer thickness, so that either small thickness or planarization of the spinel prior to the silicon epitaxy step is required.

8.2 Doping by Ion Implantation and Diffusion

In this section we add to the discussion of diffusion under the conditions of thermal oxidation information on the coupled diffusion of impurities and point defects in solids. Under steady-state conditions, impurities diffuse according to Fick's first law [Eq. (8.1)], generally with a temperature- and concentration-dependent diffusion coefficient $D(c_j, T)$. The temperature dependence of D is described by

$$D = D_0 \exp\left(-\frac{E_a}{kT}\right) \tag{8.13}$$

where D_0 is a frequency factor and E_a is the activation energy. Table 8.3 shows the values of D_0 and E_a for selected impurities in Si.

Under dynamic conditions, conservation of matter requires that the change in the concentration with time must equal the local change in the diffusive flux,

$$\frac{\partial c}{\partial t} = -\nabla \mathbf{J}_D = -\nabla D \nabla c \tag{8.14}$$

which simplifies for dilute concentrations where D is a constant to

$$\frac{\partial c}{\partial t} = -D \Delta c \tag{8.15}$$

For linear diffusion to a planar interface

$$\frac{\partial c}{\partial t} = D \frac{\partial^2 c}{\partial x^2} \tag{8.16}$$

Solving Eq. (8.16) with the initial condition

$$c(x, t = 0) = 0 \tag{8.17}$$

Table 8.3 Frequency factor D_0 and activation energy E_a for selected impurities in silicon (after Tsai [21]).

Diffusing atom	$D_0(\mathrm{cm^2\,s^{-1}})^a$	$E_a(\mathrm{eV})^a$
B	0.76	3.46
P	3.85	3.66
pAs	24	4.08
Sb	0.21	3.65

[a] The values of these constants depend somewhat on the method of analysis and of the diffusion conditions.

and the boundary conditions

$$c(x = 0, t) = c_s \quad \text{and} \quad c(x = \infty, t) = 0 \tag{8.18}$$

results in

$$c(x, t) = c_s \operatorname{erfc}\left(\frac{x}{2\sqrt{Dt}}\right) \tag{8.19}$$

This is the case of diffusion of a dopant held at constant surface concentration c_s into an undoped semiconductor wafer. However, frequently a fixed amount a_d of dopant is supplied to the surface and is allowed to distribute into the semiconductor. The initial condition remains as in the case of constant surface concentration, but the boundary conditions are

$$\int_0^\infty c(x, t)\, dx = a_d \tag{8.20}$$

$$c(x, t = \infty) = 0 \tag{8.21}$$

With these boundary conditions, Eq. (8.16) has the solution

$$c(x, t) = \frac{a_d}{\sqrt{\pi D t}} \exp\left(-\frac{x^2}{4Dt}\right) \tag{8.22}$$

Although Eqs. (8.19) and (8.22) provide a useful basis for semiquantitative assessments of the diffusion profile in simple cases, generally the diffusion mechanism is more complicated since it involves the coupled diffusion of several impurities and native point defects. Figure 8.17(a) shows a selection of models of impurity–native defect interactions in diffusive transport. If the concentration of one particular defect is much larger than the concentration of other native defects, the diffusion of a dominant extrinsic impurity can be described in terms of a simplified mechanism. For example, the emitter push effect has been explained by a simplified model based on the generation of excess vacancies in the phosphorus-doped region by dissociation of p^+V^{2-} pairs. These pairs form in the highly doped surface region and dissociate in the more lightly doped interior regions [42]. In the simple model of [59], a constant generation rate in the p-doped region is assumed, extending to the break in the diffusion profile. The vacancies diffuse into the interior and toward the surface, where conditions close to equilibrium are maintained. Thus a vacancy distribution is obtained, as shown in Figure 8.17(b). An enhancement of diffusion in the crystal region of maximum excess vacancy concentration explains the tail in the phosphorus diffusion profile and the concomitant increase in the boron diffusion rate in this portion of the crystal. Note that this enhanced diffusion is not merely a function of the dilation of the lattice, but depends on the charge of the vacancies as well. Due to Jahn–Teller distortions (see Section 2.5), the charging of the vacancies may affect the local structure of the defect, and the diffusivities thus depend on the position of the Fermi level relative to its intrinsic position. Also, there are generally more than one native

(a)

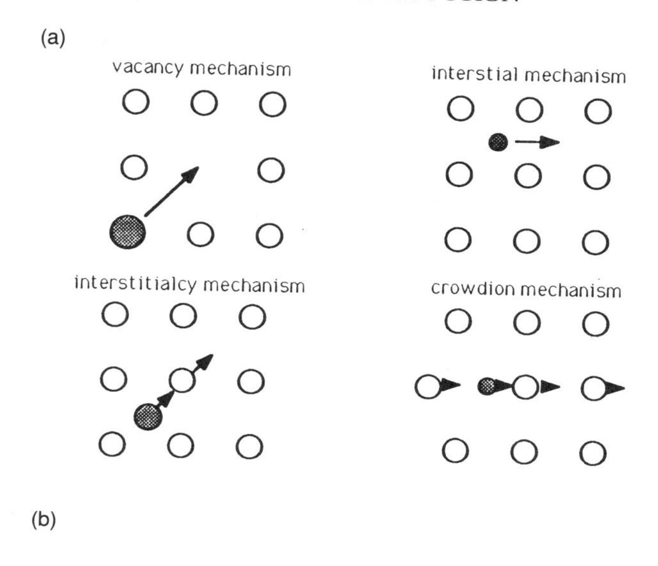

(b)

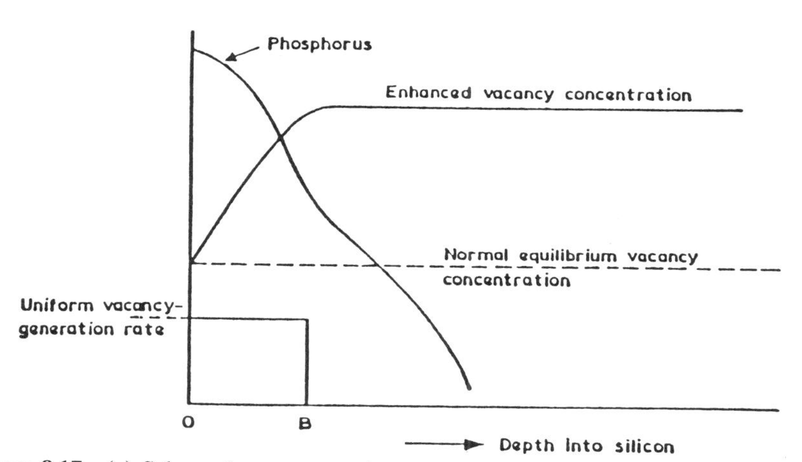

Figure 8.17 (a) Schematic representation of various mechanisms of point-defect diffusion. (b) Simplified model explaining the emitter push effect on the basis of a vacancy-enhanced diffusion model. After Lee [65] of Chapter 7; copyright © Philips Research Laboratories, Eindhoven.

defect involved in the diffusion mechanism. Although the concentrations of these defects are interrelated by the mass action law, deviations from equilibrium may exist, complicating the interpretation of the diffusion profiles.

In general, the impurities exist in electrically active form at concentration C_j and clustered form at concentration C_j^{clust}. Denoting the intrinsic diffusivity of impurity j by D_j, the interstitial and vacancy diffusivities by D_{I} and D_{V}, the fractional interstitial and vacancy diffusion components involving impurity j by f_j and $\bar{f}_j$, the bulk interstitial–vacancy recombination rate by k_{iv}, the average sink efficiency in the presence of structural defects (e.g., dislocation loops that expand or shrink upon interaction with the native point defects) by α_s, and the interstitial generation rate by β_l, Eq. (8.14) takes on the forms

$$\frac{\partial C_j}{\partial t} = \nabla\left(D_j f_j \frac{C_1}{C_1^{eq}} \nabla C_j + D_j C_j \frac{f_j}{C_1^{eq}} \nabla C_l\right) + D_j \bar{f}_j \frac{C_j}{C_V^{eq}} \nabla C_V$$

$$+ \left[Z_j C_j D_j \frac{q}{kT}\left(f_j \frac{C_1}{C_1^{eq}} + \bar{f}_j \frac{C_j}{C_V^{eq}}\right)\nabla\Psi\right] - k_d^j C_j^{mj} c^k + k_D^j C_j^{clust.} \tag{8.23}$$

$$\frac{\partial C_j}{\partial t} = k_d^j C_j^{mj} c^k + k_D^j C_j^{clust.}, \qquad j = 1, \ldots, N \tag{8.24}$$

$$\frac{\partial C_1}{\partial t} = \nabla\left[\left(D_1 + \sum_{j=1}^{N} f_j D_j \frac{C_j}{C_1^{eq}}\right)\nabla C_1\right] + \sum_{j=1}^{N} f_j D_j \frac{C_1}{C_1^{eq}} \nabla C_j$$

$$+ \left(\sum_{j=1}^{N} f_j Z_j D_j C_i \frac{C_1 q}{C_1^{eq} kT} \nabla\Psi\right) + k_{iv}(C_1 C_V - C_1^{eq} C_V^{eq})$$

$$+ \alpha_s D_1 C_1 g(x) - \sum_{j=1}^{N} \beta_j \frac{\partial C_j^{clust.}}{\partial t} \tag{8.25}$$

$$\frac{\partial C_V}{\partial t} = \nabla\left[\left(D_V + \sum_{j=1}^{N} \bar{f}_j D_j \frac{C_j}{C_V^{eq}}\right)\nabla C_V\right] + \sum_{j=1}^{N} \bar{f}_j D_j \frac{C_V}{C_V^{eq}} \nabla C_j$$

$$+ \left(\sum_{j=1}^{N} \bar{f}_j Z_j D_j C_i \frac{C_V q}{C_V^{eq} kT} \nabla\Psi\right) + k_{iv}(C_1 C_V - C_1^{eq} C_V^{eq}) \tag{8.26}$$

which can be used as a basis for the calculation of the diffusion profiles [43], [44].

Figure 8.18(a) shows a comparison of experimental data and a simulation of the coupled diffusion of boron and interstitials in silicon with and without amorphization of the Si subsurface region by ion implantation [43]. Boron diffusion is associated with interstials, and the effect of extended defects in the damaged region is to act as sinks for the interstitials, lowering the interstitial concentration in the bulk. This is incorporated into the simulation by turning on the parameter α_s at the depth of the damage that can be checked experimentally by TEM.

Figure 8.18(b) shows a similar simulation for phosphorus diffusion with and without amorphization and its implications regarding the emitter push effect

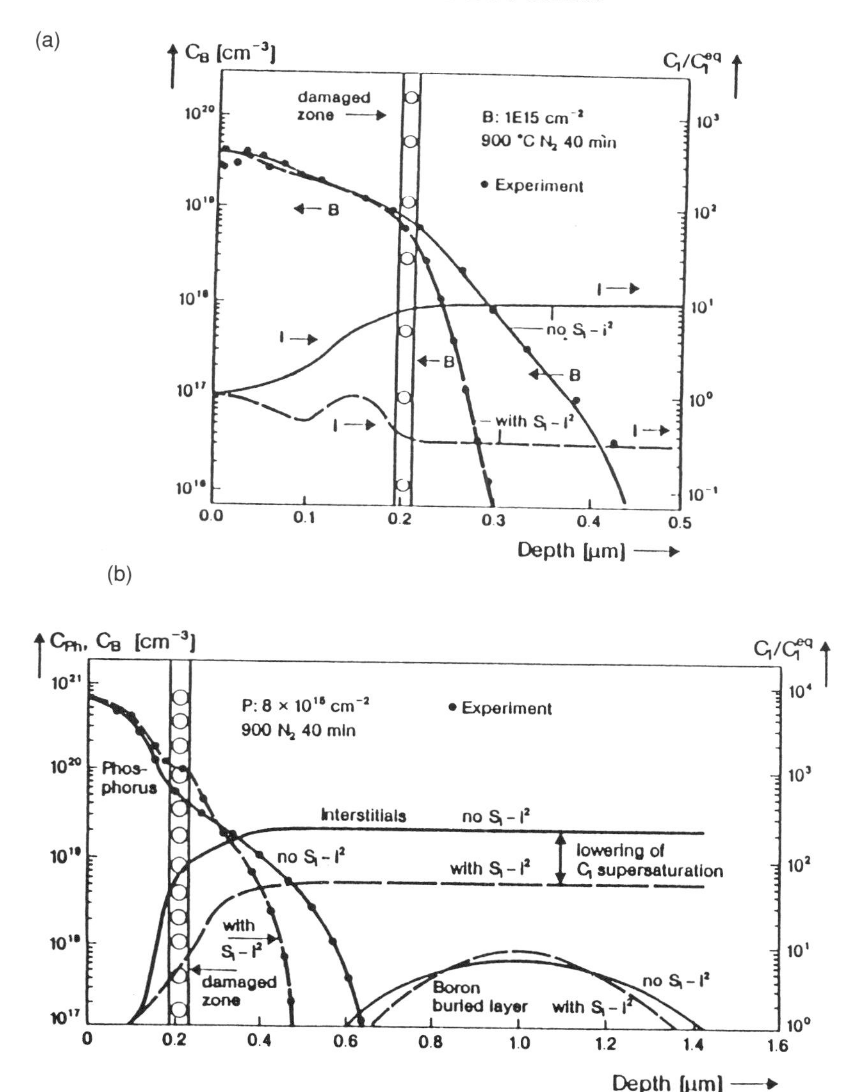

Figure 8.18 Simulation and experimental data for the coupled diffusion of (a)boron and interstitials and (b) phosphorus and interstitials in silicon with and without amorphization. After Orlowski [43]; copyright © 1988, IEDM, New York.

on a buried layer, which is retarded. Access to reliable simulations is essential for the design of ULSI processes where the tight tolerances demand intimate before-the-fact knowledge of the implications of changes in the process parameters. An example of the utility of this modeling for devising processing strategies is the generation of shallow p^+n junctions in Si [44]. Simulations show that the diffusivity of B in Si is retarded in the presence of Ga. Thus preamorphization by Ga implantation is utilized to control the B diffusion profile.

In view of the generation of point defects, such as Si interstitials at the Si/SiO_2 interface and upon ion implantation, the diffusion mechanism at interfaces and in implanted regions frequently differs from the mechanism that is valid for bulk material. Also, in heterostructures, the interfacial region generally provides for a unique chemical bonding and strain environment and must be treated as a third phase communicating with the two phases connected to the interface by diffusion. In particular, dopant segregation has been observed at the Si/SiO_2 interface, as illustrated in Figure 8.19 for phosphorus. This excess phosphorus may be bonded, in part, to oxygen and, in part, may be preferentially incorporated simply to relieve the interfacial strain. Denoting the total density of dopant traps at the interface by T_t and the concentrations of phosphorus on both sides of the interface as C_1 and C_2, the density of filled traps T_d can be calculated according to

$$T_d = T_t(C_1 + \alpha_t C_2)/(C_1 + \alpha_t C_2 + \beta_e) \tag{8.27}$$

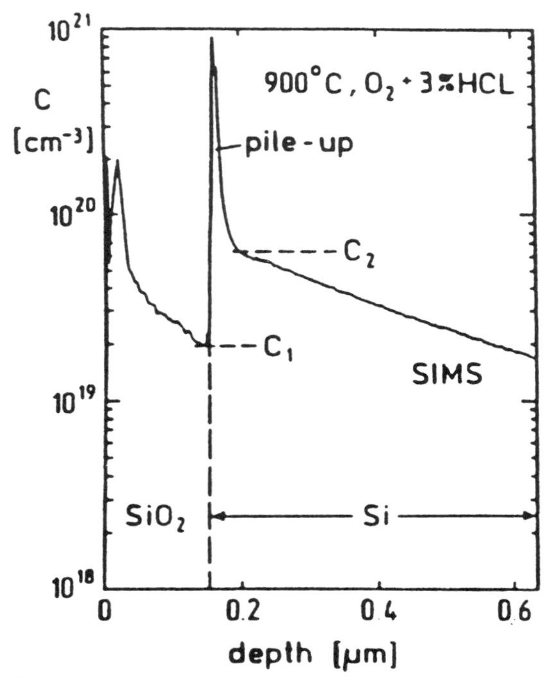

Figure 8.19 Phosphorus concentration C as a function of depth measured from the surface into an ion-implanted and oxidized silicon wafer. The P dose is $2.5 \times 10^{15}\ cm^{-2}$, and the dry oxidation is carried out at 900°C in an HCl-containing ambient for 450 min. After Lau et al. [45]; copyright © 1989, Springer Verlag, Berlin.

with

$$\alpha_t = t_1/t_2 \tag{8.28}$$

$$\beta_e = (e_1 + e_2)/t_1 \tag{8.29}$$

where t_1, t_2 are the trapping probabilities and e_1, e_2 are the emission probabilities, respectively, of dopant atoms on the two sides of the junction [45].

Ion implantation is accomplished by the exposure of the surface of a solid to a rastered beam of energetic ions, typically at energies $\leqslant 500\,\text{keV}$ for singly charged ions. It permits convenient access to a wide variety of dopants and allows the precise dosing of the number of dopant atoms implanted into the subsurface region. The number of ions deposited per unit area is called the *ion dose* Φ_i. Figure 8.20 shows a schematic representation of the cross section through an ion implanter. The ions are generated by electron impact and are extracted by an electrostatic field into a magnetic field that in conjunction with a slit selects the mass of the implanted ions. After further acceleration and focusing the ions pass through the scanning plates that deflect the ion beam, thus rastering the target surface. There the ions collide with the atoms of the lattice and come to rest at a characteristic depth called the *range* R_i of the ion implantation process. In random scattering of energetic ions, the displacement of the secondary recoils from their lattice positions causes considerable damage to the perfection of the lattice about the primary ion path. Figure 8.21 shows as an example an HREM image of the damage plume of a B^+ implant in silicon.

The projection of R_i onto the normal to the target surface is called the *projected range*. The statistical variations in R_i impose onto R_p the straggles $\Delta R_{\parallel}$ and $\Delta R_{\perp}$ parallel and transverse to the normal to the target surface, respectively. The transverse straggle $\Delta R_{\perp}$ is important, as it undercuts the edges

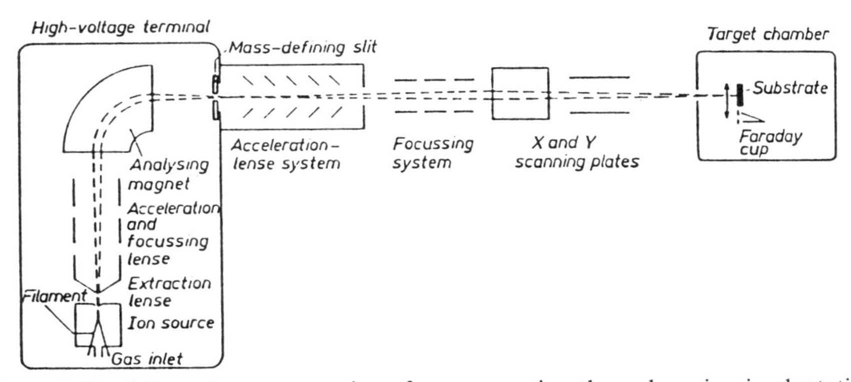

Figure 8.20 Schematic representation of a cross section through an ion implantation apparatus. After Hofker [52]; copyright © 1975, Philips Research Laboratories, Eindhoven.

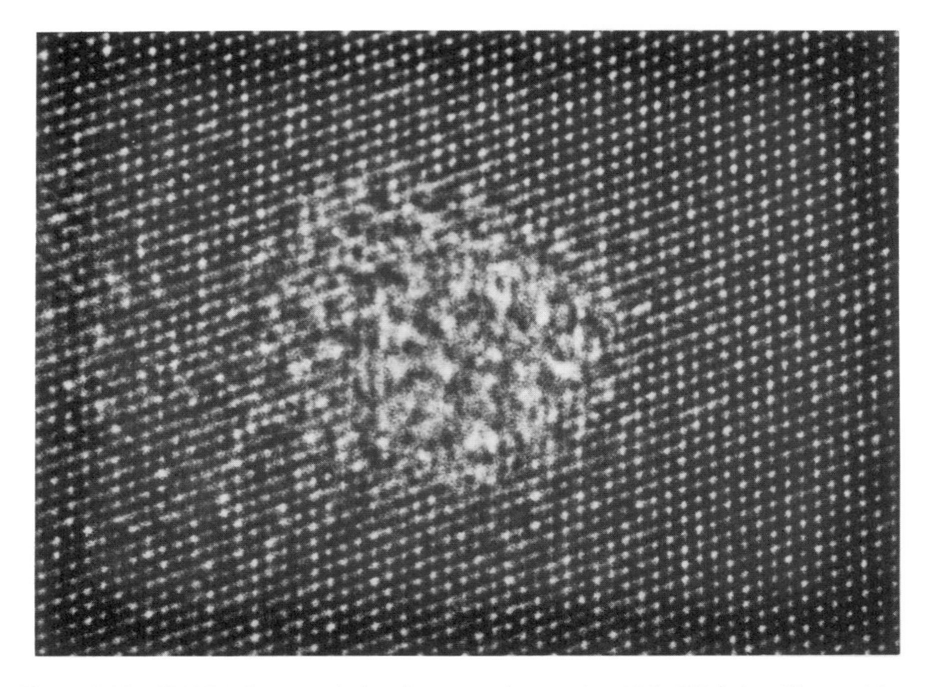

Figure 8.21 HREM image of the damage plume of a 70 keV B^+ in silicon. After Narayan and Holland [51]; copyright © 1988, The Electrochemical Society, Pennington, NJ.

of the mask. Figure 8.22 shows the ion concentration as a function of lateral coordinate y for a gate mask of width $a_m \gg \Delta R_\perp$ (top) and iso-ion concentration contours (bottom) for 70 keV B^+ implantation into Si [46]. The projected range and straggles are $R_p = 2710$ Å, $\Delta R_\perp = 1006$ Å, and $\Delta R_\parallel = 824$ Å.

A method for the calculation of the ion range as a function of the ion energy

$$R_i(E) = \frac{1}{N_t} \int_0^E \frac{dE}{S_{ion}(E)} \tag{8.30}$$

has been provided by the Linhart–Scharff–Schiott theory [47], which assumes that the energy losses of the ions due to nuclear collisions and electronic effects are independent and add. The stopping power

$$S_{ion} = \frac{dE}{dx} = S_{nuclear} + S_{electronic} \tag{8.31}$$

defined as the decrease in ion energy with increasing path length x in the solid, is thus separable into two terms. It can be shown that the nuclear stopping power for an incident ion of mass M_i and atomic number Z_i by collisions with

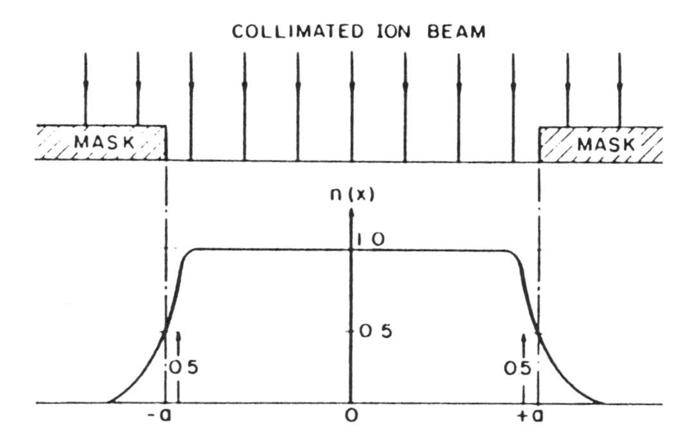

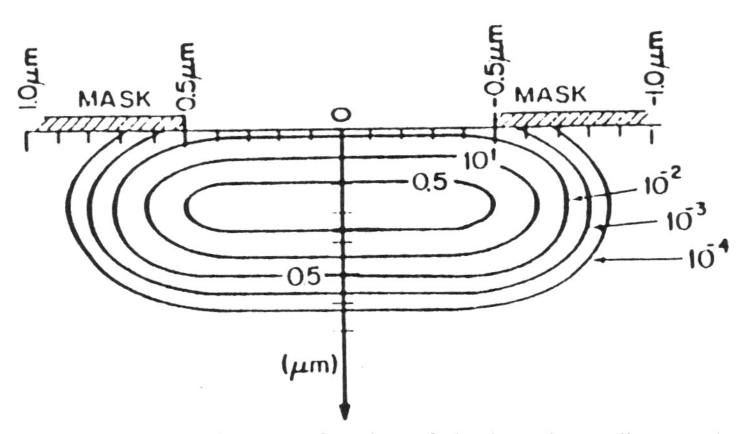

Figure 8.22 Ion concentration as a function of the lateral coordinate *y* (top) and iso-ion concentration contours for the implantation of 70 keV B^+ into Si through a gate mask. After Furukawa and Ishiwara [46]; copyright © 1972, The Japanese J. Appl. Phys., Tokyo

target atoms of mass M_t and atomic number Z_t can be expressed as

$$S_{\text{nuclear}} = \int_0^{E_{tm}} E_t \, d\sigma_s \tag{8.32}$$

where

$$E_t = \frac{4M_i M_t}{(M_i + M_t)^2} \sin^2 \frac{\theta}{2} E_i \tag{8.33}$$

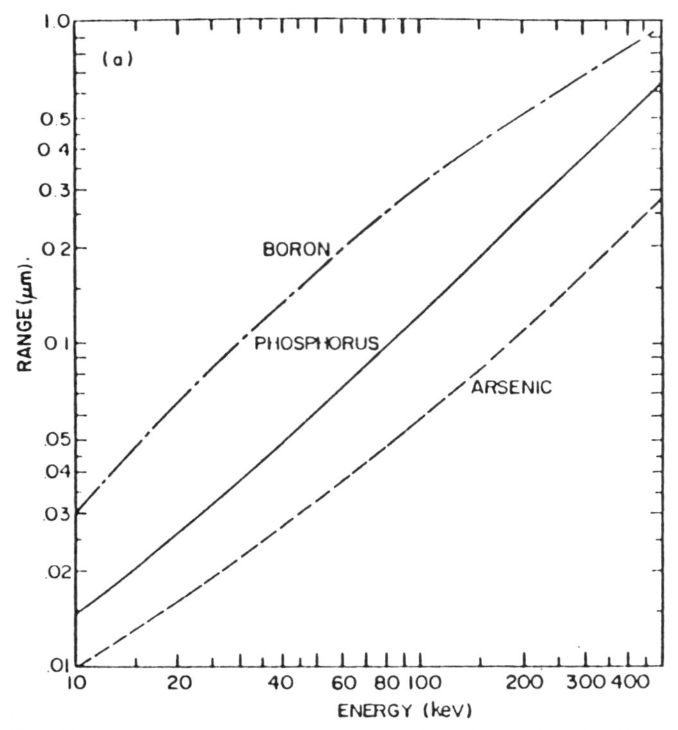

Figure 8.23 The ranges of boron, arsenic, and phosphorus ions implanted into silicon. After Pickar [48]; copyright © 1975, Academic Press, New York.

is the energy transferred in the scattering event, θ is the collision angle, and $d\sigma_s$ is the differential scattering cross section. The electronic stopping power is

$$S_{\text{electronic}} = f_e\{Z_i, Z_t, M_i, M_t\}\sqrt{E_i} \tag{8.34}$$

where $f_e\{Z_i, Z_t, M_i, M_t\}$ is an insensitive function of the variables. For light ions, such as boron, the electronic part of the stopping power dominates, while for heavy ions, such as As, the nuclear part of the stopping power dominates, at least at moderate ion energies. Figure 8.23 shows the projected ranges for boron, phosphorus, and arsenic as a function of the ion energy.

The distribution of the implanted dopants may be approximated by a Gaussian distribution; that is, for implantation through an infinitely long slit along the z direction of width $2a_m$ parallel to the y axis

$$N(x, y) = \frac{\Phi}{\sqrt{2\pi}\,\Delta R_{\parallel}} \exp\left[-\left(\frac{x - R_p}{\sqrt{2}\,\Delta R_{\parallel}}\right)^2\right] \frac{1}{\sqrt{\pi}} \operatorname{erfc}\left(\frac{y - a_m}{\sqrt{2}\,\Delta R_{\perp}}\right) \tag{8.35}$$

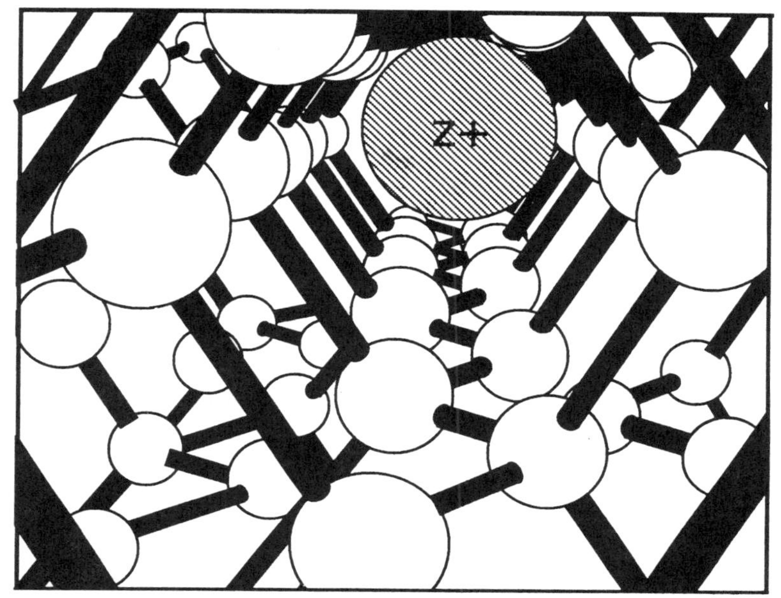

Figure 8.24 ⟨110⟩ channeling of an ion in a diamond structure lattice.

The depth distribution in the direction x is thus approximated by

$$N(x) = \frac{\Phi}{\sqrt{2\pi}\,\Delta R_{\parallel}} \exp\left(-\frac{x - R_{\mathrm{p}}}{\sqrt{2}\,\Delta R_{\parallel}}\right)^2 \tag{8.36}$$

which holds to a falloff of approximately an order of magnitude below the peak implanted ion concentration $N_{\max} = \Phi/\sqrt{2\pi}\,\Delta R_{\parallel}$. The wings in the ion distribution beyond this range are associated with ion channeling and, at elevated temperatures, diffusion (see ref. [49] for a more detailed discussion).

Ion channeling occurs in lattice directions where open channels exist, as illustrated in Figure 8.24 for the ⟨110⟩ channels in a diamond structure lattice. Ions that enter into these channels of open space are guided by glancing collisions, with the lattice atoms forming the channel walls. In view of the $\sin^2 \theta/2$ term in Eq. (8.33), these collisions cause only a relatively small energy loss and consequently result in a far larger range than ions that are not guided in lattice channels. It also results in a much smaller yield of backscattered ions, which is detected in Rutherford backscattering spectroscopy (RBS), which is discussed in more detail in the following.

In Si IC technology annealing at >800°C typically is required to reestablish crystalline order after ion implantation. Note that while the peak position in the distribution under these annealing conditions does not change, the wing structure is strongly affected even at 800°C. The annealing time can be substantially reduced by rapid annealing techniques, such as flash lamp annealing, that deposit the thermal energy into the near-surface region, where the damage is done, improving the thermal budget and the cleanliness of the operation. Because the implanted ions knock out atoms from substitutional into instertitial sites, the implantation of a particular type of ion effects the distribution of all other dopants that were introduced into the lattice previously [52].

As discussed in Section 8.1, for submicrometers CMOS technology high-quality silicon films of ~100 nm thickness are required, preferably on a dielectric substrate to suppress latchup between the *n*- and *p*-channel transistors. Ion implantation is of considerable utility in this context. For example, the currently most mature technology for providing dielectrically isolated thin films is the deposition of silicon on sapphire (SOS) substrates by CVD. Since sapphire is not lattice matched by silicon, a substantial density of defects exists at the silicon–sapphire interface. This defect density can be significantly reduced by a double solid-phase epitaxy (DSPE) method [53]. It is based on amorphization of the as-grown SOS film by ^{28}Si implantation at 170 keV and a dose of $1 \times 10^{15}\,\mathrm{cm}^{-2}$ followed by annealing which recrystallizes the film. A second ion implantation cycle at $2 \times 10^{15}\,\mathrm{cm}^{-2}$ dose and 100 keV energy followed by annealing is then added and essentially removes the twinning observed for the as-grown films. This leads to substantial improvements in both the channel mobility and transconductance of *n*- and *p*-channel transistors made after application of the DSPE process, as compared to the as-grown SOS material [54]. However, the lower dielectric constant of SiO_2 as compared to Al_2O_3 reducing parasitics and the relatively high cost of sapphire substrates currently motivates research on alternative structures that make use of the excellent quality of thermal oxide films for the isolation of complementary devices. One of these alternatives, ELO technology, already has been discussed in Section 8.1.

Another alternative, SIMOX technology, relies on a heavy ion implant of oxygen at $10^{18}\,\mathrm{cm}^{-2}$ dose using ~150 keV energy to drive the peak oxygen distribution several thousand Å below the surface [55]. Dissolved oxygen in silicon is located in interstitial positions about midway on the Si–Si bonds. The solubility of oxygen in these interstitial positions in the silicon lattice is given by the relation

$$c_{\mathrm{Oi}} = 1.5 \times 10^{21} \exp\left[\frac{-1.03\,\mathrm{eV}}{kT}\right) \tag{8.37}$$

and is exceeded at the high implant dose. Therefore, upon annealing a silicon dioxide film is formed under a surface layer of high-quality silicon that is typically 0.25 μm thick, that is, one suitable for CMOS processing.

The motion of oxygen implanted into silicon has been a puzzle for a long time, since at low temperatures and unusually fast-moving species is involved. This behavior is also exhibited in the out-diffusion of oxygen at the surface of

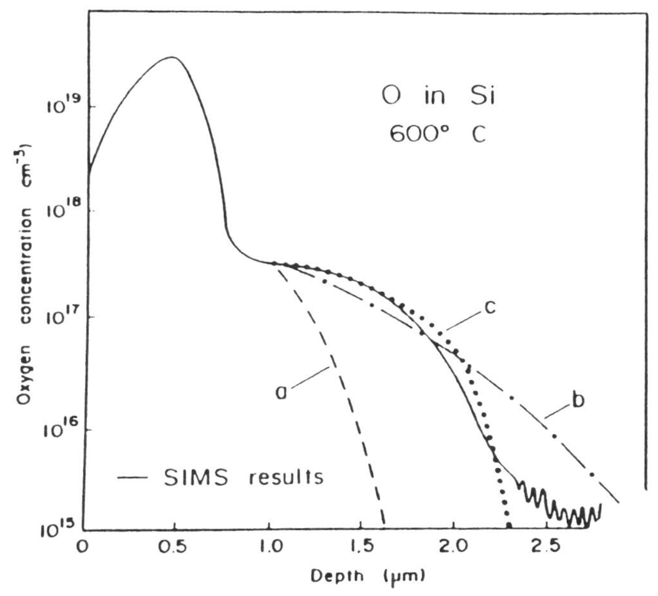

Figure 8.26 SIMS profile of ^{18}O concentration in silicon after ion implantation and subsequent annealing at 600°C. After Gösele et al. [56]; copyright © 1989, Springer Verlag, Berlin.

Czochralski pulled silicon wafers. The high-temperature oxygen diffusion data for silicon fit very well an Arrhenius plot [57], resulting in a diffusion coefficient

$$D_{\mathrm{Oi}} = 0.07 \exp\left(\frac{-2.44\,\mathrm{eV}}{kT}\right) \tag{8.38}$$

However, at $T < 700°\mathrm{C}$, the diffusion behavior is no longer determined by a simple thermally activated motion between interstitial positions.

Figure 8.26 shows the diffusion profile of oxygen implanted into silicon and annealed at 600°C. The dashed and dashed-dotted lines in this figure refer to attempts to fit the diffusion profile with a simple thermally activated diffusion process, as represented by Eq. (8.38) and with a diffusion constant $10\,D_{\mathrm{Oi}}$, respectively. The dotted line was obtained in a calculation employing a concentration-dependent diffusivity based on the following model reaction

$$\mathrm{O_i} + \mathrm{O_i} \rightleftarrows \mathrm{O_2^m} \tag{8.39}$$

$$\mathrm{O_2^m} + \mathrm{Si^x} \rightarrow \mathrm{O_2^{grec}} + \mathrm{Si_i} \tag{8.40}$$

where $\mathrm{O_i}$ and $\mathrm{Si_i}$ are oxygen and silicon atoms in interstitial positions, $\mathrm{Si^x}$ denotes a lattice silicon atom, and $\mathrm{O_2^{grec}}$ and $\mathrm{O_2^m}$ denotes oxygen dimers in

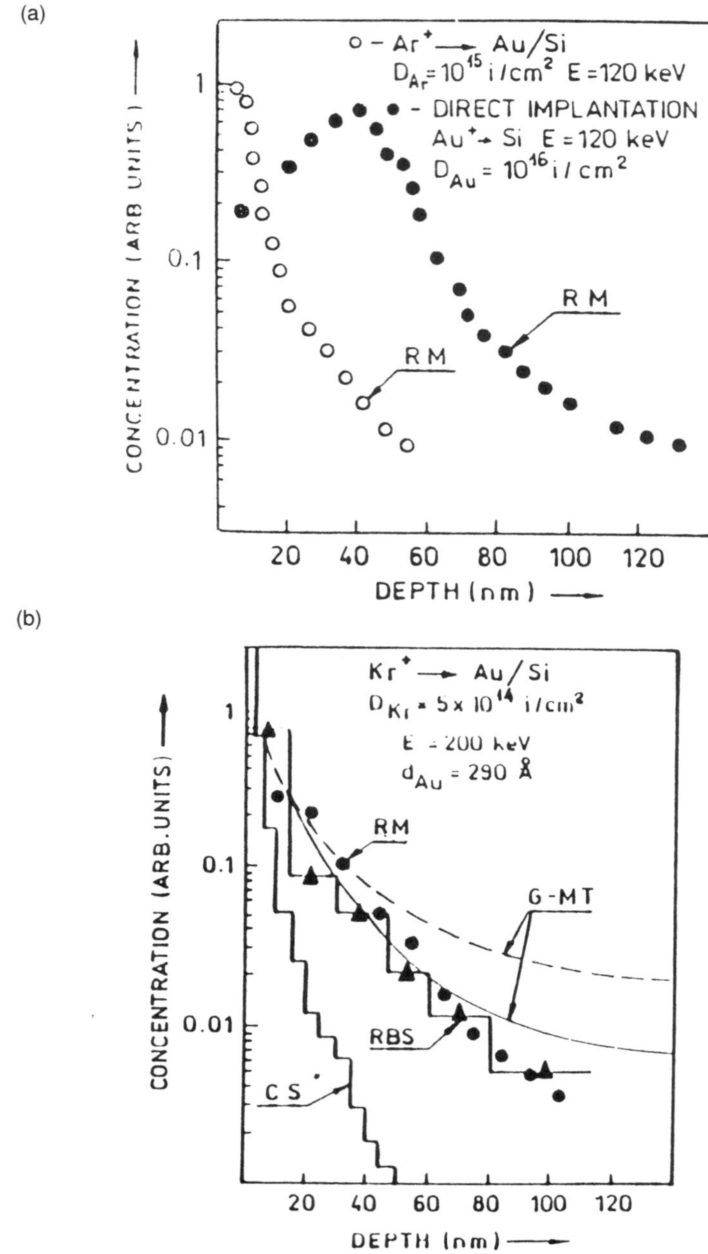

Figure 8.27 (a) Depth distributions of Au in Si produced by direct (full circles) and recoil ion implantation (open circles). (b) Comparison of the RM data (full circles) and RBS data (full triangles) for recoil implantation with the results of the Gras–Marti model (G-MT) and Monte Carlo simulations (step functions). After Paprocki, Brylowskam, and Syszko [58]; copyright © 1988, Springer Verlag, Berlin.

precipitated and mobile forms, respectively [56]. The understanding of the low-temperature diffusion of oxygen is important in the context of both thermal donors, which are formed at ~450°C and are thought of as agglomerates of oxygen, starting with the species O_2^{grec}, and the formation of SiO_2 precipitates in silicon that is supersaturated with regard to oxygen. Also, it is necessary for the understanding of intrinsic gettering by such precipitates and of the formation of the SIMOX structure.

Shallow implants are important because of the device scaling that in ULSI requires not only shrinking lateral, but also decreasing vertical, dimensions of the devices. Implants with a range below 2000 Å depth are obtained by recoil implantation. This method relies on the recoil of atoms from a thin surface film into the subsurface region of the solid upon bombardment with inert ions. Figure 8.27(a) shows a comparison of the depth distributions obtained by the radioactive tracer method for direct implantation of Au into Si and for recoil implantation of Au from a 250–300 Å thick Au film by bombardment with noble gas ions [58]. Figure 8.27(b) shows a comparison of the experiment depth distribution obtained by radioactive tracer experiments (RM) and Rutherford backscattering (RBS) [59] with the results of Monte Carlo simulations and with two analytical expressions derived on the basis of a particular model [60] for primary recoil (dashed line) and for cascade recoil (full line), respectively.

The experimental arrangement of RBS is illustrated in the inset to Figure 8.28. ^{4}He ions of 1–3 MeV energy incident at an angle of 45° to the surface normal are scattered by the lattice atoms and the flux and energy of the backscattered ions is measured by an energy-dispersive detector. Each channel corresponds to a window of energy $[E_1, E_2]$. Since the ions lose energy upon backscattering, the energy loss

$$\Delta E = \sigma_s \rho_a \Delta x \tag{8.41}$$

provides information on the scattering depth Δx if the scattering cross section σ_s and the atom density ρ_a of the sample are known. The number of counts is

$$N_c = \rho_a \Delta x J_i \Delta\Omega \frac{d\sigma_s}{d\Omega} \tag{8.42}$$

where J_i and $\Delta\Omega$ are the beam current and the acceptance angle of the detector, respectively. Since the differential scattering cross section

$$\frac{d\sigma_s}{d\Omega} = k_R Z^2 \tag{8.43}$$

is proportional to Z^2, RBS is particularly sensitive to heavy atoms.

When the incident $^4He^+$ beam enters the solid in a channeling direction, the

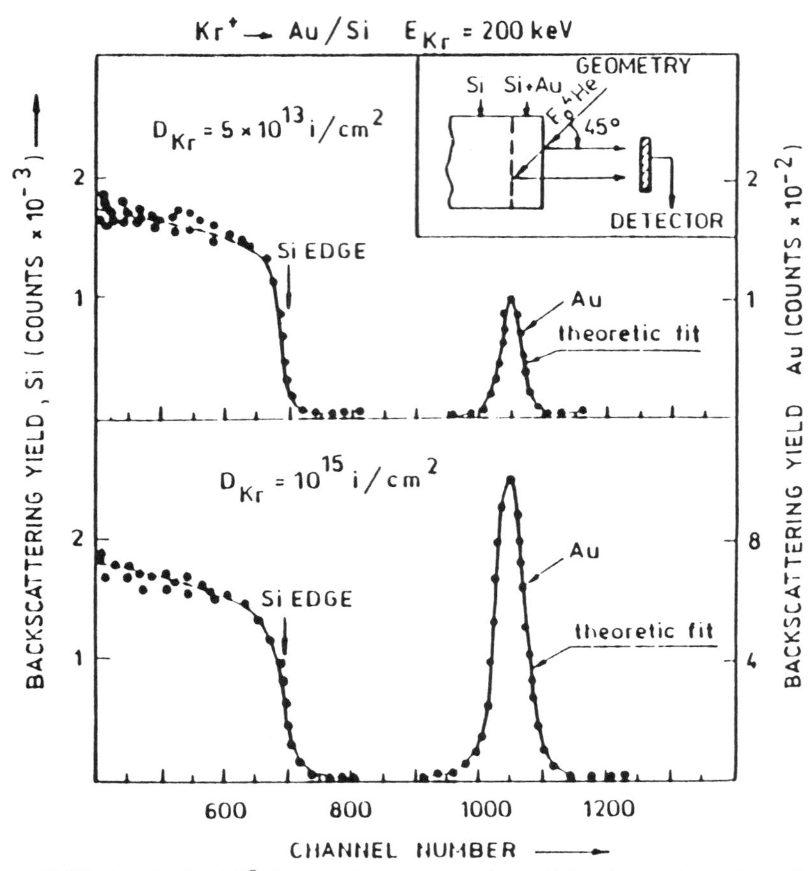

Figure 8.28 Rutherford backscattering spectra of recoil implants of Au into Si (top) and Au distribution measured by the RM method (bottom). After Paprocki, Brylowskam, and Syszko [58]; copyright © 1988, Springer Verlag, Berlin.

backscattering yield drops sharply as compared to a random direction. Note that as atoms in interstitial positions are generated (e.g., by ion implant damage), the backscattered yield in the channeling direction increases, which allows the quantitative determination of the number of atoms in interstitial spaces in the crystal. For example, Figure 8.29 shows the RBS spectra of As implanted into Si, revealing the disorder in the as-implanted silicon by the nearly random yield for ⟨100⟩ channeling and the recovery of the lattice perfection upon laser annealing using Nd:YAG laser light.

The damage created by ion implantation may be utilized for gettering since the defects that form in the damaged region can act as sinks for impurities. Also, the coupled diffusion of an implanted species and other impurities may be utilized to generate a zone that is denuded of unwanted residual impurities ahead of the implanted region. This ion pairing effect is illustrated in Figure

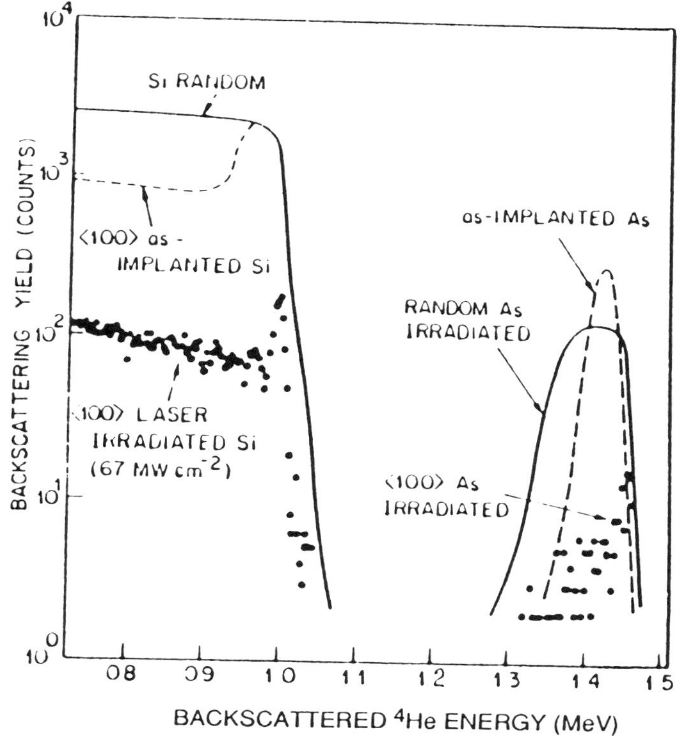

Figure 8.29 RBS spectra of As-implanted Si prior (solid and dashed curves) and after (full circles) laser annealing. After Celler, Poate, and Kimerling [61]; copyright © 1978, American Institute of Physics, New York.

8.30, which shows the concentration profiles of deliberately in-diffused phosphorus and of residual gold in Si [62]. Since these two species diffuse as P^+ and Au^-, they interact so that their diffuson profiles match. However, gettering by backside diffusion is usually less effective than gettering by backside damage.

Additional information on the energy position of traps in semiconductors and on the kinetics of capture and emission of carriers by such traps is obtained by deep-level transient spectroscopy (DLTS), which was introduced in 1974 [63]. DLTS is based on the charging of the traps by a voltage pulse. For example, upon switching from reverse voltage V_{r1}, corresponding to the depletion layer width W_{r1}, to a smaller reverse voltage V_{r2}, the depletion layer width shrinks according to Eq. (4.16) to $W_{r2} < W_{r1}$. This is illustrated in Figure 8.31(a) and (b) for a Schotty contact to an n-type semiconductor. Traps that are initially empty to a depth x_{ir1} are filled to a depth x_{ir2}. Upon switching the voltage back to V_r the depletion layer spreads back toward W_{r1}, and the excess filled traps are discharged, producing a capacitance transient as illustrated in Figure 8.31(d) [65]. If the pulse toward forward bias leads to the charging and

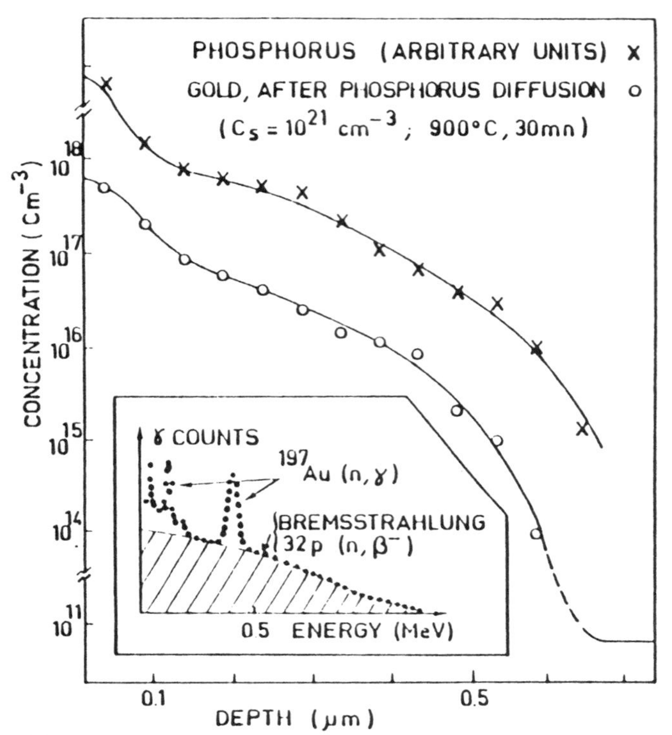

Figure 8.30 Gettering by backside diffusion. After Lecrosnier et al. [62]; copyright © 1980, The American Institute of Physics, New York.

discharging of minority carrier traps, a capacitance transient of opposite sign may result, as illustrated in Figure 8.31(e). Alternatively the reemission process may be observed by measuring the associated current [66]. Thus both minority and majority traps can be evaluated and distinguished by DLTS. In the presence of several traps located at different energies in the bandgap and being characterized by different capture cross sections, a superposition of several decay processes with different time constants must be analyzed.

A frequently used experimental implementation of this method is shown schematically in Figure 8.32(a) [64]. The decay of the capacitance after the charging pulse is measured at two times, t_1 and t_2. The associated

$$\Delta C = C(t_1) - C(t_2) \tag{8.44}$$

goes through a maximum if these times correlate with the characteristic time constant of the decay process as

$$\tau = \frac{(t_1 - t_2)}{\ln(t_1/t_2)} \tag{8.45}$$

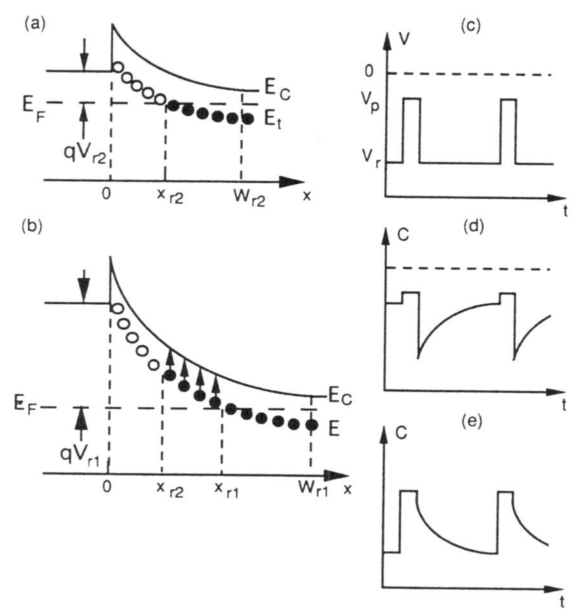

Figure 8.31 (a) and (b) Schematic representation of the charging and discharging of traps during the time intervals $0 \leqslant t \leqslant t_0$ and $t \geqslant t_0$ associated with the switching sequence of the bias voltage (c). (d) and (e) Capacitance versus time traces associated with majority and minority carrier traps, respectively. After Lefevre [67].

At a given choice of t_1 and t_2 this translates into a plot of ΔC versus T having a maximum at a characteristic temperature T_{max} where the condition Eq. (8.45) is satisfied. This is illustrated in Figure 8.32(b).

The energy position of the trap E_t, relative to the position of the conduction-band edge E_c is obtained from Arrhenius plots $\ln(1/\tau T^2]$ versus $1/T$ and may be estimated from measurements with two or more different windows τ_1 and τ_2 corresponding to $T_{max\,1}$ and $T_{max\,2}$, respectively. In this case,

$$E_c - E_t = k\frac{T_{max\,1}T_{max\,2}}{T_{max\,1} - T_{max\,2}} \ln\left[\frac{\tau_2}{\tau_1}\left(\frac{T_{max\,2}}{T_{max\,1}}\right)\right] \tag{8.46}$$

Figure 8.33 shows the DLTS spectra for hole traps in n-type GaAs for four time windows. If there are several traps, the ΔC versus T spectrum is a convolution on the individual curves, which may be resolved in favorable cases into several maxima at distinct temperatures. Since for durations t_p of the charging pulse that are small compared to the time constant of the capture process, the traps are only incompletely filled, and the amplitude of the signal

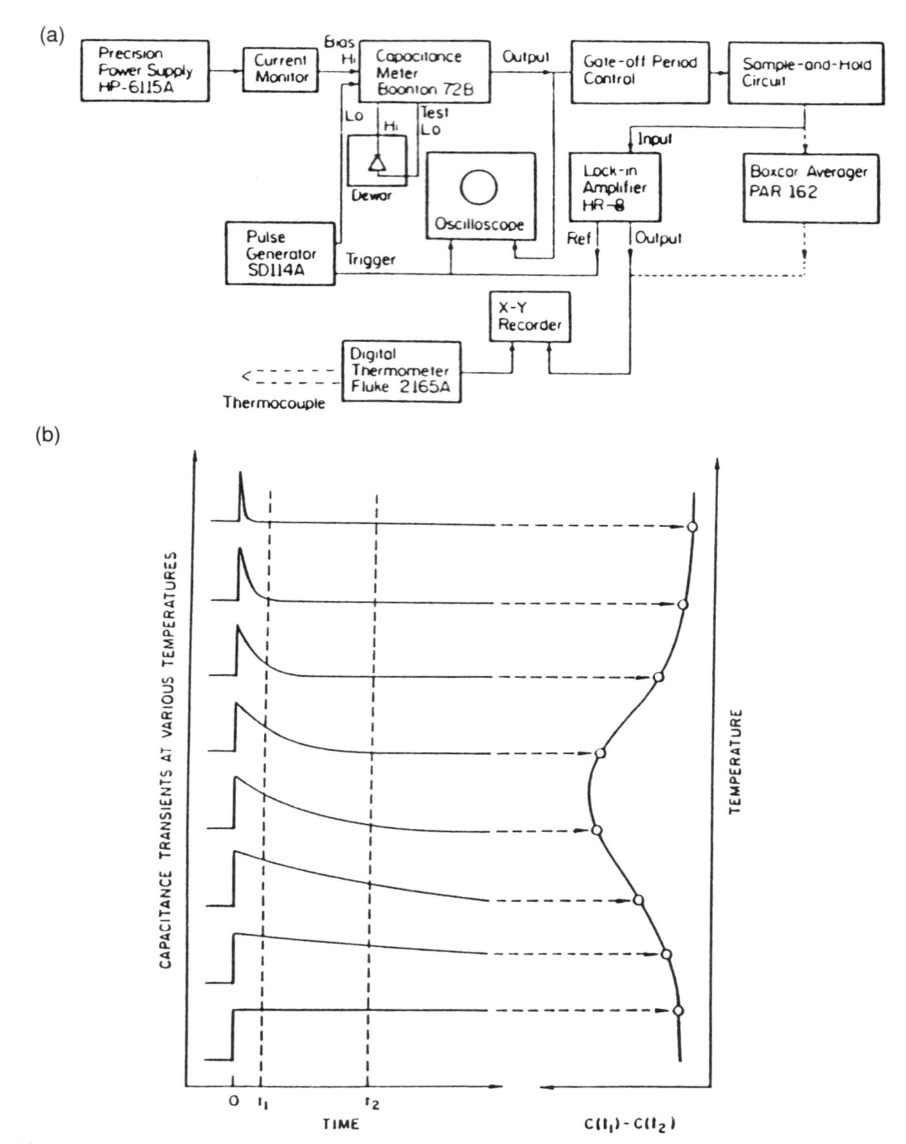

Figure 8.32 (a) Block diagram for the electronics required for DLTS measurements. (b) illustration of the use of a rate window in creating from the capacitance transient a DLTS spectrum. After Lang (a): [64] and (b) [63]; copyrights (a) © 1979 and (b) © 1974, The American Institute of Physics, New York.

$\mathscr{A}(t_p)$ depends exponentially on t_p

$$\mathscr{A}(t_p) = (n_{tf} - n_{ti})[1 - \exp(-v_{tn}t_p)] \tag{8.47}$$

where $n_{tf} - n_{ti}$ is the difference between the final and initial occupations of the

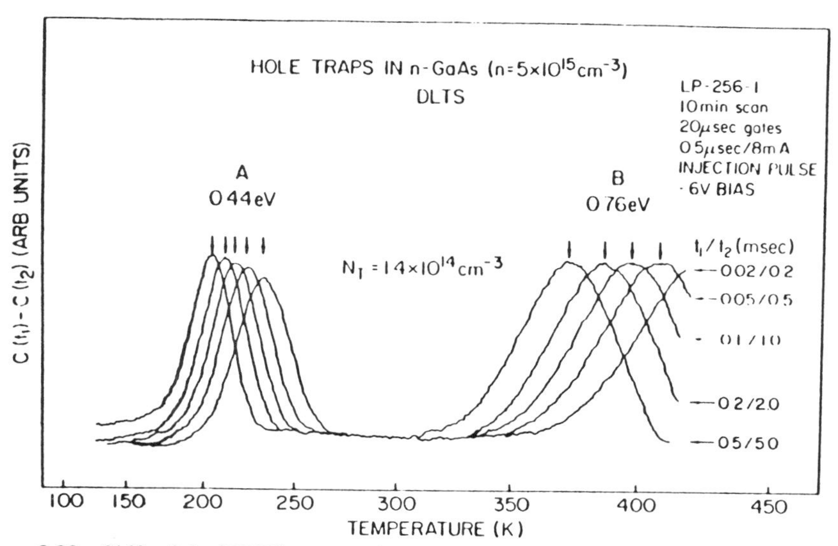

Figure 8.33 Shift of the DLTS spectra for hole traps in *n*-type GaAs. After Lang [63]; copyright © 1974, American Institute of Physics Society, New York.

traps by electrons for a change in the dc potential V_{r1} to V_{r2}. This offers an opportunity for the determination of the capture rate, which in the above-considered case of electron traps by the aid of Eqs. (3.84) and (8.47) provides for the determination of the capture cross section σ_n. Similar relations for hole traps can be worked out. The Arrhenius plot associated with the two hole traps revealed in Figure 8.33 is shown in Figure 8.34. Caution in the oversimplified interpretation of such data is needed, since it relies on the assumption of constant activation energies and capture cross sections, which is not always justified.

For the measurement of the concentration of traps, the double-correlation deep-level transient spectroscopy (DDLTS) method is particularly useful because it allows the profiling of trap concentrations as a function of the depth inside the depletion layer [67], [68]. The method is based on the filling of traps by two charging pulses of different height to two different distances from the junction into the depletion layer x_p and x_p'. Both pulses result in emission after the end of the pulse as the depth to which traps are charged changes to x_r. Forming the difference $\Delta C_{\text{DDLTS}} = \Delta C' - \Delta C$ thus probes the trap density in the spatial window $x_p - x_{p'}$. For closely spaced x_p and $x_{p'}$, the use of an average doping concentration $N_D(x_p)$ in this range is permissible even for inhomogeneous doping, resulting in the normalized trap concentration

$$\frac{N_t(x_p)}{N_D(x_p)} = \exp\left(\frac{t_1}{\tau}\right) \varepsilon q A^2 N_D(x_r) \frac{\Delta C(t_1)}{C_r(t_1)^3} \frac{1}{F(V_p, V_{p'})} \tag{8.48}$$

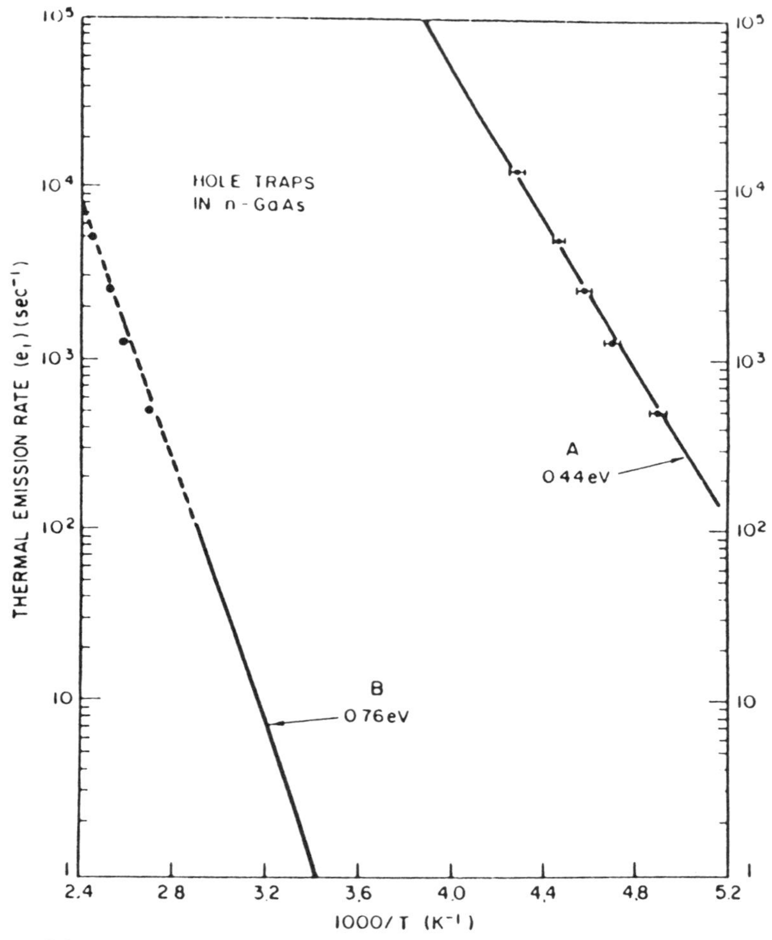

Figure 8.34 Arrhenius plot for the hole traps in the *n*-type GaAs of Figure 8.33. After Lang [63]; copyright © 1974, American Physical Society, New York.

where

$$F(V_{\mathrm{p}}, V_{\mathrm{p'}}) = V_{\mathrm{p}} - V_{\mathrm{p'}} + 2\left[\frac{E_{\mathrm{F}} - E_{\mathrm{t}}}{q}\right]^{1/2}\left[\left(V_{\mathrm{D}} - V_{\mathrm{p}} - \frac{kT}{q}\right]^{1/2} - \left(V_{\mathrm{D}} - V_{\mathrm{p'}} - \frac{kT}{q}\right)^{1/2} \qquad (8.49)$$

Thus by the choice of V_{r}, V_{p}, and $V_{\mathrm{p'}}$ depth profiling for traps is possible.

Thus far we have focused on radiation damage to the semiconductor lattice during IC processing. However, equally important are damage phenomena at

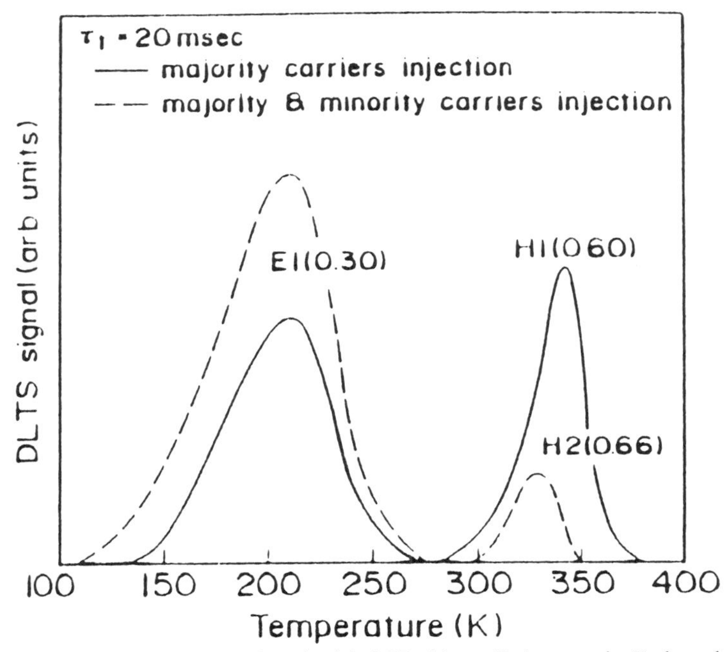

Figure 8.35 DLTS spectra associated with RIE side-wall damage in B-doped Si MIS trench capacitors. After Hamamoto [70]; copyright © 1991, The American Institute of Physics, New York.

the semiconductor–dielectric interface and in the dielectrics themselves by energetic charged particles and photons [69]. For example, the characterization of the chemical and physical damage created during RIE machining of trench capacitors and multidimensionally confined heterostructures is of considerable interest. Figure 8.35 shows the DLTS spectra of deep levels associated with the side walls of trench MIS capacitors made by magnetron etching with Cl_2 and $SiCl_4$ of silicon. Two major peaks are observed that are associated with an electron trap at $E_c - 0.30\,eV$ and hole trap at $E_c - 0.60\,eV$. The deep hole trap acts as a recombination center, that is, can trap both electrons and holes. It disappears upon electron injection and is replaced by a second hole trap at $E_c - 0.66\,eV$. The concentration of this trap increases strongly upon annealing at $T > 1220\,K$ with a concomitant strong reduction in the concentration of the recombination center at $E_c - 0.60\,eV$.

Also, x rays are known to generate damage in the gate and field insulators of MOSFETs [71] (e.g., during x-ray lithography), leading to the incorporation of neutral hole traps, neutral electron traps, fixed positive charge and fixed negative charge. Neutral hole traps exist in relatively high concentration (i.e., $\sim 10^{13}\,cm^{-3}$) in the oxide of unirradiated devices. They have a capture cross section for holes of $10^{-13}\,cm^{-2}$, corresponding to a large trapping probability [72]. Charged hole traps represent a fixed positive charge that is associated with a field, shifting the turnon voltage to a negative direction. The positive voltage required for turning an n-channel MOSFET on is thus decreased, and the negative voltage required for turning a p-channel MOSFET on is increased. Upon irradiation, the concentration of neutral hole traps has been reported to decrease [73]. Neutral traps exist in unirradiated oxide only at very low concentration (i.e., $\sim 10^{11}\,\text{cm}^{-3}$), which is increased by orders of magnitude upon exposure to ionizing radiation [74]. Their capture cross sections vary in the range between 10^{-19} and $10^{-15}\,\text{cm}^{-2}$. Hot electrons that are captured by these traps represent a fixed negative charge, shifting the turnon voltage to a positive direction, opposing the applied positive bias on n-channel and complementing the applied negative bias in p-channel MOSFETs. Since the threshold voltage shifts associated with radiation damage of the dielectric are unacceptable, it is necessary either to remove the damage as part of the processing sequence or to limit its extent. Since the concentrations of the three types of defects all decrease with decreasing exposure, one possible avenue to limiting the damage is the utilization of multilayer resists that keep the exposure of the dielectric to $\leqslant 10^6$ rad [75]. However, since multilayer resist processing is costly, alternative methods are needed. Annealing of the radiation damage would be a preferred choice. However, temperatures in excess of 550°C are required to anneal out the neutral electron traps and fixed positive charge in hydrogen [76]. This represents an undesirably high thermal load in the fabrication of submicrometer ICs. Advances in controlling the x-ray damage may become possible by annealing at elevated pressure, which has been shown

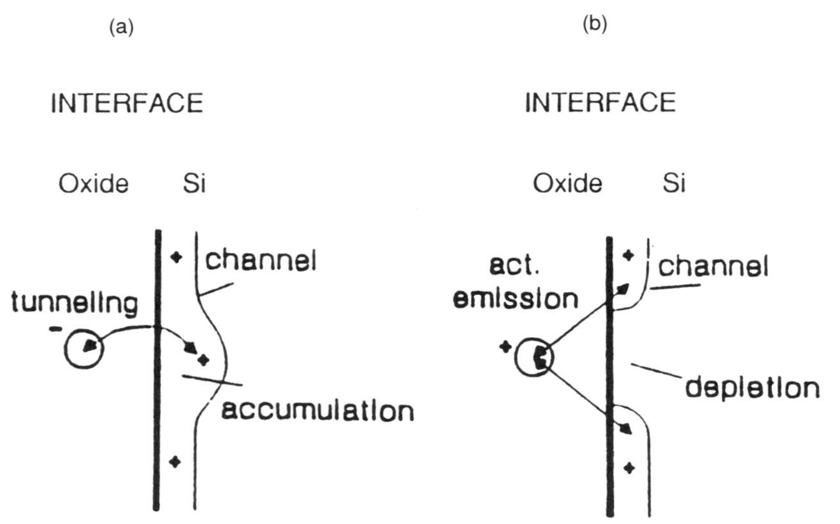

Figure 8.36 Schematic representation of the charge distribution in the vicinity of (a) a Coulomb-attractive and (b) a Coulomb-repulsive center in the oxide near the GOX/silicon interface of a MOSFET. After Schulz and Karmann [78]; copyright © 1991, Springer Verlag, Berlin.

to provide for the possible removal of neutral electron traps and positive fixed charge at $\leqslant 400°C$ [77].

Further details on the emission and capture of carriers by traps in the oxide are obtained from the switching characteristics of MOSFET devices. Figure 8.36 shows a schematic representation of the exchange of holes between a Coulomb-attractive center in the oxide near the interface and a hole in the channel of a *p*-channel MOSFET [78]. If the hole is free, its capture by the center in the oxide changes merely the number density of carriers in the channel. If the hole is bound to a trap in the channel, it does not contribute to the conduction in either state. However, its charge is screened when it resides at the center of the oxide, but is not when it resides on the hole trap in the channel. Therefore, the hole contributes to the scattering only while it resides in the channel, causing a change in the channel mobility upon hole transfer between the two states. Thus the individual transfer events are associated with telegraph noise in the source–drain conductance. Figure 8.36(b) shows the conditions for a Coulomb–repulsive center. Since the channel is depleted in the vicinity of this center and the tunneling probability is a strong function of the barrier width, thermal activation is needed. Also, the channel mobility is not affected by the switching because the hole in the channel is free. Therefore, the $1/f$ noise of the transistor is dominated by the superposition of the telegraph noise signals associated with individual switching events at Coulomb-attractive centers [9], [0].

Figure 8.37(a) shows the quantized transients in the source–drain voltage at constant current after a filling pulse that are associated with the re-emission of holes from individual defect centers in a *p*-channel MOSFET (a). The emission and capture processes at these defect centers are marked n1, a1, and a2, respectively. Note that the steps associated with emission and capture at a1 and a2 are anomalous as emission increases and capture decreases the voltage. Figure 8.37(b) shows a schematic representation of the energy positions of the hole E_1 and E_2 bound to the traps in the oxide and the channel, respectively. The Fermi energy position is related to E_1 and E_2 according to

$$E_F = \frac{E_1 + E_2}{2} + \tfrac{1}{2}kT \ln g \qquad (8.50)$$

where g accounts for the favoring of the occupancy of one level (e.g., E_2) over the other because of degeneracies. Since the two-level system is occupied by a single hole, one state is filled while the other is empty. Therefore, the Fermi occupation probabilities are

$$f_2 = 1 - f_1 = \frac{1}{1 + g^{-1/2}\exp(U/kT)} = \frac{1/\tau_e}{1/\tau_c + 1/\tau_e} \qquad (8.51)$$

where $1/\tau_c$ and $1/\tau_e$ are the rate constants for capture and emission, respective-

(a)

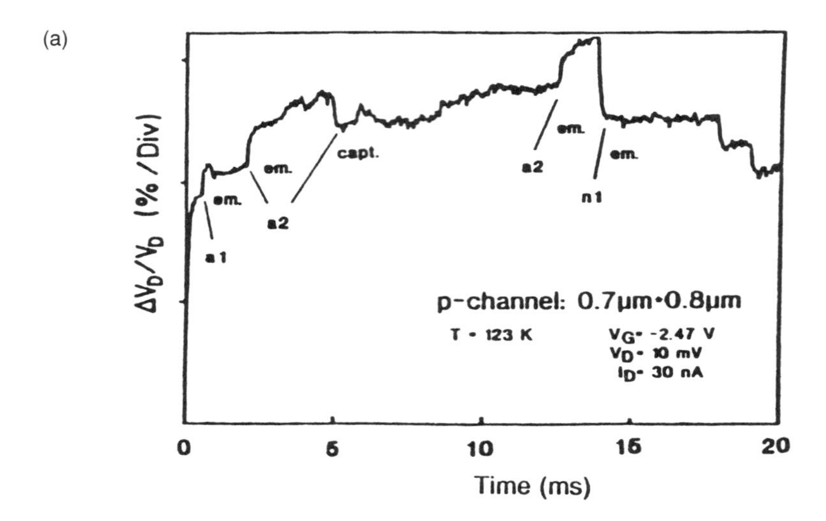

(b)

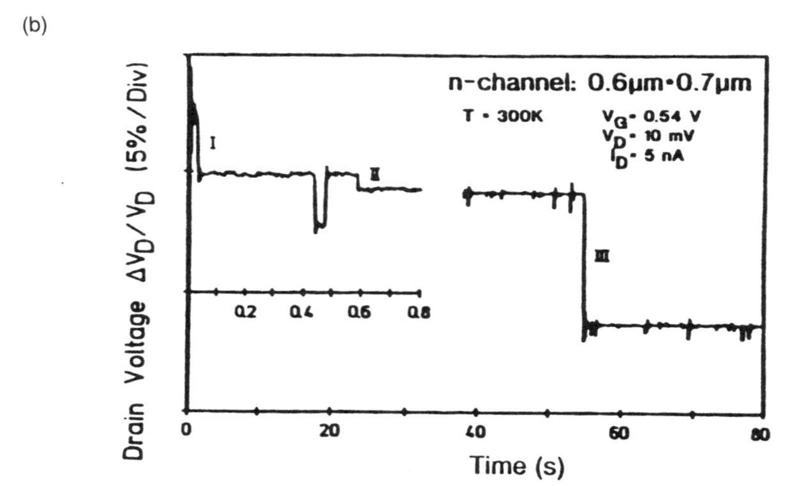

Figure 8.37 Telegraphic noise in the drain voltage at constant current after a gate voltage filling pulse. After Schulz and Karmann [78]; copyright © 1991, Springer Verlag, Berlin.

ly, and $U = (E_1 + E_2)/2$ is the offset of the two levels from the center of energy. Solving for the rate constants

$$1/\tau_c = \mathscr{P}(V_G, T)g^{-1/4}\exp(U/2kT) \tag{8.52}$$

$$1/\tau_e = \mathscr{P}(V_G, T)g^{1/4}\exp(-U/2kT) \tag{8.53}$$

$$\mathscr{P}(V_G, T) = \left(\frac{1}{\tau_c}\frac{1}{\tau_e}\right)^{1/2} \tag{8.54}$$

is the transfer probability, which depends on both the gate voltage V_G and the temperature. From the measurement values for $1/\tau_c$ and $1/\tau_e$ as a function of V_G at constant T shown in Figure 8.38(a), the functions $\mathscr{P}(V_G)$ and $U(V_G)$ can be determined and are hown in Figure 8.38(b) and (c). Also, since the gate

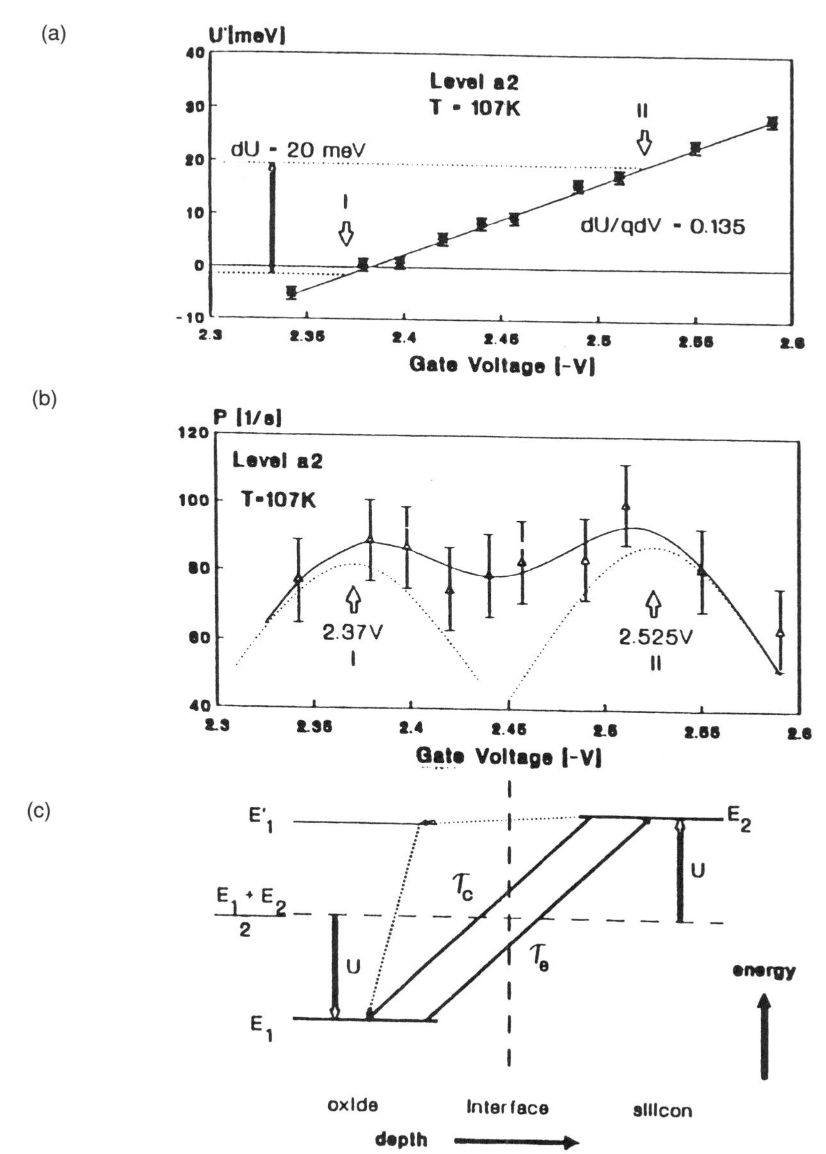

Figure 8.38 Plots of the (a) average emission and capture rates, (b) transfer probability, and (c) level offset versus the gate voltage of a *p*-channel MOSFET. After Schulz and Karmann [78]; copyright © 1991, Springer Verlag, Berlin.

voltage falls off across the gate oxide of thickness t_{ox}, the slope of $U(V_G)$ is related to the depth of the center in the oxide measured from the interface t_C according to

$$t_C = t_{ox}\frac{1}{q}\frac{dU}{dV_G} \tag{8.55}$$

resulting in an average value of 2.4 nm for this particular set of data.

8.3 Native Oxides on III–V Surfaces

Since the oxidation of compound semiconductors results in complex chemical changes at the interface and strongly affects the interfacial properties, extensive studies have been performed by a variety of surface analytical methods, such as EXAFS, XPS, UPS, and AES. These methods pertain not only to the understanding of compound semiconductor MIS devices, but also to ohmic contacts and Schottky barriers on compound semiconductors and are further discussed in Section 8.4.

Under the conditions of AES, the excitation of a core electron by bombardment with incident electrons of high energy generates a hole on an inner shell that is filled from a higher-lying shell. In an Auger process, the energy of this transition is utilized in the emission of one or several electrons, as illustrated in Figure 8.39(a)–(c). The labels $KL_\nu L_\mu$ for the Auger processes considered in this example signify that a hole in the K shell is filled by an electron in the state $\langle\nu|$ of the L shell, expelling an Auger electron from state $\langle\mu|$ of the L shell. Since the Auger electrons possess a well-defined kinetic energy

$$E = E_K - E_{L\nu} - E_{L\mu} \tag{8.56}$$

they are observed as structures in the secondary electron yield upon bombardment of a solid by high-energy electrons. Light elements are particularly well suited for AES because the ratio of the probability of an Auger process to the probability of an x-ray emission process increases with decreasing nuclear charge, becoming approximately equal at $Z = 30$ (Zn). Since the selection rules for the electronic transitions provide for elements specific signatures, the composition of the surface under investigation can be determined. Although quantitative analysis is possible based on a measurement of the peak-to-peak value of the AES signal, absolute measurements of the concentration of a particular element is difficult due to matrix effects. However, where calibration against a series of standards in the range of compositions under test is possible, AES provides reliable quantitative information for concentrations at a 1% level and above. Because the number of Auger electrons emitted is small compared to the background of backscattered secondary electrons, sensitive detection is an essential requirement of AES. Also, in order to reveal small chemical shifts in the energy positions of the Auger peaks, high resolving power is important,

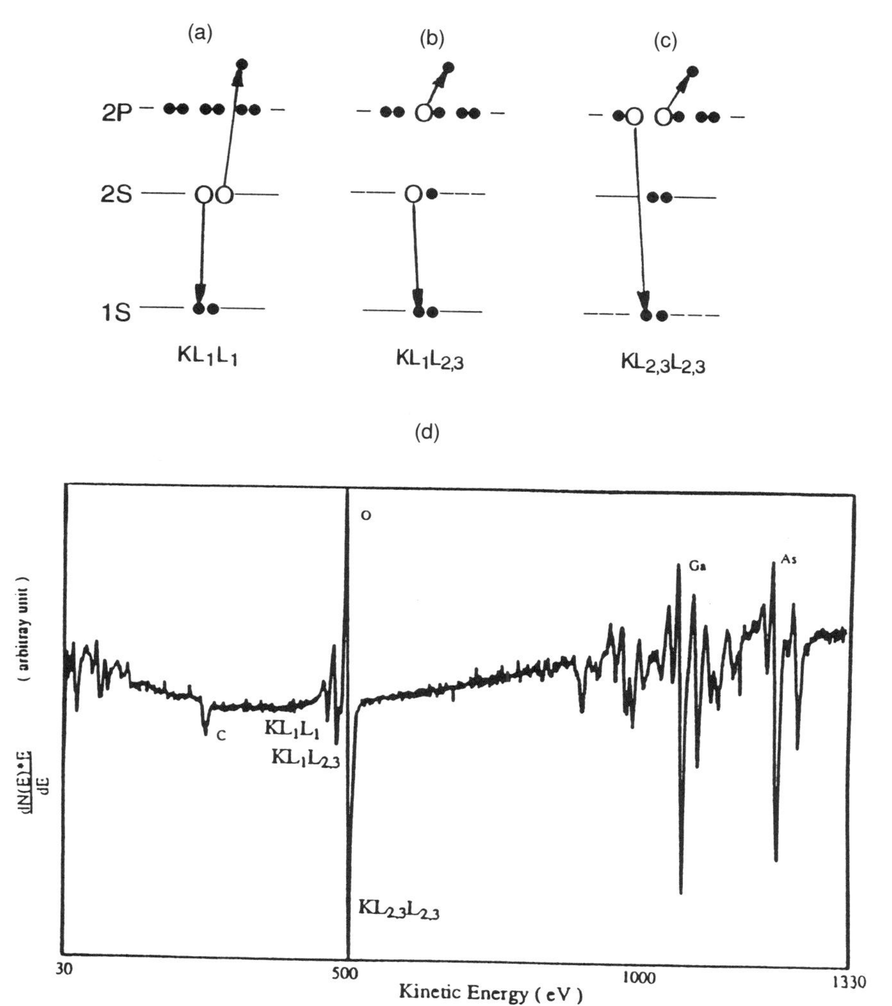

Figure 8.39 (a)–(c) Schematic representation of the *KLL* transitions of oxygen. (d) AES spectrum of an oxidized GaAs surface.

which mandates careful selection of the type of electron energy analyzer employed in AES analysis.

Magnetic electron spectrometers select the electron trajectory matched to the position of the detector by controlling the magnetic field. The fraction of the solid angle of electron emission utilized by the detector in magnetic sector instruments is $\Omega/2\pi = 0.03$–0.12, and the resolving power is $E/\Delta E = (2$–$8) \times 10^3$. The electrostatic retarding potential analyzers achieve $\Omega/2\pi = 0.1$–0.4 but have significantly smaller resolving power than magnetic sector instruments. Electrostatic cylindrical mirror analyzers (CMA) (see Figure 8.45) and electrostatic concentric hemispherical analyzers (CHA) are characterized by similar values $0.01 \leqslant \Omega/2\pi \leqslant 0.4$. However, the CHA achieves somewhat higher $E/\Delta E$ values, which are comparable to magnetic sector instruments. For more detailed discussions and a concise review of the experimental methods of surface physics, refer to Ref. [81].

The depth under the surface from which electrons can be collected depends on their interactions with the solid in inelastic scattering events during their passage toward the surface, which is critically affected by their energies. At low electron energies, the mean free path is long because of a lack of excitations. At high electron energies, the mean free path also becomes large due to the decreasing cross sections of the excitations. Therefore, the escape depth versus kinetic energy plot for the secondary electrons has a U shape with a minimum at ~100–200 eV corresponding to a few Å escape depth, as illustrated in Figure 8.40. For example, at ~500 eV energy, the Auger electrons associated with oxygen for an oxidized GaAs surface as shown in Figure 8.39(d) have an escape depth of $\leqslant 10$ Å.

Instead of plotting the number of electrons versus energy, in Figure 8.39(d), the first derivative spectrum is plotted in order to enhance the Auger electron related features relative to the secondary electron background that decreases monotonously with increasing energy. The energy position of the Auger lines depends weakly on the environment. For example, the Ga $L_3M_{45}M_{45}$ Auger lines for GaAs and GaP occurs at 1066.1 and 1065.7 eV, respectively. A

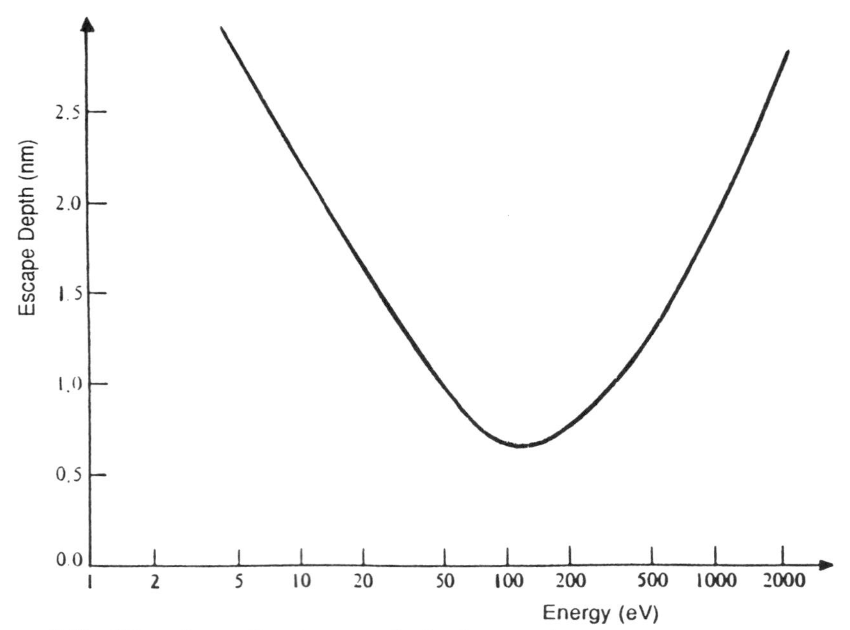

Figure 8.40 Mean free path prior to an inelastic scattering event versus kinetic energy.

substantially larger chemical shift of the Ga *LMM* line is observed upon transferring Ga from a GaAs environment into a Ga_2O_3 environment, which may be utilized to follow the native oxidation process. For the native oxide on GaAs, produced by exposing a clean cleaved GaAs surface at room temperature for 2 h to molecular oxygen at 760 Torr, this chemical shift is 3.3 eV, which is well above the resolution of modern Auger spectrometers. Also the As $L_3M_{45}M_{45}$ Auger line, which for GaAs is located at 1225.4 eV, exhibits a chemical shift of 4.4 eV upon exposure to oxygen, indicating participation of both elements in the surface oxidation process under the above-described conditions. The features near 500 eV are the *KLL* transitions of oxygen.

Further information on the interaction of molecular oxygen with GaAs surfaces is obtained from EXAFS studies. The existence of fine structure in the x-ray absorption coefficient above threshold was discovered by Kossel in 1920 [82] and was further investigated by Kronig in 1932 [83], extending to the range >20 eV above the absorption edge. The utility of EXAFS for local structure determination was recognized in 1971 [84]. Due to the ready access to synchrotron high-intensity x-ray sources, EXAFS analysis is applicable today even to dilute systems, such as adsorbates on crystal surfaces. The basis for the observation of EXAFS is the backscattering of photoelectrons generated in the x-ray absorption process by nearest-neighbor atoms interfering with the outgoing photoelectron wave. This modulates the absorption cross section, causing structure in the absorption coefficient and in the photoelectron yield above the inner-core absorption edge energy. Since the scattering in the range $\geqslant 50$ eV above the threshold for inner-core photoexcitations is weak, the interpretation of EXAFS spectra can be based on short-range single-scattering theory [86]. It provides structural information about the local environment of the absorbing atom in terms of radial distribution functions for the nearest-neighbor shell, but requires modification if more than one shell is involved in the scattering [87]. Multiple-scattering theory becomes mandatory in the analysis of the fine structure in the 10–50 eV range, which carries the same burden as the interpretation of LEED, but provides in return more detailed information on bond angles and on the distances and geometry of second- and higher-order neighbor shells about the absorbing atom than the higher-energy spectra [88]. To distinguish this region from EXAFS, it has been named *x-ray absorption near-edge structure* (XANES).

Since the absorption of x rays generates holes in the inner shells of the absorbing atoms, the modulation of the absorption coefficient by the scattering events leading to EXAFS must be reflected in the Auger and secondary electron yields. In view of the small escape depth of these electrons, there exists an opportunity to adapt EXAFS to surface studies (SEXAFS).

Figure 8.41 shows the Auger yield spectrum and the SEXAFS portion above threshold for a silicon (111) surface covered by chlorine. Figure 8.42 shows (a) the SEXAFS spectrum for oxygen on a GaAS(110) cleavage face and (b) the Fourier transformation of the signal of (a). The smooth line drawn in (a) is the Fourier-filtered signal corresponding to peak A only. Analysis of this peak in

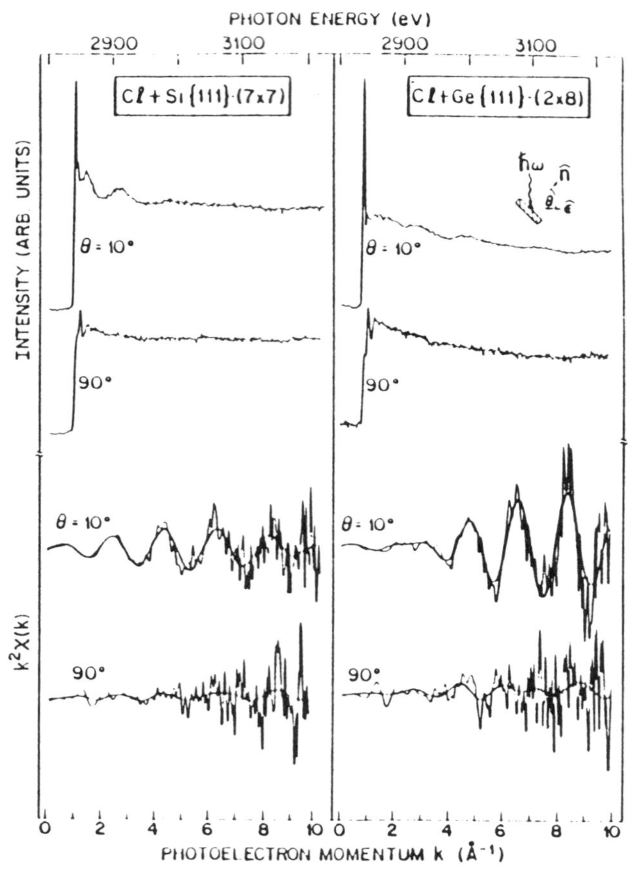

Figure 8.41 Top: Auger yield spectrum at the Cl K edge measured with polarization vector parallel to a silicon (111) surface covered with chlorine to saturation. Bottom: Background-substracted SEXAFS and filtered data. After Citrin, Rowe, and Eisenberger [85]; copyright © 1983, American Physical Society, New York.

terms of the phase $\varphi(\mathbf{k})$ of the EXAFS signal and the scattering phase shift of the photoelectron $\Phi(\mathbf{k})$ results in a nearest-neighbor distance

$$R = \frac{\varphi(\mathbf{k}) - \Phi(\mathbf{k})}{2|\mathbf{k}|} \tag{8.57}$$

of 1.70 ± 0.05 Å, which is too large for an O–O distance, leading to the conclusion that molecular oxygen adsorbs on the GaAs(110) surface dissociatively and that the nearest-neighbor distance measured by EXAFS corresponds to the oxygen bonded to the surface.

Fine structure due to backscattering of reflected electrons by neighboring

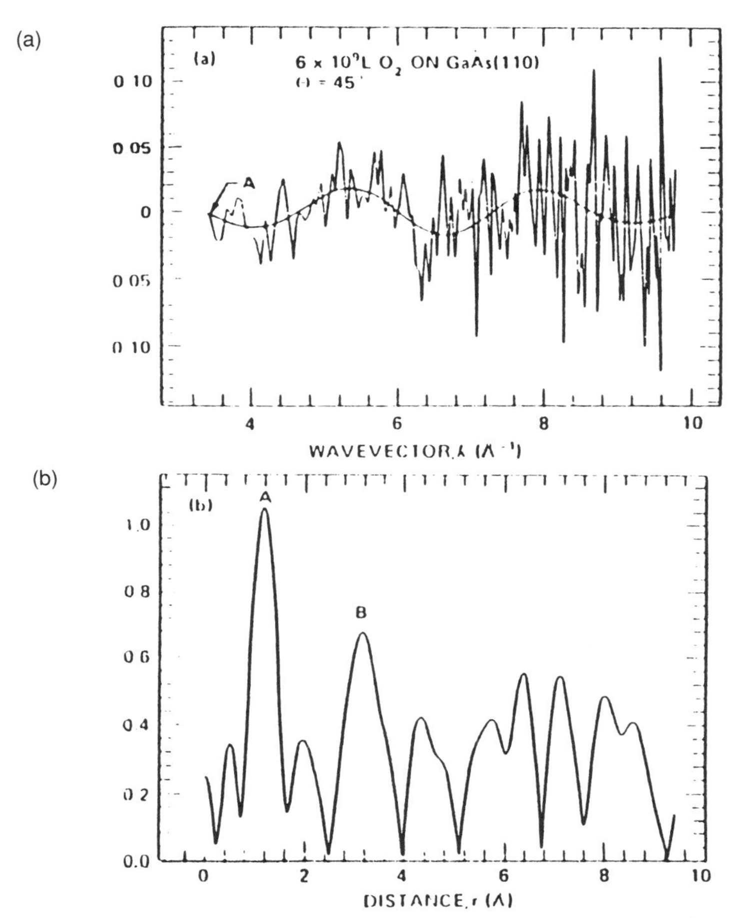

Figure 8.42 SEXAFS spectra of a cleaved GaAs(110) surface after exposure to molecular oxygen: (a) experimental data; (b) Fourier transform of these data. After Stöhr et al. [898], copyright © 1979, American Institute of Physics, New York.

surface atoms also introduces fine structure into the electron energy loss spectra. Although the interpretation of electron energy loss fine structure (EELFS) is more complex than the interpretation of EXAFS, it may become of utility in studies of growth processes under the conditions of UHV, since it is conveniently implemented with standard surface analysis components incorporated into such systems for in situ characterization. For example, EELFS has been applied in investigations of the nitridation of a silicon surface [89], the adsorption of oxygen on metals [90], [91], as well as compound semiconductors and quantum-well heterostructures [92].

UPS and XPS probe the joint density of states between filled valence-band states and empty states in the conduction bands. In UPS, the excitation is accomplished by UV light usually obtained from a He discharge lamp at either

21.2 or 40.8 eV energy, which results in the emission of electrons of kinetic energy.

$$E = h\nu - E_B - \Phi_s \tag{8.58}$$

where E_B is the binding energy of the photoemitted electron measured from the Fermi level and Φ_s is the spectrometer work function. Thus the deepest valence state that can be excited to become a free electron is located at energy $h\nu$ below the vacuum level, and the highest conduction band state into which a photoelectron can be excited lies above the valence-band maximum of the semiconductor, by the energy $h\nu$. The energy distribution curves associated with the structure in the joint density of states within this window thus tend to zero at the energy of the valence-band maximum. Therefore, an independent measurement of the position of the Fermi level of a metal standard, connected

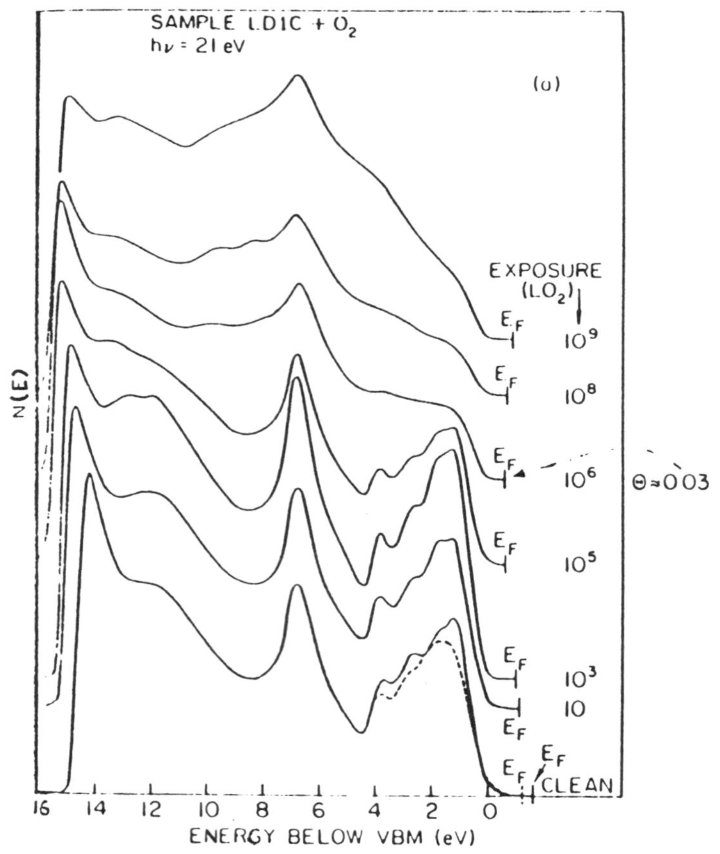

Figure 8.43 UPS spectra obtained by excitation of the GaAs(110) surface by 21 eV He light for various stages of oxidation. After Spicer et al. [93]; copyright © 1979, American Vacuum Society, New York.

by an ohmic contact to the semiconductor, provides for the experimental determination of the relative position of the Fermi level at the surface of the semiconductor with respect to the valence-band edge.

Figure 8.43 shows the UPS spectra of GaAs excited with 21 eV light of a He lamp, revealing the Fermi-level shift toward the midgap position with increasing exposure to oxygen. Alternatively, if a core level that does not exhibit a significant chemical shift is excited simultaneously, it can be utilized as a reference to determine the surface Fermi level (SFL) movement upon oxidation. In the case of GaAs, by means of unfiltered light from a He lamp, the line at 40.8 eV suffices to excite the Ga 3*d* core level as a reference. Figure 8.44 shows the position of the surface Fermi level as a function of the oxygen coverage for Si and GaAs. For GaAs, submonolayer coverage by oxygen moves the SFL toward midgap position, while for Si the SFL moves toward the majority carrier band edge.

Further information on the early stages of oxidation of Si, GaAs, and other compound semiconductors is obtained by XPS. XPS utilizes soft x rays for the photoexcitation of, for example, Mg $K\alpha$ or Al $K\alpha$ radiation at 1254 and 1487 eV, respectively. This energy is sufficient for the excitation of core levels. A typical arrangement of an XPS spectrometer employing a CMA for analysis

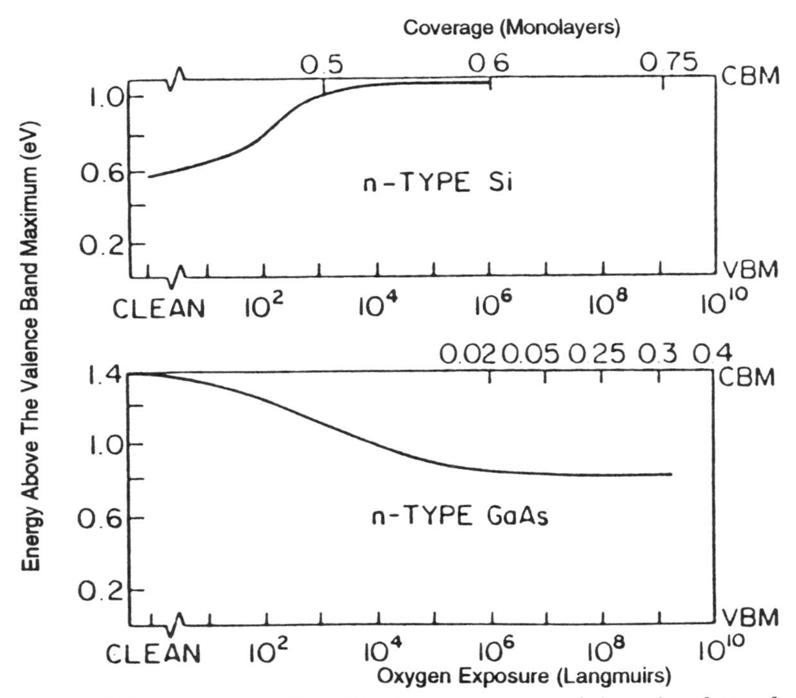

Figure 8.44 Shift of the surface Fermi level upon exposure of cleaved surfaces of *n*-type Si and GaAs to oxygen. After Spicer et al [93]; copyright © 1979, American Institute of Physics, New York.

of the energy of the photoemitted electrons is presented in Figure 8.45. In this particular instrument a sputter ion gun and an electron source are added to permit profiling and AES as a supplemental surface analysis method.

Figure 8.46 shows the XPS As 3*d* and Ga 3*d* core-level peaks for a clean UHV cleaved GaAs(110) surface, and Figure 8.47 shows the P 2*p* core-level peak, after exposure at room temperature to molecular oxygen at 760 Torr. In all cases the appearance of additional peaks at the high-energy side of the bulk core-level peaks is observed. They are considerably enhanced upon raising the temperature and prolonging the oxidation. However, while for GaAs the ratio of the Ga 3*d* to As **3d** peaks varies in accord with an arsenic loss/burial at the surface, for GaP the ratio of the Ga 3d to P 2*p* peaks stays approximately constant, as would be expected for the formation of gallium phosphate [94]. Analysis of the XPS and Auger data suggests that the initial stage of the native oxide formation on GaAs and GaP is represented by two monolayers of the compound with all gallium atoms bonded to oxygen. For GaP, all phosphorus atoms in this layer are also bonded to oxygen, but for GaAs only half of the arsenic atoms are bonded to oxygen. Free arsenic at the interface has been found by Raman spectroscopy of anodized GaAs [95], and amorphous arsenic has been found by spectroscopic ellipsometry to be mixed into plasma oxide grown on the surface of GaAs [96]. However, in the latter study the excess arsenic represented only a minor fraction of the oxide, consisting primarily of Ga_2O_3 and As_2O_3. The concentration of amorphous arsenic depended on the rate of the plasma oxidation process. On the basis of bulk thermochemical data, the equilibrium

$$As_2O_3 + 2GaAs \rightleftarrows Ga_2O_3 + 4As \tag{8.59}$$

at room temperature results primarily in products at the right side of Eq. (8.59)

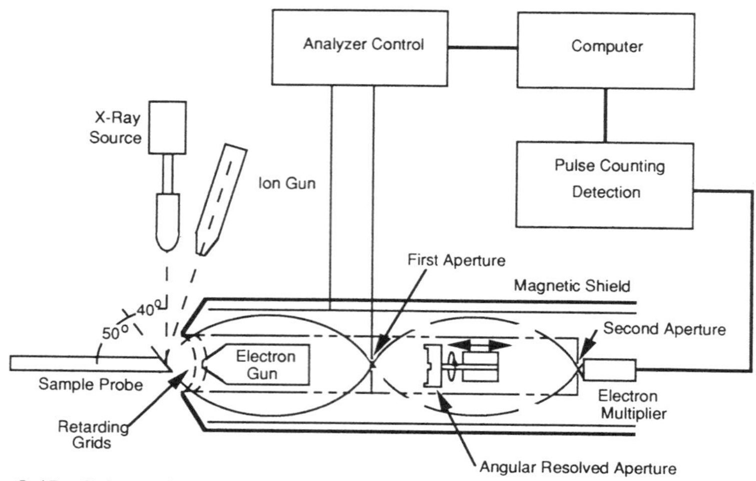

Figure 8.45 Schematic representation of an XPS spectrometer.

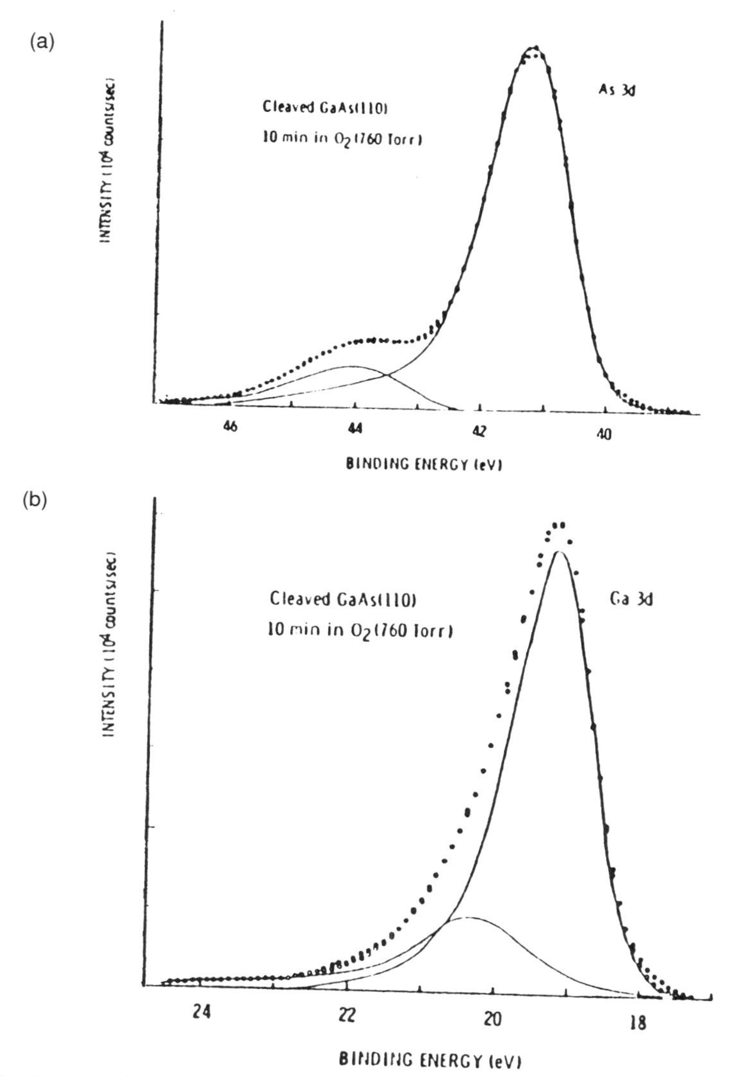

Figure 8.46 Expanded XPS spectra of (a) the Ga 3*d* level; (b) the As 3*d* level of GaAs oxidized by molecular oxygen at room temperature. After Iwasaki et al. [94]; copyright © 1978, Jpn. J. Appl. Phys, Tokyo.

[97]. These experimental results suggest a considerable barrier to the nucleation of elemental arsenic precipitates at the GaAs/native oxide interface from supersaturated solid solutions of free arsenic generated in reaction (8.59), so that, for kinetic reasons, the reaction of GaAs with oxygen results in a substantial formation of As_2O_3. As a consequence of this behavior, the native oxide GaAs formed at room temperature upon contact with molecular oxygen is inhomogeneous, which is confirmed by recent STM imaging and poten-

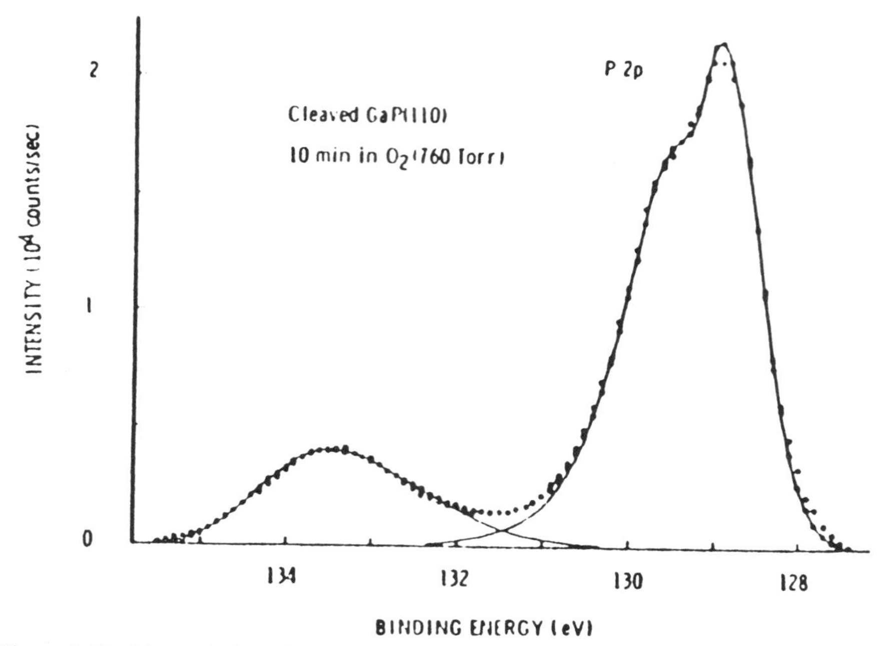

Figure 8.47 Expanded XPS spectra of the P $2p$ level of GaP oxidized by molecular oxygen at room temperature. After Iwasaki et al. [94]; copyright © 1978, Japanese J. Appl. Phys., Tokyo.

tiometry measurements [98]. The movement of the SFL is unequivocally related to the presence of excess As on the surface by the As $3d$ core-level shift measured relative to the XPS Ga $3d$ core-level peak upon exposure of a clean GaAs(110) surface to arsenic vapor. As shown in Figure 8.48, annealing of the arsenic exposed surface, leading to a loss of arsenic, restores the relative XPS peak positions of the clean surface.

Several models have been explored to explain the Fermi level pinning upon oxidation. The effective work function model of [99] explains the Fermi-level pinning by a modification of the effective work function of the semiconductor surface due to the presence of elemental arsenic, which, in the crystalline state, is a semimetal. Alternatively, the pinning of the SFL has been explained as a consequence of changes in the native point-defect chemistry in the vicinity of the semiconductor/oxide interface, specifically attributing the Fermi-level shift upon oxidation of GaAs to the formation of As_{Ga} antisite defects [94], [100]. As a third alternative the introduction of acceptor states by the adsorption of elements possessing high electronegativity values has been proposed and indeed explains the strong shift of the Fermi level upon chlorine adsorption on n-GaAs, while on p-GaAs chlorine adsorption leaves the SFL relatively unperturbed [101].

Figure 8.49 shows the movement of the Fermi-level position on the (110) surface of GaAs as a function of the exposure to molecular oxygen for various temperatures in the range $100 \leqslant T \leqslant 300$ K. At low temperature the oxygen is

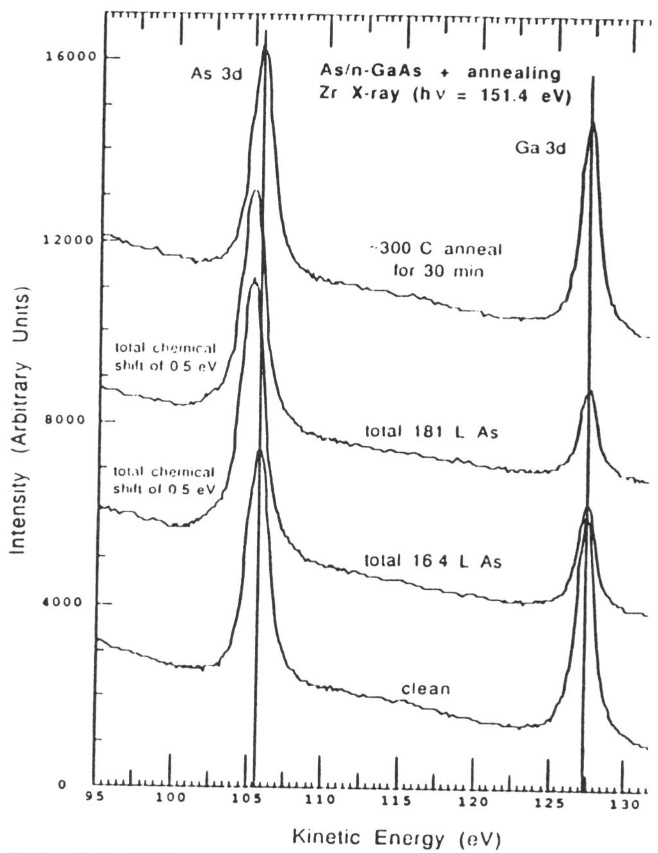

Figure 8.48 Shift of the SFL of a GaAs(110) surface upon exposure to arsenic vapor and annealing. After Chiang and Spicer [93]; copyright © 1989, American Institute of Physics, New York.

adsorbed without dissociation, and the expected behavior of the adsorbate-induced acceptor model is observed. This is further confirmed by STM studies that reveal the formation of a depletion layer about the oxygen adsorbate on *n*-GaAs, which does not exist on *p*-GaAs [103]. At room temperature the surface Fermi level pins at near midgap position for both the *n*- and *p*-type material due to bond-breaking surface reactions, which are likely to affect the point-defect chemistry in the subsurface region and generate an excess of arsenic at the interface. However, thus far no direct proof for the identification of the pinning defects as antisite defects exists, and further research is necessary to clarify the nature of the pinning process.

Although the air-grown native oxide on bulk GaAs is inhomogeneous and pins the surface Fermi level, it plays a beneficial role in GaAs processing as it protects the surface from contamination [104]. Similar experience exists for silicon, which rapidly picks up contaminants from the environment for ox-

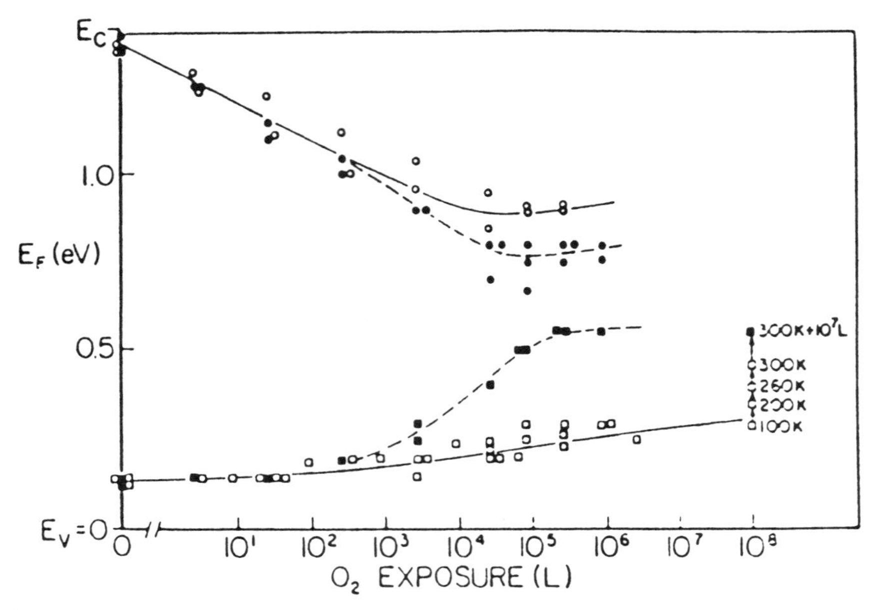

Figure 8.49 Position of the Fermi level on the GaAs(110) surface as a function of the oxygen dose for selected temperatures as paramenter. After Stiles, Mao, and Kahn [102]; copyright © 1988, American Vacuum Society, New York.

ide-stripped surfaces as compared to the thin oxide-covered surface [105]. Also, the H_2SO_4:H_2O_2:H_2O etch followed by a DI water rinse, usually employed as a surface-cleaning step prior to MBE of GaAs, creates such a protective oxide. It is quantitatively removed in UHV by thermal desorption at 525–535°C, which is well below the limit for surface decomposition [106]. Oxide stripping and simultaneously removing carbon contamination from the surface are obtained at even lower temperatures in a hydrogen plasma [107] and in a beam of atomic hydrogen, while annealing at ~600°C is required to remove the oxide in a stream of pure hydrogen [108].

For InP, the formation of a P-rich oxide is far less detrimental to the interfacial recombination velocity and device performance than the oxidation of GaAs [109], [110]. The shift of the SFL of InP upon exposure to oxygen is shown in Figure 8.50. While for *n*-type material the SFL pins close to the majority carrier band edge, for *p*-type material the SFL moves towards the conduction band edge.

In accord with the observed shift in Fermi level of GaAs upon oxidation, the analysis of the frequency dispersion of the C–V and G–V characteristics of GaAs MIS capacitors results in a Fermi-level position at 0.8–0.9 eV below the CBE and 0.7–0.8 eV above the VBE for *n*-type and *p*-type material, respectively [111]. The density of interface states has an approximately U-shaped dependence of the surface potential with a minimum at $\sim 10^{12}\,cm^{-2}\,eV^{-1}$ [111]. Due

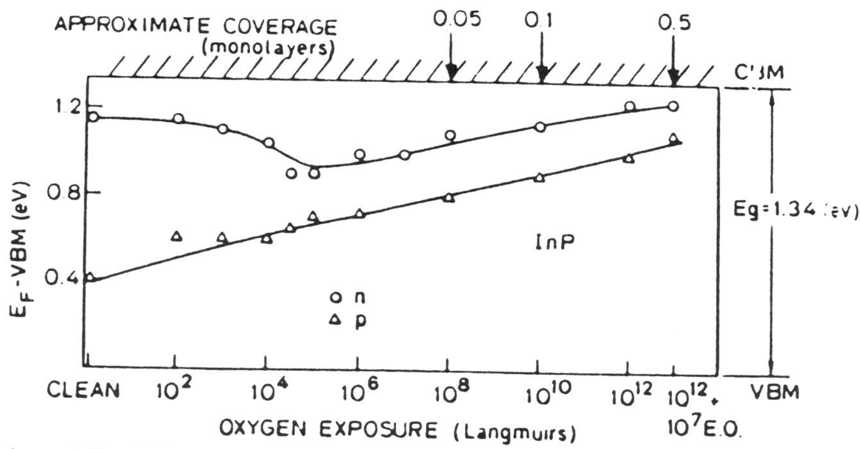

Figure 8.50 Shift of the SFL of an InP(110) surface exposure to oxygen. After Spicer et al. [93]; copyright © 1979, American Institute of Physics, New York.

to this high interface state density, the available range in the surface potential swing for such interfaces corresponds to about $\frac{1}{3}E_g/q$. This is insufficient to reach either flatband or accumulation conditions at a dc gate bias voltage below breakdown. Therefore, the utilization of FETs in the development of GaAs digital ICs has been limited thus far to JFET and MESFET technologies.

The C–V characteristics of InP MIS capacitors [113] are more well behaved, permitting a voltage swing from accumulation to inversion and consequently the utilization of InP for the construction of MISFET circuits. However, thus far InP MISFETs suffer from drift in the current-density–voltage characteristics and a lack of uniformity that seriously impede their development and applications [114]. RPCVD has been successful in controlling the drain current drift in InP MISFETs if the surface was first covered by a thin film of phosphorus [115].

Alternatively the surface of InP can be passivated by a layer of sulfur that sublimes off at ~110°C. If the in situ sublimation stage is followed immediately by RPCVD of SiO_2 at 250°C, excellent interfacial properties result, which is illustrated in Figure 8.51. It shows the drain current drift for enhancement-mode InP MISFETs having a channel width and length of 50 and 40 μm, respectively, and employing (a) deposition of plain SiO_2, (b) deposition of SiO_2 after surface sulfurization, (c) deposition and annealing of SiO_2 after surface sulfurization, (d) deposition of SiO_2 after surface coverage by phosphorus, (e) deposition of SiO_2 after in-chamber surface sulfurization. In addition to the pronounced differences in the drain current drift, the conditioning of the InP/gate dielectric interface results in distinct improvements in the transconductance of the FETs. For example, the devices corresponding to curves (e) and (d) have transconductances of 12.1 and 13.86 mS/mm as compared to

(a)

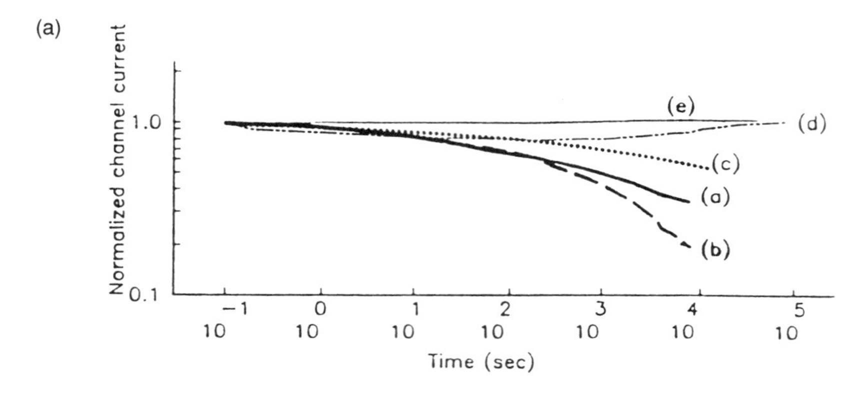

(b)

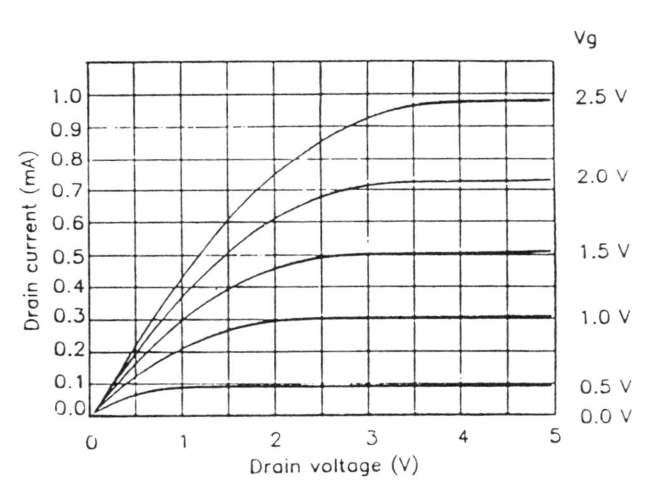

Figure 8.51 (a) Drain current drift for enhancement-mode InP MISFETs subjected to different surface treatments prior to and after the deposition of the SiO_2 gate isolation layer; (b) output dc J–V characteristics of a device employing sulfurization of the InP surface prior to SiO_2 gate dielectric deposition. After lyer, Chang, and Lile [112]; copyright © 1988, American Institute of Physics, New York.

2.3 mS/mm for the InP coated by plain SiO_2, all normalized to a gate capacitance of 1040 pF [115].

Figure 8.52(a) and (b) show the C–V characteristics and a plot of the interface state density versus the surface potential. N_{ss} values in the low $10^{-10}\,cm^{-2}\,eV^{-1}$ range are obtained at a surface potential of $-0.4\,eV$ [112]. An explanation for the beneficial behavior of sulfurized InP surfaces was provided by photoluminescence studies that reveal higher stability to the

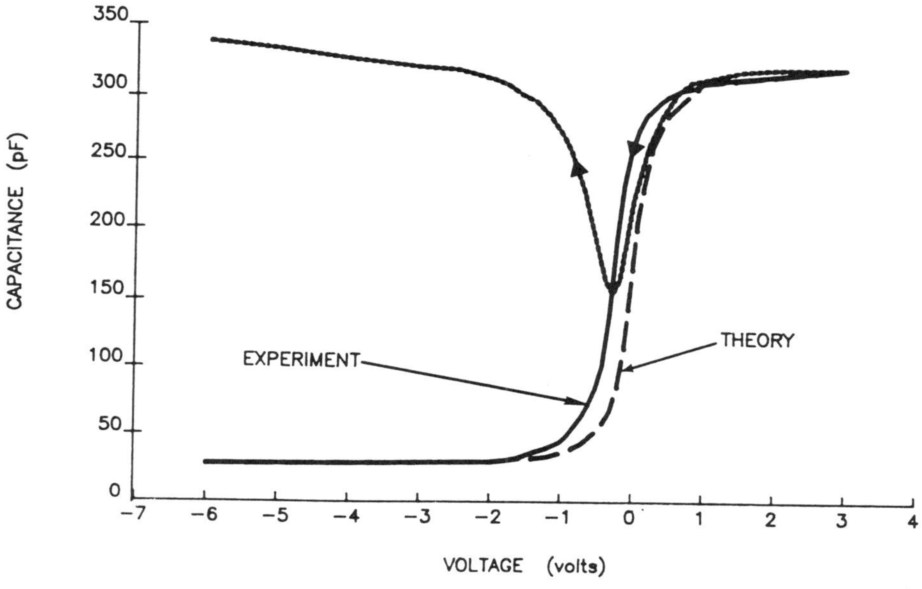

Figure 8.52 High-frequency and quasistatic C–V characteristics (a) and density of interface traps (b) of sulfurized n- InP/SiO_2 MIS capacitors. After Iyer, Chang, and Lile [112]; copyright © 1988, American Institute of Physics, New York.

thermal decomposition of the InP surface upon partial substitution of P by S [116]. Similar N_{ss} values as for sulfur-treated surface have been obtained for phosphorus nitride passivated InP surfaces. This lends additional supports to the interpretation of the above-mentioned drift problem as an effect of phosphorus loss from the surface during processing [117].

Further advances may become possible by replacing SiO_2 by alternative gate dielectrics, such as $Sr_xBa_{1-x}F_2$ (see Figure 8.16). F binds strongly to the InP surface, and MIS capacitors employing BaF_2 or its solid solutions with SrF_2 have achieved N_{ss} values in the mid-$10^{10}\,cm^{-2}\,eV^{-1}$ range without prior passivation of the InP surface [118]. Yet another interesting possibility for improving the properties of MIS devices of materials, exhibiting Fermi-level pinning at the native oxide interface, is provided by interlayers that seal the surface from harmful oxidation. For example, no hysteresis is observed in the high-frequency C–V curve of $SiO_2/Si/Ge$ MIS capacitors, and the interface state density is lowered from the typical value of $10^{11}\,cm^{-2}\,eV^{-1}$ for Ge/SiO_2 MIS capacitors to $5 \times 10^{10}\,cm^{-2}\,eV^{-1}$ [119]. Although there may thus exist an opportunity for the realization of GaAs MISFETs, at present InP seems to be a superior materials choice.

8.4 Schottky Barriers

As interactions of oxygen with the surface of semiconductors induce movement of the surface Fermi level, so do interactions with metal atoms, condensing on the surface during the early stages of contact formation, until the SFL locks into a final position. Ideally, the barrier height at the metal contact thus established depends on the work function difference between the metal and the semiconductor. If in equilibrium the surface is depleted, as in Figure 8.51(a) and (b), the contact is rectifying (Schottky contact). On the other hand, if the surface is in accumulation, the contact resistance is very small (Ohmic contact). This is illustrated in Figure 8.53(c) and (d). In practice, considerable deviations from ideal band bending at metal contacts are observed; that is, the surface Fermi level pins due to the charging of interface states to a position inside the bandgap. This renders the predictions of ideal Schottky barrier theory invalid. In this section, we discuss various materials and metal–semiconductor FETs (MESFETs) that employ a Schottky contact at the gate. Since the voltage swing at the gate of the MESFET device must cover the full range from depletion into accumulation, it is important to understand the reasons for Fermi-level pinning and to find ways of maximizing the barrier height.

The detailed modeling of the metal-induced SFL motion requires careful studies of the metal deposition process, which generally causes extensive reconstruction of the atomic arrangement at the interface. For example, Figure 8.54 shows the Fourier transforms of the SEXAFS signal for (a) $\frac{1}{3}$, (b) 1, (c) 2.5 monolayer coverage of Ag on the Si(111) surface and (d) for bulk silver metal

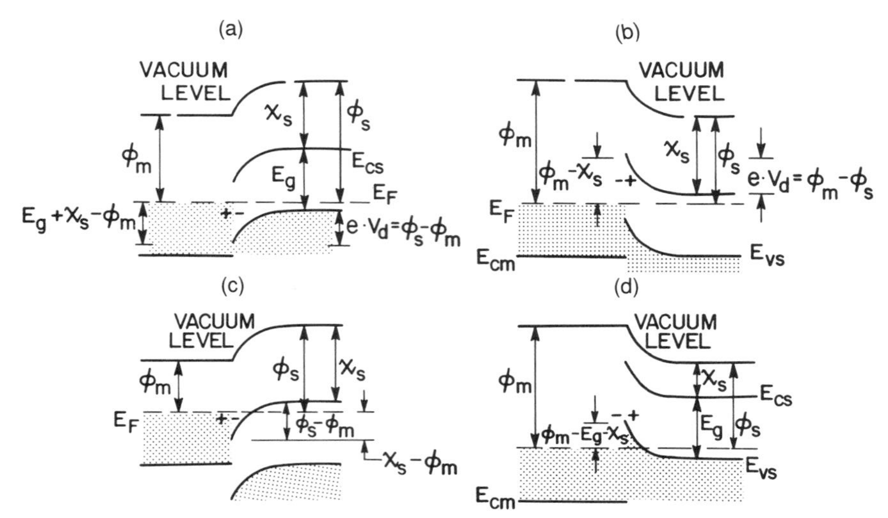

Figure 8.53 Idealized energy-band diagrams of metal–semiconductor contacts: (a) p-type: $\phi_M < \phi_S$, (b) n-type: $\phi_M > \phi_S$; (c) p-type: $\phi_M > \phi_S$; (d) n-type: $\phi_M < \phi_S$.

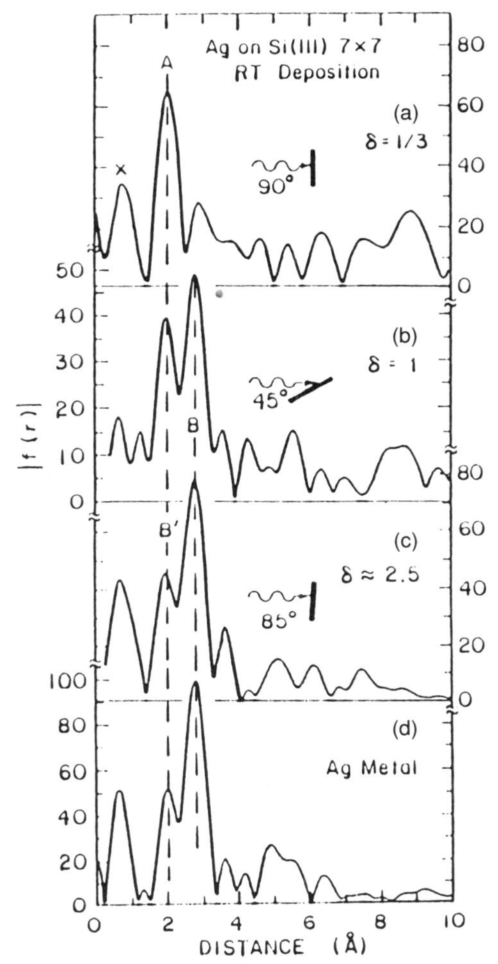

Figure 8.54 Fourier transforms of the EXAFS signal for Ag on Si(111). After Stöhr et al. [120]; copyright © 1983, North Holland Publishing Company, Amsterdam.

[120]. Upon exposure to Ag atoms, the surface transforms from its original 7×7 reconstruction into a $(\sqrt{3} \times \sqrt{3})R30^\circ$ surface mesh, which is established at both coverages (a) and (b). The peak A corresponds to the Ag–Si nearest-neighbor distance, while peak B corresponds to Ag–Ag nearest-neighbor distance. Both peaks B and B′ are observed in the SEXAFS spectra of Ag metal. The absence of these peaks at $\frac{1}{3}$ML coverage thus indicates atomic adsorption with coordination of the silver atoms by silicon surface atoms, while their presence at 2.5 ML coverage indicates clustering of the Ag atoms. While for low coverage the deposition process is two dimensional, at a critical coverage, the metal atoms form clusters followed by three-dimensional growth of metal blocks on the semiconductor surface. This Stranski–Krastanov

mechanism is linked to the buildup of strain in the joining of non-commensurate lattices. Up to a critical value of coverage, the strain is accommodated by pseudomorphic growth, while above this critical coverage the growth mechanism kicks over into homoepitaxy on already deposited clusters, minimizing the strain and leading to island deposition.

Figure 8.55(a) and (b) shows a model of the Ag/Si(111)-$\sqrt{3} \times \sqrt{3}R30°$ surface and schematically the transition from two-dimensional growth to bulk metal deposition. STM studies have revealed the boundary between surface domains on a Si(111) surface associated with the 7×7 silicon/UHV interface and Ag-atom-induced $\sqrt{3} \times \sqrt{3}R30°$ reconstruction, corresponding to an intermediate stage of the two-dimensional growth phase. The SFL shifts with increasing coverage gradually to its final position corresponding roughly to the coverage at which the transition from two-dimensional to island growth is observed [122]. For group III atoms (e.g., Al) deposited at room temperature on silicon, typical reconstructions are $\sqrt{3} \times \sqrt{3}$ for (111) and 2×2 for (100). The latter structure is derived by forming Al bridges in between the rows of Si dimers [124]. The mechanism of metal–semiconductor interactions in the initial phase of Schottky barrier formation is thus intimately linked to changes in the reconstruction of the semiconductor surface.

In the analysis of the complex evolution of the final Fermi level position upon metal deposition on a semiconductor surface, a distinction must be made between changes in the interface state density due to bond-breaking/forming chemical surface reactions and effects that are intrinsic to the presence of the metal on the surface of the semiconductor, that is, band bending due to the

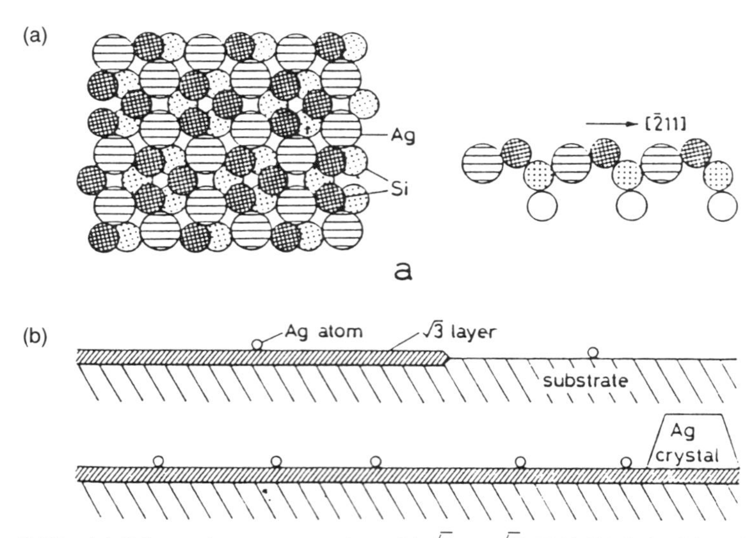

Figure 8.55 (a) Schematic representation of $(\sqrt{3} \times \sqrt{3})R30°$ (111) Ag/Si surface. After Saitoh, Shoji, Oura, and Hanawa [121]; copyright © 1981, North Holland Publishing Company, Amsterdam.

work function differences and the possible pinning of the SFL by metal induced gap states (MIGS) [126]. In order to explain the concept of MIGs, in Fig. 8.56(a) and (b) the electron density contours in the silicon and the total valence charge density are shown, averaging parallel to the interface plotted along a direction perpendicular to the interface for an aluminum/(111)-silicon junction [123b]. In the metal, this valence charge is constant while in the semiconductor it is periodic due to the concentration of charge into the bonds between the silicon atoms [127].

Figure 8.57 shows the local density of states for these layers. Since the electronic states associated with the bulk metal decay exponentially with increasing distance into the semiconductor, the density of states in the mid-gap region is raised at the interface and decays rapidly into the semiconductor. Figure 8.56(c) shows the charge density associated with the interface states

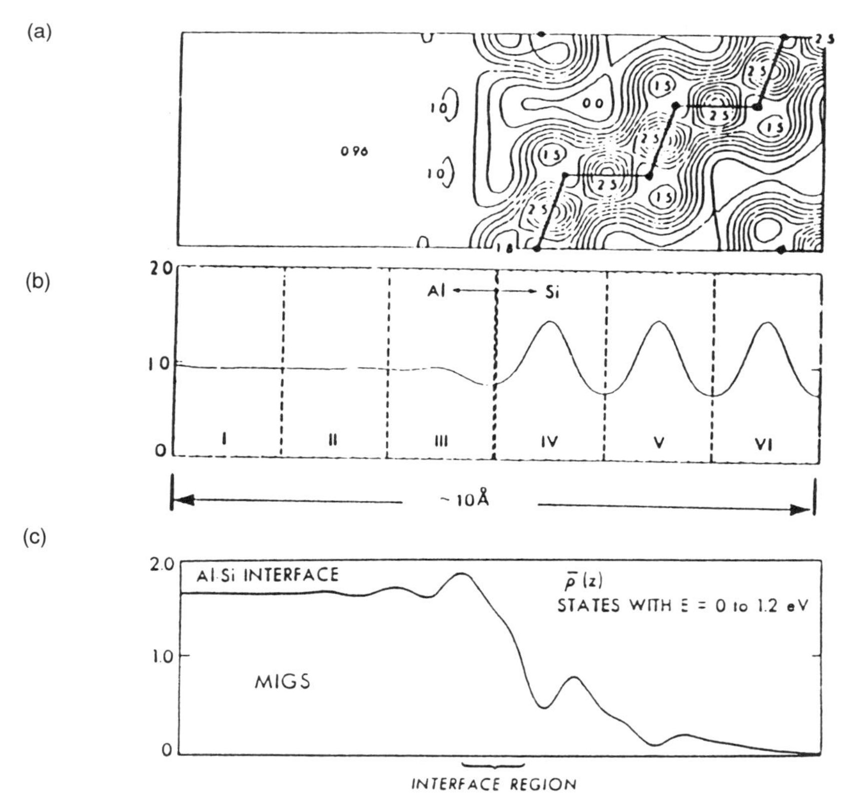

Figure 8.56 (a) Electron density contours, (b) total valence charge density in a six-layer stack containing the interface between and Al contact and Si. (c) Charge density associated with the interface states located in energy between 0 and 1.2 eV. (a), (b) After Louie and Cohen [123a] and (c) After Cohen [123b]. Copyrights (a), (b) © 1975, American Institute of Physics, New York, NY; (c) © 1980, Academic Press, Inc., New York.

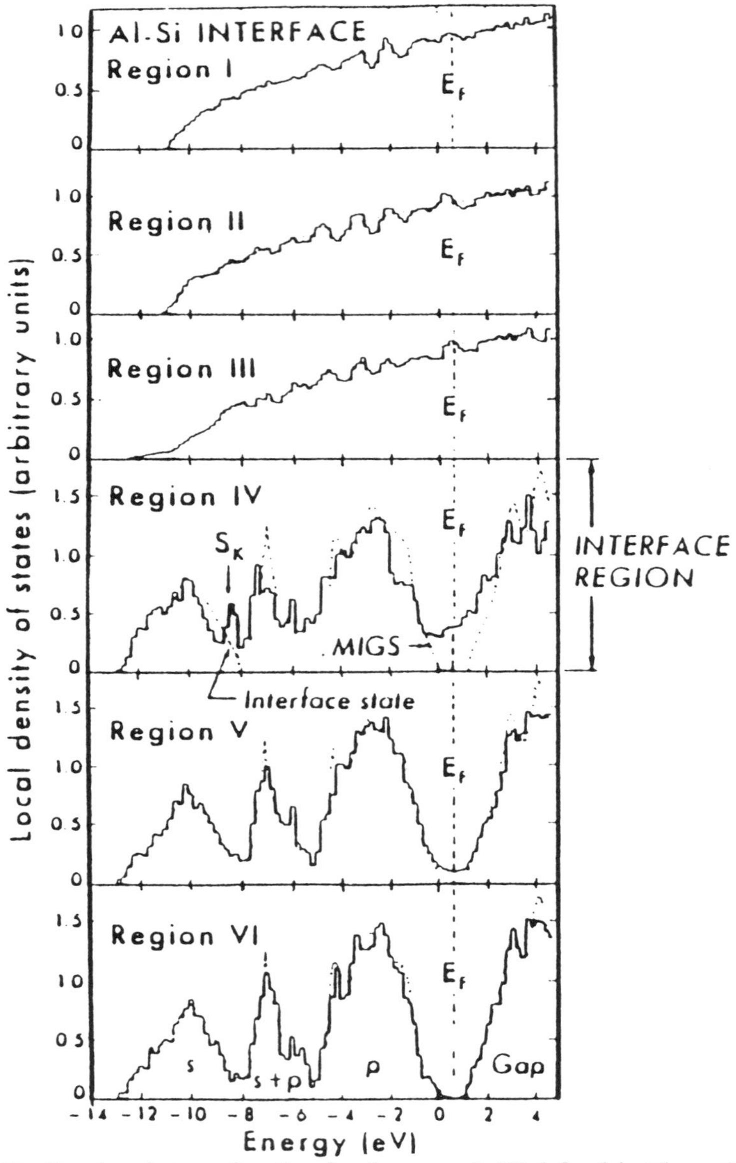

Figure 8.57 Density of states function for the zones I–VI defined in Figure 8.56(a). After Louie and Cohen [123b]; copyright © 1980, Academic Press, Inc., New York.

between 0 and 1.2 eV, averaged parallel to the interface and plotted along a direction perpendicular to the interface, revealing the exponential decay of these MIGS in the semiconductor, corresponding to charge-transfer between the metal and the semiconductor.

Strong chemical interactions are observed on the surface of silicon upon deposition of transition metals that form silicide intermetallic compounds, which are important in the context of both Schottky barriers and ohmic contacts. Attempts have been made at explaining the barrier height of transition metal Schottky contacts on Si in terms of their bulk properties. For example, the barrier height of metal silicides depends linearly on their heats of formation, suggesting that it may be explained on the basis of chemical bonding models [128]. However, inconsistencies exist in the behavior of *n*- and *p*-type semiconductors, and the correlations of bulk properties, such as the enthalpy of formation ΔH_f, to thin film properties may be affected by variations in the stoichiometry of the films. Such correlations thus cannot provide for the detailed understanding of Schottky contacts that we seek; that is, detailed investigations of their structure and interfacial chemistry are needed.

Two silicides that have been carefully evaluated are nickel and cobalt silicide, since they form nearly lattice-matched heterointerfaces with silicon [130], [131]. HREM studies show that the interfacial structure and properties depend critically on their processing. Nickel silicide forms spontaneously upon room-temperature deposition of Ni onto the Si(111) surface by solid-state epitaxy [132]. An analysis of the SEXAFS spectra and the associated Fourier transforms of $NiSi_2$ at several coverages of nickel on Si(111) reveals that the room-temperature deposition of Ni on Si starts out with the spontaneous formation of $NiSi_2$, which at 0.5 ML coverage results in a Ni–Si bond length of 2.37 ± 0.03 Å. This compares to 2.336 Å for bulk $NiSi_2$ [133]. The effective coordination number derived from the *K*-edge amplitude function is 7.4 ± 0.9 and thus agrees, within the error limits, with the eightfold coordination of the Ni in $NiSi_2$. The best site in the silicon surface lattice that would be consistent with this result is the initial insertion of the Ni atoms into sixfold-coordinated sites between the first and second silicon layer with two missing bonds. This initial reaction at low coverage is followed by further Ni substitution of Si atoms within the $NiSi_2$ overlayer structure, leading to a monotonic reduction of the SEXAFS amplitude.

The thickening mechanism that leads to the evolution of a 3D epitaxial $NiSi_2$ overlayer on the Si is still not completely understood, and different routes of the surface reaction may indeed be taken, depending on the details of the processing conditions. For example, the results of XPS and RHEED studies of the formation of nickel silicide by solid-phase epitaxy suggest that coverage of the Si(111) surface by a Ni film of a few Å thickness first replaces the 7×7 surface structure by a 1×1 structure nickel-rich layer of NiSi [129]. This disordered film converts at 160°C into a $\sqrt{3} \times \sqrt{3}$ structure, corresponding to epitaxial hexagonal NiSi on Si. Only at a temperature $\geqslant 300°C$ is an epitaxial film of $NiSi_2$ formed with a 1×1 surface mesh. $CoSi_2$, as $NiSi_2$, crystallizes in the CaF_2 structure with a fairly close lattice match to Si ($\Delta a/a = 1.2\%$). Since it has a lower resistivity than $NiSi_2$, $CoSi_2$ has been explored as a possible, but currently not proven, epitaxial contact material for ICs (see Section 8.5).

For compound semiconductors, generally the chemical interactions of the metal with the surface are more complex than for silicon and are thus even more difficult to unravel. However, the SFL shift upon low-temperature metallization of carefully stabilized GaAs(001) surfaces, which show no pinning of the SFL for the free surface, is quite simple [177]. Figure 8.58 shows the motion of the SFL associated with the deposition of Al and Au on such unpinned GaAs surfaces made by MBE. For coverages $\leqslant 1$ monolayer, little change in the SFL is observed, and at coverages >1 monolayer the Fermi level locks into the positions mandated by the work functions of the two metals

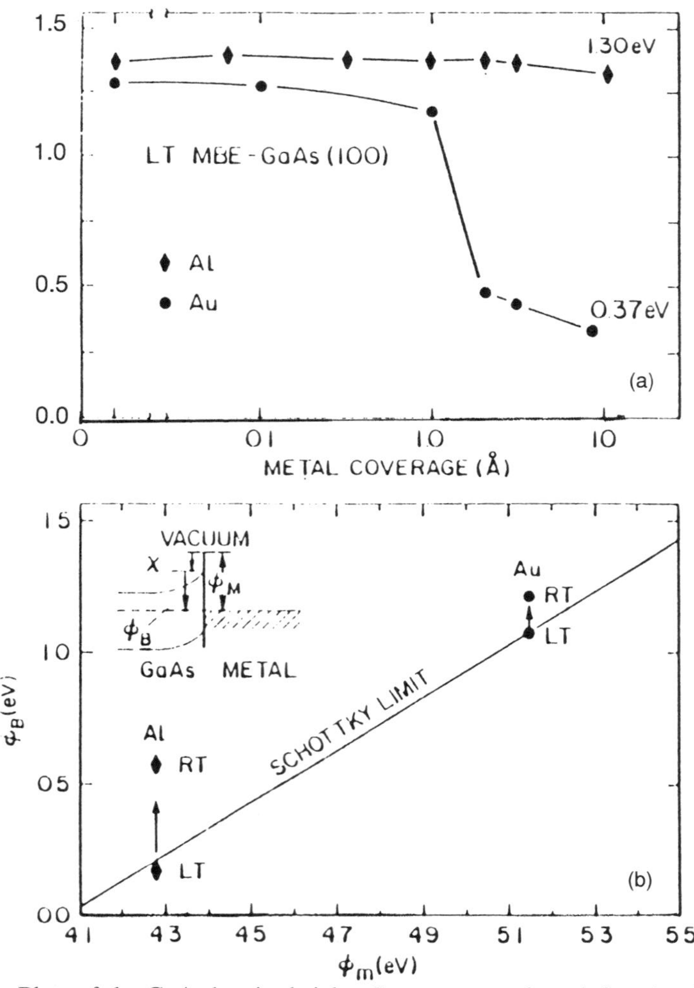

Figure 8.58 Plots of the GaAs barrier heights Φ_B versus metal work function Φ_m and E_F positions below the vacuum level (Φ_B + electron affinity χ) versus coverage. After Brillson et al. [134]; copyright © 1989, American Institute of Physics, New York.

relative to the work function of the GaAs in accord with ideal Schottky barrier theory. On the other hand, upon deposition of the same metals on analogously prepared unpinned GaAs(001) surfaces at room temperature, Fermi-level pinning is observed. The absence of pinning for the low-temperature depositions indicates that, at least for Al and Au on unpinned (001) surfaces of GaAs, MIGS play no significant role. However, chemical interactions and interdiffusion clearly affect the Fermi-level position when the contact is formed at and above room temperature. For example, in the case of Al deposited onto a cleaved GaAs(110) surface at room temperature, a dynamical analysis of the LEED current-voltage curves reveals exchange of Ga by Al, starting at low coverage with the second Ga layer and extending to a depth of several monolayers at higher coverage [136]. The first Ga layer is resilient to Al substitutions because substitution on subsurface layers maximizes the number of Al–As bonds and is thus energetically preferred.

In view of the complex point-defect chemistry, with few exceptions, the nature of the pinning states, formed in compound semiconductors upon metallization at and above room temperature, is unknown. The reason for this is that the most frequently used methods of Schottky barrier investigations primarily provide information on energy positions. For example, photoluminescence (PL) and cathodoluminescence (CL) experiments reveal metal-induced features that can be related to defects at specific energy positions in the bandgap of GaAs [135]. Figure 8.59 shows the results of CL measurements that exhibit intensity at 0.8 and 1.25 eV for the difference spectra between metal-exposed surfaces and clean surfaces. These features are related to transitions between defect and valence-band states at 0.8 eV and defect and conduction-band states at 1.25 eV, respectively, placing the defect states at 0.8 and 0.2 eV relative to the VBE. Similar energy positions are indicated by the results of STP [137]. However, although these and other investigations establish the energy position of the pinning states, they do not reveal their chemical nature.

Photoemission studies have been applied to the problem of Fermi-level pinning on metal contacts to semiconductors for many years since they permit the evaluation of SFL shifts. Figure 8.60 shows the movement of the SFL, measured by UPS as a function of coverage for Au, Ag, and Cu on clean InP(110) surfaces for low-temperature (80 K) and room-temperature metal deposition [138]. For Au and Ag, the low-temperature behavior is in accord with the physical models of atomic virtual gap states and MIGS at high coverage. However, for Cu, the observed movements of the SFL are not consistent with purely physical models, and the final position is the same for 80 K and room-temperature deposition, respectively. As in the case of PL and CL measurements, photoemission studies of the interactions of metals with compound semiconductor surfaces do not reveal the chemical nature of SFL pinning states. Since the photoexcitation to energies well above the lowest conduction-band minimum is inherently accompanied by minority carrier generation with concomitant shifts in the SFL that cannot be assessed with

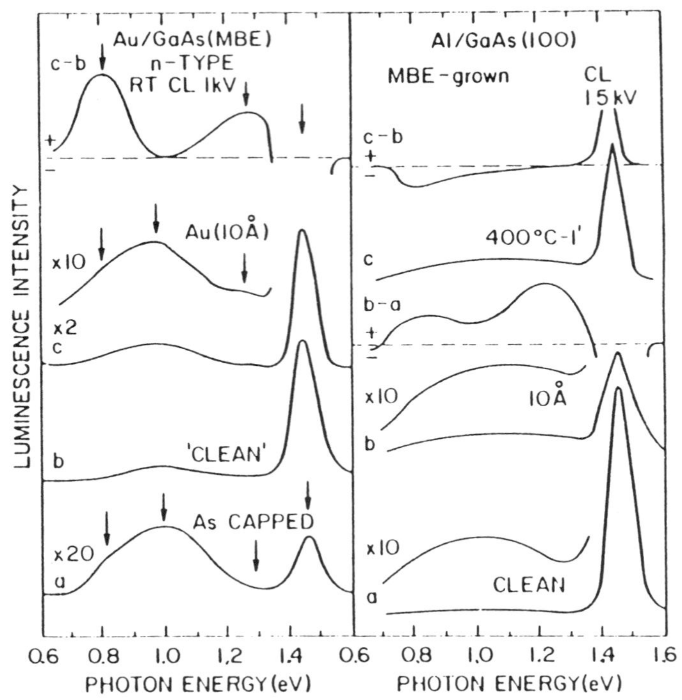

Figure 8.59 Cathodoluminescence spectra of Au- and Al-covered (100) surfaces of GaAs grown by MBE. Vituro, Slade, and Brillson [135]; copyright © 1986, The American Institute of Physics, New York.

high accuracy, photoemission experiments are also not a reliable choice for measurements of the barrier heights of low-gap semiconductors.

Interesting recent techniques that probe the Schottky barrier height on a local scale are ballistic electron emission microscopy (BEEM) [139] and ballistic electron emission spectroscopy (BEES) [140]. Figue 8.61 shows (a) a schematic representation of the experimental arrangement and (b) of the tunneling of electrons from the tip of an STM into states in a semiconductor, covered by a thin overlayer of a metal. Just above the Fermi energy, the mean free path of the electrons in the metal is sufficiently large to permit ballistic transport of the electrons into the conduction band of the semiconductor if the tip voltage exceeds the height of the Schottky barrier. The BEEM current is

$$I_{\mathrm{B}} \propto \int_{-\infty}^{\infty} \int_{0}^{4\pi} f(E - V_{\mathrm{T}}, T)\mathscr{P}_{\mathrm{t}}(E, V_{\mathrm{T}}, \mathbf{u}) \exp\left(-\frac{d(\mathbf{u})}{\lambda(E)} \mathscr{P}_{\mathrm{tr}}(E, \mathbf{u})\right) d\Omega dE \tag{8.61}$$

where $d(\mathbf{u})$ and $\lambda(E)$ refer to the electron path length and mean free path respectively, $f(E - V_{\mathrm{T}}, T)$ is the Fermi function and $\mathscr{P}_{\mathrm{t}}(E, V_{\mathrm{T}}, \mathbf{u})$ and $\mathscr{P}_{\mathrm{tr}}(E, \mathbf{u})$ are the probabilities of tunneling from the tip into the metal and transmission,

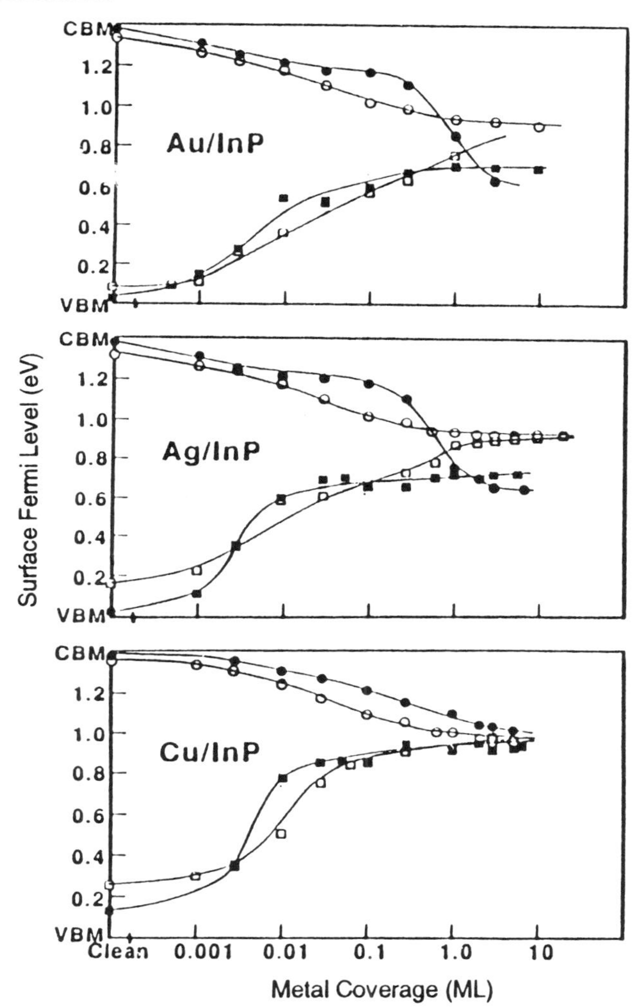

Figure 8.60 Surface Fermi-level positions as a function of coverage of clean InP(100) surfaces by Au, Ag, and Cu deposited at 80 K (full symbols) and room temperature (open symbols). After Cao et al. [138]; copyright © 1989, The American Institute of Physics, New York.

respectively, into the solid angle segment $d\Omega$ with direction $\mathbf{u}$ [141]. This can be further simplified to

$$I_B \propto \int_{\Phi_B}^{\infty} f(E - qV_T, T)(E - \Phi_B)^{3/2}\, dE \tag{8.61}$$

and at $T = 0$

$$I_B \propto (qV_T - \Phi_B)^{5/2} \tag{8.62}$$

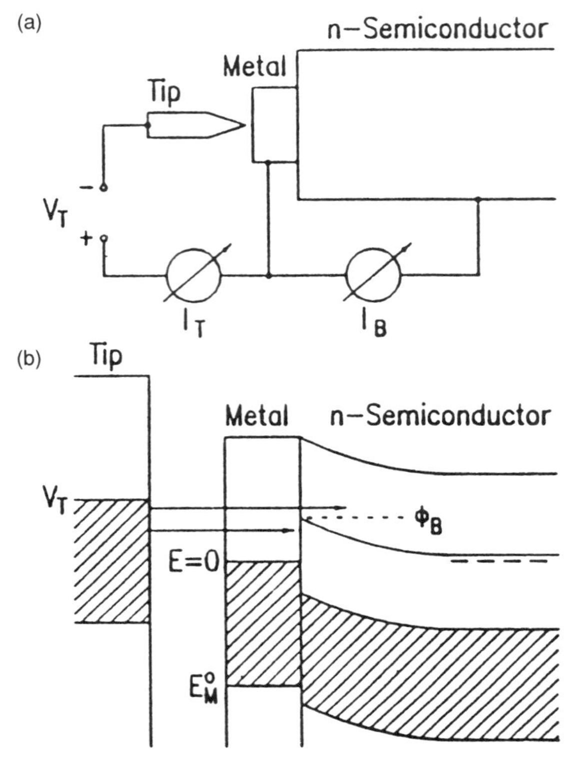

Figure 8.61 Schematic representation of (a) the experimental arrangement of BEEM and (b) the tunneling of electrons from the tip of an STM into states in the metal and the semiconductor. After Prietsch and Ludeke [141]; copyright © 1991, Elsevier Science Publishers B.V., Amsterdam.

The variation of I_B upon scanning the surface can be utilized for imaging, and its voltage dependence provides for spectroscopic evaluations of Schottky contacts and provides for excellent accuracy, provided that contamination of the tip is studiously precluded.

For example, Figure 8.62 shows the BEEM spectra for GaP(110) covered by thin films of Au, Ag, Cu, Mg, and Ni [141]. Within 0.4 V above threshold the $\frac{5}{2}$ power law is in excellent agreement with the experimental data and barrier heights are measured with a precision of ± 20–60 meV, depending on the metal. At higher voltages the energy dependence of the mean free path in the metal requires corrections, which are particularly large for Mg, as illustrated in Figure 8.63 [142].

At present the most important application of Schottky contacts is the manufacturing of GaAs MESFETs, which are widely used in low-noise microwave amplification and high-speed digital applications. There exist a variety of device structures and substrate technologies for GaAs MESFET

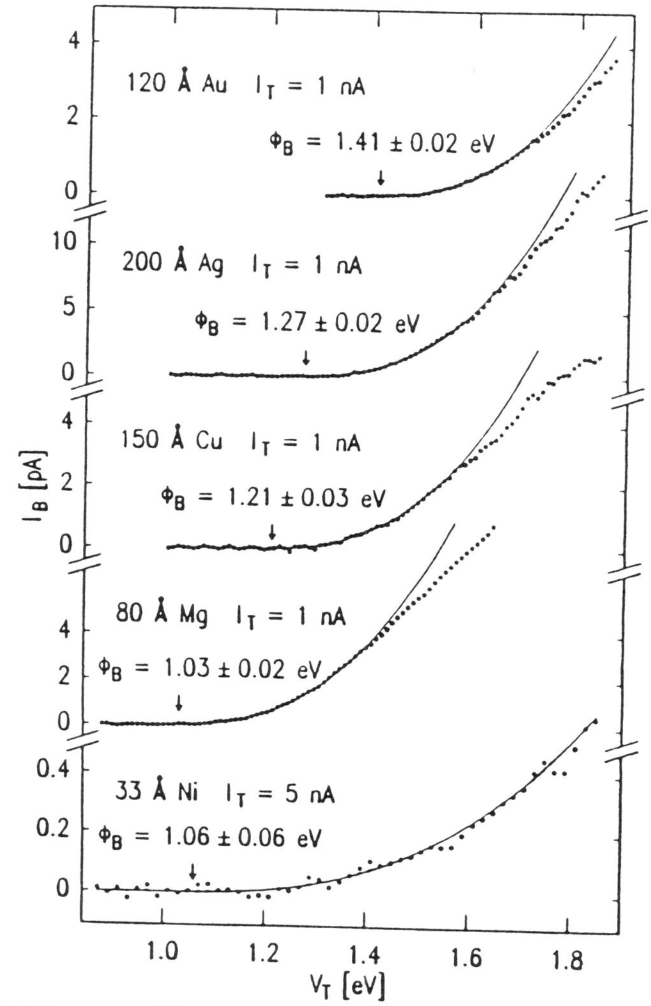

Figure 8.62 BEES spectra of GaP(110) covered by selected metals. After Prietsch and Ludeke [141]; copyright © 1991, Elsevier Science Publishers B.V., Amsterdam.

circuits. For example, they are built either in mesa-etched high-purity epitaxial films or are produced by direct ion implantation into semi-insulating GaAs substrates [144]. Also, GaAs directly grown on Si substrates and GaAs-on-insulator devices employing $Ca_xSr_{1-x}F$ as the insulating layer [145] have been used [146]. Figure 8.64 shows schematic cross section of a planar GaAs MESFET made by ion implantation that permits the manufacturing of both enhancement-mode (E) (i.e., normally off) and depletion mode (D) (i.e., normally on) devices [147]. The Be implanted P^--region is provided to counteract backgating effects [148]. Mesa structure 0.5 μm E-MESFETs with

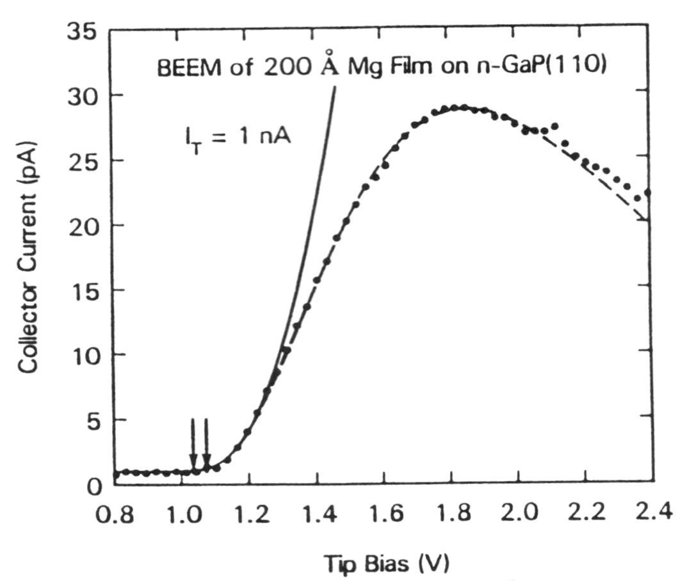

Figure 8.63 BEES spectrum of GaP(110) covered by 200 Å Mg. The dashed and solid lines represent fits of the data with and without inclusion of an energy-dependent electron mean-free path, respectively. After Ludeke, Prietsch, and Samsavar [142]; copyright © 1991, American Institute of Physics, New York.

a very shallow 25 nm channel have been made by MBE [147] and achieve at present a maximum transconductance of 400 mS/mm at $V_{DS} = 2V$ and $V_G \approx 0.7V$. Note that as in the case of Si IC technology a multilayer metallization is imployed to optimize the ohmic contacts at the source and drain and the Schottky contact (see Section 8.5).

Digital GaAs MESFET ICs achieve at present power-delay products of 214 fJ at 6.9 ps delay per gate [150]. Below 0.25 μm gate length, simulations of

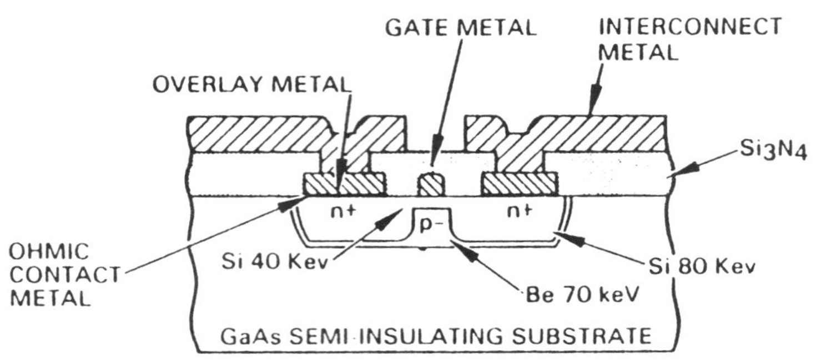

Figure 8.64 Schematic cross sections through a planar GaAs MESFET. After Sato et al. [143]; copyright © 1988, Institute of Electrical and Electronics Engineers, New York.

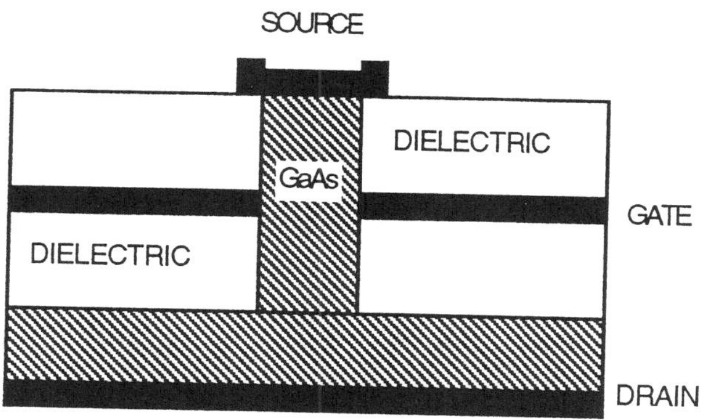

Figure 8.65 Schematic cross section through a GaAs permeable base transistor.

the performance of short channel GaAs MESFETs predict saturation in the values for the transconductance, maximum frequency of oscillation, and cutoff frequency [151]. In microwave applications, discrete GaAs MESFETs achieve at present conversion efficiencies of 25–30% in the 10–15 GHz range at a power output of ~13W [150]. An alternative simple device design with potential microwave applications is the GaAs permeable base transistor depicted schematically in Figure 8.65. It consists of a narrow GaAs plug that is surrounded by a ring-shaped Schottky barrier gate, which is separated by dielectric isolation layers (e.g., SiO_2) from the source and drain contacts on its ends. Although higher cut-off frequencies have been predicted for this device, at present, difficulties with controlling the contact metallurgy limit the devices at present to $f_t \approx 50$GHz.

8.5 Ohmic Contacts and Interconnects

As mentioned in Chapter 4, ohmic contacts are obtained either by highly doping the contact areas as part of the epitaxial steps in the device fabrication, or by driving in a heat treatment step dopants from a metal alloy into the semiconductor (alloyed contact). Thus, narrow, shallow barriers are formed under the contact that are either overcome by thermionic emission or are tunneled through. If the contact is abrupt and is dominated by thermionic emission, the contact resistance is given by

$$R_c = \frac{k_B}{C_R T} \exp\left[\frac{q\Phi_B}{k_B T}\right] \tag{8.63}$$

where C_R is the Richardson constant and Φ_B is the barrier height. If the current is dominated by tunneling through a very narrow barrier, the contact resistance depends only weakly on the temperature. Therefore, the form of the $R_c T$ versus

$1/T$ plot permits the determination of the current mechanism. Note that sharp interfaces are rarely established below the contacts, i.e., the contact resistance depends on the contact microstructure [49].

The contact and interconnect metallization of integrated circuits must satisfy the following conditions:

1. Stable, high conductivity of the metal;
2. Good step coverage by the interconnects;
3. High selectivity of the contact metal deposition;
4. Minimum chemical interactions with the underlying device regions;
5. Minimum mechanical stress;
6. Minimum contribution of the processing to the thermal budget;

Usually a multilayer contact structure is needed to realize all these requirements including diffusion barriers that allow the use of reactive metals to meet conditions 1–3 and yet satisfy condition 4. For example, the VLSI implementation of an inverter circuit shown in Figure 4.20(a) has two levels of metallization. In this case, aluminum is the metallization of choice. Even for the design rules for VLSI the current densities in the metal interconnecting network are already high enough to cause concern with regard to electromigration. Electromigration refers to the transport of matter that results in a conductor upon passage of a dc current because of momentum transfer from the electrons onto the ionic cores. The directional transport of the metal at the regions of highest current density thus induced leads to separation of the thin lines of the metal interconnect network of ICs at a characteristic mean time to failure τ_f, which is inversely proportional to the current density. Since void formation is initiated at the surface of the interconnect line and progresses along grain boundaries, τ_f depends on the grain size distribution. In addition to the effects of the grain size distribution, the texture of the grains in the metal lines also affects τ_f [154]. For single crystalline aluminum lines on silicon, τ_f is significantly extended [155]. Although alloys of aluminum containing Cu and Si additions are more resistant to electromigration than pure aluminum, intermetallic particles formed in such alloy interconnect have been shown to migrate upon current flow, thus contributing to the degradation [156].

Since the local equilibrium vacancy concentration increases in the presence of a hydrostatic tensile stress, and vacancies participate in the formation of voids, the stress induced due to thermal expansion coefficient mismatch during processing is an important factor in the failure of contacts and interconnects. As illustrated in Tables 8.4 and 8.5, the thermal expansion coefficients α of metals are generally much larger and the thermal expansion coefficients of dielectrics are smaller in comparison to silicon. For a film of a material of uniform thickness, the rate of stress change with temperature $d\sigma/dT$ is related to the difference of the expansion coefficients of the film and the substrate $\Delta\alpha$, Young's modulus E, and the Poisson ratio υ of the film according to

$$\frac{d\sigma}{dT} = \Delta\alpha \frac{E}{1-\upsilon}. \tag{8.64}$$

Table 8.4 Mechanical properties of selected metals in comparison to silicon (after Flinn et al. [157]).

Material	c_{11}(GPa)	c_{11}(GPa)	c_{11}(GPa)	$E/(1-\upsilon)$ (GPa)	$\alpha(10^{-6})$	$d\sigma/dT$ (MPa/K)
Aluminum	108	62	23.3	113	23.0	−2.3
Copper	169	122	75.3	262	16.6	−3.6
Silver	123	92	45.3	170	19.0	−2.7
Gold	190	161	42.3	189	14.2	−2.1
Silicon	166	64	79.6	229	3.0	

To account for elastic anisotropy of crystalline materials, the values $E/(1-\upsilon)$ presented in Table 8.4 refer to dominant (111) orientation. Based on this simplified estimate, stresses of several hundred MPa are expected to result in the range of typical processing temperatures. Since the failure of contacts and lines by the formation of voids expanding along grain boundaries depends on the surface and grainboundary energies, the decoration of grainboundaries by impurities and the surface cleanliness have significant effects on the failure rate. Also, interfacial reactions that alter the point defect concentrations in the vicinity of the contact and the embrittlement of the metal by bulk impurities must be considered. Since silicon processing frequently is carried out in a hydrogen containing ambient, hydrogen embrittlement is of particular concern. Thus, the reliability of contacts and interconnects depends on both the thermal and chemical history of the processing sequence. For a more detailed review of this topic, the reader is referred to Ref. [157].

Because of the reduced line width and concomitantly higher current densities in the interconnecting network under the conditions of ULSI, there is certainly a limit of the design rules, below which electromigration problems eliminate aluminum metallization as a viable option. In addition, the RC time constant of an aluminum-based interconnecting network limits the gain in the power-delay product provided by ULSI, and makes its replacement by metals with higher conductivity desirable.

Table 8.5 Mechanical properties of selected metals in comparison to silicon. After Flinn et al. [157].

Material	$E/(1-\upsilon)^2$ (GPa)	$E/(1-\upsilon)$ (GPa)	$\alpha(10^{-6})$	$d\sigma/dT$ (MPa/K)
CVD silicon oxide	120	150	1.5	0.22
PECVD TEOS	100	125		
PECVD silicon oxynitride	160	200	2.0	0.12
PECVD silicon nitride	170	210	2.0	0.12
Sintered silicon nitride		210		
Silicon		229	3.0	

Table 8.6 Resistivities of selected transition metals and transition metal silicides

Material:	Al	Cu	Ti	Ta	Mo	W	Si[a]	$NiSi_2$
Resistivity ($\mu\Omega$ cm):	2.65	1.68	60	70	10	8	> 500	50
Material:	$TiSi_2$	$TaSi_2$	$MoSi_2$	WSi_2	$CoSi_2$	PtSi		Pd_2Si
Resistivity ($\mu\Omega$ cm):	13–16	40–60	40–200	20–100	18–30	25–35		25–30

[a]Doped polysilicon.

Table 8.6 shows the resistivities of selected metals and metal silicides. Copper is a possible alternative to aluminum since, its resistivity is significantly smaller. Both evaporation and sputtering are suitable for the deposition of a blanket Cu film that is subsequently patterned by lithographic processes. Also, CVD of Cu is of interest because it can be done selectively on tungsten pads. Organometallic complexes of copper ions employing oxygen-containig ligands derived from acetylacetone (acac) [158] or tert.-butoxide [159], [160], as well as oxygen-free (trialkylphosphine)cyclopentadienyl-copper(l) complexes [161]-[163] may be utilized as source materials. The dionate complexes tend to form adducts in the gas phase, while the (trialkylphosphine)cyclopentadienyl complexes are well behaved. Good uniformity of the overgrowth of a patterned Si/SiO_2 structure is achieved with this source [164]. Table 8.7 shows the activation energies and processing windows in temperature for various Cu source materials and processes. The deposition temperatures are low enough to suit the low thermal budget requirement of ULSI technology.

Figure 8.66 shows (a) the SEM image of an oxide trench with a 0.1 μm TiW barrier metal coated by a conformal copper deposit; (b) a simulation of the trench filling assuming a sticking coefficient of 0.015 and (c) a Cu coated cantilever structure that reveals the excellent step-coverage of the CVD

Table 8.7 Processing temperature ranges and activation energies for the decomposition of various organometallic source materials for copper deposition. After Kaloyeros and Fury

Source material	Temperature Range (°C)	Ativation Energy (kcal/mole)	Process
$Cu^{II}(hfac)_2$[a]	300–390	15	CVD
	150–250	9	PACVD
$(hfac)Cu^{I}(COD)$[b]	160–180	55	CVD
(hfac)Cu(tmvs)[c]	160–190	13	CVD
	150–250	7	PACVD

[a]hfac = 1,1,1,5,5,5-hexafluoroacetylacetone,
[b]COD = 1,5-cyclooctadiene,
[c]tmvs = vinyltrimethylsilane.

process. The drawn-in contures are the result of a simulation taking the same sticking coefficient as in the simulation of Figure 8.66(b). A problem of Cu interconnect technology is the difficulty of etching because of the low volatility of copper halides. However, etch rates as high as 1 μm/min have been obtained by either simultaneous or sequential feeding of Cl_2 and $P(C_2H_5)_3$ in the chlorine etching of copper, forming a volatile complex according to the reaction.

$$CuCl + :P(C_2H_5)_3 \rightarrow ClCu(P(C_2H_5)_3)_2 \tag{8.65}$$

Electroless plating of copper has been considered as an alternative to vapor deposition in the context of VLSI applications. It generally employs the

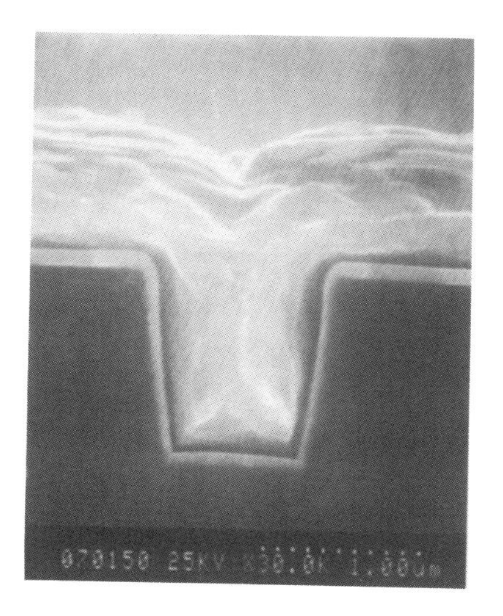

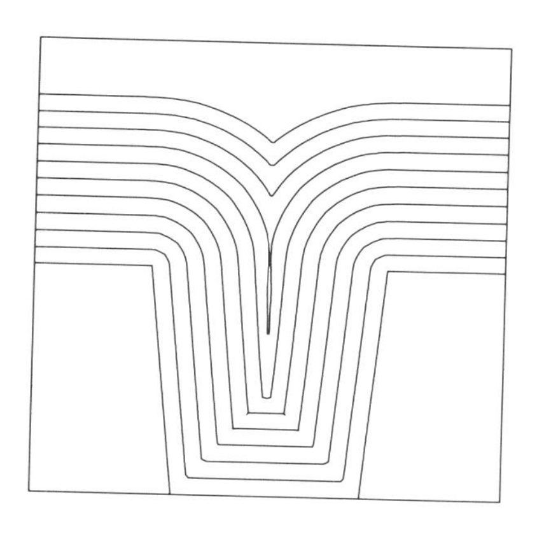

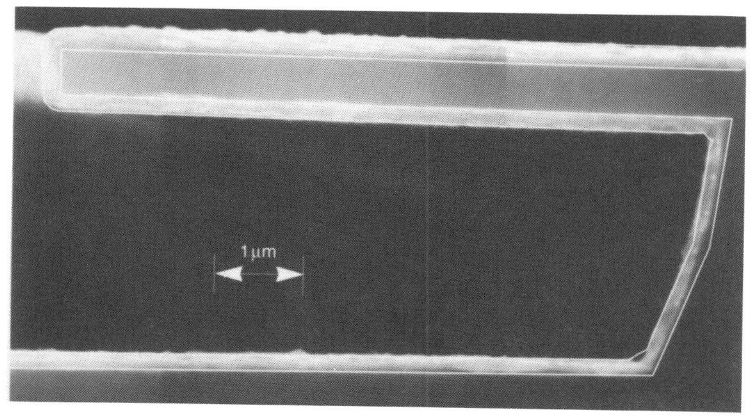

Figure 8.66 (a) Oxide trench with 0.1 μm TiW barrier filled with copper; (b) SPEEDIE simulation of the filling process; (c) Cu coated cantilever revealing the excellent step coverage by CVD copper. After Cho [166]; copyright © 1992, Institute of Electrical and Electronics Engineers, New York.

interaction between a reducing component of the solution C_r that is oxidized by copper ions to deposit copper metal according to

$$mC_r + Cu^{2+} \rightarrow mC_{ox} + Cu^0 \tag{8.66}$$

This implies the simultaneous occurrence of a redox reaction and a metal–metal-ion reaction, for example,

$$2HCHO + 4OH^- \rightarrow H_2 + 2HCOO^- + 2H_2O + 2e^- \tag{8.67a}$$

$$Cu^{2+} + 2e^- \rightarrow Cu^0 \tag{8.67b}$$

$$2HCHO + 4OH^- + Cu^{2+} \rightarrow H_2 + 2HCOO^- + 2H_2O + Cu^0 \tag{8.68}$$

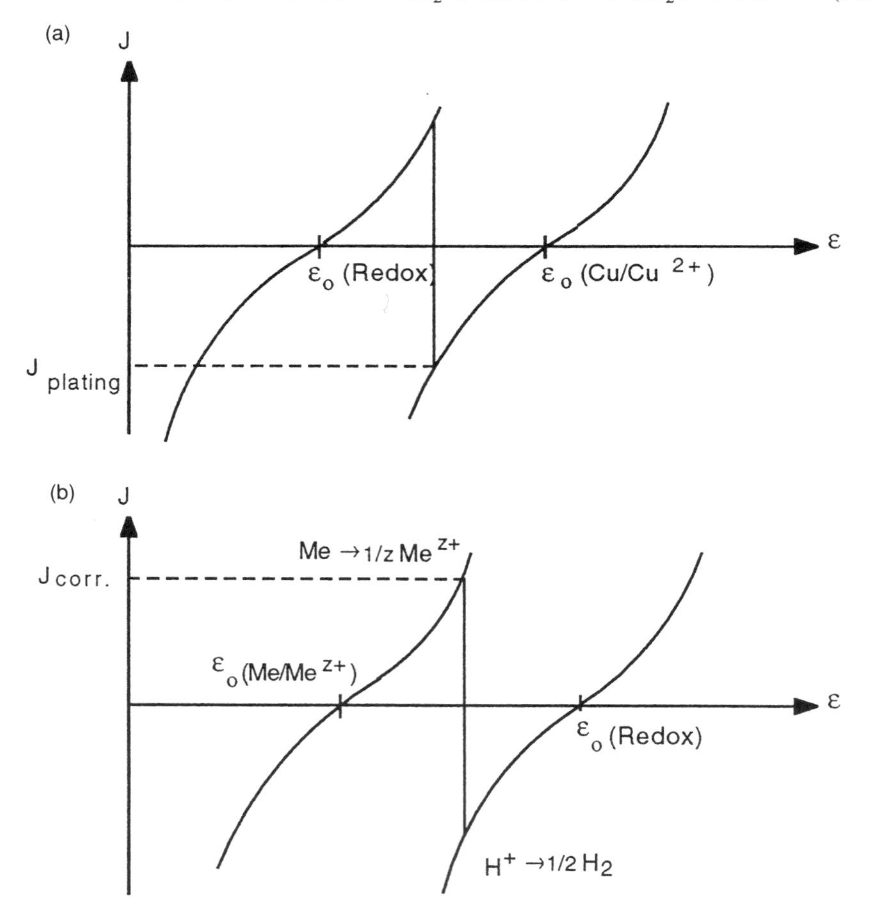

Figure 8.67 Schematic representation of mixed potentials under the conditions of (a) electroless plating, where the cathodic current for the plating at zero external current is compensated by an anodic current due to an oxidative redox reaction, and (b) corrosion, where the anodic corrosion current at zero external current is compensated by a cathodic current due to a reductive redox reaction (e.g., hydrogen evolution) or a sacrificial metal/metal ion reaction.

(a)

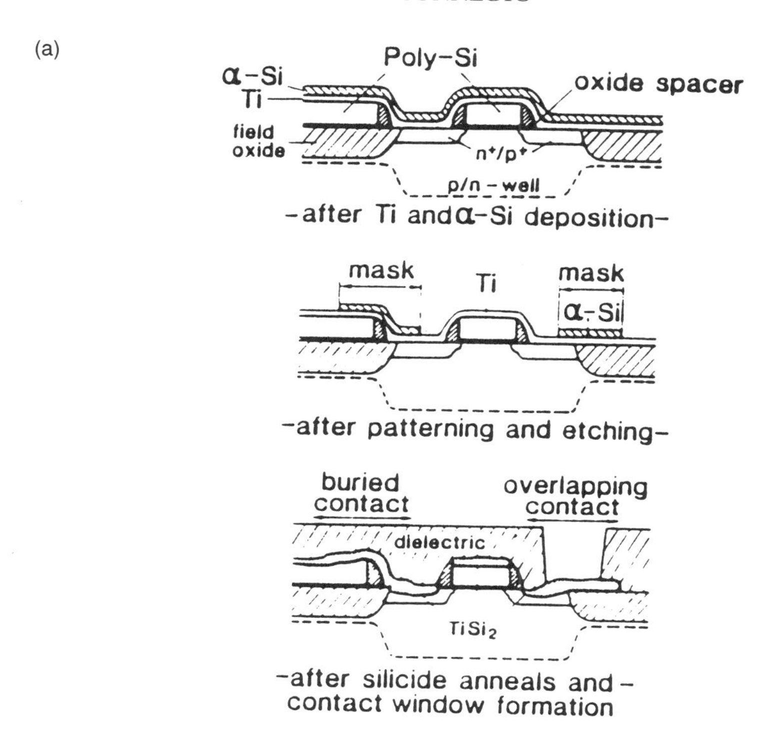

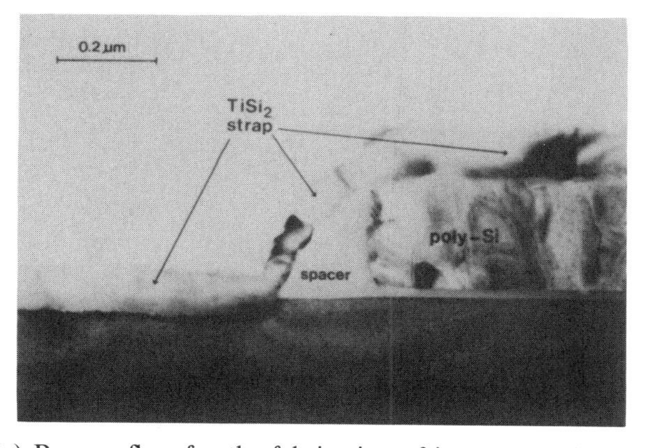

Figure 8.68 (a) Process flow for the fabrication of interconnecting $TiSi_2$ straps from active device regions over SiO_2 to polysilicon. (b) Cross-sectional TEM image of such a $TiSi_2$ interconnect. After Kramer [19 of Chapter 4]; copyright © 1989, The Electrochemical Society, Pennington, NJ.

Figure 8.67 shows schematically the current potential curves for the redox reaction Eq. (8.67a) and the metal–metal-ion reaction Eq. (8.67b), resulting at the mixed potential ε_m, where the anodic current density associated with the oxidation of formaldehyde matches the cathodic reduction current density of the metal deposition reaction so that no external current flow is observed. We note that if the equilibrium potential of the redox reaction were positive with respect to that of the metal–metal-ion reaction, corrosion of the metal would ensue, which requires careful protection of the electronic circuit with regard to contact with detrimental electrolytes.

Transmission metal silicides are widely used in the metallization of ULSI

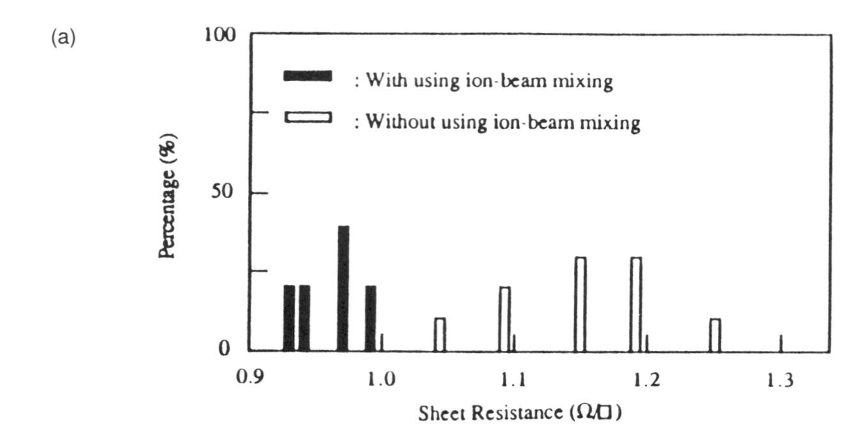

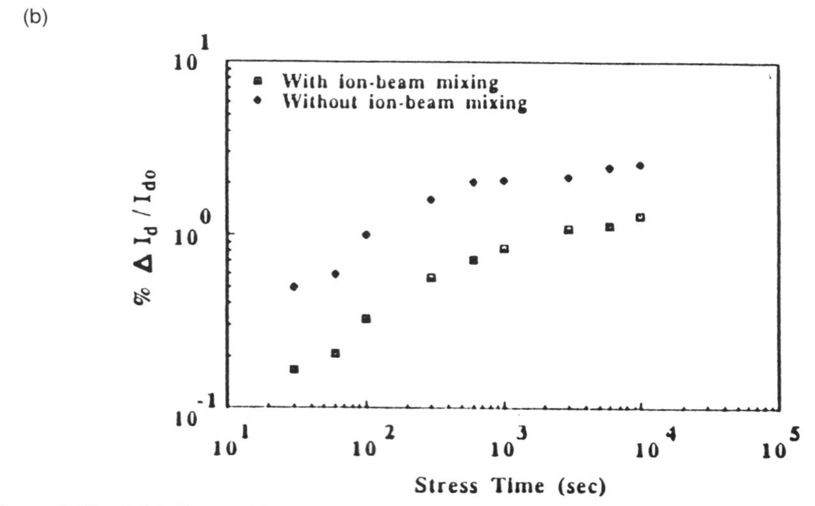

Figure 8.69 (a) Effects of ion-beam mixing on the sheet resistance of Ti salicide for a Ti thickness of 750 Å. (b) Degradation of the drain current in the linear region of Ti salicide LDDNMOSFETS. After Ku, Lee, and Kwong [166]; copyright © 1989, The Electrochemical Society, Pennington.

circuits. Titanium silicide, which exists in a wide range of metastable compositions about Ti:Si = 1:2, is of particular interest as a contact metal for ULSI circuits because it has in its C54 phase the lowest resistivity of all silicides studied to date, that is, 13–16 $\mu\Omega$ cm (see Table 8.3). Also, it has the virtue of reducing the interfacial native oxide [164] and provides for excellent ohmic contacts to both n^+- and p^+-type material. In view of the tight tolerances of ULSI processing, self-alignment of the contacts to the source, gate, and drain regions is extremely important and is best achieved by self-aligned silicide (salicide) contact formation. In the salicide process, first a blanket deposition of Ti metal by electron-beam evaporation is provided, followed by a ~20 s rapid thermal anneal in nitrogen at 650°C. This temperature and time are sufficiently small to prevent reaction of the Ti with the SiO_2, but establish the C49 phase of $TiSi_x$ in the windows defined by lithographic techniques and a passivating film of TiN elsewhere. The TiN is removed by a selective etch in $H_2O{:}H_2O_2{:}NH_4OH$ = 4:1:1. Then the low-resistivity C54 phase is formed in a second anneal at 800°C.

In the fabrication of the CMOS-SRAM circuit shown in Figure 4.20(b), connections between the active device regions and polysilicon are made by $TiSi_2$ straps, as illustrated in Figure 8.68(a). The silicon needed for $TiSi_2$ formation over the SiO_2 isolation is provided by amorphous silicon deposition

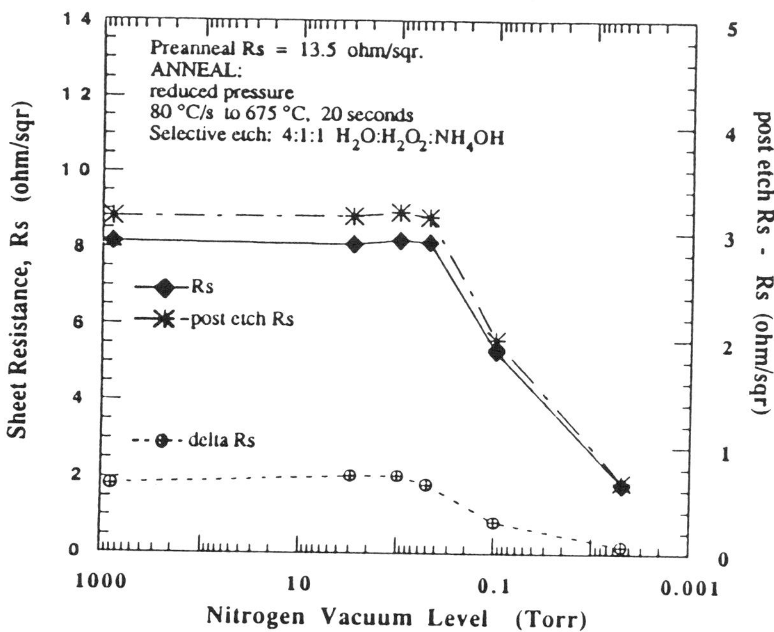

Figure 8.70 Sheet resistance of TiN/$TiSi_x$ metallization layers as a function of the nitrogen pressure employed during annealing. After Nulman [165]; copyright © 1990, Materials Research Society, Pittsburgh, PA.

in appropriately masked regions of the initially deposited Ti film. Figure 8.68(b) shows an XTEM image of the $TiSi_2$ strap running from the source region on the wafer surface pover SiO_2 to an isolated polysilicon layer.

Since the silicide formation is retarded by high As doping at the Ti/Si interface, different amounts of silicon consumption and differences in the dopant redistribution result during RTA at the As-doped n^+ and B-doped p^+ contact regions of CMOS devices during the silicide formation. Thus an ion mixing step at the interfaces prior to RTA by ^{28}Si implantation may be utilized to improve the contact properties [165]. Also, ion implantation of impurities into the silicide followed by an RTA drive-in step has been used to form shallow n^+ and p^+ junctions under the contact avoiding the implantation damage of the silicon. Figure 8.69 shows the improvement of the sheet resistance and the hot carrier degradation of the drain current in the linear region of n-channel salicide LDD MOSFETs fabricated with and without ion mixing. Also, the gate voltage shift upon constant current stress is smaller for devices made with ion mixing as compared to conventional processing. The improvements in the device performance and reliability are primarily due to the reduction of stress by the ion mixing, leading to improvements of the interface state density at the Si/SiO_2 interface.

Since the sheet resistance of the $TiSi_x$ metallization increases upon a temporary exposure of the Ti to air prior to the rapid thermal annealing step, $TiSi_x$ salicide processing benefits from an integrated thermal processing approach where the wafer is not removed from a low-pressure inert gas environment for the entire process. The sheet resistance obtained depends on the inert gas pressure as shown for nitrogen in Figure 8.70. Good selectivity in the deposition of Ti on Si and SiO_2 has been accomplished by the use of $TiCl_4/GeH_4$ PECVD; that is, the $TiSi_x$ forms in the window regions, but does not deposit on the oxide. Thus the wet etching step that is necessary after blanket metallization may become avoidable in integrated IC manufacturing schemes employing $TiCl_4$ reduction by GeH_4 [167].

In $TiSi_x$ gate metallization the degradation of the gate oxide beneath the $TiSi_x$ during thermal treatments is of concern [168]. Therefore, underlayers of polysilicon or amorphous silicon must be provided to separate the $TiSi_x$ from the gate dielectric [169]. Doped polysilicon is deposited by pyrolysis of silane in an LPCVD reactor at 600–650°C at a rate of 100–200 Å/min [172]. It is the proven gate electrode of VLSI technology serving both as gate contact and as local interconnect metal. The doping can be either done ex situ by diffusion or ion implantation or in situ by admixtures of phosphine, arsine, or diborane to the SiH_4. In situ doping requires for optimum resistivity deposition at temperatures $\geqslant 525$ and $\geqslant 625°C$ for boron and phosphorus doping, respectively, corresponding to columnar growth morphology. The dopant concentration reaches $10^{21}\,cm^{-3}$ under these conditions, but the mobility is low, typically $10–30\,cm^2\,V^{-1}\,s^{-1}$, due to scattering at impurity atoms and at the grain boundaries [173]. The resistivity of polysilicon is thus $\geqslant 500\,\mu\Omega\,cm$ and improvements, for example, by polycide formation, are desirable in the context

(a)

Si

$CoSi_2$

Si

50 nm

(b)

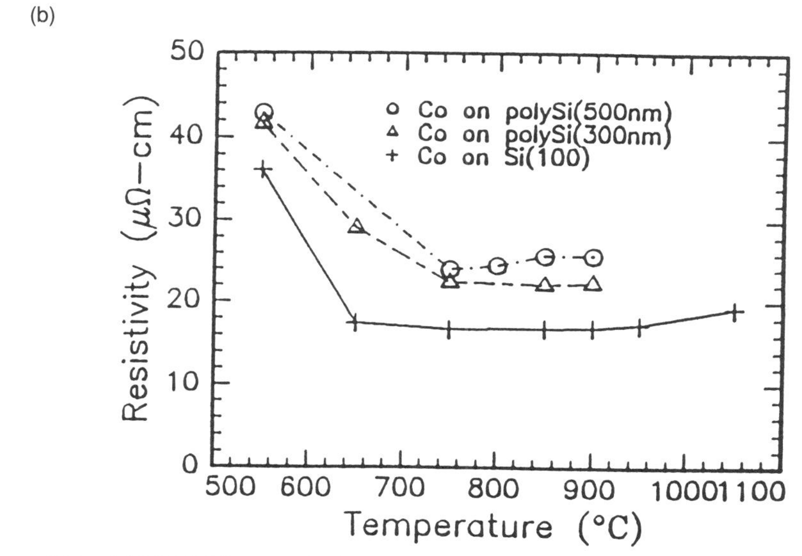

Figure 8.71 (a) XTEM image of a buried epitaxial $CoSi_2$ layer in silicon produced by ion implantation and annealing. (b) Resistivity of CoSi films as a function of the annealing temperature. After (a) Maex [170] and (b) Phillips et al. [174]; copyright © 1990 Materials Research Society, Pittsburgh.

of ULSI processing. At sufficient thickness of the polysilion buffer ($\sim$200 nm), polysilicide gates maintain the excellent stability of polysilicon/thermal SiO_2 gate structures even at gate oxide thicknesses $\leqslant$20 nm.

In addition to $TiSi_2$, $CoSi_2$ has been evaluated as a potential metallization material because of its close lattice match to silicon, relatively low specific bulk resistivity, and inertness with regard to several wet and dry etching agents. Figure 8.71(a) shows an HREM image of a $CoSi_2$/Si interface produced by ion implantation of Co into Si and annealing. Figure 8.71(b) shows the resistivity of $CoSi_2$ formed by reaction of a blanket Co layer on poly- and single-crystalline silicon as a function of the annealing temperature. Although these data show that resistivities close to the bulk value can be obtained, there are still questions remaining with regard to fundamental limitations of $CoSi_2$ as a contact material that require further evaluation [174].

In order to interconnect the various levels of metallization of ICs, vias must be opened in the separating dielectric layers and filled with metal. Figure 8.72 shows a schematic representation of a cross section through a set of vias opened after the planarizing etch (step 38 in Table 4.3) in the dielectric layer separating the first and second metal of an IC. Tungsten deposition by hydrogen reduction of WF_6 is well suited for the filling of vias since it is highly selective, that is, does not occur on the areas masked by SiO_2. This is illustrated in Figure 8.73, which shows (a) a selectivly deposited tungsten plug, connecting through a SiO_2 layer to a molybdenum layer, and (b) a tungsten contact plug. Note the bird's beak in the field oxide to both sides of the contact. The WF_6 reduction by hydrogen proceeds according to the reactions

$$2WF_6 + 3Si \rightarrow 2W + 3SiF_4 \tag{8.69}$$

$$WF_6 + 3H_2 \rightarrow W + 6HF \tag{8.70}$$

[176]. Reaction (8.69) is very fast and stops due to mass transport limitations

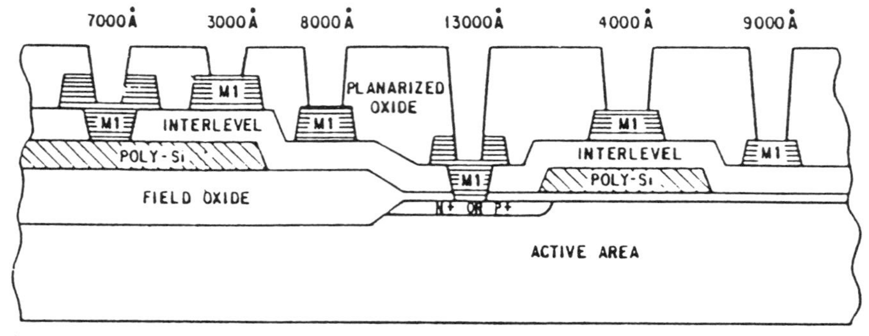

Figure 8.72 Schematic representation of vias opened in the dielectric separating the first and second metal of an IC after planarization. After Brown et al. [178]; copyright © 1987, Institute of Electrical and Electronics Engineers, New York.

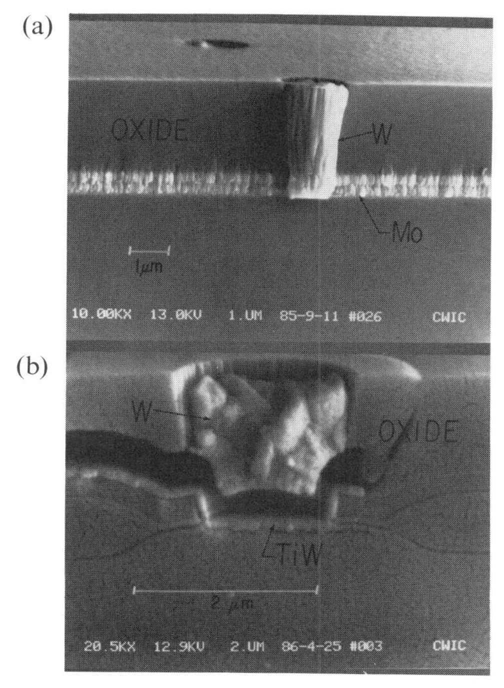

Figure 8.73 (a) Tungsten plug formed by hydrogen reduction of WF_6 in a via in a SiO_2 film making a connection to a Mo layer on a silicon substrate. (b) Tungsten contact plug. After Brown et al. [178]; copyrights © 1987, Institute of Electrical and Electronic Engineers, New York.

of W to the interface at a film thickness of 100–200 Å. It is responsible for the selectivity since it nucleates the tungsten plug on the exposed silicon surface. Once a tungsten film is deposited, the growth of tungsten on tungsten is favored over the growth of tungsten on SiO_2 and continues selectively by the second mechanism. The rate of reaction (8.70) is limited by the dissociation of H_2 at the substrate surface and depends strongly on the temperature of deposition, which is typically in the range of 300–450°C. The resistivity of W films of >3000 Å thickness deposited at 400°C is typically $\sim 12\,\mu\Omega\,\text{cm}$, as compared to $5.7\,\mu\Omega\,\text{cm}$ for bulk tungsten but increases for thinner films due to gettering effects in the initial phases of film growth and increased scattering [177]. The stress in a thick W film deposited at 400°C is typically $6 \times 10^9\,\text{dynes/cm}^2$ and decreases with increasing deposition temperature. Above 650°C the W/Si interface becomes unstable due to the formation of WSi_x. This problem is removed by the deposition of a WSi_x diffusion barrier by either cosputtering of Si and W [178] or the dichlorosilane reduction of WF_6 [179]. $W/WSi_{2.1}/Si$ contacts made by cosputtering of the silicide barrier have been shown to be stable to at least 800°C. The transition metals W, Mo, and Ta are also useful as diffusion barriers in multilevel contacts to aluminum and copper. Cu reacts at $\geqslant 300°C$ with silicon to form Cu_3Si, which degrades in air. However, $Ta/Cu/Ta/W/TiN/TiSi_x$ contacts are stable with regard to interdiffusion up to 600°C [180].

References

1. B. E. Deal and A. S. Grove, *J. Appl. Phys.* **36**, 3770 (1965).
2. Y. -L. Chiou, C. H. Sow, and K. Ports, *IEEE Electron Dev. Lett.* **10**, 1 (1989).
3. E. A. Irene and Y. J. Vander Meulen, *J. Electrochem. Soc.* **123**, 1380 (1976).
4. W. A. Tiller, *J. Electrochem. Soc.* **128**, 689 (1981); A. Lin, R. W. Dutton, D. A. Antoniades, and W. A. Tiller, *J. Electrochem. Soc.* **128**, 1121 (1981).
5. N. Tsubouchi, H. Miyoshi, and H. Abe, *J. Appl. Phys.* **17**, 223 (1978).
6. E. Nicollian, in *Semiconductor 1986*, H. R. Huff, T. Abe, and B. Kolbesen, eds., The Electrochemical Society, Pennington, NJ 1986, 437.
7. D. W. Hess and B. E. Deal, *J. Electrochem. Soc.* **124**, 735 (1977).
8. Y. Miyai, K. Yoneda, H. Oishi, H. Uchida, and M. Inoue, *J. Electrochem. Soc.* **135**, 150 (1988).
9. K. L. Brower, *Appl Phys. Lett.* **43**, 111 (1988); *Phys. Rev. B* **38**, 9657 (1988).
10. K. Makihara, A. Teramoto, K. Nakamura, M. Y. Kwon, M. Morita, and T. Ohmi, *Jpn. J. Appl. Phys.* **32**, 294 (1993).
11. K. L. Brower and S. M. Myers, *Appl. Phys. Lett.* **57**, 162 (1990).
12. L. Do Thanh and P. Balk, *J. Electrochem. Soc.* **135**, 1797 (1988).
13. A. H. Edwards, *Phys. Rev. B* **44**, 1832 (1991).
14. H. Ibach, K. Horn, R. Dorn, and H. Lüth, *Surface Sci.* **38**, 433 (1973).
15. A. Koma and R. Ludeke, *Phys. Rev. Lett.* **35**, 107 (1975).
16. J. P. Pelz and R. H. Koch, *Phys. Rev. B* **42**, 3761 (1990).
17. A. J. Schell-Sorokin and J. E. Demuth, *Surf. Sci.* **157**, 273 (1985).
18. M. Udagawa, M. Niwa, and I. Sumita, *Jpn. J. Appl. Phys.* **32**, 282 (1993).
19. T. Klitsner, R. S. Becker, and J. S. Vickers, *Phys. Rev. B* **44**, 1817 (1991).
20. A. Ourmazd, D. W. Taylor, and J. A. Rentschler, *Phys. Rev. Lett.* **59**, 213 (1987).
21. J. C. C. Tsai, in *VLSI Technology*, S. Sze, ed., McGraw Hill Book Company, New York, 1983.
22. J. Halbritter, *J. Mater. Sci.* **3**, 506 (1988).
23. J. A. Appels, E. Kooi, J. M. Paffen, J. J. H. Schatorje, and W. H. C. G. Verkuijlen, *Philips. Res. Rep.* **25**, 118 (1970).
24. M. Kishimoto, A. Kajiya, Y. Todokoro, and M. Inoue, in *ULSI Science and Technology*, C. Osburn and J. M. Andrews, eds., The Electrochemical Society, Pennington, NJ, 1989, p. 67.
25. J. Hui, T. Chiu, S. Wong, and W. Oldham, *IEDM Trans. ED-29*, 554 (1982).
26. H. -H. Tsai, C. -L. Yu, and C. -Y. Wu, *IEEE Electron Dev. Lett.* **10**, 307 (1989).

27. N. Shimizu, Y. Naito, Y. Itoh, Y. Shibata, K. Hashimoto, M. Nishio, A. Asai, K. Ohe, H. Umimoto, and Y. Hirofuji, *IEDM Techn. Dig.*, San Francisco, 1992, p. 279.

28. S. Onishi, H. Takeoka, K. Tanaka, and K. Sakiyama, in *ULSI Science and Technology*, C. Osburn and J. M. Andrews, eds., The Electrochemical Society, Pennington, 1989, p. 734.

29. Y. Tamaki, *Jpn. J. Appl. Phys.* **37–40** (1981).

30. S. F. Chuang, S. D. Collins, and R. L. Smith, *Appl. Phys. Lett.* **55**, 675 (1989).

31. K. Barla, G. Bomchil, R. Herino, and J. C. Pfister, *J. Crystal Growth* **68**, 721 (1984).

32. S. Konaka, M. Tabe, and T. Sakai, *Appl. Phys. Lett.* **41**, 86 (1982).

33. G. Bomchil, R. Herino, K. Barla, and J. C. Pfister, *J. Electrochem. Soc.* **130**, 1161 (1983).

34. P. Gupta, A. C. Dillon, A. S. Bracker, and S. M. George, *Surface Sci.* **245**, 360 (1991).

35. L. Jastrzebski, *J. Crystal Growth* **63**, 493 (1983).

36. D. P. Vu, M. Haond, D. Bensahel, and M. Dupuy, *J. Appl. Phys.* **54**, 437 (1983).

37. L. Jastrzebski, A. C. Ipri, and J. F. Corboy, *IEEE Electron. Dev. Lett.* **EDL-4**, 32 (1983).

38. Y. Hokari, M. Mikami, K. Egami, H. Tsuya, and M. Kanamori, *IEEE J. of Solid State Circuits* **SC-20**, 173 (1985).

39. H. Zogg, S. Blunier, and J. Masek, *J. Electrochem. Soc.* **136**, 775 (1989).

40. H. Ishiwara and T. Asano, *Appl. Phys. Lett.* **40**, 66 (1982); M. Ihara, Y. Arimoto, M. Jifuko, T. Kimura, S. Kodama, H. Yamawaki, and T. Yamaoka, *J. Electrochem. Soc.* **129**, 2569 (1982).

41. K. Egami, M. Mikami, and H. Tsuya, *Appl. Phys. Lett.* **43**, 757 (1983).

42. R. J. Peart and R. C. Newman, *Inst. Phys. Conf. Ser.* 16, paper 19, The Institute of Physics, London, 1973.

43. M. Orlowski, *IEDM Techn. Digest* (1988); *Phys. Lett. A* **137**, 115 (1989).

44. L. Mader, M. Orlowski, and I. Weitzel, in *ESSDERC '89*, A. Heuberger, H. Ryssel, and P. Lange, eds., Springer Verlag, Berlin, 1989, p. 41.

41. F. Lau, L. Mader, C. Mazure, Ch. Werner, and M. Orlowski, *Appl. Phys. A* **49**, 671 (1989).

46. S. Furukawa, H. Matsumura, and H. Ishiwara, *Jpn. J. Appl. Phys.* **11**, 134 (1972).

47. J. Linhard, M. Scharff, and H. Schiott, *Mat. -fys. Med., Dan. Vid. Selks* **33**, 1 (1963).

48. K. A. Pickar, *Appl. Solid State Sci.* **5**, 151 (1975).

49. T. E. Seidel, in *VLSI Technology*, S. M. Sze, ed., McGraw-Hill Book Company, New York, 1983, p. 219.

50. W. S. Johnson and J. F. Gibbons, *Projected Range Statistics in Semiconductors*, Stanford University Bookstore, Palo Alto, CA (1969).

51. J. Narayan and O. W. Holland, *J. Electrochem. Soc.* **131**, 2651 (1988).

52. W. K. Hofker, *Philips Res. Repts. Suppl.* **8** (1975).

53. T. Yoshii, S. Taguchi, T. Inoue, and H. Tango, *Jpn. J. Appl. Phys.* **21**, *Suppl.* **21-1**, 175 (1982).

54. R. E. Reedy and G. A. Garcia, *Mat. Res. Soc. Symp. Proc.* **107**, 365 (1988).

55. H. W. Laen and R. F. Pinizzotto, *J. Crystal Growth* **63**, 554 (1983).

56. U. Gösele, K. -Y. Ahm, B. R. P. Marionton, T. Y. Tan, and S. -T. Lee, *Appl. Phys.* 48, 219 (1989).

57. S. -T. Lee, P. Fellinger, and S. Chen, *J. Appl. Phys.* **63**, 1924 (1986).

58. K. Paprocki, I. Brylowskam, and W. Syszko, *Appl. Phys. A* **45**, 109 (1988).

59. W. -K. Chu, J. W. Mayer, and M. -A. Nicolet, *Backscattering Spectroscopy*, Academic Press, New York, 1978.

60. A. Gras-Marti, *Phys. Stat. Sol. (a)* **76**, 621 (1983); I. Abril, R. Garcia-Molina, and A. Gras-Marti, *ibid.* **95**, 766 (1986).

61. G. K. Celler, J. M. Poate, and L. C. Kimerling, *Appl. Phys. Lett.* **32**, 464 (1978).

62. D. Lecrosnier, J. Paugam, F. Richow, G. Pepous, and F. Berniere, *J. Appl. Phys.* **51**, 1036 (1980).

63. D. V. Lang, *J. Appl. Phys.* **45**, 3014 (1974).

64. D. V. Lang, *J. Appl. Phys.* **50**, 5093 (1979).

65. G. L. Miller, D. V. Lang, and L. C. Kimerling, *Ann. Rev. Mater. Sci.* 377 (1977).

66. H. G. Grimeis, *Ann. Rev. Mater. Sci.* 341 (1977).

67. H. Lefevre, *thesis*, Friedrich-Alexander Universität Erlangen-Nürnberg 1981.

68. H. Lefevre and M. Schulz, *Trans. IEEE Electron Dev.* **ED-24**, 973 (1977).

69. E. H. Snow, A. S. Grove, and D. J. Fitzgerald, *Proc. IEEE* **55**, 1168 (1967).

70. T. Hamamoto, *Appl. Phys. Lett.* **58**, 2942 (1991).

71. J. M. Aitken and D. R. Young, *J. Appl. Phys.* **47**, 1196 (1976).

72. T. H. Ning, *J. Appl. Phys.* **47**, 1079 (1976).

73. L. Lipkin, A. Reisman, and C. K. Williams, *MCNC Techn. Rep. TR90-27*, Research Triangle Park, NC, 1990.

74. A. Reisman, C. J. Merz, J. R. Maldonado, and W. W. Molzen, Jr., *J. Electrochem. Soc.* **133**, 628 (1984); A. Reisman, C. K. Williams, and J. R. Maldonado, *J. Appl. Phys.* **62**, 868 (1987).

75. J. R. Maldonado, A. Reisman, H. Lezec, C. K. Williams, and S. S. Eyer, *J. Electrochem. Soc.* **133**, 628 (1986); *J. Vac. Sci. Technol.* **B5**, 248 (1987).

76. J. M. Aitken, D. R. Young, and K. Pan, *J. Appl. Phys.* **49**, 3386 (1978).

77. A. Reisman and C. J. Merz, *J. Electrochem. Soc.* **130**, 1384 (1983).

78. M. Schulz and A. Karmann, *Appl. Phys. A* **52**, 104 (1991).

79. A. VanderZiel, *Adv. Electron. Electron Phys.* **49**, 225 (1979).

80. M. J. Kirton and M. J. Uren, *Adv. Phys.* **38**, 367 (1989).

81. M. Prutton, *Surface Physics*, Clarendon Press, Oxford, 1982.

82. W. Kossel, *Z. Phys.* **1**, 119 (1920).

83. R. de L. Kronig, *Z. Phys.* **70**, 317 (1931); **75**, 468 (1932).

84. D. E. Sayers, E. A. Stern, and F. W. Lyttle, *Phys. Rev. Lett.* **27**, 1204 (1971).

85. P. H. Citrin, J. E. Rowe, and P. Eisenberger, *Phys. Rev. B* **28**, 2299 (1983).

86. C. A. Ashley and S. Doniach, *Phys. Rev. B* **11**, 1279 (1975).

87. B. K. Teo, *J. Am. Chem. Soc.* **103**, 3990 (1981).

88. J. B. Pendry, The Transition between XANES and EXAFS in *EXAFS and Near Edge Structure*, A. Bianconi, L. Incoccia, and S. Stipcich, eds., Springer Verlag, Berlin, 1983, p. 4.

89. J. Stöhr, R. S. Bauer, J. C. Menamin, L. I. Johansson, and S. Brennan, *J. Vac. Sci. Technol.* **16**, 1195 (1979).

90. M. Nishijima, H. Kobayashi, K. Edamoto, and M. Onchi, *Surface Sci.* **137**, 473 (1984).

91. E. Chainet, M. De Crescence, J. Derrien, T. T. A. Nguyen, and R. C. Cinti, *Surf. Sci.* **168**, 801 (1986).

92. F. D. Schonwald and F. J. Gunthaner, *J. Vac. Sci. Technol. B* **6**, 1368 (1988).

93. W. E. Spicer, P. W. Chye, P. R. Skeath, C. Y. Su, and I. Lindau, *J. Vac. Sci. Technol.* **16**, 1422 (1979); T. T. Chiang and W. E. Spicer, *J. Vac. Sci. Technol. A* **7**, 724 (1989).

94. H. Iwasaki, Y. Mizokawa, R. Nishitani, and S. Nakamura, *Jpn. J. Appl. Phys.* **17**, 1925 (1978).

95. G. P. Schwartz, J. E. Griffith, and B. Schwartz, *J. Vac. Sci. Technol.* **16**, 1383 (1979).

96. D. E. Aspnes, J. B. Theeton, and R. P. H. Chang, *J. Vac. Sci. Technol.* **16**, 1374 (1979).

97. G. P. Schwartz, G. J. Gualtieri, J. E. Griffith, C. D. Thurmond, and B. Schwartz, *J. Electrochem. Soc.* **127**, 2488 (1980).

98. J. A. Stroscio, R. M. Feenstra, D. M. Newns, and A. P. Fein, *J. Vac. Sci. Technol. A* **6**, 499 (1988).

99. J. L. Freeouf and J. M. Woodall, *Appl. Phys. Lett.* **39**, 727 (1981).

100. E. R. Weber and J. Schneider, *Physica B* **116**, 398 (1983).

101. D. Troost, L. Koenders, L. -Y. Fan, and W. Mönch, *J. Van. Sci. Technol. B* **5**, 1119 (1987).

102. K. Stiles, D. Mao, and A. Kahn, *J. Vac. Sci. Technol. B* **6**, 1170 (1988).

103. R. M. Feenstra and J. A. Stroscio, *J. Vac. Sci. Technol. B* **5**, 923 (1987).

104. A. Y. Cho and J. R. Arthur, *Progr. Solid State Chem.* **10**, 157 (1975).

105. M. Gunder and H. Jacob, *Appl. Phys. A* **39**, 73 (1987).

106. A. Y. Cho and J. C. Tracy, Jr., U.S. Patent No. 3,969,164 (1976).

107. A. Takamori, S. Sugata, K. Asakawa, E. Miayuchi, and H. Hashimoto, *Japan. J. Appl. Phys.* **26**, L142 (1987).

108. S. I. J. Ingrey, *J. Vac. Sci. Technol.* **7**, 1554 (1989).

109. K. J. Bachmann, H. Schreiber, Jr., W. R. Sinclair, F. A. Thiel, E. G. Spencer, G. Pasteur, W. L. Feldman, and K. Sreeharsha, *J. Appl. Phys.* **50**, 3441 (1979).

110. R. Iyer, R. R. Chang, A. Dubey, and D. L. Lile, *J. Vac. Sci. Technol. B* **6**, 1174 (1988).

111. L. G. Meiners, *Thin Solid Films* **56**, 201 (1979).

112. R. Iyer, R. R. Chang, and D. L. Lile, *Appl. Phys. Lett.* **53**, 134 (1988).

113. H. H. Wieder, *J. Vac. Sci. Technol.* **15**, 1498 (1978).

114. M. Okamura and T. Kobayashi, *Electron. Lett.* **17**, 941 (1981).

115. K. P. Pande, M. A. Fathimula, D. Guitierrez, and L. Messick, *IEEE Electron. Dev. Lett.* **EDL-7**, 407 (1986).

116. R. Leonelli, C. S. Sudararaman, and J. F. Curie, *Appl. Phys. Lett.* **57**, 2678 (1990).

117. Y. -H. Jeong, J. -H. Lee, Y. -H. Bae, and Y. -Y. Hong, *Appl. Phys. Lett.* **57**, 2680 (1990).

118. T. K. Paul and D. N. Bose, *J. Appl. Phys.* **67**, 3744 (1990).

119. S. V. Hattangady, G. G. Fountain, R. A. Rudder, M. J. Mantini, D. J. Vitkavage, and R. J. Markunas, *Appl. Phys. Lett.* **57**, 581 (1990).

120. J. Stöhr, R. Jaeger, G. Rossi, T. Kendlewicz, and I. Lindau, *Surface Sci.* **134**, 813 (1983).

121. M. Saitoh, F. Shoji, K. Oura, and T. Hanawa, *Surface Sci.* **112**, 306 (1981).

122. D. G. Cahill and R. J. Hamers, *Phys. Rev. B* **44**, 1387 (1991).

123. (a) S. G. Louie and M. L. Cohen, *Phys. Rev. Lett.* **35**, 866 (1975); (b) M. L. Cohen, *Adv. Electron. Electron Phys.* **51**, 1 (1980).

124. T. Ide, *Surface Sci.* **209**, 334 (1989).

125. K. Stiles and A. Kahn, *Phys. Rev. Lett.* **60**, 440 (1988).

126, J. Tersoff, *Phys. Rev. Lett.* **52**, 465 (1984); *Phys. Rev. B* **32**, 6968 (1985).

127. J. Ihm and J. D. Joannnopoulos, *J. Vac. Sci. Technol.* **21**, 340 (1982).

128. J. M. Andrews and J. C. Philips, *Phys. Rev. Lett.* **35**, 56 (1975).

129. H. von Känel, T. Graf, J. Henz, M. Ospelt, and P. Wachter, *J. Crystal Growth* **81**, 470 (1987).

130. S. Saitoh, H. Ishiwara, and S. Furakawa, *Appl. Phys. Lett.* **37**, 203 (1980).

131. D. Cherns, G. R. Antsis, J. L. Hutchinson, and J. C. H. Spence, *Phil. Mag. A* **46**, 849 (1982).

132. R. T. Tung, J. M. Gibson, and J. M. Poate, *Appl. Phys. Lett.* **42**, 888 (1983).

133. F. Comin, J. E. Rowe, and P. H. Citrin, *Phys. Rev. Lett.* **51**, 2402 (1983).

134. L. J. Brillson, R. E. Viturro, C. Mailhiot, J. L. Shaw, N. Tache, J. McKinley, G. Margaritondo, J. M. Woodall, P. D. Kirchner, G. D. Pettit, and S. L. Wright, *J. Vac. Sci. Technol. B* **6**, 1263 (1989).

135. R. E. Vituro, M. L. Slade, and L. J. Brillson, *Phys. Rev. Lett.* **57**, 487 (1986); *J. Vac. Sci. Technol. A* **5**, 1516 (1987).

136. A. Kahn, D. Kanani, J. Carelli, J. L. Yeh, C. B. Duke, R. J. Meyer, A. Patton, and L. Brillson, *J. Vac. Sci. Technol.* **18**, 792 (1981).

137. R. M. Feenstra and P. Martensson, *Phys. Rev. Lett.* **61**, 447 (1988).

138. R. Cao, K. Miyano, I. Lindau, and W. E. Spicer, *J. Vac. Sci. Technol. A* **7**, 861 (1989).

139. W. J. Kaiser, and L. D. Bell, *Phys. Rev. Lett.* **60**, 1406 (1988).

140. M. D. Stiles and D. R. Hamann, *Phys. Rev. B* **40**, 1349 (1989).

141. M. Prietsch and R. Ludeke, *Surface Sci.* **251/252**, 413 (1991).

142. R. Ludeke, M. Prietsch, and A. Samsavar, *J. Vac. Sci. Technol. B* **9**, 2342 (1991).

143. R. N. Sato, M. Sokolich, N. Doudoumopoulos, and J. R. Duffy, *IEEE Electron Dev. Lett.* **EDL-9**, 238 (1988).

144. S. T. Long, B. M. Welch, R. Zucca, P. M. Asbeck, C. -P. Lee, C. G. Kirkpattrick, F. S. Lee, G. R. Kaelin, and R. C. Eden, *Proc. IEEE* **70**, 35 (1982).

145. K. Tsutsui, T. Nakazawa, T. Asano, H. Ishiwara, and S. Furukawa, *IEEE Electron Dev. Lett.* **EDL-8**, 277 (1987).

146. R. Van Tuyl and C. A. Liechti, *IEEE J. Solid-State Circuits* **SC-9**, 269 (1977).

147. B. J. Van Zeghbroeck, W. Pattrick, H. Meyer, and P. Vettiger, *IEEE Electron Dev. Lett.* **EDL-8**, 118 (1987).

148. C. Kocot and C. Stolte, *IEEE Trans. Electron Dev.* **ED-29**, 1059 (1982).

149. A. Katz, S. N. G. Chu, B. E. Weir, W. Savin, D. W. Harris, W. C. Dautremont-Smith, R. A. Logan, and T. T. Tabun-Ek, *J. Appl. Phys.* **68**, 4141 (1990).

150. W. Keller, GaAs Electronic Devices, in *ESDERC'89*, A. Heuberger, H. Ryssel, and P. Lange, eds., Springer Verlag, Berlin, 1989, p. 3.

151. N. Bannov, K. Valiev, V. Ryzhhii, and G. Khrenov, Scaling-down of submicrometer GaAs MESFETS, in *ESDERC'89*, A. Heuberger, H. Ryssel, and P. Lange, eds., Springer Verlag, Berlin, 1989, p. 81.

152. S. K. Wang, *IEEE Trans. Electron Dev.* **ED-32**, 2766 (1985).

153. B. Kim, M. Wurtele, H. D. Shih and H. Q. Tseng, *IEEE Electron Dev. Lett.* **EDL-9**, 57 (1988).

154. S. Vaida, D. B. Fraser, and A. K. Sinha, *Proc. 18th Reliability Physics Symp.*, IEEE, New York, 1980, p. 165.

155. S. Yokoyama and K. Okamoto, *Jpn. J. Appl. Phys.* **30**, 239 (1991).

156. S. Shigubara, N. Nishida, A. Fukukawa, H. Sakaue, and Y. Horiike, *Proc. 8th IEEE VLSI Multilevel Interconection Conf.*, (1991), p. 265.

157. P. A. Flinn, A. Sauter Mack, P. R. Besser, and T. N. Marieb, *MRS Bull.* **18**, 26 (1993).

158. R. L. van Hermert, L. B. Spendlove, and R. B. Sievers, *J. Electrochem. Soc.* **112**, 1123 (1965).

159. P. M. Jeffries and G. S. Girolami, *Chem. Mater.* **1**, 8 (1989).

160. M. J. Hampden-Smith, T. T. Kodas, M. Paffett, J. D. Farr, and H. -K. Shin, *Chem. Mater* **2**, 636 (1990).

161. C. G. Dupuy, D. B. Beach, J. E. Hurt, and J. M. Jasinki, *Chem. Mater.* **1**, 16 (1989).

162. D. B. Beach, F. K. LeGoues, and C. K. Hu, *Chem. Mater.* **2**, 216 (1990).

163. A. L. Kaloyeros, A. N. Saxena, K. Brooks, S. Gosh, and E. Eisenbraun, *Advanced Metallizations in Microelectronics*, A. Katz, S. P. Murarka, and A. Appelbaum, eds., MRS Symp. Proc. Vol. 181, Mater. Res. Soc., Pittsburgh, 1990, p. 79.

164. D. B. Beach, W. F. Kane, F. K. Legues, and C. J. Knors, in *Advanced Metallizations in Microelectronics*, A. Katz, S. P. Murarka, and A. Appelbaum, eds., MRS Symp. Proc. Vol. 181, Mater. Res. Soc., Pittsburgh, 1990, p. 73.

165. A. E. Kayoleros and M. A. Fury, *Mater. Res. Bull.* **18**, 22 (1993).

166. J. S. H. Cho, H. -K. Kang, I. Asano, and S. S. Wong, *Proc. IEDM 92*, p. 297.

167. M. E. Alperin, T. C. Holloway, R. A. Haken, C. D. Grosmeyer, R. V. Karnaugh, and W. D. Parmantie, *IEEE Trans. Electron. Dev.* **ED-32**, 141 (1985).

168. J. Nulman, in *Advanced Metallizations in Microelectronics*, A. Katz, S. P. Muraka, and A. Appelbaum, eds., MRS Symp. Proc. Vol. 181, Mater. Res. Soc., Pittsburgh, 1990, 123.

169. Y. H. Ku, S. K. Lee, and D. L. Kwong, in *ULSI Science and Technology*, C. Osburn and J. M. Andrews, eds., The Electrochemical Society, Pennington, 1989, p. 144.

170. Van der Jeunge, *Appl. Phys. Lett.* **57**, 354 (1990).

171. R. J. Gieske, J. J. McMullen, and L. F. Donaghey, in *Proc. 6th Int. Conf. on Chemical Vapor Deposition*, L. F. Donaghey, P. Ray- Choudhury, and R. N. Tauber, eds., The Electrochemical Society, Princeton, NJ, 1977, p. 183.

172. R. S. Rosler and G. M. Engle, *J. Vac. Sci. Technol. B* **4**, 723 (1986).

173. K. Maex, in *Advanced Metallizations in Microelectronics*, A. Katz, S. P. Murarka, and A. Appelbaum, eds., MRS Symp. Proc. Vol. 181, Materials Research Society, Pittsburgh, 1990, 111.

174. J. R. Phillips, P. Revesz, J. O. Olowolafse, and J. W. Mayer, in *Advanced Metallizations in Microelectronics*, A. Katz, S. P. Murarka, and A. Appelbaum, eds., MRS Symp. Proc. Vol. 181, Materials Research Society, Pittsburgh, 1990, 159.

175. C. H. J. Van Den Brekel and L. J. M. Bollen, *J. Crystal Growth* **54**, 310 (1981).

176. M. Kuisl and W. Langheinrich, in *Proc. 5th Int. Conf. on Chemical Vapor Deposition*, J. M. Blocher, Jr., H. E. Hinterman, and L. H. Hall, eds., The Electrochemical Society, Princeton, NJ, p. 380.

177. M. D. Stiles and D. R. Hamann, *J. Vac. Sci. Technol. B* **9**, 2394 (1991).

178. D. M. Brown, B. Gorowitz, P. Piacente, R. Saia, R. Wilson, and D. Woodruff, *IEEE Electron. Lett.* **EDL-8**, 55 (1987).

179. E. K. Broadbent and C. L. Ramiller, *J. Electrochem. Soc.* **131**, 1427 (1984).

180. C. Yang, S. Mehta, and P. Davis, in *Science and Technology of Microfabrication*, R. E. Howard, E. L. Hu, S. Namba, and S. W. Pang, eds., Materials Research Society, Pittsburgh, 1989, 247.

181. T. S. Cale, G. B. Raupp, and Manoj K. Jain, in *Advanced Metallizations in Microelectronics*, A. Katz, S. P. Murarka, and A. Appelbaum, eds., MRS Symp. Proc. Vol. 181, Materials Research Society, Pittsburgh, 1990, 517.

182. C. A. Chang and C. K. Hu, *Appl. Phys. Lett.* **67**, 617 (1990).

CHAPTER 9

Optical Electronics

Although semiconductor microelectronics currently holds the largest share of the solid-state electronics market, there are important additional components to this market, such as optical electronics, that have developed in parallel to microelectronics. Devices that have been pursued in the past in the context of optical electronics are semiconductor light sources and detectors, solar cells, electro-optic switches and modulators, nonlinear-optical components for harmonic generation and optical amplification, photorefractive memory, and magneto-optic isolators. In the past, many of these devices of optical electronics have been developed independently of microelectronics. However, at present the integration of optical electronics and microelectronics is driven from both sides. Examples of current goals of these developments are the integration of storage and signal processing circuits with detector arrays in sensors, the integration of transistorized driver circuits and semiconductor lasers on the same chip, the provision of optical input/output capability for ICs, and the integration of photonic and microelectronic elements in hybrid parallel computers. Also, ferroelectric memory and superconducting switching circuits provide in special applications alternatives to semiconductor microelectronics, and both magnetic and optical mass storage are already indispensible backups to semiconductor memory in modern computers. A detailed discussion of the underlying physics, the optimum materials choices, and processing methods in this field is outside the scope of this book (for a brief encyclopedic review, see Ref. [2].

Figure 9.1 shows a schematic representation of an optical communications system consisting of a light source, such as a semiconductor laser or light-

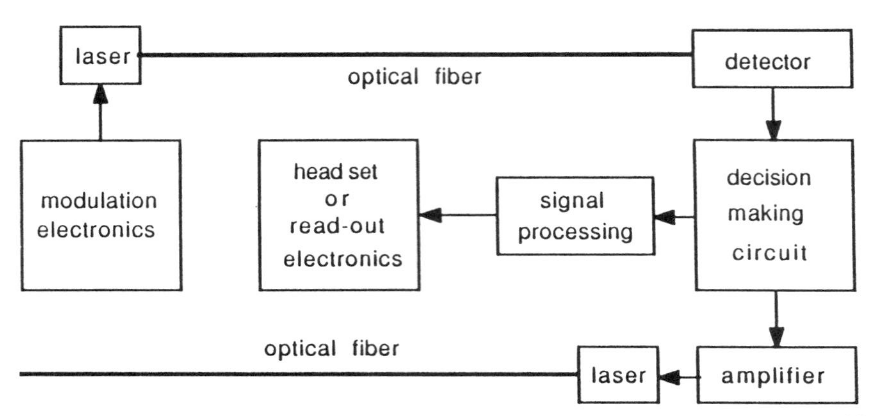

Figure 9.1 Schematic representation of an optical communications system. After Midwinter [1].

emitting diode (LED), that is fast enough to lend itself to direct digital modulation by an appropriate electronic circuit. The modulated light output of these devices is launched into an optical fiber that guides the light to a detector at the other end. If the detector is part of a repeater, its output is amplified and fed via a second light source into another length of optical fiber, repeating this step until the final destination is reached, at which point the optical signal is demodulated and read out.

The optical fiber is a critical component in such a system because its properties determine the repeater distance and with it the viability of the system as well as the materials choice for the optoelectronic components of the system. In the early 1970s, it was still unknown as to whether or not the break-even point of 20 dB/km in the total loss of the fiber could be reached at which fiber optical transmission becomes comparable in cost to the conventional transmission choices. Today total losses of less than 1 dB/km are routinely achieved. This is a remarkable achievement, which has revolutionized the communications industry. It was made possible by the preceding materials purification effort and the development of CVD methods in the context of silicon CVD.

In fiber manufacturing, phophorus- or germania-doped silica layers are deposited inside a fused silica tube, as illustrated in Figure 9.2(a). This tube is subsequently collapsed to form a preform rod for fiber drawing. Alternatively, a boron-doped layer of fused silica may be deposited from the vapor phase on the outside of a fused silica rod. The fiber drawing process is illustrated in Figure 9.2(b).

Figure 9.3 shows (a) a schematic representation of the propagation of light rays in an optical fiber and (b), (c) refractive index profiles perpendicular to the axis of an optical fiber. Due to a controlled variation in composition, the core of the fiber has a higher refractive index than the cladding, guiding the light inside the core region with an exponential decay of the evanescent wave

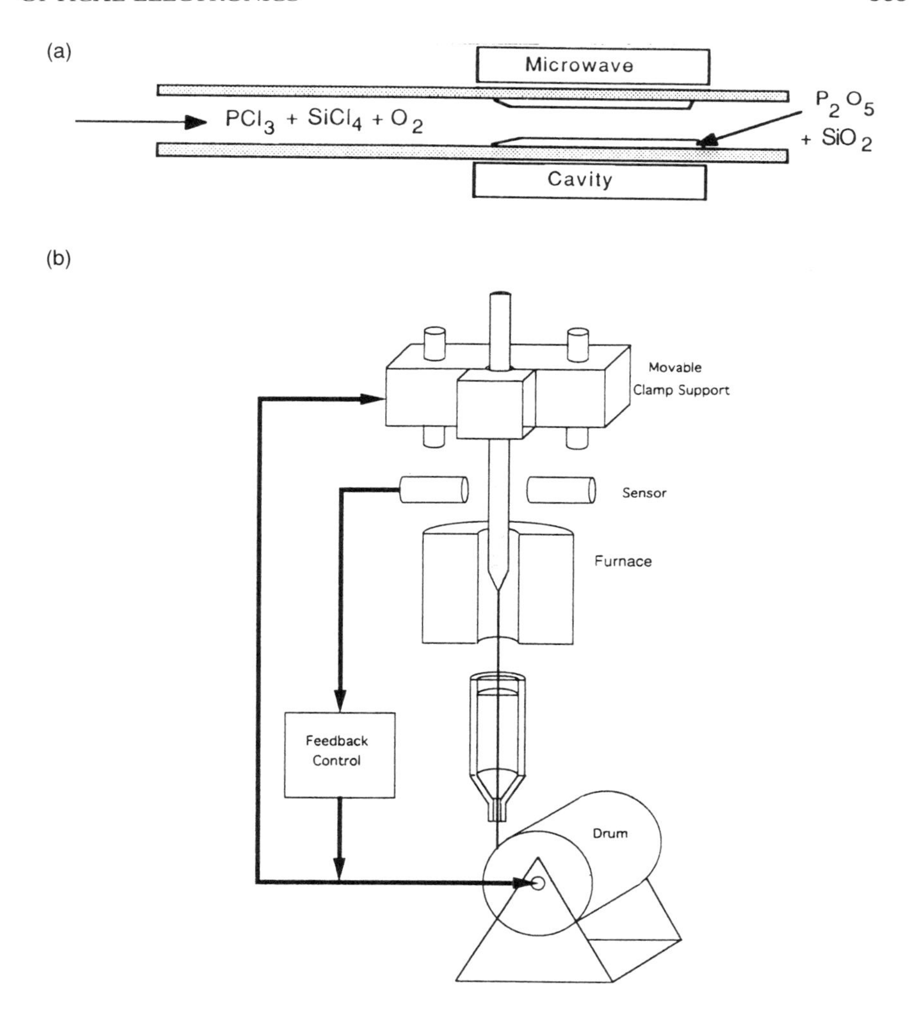

Figure 9.2 Schematic representations of (a) microwave plasma deposition of the doped silica layers and (b) fiber drawing. (b) After Midwinter [1].

into the cladding. The cladding must be thick enough to prevent leakage to the outside. It can be shown [1] that the modes of propagation supported by such a fiber form a discrete set. For the step index profile shown in Figure 9.2(b), the higher-order modes travel along a longer optical path than the low-order modes. This causes the broadening of an initially square light pulse, degrading the modulation. However, this mode dispersion can be removed by appropriate shaping of the refractive index profile.

From a materials processing point of view, this demands the deposition of multilayers of doped silica with well-defined changes in the doping concentration, similar to the planar doping of semiconductors under the conditions of

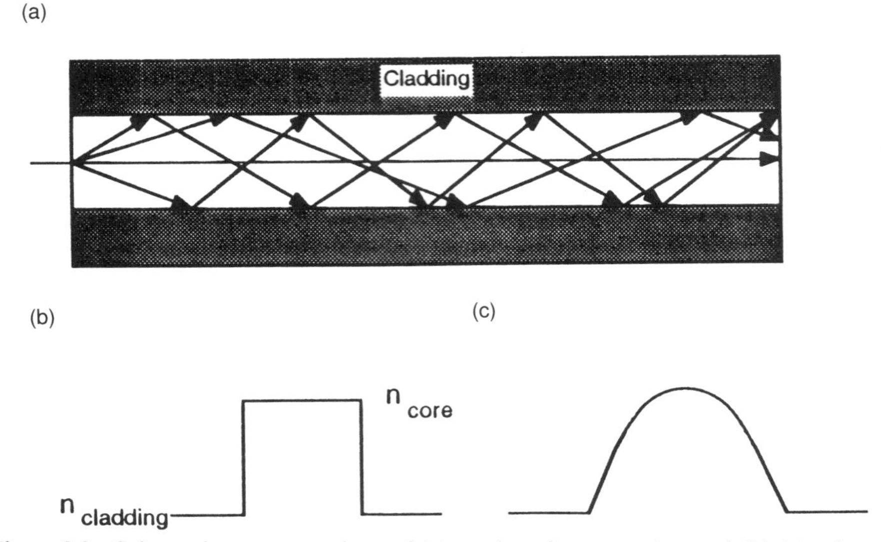

Figure 9.3 Schematic representations of (a) modes of propagation and (b), (c) refractive index profiles for (a) an axial cut and (b), (c) cross sections, respectively, through optical fibers. (b) Step index profile; (c) parabolic index profile.

CVD discussed in Chapter 6. The controlled interdiffusion of this digitally doped core region during preform fabrication and fiber drawing produces the desired refractive index profile. Another cause for dispersion is the wavelength dependence of the refractive index, which is important since semiconductor light sources emit over a more or less extended spectral region, depending on the design of the device. Fortunately, this materials dispersion, for silica-based fibers, becomes zero near 1.3 μm wavelength [3].

LEDs and lasers emitting in this range as well as detectors covering the wavelength region $0.95 \leqslant \lambda \leqslant 1.65\,\mu$m can be made in the quaternary III–V alloy system $Ga_xIn_{1-x}P_yAs_{1-y}$ using compositions that are exactly lattice matched to InP substrates [4]. This is illustrated in Figure 8.16(b), which shows a plot of the approximate bandgaps versus the lattice parameters for selected III–V alloy system. The continuous range of compositions of quaternary alloys that lattice match InP is represented by a vertical line starting at the bandgap of InP and ending on the GaAs–InAs pseudobinary at composition $Ga_{0.47}In_{0.53}As$. Note that the linear approximations to the energy gaps as a function of composition on the pseudobinaries represent a gross simplification

Figure 9.4(b) shows a schematic representation of a heterojunction detector, employing a wide-gap window material, such as *n*-type InP, and a narrow-gap absorber, such as *p*-type $Ga_{0.47}In_{0.53}As$. The detection of light in the detector is based on the absorption of photons of above-bandgap energy under the generation of electron–hole pairs that are separated in the field of the junction. Since the absorption coefficient of III–V compounds and alloys rises at the absorption edge to typically $10^4\,cm^{-1}$, corresponding to an absorption length of the order of micrometers, while the depletion layer width is of the order of

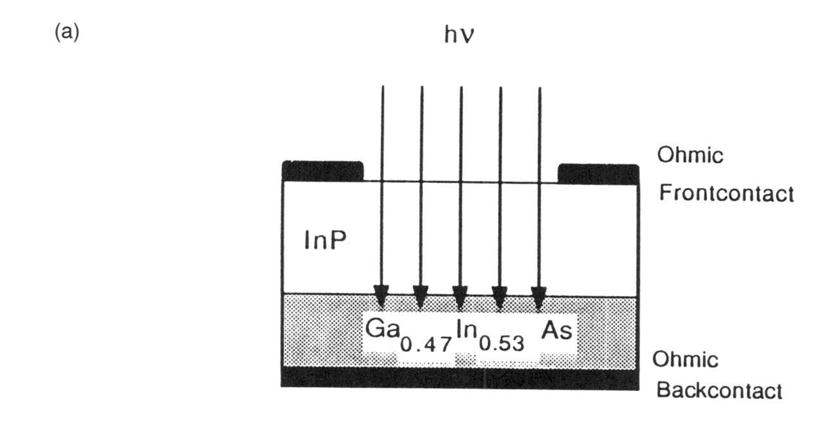

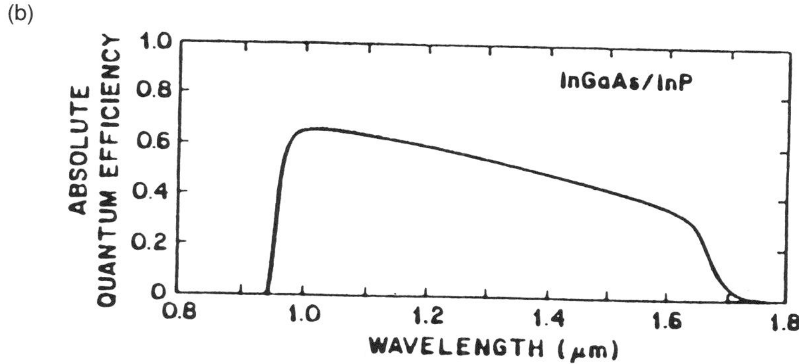

Figure 9.4 (a) Schematic representation of an InP/$Ga_{0.47}In_{0.53}As$ detector. (b) Spectral response of an early InP/$Ga_{0.47}In_{0.53}As$ detector. (b) After Bachmann and Shay [5]; copyright © 1978, American Institute of Physics, New York.

0.1 μm, a large fraction of the photogenerated minority carriers must diffuse to the junction and is endangered during this process by recombination events.

The collection efficiency of such a detector is defined as the ratio of the number of minority carriers collected per unit area and time and the number of incident photons per unit area and time. Figure 9.4(c) shows the spectral response of the InP/$Ga_{0.47}In_{0.53}As$ detector, that is, the wavelength dependence of the collection efficiency. The detector is illuminated through the InP, which has a bandgap of 1.25 eV at room temperature, that is, begins to transmit at 0.95 μm. The photons transmitted by the InP are absorbed in the lattice-matched $Ga_{0.47}In_{0.53}As$ layer. This results in the photogeneration of carriers and the onset of current flow at 0.95 μm. The $Ga_{0.47}In_{0.53}As$ becomes transparent to photons at wavelengths > 1.65 μm which explains the long-wavelength cutoff of the spectral response. The $Ga_xIn_{1-x}As$/InP system thus covers the ranges near 1.3 μm, which is preferred for LED-driven systems and near 1.55 μm wavelength where optical fibers have minimum total loss.

This is illustrated in Figure 9.5, which shows the superposition of the tail of the uv adsorption edge and the onset of the infrared absorption edge, creating

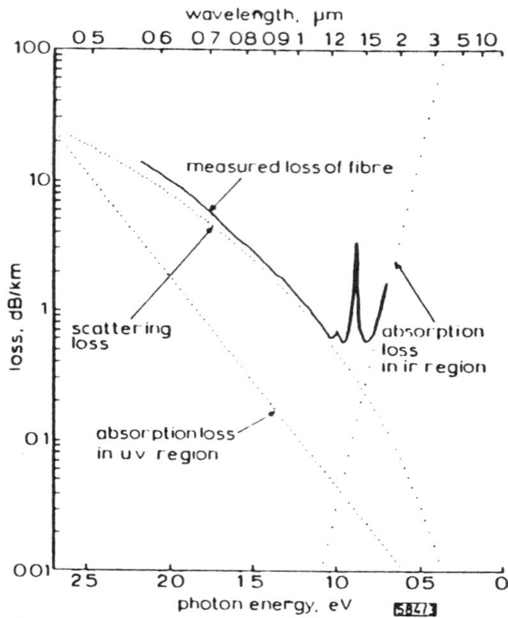

Figure 9.5 Total loss of GeO_2- and P_2O_5-doped fused silica fibers. After Osonai et al. [6]; copyright © 1976, Institute of Electrical and Electronics Engineers, New York.

a window of minimum absorption at 1.55 μm. The uv edge is associated with band-to-band electronic transitions, and the infrared edge is caused by the excitation of vibrational transitions in the SiO_2, including the O–H stretching vibration and harmonics thereof. Research on alternative fiber materials, such as fluorides, is underway to shift this window of minimum adsorption towards longer wavelength. This is desirable because the ultimate loss of the fiber is dominated by scattering, which, for Rayleigh scattering, varies as $1/\lambda^4$. However, because of the superior mechanical and chemical properties of fused silica and the advanced state of the art of materials purification and fiber processing, these alternative fiber materials have at present little economic significance.

Laser of narrow wavelength distribution near 1.55 μm can be fabricated from $Ga_xIn_{1-x}P_yAs_{1-y}/InP$ [7]. Since for the operation of semiconductor lasers, it is necessary to provide for population inversion of carriers and optical feedback, heterostructures achieving the appropriate profile of the refractive index and of the band edges are important. The selection of particular modes of propagation for amplification may be implemented by mirror surfaces forming a Fabry–Perot cavity or by feedback schemes that involve periodic refractive index variations at the interface of the active region (DFB laser) or adjacent to it (DBR laser) [8]. There exists considerable interest at present in surface-emitting lasers, which require the growth of high-quality multilayer mirror stacks. The accuracy needed in the surface structuring and epitaxial crystal growth is thus comparable to that of ULSI processing. Instead of tailoring the bandgap by the choice of the $Ga_xIn_{1-x}P_yAs_{1-y}$ alloy composition, the option of controlling the wavelength of optical transitions by the choice of the width and depth of quantum-confined heterostructures is widely exercised in the construction of advanced semiconductor lasers [9]. Figure 9.6(a) shows the light output as a function of current density for a low-threshold MQW laser diode [10]. The advances in epitaxial technologies and pattern definition, discussed in Chapters 6 and 7 in the context of advanced

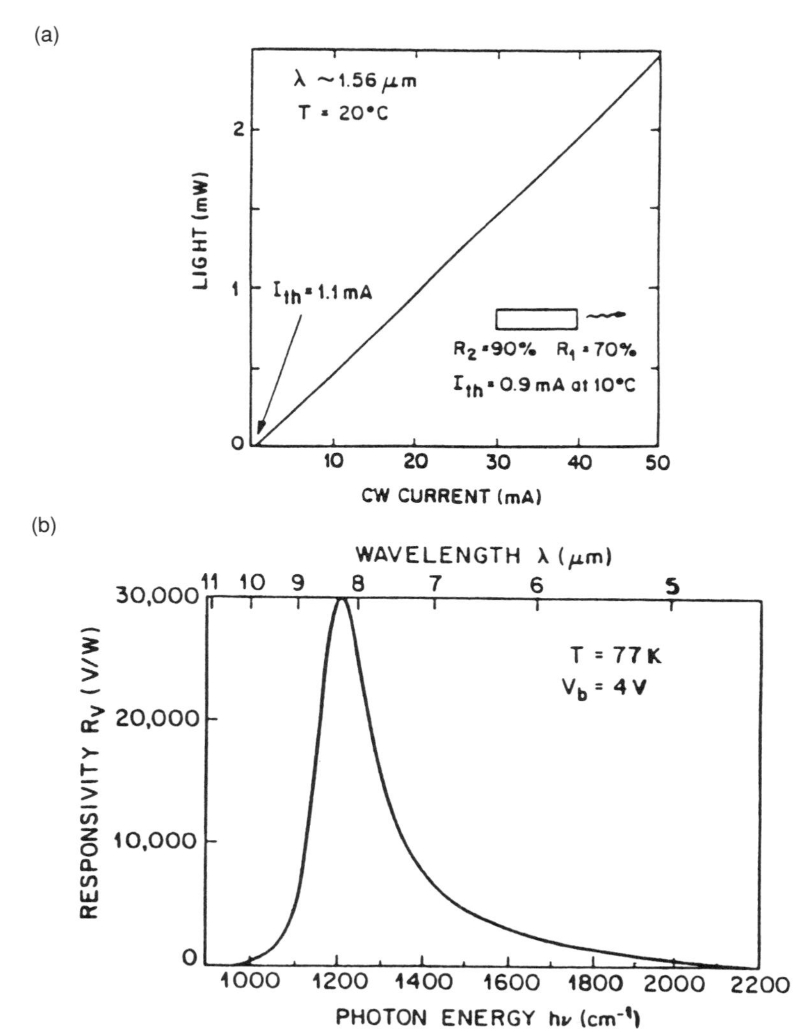

Figure 9.6 (a) Light output versus current for a GaInAs/InP MQW heterostructure laser. (b) Responsivity of a AlGaAs/GaAs MQW infrared detector [18]. After (a) Temkin et al. [10] and (b) Levine et al. [11]; copyrights (a) © 1990, and (b) © 1988, American Institute of Physics, New York.

microelectronic devices, apply directly to the optimization of light sources for optical communications and other applications in optical electronics.

Radiative transitions between sub-bands in MQW heterostructures also have been utilized successfully for the detection of infrared radiation, which is illustrated in Figure 9.6(b), showing the voltage responsivity for an $Al_xGa_{1-x}As/GaAs$ MQW detector [11]. The detectivity of this device is comparable to photoconductive $Cd_xHg_{1-x}Te$ detectors, which are the currently preferred choice in conventional sensor designs.

Solar cells, invented in 1954 by Chapin, Fuller, and Pearson [12], have been explored vigorously in the past, at first in the context of space applications and provision of power for applications in remote areas, and more recently in the context of terrestrial power generation. For space applications, efficient and highly reliable single crystal $Al_xGa_{1-x}As/GaAs$ and Si solar cells are available that are supplemented by even more efficient multijunction devices employing combinations of semiconductor alloys. Single crystal silicon solar cells achieve at present power conversion efficiencies of 21% and 17% with and without concentration of the incident solar radiation by passive collectors, respectively [13]. The progress in the area of high-efficiency solar cells relies to a large extent on discoveries made in the context of the other semiconductor devices discussed and is based on essentially the same process technology.

A completely different situation exists with regard to large-scale terrestial power generation, where the targeted production volume and the economical constraints, imposed by the competing energy technologies, requires an inventive approach. A number of new technologies, such as amorphous hydrogenated Si and $CuInSe_2$ thin film cells as well as low-cost crystalline Si cells, have been introduced in this context. They represent viable options for electricity generation in remote areas and for powering consumer goods, but are currently inadequate in both efficiency and cost to make an impact on the utility market. However, in contrast to the established energy technologies, photovoltaic installations are modular and offer considerable adaptability to changing needs. Although the present contribution of photovoltaic solar energy conversion to the world's energy production is below the expectations raised in the mid-1970s, in the opinion of the author, research in this field is still a worthwhile investment for the future. Innovative thinking rather than the incremental improvement of not even marginally competitive known devices and fabrication technologies is needed for making an impact.

Ternary semiconductors of the type I–III–VI_2 and II–IV–V_2, such as $ZnGeP_2$, are of interest in the context of phase-matched nonlinear optical applications, such as harmonic generation and optical parametric oscillators. They compete in this field with a variety of alternative inorganic and organic materials that have high figures of merit and provide great opportunities for the engineering of nonlinear properties. Recent reviews of the literature in this field are provided in [14] and [15].

Since in present transoceanic fiber-optical communications both the fiber and the required repeaters are located on the ocean floor, the lifetime of the components and options for corrective action upon device failure, which do not require hoisting the cable up for repair, are important. For example, the provision of a redundant number of laser sources in a repeater, in the case of failure of a particular device, permits replacement by another laser. However, the routing of its output to the fiber in need of a new source is required in this case, which may be accomplished by electrical signals fed into electro-optic switches. The electro-optic effect is the change of the refractive index of a material (e.g., $LiNbO_3$) in an electric field. Closely related to electro-optic materials are photorefractive materials. They contain traps that are charged by photogenerated carriers, resulting in local refractive index variations in a photorefractive crystal, replicating the incident light intensity distribution. Optical computing schemes have been proposed that use arrays of selectivity switched surface-emitting lasers for logic and addressable optical storage of information in photorefractive media to replace electronic computers by entirely photonic systems. As the optical power handled in integrated optical circuits increases, it is necessary to introduce components that protect vulnerable devices (e.g., laser diodes) from reflected light. For example, magneto-optic materials that rotate the plane of polarization of a laser beam passing through it allow the construction devices that pass the laser radiation in only one direction and are thus suitable for the realization of optical isolation.

In view of the advanced state of semiconductor light sources and detectors, and the clearly established limitations to the progress of microelectronics due to electromigration and delay problems associated with metal interconnects, it appears to be reasonable to explore at this time the partial replacement of electrical interconnects by optical interconnects. Figure 9.7 shows schematic representations of (a) the layout of the optically interconnected common memory (OICM) that is of considerable potential for the implementation of parallel computing, and (b) a cross section through the light sources and detectors that are located at neighboring memory planes [17]. The technology used at present is based on a single mask set that achieves coincidence of the sources and detectors upon flipping the chip of plane $n + 1$ by 180° with regard to plane n about a symmetry axis contained in this plane, and bonding by a transparent layer of glue. For large in-plane diode spacings this technology is adequate, but alignment problems force a transition to a monolithic epitaxial multiheterostructure as the density of optical elements increases.

Two possible avenues to accomplishing this task are available: (1) The fabrication of the light emitting and detecting devices in silicon layers separated by epitaxial dielectrics; (2) the fabrication of the light emitting and detecting devices in epitaxial compound semiconductor on silicon heterostructures. In the opinion of the author, choosing closely lattice-matched materials combinations is essential in the latter case to achieve high reliability and performance, particularly of lasers that are required when high speed data transfer is desired. Because of the very short distance of the information transfer, the limitations to the laser power by the limited cavity length of vertically emitting laser diode arrays is not a significant factor. However, the power dissipation in the vicinity

(a)

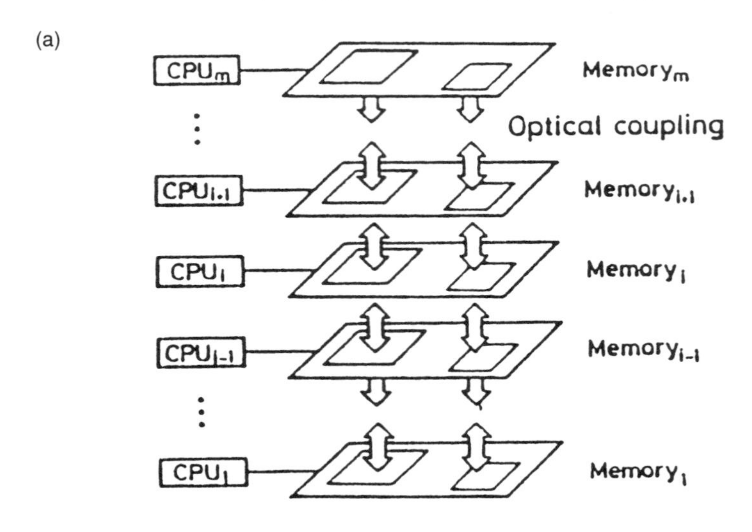

(b)

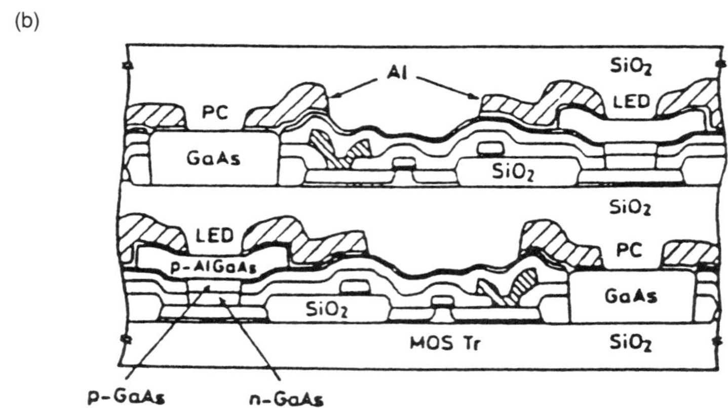

Figure 9.7 Schematic representations of (a) the implementation of optically interconnected common memory for parallel computing tasks and (b) the coupling of light emitting diodes and detectors on neighboring memory planes. After Koyanagi [16], copyright © 1991 Elsevier Science Publishers B.V., Amsterdam.

of the location of the light sources is a problem that requires inventive designs and materials research.

Superconductors, which exhibit ideal diamagnetic behavior and infinite conductivity below a critical temperature T_c, current density $\mathbf{J}_c$, and magnetic field $\mathbf{H}_c$ represent another approach to further reducing the power delay of product switching devices. It has been shown [17] that the superconducting state is due to electron–electron correlations that lower the energy of the system upon pair formation (Cooper pairs). Tunneling across an insulating barrier between two superconducting regions, in equilibrium, establishes coherency between the Cooper pairs on both sides of the junction. A phase lag $\Delta\varphi$

across the junction produces a current density $\mathbf{J} = \mathbf{J}_c \sin(\Delta\varphi)$, which represents the dc Josephson effect [18]. In addition, a dc voltage V that shifts the respective Fermi levels on both sides of a Josephson junction relative to each other by $2qV$ produces an alternating current of frequency $\nu = 2qV/h = 483\,\mathrm{THz/V}$. A variety of quantum interference devices have been developed that provide for extremely fast switching and small power dissipation and thus have potential for microelectronics applications. For example, delay times of 7 ps and power dissipations as low as 4 μW per gate have been reported for conventional 2.5 μm square 4 JL circuits fabricated from $Nb/Al_2O_3/Nb$ junctions in the early 1980s and superconductor/semiconductor/superconductor FET circuits are expected to achieve nonvolatile storage at extremely low power consumption for packing densities of 10^6 transistors/cm^2 [19]–[21]. Recently, the discovery of novel superconducting ceramics [22] with critical temperatures well above liquid nitrogen temperature [23] has shown that significant advances still can be made by rather simple exploratory materials research, which is particularly important in applications, where the established semiconductor technologies are either inadequate or are reaching the limits of performance and reliability.

Vigorous efforts are devoted at present to the replacement of silicon in microelectronics applications by wide gap semiconductors, notably diamond, silicon carbide, and the $Al_xGa_{1-x}N$ system, which are strongly bonded and offer potential advantages in applications in hostile radiative, chemical, and thermal environments. Unfortunately, the realization of epitaxial growth, shallow doping, and contact formation are far more difficult for wide gap materials than for the conventional semiconductors. Therefore, their success in the context of microelectronics applications is by no means assured. However, in view of the unique thermal conductivity, chemical inertness, and mechanical properties of diamond, the development of vapor deposition methods for this material is likely to pay off even if its application as active semiconductor in microelectronics remains elusive. Also, the economic potential of blue light-emitting devices provides a continuing driving force for research on wide bandgap materials, because none of the materials that are being pursued at present has resulted in a reliable technology or is sufficiently free of serious problems to establish an optimum choice.

In view of the increasing capability of nanoscale fabrication methods, the materials science of silicon also remains an open-ended and vital area of research. Simulations of the scaling of MOSFETs, keeping the field in the devices constant, and adjusting the threshold voltage in proportion to the circuit voltage reveal serious degradation of the logic swing normalized to the circuit voltage at feature sizes below 50 nm for CMOS inverters. As discussed in Section 4.3, DRAM memory is limited at present to a feature size of $\sim 0.25\,\mu$m, but could possibly shrink to feature sizes of 30–50 nm, which represents a substantial potential gain in complexity [24].

Primary challenges in materials processing are the lack of a low thermal budget dielectric isolation technology, of proven alternatives to the present interconnect technology, and of a reliable, high throughput pattern definition method that fits the requirements of single wafer processing for feature sizes below 180 nm. Ideas how to solve these problems exist, but, even if they are realized, it is currently not clear as to whether or not the high cost of supermodern ULSI processing facilities will prevent their economic exploitation. Therefore, new cost-saving processing methods must be developed, and novel approaches must be sought to the efficient distribution and the management of information. The development of new processing methods and their introduction into manufacturing are facilitated by the replacement of very large fabrication lines by more flexible cluster tools.

In the few already-established fields of exploratory materials research that may lead to novel approaches to information transfer and management, future advances clearly demand a greater commitment to long term fundamental research. For example, structures of nanoscale dimensions are presently only superficially understood. From classical models we may infer that the surface contributions to the chemical potential become comparable to the bulk contributions and must be incorporated into model calculations. However, recent studies of nanocrystals have shown that the bulk properties themselves depend on size [25], and the properties of nanoscale surfaces are likewise expected to deviate from those of macroscopic dimensions. Therefore, at present we lack the input parameters for quantitative assessments of the changes of the chemical reactivity and physical properties with size in the range where first principles calculations are not yet a viable avenue for the generation of such data. Fortunately, the access to advanced methods of patterning, crystal growth, and characterization with atomic resolution, developed in the context of microelectronics, provides excellent opportunities for gaining experimental data that will rectify this situation and will stimulate and assist theoretical efforts. Although the challenges are formidable, there is good reason for optimism with regard to future progress in this and other advanced areas of materials research that aid the development of solid state electronics.

References

1. John E. Midwinter, *Optical Fibers for Transmission*, John Wiley & Sons, Inc., New York, 1979.
2. D. Bloor, R. J. Brook, M. C. Fleming, and S. Mahajan, *Encyclopedia of Advanced Materials*, Pergamon Press, Oxford, 1994, in print.
3. D. Payne and W. A. Gambling, *Electron. Lett.* **11**, 176 (1975).
4. L. M. Dolginov, N. Ibrakhimov, M. G. Mil'vidskii, V. Yu. Rogulin, and E. G. Shevchenko, *Fiz. Tekh. Poloprovodn.* **9**, 1282 (1975).
5. K. J. Bachmann and J. L. Shay, *Appl. Phys. Lett.* **32**, 446 (1978).
6. H. Osonai, T. Shioda, T. Moriyama, S. Araki, M. Horiguchi, T. Izawa, and H. Takata, *Electron. Lett.* **12**, 550 (1976).
7. J. J. Hsieh, J. A. Rossi, and J. P. Donelly, *Appl. Phys. Lett.* **28**, 283 and 709 (1976).

8. H. C. Casey and M. B. Panish, *Heterostructure Lasers*, Academic Press, New York, (1978).

9. W. T. Tsang, in *Applications of Quantum Wells, Selective Doping and Superlattices*, R. Dingle, ed., *Semiconductors and Semimetals*, Vol. 24, Academic Press, Orlando, 1987, p. 397.

10. H. Temkin, N. K. Dutta, T. Tanbun-Ek, R. A. Logan, and A. M. Sergent, *Appl. Phys. Lett.* **57**, 1610 (1990).

11. B. F. Levine, C. G. Bethea, G. Hasnain, J. Walker, and R. J. Malik, *Appl. Phys. Lett.* **53**, 296 (1988).

12. D. M. Chapin, C. S. Fuller, and G. L. Pearson, *J. Appl. Phys.* **25**, 676 (1954).

13. M. A. Green, S. R. Wenham, and A. W. Blakers, *Conf. Records of the 19th IEEE Photovoltaic Specialists Conference*, New Orleans, 1987, p. 6.

14. K. J. Bachman, in *A Concise Encyclopedia of Semiconducting Materials and Related Technologies*, S. Mahajan and L. C. Kimerling, eds., Pergamon Press, Oxford, 1992, p. 27.

15. A. M. Glass, in *A Concise Encyclopedia of Semiconducting Materials and Related Technologies*, S. Mahajan and L. C. Kimerling, eds., Pergamon Press, Oxford, 1992, p. 327.

16. M. Koyanagi, in VLSI91, A. Halaas and P. B. Denyer, eds., Elsevier Science Publishers B.V., Amsterdam, 1992.

17. J. Bardeen, L. N. Cooper and J. R. Schrieffer, *Phys. Rev.* **108**, 1175 (1957).

18. B. D. Josephson, *Phys. Lett.* **1**, 251 (1962); *Adv. Phys.* **14**, 419 (1965).

19. S. Takada, I. Kurosawa, H. Nakagawa, M. Aoyagi, S. Kosaka, and F. Shinoki, in *Superconductivity Electronics*, Ko Hara, ed., Ohmsha Ltd., Tokyo, 1987, p. 2.

20. S. Yano, Y. Hatano, and U. Kawabe, in *Superconductivity Electronics*, Ko Hara, ed., Ohmsha Ltd., Tokyo, 1987, p. 34.

21. M. Mück and Th. Becker, *Applied Physics A* **54**, 47 (1992).

22. J. G. Bednorz and K. A. Muller, *Z. Physik B* **64**, 189 (1986).

23. M. K. Wu, J. R. Ashburn, C. T. Torng, P. H. Hor, R. L. Meng, L. Gao, Z. J. Huang, Y. Q. Wang, and C. W. Chu, *Phys. Rev. Lett.* **58**, 908 (1987).

24. C. Mead, *The Journal of VLSI Signal Processing*, Kluwer Academic Publishers, 1994, in print.

25. A. N. Goldstein, C. M. Echer, and A. P. Alivisatos, *Science* **256**, 1425 (1992).

APPENDIX

A

List of Abbreviations

Abbreviation	Explanation	pages
AFM	atomic force microscopy	
ALE	atomic layer epitaxy	
ALU	arithmetic logic unit	
APW	augmented plane-wave method	
BEEM	ballistic electron emission microscopy	
BEES	ballistic electron emission spectroscopy	
BICFET	bipolar inversion channel field-effect transistor	
CARS	coherent anti-stokes Raman scattering	
CBE	chemical beam epitaxy	
CBE	conduction-band edge	
CCD	charge-coupled device	
CMOS	complementary metal-oxide silicon	
CPU	central processing unit	
CVD	chemical vapor deposition	
DCFL	direct-coupled FET logic	
DDLTS	double-correlation DLTS	
DLTS	deep-level transient spectroscopy	
DOF	depth of focus	
DRAM	dynamic random-access memory	
ECL	emitter-coupled logic	
ECR	electron cyclotron resonance	
EDTA	ethylenediamine-tetra-acetic acid	

EELFS	electron energy-loss fine structure
ELO	epitaxial layer overgrowth
EMT	effective mass theory
EPTR	epoxy novolac-type resist
EXAFS	extended x-ray absorption fine structure
FET	field-effect transistor
FIM	field ion microscopy
FOX	field oxide
GTO	Gaussian-type orbital
HBT	heterstructure base transistor
HEMT	high-electron-mobility transistor
HOMO	highest occupied orbital
HREELS	high-resolution electron energy loss spectroscopy
HREM	high-resolution electron microscopy
HSG-Si	hemispherical grained silicon
IC	integrated circuit
ICCBE	interrupted-cycle chemical beam epitaxy
IMPATT	impact ionization avalanche transit time
JFET	junction field-effect transistor
LCAO	linear combination of atomic orbitals
LEC	liquid-encapsulated Czochralski
LEED	low-energy electron diffraction
LED	light-emitting diode
LEIS	low-energy ion scattering
LIMIS	liquid metal ion source
LPCVD	low-pressure chemical vapor deposition
LUMO	lowest unoccupied orbital
MBE	molecular beam epitaxy
MCSCF	multiconfiguration self-consistent field
MEE	migration-enhanced epitaxy
MESFET	metal–semiconductor field-effect transistor
MIGS	metal-induced gap states
MIS	metal–insulator–semiconductor
MMIC	monolithic microwave integrated circuit
MODFET	modulation-doped field-effect transistor
MOSFET	metal–oxide–silicon field effect transistor
MQW	multiple quantum well
MSG	maximum stable power gain
NOLOCOS	nitride–oxide local oxidation of silicon
NTD	neutron transmutation doping
OPW	orthogonalized plane wave
PECVD	plasma-enhanced chemical vapor deposition
OSF	oxidation-induced stacking fault
PGMA	propyleneglycol–methylether acetate
PPMA	polymethylmethacrylate

RBS	Rutherford backscattering
RHEED	reflection high-energy electron diffraction
ROM	read-only memory
RTA	rapid thermal annealing
RTBT	resonant tunneling bipolar transistor
RTD	resonant tunnel diode
RTO	rapid thermal oxidation
SALICIDE	self-aligned silicide
SAO	symmetry-adapted orbital
SCF	self-consisting field
SDS	sawtooth-doping superlattice
SET	single-electron transistor
SEXAFS	surface extended x-ray absorption fine structure
SIMS	secondary ion mass spectroscopy
SIRIS	sputter-induced resonance ionization spectroscopy
SRAM	static random-access memory
SSMS	spark-source mass spectrometric analysis
STM	scanning-tunneling microscope
STO	Slater-type orbital
TED	transferred electron device
TED	transmission electron diffraction
TEM	transmission electron microscope
TEOS	tetraethylorthosilicate
TDS	thermal desorption spectroscopy
TDSHF	time-dependent screened Hartree–Fock
TWA	maximum allowed workplace concentration
ULSI	ultra-large-scale integration
VLSI	very large-scale integration
UPS	ultraviolet photoemission spectroscopy
XANES	x-ray-absorption near-edge structure
XPS	x-ray photoemission spectroscopy

APPENDIX B

Tables of Physical Constants, Properties, Conversions, Correlations, and Standard Potentials

1. Physical Constants

Constant	Symbol	Value (cgs)	Value (SI)
Avogaro's number	N_A	6.02252×10^{23} mole^{-1}	6.02252×10^{23} mole^{-1}
Boltzmann constant	k_B	1.38054×10^{-16} erg K^{-1}	1.38054×10^{-23} joule K^{-1}
Bohr radius	r_B	5.29167×10^{-9} cm	5.29167×10^{-11} m
Bohr magneton	μ_B	9.274×10^{-21} oersted cm^3	9.274×10^{-24} joule tesla^{-1}
Electron radius	r_e	2.81777×10^{-13} cm	2.81777×10^{-15} m
Electron rest mass	m_o	9.1091×10^{-28} g	9.1091×10^{-31} Kg
Elementary charge	q	4.80298×10^{-10} cm$^{3/2}$g$^{1/2}$s^{-1}	1.60099×10^{-19} Coulomb
Gas constant	R	8.31466×10^{7} erg mole^{-1}K^{-1}	8.3166 joule mole^{-1}K^{-1}
Gravitational constant	γ	6.670×10 cm^{3}g^{-1}s^{-2}	6.670×10^{-11}newton m^{2}Kg^{-2}
Neutron rest mass	m_n	1.6747×10^{-24} g	$.6747 \times 10^{-27}$ Kg
Permeability	μ_o	-	1.257×10^{-6} henry m^{-1}
Permitivity	ε_o	-	8.859×10^{-12} farad m^{-1}
Planck's constant	h	6.626×10^{-27} erg s	6.626×10^{-34} joule s
Proton rest mass	m_p	1.67252×10^{-24} g	1.67252×10^{-27} Kg
Rydberg constant	R_∞	1.0973731×10^{5} cm^{-1}	1.0973731×10^{7} m^{-1}
Speed of light	c	2.997925×10^{10} cm s^{-1}	2.997925×10^{8} m s^{-1}

2. Conversion of energy units

	erg	joule	eV	cal_{th}*
1erg =	1	10^{-7}	$6.2418x10^{11}$	$2.3901x10^{-8}$
1 joule =	10^{7}	1	$6.2418x10^{18}$	0.23901
1eV =	$1.6021x10^{-12}$	$1.621x10^{-19}$	1	$3.8291x10^{-20}$
1cal_{th}* =	$4.1840x10^{7}$	4.1840	$2.6116x10^{19}$	1

* cal_{th} = thermochemical calory

3. Conversion of SI and cgs units

	SI	esu	emu
Length	meter	10^2 centimeter	10^2 centimeter
Mass	kilogram	10^3 gram	10^3 gram
Time	second	second	second
Force	newton	10^5 dyne	10^5 dyne
Energy	joule	10^7 erg	10^7 erg
Power	watt	10^7 erg s^{-1}	10^7 erg s^{-1}
Electrical charge	coulomb	$3x10^9$ statcoul	10^{-1} abcoul
current	ampere	$3x10^9$ statamp	10^{-1} abamp
displacement	coulomb $meter^{-2}$	$12\pi x10^5$ statcoul cm^{-1}	$4\pi x10^{-5}$ abcoul cm^{-2}
field	volt $meter^{-1}$	$1/(3x10^4)$ statvolt cm^{-1}	10^6 abvolt cm^{-1}
polarization	coulomb $meter^{-2}$	$3x10^5$ statcoul cm^{-2}	10^{-5} abcoul cm^{-2}
potential	volt	1/300 statvolt	10^8 abvolt
resistance	ohm	$1/(9x10^{11})$statohm	10^9 abohm
Capacitance	farad	$9x10^{11}$statcoul $statvolt^{-1}$	10^{-9} abcoul $abvolt^{-1}$
Magnetic field	ampere $meter^{-1}$	$12\pi x10^7$ statamp cm^{-1}	$4\pi x10^{-3}$ oersted
flux	weber	$1/300\ cm^{3/2}g^{-1/2}s^{-1}$	10^8 maxwell
induction	tesla	$1/(3x10^6)\ g^{1/2}cm^{-1/2}s^{-1}$	10^4 gauss
Inductance	henry	$\frac{1}{9x10^{11}} \frac{\text{statvolt s}}{\text{statamp}}$	$10^9 \frac{\text{abvolt s}}{\text{abamp}}$

1 (stat amp) = 1 $(cm^{3/2}g^{1/2}s^{-2})$ = 1/c (abamp), 1 (statcoul) = 1 $(cm^{3/2}g^{1/2}s^{-1})$ = 1/c (abcoul),

1 (statohm) = 1 (s cm^{-1}) = c (abohm), 1 (statvolt) = 1 $(cm^{1/2}g^{1/2}s)$ = c (abvolt)

In the above conversion table the speed of light has been set to: c = $3x10^{10}$ cm/s

4. Correlation tables for O_h, T_d and D_{3d}

O_h	O	T_d	T_h	D_{4h}	D_{3d}
A_{1g}	A_1	A_1	A_g	A_{1g}	A_{1g}
A_{2g}	A_2	A_2	A_g	B_{1g}	A_{2g}
E_g	E	E	E_g	$A_{1g} + B_{1g}$	E_g
T_{1g}	T_1	T_1	T_g	$A_{2g} + E_g$	$A_{2g} + E_g$
T_{2g}	T_2	T_2	T_g	$B_{2g} + E_g$	$A_{1g} + E_g$
A_{1u}	A_1	A_2	A_u	A_{1u}	A_{1u}
A_{2u}	A_2	A_1	A_u	B_{1u}	A_{2u}
E_u	E	E	E_u	$A_{1u} + B_{1u}$	E_u
T_{1u}	T_1	T_2	T_u	$A_{2u} + E_u$	$A_{2u} + E_u$
T_{2u}	T_2	T_1	T_u	$B_{2u} + E_u$	$A_{1u} + E_u$

T_d	T	D_{2d}	C_{3v}	S_4
A_1	A	A_1	A_1	A
A_2	A	B_1	A_2	B
E	E	$A_1 + B_1$	E	$A + B$
T_1	T	$A_2 + E$	$A_2 + E$	$A + E$
T_2	T	$B_2 + E$	$A_1 + E$	$B + E$

D_{3d}	D_3	C_{3v}	S_6	C_3	C_{2h}
A_{1g}	A_1	A_1	A_g	A	A_g
A_{2g}	A_2	A_2	A_g	A	B_g
E_g	E	E	E_g	E	$A_g + B_g$
A_{1u}	A_1	A_2	A_u	A	A_u
A_{2u}	A_2	A_1	A_u	A	B_u
E_u	E	E	E_u	E	$A_u + B_u$

5. Selected Properties of Semiconductors at Room Temperature

Material	E (eV)	χ (eV)	a_o (Å)	m_e^*/m_o	m_h^*/m_o	μ_e (cm^2/Vs)	μ_h (cm^2/Vs)	ε_{static}
Diamond	5.47 d	-	3.567	0.2	0.25	1800	1200	5.5
Si	1.17 i	4.01	5.431	0.97 (l) 0.19 (t)	0.16 (lh) 0.5 (hh)	1350	480	11.8
Ge	0.66 i	4.13	5.658	1.6 (l) 0.082 (t)	0.04 (lh) 0.3 (hh)	3600	1800	16.0
α-SiC	2.99 i	-	3.082 w	0.60	1.00	200	180	10.2
GaP	2.26 i	4.3	5.451	0.82	0.60	110	75	10
InP	1.29 d	4.38	5.869	0.07	0.4	4600	150	12.1
AlAs	2.15 i	-	5.661	0.11	0.22	280	-	10.1
GaAs	1.42 d	4.07	5.654	0.067	0.082	8500	400	11.5
InAs	0.36 d	4.9	6.058	0.023	0.40	33000	460	12.5
AlSb	1.6 i	3.65	6.136	0.3	0.4	900	400	10.3
GaSb	0.68 d	4.06	6.095	0.068	0.5	4000	1400	14.8
InSb	0.17 d	4.59	6.479	0.014	0.40	78000	750	15.9
ZnS	3.58 d	3.9	3.814 w	0.28	1.4	120	-	8.3
ZnSe	2.67 d	4.09	5.667	0.17	0.6	530	-	9.1
ZnTe	2.26 d	3.5	6.103	0.09	1.5	530	130	10.1
CdS	2.42 d	4.5	4.137 w	0.17	0.6	340	-	10
CdSe	1.7 d	4.95	4.298 w	0.13	0.45	600	-	9.9
CdTe	1.44d	4.28	6.477	0.11	~2	700	65	9.6

d = direct bandgap, i = indirect bandgap
w = wurzite structure a-axis lattice parameter, equivalent a-axis lattice parameters: $a_{zb} = a_w\sqrt{2}$
l = longitudinal effective mass, t = transverse effective mass, lh = light hole mass, hh = heavy hole mass

6. Standard Potentials of Selected Metal/Metal Ion Electrodes at 25 °C

Electrode Reaction	E_o (V)	Electrode Reaction	E_o (V)
$Au \leftrightarrows Au^{3+} + 3e^-$	+1.50	$Ga \leftrightarrows Ga^{3+} + 3e^-$	-0.53
$Pd \leftrightarrows Pd^{2+} + 2e^-$	+0.987	$Cr \leftrightarrows Cr^{3+} + 3e^-$	-0.74
$Ag \leftrightarrows Ag^{+} + e^-$	+0.7991	$Zn \leftrightarrows Zn^{2+} + 2e^-$	-0.763
$Hg \leftrightarrows Hg_2^{2+} + 2e^-$	+0.789	$V \leftrightarrows V^{2+} + 2e^-$	-1.18
$Cu \leftrightarrows Cu^{+} + e^-$	+0.521	$Mn \leftrightarrows Mn^{2+} + 2e^-$	-1.18
$Cu \leftrightarrows Cu^{2+} + 2e^-$	+0.337	$Zr \leftrightarrows Zr^{4+} + 4e^-$	-1.53
$H_2 \leftrightarrows H^{+} + e^-$	0	$Ti \leftrightarrows Ti^{2+} + 2e^-$	-1.63
$Fe \leftrightarrows Fe^{3+} + 3e^-$	-0.036	$Al \leftrightarrows Al^{3+} + 3e^-$	-1.66
$Pb \leftrightarrows Pb^{2+} + 2e^-$	-0.126	$Be \leftrightarrows Be^{2+} + 2e^-$	-1.85
$Sn \leftrightarrows Sn^{2+} + 2e^-$	-0.136	$Mg \leftrightarrows Mg^{2+} + 2e^-$	-2.37
$Ni \leftrightarrows Ni^{2+} + 2e^-$	-0.250	$Ce \leftrightarrows Ce^{3+} + 3e^-$	-2.48
$Co \leftrightarrows Co^{2+} + 2e^-$	-0.277	$Na \leftrightarrows Na^{+} + e^-$	-2.714
$Tl \leftrightarrows Tl^{+} + e^-$	-0.336	$Ca \leftrightarrows Ca^{2+} + 2e^-$	-2.87
$In \leftrightarrows In^{3+} + 3e^-$	-0.342	$Ba \leftrightarrows Ba^{2+} + 2e^-$	-2.90
$Cd \leftrightarrows Cd^{2+} + 2e^-$	-0.403	$K \leftrightarrows K^{+} + e^-$	-2.925
$Fe \leftrightarrows Fe^{2+} + 2e^-$	-0.440	$Li \leftrightarrows Li^{+} + e^-$	-3.045

7. Standard Potentials of Selected Redox Electrodesat 25 °C

Electrode Reaction	E_o (V)	Electrode Reaction	E_o (V)
$2F^- \leftrightarrows F_2 + 2e^-$	+2.87	$NO + 2H_2O \leftrightarrows NO_3^- + 4H^+ + 3e^-$	+0.96
$Ag^+ \leftrightarrows Ag^{2+} + e^-$	+1.98	$Hg_2^{2+} \leftrightarrows 2Hg^{2+} + 2e^-$	+0.92
$Co^{2+} \leftrightarrows Co^{3+} + e^-$	+1.82	$Fe^{2+} \leftrightarrows Fe^{3+} + e^-$	+0.771
$2\,H_2O \leftrightarrows H_2O_2 + 2H + 2e^-$	+1.77	$H_2O_2 \leftrightarrows HNO_2 + H^+ + e^-$	+0.682
$MnO_2 + 2H_2O \leftrightarrows MnO_4^- + 4H^+ + 3e^-$	+1.695	$HAsO_2 + 2H_2O \leftrightarrows H_3AsO_4 + 2H^+ + 2e-$	+0.559
$PbSO_4 + 2H_2O \leftrightarrows PbO_2 + 2H_2SO_4 + 3e^-$	+1.685	$2J^- \leftrightarrows J_{2} + 2e^-$	+0.5355
$Ce^{3+} \leftrightarrows Ce^{4+} + e^-$	+1.61	$Fe(CN)_6^{4-} \leftrightarrows Fe(CN)_6^{3-} + e^-$	+0.36
$Mn^{2+} + 4H_2O \leftrightarrows MnO_4^- + 8H^+ + 5e^-$	+1.51	$As + 2H_2O \leftrightarrows HAsO_2 + 3H^+ + 3e^-$	+0.247
$Pb^{2+} + 2H_2O \leftrightarrows PbO + 4H + 2e^-$	+1.455	$Cu^+ \leftrightarrows Cu^{2+}$	+0.153
$2Cl^- \leftrightarrows Cl_2 + 2e^-$	+1.3595	$Sn^{2+} \leftrightarrows Sn^{4+} + 2e^-$	+0.15
$2Cr^{3+} + 7H_2O \leftrightarrows Cr_2O_7^{2-} + 14H^+ + 6e^-$	+1.33	$Ti^{3+} + H_2O \leftrightarrows TiO + 2H + e^-$	+0.1
$Mn^{2+} + 2H_2O \leftrightarrows MnO_2 + 4H^+ + e^-$	+1.23	$H_2 \leftrightarrows 2H^+ + e^-$	0
$2H_2O \leftrightarrows O_2 + 4H^+ + 4e^-$	+1.229	$V^{2+} \leftrightarrows V^{3+}$	-0.255
$1/2J_2 + 3H_2O \leftrightarrows JO_3^- + 6H^+ + 5e^-$	+1.195	$Ti^{3+} \leftrightarrows Ti^{4+} + e^-$	-0.37
$2Br^- \leftrightarrows Br_2(l) + 2e^-$	+1.0652	$Cr^{2+} \leftrightarrows Cr^{3+} + e^-$	-2.925
$NO + H_2O \leftrightarrows HNO_2 + H^+ + e^-$	+1.00	$Fe(OH)_2 + OH^- \leftrightarrows Fe(OH)_3 + e^-$	-3.045

Index